RSMeans Cost Data, Student Edition

RSMeans
A division of Reed Construction Data
Construction Publishers & Consultants
700 Longwater Drive
Norwell, MA 02061
USA
1-800-334-3509

Printed in the United States of America

JOHN WILEY & SONS, INC.

Senior Editor
Marilyn Phelan, AIA

Contributing Editors
Christopher Babbitt
Ted Baker
Barbara Balboni
Robert A. Bastoni
John H. Chiang, PE
Gary W. Christensen
David G. Drain, PE
Cheryl Elsmore
Robert J. Kuchta
Robert C. McNichols
Melville J. Mossman, PE
Jeannene D. Murphy
Stephen C. Plotner
Eugene R. Spencer
Marshall J. Stetson
Phillip R. Waier, PE

President & CEO
Iain Melville

Vice President, Product & Development
Richard Remington

Vice President of Finance & Operations
David C. Walsh

Chief Customer Officer
Lisa Fiondella

Director, RSMeans
Daimon Bridge

Vice President, Marketing & Business
Steven Ritchie

Engineering Manager
Bob Mewis, CCC

Product Manager
Andrea Sillah

Production Manager
Michael Kokernak

Production Coordinator
Jill Goodman

Production
Paula Reale-Camelio
Sheryl A. Rose
Jonathan Forgit
Debbie Panarelli
Mary Lou Geary
Caroline Sullivan

Technical Support
Chris Anderson
Tom Dion
Gary L. Hoitt
Genevieve Medeiros
Kathryn S. Rodriguez
Sharon Proulx

Your purchase includes one year of access to RSMeans Online.

This includes the book's companion cost data in convenient online form, as well as the book's problem sets and sample building plans.

Please use the access code found on the inside back cover of this book for access. If the access code is missing or has already been used you can gain access to the online materials by registering here: https://rsmeansonline.com/academic

This book is printed on acid-free paper.

Published by John Wiley & Sons, Inc., Hoboken, New Jersey.

Published simultaneously in Canada.

For general information on our other products and services, or technical support, please contact our Customer Care Department within the United States at 800-762-2974, outside the United States at 317-572-3993, or fax 317-572-4002.

Wiley publishes in a variety of print and electronic formats and by print-on-demand. Some material included with standard print versions of this book may not be included in e-books or in print-on-demand. If this book refers to media such as a CD or DVD that is not included in the version you purchased, you may download this material at http://booksupport.wiley.com.

For more information about Wiley products, visit our Web site at http://www.wiley.com.

ISBN 978-1-118-33590-1 (pbk.); 978-1-118-34844-4 (ebk); 978-1-118-33753-0 (ebk); 978-1-118-33754-7 (ebk)

Printed in the United States of America

SKY10042311_020223

Table of Contents

Introduction

The purpose of this book is to provide students with reliable information to assist in determining quantities of materials, as well as labor required for construction projects. Construction estimating can be separated into two basic components: how many units are required and reasonable costs for those units. The determination of quantities is commonly called the quantity takeoff. It is, simply, the tabulation of the physical units of materials needed for the construction of the project. While this may sound like a straightforward task, any estimator knows that the process can be quite involved. For example, when estimating a concrete footing (excavation and backfill not included), the items necessary include formwork (forms, ties or spreaders, keyway, inserts, form oil, delivery, erecting, stripping, and cleaning); reinforcing (delivery, cutting, bending, ties, splices, chairs, and other accessories); and the concrete itself (placing, consolidating, finishing, curing, protecting, and patching). A working knowledge of construction materials and methods is a must for a successful estimator. This knowledge helps to ensure that all items are accounted for and tabulated correctly—a sound basis for a good estimate.

The determination of "reasonable" costs for tabulated quantities is one of the main reasons why estimates vary. What may be reasonable for one job may be unrealistic for another. Labor rates (and productivity) will vary from region to region. Costs for the same materials also vary—from city to city and even from supplier to supplier within the same town. The experienced estimator evaluates each project individually, investigates fluctuations in costs, and uses that knowledge as an advantage.

Estimating for building construction is certainly not as simple a task as many people believe. Before one can obtain a quantity and apply a cost per unit to that quantity, many hours, days, or even weeks of hard work may be put into a detailed project before arriving at that one "magic number."

Electronic access to Means CostWorks can be found at **https://rsmeansonline.com/academic**. There you will find the drawings referred to in the chapter exercises as well as the electronic version of the database. Estimating requires practice, consistency, and attention to detail. By challenging yourself with the exercises at the end of each chapter, you'll be well on your way to developing sound estimating skills.

*Note: The cost data included in this book has been abbreviated. Access to the complete data set can be obtained at **https://rsmeansonline.com/academic**

Overview

Types of Estimates

Several different levels of estimates are used to project construction costs. Each has a different purpose. The various types may be referred to by different names, some of which may not be recognized by all as necessary or definitive. Most estimators will agree to several basic levels, each of which has its place in the construction estimating process.

RSMeans, through its publications and seminars, has traditionally presented the following four levels of estimates:

Order of Magnitude Estimates: The order of magnitude estimate could be loosely described as an educated guess. It can be completed in a matter of minutes. Accuracy is –30% to +50%.

Square Foot and Cubic Foot Estimates: This type of estimate is most often used when only the proposed size and use of a planned building are known. Accuracy is –20% to +30%.

Assemblies (Systems) Estimate: An assemblies estimate is best used as a budgetary tool in the planning stages of a project. Accuracy is expected at –10% to +20%.

Unit Price Estimate: Working plans and full specifications are required to complete a unit price estimate. It is the most accurate of the four types, but is also the most time-consuming. Used primarily for bidding purposes, accuracy is –5% to +10%.

Figure 0.1 *Estimating Time vs. Accuracy Curve*

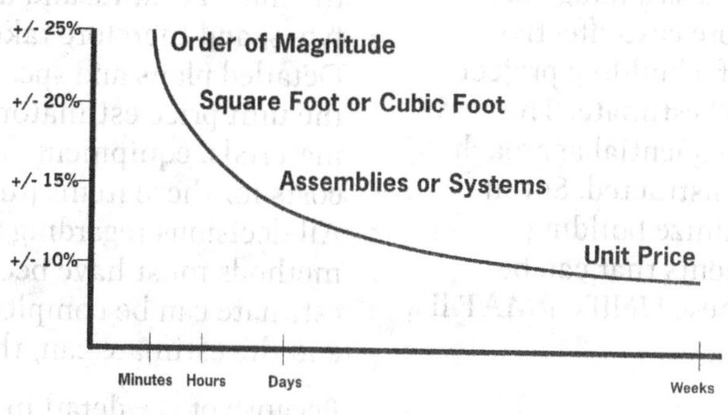

Order of Magnitude Estimates:

The order of magnitude estimate can be completed when only minimal information is available. The use and size of the proposed structure should be known. The "units" can be general. For example: "A small office building for a service company in a suburban industrial park will cost about $1,000,000." This type of statement (or estimate) can be made after a few minutes of thought, drawing on experience, and by making comparisons to similar projects. End product units estimating methods may also be used. Instead of expressing costs in dollars per square foot or dollars per cubic foot, costs are expressed in units related to the particular type of project, such as dollars per bed for hospitals, per student for schools, or per car for parking garages. While this rough figure might be appropriate for a project in one region of the country, an adjustment may be required for a change of location and for cost changes over time (price changes, inflation, etc.).

Square Foot and Cubic Foot Estimates: The use of these types of estimates is most appropriate prior to the preparation of plans or preliminary drawings, when budgetary parameters are being analyzed and established. Costs may be broken down into different components or elements, and then according to the relationship of each component to the project as a whole, in terms of costs per square foot. This breakdown enables the designer, planner, or estimator to adjust certain components according to the unique requirements of the proposed project.

Historical data for square foot costs of new construction are plentiful. However, the best source of square foot costs is the estimator's own cost records for similar projects, adjusted to the parameters of the current project. While helpful for preparing preliminary budgets, square foot and cubic foot estimates can also be useful as checks against other, more detailed estimates. While slightly more time is required than with order of magnitude estimates, a greater accuracy is achieved due to more specific definition of the project.

Assemblies (or Systems) Estimates: Ever-increasing design and construction costs make budgeting and cost efficiency increasingly important in the early stages of designing building projects. Unit price estimating, because of the time and detailed information required, is not suited as a budgetary or planning tool. A faster and more cost-effective method for the planning phase of a building project is the "assemblies," or "systems" estimate. The assemblies method is a logical, sequential approach that reflects how a building is constructed. Seven "UNIFORMAT II" elements organize building construction into major components that can be used in assemblies estimates. These UNIFORMAT II elements are listed below.

A— Substructure
B — Shell
C— Interiors
D— Services
E — Equipment & Furnishings
F — Special Construction
G— Building Site Work

Each element is further broken down into individual assemblies. Each individual assembly incorporates several different items into a system that is commonly used in building construction.

In the assemblies format, a construction material may appear within more than one element. For example, concrete is found in Substructure, as well as in Shell and Equipment. Conversely, each element may incorporate many different areas of construction and the labor of different trades.

A great advantage of the assemblies estimate is that the estimator/designer is able to substitute one system for another during design development and can quickly determine the cost differential. The owner can then anticipate accurate budgetary requirements before final details and dimensions are established.

Unlike unit price estimates, the assemblies method does not require final design details, but the estimators who use it must have solid background knowledge of construction materials and methods, code requirements, design options, and budgetary restrictions.

The assemblies estimate should not be used as a substitute for the unit price estimate. While the assemblies approach can be an invaluable tool in the planning stages of a project, it should be supported by unit price estimating when greater accuracy is required.

Unit Price Estimates: The unit price estimate is the most accurate and detailed of the four estimate types, and therefore takes the most time to complete. Detailed plans and specifications must be available to the unit price estimator to determine the quantities of materials, equipment, and labor. Current and accurate costs for these items (unit prices) are also necessary. All decisions regarding the building's materials and methods must have been made before this type of estimate can be completed. There are fewer variables, and the estimate can, therefore, be more accurate.

Because of the detail involved and the need for accuracy, unit price estimates require a great deal of time and effort to complete properly. For this reason, unit price estimating is often used for construction bidding. It can also be effective for determining certain detailed costs in conceptual budgets or during design development.

Most construction specification manuals and cost reference books, such as *RSMeans Building Construction Cost Data*, divide all unit price estimating information into the 50 CSI MasterFormat divisions.

Before Starting the Estimate

In recent years, plans and specifications have become massive volumes containing a wealth of information. It is of utmost importance that the estimator read all contract documents thoroughly. These documents exist to protect all parties involved in the construction process. The contract documents are prepared so that the estimators will be bidding equally and competitively, ensuring that all items in a project are included. The contract documents protect the designer (the architect or engineer) by ensuring that all work is supplied and installed as specified. The owner also benefits from thorough and complete construction documents, being guaranteed a measure of quality control and a complete job. Finally, the contractor benefits because the scope of work is well defined, eliminating the gray areas of what is implied, but not stated. "Extras" are more readily avoided. Change orders, if required, are accepted with less argument if the original contract documents are complete, well stated, and most important, read by all concerned parties.

During the first review of the specifications, all items to be estimated should be identified and noted. The General Conditions, Supplemental Conditions, and Special Conditions sections of the specifications should be examined carefully. These sections describe the items that have a direct bearing on the proposed project, but may not be part of the actual, physical construction. An office trailer, temporary utilities, and material testing are examples.

While analyzing the plans and specifications, the estimator should evaluate the different portions of the project to determine which areas warrant the most attention. For example, if a building is to have a steel framework with a glass and aluminum frame skin, then more time should be spent estimating Division 05, Metals; and Division 08, Openings; than Division 06, Wood, Plastics & Composites.

The estimator should always visit the project site to ascertain the overall scope and to address specific conditions that may not be apparent from review of the plans. Items to look for include site access, proximity to resources, including utilities, and adequacy of space for storage and equipment.

Figures 0.2 and 0.3 are charts showing the relative percentage of different construction components by MasterFormat Division and Assemblies (UNIFORMAT II) Element, respectively. These charts have been developed to represent the average percentages for new construction as a whole. All commonly used building types are included. The estimator should determine, for a given project, the relative proportions of each component, and estimating time should be allocated accordingly. More time and care should be given to estimating those areas that contribute more to the cost of the project.

The Quantity Takeoff

The quantity takeoff can be thought of as two processes: quantifying and tabulating. In the quantifying process, materials and work items are scaled, counted, and calculated. In the tabulating process, the resultant quantities are tabulated in a way that costs can be assigned to them.

Quantifying can be performed manually, electronically assisted, or digitally. The manual method is done with pencil and paper, architect's scales, and calculators. The electronically assisted method typically employs the use of a device called a "digitizer." The digitizer allows the quantity surveyor to save time because the scaling operation is much more efficient. Using a pointing device, the quantity surveyor simply points to the beginning and end points of a line on a plan, and it is automatically scaled. Similarly, square footages of areas can be determined, and items counted. The digital method is used with CAD (computer assisted design) plans or BIM (building information modeling) based projects. No paper is required since CAD plans are viewed on a computer monitor, and dimensions can be determined by highlighting items or through an electronically generated report. With the necessary programming, not only quantities, but descriptions, labor-hours, and raw costs, can be linked to items on the plan. This information can be used to start a quantity takeoff and estimate within the software, or exported into a separate estimating software package. The estimating information must be modified utilizing the estimator's knowledge and experience.

For the tabulation process, the resultant calculations from any of the quantification methods are transferred, tabulated, and stored—either manually on paper (columnar sheets or preprinted forms) or electronically (using a spreadsheet or database program). An advantage of spreadsheet and database programs is that quantification and tabulation are performed concurrently. In addition, multiple codes can be assigned to each takeoff item so that they can be easily sorted or filtered by project phase, location, expected order of work, or any user-defined criteria.

When working with the plans during the quantity takeoff, consistency is the most important consideration. If each job is approached in the same manner, a pattern will develop, such as moving from the lower floors to the top, clockwise, or counterclockwise. Consistency is more important than the method used and helps to avoid duplications, omissions, and errors. Preprinted forms and spreadsheet templates create documents that provide an excellent means for developing consistent patterns.

Figure 3.1 is an example of a pre-printed form. Preprinted forms can also be used as guides for creating customized spreadsheet templates. Templates are created ahead of time by the user and include all key labels, formulas, and formatting. Only the variables need to be input when estimates are produced.

Costing the Estimate

When the quantities have been determined, then prices, or unit costs, must be applied to determine total costs. No matter which source of cost information is used, the system and sequence of pricing should be the same as those used for the quantity takeoff. This consistent approach should continue through both accounting and cost control during construction of the project. Refer to Figure 5.1 for a sample Cost Analysis form.

Unit price estimates for building construction are typically organized according to the 50 divisions of the CSI MasterFormat. However, other organizational structures can be used. Within each division, the components or individual construction items are identified, listed, and assigned costs. This kind of definition and detail is necessary to complete an accurate estimate. In addition, each "item" can

Figure 0.2 *Cost Distribution by MasterFormat Division*

NO.	DIVISION	%
015433	CONTRACTOR EQUIPMENT*	6.6%
3120,3130	Earth Moving & Earthwork	4.1
3160,3170	LB Elements & Tunneling	0.1
3210	Bases, Ballasts & Paving	0.3
3310,3330,3340,3350	Utility Services & Drainage	0.2
3230	Site Improvements	0.1
3290	Planting	0.6
0241,31-34	**SITE & INFRASTRUCTURE, DEMO**	**5.4**
0310	Conc. Forming & Accessories	2.8
0320	Concrete Reinforcing	2.3
0330	Cast-in-Place Concrete	5.2
0340	Precast Concrete	2.1
03	**CONCRETE**	**12.4**
0405	Basic Masonry Mat. & Methods	0.6
0420	Unit Masonry	6.0
0440	Stone Assemblies	0.4
04	**MASONRY**	**7.0**

NO.	DIVISION	%
0510	Structural Metal Framing	3.9%
0520, 0530	Metal Joists & Decking	9.2
0550	Metal Fabrications	1.0
05	**METALS**	**14.1**
0610	Rough Carpentry	0.7
0620	Finish Carpentry	0.3
06	**WOOD & PLASTICS & Composites**	**1.0**
0710	Dampproofing & Waterproofing	0.1
0720, 0780	Thermal, Fire & Smoke Protection	1.4
0740, 0750	Roofing & Siding	1.2
0760	Flashing & Sheet Metal	0.3
07	**THERMAL & MOISTURE PROT.**	**3.0**
0810, 0830	Doors & Frames	6.1
0840, 0880	Glazing & Curtain Walls	1.0
08	**Openings**	**7.1**

NO.	DIVISION	%
0920	Plaster & Gypsum Board	1.9%
0930, 0966	Tile & Terrazzo	2.3
0950, 0980	Ceilings & Acoustical Treatment	2.9
0960	Flooring	1.9
0970, 0990	Wall Finishes & Painting/Coating	0.9
09	**FINISHES**	**9.9**
Covers	**Divs 10-14, 25,28,41,43,44**	**6.0**
2210, 2230, 2240, 2320	Piping, Pumps, Plumbing Equipment	13.4
2113	Fire Suppression Sprinkler Systems	3.0
2350	Central Heating Equipment	2.6
2330, 2340 ,2360 ,2370, 2380	Air Conditioning & Ventilation	3.1
21, 22, 23	**FIRE SUPPRESS., PLUMB. & HVAC**	**22.1**
26, 27, 3370	**ELECTRICAL, COMMUN. & UTIL.**	**12.0**
	TOTAL (Div. 1-16)	**100.0%**

* Percentage for contractor equipment is spread among divisions and included above for information only

Figure 0.3 *Cost Distribution by UNIFORMAT II Elements (Assemblies)*

Division No.	Building System	Percentage	Division No.	Building System	Percentage
A	Substructure	6.3%	D10	Services: Conveying	3.9%
B10	Shell: Superstructure	19.8	D20-40	Mechanical	22.1
B20	Exterior Closure	11.5	D50	Electrical	12.0
B30	Roofing	2.9	E10	Equipment & Furnishings	2.1
C	Interior Construction	15.5	G	Site Work	3.9
				Total (Div. A-G)	100.0%

be broken down further into material, labor, and equipment components.

Types of Costs: All costs included in a unit price estimate can be divided into two types: direct and indirect. Direct costs are those directly linked to the physical construction of a project, those costs without which the project could not be completed. The material, labor, and equipment costs mentioned above, as well as subcontract costs, are all direct costs. These may also be referred to as "bare," or "unburdened" costs.

Indirect, or overhead, costs are those costs which are incurred in achieving project completion, but not applicable to any specific task. They may include items such as supervision, temporary facilities, insurance, professional services, and contingencies. Indirect costs are separated into two groups: job site overhead, and main office overhead. Job site overhead costs are those indirect costs associated with a job site. They can be estimated in detail, but are also calculated as a percentage of direct costs and included in CSI MasterFormat Division 1 of the estimate. Main office overhead costs are costs associated with the operation of the contractor's home (or main) office. Overhead costs are typically calculated as a percentage of the total project cost and added at the end of the estimate. Some costs, such as professional services, may be either project overhead or main office overhead cost, depending on how the resource is used.

Compiling the Data: At the costing stage of the estimate, there is typically a large amount of data that must be assembled, analyzed, and organized. Generally, the information contained in these documents falls into the following major categories, and can exist in paper or digital form.

- Quantity takeoffs for all general contractor items
- Material supplier written quotations and published prices
- Material supplier telephone quotations
- Subcontractor quotations
- Equipment supplier quotations and published prices
- Cost analysis
- Historic cost data from previous projects
- Costs and historic data from independent sources

A system is needed to efficiently handle this mass of data and to ensure that everything will get transferred (and only once) from the quantity takeoff to the cost estimate. Some general rules for this procedure are:

- Code each document with a division number in a consistent place.
- Use telephone quotation forms and templates for uniformity when recording telephone quotes.
- Document the source of every quantity and price.
- Use a logical, consistent directory filing system and file naming convention.
- Back up all important data.

All subcontract costs should be properly noted and listed separately. These costs contain the subcontractors' markups, and will be treated differently from other direct costs when the estimator calculates the project overhead, profit, and contingency allowance.

Additional Tips: When estimating manually, follow these guidelines:

- Write on only one side of a page if possible.
- Keep each type of document in its pile (quantities, material, subcontractors, equipment) filed in order by division number.
- Keep the entire estimate file in one or more compartmented folders.

When using electronic spreadsheets:

- Save commonly used templates in a separate folder.
- Spot-check important formula results with manual calculations to verify results.
- Frequently save spreadsheets to avoid losing work.
- Combine related spreadsheets into workbooks.
- Use a naming convention that indicates whether a spreadsheet is in progress or complete.
- When entering quantities into a spreadsheet cell, include useful numerical information. For example, *if 10% is added to a quantity of 300, instead of simply entering 330, enter =300*1.1*

Since over 50% of the work on a typical building project is performed by subcontractors, two aspects of bid preparation deserve special attention. The first is the subcontractor's scope of work, a clear

Figure 0.4 *Sample Estimate*

How RSMeans Data Works

Sample Estimate

This sample demonstrates the elements of an estimate, including a tally of the RSMeans data lines, and a summary of the markups on a contractor's work to arrive at a total cost to the owner. The RSMeans Location Factor is added at the bottom of the estimate to adjust the cost of the work to a specific location.

Work Performed: The body of the estimate shows the RSMeans data selected, including line number, a brief description of each item, its take-off unit and quantity, and the bare costs of materials, labor, and equipment. This estimate also includes a column titled "SubContract." This data is taken from the RSMeans column "Total Incl O&P," and represents the total that a subcontractor would charge a general contractor for the work, including the sub's markup for overhead and profit.

General Requirements: This item covers project-wide needs provided by the general contractor. These items vary by project, but may include temporary facilities and utilities, security, testing, project cleanup, etc. For small projects a percentage can be used, typically between 5% and 15% of project cost. For large projects the costs may be itemized and priced individually.

Bonds: Bond costs should be added to the estimate. The figures here represent a typical performance bond, ensuring the owner that if the general contractor does not complete the obligations in the construction contract the bonding company will pay the cost for completion of the work.

Location Adjustment: RSMeans published data is based on national average costs. If necessary, adjust the total cost of the project using a location factor from the "Location Factor" table or the "City Cost Index" table. Use location factors if the work is general, covering multiple trades. If the work is by a single trade (e.g., masonry) use the more specific data found in the "City Cost Indexes."

Project Name:	Pre-Engineered Steel Building	Architect: As Shown		
Location:	Anywhere, USA			
Line Number	Description	Qty	Unit	Material
03 30 53.40 3950	Strip footing, 12" x 36"	33.33	C.Y.	$4,232.91
03 30 53.40 3940	Strip footing, 12" x24"	14.8	C.Y.	$1,953.60
03 31 05.35 0300	Concrete ready mix, 4000 psi, for slab	185.2	C.Y.	$19,075.60
03 31 05.70 4300	Placing concrete, slab on grade	185.2	C.Y.	$0.00
03 11 13.65 3000	Concrete forms	500	L.F.	$135.00
03 22 05.50 0200	WWF	150	C.S.F.	$3,225.00
03 39 23.13 0300	Curing	150	C.S.F.	$840.00
03 35 29.30 0250	Concrete finishing	15000	S.F.	$0.00
03 35 29.35 0160	Control joints	650	L.F.	$65.00
Division 03	Subtotal			$29,527.11
08 33 23.10 0100	10' x 10' OH Door	8	Ea.	$14,200.00
08 33 23.10 3300	Add for enamel finish	800	S.F.	$1,320.00
Division 8	Subtotal			$15,520.00
13 34 19.50 1000	Pre-Engineered Steel Building	15,000	SF Flr.	
13 34 19.50 6050	Framing for doors, 3' x 7'	4	Opng.	$0.00
13 34 19.50 6100	Framing for doors, 10' x 10'	8	Opng.	$0.00
13 34 19.50 6200	Framing for windows, 3' x 4'	6	Opng.	$0.00
13 34 19.50 5750	PESB doors	4	Opng.	$2,240.00
13 34 19.50 7750	PESB windows	6	Opng.	$2,130.00
13 34 19.50 6550	Gutters	380	L.F.	$2,128.00
13 34 19.50 8650	Roof vent	15	Ea.	$360.00
13 34 19.50 6900	Insulation	25,000	S.F.	$10,500.00
Division 13	Subtotal			$17,358.00
Estimate Subtotal				$62,405.11
			Gen. Requirements	4,368.36
			Sales Tax	3,120.26
			Subtotal	69,893.72
			GC O & P	6,989.37
			Subtotal	76,883.10
			Contingency	@5%
			Subtotal	
			Bond	$12/1000 +10% O&P
			Subtotal	
			Location Adjustment	Factor
			Grand Total	

Figure 0.4 Sample Estimate (continued)

This example shows the cost to construct a pre-engineered steel building. The foundation, doors, windows, and insulation will be installed by the general contractor. A subcontractor will install the structural steel, roofing, and siding.

	01/01/12			
Labor	**Equipment**	**SubContract**	**EstimateTotal**	
$1,983.14	$13.00	$0.00		
$1,102.60	$7.25	$0.00		
$0.00	$0.00	$0.00		
$2,240.92	$77.78	$0.00		
$795.00	$0.00	$0.00		
$3,000.00				
$667.50	$0.00	$0.00		
$6,300.00	$300.00	$0.00		
$208.00	$52.00	$0.00		
$16,297.16	**$450.03**	**$0.00**	**$46,274.30**	
$3,320.00	$0.00	$0.00		
$0.00	$0.00	$0.00		
$3,320.00	**$0.00**	**$0.00**	**$18,840.00**	
		$279,010.00		
$0.00	$0.00	$1,799.82		
$0.00	$0.00	$7,519.60		
$0.00	$0.00	$2,669.86		
$464.00	$0.00	$0.00		
$270.00	$50.40	$0.00		
$691.60	$0.00	$0.00		
$2,190.00	$0.00	$0.00		
$5,750.00	$0.00	$0.00		
$9,365.60	**$50.40**	**$290,999.93**	**$3[illegible]7,773.93**	
$28,982.76	**$500.43**	**$290,999.93**	**$382,888.2[illegible]**	
2,028.79	**35.03**	**20,370.00**		**Gen. Requirements**
	25.02	**7,275.00**		**Sales Tax**
31,011.55	**560.48**	**318,644.92**		**Sub Total**
20,901.79	**56.05**	**31,864.49**		**GC O & P**
51,913.34	**616.53**	**350,509.42**	**$479,922.38**	**Subtotal**
			23,996.12	**Contingency**
			$503,918.50	**Subtotal**
			6,651.72	**Bond**
			$510,570.22	**Subtotal**
102.30			**11,743.12**	**Location Adjustment**
			$522,313.34	**Grand Total**

Sales Tax: If the work is subject to state or local sales taxes, the amount must be added to the estimate. Sales tax may be added to material costs, equipment costs, and subcontracted work. In this case, sales tax was added in all three categories. It was assumed that approximately half the subcontracted work would be material cost, so the tax was applied to 50% of the subcontract total.

GC O&P: This entry represents the general contractor's markup on material, labor, equipment, and subcontractor costs. RSMeans' standard markup on materials, equipment, and subcontracted work is 10%. In this estimate, the markup on the labor performed by the GC's workers uses "Skilled Workers Average" shown in Column F on the table "Installing Contractor's Overhead & Profit," which can be found on the inside-back cover of the book.

Contingency: A factor for contingency may be added to any estimate to represent the cost of unknowns that may occur between the time that the estimate is performed and the time the project is constructed. The amount of the allowance will depend on the stage of design at which the estimate is done, and the contractor's assessment of the risk involved. Refer to section 01 21 16.50 for contigency allowances.

understanding of which is essential, not only to compare the bids of competing subs, but to ensure that the estimator has included all the items that the subcontractors may have excluded (such as cutting and patching, temporary protection, scaffolding, hoisting, etc.). The second is that subcontractor prices typically do not arrive until bid day, which leaves little time to analyze competing bids, and makes last-minute gaps in coverage difficult to address if the estimator has not done the necessary coordination with these subs beforehand.

The Estimate Summary

When the pricing of all direct costs is complete, the estimator can transfer the total costs for each subdivision to the Estimate Summary sheet and apply all indirect overhead costs to this document. This step should be double-checked, since an error of transposition may easily occur. Refer to Figure 0.4 for a sample and Figure 5.2 for a blank summary sheet. As items are listed in the proper columns, each category is added, and appropriate markups applied to the total dollar values. Generally, the percentages for each of the following categories of indirect costs are added to the sum of each column at the end of the estimate:

- General Requirements
- Sales Tax
- General Contractor Overhead & Profit
- Contingency
- Bonds

Chapter 1 The Contract Documents

Estimating projects requires fluency in the language and symbols used in construction drawings and a familiarity with the various components included in the project manual. This chapter provides an overview of the contract documents: the working drawings and project manual which includes the technical specifications and construction contract. It does not offer detailed instruction in plan reading, but will review the organization of the drawings and specifications and highlight the various parts of the project manual—essential information for estimating a construction project.

The working drawings are the graphic representation or illustration of the project, and comprise the lines, symbols, and abbreviations printed on paper that represents the owner's wishes, as interpreted by the architect. They convey quantitative information such as how many doors and where they are located. The technical specifications however, convey information that is more qualitative in nature such as specifying the quality of the materials and workmanship.

Together, the drawings and specifications make up the contract documents and form the basis of the contract for construction

Working Drawings

Most drawings develop over several generations of review and modification as a result of owner input, coordination with other design disciplines, building code compliance, and general fine-tuning. This process is referred to as design development and occurs before the release of the final version of drawings, called the working drawings. Working drawings are the completed design—a code-compliant representation of the project, ready for bidding and, ultimately, construction. They will be the focus of this section and are the prerequisite for preparing a detailed unit price estimate. (Note: "Preliminary" drawings created early in the design process may be used as a basis for budget estimates. Budget estimating, though, requires specific skills that are covered in chapters 7, 8 and 9.)

The completed drawings become a "set," which incorporates all adjustments, changes, and refinements made by the architect or engineer as the final step in design development. Working drawings should comply with all applicable building codes, including any local ordinances having jurisdiction. Drawings should include all the information needed to prepare a detailed estimate and eventually build the project. The set of working drawings consists of various disciplines, including architectural, showing the layout of the building and its use of space; structural engineering, design of the structure to ensure that it will support the imposed loads; and mechanical and electrical design, to make the space habitable.

Other drawings in the set include designs that are less concerned with the structure itself than with support services, such as utilities that will be provided to the structure. These *civil* or *site drawings* include grading and drainage plans, which indicate how surface water will be channeled away from the structure; landscaping and irrigation design, paving, and curbing layout. Fire protection, equipment plans, and furniture plans are examples of other drawing types that may also be included in a working drawing set.

Organization of Working Drawings

There is a distinct organizational structure to the working drawings, which is almost universally accepted, and is as follows:

The Cover Sheet

The cover sheet, although very basic in nature, is one of the most important pages in a set of drawings. It lists information such as the name of the project, the location, and the names of the architects, engineers, owners, and other consultants involved in the design. The cover sheet also lists the drawings that comprise the set in the order they will appear. The drawing list is organized by the number of each drawing and the title of the page on which it appears. The cover sheet may also list information specifically required by the building code, including the total square foot area of the building, the use group, and the type of construction.

Another important element on the cover sheet is a list of abbreviations or graphic symbols used in the drawing set. There is often a section that contains "general notes," such as "All dimensions shall be verified in the field," or "All dimensions are to face of masonry." These notes help set the standards for background information that you will encounter throughout the drawings. In the absence of a separate set of bound specifications (most common in the residential market, where separate specs are not often written), the cover sheet may list the general technical specifications that will govern the quality of materials used in the work. Optional information, such as a locus plan locating the project with respect to local landmarks or roadways or an architectural rendering of the building may be included in the cover sheet.

Title Block

The title block is typically located in the lower right-hand corner of each sheet in the drawing set and should include the following information:

- The prefixed letter and number of the sheet (the letter designates the discipline of the drawing followed by the sheet number)
- The name of the drawing (e.g., "First Floor Plan")
- The date of the drawing
- The initials of the draftsperson
- Any revisions to the final set of drawings

The date and scope of the revisions should be noted within the title block. If there is not enough space available, the revisions should be noted close to it. The title block should specify whether the entire drawing is one scale, or whether the scale varies per detail, as in the case of a sheet of details. Sets of drawings for commercial projects require a stamp (and usually a signature) of the architect or engineer responsible for the design.

Revisions

Often, after the set of working drawings has been completed, recommendations are made for correction or clarification of a particular detail, plan, or elevation. While major changes may require redrafting an entire sheet, smaller changes are shown as a revision of the original. All changes must be clearly recognizable. They are indicated with a revision marker, which encloses the revised detail within a scalloped line that resembles a cloud. Tied to the revision marker is a triangle that encloses the number of the revision. Revisions are noted in the title block, or close to it, by date and number. This procedure provides a mechanism for identifying the latest version of drawings.

Drawings, Major Disciplines

Site Plan (SI)

The main purpose of the site plan is to locate the structure within the confines of the building lot. Even the most basic site plans clearly establish the building's dimensions, usually by the foundation's size and the distance to property lines. The latter, called the *setback* dimensions, are shown in feet and hundredths of a foot, versus feet and inches on architectural drawings. For example, the architectural dimension of 22′-6″ would be 22.50′ on a site plan. This decimal system is used because it is the basis of measurement for the land surveyor, the predominant engineer responsible for laying out the site. Site drawings are prefixed with the letters "SI" and numbered sequentially.

Civil Drawings (C)

The grouping of different types of site drawings, such as utility and drainage, grading, site improvement, and landscaping plans, is known under the general classification of *civil drawings.* Civil drawings encompass all the work that serves the building and is outside the structure itself. The most obvious difference between civil drawings and architectural drawings is the scale of the drawing and use of the engineer's scale.

Architetural drawings (A)

Architectural drawings are considered core drawings, since all the other drawings in the set support them. They show the layout of the building and its use of space. Architectural drawings convey the aesthetics of the building, denote the construction type, and indicate the dimensions and placement of all key features. The first architectural drawings in a set generally show large areas in less detail. As one progresses through the set, the level of detail increases. These drawings are prefixed by the letter "A" and sequentially numbered.

Structural drawings (S)

Structural drawings illustrate how the various load-carrying systems transmit both the live and dead loads of the structure to the earth below. Structural design is based on the architectural features, and is designed around the core drawings. (For example, columns and beams are designed to avoid interrupting a space.) Structural drawings are prefixed by the letter "S" and are sequentially numbered.

Mechanical drawings (M, P, FP)

Mechanical drawings illustrate the environmental systems of a building, such as plumbing, fire suppression/protection, and HVAC (heating, ventilating, and air conditioning) systems. These drawings may be prefixed by the letter "M" for mechanical, "P" for plumbing, or "FP" for fire protection. Building services drawings, as they are often named, are shown mainly in plan view and are sequentially numbered.

Electrical drawings (E)

Electrical drawings illustrate the electrical requirements of the project, including power distribution, lighting, and low-voltage specialty wiring, such as for fire alarms, telephone/data, and technology wiring. They often show the provision for power wiring of equipment illustrated on other types of drawings. They are prefixed by the letter "E" and are sequentially numbered.

Specialty drawings

Specialty drawings illustrate the unique requirements of various spaces' special uses (such as kitchens, libraries, retail, and home theater spaces). They define the coordination among other building systems, most commonly the mechanical and electrical systems. The drawings are sequentially numbered, and named according to the type of drawings. For example, "K" might be used for kitchen drawings, "F" for fixture drawings, and so forth.

Graphic Formats Used in Working Drawings

There are accepted standards or methods that architects and engineers use to present graphic information. Different views ensure that all required information is visible on the drawings. There are six main graphic formats:

- Plan views
- Elevations
- Sections
- Details
- Schedules
- Diagrams

Plans, elevations, and sections belong to a family of drawings called orthographic projections. Each represents a specific view or aspect of a building. Each drawing describes a limited part of the building, but when seen together, accurately describes the entire building. These types of drawings are commonly used for working drawings because they are true to size and shape. In other words, they are quantifiable.

Plan Views

The most common graphic view, the plan view, is presented as if looking down on the space. It is the view seen as if an imaginary horizontal cut is made several feet above the floor plane. In this sense, it is actually a section view looking down. Plan views form the basis of the project, and often provide the most complete view. The most common plan view is the *architectural floor plan*, which shows the shape of the building or building footprint, doors, windows, walls, and partitions. Other types of plan views include *reflected ceiling plans* and *partial plan views*, which illustrate a particular area and enlarge it for clarity. Partial views are most often used in areas of high congestion or detail. *Demolition plans* show proposed changes to the existing floor plan. *Roof plans* show the roof layout as would be seen from overhead. Plan views provide dimensions, which are needed to calculate areas. Dimensions should be accurate, clear, and complete, showing both exterior and interior measurements of the space. Plan views are also a starting point from which the architect directs the reader to other drawings for more information. They are also used extensively in other disciplines throughout the drawing set including *structural*, *fire suppression*, *plumbing*, *HVAC*, and *electrical plans*. Each shows the work of the respective trades in plan view as they fit into the architectural floor plan.

Elevations

Elevations provide a pictorial view of the walls of the building, similar to a photograph. They are used to depict both exterior and interior wall surfaces. Building surfaces that are perpendicular to the viewer's line of sight (principal wall surface in an elevation) are true to size and shape and are therefore quantifiable. Angled surfaces or those that are not perpendicular to the line of sight are distorted and cannot be measured.

Exterior elevations may be titled based on their location with respect to the headings of a compass (north, south, east, or west elevation), or their physical location (front, rear, right side, or left side elevation). The scale of the elevation should be noted either in the title block or under the title of the elevation. Interior elevations provide views of the walls on the inside of a building. They illustrate architectural features, such as casework, standing and running trims, fixtures, doors, and windows. Exterior elevations provide a clear depiction of the exterior shape of the building, the doors, windows and exterior features. Doors and windows are often tagged with numbers or letters in circles to show types that correspond to information provided in door and window schedules. In addition, elevations show the surface materials of walls, and any changes within the plane of the elevation or facade. While the floor plan shows measurements in two dimensions along a horizontal plane, elevations provide two dimensional measurements in a vertical plane. These dimensions provide the vertical measure of floor-to-floor heights, windowsill or head heights, floor-to-plate heights, roof heights, ceiling heights, or a variety of dimensions from a fixed horizontal surface. Since elevations are true to size and shape, these measurements can be used to calculate quantities of materials needed.

Building Sections

The building section, is a "vertical slice" or cut-through of a particular part of the building. It offers a view through a part of the structure not found on other drawings. Several different sections may be incorporated into the drawings. Sections taken from a plan view are called *building-sections*; and can be both longitudinal, cut through the building lengthwise, or transverse. Sections taken from an elevation are commonly referred to as *wall sections.* Wall sections provide an exposed view of the materials and construction of the wall and their connection to other horizontal components of the building such as floor plates and roofs. By referring to sections in conjunction with floor plans and elevations, a complete understanding of the size, shape, and construction of a building is possible.

Details

For greater clarification and understanding, certain areas of a floor plan, elevation, section, or other part of the drawing may need to be enlarged. This enlargement provides information that is critical to a part of the building item that may otherwise not be available in another view. Enlargements are drawn to a larger scale and are referred to as *details.* Details can be found either on the sheet where they are first referenced, or grouped together on a separate detail sheet included in the various disciplines they reference. The detail is shown in larger scale to provide additional information and space for dimensions and notes. Details are not limited to architectural drawings, but can be used in structural and site plans and, to a lesser extent, in mechanical or electrical plans.

Schedules

In an effort to keep drawings from becoming cluttered with too much printed information or too many details, architects have devised a system to organize all types of repetitive information in an easy-to-read table, known as a *schedule.* Schedules list information pertaining to a similar group of items, such as doors, windows, room finishes, columns, trusses, and electrical or plumbing fixtures. The most common schedules are door, window, and room finish schedules. However, information of a repetitive nature of any item type can be assembled into a matrix and incorporated in a set of drawings. Schedules are not limited to architectural drawings, but can be found in any discipline included within the set. A typical door schedule lists each door by number, or *mark*, and provides information on size and type, thickness, frame material, composition, and hardware. In addition, the door schedule will provide specific instructions or requirements for an individual door, such as fire ratings, undercutting, or weather-stripping. Door schedules are accompanied by door elevations which indicate panel configurations or size and number of vision panels. In the "remarks" portion of the schedule, the architect lists any non-standard requirements or special notes to the installer.

Diagrams

Some of the information presented in the set of drawings is more diagrammatic than pictorial. A *diagram* illustrates how the various components of a system are configured, and is often provided for purposes of coordination. Diagrams are commonly used for mechanical and electrical drawings, because of the complex nature of the work. Common examples include diagrams for fire alarm risers, waste and vent piping risers, and fire protection.

Drawing Conventions

Certain conventions have been adopted to provide a standard for drawings—from one design firm to another. The most common graphic features are lines, in-fill techniques, and shading, which can often contain subtle, but very important information relative to the detail shown. While most of these conventions are widely accepted and practiced, there are always minor deviations based on local practices. This is most apparent in the use of abbreviations and symbols. In many cases, unfamiliar symbols and abbreviations usually become clear after studying the drawings.

Line Types

Drawings must convey a great deal of information in a relatively small space, where there is no room for a lot of text. Consequently, different types of lines are used to communicate information. The most common ones are discussed below.

Main object line: A thick, heavy, unbroken line that defines the outline of the structure or object. Used for the main outlines of walls, floors, elevations, details, or sections.

Dimension line: A light, fine line with arrowheads or "tic" marks at each end, used to show the measurements of the building or main objects. The arrowheads fall between extension lines that extend from the main object to show the limits of the item drawn. The number that appears within the break or above the dimension line is the required measurement between extension lines.

Extension line: A light line that extends from the edge or end of the main object line, touching the arrowheads. Used together with dimension lines to help determine the limits of a particular feature.

Hidden or invisible line: A light dashed line of equal segments that indicates the outlines of an object hidden from view, under or behind some other part of the structure, such as a foundation shown in elevation that is below grade.

Center line: A light line of alternating long and short segments that indicates the center of a particular object. Frequently labeled with the letter "C" superimposed over the letter "L."

Material Indication Symbols and Shading

In-filling techniques are used on drawings to help convey their content or composition. They indicate the type of material used and are named *material indication symbols*. Because of the various views of drawings, different materials must be recognizable in each view, from plan to section to elevation. As with abbreviations, material indication symbols are subject to change based on specific materials used in various parts of the country.

Shading

Architects and engineers convey information in a subtler manner by changing the intensity of a particular feature. This effect, called *shading*, increases or decreases the focus on the item, merely by its intensity. Items in the foreground or focus are often drawn darker or thicker. Objects in the background are lighter, and drawn less sharply. Shading is often used to differentiate between proposed and existing work on renovation projects. Using grayed-out architectural base drawings is a technique that is often used by various disciplines such as mechanical or electrical to distinguish their specific work while showing it relative to the architectural plans.

Graphic Symbols

Graphic symbols are another means of providing a standardized way to recognize information and depict repetitive information on drawings. *Section markers* indicate where a section is cut through an object, and can be directional or non-directional. *Elevation symbols* on a floor plan direct the reader to the sheet that contains the noted elevation. Another type of elevation symbol indicates differences in vertical height, such as the distance between floors, and provides a reference point used to calculate the height of components in walls or partitions.

Frequently, the design professional draws a feature, and, to save drawing space on the page, uses a *break line*. This symbol conveys that the feature is not

Figure 1.1 Graphic Symbols

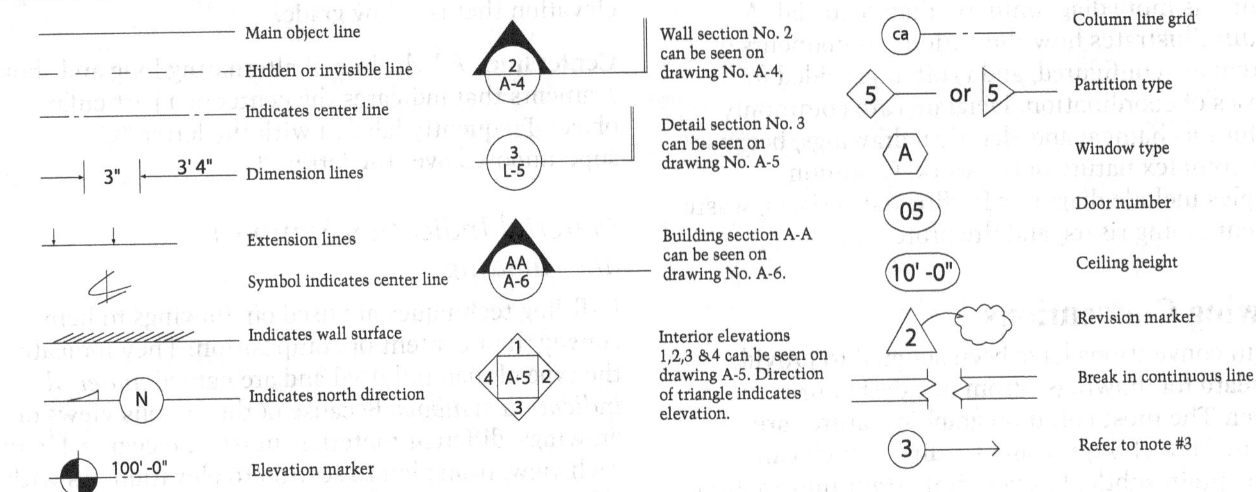

drawn in its entirety. Geometric shapes with letters, numbers, or dimensions within them define certain features or main objects. This graphic symbol is frequently used to name windows, doors, rooms, partition types, and ceiling heights. The important information is within the shape, not the shape itself. Different shapes are often used to differentiate the types of items named.

Trade-Specific Symbols

Like graphic symbols, trade-specific symbols depict items that are common to the various trades. Because of the highly diagrammatic nature of mechanical and electrical drawings, there is an abundance of unique, trade-specific symbols used on these drawings. Engineers typically provide legends that define the symbols used. Some, such as for a water closet or toilet, are highly recognizable because they mirror the feature in real life.

Abbreviations

Abbreviations are used to save design professionals time, as well as space, on drawings. There is a wide and varied selection of abbreviations used in daily practice. It is not necessary to memorize each abbreviation. Standard practice is to list the abbreviations on the cover sheet of the set of drawings. This compilation of abbreviations saves time by locating the meaning of each abbreviation in a central location

Scale

Since there are obvious physical limitations to drawing a building's actual size on a piece of paper, drawings retain their relationship to the actual size of the building using a ratio, or scale, between full size and what is seen on the drawings. There are two major types of scales: the *architect's scale* and the *engineer's scale*. Occasionally, architects and engineers include a detail strictly for visual clarification. These details are labeled "NTS," meaning "Not to Scale." This lets the reader know that the details are not for determining quantities and measurements, but for illustrating a feature that would otherwise be unclear. Diagrams are also typically not drawn to scale.

Architect's Scale

The architect's scale is used for the architectural drawings and the engineering drawings as well. The actual architect's scale may be flat, like a ruler, or three-sided. The three-sided architect's scale has ten separate scales: 1/8″ and 1/4″, 1″ and 1/2″, 3/4″ and 3/8″, 3/16″ and 3/32″, and 1-1/12″ and 3″. The one remaining side is in inches, similar to a ruler. Each noted scale is equal to one foot. For example, when used on a floor plan that is drawn at 1/4″ scale, each 1/4″ increment represents one foot. The same rule applies for 1/8″ scale; each 1/8″ segment on the drawing represents 1′-0″ of actual size. The first foot on each scale is further divided into inches. There is no strict convention that states which scale should be used on which drawings, although generally, floor

Figure 1.2 *Material Indication Symbols*

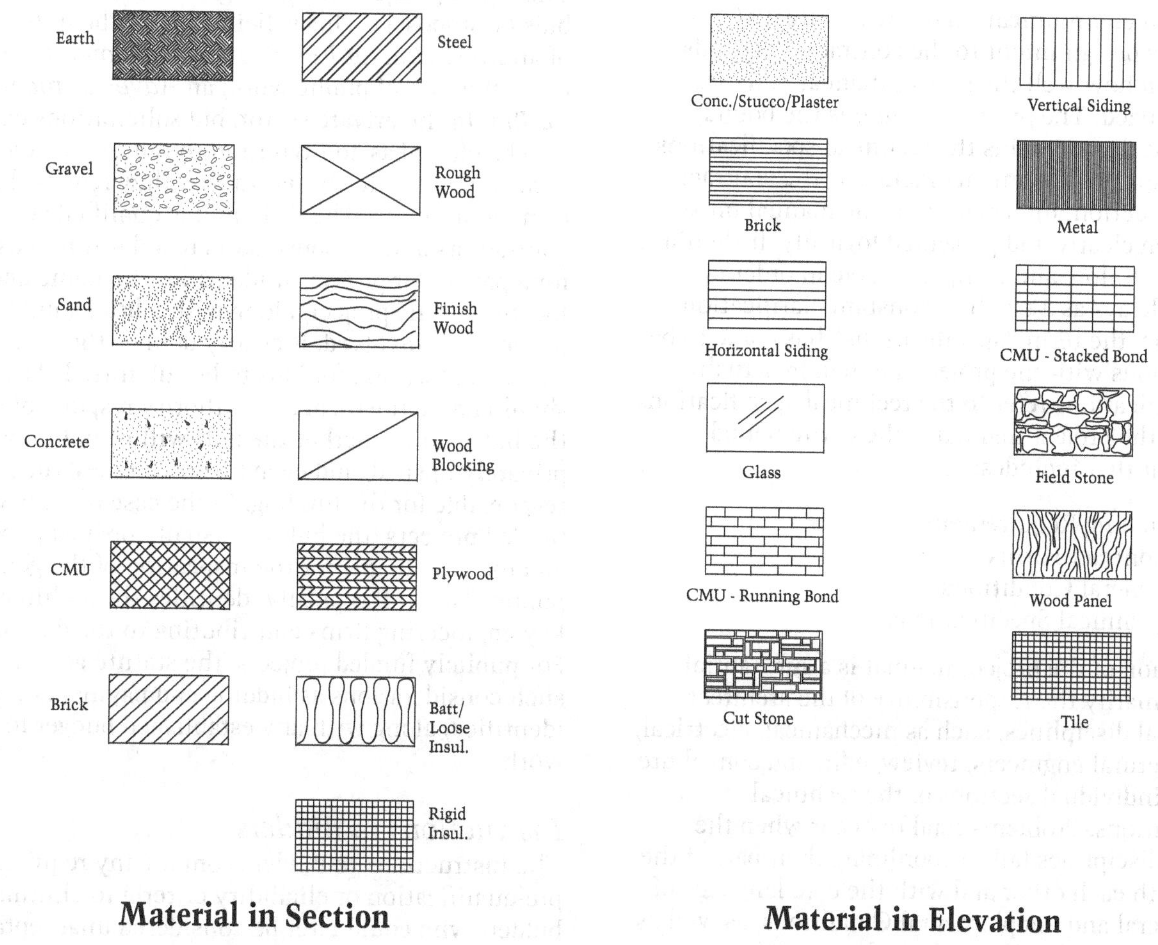

plans are drawn at 1/4″ or 1/8″ scale. Details are better illustrated in 1/2″ or 3/4″ scale for clarity.

Engineer's Scale

The engineer's scale is similar to the architect's scale and is typically (though not exclusively) used to prepare civil drawings. The difference is the size of the increments on the sides of the scale. The engineer's scale has six scales: 10, 20, 30, 40, 50, and 60. For example, the 10 scale refers to 10 feet per inch; the 20 scale is 20 feet per inch, and so on. The engineer's scale is further divided into hundredths of a foot. Other specialty scales are divided into even smaller increments, such as 100. The engineer's scale is used to measure distance on site plans, when it is greater than would be encountered in the plans of the building

Working Drawing Review

At this point we have reviewed the different types of plans and drawing elements that together comprise a full set of working drawings. It is essential to become familiar with the drawings prior to the site inspection and quantity takeoff. A thorough review of the drawings may reveal discrepancies or omissions and will help formulate questions posed to the architect before starting the quantity takeoff. It should also be noted that the various views should be used together. Information located on one drawing can often be corroborated on another. This checks and balances process is fundamental in estimating.

The Project Manual

Successful communication of the architect's/engineer's design intent to the contractor depends heavily on how well the project manual is written and organized. The *project manual* is the bound document that contains the technical specifications and the upfront documents including the contract for construction. Information in the manual must be written clearly and presented logically. It should be easy to follow and comprehensive in order to prevent delays as a result of constant clarification. Over time, the term "specifications" has come to be synonymous with the project manual. In actuality, the specifications refer to the technical specifications, whereas the project manual is the entire bound document that includes:

- Bidding Requirements
- Contract Forms
- General Conditions
- Technical Specifications

Preparation of the project manual is a substantial task, primarily the responsibility of the architect. Individual disciplines, such as mechanical, electrical, and structural engineers, review, edit, and contribute to their individual sections of the technical specifications. Problems tend to occur when the various disciplines fail to coordinate their part of the work with each other and with the core language of the General and Supplemental Conditions, as well as with the drawings. The technical writer must create a complete document using very specific language that will guide the contractor in the bidding and building processes, and will also serve as a powerful tool to enforce the contract. It takes a skillful use of language, a high level of proficiency in understanding and coordinating technical information, and the ability to process that data into usable information. Over the next several pages, we will review the structure of the project manual and its four main components.

Bidding Requirements

The Bidding Requirements are composed of the following items:

- Bid Solicitation
- Instructions to Bidders
- Information Available to Bidders
- Bid Forms and Supplements

Bid Solicitation

The bidding requirements begin with a solicitation for bids or proposals. This solicitation can be in the form of an *Invitation for Bid, Request for Proposals (RFP)*, or, in the case of public work, an *Advertisement for Bid*. In the private sector, bid solicitations can also be offered as an Invitation to Bid to selected firms only. All are similar in that they request bids from contractors. The RFP invites qualified general contractors and subcontractors to submit proposals for a particular project. It identifies the name and location of the project, along with a brief summary of the work involved. It clearly defines the date, time, and location for bids to be submitted. The RFP should name the owner or authority responsible for the bid award, whether the bids will be publicly or privately opened, and even (in some cases) the lender responsible for the funding. In the case of taxpayer-funded projects, the bids are usually opened publicly and made available for the inspection of the general public. The RFP typically identifies the architect and key engineering firms contributing to the design. For publicly funded projects, the statute governing such considerations as bidding and payments is also identified, along with any established budget for the work.

Instructions to Bidders

The Instructions to Bidders contain any required pre-qualification or eligibility criteria to eliminate bidders who could later be considered unacceptable. In the case of private bidding, the Invitation to Bid may be all that is required. In some states, publicly bid projects require formal qualification forms and a summary of the contractor's performance record. If a pre-bid conference or site inspection is scheduled, the date, time, and location are also stated. Additionally, the Instructions to Bidders defines the various forms and amount of bid security or bid bond that will be required. It states any liquidated damages that may be part of the contract, times for the commencement and completion of the work, and addenda or rules governing interpretation of the documents.

The Instructions to Bidders portion of the Bidding Requirements indicate the date, time, and location for procuring a set of contract documents and the cost, if any, to bidders. Other pertinent information, such as the time frame for award or rejection, special wage rates, tax-exempt status, or legal rights of the awarding authority to accept or reject proposals, is also provided.

Information Available to Bidders

The Information Available to Bidders provides locations where bidders can obtain copies of additional documents helpful in the bidding process. These documents could include geotechnical reports or subsurface investigation, property surveys and record drawings, conservation commission reports or directives, and hazardous materials management reports.

Bid Forms and Supplements

This section contains the forms developed by the architect for use by the contractors submitting bids, as well as bid security forms. Bid forms are used to keep proposals uniform in appearance and content. They provide the owner and architect with a mechanism for comparing "apples to apples." Bid forms provide the language of the proposal with blanks for the contractor to fill in. Space is provided to acknowledge addenda; add or deduct alternates; enter unit prices; record the name, address, and signatory party of the bidding contractor; and write the dollar amount. "Non-responsive" is the term applied to a bidder who has incorrectly filled out or inadvertently left out information on the bid form, thereby rendering that bidder ineligible for award.

Contract Forms

The most important contract form is the agreement between the owner and contractor, more commonly referred to as the *Contract for Construction*. This is a legal instrument supported by all of the contract documents. The Contract for Construction must contain the following basic items in order for it to be considered a functional document:

- Clear identification of the parties to the agreement
- Clear identification of the project
- Rights and responsibilities of each party
- Basis and terms of compensation

This agreement is incorporated within the project manual so that prospective bidders can carefully review the contract that will be executed when the project is awarded.

Performance and Payment Bonds

Other contract forms include performance and payment bond forms. A *performance and payment bond* is a type of contractual guarantee offered by a contractor to the owner of a property for a specific project that the contractor is willing to do. The bond ensures that the contractor will complete the project as specified, or face serious default penalties. Many organizations, including the government, require performance bonds when they choose a contractor to work on projects.

Certificates

The last section in the contract forms section of the specifications contains forms for insurance required for the project. This document defines the dollar limits for the various policies required.

General Conditions of the Contract

Although the Contract for Construction is the primary legal instrument in the project manual, it is insufficient on its own. Because of its complexity, a separate set of guidelines is necessary, called the *General Conditions of the Contract for Construction*. The General Conditions are meant to complement the Contract for Construction, defining the complex relationships between the owner, architect, and contractor and the mutual responsibilities and rights of the signatory parties. The General Conditions include the definitions of key terms and provide procedures and mechanisms for resolving disputes or clarifying information provided on the drawings or specifications.

Supplements to the General Conditions of the Contract

As noted previously, the General Conditions of the Contract address specific issues (in a general format) that could be considered applicable to the industry as a whole. Often projects have specific needs or unique conditions that require an amendment to the General Conditions. Because the General Conditions document is intended to interface with a whole series of other legal documents, any modifications can have serious legal ramifications. For this reason, a separate document is added, called the *Supplemental Conditions of the Contract*, or the *Supplementary General Conditions*. This custom-tailored document allows the author of the project manual great flexibility in meeting the specific needs of the individual client or project without risking the loss of continuity that the General Conditions provide. They are often presented in a way that a dollar value can be established against their impact. A classic example is insurance requirements. While the General Conditions describe the type and extent of insurance coverage, the Supplementary Conditions establish its limits. Using this information, you can establish the increased insurance policy dollars and thereby include the difference in the appropriate category of the estimate.

Technical Specifications

The last of the four categories is called the *Technical Specifications*, which define the scope, products, and execution of the work. This is the "meat and potatoes" section, providing the estimator with the necessary information (in a highly organized and industry-accepted format) to accurately price and build the structure. The technical sections provide the following information for each activity:

- Administrative (submittal) requirements
- Quality or governing industry standards
- Products and accessories
- Installation or application procedures
- Workmanship requirements

The specifications, or specs, as they are commonly referred to, are part of the contract documents, along with the working drawings. They are written documents included in the project manual that define in detail the processes and materials of the project. While the drawings define "how much and how many," the technical specs define the quality of materials and workmanship. For most light commercial and many upscale residential projects, working drawings are issued with a separate set of specifications. Even the simplest projects have some specifications, whether incorporated on the drawings or issued as a separate document to guide the contractor and subcontractors.

As owners become more informed and technically savvy, they are no longer satisfied by the term, "industry standard" when defining the quality of materials or workmanship to be included in a project. As a result, many contractors, especially in the high-end residential market, use technical specifications to establish the quality level for owners and as a guideline for subcontractors. Over the last decade, in fact, specifications have become increasingly popular as *the* standard of measurement for quality.

The specs perform a variety of functions, including:

- Defining the quality or grade of materials to be used in the project
- Defining the acceptable workmanship or providing standards to judge workmanship
- Providing a basis for accurately estimating cost
- Complementing the graphic portion of the project, the drawings

The plans and specifications are of equal importance to the contract documents, each documenting different aspects of the project. The specs are intended to be used in conjunction with the drawings. If the drawings are the *quantitative* representation of the project shown in a graphic format, then the specs are the *qualitative* requirements of the project described in a written document. Technologies, processes, and products are continually evolving in the construction industry, and architects and engineers incorporate these advancements more frequently into their designs. As a result, highly specific information is needed in the specifications. The materials and processes are described in such detail that the intent of the designer, as well as the product or system, can be upheld in case of a dispute or if products are installed incorrectly.

The specifications serve as a basis for bidding and performing the work. The person preparing the specifications, sometimes called a *specification or technical writer*, makes every effort to cover all of the items or segments of work shown on the working drawings. In the past, if there was a discrepancy between the specifications and the drawings, the specifications generally took precedence. This is no longer always the case. Many specifications now state that when there is a discrepancy between the plans and specifications, whichever results in the greater quantity, is more expensive, or is of greater benefit to the project will supersede.

Organization of Specifications

CSI MasterFormat is the most widely accepted system for arranging construction specifications and estimates. Developed by the Construction Specifications Institute, the MasterFormat system is also used for classifying data and organizing manufacturers' literature for construction products and services.

Division 01 — General Requirements
Division 02 — Existing Conditions
Division 03 — Concrete
Division 04 — Masonry
Division 05 — Metals
Division 06 — Wood, Plastics, and Composites
Division 07 — Thermal and Moisture Protection
Division 08 — Openings

Division 09 — Finishes
Division 10 — Specialties
Division 11 — Equipment
Division 12 — Furnishings
Division 13 — Special Construction
Division 14 — Conveying Equipment
Division 21 — Fire Suppression
Division 22 — Plumbing
Division 23 — Heating Ventilation and Air Conditioning
Division 25 — Integrated Automation
Division 26 — Electrical
Division 27 — Communications
Division 28 — Electronic Safety and Security
Division 31 — Earthwork
Division 32 — Exterior Improvements
Division 33 — Utilities
Division 34 — Transportation
Division 35 — Waterway and Marine
Division 40 — Process Integration
Division 41 — Material Processing and Handling Equipment
Division 42 — Process Heating, Cooling, and Drying Equipment
Division 43 — Process Gas and Liquid Handling, Purification and Storage Equipment
Division 44 — Pollution and Waste Control Equipment
Division 45 — Industry-Specific Manufacturing Equipment
Division 46 — Water and Wastewater Equipment
Division 48 — Electrical Power Generation

Format of Individual Specification Sections

In addition to the overall organization of the specifications, each spec section follows a specific format. The three-part format of each technical section provides a consistent organizational system for locating pertinent information quickly and efficiently:

Part 1 — General
Part 2 — Products
Part 3 — Execution

Part 1, General

Part 1, the general section of the specifications, provides a summary of the work included within that particular section. It ties the technical section to the General Conditions and Supplementary General Conditions of the Contract, an essential feature in maintaining continuity between the general contractor and subcontractors. Part 1 identifies the applicable agencies or organizations by which quality assurance will be measured. It defines the scope of work that will be governed by this technical section including, but not limited to, items to be furnished by this section only, or furnished by others and installed under this section. It also identifies other technical sections that have potential coordination requirements with this section, and defines the required submittals or shop drawings for the scope of work described in this section. Part 1 also establishes critical procedures for the care, handling, and protection of work within this section, including such ambient conditions as temperature and humidity. If applicable, it addresses inspection or testing services required for this scope of work.

Part 2, Products

Part 2 deals exclusively with the products and materials to be incorporated within this technical section of the work. For products that are directly purchased by the contractor from a manufacturer or supplier, the items can be identified using one of three methods:

- Proprietary Specification
- Performance Specification
- Descriptive Specification

Proprietary Specifications: These specifications spell out a product by name and model number. Proprietary specifications have the unique advantage of allowing the architect or owner to select a product they desire or have used successfully on prior projects. The advantage of requiring specific products is the level of reliability they provide. The disadvantage is that they eliminate open competition.

Performance Specifications: An alternate method of specifying products and materials is based less on makes and models and more on the ability to satisfy a design requirement or perform a specific function. This type of specifying is called a *performance specification*. In lieu of specifying a particular product by name, the architect opens competition to all products or materials that can perform the specific functions required to complete the design. This approach allows healthy competition among various manufacturers that have a similar line of products. It ensures competitive pricing and more aggressive delivery schedules. Performance specs can identify products by characteristics, such as size, shape, color, durability, longevity, resistivity, and an entire host of other requirements. For a specification section that involves custom-fabricated work, the language might be a mixture of proprietary and performance specifications.

Descriptive Specifications: The last method of specifying a product or process is by using *descriptive specifications*, which are written instructions or details for assembling various components to comprise a system or assembly. Most often, descriptive specifying is used for generic products such as mortar or concrete. Frequently, no manufacturers' or proprietary names are mentioned or needed.

Part 3, Execution

Part 3, called the *execution*, deals exclusively with the method, techniques, and quality of the workmanship. This section makes clear the allowable tolerances of the workmanship. The term "tolerances" refers to plumb, straight, level, or true. The Execution section should also describe any required preparation to the existing surfaces in order to accommodate the new work, as well as a particular technique or method for executing the work. This method or technique should be considered when developing an estimate since various work tasks require more or less labor and will ultimately affect the cost of the project. Part 3 also addresses issues such as fine-tuning or adjustments to the work after initial installation, general cleanup of the debris generated, final cleaning, and protection of the work once it is in place. Some sections of Part 3 may identify any ancillary equipment or special tools required to perform the work, such as staging or scaffolding.

Modifications to the Contract Documents

Addenda

The bidding process often produces questions that require clarification from the architect or engineer. Any changes to the contract documents made during the bidding period (the time period beginning on the date the drawings are issued and ending on bid day) in the form of modifications, clarifications, or revisions, are called *addenda*. Addenda, or an addendum (singular), can be issued only by the architect. Addenda must be issued in writing and will automatically become part of the contract documents, complete with all of the benefits of the Contract for Construction and the General Conditions of the Contract. Addenda should, at a minimum, contain the following information:

- Number of the addendum and date of issue
- Name and address of the architect and/or engineer
- Project name and location
- Bidders' names to whom the addendum is addressed
- Contract documents that are to be modified
- Explanation of the addendum's purpose

Bid forms often include an area for bidders to acknowledge addenda. Failure to do so could render the bid non-responsive. Since each addendum affects the bid price, each one should be carefully evaluated. Since addenda can affect the bids of all parties involved, subcontractors and materials suppliers should be made aware of any addenda, so they can adjust their bids accordingly.

Alternates

Often, owners want to see how a change in materials, method of construction, or addition or subtraction of work will affect the project's price. This information is presented in the form of additions or deletions to the base price, called *alternates*. Typically, the alternate is listed at the end of the specification section that is affected by it, and also in Division 1 under the section, "Alternates." Any increase or decrease in the cost for the work, including all taxes, labor burden, and overhead—both direct and indirect costs and profit should be included.

> For example, the total consequence of an alternate might look like: *Alternate 1: Delete door frame and hardware for Door #3 in its entirety. DELETE $500.*

In this example, you would not only delete the cost of the door, frame, and hardware, but must also include the additional wall materials, wood, gypsum board, paint, etc., to fill in the area Door #3 originally occupied in the base bid. The actual price of the alternate is the difference between the two. In some cases, the addition or deletion of large scopes of work by alternates can have a tremendous effect on the project's duration, thereby increasing or decreasing overhead and other time-sensitive costs of the project. Projects with limited budgets often include a series of alternates as a way of choosing how to most effectively use the budget.

Allowances

Occasionally, as the contract documents are ready to be issued, certain items have yet to be finalized and are not ready for inclusion in the bid set. Rather than leaving the item out altogether, the designer includes a cash allowance. The allowance is a fixed lump sum, such as "$10,000 for the purchase and delivery of sod and plantings." The allowance can also be in the form of a unit price, such as, "an allowance of $450 per M (thousand) for brick, including delivery to the job site." Typically, it is clearly stated what the allowance is for: materials, furnished and delivered only; materials and labor; or the entire scope of work. If there is any doubt, request clarification. At the completion of the project, the actual cost is computed for items included as allowances, savings are returned to the owner, and overages are added to the contract price. Typically, allowances are listed in Division 1 under the section, "Allowances."

Unit Prices

In the course of design for some projects, architects or engineers are sometimes unable to provide sufficient detail to the drawings so that the estimator can determine an exact quantity of a certain task or activity. An example of this is excavation of rock or unsuitable fill materials. The architect or engineer may be aware of what needs to be done and the techniques or quality required, but is unable to determine the exact amount of rock or materials to be removed. In an effort to at least establish the cost of this work for post-bid purposes, unit prices are requested and submitted as part of the bid form or proposal. Unit prices are included on the bid for each item by a unit of measure, such as excavation of unsuitable materials at $45 per CY. The unit price should always include markups for taxes, insurance, overhead, and profit.

Frequently, the unit price may be tiered-based on stipulated quantities since unit prices tend to decrease as the quantity increases. For example:

Excavation of unsuitable materials $75 per CY for 1 to 50 CY quantities

Excavation of unsuitable materials $65 per CY for 51 to 200 CY quantities

Excavation of unsuitable materials $45 per CY for 201 CY and over quantities

Conclusion

As you can see, all aspects of the contract documents affect the cost of a building project. Once the plans, specifications, and addenda have been documented and, if appropriate, a site visit has been conducted, then the quantity survey or takeoff can begin. (This process will be discussed in chapter 3.) Prior to beginning the takeoff, it may be helpful to review basic area and volume calculations, covered in chapter 2 since performing an accurate quantity takeoff requires careful area calculations and consistent use of units.

Chapter 1 Exercises

1. Why is it important to study the contract documents prior to preparing a bid?
2. Why are scales used in drawings, and what two types are commonly used in working drawing sets?
3. An abbreviation commonly found in working drawings is NTS. What does it mean and where do you find definitions for abbreviations used in the working drawing set?
4. On which drawing type would you typically find floor to floor heights?
5. What is a schedule? What type of information is commonly depicted in one?
6. Which scale is commonly used on a site plan? Why?
7. What is a performance and payment bond? Why is it important information for an estimator to know while preparing a bid?
8. What are the technical specifications?
9. Addenda are often part of a bid package. What are they and why should an estimator familiarize him/herself with them before preparing a bid?

10. How are the technical specifications organized?
11. Define the term, unit price. Where are unit prices typically found in a construction document set?
12. Graphic conventions are used in drawings to convey specific information about a building. What does a hidden line represent?
13. What is an allowance? How is it used in a construction document set?
14. What is an alternate? Provide an example of how it is used in a construction document set.
15. What is the graphic convention used to indicate a revision to a drawing? Why is this important for an estimator to know?

Chapter 2

Calculating Linear Measure, Area & Volume

To perform even the most basic quantity takeoff, estimators should be well-versed in the calculations of linear measurements, area, and volume. This chapter will review the basic formulas and relationships needed to perform these calculations. The formulas are fairly simple, and it is not necessary to memorize them. However, it is essential to know which formula to use in the proper application, and where it can be found.

Units of Measure

The most fundamental rule is to use the correct units of measure for area and volume. Area is always expressed in square units, most often square feet (SF) or square yards (SY). Volume is always in cubic units, the most common of which are cubic feet (CF) or cubic yards (CY). Another important point to remember is to be consistent and keep the units the same. For example, feet multiplied by feet results in square feet, yards multiplied by yards results in square yards, and so on. Multiplying a dimension in feet by a dimension in inches leads to an erroneous value. It is common to find different dimensions used in various parts of the drawings. Be sure to convert the dimensions given into the same units. Often, a dimension on the drawings is given in both feet and inches.

Decimal Equivalents

To calculate the area of a space that is 24′-6″ × 20′-3″, the dimensions must be converted to their decimal equivalents. The feet-and-inches dimensions are changed to feet in order to arrive at a measurement in square feet for the area. This is a fairly simple process to convert the inches portion of the dimension to its decimal form and then add it to the whole number. Once this has been done, the values can be easily entered into a calculator.

To convert inches to decimals of a foot, divide the inches by 12 (12 inches in 1 foot) Two-place accuracy after the decimal point is sufficient for estimating purposes in most cases. (If the decimal form is 0.166666 of a foot, the value is rounded up to 0.17.) 24′-6″ becomes 24.5′ and 20′-3″ becomes 20.25′. The area is 496.125 SF or 496.13 SF.

Round up total quantities in a takeoff. For example, if the total quantity of concrete in a takeoff is 34.22 CY, round to 35 CY for use in the cost analysis portion of the estimate.

To calculate the area of a room that is 24′-6″ × 20′-3-5/8″, first determine that the decimal equivalent is 24.5′ × 20.30′, which equals 497.35 SF. Decimal equivalents can be found using a calculator. Decimal equivalents of fractions of an inch are as easy as pushing a few buttons. For example: 3-1/2″ = 3.5″; 3.5″ divided by 12″/ft. = 0.29′. The remainder of this chapter will be divided into three sections:

- Linear Measurement: the measurement of lines in a single dimension.
- Area: the measurement of surfaces in two dimensions—length by width.
- Volume: the measurement in three dimensions—length, width, and height or thickness.

Linear Measurement

Perimeter

If, for a moment, we imagine the floor of a building as a simple planar surface with no depth, the sum of the sides of that planar surface is called its perimeter. Since perimeter is a linear measurement, and the dimensions of the sides are added, its units are also linear, most often LF. The perimeter of a rectangular surface can be found by adding the length to the width and multiplying it by 2, or by adding the length and width of all the sides. The perimeter of a circle or circumference can be found using the formula: $C = \pi D$, where C = circumference, $\pi = 3.14$ and D = diameter.

Knowing how to find the perimeter is helpful in determining the length of various items you would find in a building, such as baseboard and other running trims within various rooms.

Angles

If a rectangle with four 90° corners was divided in half by a line connecting two opposite corners, the resulting shape would be a right triangle. A right triangle has three angles that total 180°. It also has three sides: a base, an altitude, and a hypotenuse, the diagonal line connecting the end point of the base to the end point of the altitude. Calculating the length of the hypotenuse is helpful in determining the length of rafters, stair stringers, and the like. To do so, the *Pythagorean theorem* is used. The square of the hypotenuse of a right triangle is equal to the sum of the squares of the other two sides. (*See Figure 2.2.*)

Converting this to a formula: $C^2 = A^2 + B^2$, where C = length of the hypotenuse, A = length of the altitude, and B = length of the base. Using this formula, we can determine the length of a side of a right triangle, provided the lengths of the other two sides are known.

There are three types of triangles that frequently occur in a set of construction drawings. An equilateral triangle has three sides that are equal in length. An isosceles triangle has two sides that are of equal length. In a scalene triangle, none of the sides are equal in length.

Figure 2.1 *How to Calculate Rafter Length and Roof Area*

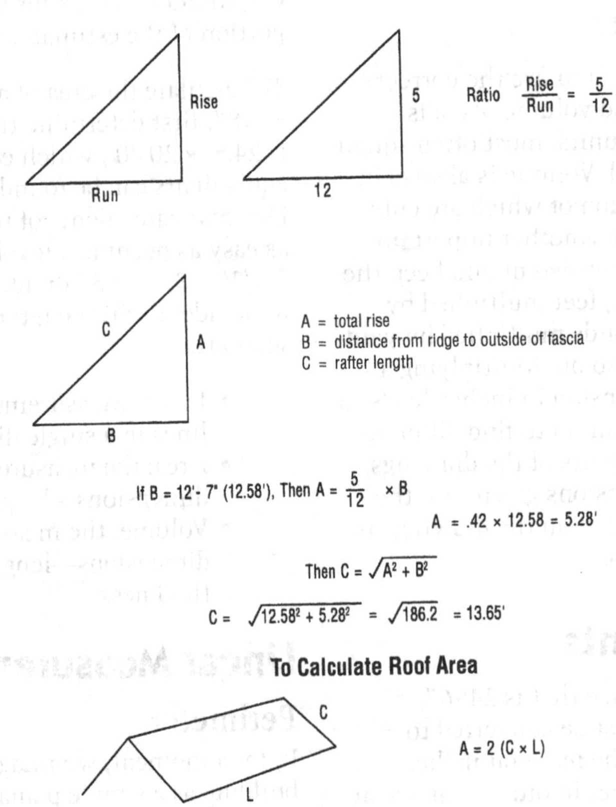

Figure 2.2 *Right Triangle*

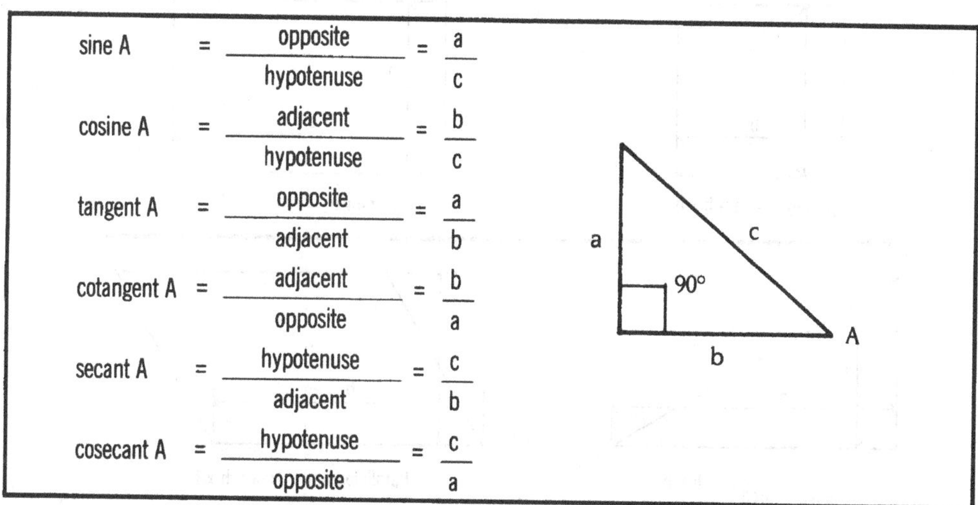

By using these relationships and a Table of Natural Trigonometric Functions, unknown angles and sides of right triangles may be found.

Area & Square Measure

Possibly the most common calculation performed by estimators is determining the area of a shape. The most common shape in the construction business is some variation of the rectangle. A rectangle, by definition, has four sides, with all angles equal to 90°. If all four sides are equal in length, it is a square. A rectangle with opposite sides only that are equal in length and parallel is a parallelogram (the angles are not 90°). A trapezoid has two opposing sides that are parallel, but not of equal length.

When no sides are equal in length, and none are parallel, the shape is a trapezium. (See Figure 2.4.)

Area calculations take into account only the surface, not the depth. Areas are expressed in square units—most commonly square feet, square yards, or square inches. The area of a rectangle or square is defined as the product of its length and width. The formula for area is:

> $A_R = L \times W$, where A_R = area of a rectangle, L = length, and W = width.

The conversion from square feet (SF) to square yards (SY): 1 SY=9 SF (3′ × 3′).

Since a right triangle is essentially a bisected rectangle, the formula for the area of a rectangle could be modified for a triangle:

> $A_T = (\frac{1}{2} B) \times A$, where A_T = area of a triangle, B = length of the base, and A = length of the altitude. This formula requires that the angle between the base and the altitude be 90°.

Irregular Shapes

Estimators may be required to calculate the areas of more complex polygons and other irregular shapes. Determining the area of a construction feature is not always one simple area calculation, and sometimes it requires additional calculations. Frequently, the area to be quantified is calculated by dividing it into several smaller areas, calculating each, then adding the results back together to arrive at the total area. (Remember to break odd-shaped features into recognizable rectangles and triangles.) The same holds true when determining a smaller portion of the whole. In this case, deduct all unwanted areas until the desired area is achieved. For most construction applications, a close approximation of the area of an irregular shape is sufficient.

Figure 2.3 Area Calculations, Regular Shapes

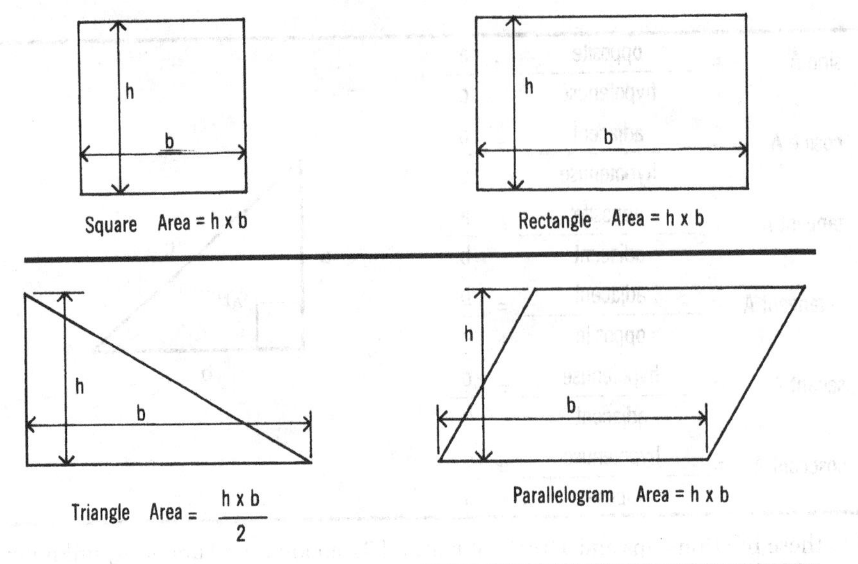

Figure 2.4 Area Calculations, Irregular Shapes

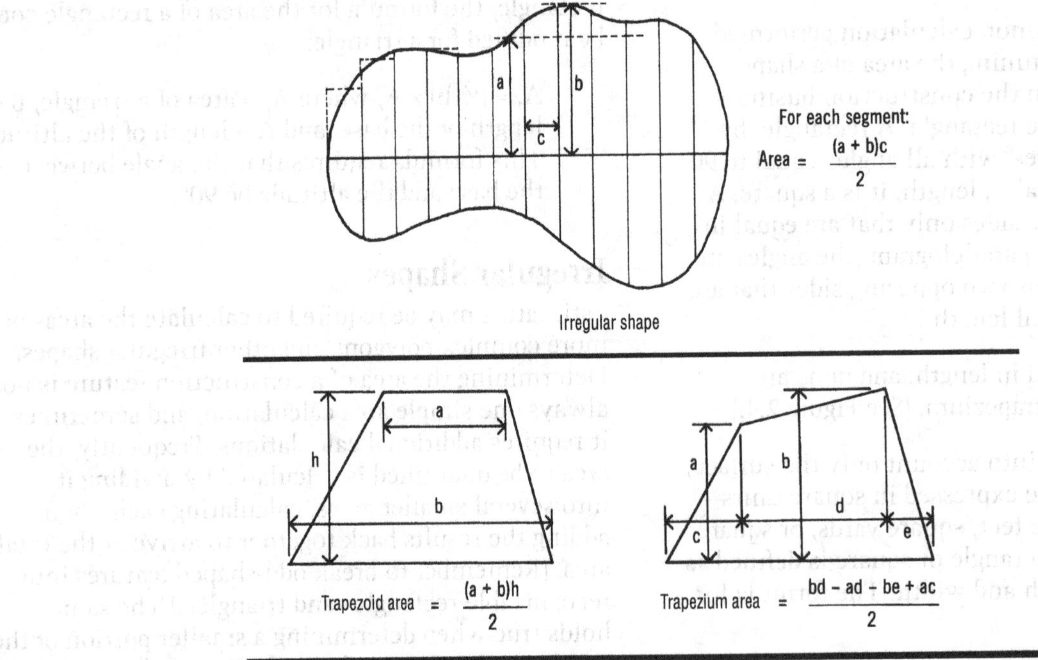

Area of a Circular Shape

Many construction elements are circular, such as brick patios and concrete piers. Before reviewing the formula for the area of a circle, it will be helpful to define various parts and some constant relationships.

The *circumference* is the perimeter of the circle. The *diameter* is a line drawn through the center of the circle, beginning and ending on the circumference. Any number of possible diameters drawn on a circle should render the two halves of that circle equal. All diameters of the same circle are also equal.

The *radius* of a circle is a line from the center point within a circle to a point on the circumference. All radii of the same circle are equal in length. The radius, by definition, is equal to one-half of the diameter. Knowing this relationship, we can establish the following formulas:

$D = 2 \times R$ or $R = 1/2 \times D$, where D = diameter and R = radius.

A constant is a number that expresses a relationship in a mathematical formula. The circumference of a circle has a constant relationship to the diameter of the same circle. That constant is the number 3.1416, or *pi*, which has a corresponding symbol, π. For most calculations in construction, π can be truncated at two places after the decimal point, or 3.14. If the area to be calculated is very large, π can be extended to a third decimal place, 3.142. The formula for the relationship between the circumference and the diameter is as follows:

$C = \pi \times D$ or $C = 2 \times \pi \times R$, where C = circumference, $\pi = 3.14$, D = diameter and R = radius.

A *chord* is a straight line connecting two points on the circumference, without passing through the center of the circle. An arc is any portion of the circumference of the circle. A circle has 360°. Therefore, if the radius and the interior angle between two radii are known, the length of the arc between them can be calculated using the formula:

$L_A = N/360 \times 2 \times \pi \times R$, where N is the central angle, $\pi = 3.14$, and R = radius.

The *tangent* of a circle is a straight line touching only one point on the circumference. A radius drawn to this point is at 90° to the tangent. The area of a circle is the radius multiplied by the radius, then multiplied by π. As a formula:

$A_c = \pi \times (R \times R)$, or $A_c = \pi \times R^2$.

An alternative method for calculating the area of a circle is to multiply the diameter by itself, then multiply the resultant area by the constant 0.7854.

Expressed as a formula:

$A_C = (D \times D) \times 0.7854$, or $A_C = D^2 \times 0.7854$, where D = diameter and 0.7854 is a constant.

It is also possible to calculate the area of a portion of a circle. If we cut a pie-shaped piece out of a circle—two radii with an angle in between, with a known radius and known angle between—we can calculate that area, which is called a *sector*. The length of the arc is a fraction of the total circumference. A similar deduction can be used to devise a formula for the area of the sector:

$A_S = N/360 \times \pi \times R^2$, where A_S = area of a sector, N = angle between the radii in degrees, $\pi = 3.14$, and R = radius.

Figure 2.5 *Parts of a Circle*

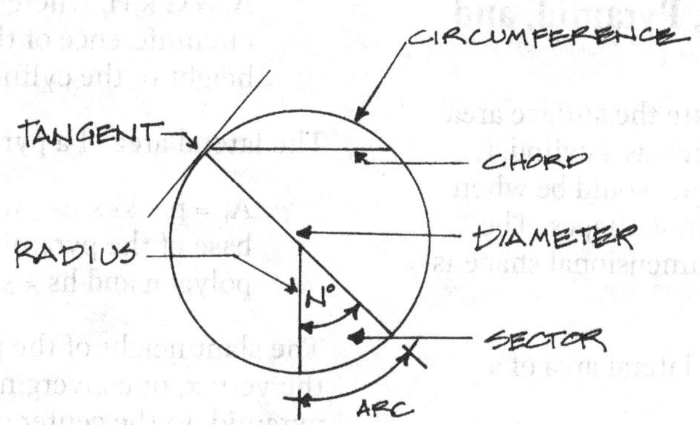

Figure 2.6 Area Calculations, Circular Shapes

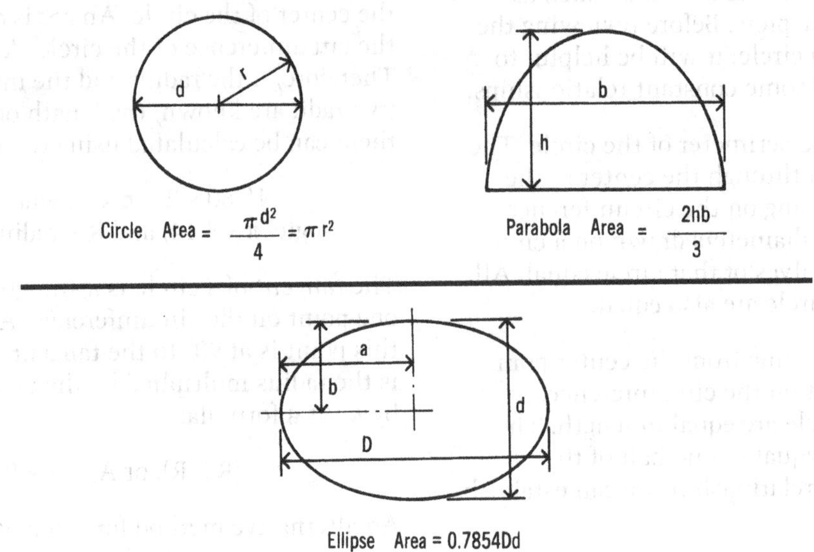

Figure 2.7 Lateral Area Diagram of Pyramid and Cone

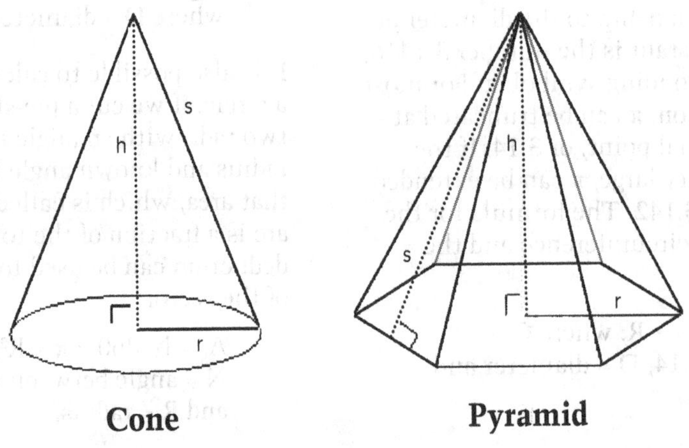

Surface Area of Cylinder, Pyramid, and Cone Shapes

Estimators often need to calculate the surface area of a three-dimensional shape, such as a cylinder, pyramid, or cone. A good example would be when estimating painting for one of these shapes. The outside surface area of a three-dimensional shape is referred to as its lateral area.

The formula for calculating the lateral area of a cylinder is:

> $A_L = C \times H$, where A_L = lateral area, C = circumference of the cylinder base and H = height of the cylinder.

The lateral area of a pyramid can be expressed as:

> $A_L = p \times \frac{1}{2} \times hs$, where p = perimeter of the base of the pyramid where the base is a regular polygon and hs = slant height of the pyramid.

The slant height of the pyramid is a line drawn from the vertex, or converging point at the top of the pyramid, to the center of any one side of the base. The lateral area of a cone is the area of its tapering side. It can be expressed as:

$A_L = C_B \times ½ \times hs$, where C_B = circumference of the base of the cone and hs = slant height of the cone.

The slant height of the cone is a straight line drawn from the vertex of the cone to the circumference. It can be calculated by solving for the hypotenuse in the Pythagorean theorem, where the altitude of the cone is its height perpendicular to the base, and the radius of the base of the cone is its base.

Volume & Cubic Measure

In contrast to area, which has only two dimensions, volume has a third dimension, depth. The depth of a shape can also be called its thickness or height. Once a shape takes on this third dimension, it is no longer planar, but becomes a solid. The term cubic refers to the volume of a solid, whereas square accounts only for its surface area. The standard units of cubic measure are cubic inches, cubic feet, and cubic yards, with the abbreviations CI, CF, and CY, respectively. There are numerous tasks encountered in construction estimating that require volume calculations. A few examples include excavation, backfill, and concrete for a form.

Volume of a Prism

If we visualize a rectangle as a surface area with a height, it would be called a prism. Prisms in construction are virtually everywhere. Examples of volume calculations of prisms that might be required include pile caps, footings, and excavations. If a prism's dimensions of length, width, and height are the same, it is a cube. This is defined in the conversion of CF to cubic yards CY , where 1 CY = 27 CF (3′ × 3′ × 3′).

If we modify the formula for the area of the planar surface by adding the new dimension, the result is the formula for the volume of a prism.

$V = A \times h$, where V = volume of the prism, A = area of the base, and h = height.

This formula applies only to shapes whose ends and opposite sides are parallel. To further expand this formula:

$V = l \times w \times h$, where l = length, w = width, and h = height.

Contractors may encounter an endless number of shapes that are variations of a prism. A triangular prism has a triangular surface area and a height. The rule of base area multiplied by height still applies:

$V = 1/2 \times l \times w \times h$, where again l = length, w = width, and h = height.

In short, it shares the same relationship that the area of a triangle has with the area of a rectangle. The volume of a triangular shape is one-half of the volume of a prism.

Volume of a Cylinder

A common, yet more sophisticated, shape is the cylinder. The formula for its volume is essential in calculating the amount of concrete in a pier or a round column form. The volume is the area of its circular base multiplied by its height. Expressed as a formula:

$V_C = \pi \times R_2 \times h$, where V_C = volume of a cylinder, $\pi = 3.14$, R = radius, and h = height.

Volume of a Pyramid

The volume of a pyramid is calculated as:

$V_P = 1/3 \times A_B \times a$, where A_B = area of the base of the pyramid provided that the base is a regular polygon, and a = the altitude of the pyramid as defined by a line drawn at 90° to the base to the vertex of the pyramid.

Volume of a Cone

Volume of a cone is very similar to that of a pyramid. It is:

$V_C = 1/3 \times A_C \times a$, where A_C = area of the round base of the cone and a = the altitude of the cone as defined by a line drawn at 90° to the base to the vertex of the cone.

Board Measure

Dimensional lumber has its own system of measurement called *board foot (BF)*. One board foot is equal to the volume of a piece of wood 1″ thick by 1′ square. When calculating BF, nominal dimensions are used. Nominal dimensions refer to the named dimension rather than the actual dressed value of the lumber. For example, a 2 × 4 piece of lumber, when dressed or planed smooth from the mill is actually

Figure 2.8 Volume Calculations

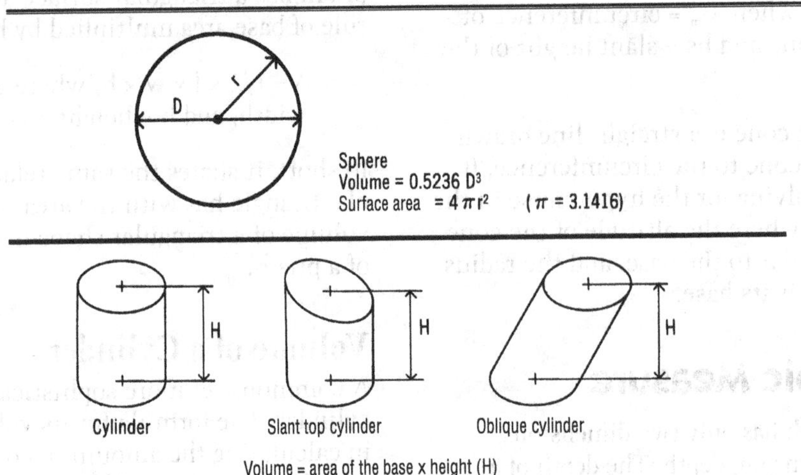

Figure 2.9 How to Calculate Board Foot Measure

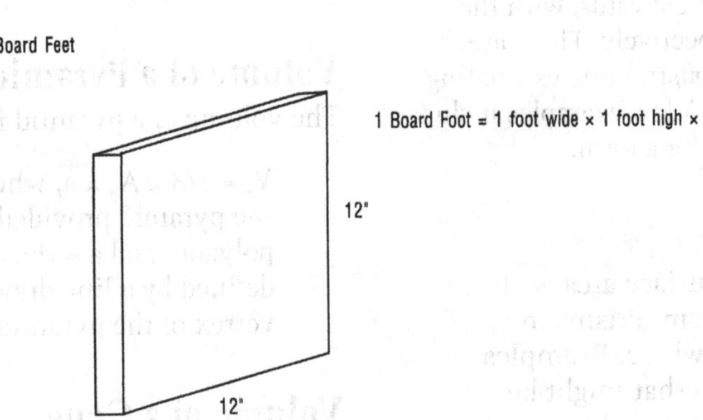

A piece of lumber 2 × 6 × 12" long also = 1 B.F.

$$\frac{L \times W \times H}{12} = \text{B.F.}$$

Where L = Feet
W = Inches
H = Inches

Example:
For a 2 × 8 – 16 feet long

$$\text{B.F.} = \frac{16 \times 2 \times 8}{12} = 21.33 \text{ B.F.}$$

Figure 2.10 Standard Weights and Measures

Linear Measure		Square Measure	
1000 mils =	1 inch	144 square inches =	1 square foot
12 inches =	1 foot	9 square feet =	1 square yard
3 feet =	1 yard	30-1/4 square yds. =	1 square rod 272-1/4 square feet
2 yards =	1 fathom 6 feet	160 square rods =	1 acre 43,560 square feet
5-1/2 yards =	1 rod 16-1/2 feet	640 acres =	1 square mile 27,878,400 square feet
40 rods =	1 furlong 660 feet	A circular mil is the area of a circle 1 mil, or 0.001 inch in diameter.	
8 furlongs =	1 mile 5280 feet	1 square inch =	1,273,239 circular mils
1.15156 miles =	1 nautical mile, or knot 6080.26 feet	A circular inch is the area of a circle 1 inch in diameter =	0.7854 square inches
3 nautical miles =	1 league 18,240.78 feet	1 square inch =	1.2732 circular inches

Dry Measure		Weight—Avoirdupois or Commercial	
2 pints =	1 quart	437.5 grains =	1 ounce
8 quarts =	1 peck	16 ounces =	1 pound
4 pecks =	1 bushel 2150.42 cubic in. 1.2445 cubic feet	112 pounds =	1 hundredweight
		20 hundredweight =	1 gross, or long ton 2240 pounds
		2000 pounds =	1 net, or short ton
		2204.6 pounds =	1 metric ton
		1 lb. of water (39.1°F) =	27.681217 cu. in.
		=	0.016019 cu. ft.
		=	0.119832 U.S. gallon
		=	0.453617 liter

1 ½″ × 3 ½″. Since the length of lumber is usually measured in feet, and the thickness and width in inches, the following formula is used to calculate board feet:

$$BF = \frac{L \times W \times H}{12} (n).$$

where BF = board foot, W = nominal thickness in inches, H = nominal width in inches, L = length in feet of the individual piece and n = number of pieces.

A common calculation required for a quantity takeoff for a wood framed building is to figure out the total quantity of board feet required of floor joists for a platform framed floor. To determine the quantity of floor joists divide the length of the floor by the joist spacing and add 1. Typically joists are spaced 12″, 16″, or 24″ on center. An additional joist is added to account for the joist at the end of the floor. Since the length of most joists is limited to a maximum span of 16 feet and the width of most buildings exceeds 16 feet, a girder is installed at the mid-point which means the number of floor joists is doubled; a row of joists on each side of the girder. For example if you have a floor that measures 24′ × 32′, and the joists are spaced 16″ on center, how many joists are needed to complete the floor plate?

$$n = 2\ [(32\ LF \div 1.33\ LF) + 1].$$

$$n = 50 \text{ floor joists.}$$

Where n = number of floor joists, 32 LF is the length of the floor, 1.33 LF is the 16″ spacing, and 1 is the additional joist at the end of the platform. This is multiplied by 2 since the width of the floor is 24′ and a girder is required at the midpoint. The total number of floor joists required is 50. Additionally, since floor joists are also required on the long dimension to complete the box frame, two additional lengths are required: 32 LF + 32 LF = 64 LF.

If each floor joist in the preceding example is a 2 × 10, how many board feet are required?

$$BF = \frac{12' \times 2'' \times 10''}{12} (50) + \frac{32' \times 2'' \times 10''}{12} (2)$$

$$= 1107 \text{ BF.}$$

RSMeans uses the unit, MBF to quantify board measure. 1107 BF = 1 MBF, so 1107BF = 1.107 MBF.

Conclusion

The shapes in this chapter are by no means all the shapes that may be encountered in a quantity takeoff, but they are the most common ones. Additional formulas for less common shapes are shown in figures 2.4, 2.6 and 2.8. Figure 2.10 is a table of conversions for linear, square, and cubic measure. Calculating quantities accurately is essential to a solid base from which to start the estimate. Now that linear, area, and volume calculations have been reviewed, we can begin the takeoff.

Chapter 2 Exercises

1. Find the area of a room that is 12′ - 7″ × 16′ - 4-1/2″.
2. Find the gross surface area of the walls in a room that is 25′ - 8-3/8″ × 16′ - 9-3/8″ with a ceiling height of 9′ - 8″.
3. Deduct the area of the following openings in the room in exercise 2 to obtain the wall area to be painted:

   (2) Doors, 3′-0″ × 7′-0″

   (4) Windows, 2′-5 3/4″ wide × 5′-6 5/8″ high
4. Find the length of vinyl baseboard you would need to purchase for a room 120′ - 7″ × 46′ - 8″. There are eight doors (each 3′ wide) in the room. Baseboard is purchased in 10′ lengths.
5. In painting the room in exercises 2 and 3, the painter needs to "cut-in" with brushwork at all inside corners and at the perimeters of all doors and windows. At the corners he will have to paint on two planes, so count the length of the corners twice. How many linear feet of cut-ins should be included in the estimate?
6. Find the length of a rafter if the plan length is 13′ - 0″ and the roof pitch is 6:12.
7. Find the area of a playground that is 120′ wide by 263′ deep.
8. Find the area of a ball field where the front of the lot is 460′; the side lines are at 90° angles from the front; the right side is 932′; and the left side is 1484′.
9. What is the circumference of a 12″ round concrete pier?
10. What is the surface area of the top of this pier?
11. The college cafeteria cuts 12″ diameter pies into 8 slices. Joe's Diner cuts their 12″ pies into 6 slices. How many extra square inches of pie do you get at Joe's?
12. What is the lateral (surface) area of a 12″ round pier that is 4′ deep (not including top and bottom)?
13. What is the total volume of concrete required for the piers in a foundation. The piers are 10″ diameter, 20′ deep, and 66 are required. Express the answer in even cubic yards.

14. What is the volume of concrete required for the footings in a building that is 140′ long and 72′ wide. There are 14 spread footings 3′ × 3′ × 16″ deep; 2 spread footings 3′ × 4′ × 20″ deep, and a continuous footing 24″ wide × 12″ deep at the perimeter. Allow 3% waste.

15. The joists used for the floor framing in a new house are 2 × 12 × 14 long, spaced 16″ O.C. The house is 42′ × 28′. What is the minimum amount of lumber required. Use the unit MBF. (Abbreviations can be found in the reference section of the book).

Chapter 3 The Quantity Takeoff

A precise and thorough quantity takeoff is the basis for a sound estimate. Errors or inaccuracies in this portion of the estimate can be compounded during the pricing phase, regardless of how reliable the unit prices are. In this chapter, we will explore the most common practices of taking off quantities for a construction project and offer suggestions for developing routine procedures that will help ensure accuracy:

Reviewing the Documentation

- A set of drawings
- The project manual which includes the technical specifications
- Any related addenda or bulletins
- Additional relevant documents, such as geotechnical reports and any special documentation or unique requirements from local authorities. Examples include conservation commission directives or conditions and/or amendments made by local inspection officials during the review of the plans.

In the absence of a formal set of specifications (bound and separate from the plans), the plans should contain, at the least, a minimal amount of general information to be used as guidelines by the individual estimating the work. As discussed in Chapter 1, the plans and specifications each contain equally important, but distinctly different, kinds of information. Both are necessary to prepare an estimate. Information illustrated on the plans should be supported by the written language in the specifications. For example, if the drawings depict a reinforced concrete footing, the specifications should define the strength and any special requirements of the concrete material and reinforcing, in such a way that the estimator is able to price the material and labor necessary to complete the task.

Reviewing the documents is essential not only to understand the structure that will be estimated, but also to become familiar with the location of various pieces of information found within the documents. This preliminary review often raises questions due to insufficient, missing, or contradictory information.

Beginning the Quantity Takeoff

The following takeoff rules are based on common sense practices that will help prevent, or at least minimize, errors. They can help make the takeoff better organized, more efficient, and more accurate.

Develop a Parameter Sheet

Develop a parameter sheet for the building. Include total square foot area, perimeter, eave height, roof pitch, number of toilets, and any other recurring information. By calculating basic information about the building correctly at the outset, the chance for error is reduced and calculated quantities are all based upon the same basic information. Remember to include in the header, the name of the project, the name of the estimator and the date.

Write Clear Task Descriptions

Descriptions should be clear and legible and should indicate the work needed and the part of the structure involved—or the location on the drawing where the quantity originates. Number each task. Be sure to include unit, quantity, dimensions and total quantities. The following is an example of a thorough task description: Form 24″ × 12″ footing including keyway along B-line as per Detail 5 on drawing S-2.

Use Industry-Accepted Units

A takeoff is not a list of materials for use in placing an order, but rather a descriptive list of construction tasks with quantities derived from the dimensions on the documents. These quantities must then be extended into accepted units for pricing. For example, concrete is estimated in cubic yards (CY), because that is how it is sold. CY is also the accepted industry standard unit.

Follow a Logical Order

The takeoff should be logical and organized. The best approach is to proceed in roughly the same order as the structure is built—from the ground up. This allows you to visualize it while performing the takeoff. The logical thought process is to consider "what is the next step?" and organize the takeoff according to industry standard classification systems such as the CSI MasterFormat classification system, discussed in Chapter 1. For example, all of the work in Division 3, Concrete, should be complete before moving on to Division 4, Masonry. Occasionally, the MasterFormat divisions may present a problem in that items that would naturally be included as part of a construction task may have components in different divisions, and one element of work may be omitted or forgotten by the time you reach the later division where the element should be listed. For example, although the vapor barrier under a slab-on-grade is placed during concrete work (Division 3), it is classified as part of Division 7, Thermal and Moisture Protection. As with every rule, there are exceptions. Place the task in the estimate where it makes the most sense. In this example, it might be best to include the vapor barrier under Division 3 work, with a simple note in Division 7 that it has been accounted for in Division 3.

Review Scales, Notes, Abbreviations, and Definitions

Review drawings and details carefully for notes and scale. Scale can change from drawing to drawing, and a general rule of thumb is that as the detail becomes smaller in focus, the scale becomes larger

Rules for Accurate Takeoffs

It is necessary to become familiar with symbols and abbreviations that typically appear on the drawings. Frequently, drawings contain legends that define material and graphic symbols, as discussed in Chapter 1. Abbreviations, such as NTS (not to scale) and TYP (typical), may be used throughout the entire set of plans. Some words in the construction industry have unique meanings, such as provide, which is defined as "to furnish and install." Carefully review any specification sections that include references or definitions.

Verify Dimensions

Wherever possible, use the dimensions exactly as they appear on the drawings. Add intermediate dimensions to arrive at total dimensions. Develop the habit of checking printed dimensions against scaled dimensions. Discrepancies should be brought to the attention of the design team or the owner. Always express dimensions in the same order, such as length × width × depth. This method avoids errors when referring to the size of certain features, such as windows or doors (which are typically dimensioned as width × height). Convert dimensions to the same units for calculation purposes: feet × feet × feet. For example, when calculating the volume of concrete in a slab with the dimensions of 10′ long × 10′ wide × 4″ thick, convert the 4″ to .33 feet so that all units are the same. Converting all dimensions to the same unit avoids arriving at erroneous values.

Be Consistent

Develop a systematic approach when working with the drawings. For instance, take measurements in a clockwise direction around a floor plan. Begin counting similar features, such as light fixtures, from the left to right and top to bottom. Whatever the procedure, it should become a standard, systematic approach.

Number Takeoff Sheets

Always number each takeoff sheet and keep them in order. Whenever possible, tasks, groups of similar tasks, or entire sheets should be identified by their location on the drawings. For example: Phase 2-B–Building Foundation Footings.

Define Units for Material, Work, and Assembly Items

Items or tasks that have no labor component are called material items. These are furnished only and will be installed under another scope of work. Good examples are lintels for masonry openings. Typically, steel lintels are furnished under Division 5, Metals and installed by a masonry contractor as the brick or block is laid up.

Items that have no material component and require labor only are referred to as work items. Examples include fine grading gravel or finishing of concrete. All tasks or items should be labeled with a unit of measure, which will be extended to the final price. For example: Calculating the volume of concrete in a footing that is 27′ long × 3′ wide × 2′ deep will result in 162 cubic feet (CF). However, concrete is priced in cubic yards (CY), so the calculated quantity of 162 CF must be converted to 6 CY by dividing 162 by 27. (1 CY equals 27 CF.) (Refer to Chapter 2 for more on calculations.)

Calculating the area of a rectangular shape results in a square area unit, most often SF or SY. Even the area of a circle is expressed in SF or SY. For example: The area of a room 12′ long by 17′ wide is 204 SF. When calculating the volume of a shape, the results should be expressed in cubic units, most often CY. For example: The volume of a prism that is 12′ long × 13′ wide × 14′ deep is 2,184 CF, or 80.88 CY.

Use Decimals

Decimals are preferable in the quantity takeoff in lieu of fractions, because they are faster, more precise, and easier to use with a calculator. Dimensions on drawings are in feet and inches. These should be converted to their decimal equivalent. To convert inches into decimals of a foot, divide the number of inches by 12. For example: A dimension of 24′-4″ should be converted to 24.33′ (4″÷12″=0.33′). In calculating the area of a room that is 24′-6″ × 24′-6″, converting both dimensions to decimals 24.5′ × 24.5′ and performing the multiplication results in an area of 600.25 SF. Always check the final units of the dimension. Adding linear dimensions results in linear feet (LF) or linear yards (LY). For example: The perimeter of a rectangle that is 12′ long × 13′ wide is 50 LF.

Verify Appropriate Level of Accuracy

While accuracy is important, over-accuracy wastes time. There is an old adage in the construction business that warns, "Don't spend ten dollars of estimating time figuring a one-dollar item . . . unless there are hundreds of them." Accuracy is relative to the task being taken off or estimated. Rarely is it necessary for a number to be calculated to more than two places after the decimal.

There are acceptable parameters for rounding, depending on the particular task being calculated. Most items can be rounded to the nearest full unit. In some cases, it is necessary to round to the nearest sales unit, if the balance of the sales unit has no inventory or future value. (Waste factors will be discussed later in the chapter.) Rounding quantities should be done when appropriate. For example: If the volume of concrete for each of 10 footings is 34.56 CF, then the total would be 345.60 CF, or 12.80 CY for all of the footings. The total concrete yardage would be rounded to 13 CY for the total amount of concrete, rather than rounding each individual footing to the nearest CY.

Calculate Net Versus Gross Quantities

Some materials require an added allowance for waste. Waste is applicable to materials only, and should not be applied to labor or confused with productivity. Before waste is added, quantities are referred to as net quantities. After an allowance for waste has been added, quantities are referred to as gross quantities. (See the "Accounting for Waste" section later in this chapter.)

Check the Takeoff

The quantity takeoff should be checked by another individual for accuracy. Ideally, it would be best to have another completely separate takeoff and estimate done as a means of checking the first. However, this may not be practical or cost-effective. Quantities derived by hand (without the use of a digitizer or computer) should be randomly checked. Select several work items or tasks throughout the estimate and recalculate their quantities. Extensions from the takeoff quantities to the final pricing units can even be checked by a reliable clerical staff person who has minimal background in estimating. The extension of quantities involves calculations that can be checked by anyone with an understanding of simple mathematics and a calculator.

Mark up the Drawings as Bid Documents

Mark the drawings using check marks, colored pencils, or highlighters to indicate the work that has already been taken off. These serve as the estimator's work papers and should be kept as a record of how quantities were derived. If original drawings must be returned, make copies and file them as records. This aspect of the takeoff is critical for projects that are bid and then go to contract. Once the project has been awarded, it will be turned over to the project manager, and the bid documents become crucial for relaying information as to how the project was bid.

Focus on the Task at Hand

Those who perform the takeoff and pricing require a high level of concentration in order to accurately do the job, and should not be subject to interruptions from all the normal distractions of the office. Phone calls, frequent drop-ins by co-workers, and any type of concentration-breaker are detrimental to accurate performance. Distractions or attempts to multi-task are often the greatest source of error. When the takeoff must be interrupted, select a natural stopping point and mark it clearly so that when work is resumed, there is no doubt as to where you left off.

Organize the Documentation

Careful organization and neatness of work papers and takeoff sheets are crucial. If supporting work papers are needed (including sketches or details as to how unusual features were estimated), they should be retained and attached to the pertinent quantity sheet. Even if the takeoff is performed using a digitizer and the estimate is done on a computer, there will still be work papers and notes.

All calculations should follow a logical and sequential process. Preprinted takeoff sheets and forms, such as shown in Figure 3.1, should be used whenever possible. Erasures should be neat and clean. Work papers, quantity sheets, and all components of the estimate should be maintained for a minimum of one year. Projects are sometimes abandoned for a number of reasons. Often, those same projects are re-started at a later point in time, due to changes in the economy, ownership, or need. Retaining the estimate and its various components is a reliable way to check for what has changed or remained the same over time.

Accounting for Waste

Quantities derived during the takeoff process are often not the same quantities that are purchased when the work is actually in process. For example, it may be determined that the area of a floor to be covered with plywood is 300 SF. However, plywood is sold in full sheets, which are each 32 SF. When the 300 SF required to cover the floor deck is divided by 32 SF per sheet, the result is 9.375 sheets. Since only full sheets are sold, 10 sheets must be included in the takeoff. The difference between what to include in the takeoff and estimate and what is actually installed is called waste. As mentioned earlier, the material quantities before waste is added are called net quantities, and gross quantities after waste has been added. Pricing is done at the gross quantities level, not on net quantities.

Waste may need to be added for any of three primary reasons:

- To adjust to the standard sales unit
- As anticipated waste resulting from handling
- To achieve a specific assembly lap, as in shingles

Other types of waste include material that is broken, damaged or defective, or lumber that is mistakenly cut shorter than desired. Defective pieces may be considered waste or they may be returned to the vendor for replacement.

Adjusting for Standard Sales Units

Materials often go through some on-site modification. The classic example is wood framing. Framing lumber is purchased in standard lengths, usually in 2′ increments, delivered to the site and then cut to exact lengths for the specific component of the frame. It is purchased in lengths as close to the in-place length as possible to minimize waste. For example, if a partition has a single top and bottom plate with a plate-to-plate height of 8′, the estimator would include studs to be purchased at 8′ long in the takeoff. The studs would then be cut down to 7′-9″ for installation between the plates. The remaining 3″, sometimes referred to as "fall-off," is the waste. It has no real value, but still needs to be accounted for in the estimate, because it is paid for.

It is important to be attentive to other types of materials with similar waste requirements. Any material with a standard sales unit larger than needed for the task qualifies as having a waste component. Construction materials sold in lengths, rolls, bundles, boxes, and sheets, and fluids sold in gallons, drums, or barrels, should be reviewed for waste.

Waste Resulting from Handling

Waste can occur as a result of handling or placement, which is fairly common. Even with careful planning and execution of a task, waste will occur. One

Figure 3.1 *Quantity Sheet*

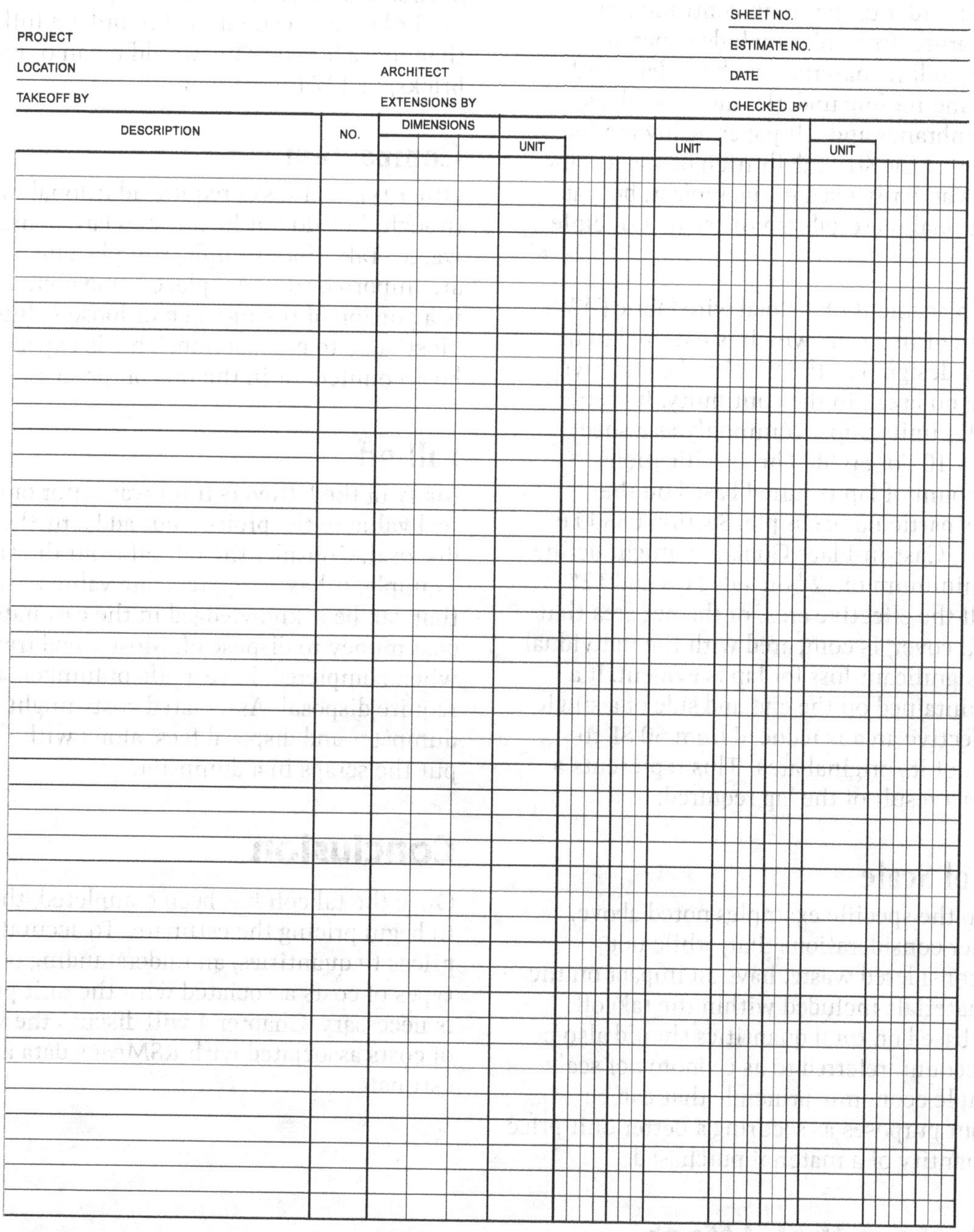

PROJECT		SHEET NO.
		ESTIMATE NO.
LOCATION	ARCHITECT	DATE
TAKEOFF BY	EXTENSIONS BY	CHECKED BY

DESCRIPTION	NO.	DIMENSIONS			UNIT	UNIT	UNIT

example is concrete with specific types of placement. Concrete placed by a chute has minimal waste, as the concrete slides directly into its final resting place, the form. There is no real handling of the concrete using this method. Concrete placed by pump, however, allows significant loss of the total amount, which needs to be accounted for in the estimate as waste.

Other examples of common materials with waste include soil, gravel, and stone delivered by the truckload. Often these materials are distributed by equipment and, due to the inaccurate nature of placing earthen materials, waste can be significant. Generally, the more the materials are handled, the more waste can be expected.

Waste Required for Lap

Often, additional materials are required to satisfy a specific lap in order to maintain continuity of a particular feature. Examples include concrete reinforcement, siding materials such as clapboard and shingles, and roofing underlayments such as bituthene membranes and felt paper. Allowing for lap does not meet the strict definition of waste, since the material is actually used in the project, but lap requires additional materials, so the same principle applies.

Consider the placement of welded wire fabric (WWF) for reinforcement in a slab, which is required to be continuous by design. For the WWF to be effective, there must be no break in the continuity. It is sold in specific sales units, most commonly as a sheet measuring 5′ × 10′, or 50 SF. The specifications define the amount of lap required based on the design. In this particular example, Section 03300 of MasterFormat, Cast-in-Place Concrete, might define the lap as a minimum of 12″ on side laps and 12″ on end laps. If the effective area, or the net area that one sheet will cover, is compared with the individual sales unit, a significant loss for lap is evident. If a 12″ lap is maintained on the end and side of a single sheet, the effective area is reduced from 50 SF to 36 SF, or 72% of its original area. This represents a 28% waste as a result of the lap required.

Economy of Scale

In addition to the specific examples noted above, there are other considerations that, while not specifically considered waste, have an impact on the amount of materials included within the takeoff. Price breaks based on total quantities should also be taken into account, referred to as economy of scale. This is a simple economic principle that can be defined for our purposes as securing a better unit price for a large quantity of a material purchased.

Other Factors that Affect Quantities

Suppliers may be able to provide price break points for certain materials. Brick, for instance, is sold by the pallet, varying in quantity based on the size and type. For example, assume that there are 500 bricks per pallet. If it is determined that 63,485 bricks, including normal waste, are needed for a particular project, it might be more economical to order a full pallet rather than loose bricks. This would mean ordering 63,500 bricks, or 127 full pallets.

Compaction

Other types of tasks require additional materials, though they do not fit the standard definition of waste. Take, for example, soil placement. When soils are imported to a site, placed, and compacted, there is a portion of the in-truck or loose volume that is "lost" due to compaction. This is expected and must be accounted for in the takeoff process.

Fall-off

Many of the leftovers from waste not only have no real value to the project, but add a further expense for disposal. Consider the fall-off from the earlier framing example. It has no appreciable value to the project that can be acknowledged in the estimate, and will cost money to dispose of. Most wood frame projects, when completed, have a pile of lumber scraps that require disposal. Associated costs might include dumpster and disposal fees, along with the labor to put the scraps in a dumpster.

Conclusion

Once the takeoff has been completed, the next step is to begin pricing the estimate. To accurately apply unit prices to quantities, an understanding of the different types of costs associated with the unit price system is necessary. Chapter 4 will discuss the various types of costs associated with RSMeans data and a unit cost estimate.

Chapter 3 Exercises

1. Develop a parameter sheet for the medical office plans found at **https://rsmeansonline.com/academic**. Include S.F. area, perimeter, eave height, roof pitch, number of toilet rooms, and any other recurring dimensions that will be used in your takeoff.
2. a. How many square feet are there in a square yard?

   b. How many cubic feet are there in a cubic yard?
3. a. How much does a 10′ long #5 rebar weigh?

   b. How many L.F. of #5 rebar are there per ton?
4. If you plan to paint a room 12′ x 15′ x 9′ high with 2 coats of latex paint that has a coverage rate of 250 SF per gallon, how much paint should you purchase?
5. You need to provide the formwork for a concrete wall that is 100′ long, 8′ high, and 12″ thick. How much formwork is required? What is the correct unit of measure?
6. How much concrete must be purchased to complete the concrete wall in Exercise 5? Assume 3% waste.
7. How many tons of #4 rebar should be purchased for the wall in Exercise 6 if bars are placed at 2′ - 0″ on center vertical and horizontal? Allow 3″ relief from bar to edge of concrete.
8. Floor plans are usually shown in 1/4″ or 1/8″ scale. The architect has failed to include a critical dimension on the plan, and you need the dimension of a partition immediately. The scale is 1/4″ and the partition measures 11-3/8″ on your ruler. How long is it?
9. What is the better solution to Exercise 8?
10. Name three cost items that will not be found in the plans, but will be clarified in the specifications.

Chapter 4 Understanding Material, Labor & Equipment Costs

Before pricing a unit cost estimate, a basic understanding of material, labor, and equipment costs is essential. In this chapter, we will explore how RSMeans data works, what is meant by bare costs, and how these costs are modified by direct and indirect costs.

Material

Material costs represent the cost paid by the contractor for material purchased at construction supply firms. The material listed in the RSMeans unit price data sheets are for average quality. They include the costs of normal methods of attachment and the small tools required for their installation. RSMeans engineers use manufacturers' recommendations, written specifications, and/or standard construction practice for size and spacing of fasteners. Costs published in the RSMeans cost books reflect typical costs that a contractor would pay if they were purchasing materials for a project costing one million dollars. They include delivery within 20 miles of the project site. Since RSMeans' costs are gathered from various suppliers located throughout the U.S., they reflect national averages. This means that to further refine a cost estimate, it is necessary to adjust the costs using a location factor. More on this will be discussed in Chapter 5. (Larger equipment is handled separately and will be discussed later under the Equipment heading.)

Several modifiers are added to the cost of materials to arrive at an accurate unit price. Sales tax is not included in the cost of materials listed in the unit price data sheet, but is added as a direct cost in the estimate summary, and will be covered in chapter 5. Another modifier that should be considered is waste. As discussed in the previous chapter, materials are often sold in standard lengths or sizes. Oftentimes a project will require additional material because of its physical configuration relative to the standard size sold. For example, carpet is manufactured in standard widths. You want to calculate the quantity of carpet required for an office space. The net area to be covered with carpet is 2,000 SY. However, the carpet is sold in 15′-0″ wide rolls and, when layout is considered, the actual quantity needed is 2,160 SY, or an additional 8%. If you increase the quantity of the unit price line for carpet to account for the waste, you will increase not only the amount of material, but the labor as well. This is incorrect because no additional labor is required, only additional material. To adjust the estimate for material waste using RSMeans' data, create a new unit price line, and adjust the material only by 8%. Sales tax will be added to the total material cost including waste in the estimate summary.

Labor

Labor costs represent the cost of labor for a typical installation during daylight hours in temperate conditions. Labor costs reflect productivity based on actual working conditions. In addition to actual installation, these figures include time spent during a normal weekday on tasks such as material receiving and handling, mobilization at site, site movement, breaks, and cleanup. Productivity data is developed over an extended period so as not to be influenced by abnormal variations, and reflects a typical average.

Installing Contractor Rate

(Column H, I, Figure 4.1)

Let's begin by defining some basic terms. RSMeans lists labor rates for forty-five trades that can be found in Figure 4.1 or on the inside back cover of the RSMeans cost book. The term,

Figure 4.1 *Labor Rates*

		A		B	C	D	E	F	G	H	I
		Base Rate Incl. Fringes		Workers' Comp. Ins.	Average Fixed Overhead	Overhead	Profit	Total Overhead & Profit		Rate with O & P	
Abbr.	Trade	Hourly	Daily					%	Amount	Hourly	Daily
Skwk	Skilled Workers Average (35 trades)	$34.05	$272.40	14.1%	16.3%	27.0%	10%	67.4%	$22.95	$57.00	$456.00
	Helpers Average (5 trades)	25.35	202.80	15.6		25.0		66.9	16.95	42.30	338.40
	Foreman Average, Inside ($.50 over trade)	34.55	276.40	14.1		27.0		67.4	23.30	57.85	462.80
	Foreman Average, Outside ($2.00 over trade)	36.05	288.40	14.1		27.0		67.4	24.30	60.35	482.80
Clab	Common Building Laborers	26.45	211.60	16.1		25.0		67.4	17.85	44.30	354.40
Asbe	Asbestos/Insulation Workers/Pipe Coverers	34.70	277.60	12.3		30.0		68.6	23.80	58.50	468.00
Boil	Boilermakers	40.30	322.40	10.3		30.0		66.6	26.85	67.15	537.20
Bric	Bricklayers	33.65	269.20	12.7		25.0		64.0	21.55	55.20	441.60
Brhe	Bricklayer Helpers	27.15	217.20	12.7		25.0		64.0	17.40	44.55	356.40
Carp	Carpenters	33.55	268.40	16.1		25.0		67.4	22.60	56.15	449.20
Cefi	Cement Finishers	31.85	254.80	8.4		25.0		59.7	19.00	50.85	406.80
Elec	Electricians	40.25	322.00	5.8		30.0		62.1	25.00	65.25	522.00
Elev	Elevator Constructors	50.55	404.40	6.1		30.0		62.4	31.55	82.10	656.80
Eqhv	Equipment Operators, Crane or Shovel	35.80	286.40	9.1		28.0		63.4	22.70	58.50	468.00
Eqmd	Equipment Operators, Medium Equipment	34.90	279.20	9.1		28.0		63.4	22.15	57.05	456.40
Eqlt	Equipment Operators, Light Equipment	33.55	268.40	9.1		28.0		63.4	21.25	54.80	438.40
Eqol	Equipment Operators, Oilers	31.05	248.40	9.1		28.0		63.4	19.70	50.75	406.00
Eqmm	Equipment Operators, Master Mechanics	36.00	288.00	9.1		28.0		63.4	22.80	58.80	470.40
Glaz	Glaziers	33.20	265.60	12.9		25.0		64.2	21.30	54.50	436.00
Lath	Lathers	30.05	240.40	8.6		25.0		59.9	18.00	48.05	384.40
Marb	Marble Setters	31.80	254.40	12.7		25.0		64.0	20.35	52.15	417.20
Mill	Millwrights	35.35	282.80	8.4		25.0		59.7	21.10	56.45	451.60
Mstz	Mosaic & Terrazzo Workers	31.30	250.40	8.5		25.0		59.8	18.70	50.00	400.00
Pord	Painters, Ordinary	29.65	237.20	11.1		25.0		62.4	18.50	48.15	385.20
Psst	Painters, Structural Steel	30.55	244.40	41.6		25.0		92.9	28.40	58.95	471.60
Pape	Paper Hangers	29.90	239.20	11.1		25.0		62.4	18.65	48.55	388.40
Pile	Pile Drivers	32.75	262.00	18.0		30.0		74.3	24.35	57.10	456.80
Plas	Plasterers	30.70	245.60	11.4		25.0		62.7	19.25	49.95	399.60
Plah	Plasterer Helpers	27.35	218.80	11.4		25.0		62.7	17.15	44.50	356.00
Plum	Plumbers	39.65	317.20	7.0		30.0		63.3	25.10	64.75	518.00
Rodm	Rodmen (Reinforcing)	36.30	290.40	19.5		28.0		73.8	26.80	63.10	504.80
Rofc	Roofers, Composition	28.15	225.20	29.2		25.0		80.5	22.65	50.80	406.40
Rots	Roofers, Tile & Slate	28.30	226.40	29.2		25.0		80.5	22.80	51.10	408.80
Rohe	Roofers, Helpers (Composition)	20.90	167.20	29.2		25.0		80.5	16.80	37.70	301.60
Shee	Sheet Metal Workers	38.20	305.60	8.9		30.0		65.2	24.90	63.10	504.80
Spri	Sprinkler Installers	39.30	314.40	7.1		30.0		63.4	24.90	64.20	513.60
Stpi	Steamfitters or Pipefitters	40.05	320.40	7.0		30.0		63.3	25.35	65.40	523.20
Ston	Stone Masons	32.85	262.80	12.7		25.0		64.0	21.00	53.85	430.80
Sswk	Structural Steel Workers	36.40	291.20	36.3		28.0		90.6	33.00	69.40	555.20
Tilf	Tile Layers	31.55	252.40	8.5		25.0		59.8	18.85	50.40	403.20
Tilh	Tile Layers Helpers	24.85	198.80	8.5		25.0		59.8	14.85	39.70	317.60
Trlt	Truck Drivers, Light	26.40	211.20	14.6		25.0		65.9	17.40	43.80	350.40
Trhv	Truck Drivers, Heavy	27.20	217.60	14.6		25.0		65.9	17.90	45.10	360.80
Sswl	Welders, Structural Steel	36.40	291.20	36.3		28.0		90.6	33.00	69.40	555.20
Wrck	*Wrecking	26.45	211.60	30.9		25.0		82.2	21.75	48.20	385.60

billing rate, or installing contractor rate refers to the cost that a subcontractor or installing contractor would charge to perform the work. The wage rates listed in this particular chart are for open shop or nonunion rates. Burdened labor cost is the term used to describe labor rates that have been modified by all taxes, insurance, and overhead components. After profit has been added, the result is the full value of the labor hour, referred to as the billing rate. This is the rate that would be charged for labor in a time and material application or the rate that the installing contractor would charge a general contractor. The installing contractor rate is the cost listed in column H, Figure 4.1 and is the cost included for the labor component in the RSMeans Unit Price data sheet column, Total Including O&P. The following is a review of the components that make up the billing rate and how those costs are calculated.

Labor Rate *(Column A, Figure 4.1)*

Labor rates are based on average open shop wages for 7 major U.S. regions. Base rates, including fringe benefits, are listed both hourly and daily. These figures are the sum of the wage rate and employer-paid fringe benefits such as vacation pay and employer-paid health costs.

Workers' Compensation Insurance

(Column B, Figure 4.1)

A type of insurance that protects employees is Workers' Compensation, provided by the employer in the event of injury, disability, or death occurring in the workplace. In most states, this is a compulsory insurance. Workers' Comp, as it is commonly referred to, is different for each classification of worker or trade, and is based on a percentage of the gross non-premium wages of the employee. Workers' Compensation rates are listed in column B. RSMeans uses the national average of state rates established for each trade.

State and Federal Taxes/Insurance

(Column C, Figure 4.1)

All states and the federal government apply a tax to wages to provide a source for unemployment benefits. The Federal Unemployment Tax Act (FUTA) and State Unemployment Tax Act (SUTA) require taxes to be paid by the employer based on the dollars earned by the employee. They are levied on the employee's gross taxable wages. These costs are set at 6.2%.

The Federal Insurance Corporation of America (FICA), or Social Security, and Medicaid taxes are assessed by the federal government; though they are more appropriately considered insurances, since they are a benefit paid by the government to retired citizens. Employees are taxed at the rate of 7.65% of gross taxable wages, deducted from the employee's paycheck. Additionally, every employer makes a matching contribution of the same percentage. It is this percentage that is recouped in the wage calculation.

Public Liability and Builder's Risk Insurance

Public liability insurance protects a contractor's interests when a third party suffers an injury, or other loss, or damage, while on the construction site. Builder's Risk Insurance indemnifies a contractor against damage to a building while under construction. These rates are set at 2.02% for Public Liability Insurance and 0.44% for Builder's Risk Insurance. They are included in column C, Average Fixed Overhead, along with the other state and federal taxes and insurances listed above. All the percentages, except for social security taxes, vary from state to state, as well as from company to company.

Assumptions

The percentages in both columns D and E are based on the assumption that the installing contractor has annual billing of $2,000,000 and up. Overhead percentages may increase if annual billing is smaller. The overhead percentages for any given contractor may vary greatly and depend on a number of factors such as the contractor's annual volume, engineering and logistical support costs, and staff requirements. The figures for overhead and profit will also vary depending on the type of job, the job location, and the prevailing economic conditions. All factors should be examined very carefully for each job.

Installing Contractor Rate with Overhead and Profit

Column F lists the total of columns B, C, D and E. Column G is the hourly base labor rate, calculated by multiplying column A, by the percentage in column F. Column H is the total of columns A and G, hourly base labor rate plus total overhead and profit. The last column, or column I is the hourly rate with overhead and profit, column H, multiplied by eight hours. The base labor rate including overhead and profit is the installing or subcontractor billing rate, the rate a subcontractor would bill a general contractor for his portion of the work. It is also the labor portion of the Total Incl. O&P (Figure 4.2). Typically in a construction project, a general contractor will hire subcontractors to execute certain portions of the work. Subcontractors are independent entities and are responsible for providing their own insurances and paying their own taxes. Although the construction contract is between the general contractor and owner, the contract assigns responsibility for the subcontractors to the general contractor. For this reason, general contractors must be compensated for the coordination, supervision, and risk assumed when hiring subcontractors. We will apply the general contractor's profit to the estimate in the estimate summary, covered in chapter 5.

Equipment

Equipment cost is an essential part of any construction project. Equipment ranging in size from a small air compressor to a huge tower crane may be rented or leased by the contractor. Although the cost of small tools associated with ordinary installation is included in the labor overhead, the cost of renting larger equipment must be considered in the cost estimate and is included in the bare equipment column. Remember, RSMeans adds an additional 10% to the cost of bare equipment to cover indirect costs. This is reflected in the last column of the unit price data sheet under the heading Total Incl O&P (total including overhead and profit).

Figure 4.2 *Cast-in-Place Concrete*

03 30 Cast-In-Place Concrete

03 30 53 – Miscellaneous Cast-In-Place Concrete

03 30 53.40 Concrete In Place			Crew	Daily Output	Labor-Hours	Unit	2011 Bare Costs Material	Labor	Equipment	Total	Total Incl O&P
0010	**CONCRETE IN PLACE**	R033053-10									
0020	Including forms (4 uses), reinforcing steel, concrete, placement,	R033053-60									
0050	and finishing unless otherwise indicated	R033105-10									
0500	Chimney foundations (5000 psi), industrial, minimum	R033105-20	C-14C	32.22	3.476	C.Y.	144	111	.73	255.73	345
0510	Maximum	R033105-50	"	23.71	4.724	"	168	151	.99	319.99	440
3540	Equipment pad (3000 psi), 3′ x 3′ x 6″ thick	R033105-70	C-14H	45	1.067	Ea.	40.50	35	.51	76.01	104
3550	4′ x 4′ x 6″ thick			30	1.600		62	52.50	.77	115.27	157
3560	5′ x 5′ x 8″ thick			18	2.667		111	87.50	1.28	199.78	270
3570	6′ x 6′ x 8″ thick			14	3.429		150	113	1.65	264.65	355
3580	8′ x 8′ x 10″ thick			8	6		320	197	2.88	519.88	690
3590	10′ x 10′ x 12″ thick		↓	5	9.600	↓	550	315	4.61	869.61	1,150
3800	Footings (3000 psi), spread under 1 C.Y.		C-14C	28	4	C.Y.	158	128	.84	286.84	390
3825	1 C.Y. to 5 C.Y.			43	2.605		185	83	.55	268.55	345
3850	Over 5 C.Y.	R033105-80	↓	75	1.493		171	47.50	.31	218.81	268
3900	Footings, strip (3000 psi), 18″ x 9″, unreinforced		C-14L	40	2.400		119	75	.58	194.58	257
3920	18″ x 9″, reinforced	R033105-85	C-14C	35	3.200		141	102	.67	243.67	330
3925	20″ x 10″, unreinforced		C-14L	45	2.133		116	66.50	.51	183.01	240
3930	20″ x 10″, reinforced		C-14C	40	2.800		134	89.50	.59	224.09	298
3935	24″ x 12″, unreinforced		C-14L	55	1.745		114	54.50	.42	168.92	217
3940	24″ x 12″, reinforced		C-14C	48	2.333		132	74.50	.49	206.99	271
3945	36″ x 12″, unreinforced		C-14L	70	1.371		111	43	.33	154.33	194
3950	36″ x 12″, reinforced		C-14C	60	1.867		127	59.50	.39	186.89	240
4000	Foundation mat (3000 psi), under 10 C.Y.			38.67	2.896		192	92.50	.61	285.11	365
4050	Over 20 C.Y.		↓	56.40	1.986	↓	169	63.50	.42	232.92	292
4520	Handicap access ramp (4000 psi), railing both sides, 3′ wide		C-14H	14.58	3.292	L.F.	278	108	1.58	387.58	490
4525	5′ wide			12.22	3.928		288	129	1.89	418.89	535
4530	With 6″ curb and rails both sides, 3′ wide			8.55	5.614		287	185	2.69	474.69	630
4535	5′ wide		↓	7.31	6.566	↓	292	216	3.15	511.15	685
4751	Slab on grade (3500 psi), incl. troweled finish, not incl. forms										
4760	or reinforcing, over 10,000 S.F., 4″ thick		C-14F	3425	.021	S.F.	1.29	.64	.01	1.94	2.46
4820	6″ thick		"	3350	.021	"	1.89	.65	.01	2.55	3.14
5000	Slab on grade (3000 psi), incl. textured finish, not incl. forms										
5001	or reinforcing, 4″ thick		C-14G	2873	.019	S.F.	1.29	.58	.01	1.88	2.37
5010	6″ thick			2590	.022		2.01	.64	.01	2.66	3.27
5020	8″ thick		↓	2320	.024	↓	2.62	.72	.01	3.35	4.07
6800	Stairs (3500 psi), not including safety treads, free standing, 3′-6″ wide		C-14H	83	.578	LF Nose	5	19	.28	24.28	38
6850	Cast on ground			125	.384	"	4.28	12.65	.18	17.11	26
7000	Stair landings, free standing			200	.240	S.F.	4.03	7.90	.12	12.05	17.75
7050	Cast on ground		↓	475	.101	"	3.29	3.32	.05	6.66	9.20

Figure 4.3 *Crews*

Crew C-14C	Hr.	Daily	Hr.	Daily	Bare Costs	Incl. O&P
1 Carpenter Foreman (out)	$35.55	$284.40	$59.50	$476.00	$31.94	$53.62
6 Carpenters	33.55	1610.40	56.15	2695.20		
2 Rodmen (reinf.)	36.30	580.80	63.10	1009.60		
4 Laborers	26.45	846.40	44.30	1417.60		
1 Cement Finisher	31.85	254.80	50.85	406.80		
1 Gas Engine Vibrator		23.20		25.52	0.21	0.23
112 L.H., Daily Totals		$3600.00		$6030.72	$32.14	$53.85

Equipment expenses are divided into two categories, rental and operating expenses. The cost to rent equipment is broken down into hourly, daily, weekly, and monthly time periods. Operating expenses include things such as fuel, oil, grease, and routine maintenance. The equipment required to perform the installation of strip footings in our example, highlighted in Figure 4.4, is an 8 horse power, gas engine vibrator, the same listed in the crew detail, Figure 4.3. For our example, use line item 01 54 33.10 3000 in the Equipment Rental, Construction Aids chart, Figure 4.4.

Equipment Cost Calculation

(based on a weekly rental)

Equipment Cost Per Day: = Rent per week ÷ 5 days per week + hourly operating cost × 8 hours/day.

Equipment Rental Cost: = $ 46 per week ÷ 5 days per week = $ 9.20/day.

Operating Cost: = $ 1.75 per hour × 8 hours per day = $ 14.00/day.

Equipment Cost Per Day: = $ 9.20 + $ 14.00 = $ 23.20 per day.

Understanding RSMeans Unit Price Data

Every unit line in the Unit Price section of this book represents a component of construction. Each construction project comprises many components, the actual number depending on the scope and complexity of the project. Before selecting a unit price line, read the entire section, and choose the unit price line that most closely resembles the construction component of your project. Each description is truncated so it is meant to be read from the selected unit price line up through the indents to include the appropriate headings. For example, the highlighted line item in Figure 4.2, item 03 30 53.40 3920, is cast-in-place concrete strip footings (3000 psi), 18″ × 9″, reinforced.

Reading across the chart, we see that the crew required to complete the work is listed as C-14C. For each line item listed in the Unit Price data sheet, RSMeans lists the crew which includes labor, and/or labor and equipment required to complete that component of construction. The abbreviation and composition of each crew can be found in the reference section at the back of the RSMeans book under Crews. (For ease of use, details of all necessary charts required to perform the calculations in this chapter are included in Figures 4.2, 4.3 and 4.4.) Looking at Figure 4.3, we see that the composition of crew C-14C is 1 Carpenter Foreman, 6 Carpenters, 2 Rodmen (reinf.), 4 Laborers, and 1 Cement Finisher. The equipment required to perform the work is 1 Gas Engine Vibrator.

Notice the boxes to the right of the unit price descriptions at the beginning of the section (*see Figure 4.2*). The numbers in the boxes (e.g. R033105-85), refer to additional reference information that is related to that section, and will assist you in your estimate. References are located in the back of the book with gray tabs that read, Reference Tables.

Figure 4.4 *Construction Aids*

01 54 | Construction Aids

	01 54 33 | Equipment Rental		Unit	Hourly Oper. Cost	Rent Per Day	Rent Per Week	Rent Per Month	Equipment Cost/Day
10	0010	**CONCRETE EQUIPMENT RENTAL** without operators R015433-10						
	0200	Bucket, concrete lightweight, 1/2 C.Y.	Ea.	.75	23	69	207	19.80
	0300	1 C.Y.		.80	28	84	252	23.20
	0400	1-1/2 C.Y.		1.00	38.50	115	345	31
	0500	2 C.Y.		1.10	45	135	405	35.80
	0580	8 C.Y.		5.70	265	795	2,375	204.60
	0600	Cart, concrete, self-propelled, operator walking, 10 C.F.		2.50	58.50	175	525	55
	0700	Operator riding, 18 C.F.		4.20	96.50	290	870	91.60
	0800	Conveyer for concrete, portable, gas, 16" wide, 26' long		9.20	127	380	1,150	149.60
	0900	46' long		9.55	153	460	1,375	168.40
	1000	56' long		9.70	162	485	1,450	174.60
	1100	Core drill, electric, 2-1/2 H.P., 1" to 8" bit diameter		1.40	59.50	179	535	47
	1150	11 H.P., 8" to 18" cores		5.10	115	345	1,025	109.80
	1200	Finisher, concrete floor, gas, riding trowel, 96" wide		8.55	148	445	1,325	157.40
	1300	Gas, walk-behind, 3 blade, 36" trowel		1.40	20.50	62	186	23.60
	1400	4 blade, 48" trowel		2.70	28	84	252	38.40
	1500	Float, hand-operated (Bull float) 48" wide		.08	13.65	41	123	8.85
	1570	Curb builder, 14 H.P., gas, single screw		11.95	253	760	2,275	247.60
	1590	Double screw		12.60	293	880	2,650	276.80
	1600	Floor grinder, concrete and terrazzo, electric, 22" path		1.62	102	305	915	73.95
	1700	Edger, concrete, electric, 7" path		.90	51.50	155	465	38.20
	1750	Vacuum pick-up system for floor grinders, wet/dry		1.35	81.50	245	735	59.80
	1800	Mixer, powered, mortar and concrete, gas, 6 C.F., 18 H.P.		6.45	120	360	1,075	123.60
	1900	10 C.F., 25 H.P.		7.90	147	440	1,325	151.20
	2000	16 C.F.		8.20	170	510	1,525	167.60
	2100	Concrete, stationary, tilt drum, 2 C.Y.		6.45	238	715	2,150	194.60
	2120	Pump, concrete, truck mounted 4" line 80' boom		23.25	925	2,775	8,325	741
	2140	5" line, 110' boom		30.70	1,250	3,770	11,300	999.60
	2160	Mud jack, 50 C.F. per hr.		6.10	128	385	1,150	125.80
	2180	225 C.F. per hr		8.15	147	440	1,325	153.20
	2190	Shotcrete pump rig, 12 C.Y./hr.		14.15	227	680	2,050	249.20
	2200	35 C.Y./hr.		16.75	243	730	2,200	280
	2600	Saw, concrete, manual, gas, 18 H.P.		4.65	43.50	130	390	63.20
	2650	Self-propelled, gas, 30 H.P.		8.90	105	315	945	134.20
	2700	Vibrators, concrete, electric, 60 cycle, 2 H.P.		.31	9	27	81	7.90
	2800	3 H.P.		.43	12	36	108	10.65
	2900	Gas engine, 5 H.P.		1.30	16.35	49	147	20.20
	3000	8 H.P.		1.75	15.35	46	138	23.20
	3050	Vibrating screed, gas engine, 8 H.P.		2.81	71.50	215	645	65.50
	3120	Concrete transit mixer, 6 x 4, 250 H.P., 8 C.Y., rear discharge		43.95	590	1,765	5,300	704.60
	3200	Front discharge		51.35	725	2,175	6,525	845.80
	3300	6 x 6, 285 H.P., 12 C.Y., rear discharge		50.45	685	2,050	6,150	813.60
	3400	Front discharge	↓	52.65	735	2,200	6,600	861.20
20	0010	**EARTHWORK EQUIPMENT RENTAL** without operators R015433-10						
	0040	Aggregate spreader, push type 8' to 12' wide	Ea.	2.25	26	78	234	33.60
	0045	Tailgate type, 8' wide		2.20	33.50	100	300	37.60
	0055	Earth auger, truck-mounted, for fence & sign posts, utility poles		11.60	515	1,545	4,625	401.80
	0060	For borings and monitoring wells		38.45	665	1,995	5,975	706.60
	0070	Portable, trailer mounted		2.45	32.50	98	294	39.20
	0075	Truck-mounted, for caissons, water wells		83.90	3,025	9,075	27,200	2,486
	0080	Horizontal boring machine, 12" to 36" diameter, 45 H.P.		20.85	200	600	1,800	286.80
	0090	12" to 48" diameter, 65 H.P.		28.90	350	1,055	3,175	442.20
	0095	Auger, for fence posts, gas engine, hand held		.45	6	18	54	7.20
	0100	Excavator, diesel hydraulic, crawler mounted, 1/2 C.Y. cap.		16.70	385	1,155	3,475	364.60
	0120	5/8 C.Y. capacity		24.50	555	1,670	5,000	530
	0140	3/4 C.Y. capacity		30.55	660	1,980	5,950	640.40
	0150	1 C.Y. capacity	↓	36.35	730	2,195	6,575	729.80

Daily output is the number of units the listed crew can install in a typical eight-hour day. It includes mobilization, layout, movement of materials, and cleanup. In our example, crew C-14C can install 35 units of 18″ × 9″ reinforced, cast-in-place strip footings (3000 psi) in one day. Daily output varies depending on a number of factors and is directly related to the productivity of the crew. Environmental conditions, such as the time of day and extremes in weather conditions, can affect crew productivity. The daily output listed in the RSMeans Unit Price data assumes ordinary working conditions, and measures average productivity over an eight hour work day.

Labor hours are the amount of labor crew C-14C requires to complete the installation of one CY (cubic yard) of the strip footing in our example. It is calculated by dividing the daily total labor hours of the crew, 112 L.H. (*Figure 4.3*), by the daily output, 35 CY per day (*Figure 4.2*). From the quantity takeoff we learned that each material is quantified by a specific unit of measure. In this example, concrete strip footing is measured in cubic yards, abbreviated as CY. If you see an abbreviation that is unfamiliar, refer to the Abbreviations section at the end of the RSMeans cost book.

Labor Hours

$$\frac{\text{Labor Hours}}{\text{Unit}} = \frac{\text{Total Crew Labor Hours}}{\text{Daily Output}}.$$

$$\frac{\text{3.2 Labor Hours}}{\text{CY}} = \frac{\text{112 Labor Hours}}{\text{35 CY / Day}}.$$

Continuing to read across the unit price line (Figure 4.2), bare costs are listed for material, labor, and equipment. As mentioned earlier, these are the costs that the installing contractor pays. They represent the cost in U.S. dollars for one unit of work. They do not include any markups for profit or labor burden.

Labor Cost

$$\frac{\text{Total Daily Bare Labor Cost}}{\text{Daily Output}} = \frac{\text{Labor Cost}}{\text{Unit}}.$$

$$\frac{\$\ 3576.80}{\text{35 CY/Day}} = \frac{\$\ 102.19}{\text{CY}}.$$

Or...

$$\frac{\text{Labor Hours}}{\text{Unit}} \times \frac{\text{Labor Cost}}{\text{Hour}} = \frac{\text{Labor Cost}}{\text{Unit}}.$$

$$\frac{\text{3.2 Labor Hours}}{\text{CY}} \times \frac{\$\ 31.94}{\text{LH}} = \frac{\$\ 102.21}{\text{CY}}.$$

Note: These numbers may vary slightly due to mathematical rounding.

Equipment Cost

$$\frac{\text{Total Daily Bare Equipment Cost}}{\text{Daily Output}} = \frac{\text{Equipment Cost}}{\text{Unit}}.$$

$$\frac{\$\ 23.20}{\text{35 CY/day}} = \frac{\$\ 0.66}{\text{CY}}$$

Or...

$$\frac{\text{Labor Hours}}{\text{Unit}} \times \frac{\text{Equipment Cost}}{\text{Labor Hours}} = \frac{\text{Equipment Cost}}{\text{Unit}}.$$

$$\frac{\text{3.2 Labor Hours}}{\text{CY}} \times \frac{\$\ 0.21}{\text{Labor Hours}} = \frac{\$\ 0.67}{\text{CY}}.$$

Formulas

Labor Hours per Unit = Crew L.H. Total ÷ Daily Output.

Labor Dollars per Unit = Daily Labor Cost (Bare) ÷ Daily Output.

Equipment Dollars per Unit = Daily Equipment Cost (Bare) ÷ Daily Output.

Crew Equipment Cost per Day = (Weekly Rental ÷ 5) + (Hours Operating × 8).

Material Cost = Material with delivery within 20 miles, no sales tax (Bare).

Total with Overhead and Profit = Material + 10% profit; plus Labor + labor burden + 10% profit; plus Equipment + 10% profit.

Conclusion

Understanding the various modifiers that apply to the material, labor and, equipment components of the unit price estimate is essential to good estimating practice. It is also a prerequisite for understanding direct overhead costs discussed in the next chapter.

Chapter 4 Exercises

1. If 2 carpenters can install 2000 board feet of 4 × 10 joists in one workday, how many hours of labor will be required to install 1 MBF of 4 × 10 joists?
2. One carpenter can install 625 SF of ceiling tile in grid on one workday. How many hours of labor are needed to install 1 S.F.?
   a. How many minutes does it take to install 1 S.F. of ceiling tile?
3. Crew Q5 can install 1.7 2-ton heat pumps in a day. How many labor hours are required to install one heat pump?
4. Crew B10B is used to backfill a trench. It can accomplish 1200 CY of backfilling per day.
   a. What is the number of labor hours per CY required?
   b. What is the bare cost of labor per CY?
   c. What is the bare cost of equipment per CY?
5. Calculate the equipment cost per day for a 200 HP dozer with the following costs. The anticipated duration of the job is (4) days.

   Daily rent: $1035
   Weekly rent: $3100
   Monthly rent: $9300
   Hourly operating cost: $55
6. If the dozer in Exercise 5 is used in Crew B10B, and can backfill 1200 C.Y. of trench in a day, what is the equipment cost per CY?
7. What is the bare labor cost to install 1,000 SF of gypsum wallboard if 2 carpenters can install 775 SF per day and their rate is $28.00 per hour?
8. Calculate total with overhead and profit for a project to install a 52 gallon electric water heater where:

   The bare material cost is $825.00.
   The bare labor cost for 1 plumber is $180.00
   No equipment is required.
9. If it takes one carpenter 0.77 minutes to install one square foot of ceiling tile, and we have 10,000 SF to install with a crew of four carpenters, how many days will this work take?
10. What is the (a) bare cost and (b) cost with overhead and profit to excavate a trench 375′ long, 4′ deep, and 3′ wide using a 1/2 C.Y. excavator? Daily output is 175 C.Y. per day. The daily rental cost is $500 per day and the hourly operating cost is $25.00. The bare rate for the equipment operator is $36.00 per hour.

Chapter 5 Pricing the Estimate/Estimate Summary

Once the takeoff has been completed and a basic understanding of material, labor, and equipment costs attained, we can begin pricing the estimate. In this chapter we will discuss how to add costs to a takeoff and apply direct costs to the estimate summary.

The Unit Price Estimate

While there are various reasons for business failures in the construction industry, a major contributor is the inability to produce an accurate, profitable, and defendable estimate. At the root of this chronic problem are poor or unprofessional practices on the part of the estimator. Estimating is a labor-intensive and costly operation that, at best, only approximates the cost of a construction project as seen through the experience and judgment of the professional estimator. Every element within the estimate, ranging from quantities to unit prices, must be substantiated in factual terms in order to be considered professional work. This is the basis for a defendable, detailed estimate.

In order to be successful, the professional estimator must:

- Be highly organized and efficient
- Understand the construction process and its materials
- Read and fully understand drawings and specifications
- Be able to visualize the project being built
- Have a working knowledge of basic mathematics

The work product of the professional estimator is the detailed, or unit price estimate. By definition, the unit price estimate consists of breaking the project down into tasks, quantifying those tasks, and then applying a price based on the units of each task. As mentioned previously, each task can have various and multiple components, e.g., materials, labor, tools, equipment, and subcontractors that make up the unit price. For example:

Task Description	*Quantity*	*Unit Cost*	*Unit Price Total*
Place and finish concrete slab	10,000 SF	@ $1.00/SF	= $10,000.

The unit price of $1.00 per SF in this example consists of the labor and equipment used in the placing and finishing of the concrete slab. The example illustrates that the $1.00 unit cost is based on the SF unit. Once the unit price has been multiplied by the quantity, the resultant cost is called the extended cost, or total cost for the item.

A correct unit price estimate has a quantity and a price for each item of work or expense identified in the bid documents. When the estimate is summarized, and profit is added, it is submitted as the offer, or the bid. Should the contractor be successful in attaining the contract, this same estimate now becomes the basis for the cost control system that will determine whether the work has been performed and the costs incurred as estimated—in short, if the project made or lost money.

The unit price estimate serves as the guideline for awarding subcontracts and purchasing materials and standard for judging productivity. It is valuable in assigning a dollar value to each category in the schedule of values that will become the basis for requisitioning payment. In essence, the estimate becomes the foundation of the project. This is why it is of such importance—it is part of virtually every aspect of the construction process. The most proficient contracting firm cannot overcome flawed estimates that represent work taken below cost, or that do not allow for an appropriate profit. At the heart of the successful estimate is the unit price. The term *unit price* can be defined as the incremental cost per unit of work.

The following are parameters that define the unit price:

- Unit prices are based on dollars, or portions of a dollar, per unit.
- The pricing unit or unit cost should equal the extended unit of each task divided by the quantity.
- Most unit prices are based on a specific time frame as a means of measuring productivity.
- Labor unit prices can be based on an individual's production or on the production of a crew.
- Unit prices can be based on historical cost data or published cost data.
- Unit prices can include all, one, or a combination of cost types (material, labor, equipment, subcontractors, etc.).

Sources of Unit Price Data

Collecting, organizing, and analyzing the data for the estimate can be a daunting task. Being able to use this data efficiently is a result of how well it has been organized. There are several sources for pricing data, but most fall into these five categories:

1. Written quotations or published prices from suppliers and vendors
2. Written quotations from subcontractors, including materials and labor
3. Written quotations from equipment suppliers or rental agencies
4. Historical cost data from your company's own previously completed projects (similar to the one currently being estimated)
5. Cost data from published sources, such as RSMeans

Categories 1 through 3 are referred to as contemporaneous pricing. They are based on a review of the project documents, with full understanding of the project's unique or special conditions. These prices can usually be assumed to be an offer or bid from a source looking to do business. Pricing by interested parties doesn't just happen. It is typically the result of time (and money) invested in soliciting pricing from various sources. Failure to establish contacts, gain the interest of the bidder, and follow up as bid day approaches can often result in gaps in the estimate on bid day. Quotes from bidders, reviewed and qualified, are usually the best source of costs.

Current pricing should be accepted only from individuals or firms that will provide prices in writing. Frequently, pricing may be verbal, with written confirmation to follow. Such quotations should be written by the person receiving the quotation on a standardized form, called a telephone quotation sheet. At the time the quotation is received, it should be qualified. This involves asking the person providing the quote what is included. An intimate knowledge of the work involved in each task is necessary in order to ask the right questions. Sometimes all that is required is to confirm that the quoting party has included what was required or requested. The qualification process can also help avoid duplication errors. For example, if a material quote includes sales tax, this should be noted so that tax is not added a second time.

Any notes or pertinent information discussed during the qualification process should be written down on the telephone quotation sheet, so that they can be compared against the written follow-up quote. At a minimum, the telephone quote must document the name of the individual delivering the quote, the date and time, and how long the price is valid. In the absence of a written quote, follow up after the bid with a "confirmation of price" fax or letter. This procedure has the benefit of documenting the price while it is fresh in both parties' minds. While most people in the construction business know that the quote is only as valid as the integrity of the firm or individual that makes it, a written quote goes a long way to ensuring compliance. A well-defined price is a big help when comparing various competing numbers.

Category 4 sources are records of actual costs of similar work previously performed by your own company. Historical costs are recorded—actual costs to perform a quantified, specific task. That means that the costs can be analyzed and compared with the project now being bid to obtain unit prices. If the total cost is divided by the total quantity, the result is the unit price. Historical data shows not only actual project costs for previously completed projects, but also the accuracy of those projects' estimates.

Historical costs are typically taken from the company's records for self-performed work. However, it is not uncommon for some general contractors to study and record the time spent and methods employed by other contractors or subcontractors to perform a defined quantity of a specific task in the hopes of gaining better estimating insight for their own future projects.

While historical costs are actual, it is rare that the unit price can be applied without making at least minor modifications to allow for different circumstances between the previously completed project and the project currently being bid. At the very least, there are considerations for cost escalations of wages and materials over time. There may be different site or weather conditions, varying productivities and supervision, learning curves, and degree of difficulty that might not be readily apparent from empirical data alone.

Historical data must be analyzed carefully, with some understanding of the unique set of circumstances under which the work was performed. All work papers from mathematical computations from the analysis should be retained for the record. Whenever possible, another individual should check the accuracy of the computations to uncover possible errors before they are incorporated into the estimate, and a bid is offered.

Category 5 costs are from published sources, such as the numerous RSMeans cost data books and software. As with any published construction cost data, there is an applied set of parameters. It is essential to understand all of the conditions that have been applied to the data (normally averages), so that you can make your own adjustments for factors such as location, skill of the crew, climate, availability of resources, and supervision.

Published data is based on normal working conditions, during regular business hours, and under average conditions. There is no accounting for unpredictable costs associated with labor resource shortages, supply and demand cycles, or travel and per diem costs. It is also important to understand price and crew size in published cost data to make adjustments based on experience and judgment.

Cost data can be presented in two ways: as bare costs without overhead and profit, and total costs, which include a markup for overhead and profit. It is important to understand whether the work will be self-performed or subcontracted. If it is self-performed, bare costs are preferable so that the markup can be applied at the end of the estimate. If the work is to be subcontracted, make sure that an allowance for the overhead and profit of the installing contractor is carried in the estimate before the estimate summary is begun.

Pricing the Quantities

After the quantity takeoff for each individual task or activity has been assembled, organized, extended into its final units, and checked, prices must be applied. As discussed previously, unit prices can be obtained from a variety of sources and should be noted within the estimate. In the case of published cost data, such as the RSMeans cost data books and software, a twelve-digit (MasterFormat) line item can be cited. In the case of historical or contemporaneous prices, the specific source should be noted and documented within the estimate. This helps establish the credibility of the unit price.

Cost Analysis

The cost analysis portion of the estimate can be recorded on preprinted forms or in estimating software applications such as CostWorks (cost data software by RSMeans). (See Figure 5.1 for a sample Cost Analysis sheet.)

Cost Analysis sheets are different from quantity takeoff sheets in that they provide space for unit prices. Unit prices can be broken into component parts, such as materials, labor, and equipment. Since each component has a different modifier, each component must be calculated and summarized separately. The format of the form allows you to apply unit prices for each component in a columnar sequence.

As each task or activity is priced, its total estimated cost is tabulated at the right of the sheet. Once each task line has been priced and totaled, then the columns are totaled from the top to the bottom of the sheet. Finally, they are totaled for each CSI MasterFormat section number or grouping of tasks.

Tasks can be grouped together for work of a similar nature. For example, the estimator might summarize and total all of the carpentry framing tasks or the entire masonry scope of work for a project. This is done so that the totals for each MasterFormat division can be brought to an Estimate Summary sheet. This represents an entire scope of work that might be subcontracted if the project bid is won. Totaling the cost of work by CSI MasterFormat division is an easy and logical way to organize the estimate. The

Figure 5.1 *Cost Analysis*

COST ANALYSIS

SHEET NO.

PROJECT | ESTIMATE NO.

ARCHITECT | DATE

TAKE OFF BY: | QUANTITIES BY: | PRICES BY: | EXTENSIONS BY: | CHECKED BY:

DESCRIPTION	SOURCE/DIMENSIONS			QUANTITY	UNIT	MATERIAL		LABOR		EQUIPMENT		SUB./TOTAL	
						UNIT COST	TOTAL	UNIT COST	TOTAL	UNIT COST	TOTAL	UNIT COST	TOTAL

total cost for materials, labor, and equipment can be summarized and brought to the Estimate Summary sheet so that you can view a single number that represents a well-defined scope of work. This allows for comparing a price generated for a division or segment of work to a price that was submitted by a subcontractor bidding the work.

Direct Overhead Costs

Direct overhead costs are costs incurred by a contractor that are directly related to one individual project. They include a wide variety of costs such as temporary facilities, trailer rental, supervision, and temporary utilities to name a few. Since these are actual costs that the contractor must pay, they must be included in the unit price estimate. They can be broken down further for estimating purposes into:

- Time-sensitive costs
- Fixed costs

Time-Sensitive Costs: Time-sensitive costs are items whose price is driven by time. The longer the particular item is on site, the higher the cost. A good example would be a project trailer. Most contractors rent trailers, and since there is a monthly fee for rental, the cost is a function of time, or time-sensitive. The longer the project goes on, the higher the accrued cost for the trailer rental. There are numerous other examples of time-sensitive costs that can be found in project overhead, such as telephone and fax line costs and the superintendent's wages. To accurately assign a dollar value to these costs, a schedule should be developed to determine how and when they apply during the term of the project. Initial schedules for determining time-sensitive costs tend to be rudimentary and develop with more information as the estimate proceeds. It is not uncommon for a project schedule to evolve through three generations before it is considered sufficient for use.

Fixed Overhead Costs: Direct overhead costs that are not affected by time are classified as fixed project overhead costs. Examples include building permit fees, registered site layout, engineering design fees, access roads or ramps, and so forth. In most circumstances, there is a single occurrence for each of these costs, independent of the project schedule.

Indirect Overhead Costs

Indirect overhead costs are referred to as main office overhead, which is any cost of a general nature that is not unique to a specific project. Indirect overhead costs are associated with being in business. Some examples include main office rent or mortgage, salaries and benefits of staff, base insurance policies, company vehicle costs, and so forth. The costs associated with indirect overhead are accumulated while work is being performed and must be recovered in each project estimate. Both subcontractors, who operate as independent contractors and general contractors must include these costs to operate their businesses profitably. The general contractor marks up not only material, labor and equipment performed in-house, but work that is accomplished by subcontractors on the project as well. Means captures these indirect overhead costs in the last column of the unit price data sheet named 'Total Incl O&P' (total including overhead and profit). 10% profit is added to each of the bare material, labor and equipment costs listed in each unit price line. In addition to the 10% profit in the labor markup, a percentage is added to the labor cost which includes such costs as employer paid fringe benefits, State and Federal Unemployment costs, Social Security Taxes, Builder's Risk Insurance and Public Liability costs. A detailed breakdown of what those individual costs are and the rates that Means uses to modify them are shown on the inside back cover of this book.

Estimate Summary

The format of the Estimate Summary sheet is straightforward. (*Figure 5.2*) The description of the tasks is in columnar format along the left side of the page, with parallel columns for materials, labor, equipment, and subcontractor costs that start adjacent to the description and move from left to right. At the far right is the total column, which contains the total value of the various components of each grouping of tasks or CSI MasterFormat division. The values of each component are added along the individual rows, which, when totaled, represent the total value for that scope of work.

As a check and balance system, each column is totaled. Then the total of each column is added and compared to the total in the lower right-hand column as a check. When all costs have been transposed from the Cost Analysis sheets to the Estimate Summary sheet, costs for project overhead are added. At this level of costs, the totals are referred to as raw costs, or unburdened costs. Modifiers are added to each category, and the indirect overhead is applied to the subtotals. This is referred to as the burdened or "real cost." In theory, if each task was performed and the

***Figure 5.2** Estimate Summary*

ESTIMATE SUMMARY

PROJECT		SHEET NO.
LOCATION	TOTAL AREA/VOLUME	ESTIMATE NO.
ARCHITECT	COST PER S.F./C.F.	DATE
PRICES BY:	EXTENSIONS BY:	NO. OF STORIES
		CHECKED BY:

DIV.	DESCRIPTION	MATERIAL	LABOR	EQUIPMENT	SUBCONTRACT	TOTAL
1.0	**General Requirements**					
	Insurance, Taxes, Bonds					
	Equipment & Tools					
	Design, Engineering, Supervision					
2.0	**Site Construction**					
	Site Preparation, Demolition					
	Earthwork					
	Caissons & Piling					
	Drainage & Utilities					
	Paving & Surfacing					
	Site Improvements, Landscaping					
3.0	**Concrete**					
	Formwork					
	Reinforcing Steel & Mesh					
	Foundations					
	Superstructure					
	Precast Concrete					
4.0	**Masonry**					
	Mortar & Reinforcing					
	Brick, Block, Stonework					
5.0	**Metals**					
	Structural Steel					
	Open-Web Joists					
	Steel Deck					
	Misc. & Ornamental Metals					
	Fasteners, Rough Hardware					
6.0	**Wood & Plastics**					
	Rough					
	Finish					
	Architectural Woodwork					
7.0	**Thermal & Moisture Protection**					
	Water & Dampproofing					
	Insulation & Fireproofing					
	Roofing & Sheet Metal					
	Siding					
	Roof Accessories					
8.0	**Doors & Windows**					
	Doors & Frames					
	Windows					
	Finish Hardware					
	Glass & Glazing					
	Curtain Wall & Entrances					
	PAGE TOTALS					

Figure 5.2 *Estimate Summary (cont.)*

DIV.	DESCRIPTION	MATERIAL	LABOR	EQUIPMENT	SUBCONTRACT	TOTAL
	Totals Brought Forward					
9.0	**Finishes**					
	Studs & Furring					
	Lath, Plaster & Stucco					
	Drywall					
	Tile, Terrazzo, Etc.					
	Acoustical Treatment					
	Floor Covering					
	Painting & Wall Coverings					
10.0	**Specialties**					
	Bathroom Accessories					
	Lockers					
	Partitions					
	Signs & Bulletin Boards					
11.0	**Equipment**					
	Appliances					
	Dock					
	Kitchen					
12.0	**Furnishings**					
	Blinds					
	Seating					
13.0	**Special Construction**					
	Integrated Ceilings					
	Pedestal Floors					
	Pre Fab Rooms & Bldgs.					
14.0	**Conveying Systems**					
	Elevators, Escalators					
	Pneumatic Tube Systems					
15.0	**Mechanical**					
	Pipe & Fittings					
	Plumbing Fixtures & Appliances					
	Fire Protection					
	Heating					
	Air Conditioning & Ventilation					
16.0	**Electrical**					
	Raceways					
	Conductors & Grounding					
	Boxes & Wiring Devices					
	Starters, Boards & Switches					
	Transformers & Bus Duct					
	Lighting					
	Special Systems					
	Subtotals					
	Sales Tax %					
	Overhead %					
	Subtotal					
	Profit %					
	Adjustments/Contingency					
	TOTAL BID					

costs realized exactly as detailed in the estimate, the real cost would be the break-even point with no profit being realized. The last step is to add general requirements, any contingencies deemed justified, taxes, G.C. profit, costs related to performance and payment bonds and, finally, a location adjustment factor. This final number rounded to the whole dollar represents the bid amount for a lump sum bid.

General Requirements cover project-specific overhead. They include the administrative costs associated with the execution of the contract and project. They are most often determined at the end of the estimate because they require a thorough understanding of the project. Comparing project requirements with the sections listed in Division 1 is a good way to itemize these costs. General Requirements vary by project but may include items such as project coordination, testing and inspection services, temporary facilities and utilities to name a few. For small projects, a percentage can be used, typically between 5% and 15% of project cost. For large projects, the costs may be itemized and priced individually to ensure all costs are covered in the estimate.

If the work is subject to state or local sales tax, this amount must be added to the estimate. Sales tax may be added to material and equipment costs and the material and equipment costs of subcontracted work if not included in the subcontractor's estimate. A rule of thumb when applying taxes to subcontracted work is to apply tax to 50% of the work, assuming approximately half the work is labor and the other half, material and equipment costs. Some taxpayer- and government-funded projects are exempt from sales tax, so there may be no need to add it. However, most residential and commercial projects are subject to a sales tax on materials and equipment. Some states also apply sales tax to labor.

As mentioned previously, the general contractor marks-up material, labor and equipment as well as subcontracted work. This mark-up is called the general contractor's overhead and profit. If the unit price estimate is thorough and complete, it will represent the cost the general contractor must pay to construct the project. But profit is the reason for doing business. It is the end result of a project done right, and the reward for risks taken. While profit is not specifically a cost, it does need to be captured in the estimate. Without it, the project would be considered a failure. Profit is most often assigned as a percentage of the total cost of the work. It is determined by a number of factors including project size, complexity and the amount of risk assumed by the contractor. RSMeans standard markup on materials, equipment and subcontracted work is 10%. To determine the markup on labor, use the average markup for the labor component in your estimate (inside back cover of the Means cost book).

Contingencies cover the cost of unknown conditions in the project. They can be used to cover undefined costs at the early stages of design in a budget estimate or can be used when there are unknown building or site conditions as may be encountered in a renovation project. Contingency allowances vary by project but typically range anywhere from 3% to 20%.

Bond costs should also be added to the estimate. They are actual costs that are borne by the contractor. A performance bond is a surety bond issued by an insurance company to guarantee satisfactory completion of a project by a contractor. Oftentimes, a job requiring a payment & performance bond will require a bid bond, to bid the job. When the job is awarded to the winning bid, a payment and performance bond is then required as security for the completion of the job. If the contractor fails to construct the building according to the specifications laid out by the contract, the client is guaranteed compensation for any monetary loss up to the amount of the performance bond. Actual rates for bonds vary from contractor to contractor and bonding company to bonding company and are dependent on a number of factors, including the contract amount and the financial strength and stability of the contractor.

Since RSMeans cost data is based on national averages, it may be necessary to adjust your total project cost by a factor, based on the specific location of the project. If the work in your estimate is general in nature and covers multiple trades, adjust your total project cost by the Location Factor listed in the reference section in the rear of the book. If the work is by a single trade, use the more specific data found in the City Cost Indexes, also located in the rear of the book.

Submitting a Bid

The bid is a by-product of the estimate—a proposal to do the work estimated. In short, a *bid* is an offer to perform work for a certain price. It is the natural progression of the estimate. In simple terms, a bid is composed of the following:

- An offer to perform a defined scope of work
- A stipulated compensation for performing the work

- Name of the party making the offer
- Name of the party to which the offer is made

Some bids or proposals are submitted from a firm on letterhead, while some are on generic preprinted proposal forms, and others are on bid forms provided by the architect or owner. In the absence of a bid form, the party making the offer has to quantify, and sometimes qualify, the exact scope of work being provided within the proposal. The bidder must identify the documents used in preparing the bid by listing the plans and specifications, the date of the documents, and any addenda issued that modified the documents. The offer should include any assumptions or qualifications on which the proposal is based. The bid should be executed by a responsible individual having the authority to do so, assigned by the company making the offer. The person who signs the bid should include his or her title. The offer should have an expiration date, and should provide a space for the receiving party to sign, date, and accept the offer.

Using letterhead or preprinted forms in lieu of a bid form has some distinct advantages to the contractor and disadvantages to the owner or architect. The contractor can include or exclude any item or scope of work chosen and can qualify or quantify any item of work within the scope without rendering the proposal null and void. The disadvantage to the owner is that it becomes difficult to compare multiple proposals from different contractors if each has noted some qualifications that affect the price.

When a bid form is provided, it alleviates some of the problems encountered with bids that are submitted on letterhead. Bid forms are printed forms specific to the project with the language of the proposal already typed in the body of the form. They are provided by the architect or owner as part of the documents and contain blank spaces for the contractor to fill in only the information required and offer no space for the contractor to qualify a bid. Using a standard bid form ensures a level of conformity so that each bidder's proposal includes the same scope, where the only difference is in the offered amount. The advantage to the architect and owner is an immediate acknowledgement of where each contractor's price lies in comparison to the prices of the other bidders.

Conclusion

A reliable and comprehensive estimate is a combination of quantities derived during the takeoff and the application of correct unit prices. Unit prices can be categorized into materials, labor, and equipment costs for self-performed work, or subcontractor costs for work performed by those individuals who are not direct employees. There are also overhead or nonproduction costs that must be applied. When all costs are summarized and profit added, the estimate is complete and can be submitted as a bid.

Chapter 5 Exercises

1. What is the difference between direct overhead costs and indirect overhead costs?
2. What is an extended cost in a unit price estimate?
3. List five sources of unit price data, in the order of timeliness and accuracy (most accurate first).
4. Many contractors purchase RSMeans unit price data, but never use the costs. What information, other than costs, is valuable?
5. What elements comprise a bid for construction?
6. What adjustments must be made to published data when using it in an estimate?
7. Define productivity in relation to a construction task—as used by RSMeans.
8. How would you calculate the amount of waste when estimating the cost of materials to frame a house?
9. What is the purpose of an estimate summary?
10. Calculate the cost of office overhead for a small general contracting firm whose annual total billing is $10,000,000. What figure should be used as a percent markup on labor in the following year to cover these expenses?

Chapter 6

Unit Cost Estimates

Unit cost estimates are the most commonly used and best known type of construction cost estimate. These estimates are time consuming and costly to create because of the quantity of information that must be researched and recorded. They look at every cost item in a project, in detail, and attach a predicted cost to it. They typically define the cost of material, labor, and equipment for every task identified in the takeoff (or work-breakdown structure).

Unit cost estimates are typically used during the bid phase of the project to establish the contract cost, or during construction to establish the cost of change orders. Practically speaking a unit cost estimate cannot be done until the design is complete and the contract documents are at least 60% complete. The more complete the documents, the more accurate the estimate.

Earlier chapters in this book have described the background needed to be able to perform an estimate, and have shown in detail how to read and use RSMeans cost data. This chapter includes the data itself. This is the complete unit cost section from *RSMeans Light Commercial Cost Data*. The labor rate used is nonunion (open shop). The data is organized using the Construction Specification Institute's MasterFormat 2010.

Chapter 6 Exercises

Using the drawings for the medical office found at **https://rsmeansonline.com/academic**, calculate the cost of the following items using RSMeans unit price data.

1. Excavation:

   Calculate the installing contractor's cost for excavating the strip footings for the Urology Clinic. Assume that the 8′-0″ basement has already been excavated.

2. Foundation:

   Calculate the total installing contractor's cost for concrete.

   Strip footings, reinforced
   Spread footings (interior)
   Slab on grade (assume 3000 psi, unreinforced)

3. Floor Plate:

   Calculate the total bare costs.

   Composite wood joists, 11-1/2″ deep, include cantilevered bays and band joists. Do not include exterior stair or porch. Add 10% for interior stair and doubled joists.
   Fiberglass blanket insulation, paper faced
   Plywood subfloor, 3/4″ thk. CDX, pneumatic nailed

4. Exterior Wall:

   Calculate the total costs including overhead and profit.

   Wall framing, use 8′ HT. including plates (assume window framing has been calculated separately). Add 15% for corners, and intersection of interior partitions.

   Fiberglass blanket, paper faced insulation.

   Exterior sheathing, CDX, pneumatic nailed.

5. Roof:

   Calculate the installing contractor's cost.

   Roof trusses, use 20′ span trusses for hip trusses.

   Insulation.

   Asphalt roof shingles, class A, 260-300lb, pneumatic nailed.

6. Exterior Windows:

   Calculate the total costs including overhead and profit.

   Basement Utility windows.

   Type A, double hung, including frame, screens and grilles, vinyl clad, premium, double insulated glass, 2′-6″ wide × 4′-0″ high.

   Type B, use cost of Type A × 2.

   Type C, use cost of Type A × 3.

   Type D, awning window including frame, screens and grilles, vinyl clad, 34″ × 22″.

7. Interior Partitions, First floor:

   Calculate the total costs of the installing contractor.

   2 × 4 studs, 8′-0″ high with single bottom and double top plates, 16″ on center, pneumatic nailed.

   Toilet room wet walls: 2 × 6 studs, 8′-0″ high with single bottom and double top plates, 16″ on center, pneumatic nailed.

   All interior partitions are finished with 1/2″ blueboard with plaster skim coat.

8. Ceilings, First floor:

   Calculate total costs including overhead and profit.

   1 × 3 furring strips, 16″ on center, pneumatic nailed.

   1/2″ blueboard with plaster skim coat.

9. A painting subcontractor is asked to submit his cost to paint interior door types 3 and 4. The system specified is to brush apply a primer with one coat of latex on each door. How much are his total costs?

10. A general contractor has asked his flooring subcontractor to submit a price to install a 12′-0″ wide broadloom carpet in the waiting area. The specifications call for a 42 oz. nylon, direct glue-down installation. Calculate the total material required including waste and the total cost of the installation. Run seams along length of room as much as possible. Assume there is no pattern repeat.

Unit Price Section

Table of Contents

How RSMeans Data Works

All RSMeans unit price data is organized in the same way.

It is important to understand the structure, so that you can find information easily and use it correctly. *Note: Data is 2012, union wage rates.*

RSMeans **Line Numbers** consist of 12 characters, which identify a unique location in the database for each task. The first 6 or 8 digits conform to the Construction Specifications Institute MasterFormat 2010. The remainder of the digits are a further breakdown by RSMeans in order to arrange items in understandable groups of similar tasks. Line numbers are consistent across all RSMeans publications, so a line number in any RSMeans product will always refer to the same unit of work.

RSMeans engineers have created **reference** information to assist you in your estimate. If there is information that applies to a section, it will be indicated at the start of the section. In this case, R033105-10 provides information on the proportionate quantities of formwork, reinforcing, and concrete used in cast-in-place concrete items such as footings, slabs, beams, and columns. The reference section is located in the back of the book on the pages with a gray edge.

RSMeans **Descriptions** are shown in a hierarchical structure to make them readable. In order to read a complete description, read up through the indents to the top of the section. Include everything that is above and to the left that is not contradicted by information below. For instance, the complete description for line 03 30 53.40 3550 is "Concrete in place, including forms (4 uses), reinforcing steel, concrete, placement, and finishing unless otherwise indicated; Equipment pad (3000 psi), 4' x 4' x 6" thick."

When using **RSMeans data**, it is important to read through an entire section to ensure that you use the data that most closely matches your work. Note that sometimes there is additional information shown in the section that may improve your price. There are frequently lines that further describe, add to, or adjust data for specific situations.

03 30 Cast-In-Place Concrete

03 30 53 – Miscellaneous Cast-In-Place Concrete

03 30 53.40 Concrete In Place

0010	CONCRETE IN PLACE	R033105-10
0020	Including forms (4 uses), reinforcing steel, concrete, placement,	R033105-20
0050	and finishing unless otherwise indicated	R033105-50
0300	Beams (3500 psi), 5 kip per L.F., 10' span	R033105-65
0350	25' span	R033105-70
0500	Chimney foundations (5000 psi), industrial, minimum	R033105-85
0510	Maximum	
0700	Columns, square (4000 psi), 12" x 12", minimum reinforcing	
3500	Add per floor for 3 to 6 stories high	
3520	For 7 to 20 stories high	
3540	Equipment pad (3000 psi), 3' x 3' x 6" thick	
3550	4' x 4' x 6" thick	
3560	5' x 5' x 8" thick	
3570	6' x 6' x 8" thick	

The data published in RSMeans print books represents a "national average" cost. This data should be modified to the project location using the **City Cost Indexes** or **Location Factors** tables found in the reference section. Use the location factors to adjust estimate totals if the project covers multiple trades. Use the city cost indexes (CCI) for single trade projects or projects where a more detailed analysis is required. All figures in the two tables are derived from the same research. The last row of data in the CCI, the weighted average, is the same as the numbers reported for each location in the location factor table.

Crews include labor or labor and equipment necessary to accomplish each task. In this case, Crew C-14H is used. RSMeans selects a crew to represent the workers and equipment that are typically used for that task. In this case, Crew C-14H consists of one carpenter foreman (outside), two carpenters, one rodman, one laborer, one cement finisher, and one gas engine vibrator. Details of all crews can be found in the reference section.

Crews

Crew No.	Bare Costs		Incl. Subs O&P		Cost Per Labor-Hour	
Crew C-14H	Hr.	Daily	Hr.	Daily	Bare Costs	Incl. O&P
1 Carpenter Foreman (outside)	$46.10	$368.80	$70.95	$567.60	$43.48	$66.89
2 Carpenters	44.10	705.60	67.85	1085.60		
1 Rodman (reinf.)	49.10	392.80	78.40	627.20		
1 Laborer	35.10	280.80	54.00	432.00		
1 Cement Finisher	42.35	338.80	62.30	498.40		
1 Gas Engine Vibrator		27.40		30.14	.57	.63
48 L.H., Daily Totals		$2114.20		$3240.94	$44.05	$67.52

The **Daily Output** is the amount of work that the crew can do in a normal 8-hour workday, including mobilization, layout, movement of materials, and cleanup. In this case, crew C-14H can install thirty 4' x 4' x 6" thick concrete pads in a day. Daily output is variable, based on many factors, including the size of the job, location, and environmental conditions. RSMeans data represents work done in daylight (or adequate lighting) and temperate conditions.

Bare Costs are the costs of materials, labor, and equipment that the installing contractor pays. They represent the cost, in U.S. dollars, for one unit of work. They do not include any markups for profit or labor burden.

Crew	Daily Output	Labor-Hours	Unit	2012 Bare Costs Material	Labor	Equipment	Total	Total Incl O&P
C-14A	15.62	12.804	C.Y.	305	565	47.50	917.50	1,275
"	18.55	10.782		320	480	40	840	1,125
C-14C	32.22	3.476		144	147	.83	291.83	385
"	23.71	4.724		168	200	1.13	369.13	495
C-14A	11.96	16.722		350	740	62	1,152	1,600
	31800	.002			.07	.02	.09	.12
	21200	.003			.10	.03	.13	.20
C-14H	45	1.067	Ea.	44	46.50	.61	91.11	121
	30	1.600		66	69.50	.91	136.41	181
	18	2.667		118	116	1.52	235.52	310
	14	3.429		159	149	1.95	309.95	405

The **Total Incl O&P column** is the total cost, including overhead and profit, that the installing contractor will charge the customer. This represents the cost of materials plus 10% profit, the cost of labor plus labor burden and 10% profit, and the cost of equipment plus 10% profit. It does not include the general contractor's overhead and profit. Note: See the inside back cover for details of how RSMeans calculates labor burden.

The **Total column** represents the total bare cost for the installing contractor, in U.S. dollars. In this case, the sum of $66 for material + $69.50 for labor + $.91 for equipment is $136.41.

The figure in the **Labor Hours** column is the amount of labor required to perform one unit of work—in this case the amount of labor required to construct one 4' x 4' equipment pad. This figure is calculated by dividing the number of hours of labor in the crew by the daily output (48 labor hours divided by 30 pads = 1.6 hours of labor per pad). Multiply 1.600 times 60 to see the value in minutes: 60 x 1.6 = 96 minutes. Note: the labor hour figure is not dependent on the crew size. A change in crew size will result in a corresponding change in daily output, but the labor hours per unit of work will not change.

All RSMeans unit cost data includes the typical **Unit of Measure** used for estimating that item. For concrete-in-place the typical unit is cubic yards (C.Y.) or each (Ea.). For installing broadloom carpet it is square yard, and for gypsum board it is square foot. The estimator needs to take special care that the unit in the data matches the unit in the take-off. Unit conversions may be found in the Reference Section.

Estimating Tips

General

- Carefully check all the plans and specifications. Concrete often appears on drawings other than structural drawings, including mechanical and electrical drawings for equipment pads. The cost of cutting and patching is often difficult to estimate. See Subdivision 03 81 for Concrete Cutting, Subdivision 02 41 19.16 for Cutout Demolition, Subdivision 03 05 05.10 for Concrete Demolition, and Subdivision 02 41 19.23 for Rubbish Handling (handling, loading and hauling of debris).
- Always obtain concrete prices from suppliers near the job site. A volume discount can often be negotiated, depending upon competition in the area. Remember to add for waste, particularly for slabs and footings on grade.

03 10 00 Concrete Forming and Accessories

- A primary cost for concrete construction is forming. Most jobs today are constructed with prefabricated forms. The selection of the forms best suited for the job and the total square feet of forms required for efficient concrete forming and placing are key elements in estimating concrete construction. Enough forms must be available for erection to make efficient use of the concrete placing equipment and crew.
- Concrete accessories for forming and placing depend upon the systems used. Study the plans and specifications to ensure that all special accessory requirements have been included in the cost estimate, such as anchor bolts, inserts, and hangers.
- Included within costs for forms-in-place are all necessary bracing and shoring.

03 20 00 Concrete Reinforcing

- Ascertain that the reinforcing steel supplier has included all accessories, cutting, bending, and an allowance for lapping, splicing, and waste. A good rule of thumb is 10% for lapping, splicing, and waste. Also, 10% waste should be allowed for welded wire fabric.
- The unit price items in the subdivisions for Reinforcing In Place, Glass Fiber Reinforcing, and Welded Wire Fabric include the labor to install accessories such as beam and slab bolsters, high chairs, and bar ties and tie wire. The material cost for these accessories is not included; they may be obtained from the Accessories Division.

03 30 00 Cast-In-Place Concrete

- When estimating structural concrete, pay particular attention to requirements for concrete additives, curing methods, and surface treatments. Special consideration for climate, hot or cold, must be included in your estimate. Be sure to include requirements for concrete placing equipment, and concrete finishing.
- For accurate concrete estimating, the estimator must consider each of the following major components individually: forms, reinforcing steel, ready-mix concrete, placement of the concrete, and finishing of the top surface. For faster estimating, Subdivision 03 30 53.40 for Concrete-In-Place can be used; here, various items of concrete work are presented that include the costs of all five major components (unless specifically stated otherwise).

03 40 00 Precast Concrete

03 50 00 Cast Decks and Underlayment

- The cost of hauling precast concrete structural members is often an important factor. For this reason, it is important to get a quote from the nearest supplier. It may become economically feasible to set up precasting beds on the site if the hauling costs are prohibitive.

Reference Numbers

Reference numbers are shown in shaded boxes at the beginning of some major classifications. These numbers refer to related items in the Reference Section. The reference information may be an estimating procedure, an alternate pricing method, or technical information.

Note: Not all subdivisions listed here necessarily appear in this publication.

03 01 Maintenance of Concrete

03 01 30 – Maintenance of Cast-In-Place Concrete

03 01 30.64 Floor Patching			Crew	Daily Output	Labor-Hours	Unit	Material	2011 Bare Costs Labor	Equipment	Total	Total Incl O&P
0010	FLOOR PATCHING										
0012	Floor patching, 1/4" thick, small areas, regular		1 Cefi	170	.047	S.F.	2.44	1.50		3.94	5.05
0100	Epoxy		"	100	.080	"	5.75	2.55		8.30	10.40

03 05 Common Work Results for Concrete

03 05 05 – Selective Concrete Demolition

03 05 05.10 Selective Demolition, Concrete			Crew	Daily Output	Labor-Hours	Unit	Material	2011 Bare Costs Labor	Equipment	Total	Total Incl O&P
0010	SELECTIVE DEMOLITION, CONCRETE	R024119-10									
0012	Excludes saw cutting, torch cutting, loading or hauling										
0050	Break up into small pieces, minimum reinforcing		B-9	24	1.667	C.Y.		45	8.15	53.15	84
0060	Average reinforcing			16	2.500			67	12.20	79.20	125
0070	Maximum reinforcing			8	5			134	24.50	158.50	252
0150	Remove whole pieces, up to 2 tons per piece		E-18	36	1.111	Ea.		40.50	29	69.50	107
0160	2 – 5 tons per piece			30	1.333			48.50	34.50	83	129
0170	5 – 10 tons per piece			24	1.667			61	43	104	161
0180	10 – 15 tons per piece			18	2.222			81	57.50	138.50	213
0250	Precast unit embedded in masonry, up to 1 C.F.		D-1	16	1			30.50		30.50	50
0260	1 – 2 C.F.			12	1.333			40.50		40.50	66.50
0270	2 – 5 C.F.			10	1.600			48.50		48.50	80
0280	5 – 10 C.F.			8	2			61		61	100

03 05 13 – Basic Concrete Materials

03 05 13.85 Winter Protection			Crew	Daily Output	Labor-Hours	Unit	Material	2011 Bare Costs Labor	Equipment	Total	Total Incl O&P
0010	WINTER PROTECTION										
0012	For heated ready mix, add, minimum					C.Y.	4.25			4.25	4.68
0050	Maximum					"	5.30			5.30	5.85
0100	Temporary heat to protect concrete, 24 hours, minimum		2 Clab	50	.320	M.S.F.	525	8.45		533.45	595
0200	Temporary shelter for slab on grade, wood frame/polyethylene sheeting										
0201	Build or remove, minimum		2 Carp	10	1.600	M.S.F.	268	53.50		321.50	385
0210	Maximum		"	3	5.333	"	325	179		504	655
0710	Electrically, heated pads, 15 watts/S.F., 20 uses, minimum					S.F.	.21			.21	.23
0800	Maximum					"	.35			.35	.39

03 11 Concrete Forming

03 11 13 – Structural Cast-In-Place Concrete Forming

03 11 13.45 Forms In Place, Footings			Crew	Daily Output	Labor-Hours	Unit	Material	2011 Bare Costs Labor	Equipment	Total	Total Incl O&P
0010	FORMS IN PLACE, FOOTINGS	R031113-40									
0020	Continuous wall, plywood, 1 use		C-1	375	.085	SFCA	5	2.54		7.54	9.75
0150	4 use		"	485	.066	"	1.63	1.96		3.59	5.10
1500	Keyway, 4 use, tapered wood, 2" x 4"	R031113-60	1 Carp	530	.015	L.F.	.18	.51		.69	1.05
1550	2" x 6"		"	500	.016	"	.26	.54		.80	1.19
5000	Spread footings, job-built lumber, 1 use		C-1	305	.105	SFCA	1.79	3.12		4.91	7.15
5150	4 use		"	414	.077	"	.58	2.30		2.88	4.48

03 11 13.65 Forms In Place, Slab On Grade			Crew	Daily Output	Labor-Hours	Unit	Material	2011 Bare Costs Labor	Equipment	Total	Total Incl O&P
0010	FORMS IN PLACE, SLAB ON GRADE	R031113-40									
1000	Bulkhead forms w/keyway, wood, 6" high, 1 use	R031113-60	C-1	510	.063	L.F.	.86	1.87		2.73	4.06
1400	Bulkhead form for slab, 4-1/2" high, exp metal, incl keyway & stakes	G		1200	.027		1.64	.79		2.43	3.13
1410	5-1/2" high	G		1100	.029		1.90	.86		2.76	3.54
1420	7-1/2" high	G		960	.033		2.31	.99		3.30	4.20

03 11 Concrete Forming

03 11 13 – Structural Cast-In-Place Concrete Forming

03 11 13.65 Forms In Place, Slab On Grade			Crew	Daily Output	Labor-Hours	Unit	Material	2011 Bare Costs Labor	Equipment	Total	Total Incl O&P
1430	9-1/2" high	G	C-1	840	.038	L.F.	2.40	1.13		3.53	4.53
2000	Curb forms, wood, 6" to 12" high, on grade, 1 use			215	.149	SFCA	1.80	4.43		6.23	9.40
2150	4 use			275	.116	"	.58	3.46		4.04	6.45
3000	Edge forms, wood, 4 use, on grade, to 6" high			600	.053	L.F.	.27	1.59		1.86	2.95
3050	7" to 12" high			435	.074	SFCA	.59	2.19		2.78	4.31
4000	For slab blockouts, to 12" high, 1 use			200	.160	L.F.	.63	4.76		5.39	8.65
4100	Plastic (extruded), to 6" high, multiple use, on grade			800	.040	"	7.90	1.19		9.09	10.70
8760	Void form, corrugated fiberboard, 4" x 12", 4' long	G		3000	.011	S.F.	2.39	.32		2.71	3.15
8770	6" x 12", 4' long		↓	3000	.011		2.81	.32		3.13	3.62
8780	1/4" thick hardboard protective cover for void form		2 Carp	1500	.011	↓	.57	.36		.93	1.23

03 11 13.85 Forms In Place, Walls			Crew	Daily Output	Labor-Hours	Unit	Material	Labor	Equipment	Total	Total Incl O&P
0010	**FORMS IN PLACE, WALLS**	R031113-10									
0100	Box out for wall openings, to 16" thick, to 10 S.F.		C-2	24	2	Ea.	23	60		83	125
0150	Over 10 S.F. (use perimeter)	R031113-40	"	280	.171	L.F.	1.94	5.15		7.09	10.75
0250	Brick shelf, 4" w, add to wall forms, use wall area abv shelf										
0260	1 use	R031113-60	C-2	240	.200	SFCA	2.06	6		8.06	12.25
0350	4 use			300	.160	"	.82	4.80		5.62	8.90
0500	Bulkhead, wood with keyway, 1 use, 2 piece		↓	265	.181	L.F.	1.69	5.45		7.14	10.95
0600	Bulkhead forms with keyway, 1 piece expanded metal, 8" wall	G	C-1	1000	.032		2.31	.95		3.26	4.13
0610	10" wall	G		800	.040		2.40	1.19		3.59	4.63
0620	12" wall	G	↓	525	.061	↓	2.88	1.81		4.69	6.20
2000	Wall, job-built plywood, to 8' high, 1 use		C-2	370	.130	SFCA	2.17	3.89		6.06	8.90
2050	2 use			435	.110		1.38	3.31		4.69	7.05
2100	3 use			495	.097		1.01	2.91		3.92	5.95
2150	4 use			505	.095		.82	2.85		3.67	5.65
2400	Over 8' to 16' high, 1 use			280	.171		2.38	5.15		7.53	11.20
2450	2 use			345	.139		1.08	4.17		5.25	8.15
2500	3 use			375	.128		.77	3.84		4.61	7.25
2550	4 use		↓	395	.122	↓	.63	3.64		4.27	6.80
7800	Modular prefabricated plywood, based on 20 uses of purchased										
7820	forms, and 4 uses of bracing lumber										
7860	To 8' high		C-2	800	.060	SFCA	.89	1.80		2.69	3.99
8060	Over 8' to 16' high		"	600	.080	"	.93	2.40		3.33	5.05

03 11 19 – Insulating Concrete Forming

03 11 19.10 Insulating Forms, Left In Place			Crew	Daily Output	Labor-Hours	Unit	Material	Labor	Equipment	Total	Total Incl O&P
0010	**INSULATING FORMS, LEFT IN PLACE**										
0020	S.F. is for exterior face, but includes forms for both faces (total R22)										
2000	4" wall, straight block, 16" x 48" (5.33 S.F.)	G	2 Carp	90	.178	Ea.	16.45	5.95		22.40	28
2010	90 corner block, exterior 16" x 38" x 22" (6.67 S.F.)	G		75	.213		18.80	7.15		25.95	32.50
2020	45 corner block, exterior 16" x 34" x 18" (5.78 S.F.)	G		75	.213		18.95	7.15		26.10	33
2100	6" wall, straight block, 16" x 48" (5.33 S.F.)	G		90	.178		16.70	5.95		22.65	28.50
2110	90 corner block, exterior 16" x 32" x 24" (6.22 S.F.)	G		75	.213		20.50	7.15		27.65	34.50
2120	45 corner block, exterior 16" x 26" x 18" (4.89 S.F.)	G		75	.213		17.80	7.15		24.95	31.50
2130	Brick ledge block, 16" x 48" (5.33 S.F.)	G		80	.200		21	6.70		27.70	35
2140	Taper top block, 16" x 48" (5.33 S.F.)	G		80	.200		19.65	6.70		26.35	33
2200	8" wall, straight block, 16" x 48" (5.33 S.F.)	G		90	.178		17.25	5.95		23.20	29
2210	90 corner block, exterior 16" x 34" x 26" (6.67 S.F.)	G		75	.213		22.50	7.15		29.65	36.50
2220	45 corner block, exterior 16" x 28" x 20" (5.33 S.F.)	G		75	.213		19.15	7.15		26.30	33
2230	Brick ledge block, 16" x 48" (5.33 S.F.)	G		80	.200		22	6.70		28.70	35.50
2240	Taper top block, 16" x 48" (5.33 S.F.)	G	↓	80	.200	↓	20	6.70		26.70	33.50

03 11 Concrete Forming

03 11 23 – Permanent Stair Forming

03 11 23.75 Forms In Place, Stairs			Crew	Daily Output	Labor-Hours	Unit	Material	Labor	Equipment	Total	Total Incl O&P
							2011 Bare Costs				
0010	FORMS IN PLACE, STAIRS	R031113-40									
0015	(Slant length x width), 1 use		C-2	165	.291	S.F.	4.65	8.70		13.35	19.70
0150	4 use	R031113-60		190	.253		1.65	7.55		9.20	14.45
2000	Stairs, cast on sloping ground (length x width), 1 use			220	.218		1.88	6.55		8.43	13
2025	2 use			232	.207		1.03	6.20		7.23	11.50
2050	3 use			244	.197		.75	5.90		6.65	10.70
2100	4 use			256	.188		.61	5.60		6.21	10.05

03 15 Concrete Accessories

03 15 05 – Concrete Forming Accessories

03 15 05.02 Anchor Bolt Accessories		Crew	Daily Output	Labor-Hours	Unit	Material	Labor	Equipment	Total	Total Incl O&P
0010	ANCHOR BOLT ACCESSORIES									
0015	For anchor bolts set in fresh concrete, see Section 05 05 23.05									
8150	Anchor bolt sleeve, plastic, 1" diam. bolts	1 Carp	60	.133	Ea.	7	4.47		11.47	15.20
8500	1-1/2" diameter		28	.286		12.40	9.60		22	29.50
8600	2" diameter		24	.333		15.45	11.20		26.65	35.50
8650	3" diameter		20	.400		28.50	13.40		41.90	53.50

03 15 13 – Waterstops

03 15 13.50 Waterstops		Crew	Daily Output	Labor-Hours	Unit	Material	Labor	Equipment	Total	Total Incl O&P
0010	WATERSTOPS, PVC and Rubber									
0020	PVC, ribbed 3/16" thick, 4" wide	1 Carp	155	.052	L.F.	.95	1.73		2.68	3.95
0050	6" wide		145	.055		1.79	1.85		3.64	5.05
0500	With center bulb, 6" wide, 3/16" thick		135	.059		1.61	1.99		3.60	5.10
0550	3/8" thick		130	.062		2.91	2.06		4.97	6.65
0600	9" wide x 3/8" thick		125	.064		4.48	2.15		6.63	8.50

03 21 Reinforcing Steel

03 21 10 – Uncoated Reinforcing Steel

03 21 10.60 Reinforcing In Place				Crew	Daily Output	Labor-Hours	Unit	Material	Labor	Equipment	Total	Total Incl O&P
0015	REINFORCING IN PLACE, 50-60 ton lots, A615 Grade 60	R032110-10										
0020	Includes labor, but not material cost, to install accessories											
0030	Made from recycled materials		G									
0500	Footings, #4 to #7		G	4 Rodm	2.10	15.238	Ton	855	555		1,410	1,900
0550	#8 to #18	R032110-20	G		3.60	8.889		810	325		1,135	1,450
0700	Walls, #3 to #7		G		3	10.667		855	385		1,240	1,625
0750	#8 to #18	R032110-25	G		4	8		855	290		1,145	1,450
0900	For other than 50 – 60 ton lots											
1000	Under 10 ton job, #3 to #7, add						Ton	25%	10%			
1010	#8 to #18, add							20%	10%			
1050	10 – 50 ton job, #3 to #7, add							10%				
1060	#8 to #18, add							5%				
1100	60 – 100 ton job, #3 to #7, deduct							5%				
1110	#8 to #18, deduct							10%				
1150	Over 100 ton job, #3 to #7, deduct							10%				
1160	#8 to #18, deduct							15%				
2400	Dowels, 2 feet long, deformed, #3		G	2 Rodm	520	.031	Ea.	.37	1.12		1.49	2.35
2410	#4	R032110-40	G		480	.033		.66	1.21		1.87	2.83
2420	#5		G		435	.037		1.03	1.34		2.37	3.46

03 21 Reinforcing Steel

03 21 10 – Uncoated Reinforcing Steel

03 21 10.60 Reinforcing In Place		Crew	Daily Output	Labor-Hours	Unit	2011 Bare Costs Material	Labor	Equipment	Total	Total Incl O&P
2430	#6 R032110-50	2 Rodm	360	.044	Ea.	1.49	1.61		3.10	4.44
2600	Dowel sleeves for CIP concrete, 2-part system									
2610	Sleeve base, plastic, for 5/8" smooth dowel sleeve, fasten to edge form R032110-70	1 Rodm	200	.040	Ea.	.57	1.45		2.02	3.15
2615	Sleeve, plastic, 12" long, for 5/8" smooth dowel, snap onto base		400	.020		1.09	.73		1.82	2.46
2620	Sleeve base, for 3/4" smooth dowel sleeve R032110-80		175	.046		.54	1.66		2.20	3.47
2625	Sleeve, 12" long, for 3/4" smooth dowel		350	.023		.97	.83		1.80	2.51
2630	Sleeve base, for 1" smooth dowel sleeve		150	.053		.64	1.94		2.58	4.07
2635	Sleeve, 12" long, for 1" smooth dowel		300	.027		1.28	.97		2.25	3.09
2700	Dowel caps, visual warning only, plastic, #3 to #8	2 Rodm	800	.020		.26	.73		.99	1.55
2720	#8 to #18		750	.021		.78	.77		1.55	2.21
2750	Impalement protective, plastic, #4 to #9		800	.020		1.41	.73		2.14	2.81

03 21 10.70 Glass Fiber Reinforced Polymer Bars

Line	Description	Crew	Daily Output	Labor-Hours	Unit	Material	Labor	Equipment	Total	Total Incl O&P
0010	**GLASS FIBER REINFORCED POLYMER BARS**									
0050	#2 bar, .043 lb./L.F.	4 Rodm	9500	.003	L.F.	.35	.12		.47	.60
0100	#3 bar, .092 lb./L.F.		9300	.003		.48	.12		.60	.75
0150	#4 bar, .160 lb./L.F.		9100	.004		.70	.13		.83	.99
0200	#5 bar, .258 lb./L.F.		8700	.004		1.05	.13		1.18	1.39
0250	#6 bar, .372 lb./L.F.		8300	.004		1.38	.14		1.52	1.76
0300	#7 bar, .497 lb./L.F.		7900	.004		1.73	.15		1.88	2.16
0350	#8 bar, .620 lb./L.F.		7400	.004		2.32	.16		2.48	2.82
0400	#9 bar, .800 lb./L.F.		6800	.005		2.92	.17		3.09	3.51
0450	#10 bar, 1.08 lb./L.F.		5800	.006		3.50	.20		3.70	4.20
0500	For Bends, add per bend				Ea.	1.25			1.25	1.38

03 22 Welded Wire Fabric Reinforcing

03 22 05 – Uncoated Welded Wire Fabric

03 22 05.50 Welded Wire Fabric

Line	Description	Crew	Daily Output	Labor-Hours	Unit	Material	Labor	Equipment	Total	Total Incl O&P
0010	**WELDED WIRE FABRIC** ASTM A185 R032205-30									
0030	Made from recycled materials G									
0050	Sheets									
0100	6 x 6 - W1.4 x W1.4 (10 x 10) 21 lb. per C.S.F. G	2 Rodm	35	.457	C.S.F.	12.50	16.60		29.10	43
0300	6 x 6 - W2.9 x W2.9 (6 x 6) 42 lb. per C.S.F. G		29	.552		21.50	20		41.50	58.50
0500	4 x 4 - W1.4 x W1.4 (10 x 10) 31 lb. per C.S.F. G		31	.516		17.90	18.75		36.65	52
0750	Rolls									
0900	2 x 2 - #12 galv. for gunite reinforcing G	2 Rodm	6.50	2.462	C.S.F.	65	89.50		154.50	227

03 23 Stressing Tendons

03 23 05 – Prestressing Tendons

03 23 05.50 Prestressing Steel

Line	Description	Crew	Daily Output	Labor-Hours	Unit	Material	Labor	Equipment	Total	Total Incl O&P
0010	**PRESTRESSING STEEL** R034136-90									
3000	Slabs on grade, 0.5-inch diam. non-bonded strands, HDPE sheathed,									
3050	attached dead-end anchors, loose stressing-end anchors									
3100	25' x 30' slab, strands @ 36" O.C., placing	2 Rodm	2940	.005	S.F.	.59	.20		.79	.99
3105	Stressing	C-4A	3750	.004			.16	.01	.17	.28
3110	42" O.C., placing	2 Rodm	3200	.005		.52	.18		.70	.89
3115	Stressing	C-4A	4040	.004			.14	.01	.15	.26
3120	48" O.C., placing	2 Rodm	3510	.005		.46	.17		.63	.79
3125	Stressing	C-4A	4390	.004			.13	.01	.14	.24

03 23 Stressing Tendons

03 23 05 – Prestressing Tendons

03 23 05.50 Prestressing Steel		Crew	Daily Output	Labor-Hours	Unit	Material	2011 Bare Costs Labor	Equipment	Total	Total Incl O&P
3150	25′ x 40′ slab, strands @ 36″ O.C., placing	2 Rodm	3370	.005	S.F.	.57	.17		.74	.93
3155	Stressing	C-4A	4360	.004			.13	.01	.14	.24
3160	42″ O.C., placing	2 Rodm	3760	.004		.49	.15		.64	.81
3165	Stressing	C-4A	4820	.003			.12	.01	.13	.22
3170	48″ O.C., placing	2 Rodm	4090	.004		.44	.14		.58	.73
3175	Stressing	C-4A	5190	.003			.11	.01	.12	.20
3200	30′ x 30′ slab, strands @ 36″ O.C., placing	2 Rodm	3260	.005		.57	.18		.75	.94
3205	Stressing	C-4A	4190	.004			.14	.01	.15	.25
3210	42″ O.C., placing	2 Rodm	3530	.005		.51	.16		.67	.86
3215	Stressing	C-4A	4500	.004			.13	.01	.14	.23
3220	48″ O.C., placing	2 Rodm	3840	.004		.46	.15		.61	.76
3225	Stressing	C-4A	4850	.003			.12	.01	.13	.22
3230	30′ x 40′ slab, strands @ 36″ O.C., placing	2 Rodm	3780	.004		.55	.15		.70	.88
3235	Stressing	C-4A	4920	.003			.12	.01	.13	.22
3240	42″ O.C., placing	2 Rodm	4190	.004		.48	.14		.62	.77
3245	Stressing	C-4A	5410	.003			.11	.01	.12	.20
3250	48″ O.C., placing	2 Rodm	4520	.004		.44	.13		.57	.70
3255	Stressing	C-4A	5790	.003			.10	.01	.11	.18
3260	30′ x 50′ slab, strands @ 36″ O.C., placing	2 Rodm	4300	.004		.52	.14		.66	.80
3265	Stressing	C-4A	5650	.003			.10	.01	.11	.19
3270	42″ O.C., placing	2 Rodm	4720	.003		.46	.12		.58	.72
3275	Stressing	C-4A	6150	.003			.09	.01	.10	.17
3280	48″ O.C., placing	2 Rodm	5240	.003		.41	.11		.52	.64
3285	Stressing	C-4A	6760	.002	↓		.09	.01	.10	.16

03 24 Fibrous Reinforcing

03 24 05 – Reinforcing Fibers

03 24 05.30 Synthetic Fibers		Crew	Daily Output	Labor-Hours	Unit	Material	Labor	Equipment	Total	Total Incl O&P
0010	**SYNTHETIC FIBERS**									
0100	Synthetic fibers, add to concrete				Lb.	5.05			5.05	5.55
0110	1-1/2 lb. per C.Y.				C.Y.	7.80			7.80	8.60

03 24 05.70 Steel Fibers		Crew	Daily Output	Labor-Hours	Unit	Material	Labor	Equipment	Total	Total Incl O&P
0010	**STEEL FIBERS**									
0150	Steel fibers, add to concrete G				Lb.	.77			.77	.85
0155	25 lb. per C.Y. G				C.Y.	19.25			19.25	21
0160	50 lb. per C.Y. G					38.50			38.50	42.50
0170	75 lb. per C.Y. G					59.50			59.50	65
0180	100 lb. per C.Y. G				↓	77			77	84.50

03 30 Cast-In-Place Concrete

03 30 53 – Miscellaneous Cast-In-Place Concrete

03 30 53.40 Concrete In Place			Crew	Daily Output	Labor-Hours	Unit	Material	2011 Bare Costs Labor	Equipment	Total	Total Incl O&P
0010	**CONCRETE IN PLACE**	R033053-10									
0020	Including forms (4 uses), reinforcing steel, concrete, placement,	R033053-60									
0050	and finishing unless otherwise indicated	R033105-10									
0500	Chimney foundations (5000 psi), industrial, minimum	R033105-20	C-14C	32.22	3.476	C.Y.	144	111	.73	255.73	345
0510	Maximum	R033105-50	"	23.71	4.724	"	168	151	.99	319.99	440
3540	Equipment pad (3000 psi), 3′ x 3′ x 6″ thick	R033105-70	C-14H	45	1.067	Ea.	40.50	35	.51	76.01	104
3550	4′ x 4′ x 6″ thick			30	1.600		62	52.50	.77	115.27	157
3560	5′ x 5′ x 8″ thick			18	2.667		111	87.50	1.28	199.78	270
3570	6′ x 6′ x 8″ thick			14	3.429		150	113	1.65	264.65	355
3580	8′ x 8′ x 10″ thick			8	6		320	197	2.88	519.88	690
3590	10′ x 10′ x 12″ thick			5	9.600		550	315	4.61	869.61	1,150
3800	Footings (3000 psi), spread under 1 C.Y.		C-14C	28	4	C.Y.	158	128	.84	286.84	390
3825	1 C.Y. to 5 C.Y.			43	2.605		185	83	.55	268.55	345
3850	Over 5 C.Y.	R033105-80		75	1.493		171	47.50	.31	218.81	268
3900	Footings, strip (3000 psi), 18″ x 9″, unreinforced		C-14L	40	2.400		119	75	.58	194.58	257
3920	18″ x 9″, reinforced	R033105-85	C-14C	35	3.200		141	102	.67	243.67	330
3925	20″ x 10″, unreinforced		C-14L	45	2.133		116	66.50	.51	183.01	240
3930	20″ x 10″, reinforced		C-14C	40	2.800		134	89.50	.59	224.09	298
3935	24″ x 12″, unreinforced		C-14L	55	1.745		114	54.50	.42	168.92	217
3940	24″ x 12″, reinforced		C-14C	48	2.333		132	74.50	.49	206.99	271
3945	36″ x 12″, unreinforced		C-14L	70	1.371		111	43	.33	154.33	194
3950	36″ x 12″, reinforced		C-14C	60	1.867		127	59.50	.39	186.89	240
4000	Foundation mat (3000 psi), under 10 C.Y.			38.67	2.896		192	92.50	.61	285.11	365
4050	Over 20 C.Y.			56.40	1.986		169	63.50	.42	232.92	292
4520	Handicap access ramp (4000 psi), railing both sides, 3′ wide		C-14H	14.58	3.292	L.F.	278	108	1.58	387.58	490
4525	5′ wide			12.22	3.928		288	129	1.89	418.89	535
4530	With 6″ curb and rails both sides, 3′ wide			8.55	5.614		287	185	2.69	474.69	630
4535	5′ wide			7.31	6.566		292	216	3.15	511.15	685
4751	Slab on grade (3500 psi), incl. troweled finish, not incl. forms										
4760	or reinforcing, over 10,000 S.F., 4″ thick		C-14F	3425	.021	S.F.	1.29	.64	.01	1.94	2.46
4820	6″ thick		"	3350	.021	"	1.89	.65	.01	2.55	3.14
5000	Slab on grade (3000 psi), incl. textured finish, not incl. forms										
5001	or reinforcing, 4″ thick		C-14G	2873	.019	S.F.	1.29	.58	.01	1.88	2.37
5010	6″ thick			2590	.022		2.01	.64	.01	2.66	3.27
5020	8″ thick			2320	.024		2.62	.72	.01	3.35	4.07
6800	Stairs (3500 psi), not including safety treads, free standing, 3′-6″ wide		C-14H	83	.578	LF Nose	5	19	.28	24.28	38
6850	Cast on ground			125	.384	"	4.28	12.65	.18	17.11	26
7000	Stair landings, free standing			200	.240	S.F.	4.03	7.90	.12	12.05	17.75
7050	Cast on ground			475	.101	"	3.29	3.32	.05	6.66	9.20

03 31 Structural Concrete

03 31 05 – Normal Weight Structural Concrete

03 31 05.35 Normal Weight Concrete, Ready Mix			Crew	Daily Output	Labor-Hours	Unit	Material	Labor	Equipment	Total	Total Incl O&P
0010	**NORMAL WEIGHT CONCRETE, READY MIX**, delivered	R033105-10									
0012	Includes local aggregate, sand, Portland cement, and water										
0015	Excludes all additives and treatments	R033105-20									
0020	2000 psi					C.Y.	91.50			91.50	101
0100	2500 psi	R033105-30					94			94	103
0150	3000 psi						99			99	109
0200	3500 psi	R033105-40					99.50			99.50	110
0300	4000 psi						103			103	113

03 31 Structural Concrete

03 31 05 – Normal Weight Structural Concrete

03 31 05.35 Normal Weight Concrete, Ready Mix

	03 31 05.35 Normal Weight Concrete, Ready Mix		Crew	Daily Output	Labor-Hours	Unit	2011 Bare Costs Material	Labor	Equipment	Total	Total Incl O&P
0350	4500 psi	R033105-50				C.Y.	106			106	116
0400	5000 psi						111			111	122
0411	6000 psi						127			127	139
0412	8000 psi						206			206	227
0413	10,000 psi						293			293	320
0414	12,000 psi						355			355	390
1000	For high early strength cement, add						10%				
1300	For winter concrete (hot water), add						4.25			4.25	4.68
1400	For hot weather concrete (ice), add						9.35			9.35	10.25
1410	For mid-range water reducer, add						4.13			4.13	4.54
1420	For high-range water reducer/superplasticizer, add						6.35			6.35	6.95
1430	For retarder, add						2.71			2.71	2.98
1440	For non-Chloride accelerator, add						4.83			4.83	5.30
1450	For Chloride accelerator, per 1%, add						3.28			3.28	3.61
1460	For fiber reinforcing, synthetic (1 lb./C.Y.), add						6.65			6.65	7.30
1500	For Saturday delivery, add					↓	8.85			8.85	9.70
1510	For truck holding/waiting time past 1st hour per load, add					Hr.	87.50			87.50	96.50
1520	For short load (less than 4 C.Y.), add per load					Ea.	112			112	124
2000	For all lightweight aggregate, add					C.Y.	45%				

03 31 05.70 Placing Concrete

	03 31 05.70 Placing Concrete		Crew	Daily Output	Labor-Hours	Unit	Material	Labor	Equipment	Total	Total Incl O&P
0010	**PLACING CONCRETE**										
0020	Includes labor and equipment to place, strike off and consolidate										
1400	Elevated slabs, less than 6" thick, pumped		C-20	140	.457	C.Y.		13	5.60	18.60	27.50
1450	With crane and bucket	R033105-70	C-7	95	.758			22	13.45	35.45	51
1500	6" to 10" thick, pumped		C-20	160	.400			11.35	4.92	16.27	24.50
1550	With crane and bucket		C-7	110	.655			18.80	11.60	30.40	44
1600	Slabs over 10" thick, pumped		C-20	180	.356			10.10	4.37	14.47	21.50
1650	With crane and bucket		C-7	130	.554			15.90	9.85	25.75	37.50
1900	Footings, continuous, shallow, direct chute		C-6	120	.400			11.05	.39	11.44	18.80
1950	Pumped		C-20	150	.427			12.15	5.25	17.40	26
2000	With crane and bucket		C-7	90	.800			23	14.20	37.20	53.50
2400	Footings, spread, under 1 C.Y., direct chute		C-6	55	.873			24	.85	24.85	41
2600	Over 5 C.Y., direct chute			120	.400			11.05	.39	11.44	18.80
2900	Foundation mats, over 20 C.Y., direct chute			350	.137			3.80	.13	3.93	6.45
4300	Slab on grade, up to 6" thick, direct chute		↓	110	.436			12.10	.42	12.52	20.50
4350	Pumped		C-20	130	.492			14	6.05	20.05	29.50
4400	With crane and bucket		C-7	110	.655			18.80	11.60	30.40	44
4900	Walls, 8" thick, direct chute		C-6	90	.533			14.75	.52	15.27	25
4950	Pumped		C-20	100	.640			18.20	7.85	26.05	38.50
5000	With crane and bucket		C-7	80	.900			26	15.95	41.95	60.50
5050	12" thick, direct chute		C-6	100	.480			13.30	.47	13.77	22.50
5100	Pumped		C-20	110	.582			16.55	7.15	23.70	35.50
5200	With crane and bucket		C-7	90	.800	↓		23	14.20	37.20	53.50
5600	Wheeled concrete dumping, add to placing costs above										
5610	Walking cart, 50' haul, add		C-18	32	.281	C.Y.		7.50	1.72	9.22	14.45
5620	150' haul, add			24	.375			10	2.29	12.29	19.25
5700	250' haul, add		↓	18	.500			13.35	3.06	16.41	26
5800	Riding cart, 50' haul, add		C-19	80	.113			3	1.15	4.15	6.30
5810	150' haul, add			60	.150			4	1.53	5.53	8.40
5900	250' haul, add		↓	45	.200	↓		5.35	2.04	7.39	11.20

03 35 Concrete Finishing

03 35 29 – Tooled Concrete Finishing

03 35 29.30 Finishing Floors		Crew	Daily Output	Labor-Hours	Unit	Material	2011 Bare Costs Labor	Equipment	Total	Total Incl O&P
0010	**FINISHING FLOORS**									
0012	Finishing requires that concrete first be placed, struck off & consolidated									
0015	Basic finishing for various unspecified flatwork									
0100	Bull float only	C-10	4000	.006	S.F.		.18		.18	.29
0125	Bull float & manual float		2000	.012			.36		.36	.58
0150	Bull float, manual float, & broom finish, w/edging & joints		1850	.013			.39		.39	.63
0200	Bull float, manual float & manual steel trowel		1265	.019			.57		.57	.92
0210	For specified Random Access Floors in ACI Classes 1, 2, 3 and 4 to achieve									
0215	Composite Overall Floor Flatness and Levelness values up to F35/F25									
0250	Bull float, machine float & machine trowel (walk-behind)	C-10C	1715	.014	S.F.		.42	.02	.44	.70
0300	Power screed, bull float, machine float & trowel (walk-behind)	C-10D	2400	.010			.30	.04	.34	.54
0350	Power screed, bull float, machine float & trowel (ride-on)	C-10E	4000	.006			.18	.06	.24	.35
0352	For specified Random Access Floors in ACI Classes 5, 6, 7 and 8 to achieve									
0354	Composite Overall Floor Flatness and Levelness values up to F50/F50									
0356	Add for two-dimensional restraightening after power float	C-10	6000	.004	S.F.		.12		.12	.19
0358	For specified Random or Defined Access Floors in ACI Class 9 to achieve									
0360	Composite Overall Floor Flatness and Levelness values up to F100/F100									
0362	Add for two-dimensional restraightening after bull float & power float	C-10	3000	.008	S.F.		.24		.24	.39
0364	For specified Superflat Defined Access Floors in ACI Class 9 to achieve									
0366	Minimum Floor Flatness and Levelness values of F100/F100									
0368	Add for 2-dim'l restraightening after bull float, power float, power trowel	C-10	2000	.012	S.F.		.36		.36	.58
1600	Exposed local aggregate finish, minimum	1 Cefi	625	.013		.21	.41		.62	.88
1650	Maximum	"	465	.017		.64	.55		1.19	1.57

03 35 29.35 Control Joints, Saw Cut		Crew	Daily Output	Labor-Hours	Unit	Material	Labor	Equipment	Total	Total Incl O&P
0010	**CONTROL JOINTS, SAW CUT**									
0100	Sawcut control joints in green concrete									
0120	1" depth	C-27	2000	.008	L.F.	.05	.25	.07	.37	.53
0140	1-1/2" depth		1800	.009		.07	.28	.07	.42	.61
0160	2" depth		1600	.010		.10	.32	.08	.50	.71
0180	Sawcut joint reservoir in cured concrete									
0182	3/8" wide x 3/4" deep, with single saw blade	C-27	1000	.016	L.F.	.07	.51	.13	.71	1.04
0184	1/2" wide x 1" deep, with double saw blades		900	.018		.14	.57	.15	.86	1.22
0186	3/4" wide x 1-1/2" deep, with double saw blades		800	.020		.30	.64	.17	1.11	1.53
0190	Water blast joint to wash away laitance, 2 passes	C-29	2500	.003			.08	.02	.10	.17
0200	Air blast joint to blow out debris and air dry, 2 passes	C-28	2000	.004			.13	.01	.14	.21
0300	For backer rod, see Section 07 91 23.10									
0340	For joint sealant, see Sections 03 15 05.25 or 07 92 13.20, if available									
0900	For replacement of joint sealant, see Section 07 01 90.81									

03 35 29.60 Finishing Walls		Crew	Daily Output	Labor-Hours	Unit	Material	Labor	Equipment	Total	Total Incl O&P
0010	**FINISHING WALLS**									
0020	Break ties and patch voids	1 Cefi	540	.015	S.F.	.03	.47		.50	.79
0050	Burlap rub with grout	"	450	.018		.03	.57		.60	.94
0300	Bush hammer, green concrete	B-39	1000	.048			1.29	.19	1.48	2.36
0350	Cured concrete	"	650	.074			1.98	.30	2.28	3.64

03 35 33 – Stamped Concrete Finishing

03 35 33.50 Slab Texture Stamping		Crew	Daily Output	Labor-Hours	Unit	Material	Labor	Equipment	Total	Total Incl O&P
0010	**SLAB TEXTURE STAMPING**									
0050	Stamping requires that concrete first be placed, struck off, consolidated,									
0060	bull floated and free of bleed water. Decorative stamping tasks include:									
0100	Step 1 - first application of dry shake colored hardener	1 Cefi	6400	.001	S.F.	.38	.04		.42	.48
0110	Step 2 - bull float		6400	.001			.04		.04	.06
0130	Step 3 - second application of dry shake colored hardener		6400	.001		.19	.04		.23	.27

03 35 Concrete Finishing

03 35 33 – Stamped Concrete Finishing

03 35 33.50 Slab Texture Stamping		Crew	Daily Output	Labor-Hours	Unit	Material	2011 Bare Costs Labor	Equipment	Total	Total Incl O&P
0140	Step 4 - bull float, manual float & steel trowel	3 Cefi	1280	.019	S.F.		.60		.60	.95
0150	Step 5 - application of dry shake colored release agent	1 Cefi	6400	.001		.08	.04		.12	.15
0160	Step 6 - place, tamp & remove mats	3 Cefi	2400	.010		1.41	.32		1.73	2.06
0170	Step 7 - touch up edges, mat joints & simulated grout lines	1 Cefi	1280	.006			.20		.20	.32
0300	Alternate stamping estimating method includes all tasks above	4 Cefi	800	.040		2.07	1.27		3.34	4.31
0400	Step 8 - pressure wash @ 3000 psi after 24 hours	1 Cefi	1600	.005			.16		.16	.25
0500	Step 9 - roll 2 coats cure/seal compound when dry	"	800	.010		.48	.32		.80	1.04

03 35 43 – Polished Concrete Finishing

03 35 43.10 Polished Concrete Floors		Crew	Daily Output	Labor-Hours	Unit	Material	Labor	Equipment	Total	Total Incl O&P
0010	**POLISHED CONCRETE FLOORS**									
0015	Processing of cured concrete to include grinding, honing,									
0020	and polishing of interior floors with 22" segmented diamond									
0025	planetary floor grinder (2 passes in different directions per grit)									
0100	Removal of pre-existing coatings, dry, with carbide discs using									
0105	dry vacuum pick-up system, final hand sweeping									
0110	Glue, adhesive or tar	J-4	1.60	15	M.S.F.	19.50	450	107	576.50	870
0120	Paint, epoxy, 1 coat		3.60	6.667		19.50	200	47.50	267	400
0130	2 coats		1.80	13.333		19.50	400	95.50	515	775
0200	Grinding and edging, wet, including wet vac pick-up and auto									
0205	scrubbing between grit changes									
0210	40-grit diamond/metal matrix	J-4A	1.60	20	M.S.F.	33	585	204	822	1,200
0220	80-grit diamond/metal matrix		2	16		33	465	163	661	975
0230	120-grit diamond/metal matrix		2.40	13.333		33	390	136	559	820
0240	200-grit diamond/metal matrix		2.80	11.429		33	335	116	484	710
0300	Spray on dye or stain (1 coat)	1 Cefi	16	.500		271	15.95		286.95	325
0400	Spray on densifier/hardener (2 coats)	"	8	1		279	32		311	355
0410	Auto scrubbing after 2nd coat, when dry	J-4B	16	.500			13.25	9.65	22.90	32.50
0500	Honing and edging, wet, including wet vac pick-up and auto									
0505	scrubbing between grit changes									
0510	100-grit diamond/resin matrix	J-4A	2.80	11.429	M.S.F.	33	335	116	484	710
0520	200-grit diamond/resin matrix	"	2.80	11.429	"	33	335	116	484	710
0530	Dry, including dry vacuum pick-up system, final hand sweeping									
0540	400-grit diamond/resin matrix	J-4A	2.80	11.429	M.S.F.	33	335	116	484	710
0600	Polishing and edging, dry, including dry vac pick-up and hand									
0605	sweeping between grit changes									
0610	800-grit diamond/resin matrix	J-4A	2.80	11.429	M.S.F.	33	335	116	484	710
0620	1500-grit diamond/resin matrix		2.80	11.429		33	335	116	484	710
0630	3000-grit diamond/resin matrix		2.80	11.429		33	335	116	484	710
0700	Auto scrubbing after final polishing step	J-4B	16	.500			13.25	9.65	22.90	32.50

03 39 Concrete Curing

03 39 13 – Water Concrete Curing

03 39 13.50 Water Curing		Crew	Daily Output	Labor-Hours	Unit	Material	Labor	Equipment	Total	Total Incl O&P
0010	**WATER CURING**									
0015	With burlap, 4 uses assumed, 7.5 oz.	2 Clab	55	.291	C.S.F.	8.55	7.70		16.25	22.50
0100	10 oz.	"	55	.291	"	15.40	7.70		23.10	30

03 39 23 – Membrane Concrete Curing

03 39 23.13 Chemical Compound Membrane Concrete Curing		Crew	Daily Output	Labor-Hours	Unit	Material	Labor	Equipment	Total	Total Incl O&P
0010	**CHEMICAL COMPOUND MEMBRANE CONCRETE CURING**									
0300	Sprayed membrane curing compound	2 Clab	95	.168	C.S.F.	5.60	4.45		10.05	13.60

03 39 Concrete Curing

03 39 23 – Membrane Concrete Curing

03 39 23.23 Sheet Membrane Concrete Curing		Crew	Daily Output	Labor-Hours	Unit	2011 Bare Costs Material	Labor	Equipment	Total	Total Incl O&P
0010	**SHEET MEMBRANE CONCRETE CURING**									
0200	Curing blanket, burlap/poly, 2-ply	2 Clab	70	.229	C.S.F.	14.85	6.05		20.90	26.50

03 41 Precast Structural Concrete

03 41 13 – Precast Concrete Hollow Core Planks

03 41 13.50 Precast Slab Planks

		Crew	Daily Output	Labor-Hours	Unit	Material	Labor	Equipment	Total	Total Incl O&P
0010	**PRECAST SLAB PLANKS** R034105-30									
0020	Prestressed roof/floor members, grouted, solid, 4" thick	C-11	2400	.023	S.F.	5.55	.81	.79	7.15	8.30
0050	6" thick		2800	.020		6.50	.69	.68	7.87	9.05
0100	Hollow, 8" thick		3200	.018		7.20	.61	.60	8.41	9.55
0150	10" thick		3600	.016		7.50	.54	.53	8.57	9.75
0200	12" thick	↓	4000	.014	↓	8	.48	.48	8.96	10.15

03 41 23 – Precast Concrete Stairs

03 41 23.50 Precast Stairs

		Crew	Daily Output	Labor-Hours	Unit	Material	Labor	Equipment	Total	Total Incl O&P
0010	**PRECAST STAIRS**									
0020	Precast concrete treads on steel stringers, 3' wide	C-12	75	.640	Riser	129	21	8.75	158.75	187
0300	Front entrance, 5' wide with 48" platform, 2 risers		16	3	Flight	460	99	41	600	715
0350	5 risers		12	4		715	132	54.50	901.50	1,075
0500	6' wide, 2 risers		15	3.200		510	106	43.50	659.50	785
0550	5 risers		11	4.364		790	144	59.50	993.50	1,175
0700	7' wide, 2 risers	↓	14	3.429		660	113	47	820	965
1200	Basement entrance stairs, steel bulkhead doors, minimum	B-51	22	2.182		1,350	58.50	7.80	1,416.30	1,600
1250	Maximum	"	11	4.364	↓	2,275	117	15.60	2,407.60	2,700

03 48 Precast Concrete Specialties

03 48 43 – Precast Concrete Trim

03 48 43.40 Precast Lintels

		Crew	Daily Output	Labor-Hours	Unit	Material	Labor	Equipment	Total	Total Incl O&P
0010	**PRECAST LINTELS**, smooth gray, prestressed, stock units only									
0800	4" wide, 8" high, x 4' long	D-10	28	1.143	Ea.	22.50	38	15.50	76	104
0850	8' long		24	1.333		57.50	44	18.10	119.60	155
1000	6" wide, 8" high, x 4' long		26	1.231		33	40.50	16.70	90.20	121
1050	10' long	↓	22	1.455	↓	88	48	19.75	155.75	198

03 48 43.90 Precast Window Sills

		Crew	Daily Output	Labor-Hours	Unit	Material	Labor	Equipment	Total	Total Incl O&P
0010	**PRECAST WINDOW SILLS**									
0600	Precast concrete, 4" tapers to 3", 9" wide	D-1	70	.229	L.F.	12.25	6.95		19.20	25
0650	11" wide		60	.267		16	8.10		24.10	31
0700	13" wide, 3 1/2" tapers to 2 1/2", 12" wall	↓	50	.320	↓	16	9.75		25.75	33.50

03 54 Cast Underlayment

03 54 13 – Gypsum Cement Underlayment

03 54 13.50 Poured Gypsum Underlayment		Crew	Daily Output	Labor-Hours	Unit	2011 Bare Costs Material	Labor	Equipment	Total	Total Incl O&P
0010	**POURED GYPSUM UNDERLAYMENT**									
0400	Underlayment, gypsum based, self-leveling 2500 psi, pumped, 1/2"	C-8	24000	.002	S.F.	.31	.07	.03	.41	.48
0500	3/4"		20000	.003		.47	.08	.04	.59	.69
0600	1"		16000	.004		.62	.10	.05	.77	.91
1400	Hand placed, 1/2"	C-18	450	.020		.31	.53	.12	.96	1.36
1500	3/4"	"	300	.030		.47	.80	.18	1.45	2.05

03 63 Epoxy Grouting

03 63 05 – Grouting of Dowels and Fasteners

03 63 05.10 Epoxy Only		Crew	Daily Output	Labor-Hours	Unit	Material	Labor	Equipment	Total	Total Incl O&P
0010	**EPOXY ONLY**									
1500	Chemical anchoring, epoxy cartridge, excludes layout, drilling, fastener									
1530	For fastener 3/4" diam. x 6" embedment	2 Skwk	72	.222	Ea.	4.95	7.55		12.50	18.10
1535	1" diam. x 8" embedment		66	.242		7.45	8.25		15.70	22
1540	1-1/4" diam. x 10" embedment		60	.267		14.85	9.10		23.95	31.50
1545	1-3/4" diam. x 12" embedment		54	.296		25	10.10		35.10	44
1550	14" embedment		48	.333		29.50	11.35		40.85	51.50
1555	2" diam. x 12" embedment		42	.381		39.50	12.95		52.45	65
1560	18" embedment		32	.500		49.50	17.05		66.55	83

03 82 Concrete Boring

03 82 16 – Concrete Drilling

	03 82 16.10 Concrete Impact Drilling	Crew	Daily Output	Labor-Hours	Unit	Material	2011 Bare Costs Labor	Equipment	Total	Total Incl O&P
0010	**CONCRETE IMPACT DRILLING**									
0050	Up to 4" deep in conc/brick floor/wall, incl. bit & layout, no anchor									
0100	Holes, 1/4" diameter	1 Carp	75	.107	Ea.	.05	3.58		3.63	6.05
0150	For each additional inch of depth, add		430	.019		.01	.62		.63	1.05
0200	3/8" diameter		63	.127		.05	4.26		4.31	7.20
0250	For each additional inch of depth, add		340	.024		.01	.79		.80	1.33
0300	1/2" diameter		50	.160		.05	5.35		5.40	9.05
0350	For each additional inch of depth, add		250	.032		.01	1.07		1.08	1.81
0400	5/8" diameter		48	.167		.07	5.60		5.67	9.45
0450	For each additional inch of depth, add		240	.033		.02	1.12		1.14	1.89
0500	3/4" diameter		45	.178		.10	5.95		6.05	10.10
0550	For each additional inch of depth, add		220	.036		.02	1.22		1.24	2.07
0600	7/8" diameter		43	.186		.13	6.25		6.38	10.60
0650	For each additional inch of depth, add		210	.038		.03	1.28		1.31	2.18
0700	1" diameter		40	.200		.14	6.70		6.84	11.40
0750	For each additional inch of depth, add		190	.042		.04	1.41		1.45	2.40
0800	1-1/4" diameter		38	.211		.22	7.05		7.27	12.05
0850	For each additional inch of depth, add		180	.044		.06	1.49		1.55	2.56
0900	1-1/2" diameter		35	.229		.34	7.65		7.99	13.20
0950	For each additional inch of depth, add		165	.048		.08	1.63		1.71	2.81
1000	For ceiling installations, add						40%			

Division Notes

			DAILY	LABOR-		2011 BARE COSTS				TOTAL
		CREW	OUTPUT	HOURS	UNIT	MAT.	LABOR	EQUIP.	TOTAL	INCL O&P

Estimating Tips

04 05 00 Common Work Results for Masonry

- The terms *mortar* and *grout* are often used interchangeably, and incorrectly. Mortar is used to bed masonry units, seal the entry of air and moisture, provide architectural appearance, and allow for size variations in the units. Grout is used primarily in reinforced masonry construction and is used to bond the masonry to the reinforcing steel. Common mortar types are M(2500 psi), S(1800 psi), N(750 psi), and O(350 psi), and conform to ASTM C270. Grout is either fine or coarse and conforms to ASTM C476, and in-place strengths generally exceed 2500 psi. Mortar and grout are different components of masonry construction and are placed by entirely different methods. An estimator should be aware of their unique uses and costs.
- Mortar is included in all assembled masonry line items. The mortar cost, part of the assembled masonry material cost, includes all ingredients, all labor, and all equipment required. Please see reference number R040513-10.
- Waste, specifically the loss/droppings of mortar and the breakage of brick and block, is included in all masonry assemblies in this division. A factor of 25% is added for mortar and 3% for brick and concrete masonry units.
- Scaffolding or staging is not included in any of the Division 4 costs. Refer to Subdivision 01 54 23 for scaffolding and staging costs.

04 20 00 Unit Masonry

- The most common types of unit masonry are brick and concrete masonry. The major classifications of brick are building brick (ASTM C62), facing brick (ASTM C216), glazed brick, fire brick, and pavers. Many varieties of texture and appearance can exist within these classifications, and the estimator would be wise to check local custom and availability within the project area. For repair and remodeling jobs, matching the existing brick may be the most important criteria.
- Brick and concrete block are priced by the piece and then converted into a price per square foot of wall. Openings less than two square feet are generally ignored by the estimator because any savings in units used is offset by the cutting and trimming required.
- It is often difficult and expensive to find and purchase small lots of historic brick. Costs can vary widely. Many design issues affect costs, selection of mortar mix, and repairs or replacement of masonry materials. Cleaning techniques must be reflected in the estimate.
- All masonry walls, whether interior or exterior, require bracing. The cost of bracing walls during construction should be included by the estimator, and this bracing must remain in place until permanent bracing is complete. Permanent bracing of masonry walls is accomplished by masonry itself, in the form of pilasters or abutting wall corners, or by anchoring the walls to the structural frame. Accessories in the form of anchors, anchor slots, and ties are used, but their supply and installation can be by different trades. For instance, anchor slots on spandrel beams and columns are supplied and welded in place by the steel fabricator, but the ties from the slots into the masonry are installed by the bricklayer. Regardless of the installation method, the estimator must be certain that these accessories are accounted for in pricing.

Reference Numbers

Reference numbers are shown in shaded boxes at the beginning of some major classifications. These numbers refer to related items in the Reference Section. The reference information may be an estimating procedure, an alternate pricing method, or technical information.

Note: Not all subdivisions listed here necessarily appear in this publication.

04 01 Maintenance of Masonry

04 01 20 – Maintenance of Unit Masonry

04 01 20.20 Pointing Masonry		Crew	Daily Output	Labor-Hours	Unit	2011 Bare Costs Material	Labor	Equipment	Total	Total Incl O&P
0010	**POINTING MASONRY**									
0300	Cut and repoint brick, hard mortar, running bond	1 Bric	80	.100	S.F.	.53	3.37		3.90	6.10
0320	Common bond		77	.104		.53	3.50		4.03	6.35
0360	Flemish bond		70	.114		.56	3.85		4.41	6.90
0400	English bond		65	.123		.56	4.14		4.70	7.40
0600	Soft old mortar, running bond		100	.080		.53	2.69		3.22	5
0620	Common bond		96	.083		.53	2.80		3.33	5.20
0640	Flemish bond		90	.089		.56	2.99		3.55	5.50
0680	English bond		82	.098		.56	3.28		3.84	6
0700	Stonework, hard mortar		140	.057	L.F.	.70	1.92		2.62	3.92
0720	Soft old mortar		160	.050	"	.70	1.68		2.38	3.53
1000	Repoint, mask and grout method, running bond		95	.084	S.F.	.70	2.83		3.53	5.40
1020	Common bond		90	.089		.70	2.99		3.69	5.70
1040	Flemish bond		86	.093		.74	3.13		3.87	5.95
1060	English bond		77	.104		.74	3.50		4.24	6.55
2000	Scrub coat, sand grout on walls, minimum		120	.067		2.77	2.24		5.01	6.75
2020	Maximum		98	.082		3.84	2.75		6.59	8.75

04 01 20.30 Pointing CMU		Crew	Daily Output	Labor-Hours	Unit	Material	Labor	Equipment	Total	Total Incl O&P
0010	**POINTING CMU**									
0300	Cut and repoint block, hard mortar, running bond	1 Bric	190	.042	S.F.	.22	1.42		1.64	2.56
0310	Stacked bond		200	.040		.22	1.35		1.57	2.45
0600	Soft old mortar, running bond		230	.035		.22	1.17		1.39	2.16
0610	Stacked bond		245	.033		.22	1.10		1.32	2.04

04 01 30 – Unit Masonry Cleaning

04 01 30.60 Brick Washing		Crew	Daily Output	Labor-Hours	Unit	Material	Labor	Equipment	Total	Total Incl O&P
0010	**BRICK WASHING**	R040130-10								
0012	Acid cleanser, smooth brick surface	1 Bric	560	.014	S.F.	.05	.48		.53	.84
0050	Rough brick		400	.020		.06	.67		.73	1.17
0060	Stone, acid wash		600	.013		.08	.45		.53	.82
1000	Muriatic acid, price per gallon in 5 gallon lots				Gal.	9.55			9.55	10.50

04 05 Common Work Results for Masonry

04 05 05 – Selective Masonry Demolition

04 05 05.10 Selective Demolition		Crew	Daily Output	Labor-Hours	Unit	Material	Labor	Equipment	Total	Total Incl O&P
0010	**SELECTIVE DEMOLITION**	R024119-10								
0200	Bond beams, 8" block with #4 bar	2 Clab	32	.500	L.F.		13.25		13.25	22
0300	Concrete block walls, unreinforced, 2" thick		1200	.013	S.F.		.35		.35	.59
0310	4" thick		1150	.014			.37		.37	.62
0320	6" thick		1100	.015			.38		.38	.64
0330	8" thick		1050	.015			.40		.40	.68
0340	10" thick		1000	.016			.42		.42	.71
0360	12" thick		950	.017			.45		.45	.75
0380	Reinforced alternate courses, 2" thick		1130	.014			.37		.37	.63
0390	4" thick		1080	.015			.39		.39	.66
0400	6" thick		1035	.015			.41		.41	.68
0410	8" thick		990	.016			.43		.43	.72
0420	10" thick		940	.017			.45		.45	.75
0430	12" thick		890	.018			.48		.48	.80
0440	Reinforced alternate courses & vertically 48" OC, 4" thick		900	.018			.47		.47	.79
0450	6" thick		850	.019			.50		.50	.83

04 05 Common Work Results for Masonry

04 05 05 – Selective Masonry Demolition

04 05 05.10 Selective Demolition		Crew	Daily Output	Labor-Hours	Unit	Material	2011 Bare Costs Labor	Equipment	Total	Total Incl O&P
0460	8" thick	2 Clab	800	.020	S.F.		.53		.53	.89
0480	10" thick		750	.021			.56		.56	.94
0490	12" thick		700	.023			.60		.60	1.01
1000	Chimney, 16" x 16", soft old mortar	1 Clab	55	.145	C.F.		3.85		3.85	6.45
1004	Chimney, flue tile, 8" x 8", soft old mortar		55	.145			3.85		3.85	6.45
1020	Hard mortar		40	.200			5.30		5.30	8.85
1030	16" x 20", soft old mortar		55	.145			3.85		3.85	6.45
1040	Hard mortar		40	.200			5.30		5.30	8.85
1050	16" x 24", soft old mortar		55	.145			3.85		3.85	6.45
1060	Hard mortar		40	.200			5.30		5.30	8.85
1080	20" x 20", soft old mortar		55	.145			3.85		3.85	6.45
1100	Hard mortar		40	.200			5.30		5.30	8.85
1110	20" x 24", soft old mortar		55	.145			3.85		3.85	6.45
1120	Hard mortar		40	.200			5.30		5.30	8.85
1140	20" x 32", soft old mortar		55	.145			3.85		3.85	6.45
1160	Hard mortar		40	.200			5.30		5.30	8.85
1200	48" x 48", soft old mortar		55	.145			3.85		3.85	6.45
1220	Hard mortar		40	.200			5.30		5.30	8.85
1250	Metal, high temp steel jacket, 24" diameter	E-2	130	.369	V.L.F.		13.50	12.50	26	38.50
1260	60" diameter	"	60	.800			29.50	27	56.50	84
1280	Flue lining, up to 12" x 12"	1 Clab	200	.040			1.06		1.06	1.77
1282	Up to 24" x 24"		150	.053			1.41		1.41	2.36
2000	Columns, 8" x 8", soft old mortar		48	.167			4.41		4.41	7.40
2020	Hard mortar		40	.200			5.30		5.30	8.85
2060	16" x 16", soft old mortar		16	.500			13.25		13.25	22
2100	Hard mortar		14	.571			15.10		15.10	25.50
2140	24" x 24", soft old mortar		8	1			26.50		26.50	44.50
2160	Hard mortar		6	1.333			35.50		35.50	59
2200	36" x 36", soft old mortar		4	2			53		53	88.50
2220	Hard mortar		3	2.667			70.50		70.50	118
2230	Alternate pricing method, soft old mortar		30	.267	C.F.		7.05		7.05	11.80
2240	Hard mortar		23	.348	"		9.20		9.20	15.40
3000	Copings, precast or masonry, to 8" wide									
3020	Soft old mortar	1 Clab	180	.044	L.F.		1.18		1.18	1.97
3040	Hard mortar	"	160	.050	"		1.32		1.32	2.22
3100	To 12" wide									
3120	Soft old mortar	1 Clab	160	.050	L.F.		1.32		1.32	2.22
3140	Hard mortar	"	140	.057	"		1.51		1.51	2.53
4000	Fireplace, brick, 30" x 24" opening									
4020	Soft old mortar	1 Clab	2	4	Ea.		106		106	177
4021	Soft old mortar		64	.125	S.F.		3.31		3.31	5.55
4040	Hard mortar		1.25	6.400	Ea.		169		169	284
4100	Stone, soft old mortar		1.50	5.333			141		141	236
4120	Hard mortar		1	8			212		212	355
4150	Up to 48" fireplace, 15' chimney and foundation		.28	28.571			755		755	1,275
4400	Premanufactured, up to 48"	2 Clab	14	1.143			30		30	50.50
5000	Veneers, brick, soft old mortar	1 Clab	140	.057	S.F.		1.51		1.51	2.53
5020	Hard mortar		125	.064			1.69		1.69	2.84
5025	Hard mortar, 12" thick		90	.089			2.35		2.35	3.94
5050	Glass block, up to 4" thick		500	.016			.42		.42	.71
5100	Granite and marble, 2" thick		180	.044			1.18		1.18	1.97
5120	4" thick		170	.047			1.24		1.24	2.08
5130	Granite and marble, 5' long x 18" wide x 2" thick		10	.800	Ea.		21		21	35.50

04 05 Common Work Results for Masonry

04 05 05 – Selective Masonry Demolition

04 05 05.10 Selective Demolition		Crew	Daily Output	Labor-Hours	Unit	Material	2011 Bare Costs Labor	Equipment	Total	Total Incl O&P
5140	Stone, 4" thick	1 Clab	180	.044	S.F.		1.18		1.18	1.97
5160	8" thick		175	.046	"		1.21		1.21	2.03
5400	Alternate pricing method, stone, 4" thick		60	.133	C.F.		3.53		3.53	5.90
5420	8" thick		85	.094			2.49		2.49	4.17
5450	Solid masonry		130	.062			1.63		1.63	2.73
5460	Stone or precast sills, treads, copings		130	.062			1.63		1.63	2.73
5470	Solid stone or precast		110	.073			1.92		1.92	3.22
5500	Remove and reset steel lintel	1 Bric	40	.200	L.F.		6.75		6.75	11.05
5600	Vent box removal	1 Clab	50	.160	S.F.		4.23		4.23	7.10
5700	Remove block pilaster for fence, 6' high		2.33	3.433	Ea.		91		91	152
5800	Remove 12" x 12" step flashing from mortar joints		240	.033	C.F.		.88		.88	1.48

04 05 13 – Masonry Mortaring

04 05 13.10 Cement

Code	Description	Crew	Daily Output	Labor-Hours	Unit	Material	Labor	Equipment	Total	Total Incl O&P
0010	**CEMENT** R040513-10									
0100	Masonry, 70 lb. bag, T.L. lots				Bag	9.45			9.45	10.35
0150	L.T.L. lots					10			10	11
0200	White, 70 lb. bag, T.L. lots					14.85			14.85	16.35
0250	L.T.L. lots					16.70			16.70	18.35

04 05 16 – Masonry Grouting

04 05 16.30 Grouting

Code	Description	Crew	Daily Output	Labor-Hours	Unit	Material	Labor	Equipment	Total	Total Incl O&P
0010	**GROUTING**									
0200	Concrete block cores, solid, 4" thk., by hand, 0.067 C.F./S.F. of wall	D-8	1100	.036	S.F.	.28	1.13		1.41	2.16
0210	6" thick, pumped, 0.175 C.F. per S.F.	D-4	720	.056		.74	1.57	.18	2.49	3.59
0250	8" thick, pumped, 0.258 C.F. per S.F.		680	.059		1.09	1.67	.19	2.95	4.14
0300	10" thick, pumped, 0.340 C.F. per S.F.		660	.061		1.44	1.72	.19	3.35	4.62
0350	12" thick, pumped, 0.422 C.F. per S.F.		640	.063		1.79	1.77	.20	3.76	5.10

04 05 19 – Masonry Anchorage and Reinforcing

04 05 19.05 Anchor Bolts

Code	Description	Crew	Daily Output	Labor-Hours	Unit	Material	Labor	Equipment	Total	Total Incl O&P
0010	**ANCHOR BOLTS**									
0015	Installed in fresh grout in CMU bond beams or filled cores, no templates									
0020	Hooked, with nut and washer, 1/2" diam., 8" long	1 Bric	132	.061	Ea.	1.36	2.04		3.40	4.84
0030	12" long		131	.061		1.51	2.06		3.57	5.05
0060	3/4" diameter, 8" long		127	.063		3.65	2.12		5.77	7.50
0070	12" long		125	.064		4.56	2.15		6.71	8.55

04 05 19.16 Masonry Anchors

Code	Description	Crew	Daily Output	Labor-Hours	Unit	Material	Labor	Equipment	Total	Total Incl O&P
0010	**MASONRY ANCHORS**									
0020	For brick veneer, galv., corrugated, 7/8" x 7", 22 Ga.	1 Bric	10.50	.762	C	9.45	25.50		34.95	52.50
0100	24 Ga.		10.50	.762		8.90	25.50		34.40	52
0150	16 Ga.		10.50	.762		25	25.50		50.50	69.50
0200	Buck anchors, galv., corrugated, 16 gauge, 2" bend, 8" x 2"		10.50	.762		49	25.50		74.50	96
0250	8" x 3"		10.50	.762		51	25.50		76.50	98
0660	Cavity wall, Z-type, galvanized, 6" long, 1/8" diam.		10.50	.762		23	25.50		48.50	67
0670	3/16" diameter		10.50	.762		21	25.50		46.50	65
0680	1/4" diameter		10.50	.762		39.50	25.50		65	85.50
0850	8" long, 3/16" diameter		10.50	.762		23.50	25.50		49	68
0855	1/4" diameter		10.50	.762		46.50	25.50		72	93
1000	Rectangular type, galvanized, 1/4" diameter, 2" x 6"		10.50	.762		67	25.50		92.50	116
1050	4" x 6"		10.50	.762		83.50	25.50		109	134
1100	3/16" diameter, 2" x 6"		10.50	.762		38	25.50		63.50	83.50
1150	4" x 6"		10.50	.762		39	25.50		64.50	85
1500	Rigid partition anchors, plain, 8" long, 1" x 1/8"		10.50	.762		109	25.50		134.50	162

04 05 Common Work Results for Masonry

04 05 19 – Masonry Anchorage and Reinforcing

04 05 19.16 Masonry Anchors		Crew	Daily Output	Labor-Hours	Unit	Material	2011 Bare Costs Labor	Equipment	Total	Total Incl O&P
1550	1" x 1/4"	1 Bric	10.50	.762	C	180	25.50		205.50	240
1580	1-1/2" x 1/8"		10.50	.762		145	25.50		170.50	202
1600	1-1/2" x 1/4"		10.50	.762		295	25.50		320.50	365
1650	2" x 1/8"		10.50	.762		188	25.50		213.50	249
1700	2" x 1/4"	↓	10.50	.762	↓	345	25.50		370.50	420

04 05 19.26 Masonry Reinforcing Bars		Crew	Daily Output	Labor-Hours	Unit	Material	Labor	Equipment	Total	Total Incl O&P
0010	**MASONRY REINFORCING BARS** R040519-50									
0015	Steel bars A615, placed horiz., #3 & #4 bars	1 Bric	450	.018	Lb.	.45	.60		1.05	1.48
0050	Placed vertical, #3 & #4 bars		350	.023		.45	.77		1.22	1.76
0060	#5 & #6 bars		650	.012	↓	.45	.41		.86	1.18
0200	Joint reinforcing, regular truss, to 6" wide, mill std galvanized		30	.267	C.L.F.	15.45	8.95		24.40	31.50
0250	12" wide		20	.400		17	13.45		30.45	40.50
0400	Cavity truss with drip section, to 6" wide		30	.267		14.80	8.95		23.75	31
0450	12" wide	↓	20	.400	↓	15.20	13.45		28.65	38.50

04 05 23 – Masonry Accessories

04 05 23.13 Masonry Control and Expansion Joints		Crew	Daily Output	Labor-Hours	Unit	Material	Labor	Equipment	Total	Total Incl O&P
0010	**MASONRY CONTROL AND EXPANSION JOINTS**									
0020	Rubber, for double wythe 8" minimum wall (Brick/CMU)	1 Bric	400	.020	L.F.	1.72	.67		2.39	2.99
0025	"T" shaped		320	.025		1.05	.84		1.89	2.54
0030	Cross-shaped for CMU units		280	.029		1.38	.96		2.34	3.10
0050	PVC, for double wythe 8" minimum wall (Brick/CMU)	↓	400	.020	↓	1.26	.67		1.93	2.49

04 21 Clay Unit Masonry

04 21 13 – Brick Masonry

04 21 13.13 Brick Veneer Masonry		Crew	Daily Output	Labor-Hours	Unit	Material	Labor	Equipment	Total	Total Incl O&P
0010	**BRICK VENEER MASONRY**, T.L. lots, excl. scaff., grout & reinforcing									
0015	Material costs incl. 3% brick and 25% mortar waste									
2000	Standard, sel. common, 4" x 2-2/3" x 8", (6.75/S.F.)	D-8	230	.174	S.F.	3.61	5.40		9.01	12.80
2020	Standard, red, 4" x 2-2/3" x 8", running bond (6.75/S.F.)		220	.182		4.19	5.65		9.84	13.85
2050	Full header every 6th course (7.88/S.F.)		185	.216		4.88	6.70		11.58	16.35
2100	English, full header every 2nd course (10.13/S.F.)		140	.286		6.25	8.85		15.10	21.50
2150	Flemish, alternate header every course (9.00/S.F.)		150	.267		5.55	8.30		13.85	19.70
2200	Flemish, alt. header every 6th course (7.13/S.F.)		205	.195		4.42	6.05		10.47	14.80
2250	Full headers throughout (13.50/S.F.)		105	.381		8.30	11.85		20.15	28.50
2300	Rowlock course (13.50/S.F.)		100	.400		8.30	12.40		20.70	29.50
2350	Rowlock stretcher (4.50/S.F.)		310	.129		2.81	4.01		6.82	9.65
2400	Soldier course (6.75/S.F.)		200	.200		4.19	6.20		10.39	14.80
2450	Sailor course (4.50/S.F.)		290	.138		2.81	4.28		7.09	10.15
2600	Buff or gray face, running bond, (6.75/S.F.)		220	.182		4.43	5.65		10.08	14.10
2700	Glazed face brick, running bond		210	.190		11.95	5.90		17.85	23
2750	Full header every 6th course (7.88/S.F.)		170	.235		13.95	7.30		21.25	27.50
3000	Jumbo, 6" x 4" x 12" running bond (3.00/S.F.)		435	.092		4.87	2.86		7.73	10.05
3050	Norman, 4" x 2-2/3" x 12" running bond, (4.5/S.F.)		320	.125		5.70	3.88		9.58	12.60
3100	Norwegian, 4" x 3-1/5" x 12" (3.75/S.F.)		375	.107		4.15	3.31		7.46	10
3150	Economy, 4" x 4" x 8" (4.50/S.F.)		310	.129		4.19	4.01		8.20	11.15
3200	Engineer, 4" x 3-1/5" x 8" (5.63/S.F.)		260	.154		3.54	4.78		8.32	11.75
3250	Roman, 4" x 2" x 12" (6.00/S.F.)		250	.160		5.75	4.97		10.72	14.50
3300	SCR, 6" x 2-2/3" x 12" (4.50/S.F.)		310	.129		5.25	4.01		9.26	12.30
3350	Utility, 4" x 4" x 12" (3.00/S.F.)	↓	450	.089	↓	4.51	2.76		7.27	9.50
3400	For cavity wall construction, add						15%			

04 21 Clay Unit Masonry

04 21 13 – Brick Masonry

04 21 13.13 Brick Veneer Masonry		Crew	Daily Output	Labor-Hours	Unit	Material	2011 Bare Costs Labor	Equipment	Total	Total Incl O&P
3450	For stacked bond, add						10%			
3500	For interior veneer construction, add						15%			
3550	For curved walls, add						30%			
04 21 13.14 Thin Brick Veneer										
0010	**THIN BRICK VENEER**									
0015	Material costs incl. 3% brick and 25% mortar waste									
0020	On & incl. metal panel support sys, modular, 2-2/3" x 5/8" x 8", red	D-7	92	.174	S.F.	8.75	4.90		13.65	17.50
0100	Closure, 4" x 5/8" x 8"		110	.145		8.75	4.10		12.85	16.20
0110	Norman, 2-2/3" x 5/8" x 12"		110	.145		8.95	4.10		13.05	16.40
0120	Utility, 4" x 5/8" x 12"		125	.128		8.95	3.61		12.56	15.60
0130	Emperor, 4" x 3/4" x 16"		175	.091		10.15	2.58		12.73	15.30
0140	Super emperor, 8" x 3/4" x 16"	↓	195	.082	↓	10.15	2.31		12.46	14.90
0150	For L shaped corners with 4" return, add				L.F.	9.25			9.25	10.20
0200	On masonry/plaster back-up, modular, 2-2/3" x 5/8" x 8", red	D-7	137	.117	S.F.	3.95	3.29		7.24	9.60
0210	Closure, 4" x 5/8" x 8"		165	.097		3.95	2.73		6.68	8.70
0220	Norman, 2-2/3" x 5/8" x 12"		165	.097		4.15	2.73		6.88	8.95
0230	Utility, 4" x 5/8" x 12"		185	.086		4.15	2.44		6.59	8.45
0240	Emperor, 4" x 3/4" x 16"		260	.062		5.35	1.74		7.09	8.65
0250	Super emperor, 8" x 3/4" x 16"	↓	285	.056	↓	5.35	1.58		6.93	8.45
0260	For L shaped corners with 4" return, add				L.F.	9.25			9.25	10.20
0270	For embedment into pre-cast concrete panels, add				S.F.	14.40			14.40	15.85
04 21 13.15 Chimney										
0010	**CHIMNEY**, excludes foundation, scaffolding, grout and reinforcing									
0100	Brick, 16" x 16", 8" flue	D-1	18.20	.879	V.L.F.	21.50	26.50		48	67.50
0150	16" x 20" with one 8" x 12" flue		16	1		34	30.50		64.50	87.50
0200	16" x 24" with two 8" x 8" flues		14	1.143		50	34.50		84.50	112
0250	20" x 20" with one 12" x 12" flue		13.70	1.168		38.50	35.50		74	101
0300	20" x 24" with two 8" x 12" flues		12	1.333		57	40.50		97.50	129
0350	20" x 32" with two 12" x 12" flues	↓	10	1.600	↓	68	48.50		116.50	155
04 21 13.18 Columns										
0010	**COLUMNS**, solid, excludes scaffolding, grout and reinforcing									
0050	Brick, 8" x 8", 9 brick per V.L.F.	D-1	56	.286	V.L.F.	5.45	8.70		14.15	20.50
0100	12" x 8", 13.5 brick per V.L.F.		37	.432		8.15	13.15		21.30	30.50
0200	12" x 12", 20 brick per V.L.F.		25	.640		12.10	19.45		31.55	45.50
0300	16" x 12", 27 brick per V.L.F.		19	.842		16.35	25.50		41.85	60
0400	16" x 16", 36 brick per V.L.F.		14	1.143		22	34.50		56.50	81
0500	20" x 16", 45 brick per V.L.F.		11	1.455		27	44		71	103
0600	20" x 20", 56 brick per V.L.F.	↓	9	1.778	↓	34	54		88	126
04 21 13.30 Oversized Brick										
0010	**OVERSIZED BRICK**, excludes scaffolding, grout and reinforcing									
0100	Veneer, 4" x 2.25" x 16"	D-8	387	.103	S.F.	6.45	3.21		9.66	12.30
0105	4" x 2.75" x 16"		412	.097		5.80	3.01		8.81	11.30
0110	4" x 4" x 16"		460	.087		5.30	2.70		8	10.30
0120	4" x 8" x 16"		533	.075		6.10	2.33		8.43	10.55
0125	Loadbearing, 6" x 4" x 16", grouted and reinforced		387	.103		9	3.21		12.21	15.15
0130	8" x 4" x 16", grouted and reinforced		327	.122		9.40	3.80		13.20	16.60
0135	6" x 8" x 16", grouted and reinforced		440	.091		8.65	2.82		11.47	14.20
0140	8" x 8" x 16", grouted and reinforced		400	.100		9.55	3.11		12.66	15.60
0145	Curtainwall/reinforced veneer, 6" x 4" x 16"		387	.103		13.20	3.21		16.41	19.75
0150	8" x 4" x 16"		327	.122		15.65	3.80		19.45	23.50
0155	6" x 8" x 16"		440	.091		13.30	2.82		16.12	19.30
0160	8" x 8" x 16"	↓	400	.100	↓	15.80	3.11		18.91	22.50

04 21 Clay Unit Masonry

04 21 13 – Brick Masonry

04 21 13.30 Oversized Brick		Crew	Daily Output	Labor-Hours	Unit	Material	Labor	Equipment	Total	Total Incl O&P
						2011 Bare Costs				
0200	For 1 to 3 slots in face, add				S.F.	15%				
0210	For 4 to 7 slots in face, add					25%				
0220	For bond beams, add					20%				
0230	For bullnose shapes, add					20%				
0240	For open end knockout, add					10%				
0250	For white or gray color group, add					10%				
0260	For 135 degree corner, add				↓	250%				
04 21 13.35 Common Building Brick										
0010	**COMMON BUILDING BRICK**, C62, TL lots, material only R042110-10									
0020	Standard, minimum				M	400			400	440
0050	Average (select)				"	430			430	475
04 21 13.45 Face Brick										
0010	**FACE BRICK** Material Only, C216, TL lots R042110-20									
0300	Standard modular, 4" x 2-2/3" x 8", minimum				M	515			515	565
0350	Maximum					685			685	750
2170	For less than truck load lots, add					15			15	16.50
2180	For buff or gray brick, add				↓	16			16	17.60

04 22 Concrete Unit Masonry

04 22 10 – Concrete Masonry Units

04 22 10.11 Autoclave Aerated Concrete Block		Crew	Daily Output	Labor-Hours	Unit	Material	Labor	Equipment	Total	Total Incl O&P
0010	**AUTOCLAVE AERATED CONCRETE BLOCK**, excl. scaffolding, grout & reinforcing									
0050	Solid, 4" x 12" x 24", incl mortar G	D-8	600	.067	S.F.	1.39	2.07		3.46	4.93
0060	6" x 12" x 24" G		600	.067		2.08	2.07		4.15	5.70
0070	8" x 8" x 24" G		575	.070		2.78	2.16		4.94	6.60
0080	10" x 12" x 24" G		575	.070		3.40	2.16		5.56	7.30
0090	12" x 12" x 24" G	↓	550	.073	↓	4.16	2.26		6.42	8.30
04 22 10.14 Concrete Block, Back-Up										
0010	**CONCRETE BLOCK, BACK-UP**, C90, 2000 psi									
0020	Normal weight, 8" x 16" units, tooled joint 1 side									
0050	Not-reinforced, 2000 psi, 2" thick	D-8	475	.084	S.F.	1.39	2.61		4	5.80
0200	4" thick		460	.087		1.58	2.70		4.28	6.15
0300	6" thick		440	.091		2.18	2.82		5	7.05
0350	8" thick		400	.100		2.49	3.11		5.60	7.85
0400	10" thick	↓	330	.121		3.12	3.76		6.88	9.60
0450	12" thick	D-9	310	.155		3.52	4.71		8.23	11.60
1000	Reinforced, alternate courses, 4" thick	D-8	450	.089		1.71	2.76		4.47	6.40
1100	6" thick		430	.093		2.31	2.89		5.20	7.30
1150	8" thick		395	.101		2.62	3.14		5.76	8.05
1200	10" thick	↓	320	.125		3.25	3.88		7.13	9.90
1250	12" thick	D-9	300	.160	↓	3.65	4.86		8.51	12
04 22 10.16 Concrete Block, Bond Beam										
0010	**CONCRETE BLOCK, BOND BEAM**, C90, 2000 psi									
0020	Not including grout or reinforcing									
0125	Regular block, 6" thick	D-8	584	.068	L.F.	2.32	2.13		4.45	6.05
0130	8" high, 8" thick	"	565	.071		2.50	2.20		4.70	6.35
0150	12" thick	D-9	510	.094		3.50	2.86		6.36	8.55
0525	Lightweight, 6" thick	D-8	592	.068	↓	2.55	2.10		4.65	6.25

04 22 Concrete Unit Masonry

04 22 10 – Concrete Masonry Units

04 22 10.19 Concrete Block, Insulation Inserts		Crew	Daily Output	Labor-Hours	Unit	Material	Labor	Equipment	Total	Total Incl O&P
						2011 Bare Costs				
0010	**CONCRETE BLOCK, INSULATION INSERTS**									
0100	Styrofoam, plant installed, add to block prices									
0200	8" x 16" units, 6" thick				S.F.	1.81			1.81	1.99
0250	8" thick					1.81			1.81	1.99
0300	10" thick					2.13			2.13	2.34
0350	12" thick				↓	2.24			2.24	2.46

04 22 10.23 Concrete Block, Decorative		Crew	Daily Output	Labor-Hours	Unit	Material	Labor	Equipment	Total	Total Incl O&P
0010	**CONCRETE BLOCK, DECORATIVE**, C90, 2000 psi									
0020	Embossed, simulated brick face									
0100	8" x 16" units, 4" thick	D-8	400	.100	S.F.	3.51	3.11		6.62	8.95
0200	8" thick		340	.118		4.83	3.65		8.48	11.30
0250	12" thick	↓	300	.133	↓	6.35	4.14		10.49	13.80
0400	Embossed both sides									
0500	8" thick	D-8	300	.133	S.F.	5.40	4.14		9.54	12.75
0550	12" thick	"	275	.145	"	6.85	4.52		11.37	14.95
1000	Fluted high strength									
1100	8" x 16" x 4" thick, flutes 1 side,	D-8	345	.116	S.F.	4.17	3.60		7.77	10.50
1150	Flutes 2 sides		335	.119		5.05	3.71		8.76	11.65
1200	8" thick	↓	300	.133		6.60	4.14		10.74	14.05
1250	For special colors, add				↓	.41			.41	.45
1400	Deep grooved, smooth face									
1450	8" x 16" x 4" thick	D-8	345	.116	S.F.	2.72	3.60		6.32	8.90
1500	8" thick	"	300	.133	"	4.68	4.14		8.82	11.95
2000	Formblock, incl. inserts & reinforcing									
2100	8" x 16" x 8" thick	D-8	345	.116	S.F.	4.76	3.60		8.36	11.15
2150	12" thick	"	310	.129	"	5.95	4.01		9.96	13.05
2500	Ground face									
2600	8" x 16" x 4" thick	D-8	345	.116	S.F.	3.27	3.60		6.87	9.50
2650	6" thick		325	.123		4.78	3.82		8.60	11.50
2700	8" thick	↓	300	.133		6.35	4.14		10.49	13.75
2750	12" thick	D-9	265	.181	↓	5.70	5.50		11.20	15.30
2900	For special colors, add, minimum					15%				
2950	For special colors, add, maximum					45%				
4000	Slump block									
4100	4" face height x 16" x 4" thick	D-1	165	.097	S.F.	3.94	2.95		6.89	9.20
4150	6" thick		160	.100		5.45	3.04		8.49	10.95
4200	8" thick		155	.103		6.05	3.14		9.19	11.80
4250	10" thick		140	.114		10.90	3.47		14.37	17.70
4300	12" thick		130	.123		11.40	3.74		15.14	18.70
4400	6" face height x 16" x 6" thick		155	.103		4.75	3.14		7.89	10.35
4450	8" thick		150	.107		6.50	3.24		9.74	12.45
4500	10" thick		130	.123		10	3.74		13.74	17.15
4550	12" thick	↓	120	.133	↓	10.35	4.05		14.40	18.05
5000	Split rib profile units, 1" deep ribs, 8 ribs									
5100	8" x 16" x 4" thick	D-8	345	.116	S.F.	3.17	3.60		6.77	9.40
5150	6" thick		325	.123		3.64	3.82		7.46	10.25
5200	8" thick	↓	300	.133		4.16	4.14		8.30	11.40
5250	12" thick	D-9	275	.175		4.93	5.30		10.23	14.15
5400	For special deeper colors, 4" thick, add					1.15			1.15	1.27
5450	12" thick, add					1.19			1.19	1.31
5600	For white, 4" thick, add					1.15			1.15	1.27
5650	6" thick, add				↓	1.14			1.14	1.26

04 22 Concrete Unit Masonry

04 22 10 – Concrete Masonry Units

04 22 10.23 Concrete Block, Decorative		Crew	Daily Output	Labor-Hours	Unit	Material	2011 Bare Costs Labor	Equipment	Total	Total Incl O&P
5700	8" thick, add				S.F.	1.20			1.20	1.32
5750	12" thick, add					1.23			1.23	1.35
6000	Split face									
6100	8" x 16" x 4" thick	D-8	350	.114	S.F.	2.82	3.55		6.37	8.90
6150	6" thick		325	.123		3.28	3.82		7.10	9.85
6200	8" thick		300	.133		3.73	4.14		7.87	10.90
6250	12" thick	D-9	270	.178		4.57	5.40		9.97	13.90
6300	For scored, add					.35			.35	.39
6400	For special deeper colors, 4" thick, add					.67			.67	.73
6450	6" thick, add					.66			.66	.72
6500	8" thick, add					.68			.68	.75
6550	12" thick, add					.70			.70	.77
6650	For white, 4" thick, add					1.14			1.14	1.26
6700	6" thick, add					1.13			1.13	1.24
6750	8" thick, add					1.16			1.16	1.28
6800	12" thick, add					1.19			1.19	1.31
7000	Scored ground face, 2 to 5 scores									
7100	8" x 16" x 4" thick	D-8	340	.118	S.F.	4.91	3.65		8.56	11.40
7150	6" thick		310	.129		5.60	4.01		9.61	12.75
7200	8" thick		290	.138		6.60	4.28		10.88	14.30
7250	12" thick	D-9	265	.181		8.20	5.50		13.70	18.10
8000	Hexagonal face profile units, 8" x 16" units									
8100	4" thick, hollow	D-8	340	.118	S.F.	3.12	3.65		6.77	9.45
8200	Solid		340	.118		4.03	3.65		7.68	10.45
8300	6" thick, hollow		310	.129		3.49	4.01		7.50	10.40
8350	8" thick, hollow		290	.138		4.73	4.28		9.01	12.25
8500	For stacked bond, add						26%			
8550	For high rise construction, add per story	D-8	67.80	.590	M.S.F.		18.30		18.30	30
8600	For scored block, add					10%				
8650	For honed or ground face, per face, add				Ea.	.38			.38	.42
8700	For honed or ground end, per end, add				"	2.98			2.98	3.28
8750	For bullnose block, add					10%				
8800	For special color, add					13%				

04 22 10.24 Concrete Block, Exterior		Crew	Daily Output	Labor-Hours	Unit	Material	Labor	Equipment	Total	Total Incl O&P
0010	**CONCRETE BLOCK, EXTERIOR**, C90, 2000 psi									
0020	Reinforced alt courses, tooled joints 2 sides									
0100	Normal weight, 8" x 16" x 6" thick	D-8	395	.101	S.F.	2.56	3.14		5.70	7.95
0200	8" thick		360	.111		3.26	3.45		6.71	9.25
0250	10" thick		290	.138		4.28	4.28		8.56	11.75
0300	12" thick	D-9	250	.192		4.29	5.85		10.14	14.30

04 22 10.26 Concrete Block Foundation Wall		Crew	Daily Output	Labor-Hours	Unit	Material	Labor	Equipment	Total	Total Incl O&P
0010	**CONCRETE BLOCK FOUNDATION WALL**, C90/C145									
0050	Normal-weight, cut joints, horiz joint reinf, no vert reinf									
0200	Hollow, 8" x 16" x 6" thick	D-8	455	.088	S.F.	2.62	2.73		5.35	7.35
0250	8" thick		425	.094		2.94	2.92		5.86	8
0300	10" thick		350	.114		3.57	3.55		7.12	9.75
0350	12" thick	D-9	300	.160		3.98	4.86		8.84	12.40
0500	Solid, 8" x 16" block, 6" thick	D-8	440	.091		2.69	2.82		5.51	7.60
0550	8" thick	"	415	.096		3.80	2.99		6.79	9.10
0600	12" thick	D-9	350	.137		5.40	4.17		9.57	12.80

04 22 Concrete Unit Masonry

04 22 10 – Concrete Masonry Units

04 22 10.32 Concrete Block, Lintels

	04 22 10.32 Concrete Block, Lintels	Crew	Daily Output	Labor-Hours	Unit	Material	2011 Bare Costs Labor	Equipment	Total	Total Incl O&P
0010	**CONCRETE BLOCK, LINTELS**, C90, normal weight									
0100	Including grout and horizontal reinforcing									
0200	8" x 8" x 8", 1 #4 bar	D-4	300	.133	L.F.	4.85	3.77	.42	9.04	12
0250	2 #4 bars		295	.136		5.05	3.84	.43	9.32	12.30
1000	12" x 8" x 8", 1 #4 bar		275	.145		6.80	4.12	.46	11.38	14.80
1150	2 #5 bars	↓	270	.148	↓	7.25	4.19	.47	11.91	15.35

04 22 10.34 Concrete Block, Partitions

	04 22 10.34 Concrete Block, Partitions	Crew	Daily Output	Labor-Hours	Unit	Material	Labor	Equipment	Total	Total Incl O&P
0010	**CONCRETE BLOCK, PARTITIONS**, excludes scaffolding									
1000	Lightweight block, tooled joints, 2 sides, hollow									
1100	Not reinforced, 8" x 16" x 4" thick	D-8	440	.091	S.F.	1.65	2.82		4.47	6.45
1150	6" thick		410	.098		2.37	3.03		5.40	7.60
1200	8" thick		385	.104		2.88	3.23		6.11	8.45
1250	10" thick	↓	370	.108		3.49	3.36		6.85	9.35
1300	12" thick	D-9	350	.137	↓	4.34	4.17		8.51	11.60
4000	Regular block, tooled joints, 2 sides, hollow									
4100	Not reinforced, 8" x 16" x 4" thick	D-8	430	.093	S.F.	1.49	2.89		4.38	6.40
4150	6" thick		400	.100		2.09	3.11		5.20	7.40
4200	8" thick		375	.107		2.40	3.31		5.71	8.10
4250	10" thick	↓	360	.111		3.03	3.45		6.48	9
4300	12" thick	D-9	340	.141	↓	3.43	4.29		7.72	10.85

04 23 Glass Unit Masonry

04 23 13 – Vertical Glass Unit Masonry

04 23 13.10 Glass Block

	04 23 13.10 Glass Block	Crew	Daily Output	Labor-Hours	Unit	Material	Labor	Equipment	Total	Total Incl O&P
0010	**GLASS BLOCK**									
0150	8" x 8"	D-8	160	.250	S.F.	15	7.75		22.75	29.50
0160	end block		160	.250		50	7.75		57.75	68
0170	90 deg corner		160	.250		49.50	7.75		57.25	67.50
0180	45 deg corner		160	.250		46	7.75		53.75	63.50
0200	12" x 12"		175	.229		21.50	7.10		28.60	35
0210	4" x 8"		160	.250		15.30	7.75		23.05	29.50
0220	6" x 8"	↓	160	.250	↓	14.70	7.75		22.45	29
0700	For solar reflective blocks, add					100%				
1000	Thinline, plain, 3-1/8" thick, under 1,000 S.F., 6" x 6"	D-8	115	.348	S.F.	17.50	10.80		28.30	37
1050	8" x 8"		160	.250		9.80	7.75		17.55	23.50
1400	For cleaning block after installation (both sides), add	↓	1000	.040	↓	.11	1.24		1.35	2.16

04 24 Adobe Unit Masonry

04 24 16 – Manufactured Adobe Unit Masonry

04 24 16.06 Adobe Brick

	04 24 16.06 Adobe Brick		Crew	Daily Output	Labor-Hours	Unit	Material	Labor	Equipment	Total	Total Incl O&P
0010	**ADOBE BRICK**, Semi-stabilized, with cement mortar										
0060	Brick, 10" x 4" x 14", 2.6/S.F.	G	D-8	560	.071	S.F.	3.49	2.22		5.71	7.50
0080	12" x 4" x 16", 2.3/S.F.	G		580	.069		4.61	2.14		6.75	8.55
0100	10" x 4" x 16", 2.3/S.F.	G		590	.068		4.37	2.11		6.48	8.25
0120	8" x 4" x 16", 2.3/S.F.	G		560	.071		3.56	2.22		5.78	7.55
0140	4" x 4" x 16", 2.3/S.F.	G		540	.074		2.56	2.30		4.86	6.60
0160	6" x 4" x 16", 2.3/S.F.	G		540	.074		2.59	2.30		4.89	6.60
0180	4" x 4" x 12", 3.0/S.F.	G	↓	520	.077	↓	2.45	2.39		4.84	6.60

04 24 Adobe Unit Masonry

04 24 16 – Manufactured Adobe Unit Masonry

04 24 16.06 Adobe Brick			Crew	Daily Output	Labor-Hours	Unit	Material	Labor	Equipment	Total	Total Incl O&P
							2011 Bare Costs				
0200	8" x 4" x 12", 3.0/S.F.	G	D-8	520	.077	S.F.	3.16	2.39		5.55	7.40

04 27 Multiple-Wythe Unit Masonry

04 27 10 – Multiple-Wythe Masonry

04 27 10.10 Cornices			Crew	Daily Output	Labor-Hours	Unit	Material	Labor	Equipment	Total	Total Incl O&P
0010	**CORNICES**										
0110	Face bricks, 12 brick/S.F., minimum		D-1	30	.533	SF Face	5.95	16.20		22.15	33
0150	15 brick/S.F., maximum		"	23	.696	"	7.15	21		28.15	42.50
04 27 10.30 Brick Walls											
0010	**BRICK WALLS**, including mortar, excludes scaffolding										
0800	Face brick, 4" thick wall, 6.75 brick/S.F.		D-8	215	.186	S.F.	4.12	5.80		9.92	14.05
0850	Common brick, 4" thick wall, 6.75 brick/S.F.			240	.167		3.33	5.20		8.53	12.15
0900	8" thick, 13.50 bricks per S.F.			135	.296		6.90	9.20		16.10	22.50
1000	12" thick, 20.25 bricks per S.F.			95	.421		10.40	13.05		23.45	33
1050	16" thick, 27.00 bricks per S.F.			75	.533		14.05	16.55		30.60	42.50
1200	Reinforced, face brick, 4" thick wall, 6.75 brick/S.F.			210	.190		4.27	5.90		10.17	14.40
1220	Common brick, 4" thick wall, 6.75 brick/S.F.			235	.170		3.48	5.30		8.78	12.50
1250	8" thick, 13.50 bricks per S.F.			130	.308		7.20	9.55		16.75	23.50
1300	12" thick, 20.25 bricks per S.F.			90	.444		10.85	13.80		24.65	34.50
1350	16" thick, 27.00 bricks per S.F.		↓	70	.571	↓	14.70	17.75		32.45	45

04 41 Dry-Placed Stone

04 41 10 – Dry Placed Stone

04 41 10.10 Rough Stone Wall			Crew	Daily Output	Labor-Hours	Unit	Material	Labor	Equipment	Total	Total Incl O&P
0011	**ROUGH STONE WALL**, Dry										
0012	Dry laid (no mortar), under 18" thick	G	D-1	60	.267	C.F.	8.80	8.10		16.90	23
0100	Random fieldstone, under 18" thick	G	D-12	60	.533		8.80	16.20		25	36
0150	Over 18" thick	G	"	63	.508		10.55	15.45		26	37
0600	Rubble stone walls, in mortar bed, up to 18" thick	G	D-11	75	.320	↓	10.65	10.05		20.70	28.50

04 43 Stone Masonry

04 43 10 – Masonry with Natural and Processed Stone

04 43 10.45 Granite			Crew	Daily Output	Labor-Hours	Unit	Material	Labor	Equipment	Total	Total Incl O&P
0010	**GRANITE**, cut to size										
0050	Veneer, polished face, 3/4" to 1-1/2" thick										
0150	Low price, gray, light gray, etc.		D-10	130	.246	S.F.	26.50	8.15	3.34	37.99	46
0220	High price, red, black, etc.		"	130	.246	"	41.50	8.15	3.34	52.99	63
0300	1-1/2" to 2-1/2" thick, veneer										
0350	Low price, gray, light gray, etc.		D-10	130	.246	S.F.	28.50	8.15	3.34	39.99	48
0550	High price, red, black, etc.		"	130	.246	"	51.50	8.15	3.34	62.99	73.50
0700	2-1/2" to 4" thick, veneer										
0750	Low price, gray, light gray, etc.		D-10	110	.291	S.F.	38	9.60	3.95	51.55	62
0950	High price, red, black, etc.		"	110	.291		62.50	9.60	3.95	76.05	88.50
1000	For bush hammered finish, deduct						5%				
1050	Coarse rubbed finish, deduct						10%				
1100	Honed finish, deduct						5%				
1150	Thermal finish, deduct					↓	18%				

04 43 Stone Masonry

04 43 10 – Masonry with Natural and Processed Stone

04 43 10.45 Granite		Crew	Daily Output	Labor-Hours	Unit	Material	2011 Bare Costs Labor	Equipment	Total	Total Incl O&P
2450	For radius under 5', add				L.F.	100%				
2500	Steps, copings, etc., finished on more than one surface									
2550	Minimum	D-10	50	.640	C.F.	91	21	8.70	120.70	144
2600	Maximum	"	50	.640	"	146	21	8.70	175.70	204
2800	Pavers, 4" x 4" x 4" blocks, split face and joints									
2850	Minimum	D-11	80	.300	S.F.	12.95	9.45		22.40	29.50
2900	Maximum	"	80	.300		28.50	9.45		37.95	47
4000	Soffits, 2" thick, minimum	D-13	35	1.371		37.50	44	12.40	93.90	128
4100	Maximum		35	1.371		90	44	12.40	146.40	185
4200	4" thick, minimum		35	1.371		62	44	12.40	118.40	155
4300	Maximum		35	1.371		117	44	12.40	173.40	215
04 43 10.55 Limestone										
0010	**LIMESTONE**, cut to size									
0020	Veneer facing panels									
0500	Texture finish, light stick, 4-1/2" thick, 5' x 12'	D-4	300	.133	S.F.	48	3.77	.42	52.19	59.50
0750	5" thick, 5' x 14' panels	D-10	275	.116		52.50	3.85	1.58	57.93	65.50
1000	Sugarcube finish, 2" Thick, 3' x 5' panels		275	.116		35	3.85	1.58	40.43	46.50
1050	3" Thick, 4' x 9' panels		275	.116		38	3.85	1.58	43.43	50
1200	4" Thick, 5' x 11' panels		275	.116		41	3.85	1.58	46.43	53
1400	Sugarcube, textured finish, 4-1/2" thick, 5' x 12'		275	.116		45	3.85	1.58	50.43	57.50
1450	5" thick, 5' x 14' panels		275	.116		53.50	3.85	1.58	58.93	66.50
2000	Coping, sugarcube finish, top & 2 sides		30	1.067	C.F.	68	35.50	14.50	118	149
2100	Sills, lintels, jambs, trim, stops, sugarcube finish, average		20	1.600		68	53	21.50	142.50	186
2150	Detailed		20	1.600		68	53	21.50	142.50	186
2300	Steps, extra hard, 14" wide, 6" rise		50	.640	L.F.	41.50	21	8.70	71.20	90
3000	Quoins, plain finish, 6" x 12" x 12"	D-12	25	1.280	Ea.	38	39		77	106
3050	6" x 16" x 24"	"	25	1.280	"	50.50	39		89.50	120
04 43 10.60 Marble										
0011	**MARBLE**, ashlar, split face, 4" + or - thick, random									
1000	Facing, polished finish, cut to size, 3/4" to 7/8" thick									
1050	Average	D-10	130	.246	S.F.	23.50	8.15	3.34	34.99	43
1100	Maximum	"	130	.246	"	55	8.15	3.34	66.49	77.50
2200	Window sills, 6" x 3/4" thick	D-1	85	.188	L.F.	8.95	5.70		14.65	19.20
2500	Flooring, polished tiles, 12" x 12" x 3/8" thick									
2510	Thin set, average	D-11	90	.267	S.F.	11.90	8.40		20.30	27
2600	Maximum		90	.267		183	8.40		191.40	215
2700	Mortar bed, average		65	.369		10.65	11.60		22.25	31
2740	Maximum		65	.369		183	11.60		194.60	220
2780	Travertine, 3/8" thick, average	D-10	130	.246		15.15	8.15	3.34	26.64	33.50
2790	Maximum	"	130	.246		37	8.15	3.34	48.49	58
3500	Thresholds, 3' long, 7/8" thick, 4" to 5" wide, plain	D-12	24	1.333	Ea.	17.15	40.50		57.65	85.50
3550	Beveled		24	1.333	"	19.90	40.50		60.40	88.50
3700	Window stools, polished, 7/8" thick, 5" wide		85	.376	L.F.	15.95	11.45		27.40	36.50
04 43 10.75 Sandstone or Brownstone										
0011	**SANDSTONE OR BROWNSTONE**									
0100	Sawed face veneer, 2-1/2" thick, to 2' x 4' panels	D-10	130	.246	S.F.	21	8.15	3.34	32.49	40
0150	4" thick, to 3'-6" x 8' panels		100	.320		21	10.60	4.35	35.95	45
0300	Split face, random sizes		100	.320		14	10.60	4.35	28.95	37.50
0350	Cut stone trim (limestone)									
0360	Ribbon stone, 4" thick, 5' pieces	D-8	120	.333	Ea.	172	10.35		182.35	206
0370	Cove stone, 4" thick, 5' pieces		105	.381		173	11.85		184.85	209
0380	Cornice stone, 10" to 12" wide		90	.444		214	13.80		227.80	258

04 43 Stone Masonry

04 43 10 – Masonry with Natural and Processed Stone

04 43 10.75 Sandstone or Brownstone		Crew	Daily Output	Labor-Hours	Unit	2011 Bare Costs Material	Labor	Equipment	Total	Total Incl O&P
0390	Band stone, 4" thick, 5' pieces	D-8	145	.276	Ea.	110	8.55		118.55	135
0410	Window and door trim, 3" to 4" wide		160	.250		93.50	7.75		101.25	116
0420	Key stone, 18" long		60	.667		98.50	20.50		119	142
04 43 10.80 Slate										
0010	**SLATE**									
3500	Stair treads, sand finish, 1" thick x 12" wide									
3600	3 L.F. to 6 L.F.	D-10	120	.267	L.F.	24	8.80	3.62	36.42	44.50
3700	Ribbon, sand finish, 1" thick x 12" wide									
3750	To 6 L.F.	D-10	120	.267	L.F.	20	8.80	3.62	32.42	40.50
04 43 10.85 Window Sill										
0010	**WINDOW SILL**									
0020	Bluestone, thermal top, 10" wide, 1-1/2" thick	D-1	85	.188	S.F.	16.05	5.70		21.75	27
0050	2" thick		75	.213	"	18.75	6.50		25.25	31
0100	Cut stone, 5" x 8" plain		48	.333	L.F.	11.90	10.15		22.05	29.50
0200	Face brick on edge, brick, 8" wide		80	.200		2.53	6.10		8.63	12.80
0400	Marble, 9" wide, 1" thick		85	.188		8.50	5.70		14.20	18.75
0900	Slate, colored, unfading, honed, 12" wide, 1" thick		85	.188		17.10	5.70		22.80	28
0950	2" thick		70	.229		24	6.95		30.95	37.50

04 51 Flue Liner Masonry

04 51 10 – Clay Flue Lining

04 51 10.10 Flue Lining		Crew	Daily Output	Labor-Hours	Unit	Material	Labor	Equipment	Total	Total Incl O&P
0010	**FLUE LINING**, including mortar									
0020	Clay, 8" x 8"	D-1	125	.128	V.L.F.	4.97	3.89		8.86	11.85
0100	8" x 12"		103	.155		7.55	4.72		12.27	16.05
0200	12" x 12"		93	.172		9.75	5.25		15	19.30
0300	12" x 18"		84	.190		18.90	5.80		24.70	30.50
0400	18" x 18"		75	.213		22.50	6.50		29	35
0500	20" x 20"		66	.242		36	7.35		43.35	51.50
0600	24" x 24"		56	.286		47	8.70		55.70	66
1000	Round, 18" diameter		66	.242		30.50	7.35		37.85	45.50
1100	24" diameter		47	.340		67.50	10.35		77.85	91

04 57 Masonry Fireplaces

04 57 10 – Brick or Stone Fireplaces

04 57 10.10 Fireplace		Crew	Daily Output	Labor-Hours	Unit	Material	Labor	Equipment	Total	Total Incl O&P
0010	**FIREPLACE**									
0100	Brick fireplace, not incl. foundations or chimneys									
0110	30" x 29" opening, incl. chamber, plain brickwork	D-1	.40	40	Ea.	535	1,225		1,760	2,600
0200	Fireplace box only (110 brick)	"	2	8	"	176	243		419	595
0300	For elaborate brickwork and details, add					35%	35%			
0400	For hearth, brick & stone, add	D-1	2	8	Ea.	197	243		440	615
0410	For steel angle, damper, cleanouts, add		4	4		138	122		260	350
0600	Plain brickwork, incl. metal circulator		.50	32		1,025	975		2,000	2,725
0800	Face brick only, standard size, 8" x 2-2/3" x 4"		.30	53.333	M	535	1,625		2,160	3,250
0900	Stone fireplace, fieldstone, add				SF Face	14.05			14.05	15.50
1000	Cut stone, add				"	15.50			15.50	17.05

04 72 Cast Stone Masonry

04 72 10 – Cast Stone Masonry Features

04 72 10.10 Coping		Crew	Daily Output	Labor-Hours	Unit	Material	2011 Bare Costs Labor	Equipment	Total	Total Incl O&P
0010	**COPING**, stock units									
0050	Precast concrete, 10" wide, 4" tapers to 3-1/2", 8" wall	D-1	75	.213	L.F.	17.55	6.50		24.05	30
0100	12" wide, 3-1/2" tapers to 3", 10" wall		70	.229		17.55	6.95		24.50	30.50
0150	16" wide, 4" tapers to 3-1/2", 14" wall		60	.267		17.10	8.10		25.20	32
0300	Limestone for 12" wall, 4" thick		90	.178		15.35	5.40		20.75	26
0350	6" thick		80	.200		23	6.10		29.10	35.50
0500	Marble, to 4" thick, no wash, 9" wide		90	.178		23	5.40		28.40	34.50
0550	12" wide		80	.200		35	6.10		41.10	48.50
0700	Terra cotta, 9" wide		90	.178		5.60	5.40		11	15
0750	12" wide		80	.200		9.20	6.10		15.30	20
0800	Aluminum, for 12" wall		80	.200		13.30	6.10		19.40	24.50

04 72 20 – Cultured Stone Veneer

04 72 20.10 Cultured Stone Veneer Components		Crew	Daily Output	Labor-Hours	Unit	Material	Labor	Equipment	Total	Total Incl O&P
0010	**CULTURED STONE VENEER COMPONENTS**									
0110	On wood frame and sheathing substrate, random sized cobbles, corner stones	D-8	70	.571	V.L.F.	12.10	17.75		29.85	42.50
0120	Field stones		140	.286	S.F.	8.80	8.85		17.65	24
0130	Random sized flats, corner stones		70	.571	V.L.F.	11.90	17.75		29.65	42
0140	Field stones		140	.286	S.F.	10.05	8.85		18.90	25.50
0150	Horizontal lined ledgestones, corner stones		75	.533	V.L.F.	12.10	16.55		28.65	40.50
0160	Field stones		150	.267	S.F.	8.80	8.30		17.10	23.50
0170	Random shaped flats, corner stones		65	.615	V.L.F.	12.10	19.10		31.20	45
0180	Field stones		150	.267	S.F.	8.80	8.30		17.10	23.50
0190	Random shaped/textured face, corner stones		65	.615	V.L.F.	12.10	19.10		31.20	45
0200	Field stones		130	.308	S.F.	8.80	9.55		18.35	25.50
0210	Random shaped river rock, corner stones		65	.615	V.L.F.	12.10	19.10		31.20	45
0220	Field stones		130	.308	S.F.	8.80	9.55		18.35	25.50
0240	On concrete or CMU substrate, random sized cobbles, corner stones		70	.571	V.L.F.	11.50	17.75		29.25	41.50
0250	Field stones		140	.286	S.F.	8.50	8.85		17.35	24
0260	Random sized flats, corner stones		70	.571	V.L.F.	11.30	17.75		29.05	41.50
0270	Field stones		140	.286	S.F.	9.75	8.85		18.60	25.50
0280	Horizontal lined ledgestones, corner stones		75	.533	V.L.F.	11.50	16.55		28.05	39.50
0290	Field stones		150	.267	S.F.	8.50	8.30		16.80	23
0300	Random shaped flats, corner stones		70	.571	V.L.F.	11.50	17.75		29.25	41.50
0310	Field stones		140	.286	S.F.	8.50	8.85		17.35	24
0320	Random shaped/textured face, corner stones		65	.615	V.L.F.	11.50	19.10		30.60	44
0330	Field stones		130	.308	S.F.	8.50	9.55		18.05	25
0340	Random shaped river rock, corner stones		65	.615	V.L.F.	11.50	19.10		30.60	44
0350	Field stones		130	.308	S.F.	8.50	9.55		18.05	25
0360	Cultured stone veneer, #15 felt weather resistant barrier	1 Clab	3700	.002	Sq.	5.65	.06		5.71	6.30
0370	Expanded metal lath, diamond, 2.5 lb./S.Y., galvanized	1 Lath	85	.094	S.Y.	2.63	2.83		5.46	7.40
0390	Water table or window sill, 18" long	1 Bric	80	.100	Ea.	10.60	3.37		13.97	17.15

Estimating Tips

05 05 00 Common Work Results for Metals

- Nuts, bolts, washers, connection angles, and plates can add a significant amount to both the tonnage of a structural steel job and the estimated cost. As a rule of thumb, add 10% to the total weight to account for these accessories.
- Type 2 steel construction, commonly referred to as "simple construction," consists generally of field-bolted connections with lateral bracing supplied by other elements of the building, such as masonry walls or x-bracing. The estimator should be aware, however, that shop connections may be accomplished by welding or bolting. The method may be particular to the fabrication shop and may have an impact on the estimated cost.

05 12 23 Structural Steel

- Steel items can be obtained from two sources: a fabrication shop or a metals service center. Fabrication shops can fabricate items under more controlled conditions than can crews in the field. They are also more efficient and can produce items more economically. Metal service centers serve as a source of long mill shapes to both fabrication shops and contractors.
- Most line items in this structural steel subdivision, and most items in 05 50 00 Metal Fabrications, are indicated as being shop fabricated. The bare material cost for these shop fabricated items is the "Invoice Cost" from the shop and includes the mill base price of steel plus mill extras, transportation to the shop, shop drawings and detailing where warranted, shop fabrication and handling, sandblasting and a shop coat of primer paint, all necessary structural bolts, and delivery to the job site. The bare labor cost and bare equipment cost for these shop fabricated items is for field installation or erection.
- Line items in Subdivision 05 12 23.40 Lightweight Framing, and other items scattered in Division 5, are indicated as being field fabricated. The bare material cost for these field fabricated items is the "Invoice Cost" from the metals service center and includes the mill base price of steel plus mill extras, transportation to the metals service center, material handling, and delivery of long lengths of mill shapes to the job site. Material costs for structural bolts and welding rods should be added to the estimate. The bare labor cost and bare equipment cost for these items is for both field fabrication and field installation or erection, and include time for cutting, welding and drilling in the fabricated metal items. Drilling into concrete and fasteners to fasten field fabricated items to other work are not included and should be added to the estimate.

05 20 00 Steel Joist Framing

- In any given project the total weight of open web steel joists is determined by the loads to be supported and the design. However, economies can be realized in minimizing the amount of labor used to place the joists. This is done by maximizing the joist spacing, and therefore minimizing the number of joists required to be installed on the job. Certain spacings and locations may be required by the design, but in other cases maximizing the spacing and keeping it as uniform as possible will keep the costs down.

05 30 00 Steel Decking

- The takeoff and estimating of metal deck involves more than simply the area of the floor or roof and the type of deck specified or shown on the drawings. Many different sizes and types of openings may exist. Small openings for individual pipes or conduits may be drilled after the floor/roof is installed, but larger openings may require special deck lengths as well as reinforcing or structural support. The estimator should determine who will be supplying this reinforcing. Additionally, some deck terminations are part of the deck package, such as screed angles and pour stops, and others will be part of the steel contract, such as angles attached to structural members and cast-in-place angles and plates. The estimator must ensure that all pieces are accounted for in the complete estimate.

05 50 00 Metal Fabrications

- The most economical steel stairs are those that use common materials, standard details, and most importantly, a uniform and relatively simple method of field assembly. Commonly available A36 channels and plates are very good choices for the main stringers of the stairs, as are angles and tees for the carrier members. Risers and treads are usually made by specialty shops, and it is most economical to use a typical detail in as many places as possible. The stairs should be pre-assembled and shipped directly to the site. The field connections should be simple and straightforward to be accomplished efficiently, and with minimum equipment and labor.

Reference Numbers

Reference numbers are shown in shaded boxes at the beginning of some major classifications. These numbers refer to related items in the Reference Section. The reference information may be an estimating procedure, an alternate pricing method, or technical information. *Note:* Not all subdivisions listed here necessarily appear in this publication.

05 01 Maintenance of Metals

05 01 10 – Maintenance of Structural Metal Framing

05 01 10.51 Cleaning of Structural Metal Framing		Crew	Daily Output	Labor-Hours	Unit	Material	2011 Bare Costs Labor	Equipment	Total	Total Incl O&P
0010	CLEANING OF STRUCTURAL METAL FRAMING									
6125	Steel surface treatments, PDCA guidelines									
6235	Com'l blast (SSPC-SP6), loose scale, fine pwder rust, 2.0#/S.F. sand	E-11	1200	.027	S.F.	.49	.81	.16	1.46	2.17
6240	Tight mill scale, little/no rust, 3.0#/S.F. sand		1000	.032		.73	.97	.20	1.90	2.76
6245	Exist coat blistered/pitted, 4.0#/S.F. sand		875	.037		.97	1.11	.22	2.30	3.30

05 05 Common Work Results for Metals

05 05 05 – Selective Metals Demolition

05 05 05.10 Selective Demolition, Metals		Crew	Daily Output	Labor-Hours	Unit	Material	Labor	Equipment	Total	Total Incl O&P
0010	SELECTIVE DEMOLITION, METALS R024119-10									
0015	Excludes shores, bracing, cutting, loading, hauling, dumping									
0020	Remove nuts only up to 3/4" diameter	1 Sswk	480	.017	Ea.		.61		.61	1.16
0030	7/8" to 1-1/4" diameter		240	.033			1.21		1.21	2.31
0040	1-3/8" to 2" diameter		160	.050			1.82		1.82	3.47
0060	Unbolt and remove structural bolts up to 3/4" diameter		240	.033			1.21		1.21	2.31
0070	7/8" to 2" diameter		160	.050			1.82		1.82	3.47
0140	Light weight framing members, remove whole or cut up, up to 20 lb		240	.033			1.21		1.21	2.31
0150	21 – 40 lb	2 Sswk	210	.076			2.77		2.77	5.30
0160	41 – 80 lb	3 Sswk	180	.133			4.85		4.85	9.25
0170	81 – 120 lb	4 Sswk	150	.213			7.75		7.75	14.80
0230	Structural members, remove whole or cut up, up to 500 lb	E-19	48	.500			18.05	21.50	39.55	56.50
0240	1/4 – 2 tons	E-18	36	1.111			40.50	29	69.50	107
0250	2 – 5 tons	E-24	30	1.067			38.50	25	63.50	98
0260	5 – 10 tons	E-20	24	2.667			96	52.50	148.50	234
0270	10 – 15 tons	E-2	18	2.667			97.50	90	187.50	281
0340	Fabricated item, remove whole or cut up, up to 20 lb	1 Sswk	96	.083			3.03		3.03	5.80
0350	21 – 40 lb	2 Sswk	84	.190			6.95		6.95	13.20
0360	41 – 80 lb	3 Sswk	72	.333			12.15		12.15	23
0370	81 – 120 lb	4 Sswk	60	.533			19.40		19.40	37
0380	121 – 500 lb	E-19	48	.500			18.05	21.50	39.55	56.50

05 05 13 – Shop-Applied Coatings for Metal

05 05 13.50 Paints and Protective Coatings		Crew	Daily Output	Labor-Hours	Unit	Material	Labor	Equipment	Total	Total Incl O&P
0010	PAINTS AND PROTECTIVE COATINGS									
5900	Galvanizing structural steel in shop, under 1 ton				Ton	460			460	505
5950	1 ton to 20 tons					400			400	440
6000	Over 20 tons					375			375	415

05 05 21 – Fastening Methods for Metal

05 05 21.15 Drilling Steel		Crew	Daily Output	Labor-Hours	Unit	Material	Labor	Equipment	Total	Total Incl O&P
0010	DRILLING STEEL									
1910	Drilling & layout for steel, up to 1/4" deep, no anchor									
1920	Holes, 1/4" diameter	1 Sswk	112	.071	Ea.	.07	2.60		2.67	5.05
1925	For each additional 1/4" depth, add		336	.024		.07	.87		.94	1.72
1930	3/8" diameter		104	.077		.08	2.80		2.88	5.45
1935	For each additional 1/4" depth, add		312	.026		.08	.93		1.01	1.87
1940	1/2" diameter		96	.083		.08	3.03		3.11	5.90
1945	For each additional 1/4" depth, add		288	.028		.08	1.01		1.09	2.02
1950	5/8" diameter		88	.091		.12	3.31		3.43	6.45
1955	For each additional 1/4" depth, add		264	.030		.12	1.10		1.22	2.23
1960	3/4" diameter		80	.100		.15	3.64		3.79	7.10
1965	For each additional 1/4" depth, add		240	.033		.15	1.21		1.36	2.47

05 05 Common Work Results for Metals

05 05 21 – Fastening Methods for Metal

05 05 21.15 Drilling Steel			Crew	Daily Output	Labor-Hours	Unit	2011 Bare Costs Material	Labor	Equipment	Total	Total Incl O&P
1970	7/8" diameter		1 Sswk	72	.111	Ea.	.19	4.04		4.23	7.90
1975	For each additional 1/4" depth, add			216	.037		.19	1.35		1.54	2.78
1980	1" diameter			64	.125		.20	4.55		4.75	8.90
1985	For each additional 1/4" depth, add			192	.042		.20	1.52		1.72	3.11
1990	For drilling up, add							40%			

05 05 23 – Metal Fastenings

05 05 23.05 Anchor Bolts			Crew	Daily Output	Labor-Hours	Unit	Material	Labor	Equipment	Total	Total Incl O&P
0010	**ANCHOR BOLTS**										
0015	Made from recycled materials	G									
0025	Single bolts installed in fresh concrete, no templates										
0030	Hooked w/nut and washer, 1/2" diameter, 8" long	G	1 Carp	132	.061	Ea.	1.36	2.03		3.39	4.89
0040	12" long	G		131	.061		1.51	2.05		3.56	5.10
0070	3/4" diameter, 8" long	G		127	.063		3.65	2.11		5.76	7.55
0080	12" long	G		125	.064		4.56	2.15		6.71	8.60
0090	2-bolt pattern, including job-built 2-hole template, per set										
0100	J-type, incl. hex nut & washer, 1/2" diameter x 6" long		1 Carp	21	.381	Set	4.50	12.80		17.30	26.50
0110	12" long			21	.381		5.10	12.80		17.90	27
0120	18" long			21	.381		6	12.80		18.80	28
0130	3/4" diameter x 8" long			20	.400		9.40	13.40		22.80	33
0140	12" long			20	.400		11.20	13.40		24.60	35
0150	18" long			20	.400		13.95	13.40		27.35	38
0160	1" diameter x 12" long			19	.421		19.65	14.15		33.80	45
0170	18" long			19	.421		23.50	14.15		37.65	49
0180	24" long			19	.421		28	14.15		42.15	54
0190	36" long			18	.444		37.50	14.90		52.40	66.50
0200	1-1/2" diameter x 18" long			17	.471		60	15.80		75.80	92.50
0210	24" long			16	.500		71.50	16.80		88.30	107
0300	L-type, incl. hex nut & washer, 3/4" diameter x 12" long			20	.400		10.70	13.40		24.10	34.50
0310	18" long			20	.400		13.10	13.40		26.50	37
0320	24" long			20	.400		15.50	13.40		28.90	39.50
0330	30" long			20	.400		19.15	13.40		32.55	43.50
0340	36" long			20	.400		21.50	13.40		34.90	46
0350	1" diameter x 12" long			19	.421		16.75	14.15		30.90	42
0360	18" long			19	.421		20.50	14.15		34.65	46
0370	24" long			19	.421		24.50	14.15		38.65	50.50
0380	30" long			19	.421		28.50	14.15		42.65	55
0390	36" long			18	.444		32.50	14.90		47.40	61
0400	42" long			18	.444		39	14.90		53.90	68
0410	48" long			18	.444		43.50	14.90		58.40	73
0420	1-1/4" diameter x 18" long			18	.444		30	14.90		44.90	58
0430	24" long			18	.444		35	14.90		49.90	64
0440	30" long			17	.471		40.50	15.80		56.30	71
0450	36" long			17	.471		45.50	15.80		61.30	77
1000	4-bolt pattern, including job-built 4-hole template, per set										
1100	J-type, incl. hex nut & washer, 1/2" diameter x 6" long	G	1 Carp	19	.421	Set	6.90	14.15		21.05	31
1110	12" long	G		19	.421		8.15	14.15		22.30	32.50
1120	18" long	G		18	.444		9.95	14.90		24.85	36
1130	3/4" diameter x 8" long	G		17	.471		16.70	15.80		32.50	45
1140	12" long	G		17	.471		20.50	15.80		36.30	49
1150	18" long	G		17	.471		26	15.80		41.80	55
1160	1" diameter x 12" long	G		16	.500		37.50	16.80		54.30	69
1170	18" long	G		15	.533		44.50	17.90		62.40	79

05 05 Common Work Results for Metals

05 05 23 – Metal Fastenings

05 05 23.05 Anchor Bolts			Crew	Daily Output	Labor-Hours	Unit	2011 Bare Costs Material	Labor	Equipment	Total	Total Incl O&P
1180	24" long	G	1 Carp	15	.533	Set	54	17.90		71.90	89
1190	36" long	G		15	.533		73	17.90		90.90	110
1200	1-1/2" diameter x 18" long	G		13	.615		118	20.50		138.50	165
1210	24" long	G		12	.667		140	22.50		162.50	193
1300	L-type, incl. hex nut & washer, 3/4" diameter x 12" long	G		17	.471		19.25	15.80		35.05	47.50
1310	18" long	G		17	.471		24	15.80		39.80	53
1320	24" long	G		17	.471		29	15.80		44.80	58.50
1330	30" long	G		16	.500		36	16.80		52.80	68
1340	36" long	G		16	.500		41	16.80		57.80	73
1350	1" diameter x 12" long	G		16	.500		31.50	16.80		48.30	62.50
1360	18" long	G		15	.533		38.50	17.90		56.40	72.50
1370	24" long	G		15	.533		47	17.90		64.90	82
1380	30" long	G		15	.533		55.50	17.90		73.40	91
1390	36" long	G		15	.533		63	17.90		80.90	99
1400	42" long	G		14	.571		76	19.15		95.15	116
1410	48" long	G		14	.571		85	19.15		104.15	126
1420	1-1/4" diameter x 18" long	G		14	.571		58	19.15		77.15	95.50
1430	24" long	G		14	.571		68.50	19.15		87.65	107
1440	30" long	G		13	.615		79	20.50		99.50	121
1450	36" long	G	↓	13	.615	↓	89.50	20.50		110	133

05 05 23.10 Bolts and Hex Nuts			Crew	Daily Output	Labor-Hours	Unit	Material	Labor	Equipment	Total	Total Incl O&P
0010	**BOLTS & HEX NUTS**, Steel, A307										
0100	1/4" diameter, 1/2" long	G	1 Sswk	140	.057	Ea.	.07	2.08		2.15	4.05
0200	1" long	G		140	.057		.08	2.08		2.16	4.06
0300	2" long	G		130	.062		.11	2.24		2.35	4.39
0400	3" long	G		130	.062		.16	2.24		2.40	4.45
0500	4" long	G		120	.067		.18	2.43		2.61	4.82
0600	3/8" diameter, 1" long	G		130	.062		.18	2.24		2.42	4.46
0700	2" long	G		130	.062		.22	2.24		2.46	4.51
0800	3" long	G		120	.067		.29	2.43		2.72	4.95
0900	4" long	G		120	.067		.36	2.43		2.79	5.05
1000	5" long	G		115	.070		.45	2.53		2.98	5.30
1100	1/2" diameter, 1-1/2" long	G		120	.067		.43	2.43		2.86	5.10
1200	2" long	G		120	.067		.49	2.43		2.92	5.15
1300	4" long	G		115	.070		.78	2.53		3.31	5.70
1400	6" long	G		110	.073		1.08	2.65		3.73	6.25
1500	8" long	G		105	.076		1.42	2.77		4.19	6.85
1600	5/8" diameter, 1-1/2" long	G		120	.067		1.01	2.43		3.44	5.75
1700	2" long	G		120	.067		1.12	2.43		3.55	5.85
1800	4" long	G		115	.070		1.61	2.53		4.14	6.60
1900	6" long	G		110	.073		2.07	2.65		4.72	7.35
2000	8" long	G		105	.076		3.07	2.77		5.84	8.70
2100	10" long	G		100	.080		3.87	2.91		6.78	9.80
2200	3/4" diameter, 2" long	G		120	.067		1.22	2.43		3.65	5.95
2300	4" long	G		110	.073		1.74	2.65		4.39	6.95
2400	6" long	G		105	.076		2.23	2.77		5	7.75
2500	8" long	G		95	.084		3.35	3.07		6.42	9.55
2600	10" long	G		85	.094		4.39	3.43		7.82	11.40
2700	12" long	G		80	.100		5.15	3.64		8.79	12.60
2800	1" diameter, 3" long	G		105	.076		3.53	2.77		6.30	9.20
2900	6" long	G		90	.089		5.25	3.24		8.49	11.90
3000	12" long	G	↓	75	.107	↓	9.55	3.88		13.43	17.90

05 05 Common Work Results for Metals

05 05 23 – Metal Fastenings

05 05 23.10 Bolts and Hex Nuts			Crew	Daily Output	Labor-Hours	Unit	Material	Labor	Equipment	Total	Total Incl O&P
							2011 Bare Costs				
3100	For galvanized, add					Ea.	75%				
3200	For stainless, add						350%				
05 05 23.15 Chemical Anchors											
0010	**CHEMICAL ANCHORS**										
0020	Includes layout & drilling										
1430	Chemical anchor, w/rod & epoxy cartridge, 3/4" diam. x 9-1/2" long		B-89A	27	.593	Ea.	8.25	17.95	4.07	30.27	43.50
1435	1" diameter x 11-3/4" long			24	.667		16.05	20	4.57	40.62	56.50
1440	1-1/4" diameter x 14" long			21	.762		29.50	23	5.25	57.75	77
1445	1-3/4" diameter x 15" long			20	.800		58	24	5.50	87.50	110
1450	18" long			17	.941		69.50	28.50	6.45	104.45	131
1455	2" diameter x 18" long			16	1		84	30.50	6.85	121.35	151
1460	24" long			15	1.067		109	32.50	7.30	148.80	182
05 05 23.20 Expansion Anchors											
0010	**EXPANSION ANCHORS**										
0100	Anchors for concrete, brick or stone, no layout and drilling										
0200	Expansion shields, zinc, 1/4" diameter, 1-5/16" long, single	G	1 Carp	90	.089	Ea.	.42	2.98		3.40	5.45
0300	1-3/8" long, double	G		85	.094		.50	3.16		3.66	5.85
0500	2" long, double	G		80	.100		1.08	3.36		4.44	6.80
0700	2-1/2" long, double	G		75	.107		1.65	3.58		5.23	7.80
0900	2-3/4" long, double	G		70	.114		2.39	3.83		6.22	9.05
1100	3-15/16" long, double	G		65	.123		4.65	4.13		8.78	12
2100	Hollow wall anchors for gypsum wall board, plaster or tile										
2500	3/16" diameter, short	G	1 Carp	150	.053	Ea.	.34	1.79		2.13	3.36
3000	Toggle bolts, bright steel, 1/8" diameter, 2" long	G		85	.094		.18	3.16		3.34	5.50
3100	4" long	G		80	.100		.23	3.36		3.59	5.85
3200	3/16" diameter, 3" long	G		80	.100		.21	3.36		3.57	5.85
3300	6" long	G		75	.107		.31	3.58		3.89	6.35
3400	1/4" diameter, 3" long	G		75	.107		.27	3.58		3.85	6.30
3500	6" long	G		70	.114		.45	3.83		4.28	6.90
3600	3/8" diameter, 3" long	G		70	.114		.67	3.83		4.50	7.15
3700	6" long	G		60	.133		1.13	4.47		5.60	8.75
3800	1/2" diameter, 4" long	G		60	.133		1.67	4.47		6.14	9.35
3900	6" long	G		50	.160		2.30	5.35		7.65	11.55
4000	Nailing anchors										
4100	Nylon nailing anchor, 1/4" diameter, 1" long		1 Carp	3.20	2.500	C	12.45	84		96.45	154
4200	1-1/2" long			2.80	2.857		15.85	96		111.85	177
4300	2" long			2.40	3.333		17.95	112		129.95	207
4400	Metal nailing anchor, 1/4" diameter, 1" long	G		3.20	2.500		15.20	84		99.20	157
4500	1-1/2" long	G		2.80	2.857		19.95	96		115.95	182
4600	2" long	G		2.40	3.333		25	112		137	215
5000	Screw anchors for concrete, masonry,										
5100	stone & tile, no layout or drilling included										
5700	Lag screw shields, 1/4" diameter, short	G	1 Carp	90	.089	Ea.	.31	2.98		3.29	5.35
5800	Long	G		85	.094		.40	3.16		3.56	5.75
5900	3/8" diameter, short	G		85	.094		.59	3.16		3.75	5.95
6000	Long	G		80	.100		.73	3.36		4.09	6.40
6100	1/2" diameter, short	G		80	.100		.99	3.36		4.35	6.70
6200	Long	G		75	.107		1.35	3.58		4.93	7.50
6300	5/8" diameter, short	G		70	.114		1.58	3.83		5.41	8.15
6400	Long	G		65	.123		2.12	4.13		6.25	9.25
6600	Lead, #6 & #8, 3/4" long	G		260	.031		.16	1.03		1.19	1.91
6700	#10 - #14, 1-1/2" long	G		200	.040		.26	1.34		1.60	2.54

05 05 Common Work Results for Metals

05 05 23 – Metal Fastenings

05 05 23.20 Expansion Anchors		Crew	Daily Output	Labor-Hours	Unit	Material	2011 Bare Costs Labor	Equipment	Total	Total Incl O&P
6800	#16 & #18, 1-1/2" long	G	1 Carp	160	.050	Ea.	.32	1.68		2	3.16
6900	Plastic, #6 & #8, 3/4" long			260	.031		.04	1.03		1.07	1.77
7000	#8 & #10, 7/8" long			240	.033		.04	1.12		1.16	1.91
7100	#10 & #12, 1" long			220	.036		.05	1.22		1.27	2.10
7200	#14 & #16, 1-1/2" long			160	.050		.07	1.68		1.75	2.89
8950	Self-drilling concrete screw, hex washer head, 3/16" diam. x 1-3/4" long	G		300	.027		.18	.89		1.07	1.70
8960	2-1/4" long	G		250	.032		.20	1.07		1.27	2.02
8970	Phillips flat head, 3/16" diam. x 1-3/4" long	G		300	.027		.18	.89		1.07	1.70
8980	2-1/4" long	G	↓	250	.032	↓	.19	1.07		1.26	2.01

05 05 23.30 Lag Screws

			Crew	Daily Output	Labor-Hours	Unit	Material	Labor	Equipment	Total	Total Incl O&P
0010	LAG SCREWS										
0020	Steel, 1/4" diameter, 2" long	G	1 Carp	200	.040	Ea.	.09	1.34		1.43	2.35
0100	3/8" diameter, 3" long	G		150	.053		.26	1.79		2.05	3.28
0200	1/2" diameter, 3" long	G		130	.062		.48	2.06		2.54	3.99
0300	5/8" diameter, 3" long	G	↓	120	.067	↓	.99	2.24		3.23	4.83

05 05 23.50 Powder Actuated Tools and Fasteners

			Crew	Daily Output	Labor-Hours	Unit	Material	Labor	Equipment	Total	Total Incl O&P
0010	POWDER ACTUATED TOOLS & FASTENERS										
0020	Stud driver, .22 caliber, buy, minimum					Ea.	104			104	114
0100	Maximum					"	370			370	405
0300	Powder charges for above, low velocity					C	7			7	7.70
0400	Standard velocity						7.70			7.70	8.50
0600	Drive pins & studs, 1/4" & 3/8" diam., to 3" long, minimum	G	1 Carp	4.80	1.667		4.02	56		60.02	98
0700	Maximum	G	"	4	2	↓	12.85	67		79.85	126

05 05 23.55 Rivets

			Crew	Daily Output	Labor-Hours	Unit	Material	Labor	Equipment	Total	Total Incl O&P
0010	RIVETS										
0100	Aluminum rivet & mandrel, 1/2" grip length x 1/8" diameter	G	1 Carp	4.80	1.667	C	6.60	56		62.60	101
0200	3/16" diameter	G		4	2		11.20	67		78.20	124
0300	Aluminum rivet, steel mandrel, 1/8" diameter	G		4.80	1.667		8.75	56		64.75	103
0400	3/16" diameter	G		4	2		14	67		81	127
0500	Copper rivet, steel mandrel, 1/8" diameter	G		4.80	1.667		7.70	56		63.70	102
0800	Stainless rivet & mandrel, 1/8" diameter	G		4.80	1.667		24	56		80	120
0900	3/16" diameter	G		4	2		39.50	67		106.50	156
1000	Stainless rivet, steel mandrel, 1/8" diameter	G		4.80	1.667		20.50	56		76.50	116
1100	3/16" diameter	G		4	2		28	67		95	143
1200	Steel rivet and mandrel, 1/8" diameter	G		4.80	1.667		7.50	56		63.50	102
1300	3/16" diameter	G	↓	4	2	↓	11.10	67		78.10	124
1400	Hand riveting tool, minimum					Ea.	55			55	60.50
1500	Maximum						257			257	283
1600	Power riveting tool, minimum						450			450	495
1700	Maximum					↓	2,450			2,450	2,700

05 05 23.70 Structural Blind Bolts

			Crew	Daily Output	Labor-Hours	Unit	Material	Labor	Equipment	Total	Total Incl O&P
0010	STRUCTURAL BLIND BOLTS										
0100	1/4" diameter x 1/4" grip	G	1 Sswk	240	.033	Ea.	1.24	1.21		2.45	3.67
0150	1/2" grip	G		216	.037		.95	1.35		2.30	3.62
0200	3/8" diameter x 1/2" grip	G		232	.034		1.75	1.26		3.01	4.32
0250	3/4" grip	G		208	.038		1.83	1.40		3.23	4.68
0300	1/2" diameter x 1/2" grip	G		224	.036		5.15	1.30		6.45	8.15
0350	3/4" grip	G		200	.040		5.60	1.46		7.06	8.95
0400	5/8" diameter x 3/4" grip	G		216	.037		8.25	1.35		9.60	11.65
0450	1" grip	G	↓	192	.042	↓	9.05	1.52		10.57	12.85

05 12 Structural Steel Framing

05 12 23 – Structural Steel for Buildings

05 12 23.10 Ceiling Supports			Crew	Daily Output	Labor-Hours	Unit	Material	2011 Bare Costs Labor	Equipment	Total	Total Incl O&P
0010	**CEILING SUPPORTS**										
1000	Entrance door/folding partition supports, shop fabricated	G	E-4	60	.533	L.F.	22.50	19.70	1.82	44.02	64.50
1100	Linear accelerator door supports	G		14	2.286		102	84.50	7.80	194.30	283
1200	Lintels or shelf angles, hung, exterior hot dipped galv.	G		267	.120		15.35	4.42	.41	20.18	26
1250	Two coats primer paint instead of galv.	G		267	.120		13.30	4.42	.41	18.13	23.50
1400	Monitor support, ceiling hung, expansion bolted	G		4	8	Ea.	355	295	27.50	677.50	985
1450	Hung from pre-set inserts	G		6	5.333		385	197	18.20	600.20	815
1600	Motor supports for overhead doors	G		4	8		181	295	27.50	503.50	795
1700	Partition support for heavy folding partitions, without pocket	G		24	1.333	L.F.	51	49	4.55	104.55	156
1750	Supports at pocket only	G		12	2.667		102	98.50	9.10	209.60	310
2000	Rolling grilles & fire door supports	G		34	.941		44	34.50	3.21	81.71	118
2100	Spider-leg light supports, expansion bolted to ceiling slab	G		8	4	Ea.	146	148	13.65	307.65	455
2150	Hung from pre-set inserts	G		12	2.667	"	158	98.50	9.10	265.60	370
2400	Toilet partition support	G		36	.889	L.F.	51	33	3.03	87.03	122
2500	X-ray travel gantry support	G		12	2.667	"	176	98.50	9.10	283.60	390

05 12 23.15 Columns, Lightweight			Crew	Daily Output	Labor-Hours	Unit	Material	Labor	Equipment	Total	Total Incl O&P
0010	**COLUMNS, LIGHTWEIGHT**										
1000	Lightweight units (lally), 3-1/2" diameter		E-2	780	.062	L.F.	6.65	2.25	2.08	10.98	13.85
1050	4" diameter		"	900	.053	"	8.10	1.95	1.80	11.85	14.50
8000	Lally columns, to 8', 3-1/2" diameter		2 Carp	24	.667	Ea.	53.50	22.50		76	96
8080	4" diameter		"	20	.800	"	64.50	27		91.50	116

05 12 23.17 Columns, Structural			Crew	Daily Output	Labor-Hours	Unit	Material	Labor	Equipment	Total	Total Incl O&P
0010	**COLUMNS, STRUCTURAL**	R051223-10									
0015	Made from recycled materials	G									
0020	Shop fab'd for 100-ton, 1-2 story project, bolted connections										
0800	Steel, concrete filled, extra strong pipe, 3-1/2" diameter		E-2	660	.073	L.F.	37	2.66	2.46	42.12	48.50
0830	4" diameter			780	.062		41.50	2.25	2.08	45.83	52
0890	5" diameter			1020	.047		49.50	1.72	1.59	52.81	59
0930	6" diameter			1200	.040		65.50	1.47	1.35	68.32	76
0940	8" diameter			1100	.044		65.50	1.60	1.47	68.57	76.50
1100	For galvanizing, add					Lb.	.20			.20	.22
1300	For web ties, angles, etc., add per added lb.		1 Sswk	945	.008		1.13	.31		1.44	1.83
1500	Steel pipe, extra strong, no concrete, 3" to 5" diameter	G	E-2	16000	.003		1.13	.11	.10	1.34	1.55
1600	6" to 12" diameter	G		14000	.003		1.13	.13	.12	1.38	1.60
2700	12" x 8" x 1/2" thk wall	G		24000	.002		1.13	.07	.07	1.27	1.45
2800	Heavy section	G		32000	.002		1.13	.05	.05	1.23	1.40
5100	Structural tubing, rect, 5" to 6" wide, light section	G		8000	.006		1.13	.22	.20	1.55	1.87
8090	For projects 75 to 99 tons, add					All	10%				
8092	50 to 74 tons, add						20%				
8094	25 to 49 tons, add						30%	10%			
8096	10 to 24 tons, add						50%	25%			
8098	2 to 9 tons, add						75%	50%			
8099	Less than 2 tons, add						100%	100%			

05 12 23.45 Lintels			Crew	Daily Output	Labor-Hours	Unit	Material	Labor	Equipment	Total	Total Incl O&P
0010	**LINTELS**										
0015	Made from recycled materials	G									
0020	Plain steel angles, shop fabricated, under 500 lb.	G	1 Bric	550	.015	Lb.	.87	.49		1.36	1.75
0100	500 to 1000 lb.	G		640	.013	"	.84	.42		1.26	1.62
2000	Steel angles, 3-1/2" x 3", 1/4" thick, 2'-6" long	G		47	.170	Ea.	12.15	5.75		17.90	23
2100	4'-6" long	G		26	.308		22	10.35		32.35	41
2600	4" x 3-1/2", 1/4" thick, 5'-0" long	G		21	.381		28	12.80		40.80	51.50
2700	9'-0" long	G		12	.667		50	22.50		72.50	92

05 12 Structural Steel Framing

05 12 23 – Structural Steel for Buildings

05 12 23.65 Plates			Crew	Daily Output	Labor-Hours	Unit	Material	2011 Bare Costs Labor	Equipment	Total	Total Incl O&P
0010	PLATES										
0015	Made from recycled materials	G									
0020	For connections & stiffener plates, shop fabricated										
0050	1/8" thick (5.1 lb./S.F.)	G				S.F.	5.75			5.75	6.30
0100	1/4" thick (10.2 lb./S.F.)	G					11.50			11.50	12.60
0300	3/8" thick (15.3 lb./S.F.)	G					17.20			17.20	18.95
0400	1/2" thick (20.4 lb./S.F.)	G					23			23	25.50
0450	3/4" thick (30.6 lb./S.F.)	G					34.50			34.50	38
0500	1" thick (40.8 lb./S.F.)	G				↓	46			46	50.50
2000	Steel plate, warehouse prices, no shop fabrication										
2100	1/4" thick (10.2 lb./S.F.)	G				S.F.	7.15			7.15	7.85

05 12 23.77 Structural Steel Projects			Crew	Daily Output	Labor-Hours	Unit	Material	Labor	Equipment	Total	Total Incl O&P
0010	STRUCTURAL STEEL PROJECTS	R050516-30									
0015	Made from recycled materials	G									
0020	Shop fab'd for 100-ton, 1-2 story project, bolted connections										
0201	Apartments, nursing homes, etc., 1 to 2 stories R050523-10	G	E-2	6.45	7.442	Ton	2,250	273	251	2,774	3,250
0701	Offices, hospitals, etc., steel bearing, 1 to 2 stories	G		6.45	7.442		2,250	273	251	2,774	3,250
3101	Roof trusses, minimum	G	↓	8.13	5.904		3,150	216	200	3,566	4,100
3200	Maximum	G	E-5	8.30	8.675		3,825	315	209	4,349	5,025
5390	For projects 75 to 99 tons, add	R051223-20					10%				
5392	50 to 74 tons, add						20%				
5394	25 to 49 tons, add	R051223-25					30%	10%			
5396	10 to 24 tons, add						50%	25%			
5398	2 to 9 tons, add	R051223-30					75%	50%			
5399	Less than 2 tons, add					↓	100%	100%			

05 12 23.79 Structural Steel			Crew	Daily Output	Labor-Hours	Unit	Material	Labor	Equipment	Total	Total Incl O&P
0010	STRUCTURAL STEEL	R050516-30									
0020	Shop fab'd for 100-ton, 1-2 story project, bolted conn's.										
0050	Beams, W 6 x 9 R050521-20	G	E-2	720	.067	L.F.	12.15	2.44	2.25	16.84	20.50
0100	W 8 x 10	G		720	.067		13.50	2.44	2.25	18.19	22
0200	Columns, W 6 x 15 R051223-10	G		540	.089		22	3.26	3	28.26	33.50
0250	W 8 x 31	G	↓	540	.089	↓	45.50	3.26	3	51.76	59.50
7990	For projects 75 to 99 tons, add	R051223-20				All	10%				
7992	50 to 75 tons, add						20%				
7994	25 to 49 tons, add	R051223-25					30%	10%			
7996	10 to 24 tons, add						50%	25%			
7998	2 to 9 tons, add	R051223-30					75%	50%			
7999	Less than 2 tons, add					↓	100%	100%			

05 15 Wire Rope Assemblies

05 15 16 – Steel Wire Rope Assemblies

05 15 16.70 Temporary Cable Safety Railing			Crew	Daily Output	Labor-Hours	Unit	Material	Labor	Equipment	Total	Total Incl O&P
0010	TEMPORARY CABLE SAFETY RAILING, Each 100' strand incl.										
0020	2 eyebolts, 1 turnbuckle, 100' cable, 2 thimbles, 6 clips										
0025	Made from recycled materials	G									
0100	One strand using 1/4" cable & accessories	G	2 Sswk	4	4	C.L.F.	191	146		337	490
0200	1/2" cable & accessories	G	"	2	8	"	400	291		691	995

05 21 Steel Joist Framing

05 21 19 – Open Web Steel Joist Framing

05 21 19.10 Open Web Joists			Crew	Daily Output	Labor-Hours	Unit	Material	2011 Bare Costs Labor	Equipment	Total	Total Incl O&P
0010	**OPEN WEB JOISTS**										
0015	Made from recycled materials	G									
0020	K series, 40-ton lots, horiz. bridging, spans to 30', shop primer, minimum	G	E-7	15	4.800	Ton	1,125	175	123	1,423	1,725
0050	Average	G		12	6		1,275	219	153	1,647	1,975
0080	Maximum	G		9	8		1,525	292	204	2,021	2,450
0130	8K1, 5.1 lb./L.F.	G		1200	.060	L.F.	3.25	2.19	1.53	6.97	9.40
0140	10K1, 5.0 lb./L.F.	G		1200	.060		3.19	2.19	1.53	6.91	9.30
0160	12K3, 5.7 lb./L.F.	G		1500	.048		3.63	1.75	1.23	6.61	8.65
0180	14K3, 6.0 lb./L.F.	G		1500	.048		3.82	1.75	1.23	6.80	8.85
0200	16K3, 6.3 lb./L.F.	G		1800	.040		4.01	1.46	1.02	6.49	8.30
0220	16K6, 8.1 lb./L.F.	G		1800	.040		5.15	1.46	1.02	7.63	9.55
0240	18K5, 7.7 lb./L.F.	G		2000	.036		4.91	1.32	.92	7.15	8.90
0260	18K9, 10.2 lb./L.F.	G		2000	.036		6.50	1.32	.92	8.74	10.65
0410	Span 30' to 50', minimum	G		17	4.235	Ton	1,125	155	108	1,388	1,625
0440	Average	G		17	4.235		1,250	155	108	1,513	1,775
0460	Maximum	G		10	7.200		1,325	263	184	1,772	2,150
0500	20K5, 8.2 lb./L.F.	G		2000	.036	L.F.	5.15	1.32	.92	7.39	9.15
0520	20K9, 10.8 lb./L.F.	G		2000	.036		6.75	1.32	.92	8.99	10.95
0540	22K5, 8.8 lb./L.F.	G		2000	.036		5.50	1.32	.92	7.74	9.55
0560	22K9, 11.3 lb./L.F.	G		2000	.036		7.05	1.32	.92	9.29	11.25
0580	24K6, 9.7 lb./L.F.	G		2200	.033		6.05	1.20	.84	8.09	9.80
0600	24K10, 13.1 lb./L.F.	G		2200	.033		8.20	1.20	.84	10.24	12.15
0620	26K6, 10.6 lb./L.F.	G		2200	.033		6.65	1.20	.84	8.69	10.45
0640	26K10, 13.8 lb./L.F.	G		2200	.033		8.65	1.20	.84	10.69	12.65
0660	28K8, 12.7 lb./L.F.	G		2400	.030		7.95	1.10	.77	9.82	11.65
0680	28K12, 17.1 lb./L.F.	G		2400	.030		10.70	1.10	.77	12.57	14.65
0700	30K8, 13.2 lb./L.F.	G		2400	.030		8.25	1.10	.77	10.12	12
0720	30K12, 17.6 lb./L.F.	G		2400	.030		11	1.10	.77	12.87	15
0800	For less than 40-ton job lots										
0802	For 30 to 39 tons, add						10%				
0804	20 to 29 tons, add						20%				
0806	10 to 19 tons, add						30%				
0807	5 to 9 tons, add						50%	25%			
0808	1 to 4 tons, add						75%	50%			
0809	Less than 1 ton, add						100%	100%			
6200	For shop prime paint other than mfrs. standard, add						20%				
6400	Individual steel bearing plate, 6" x 6" x 1/4" with J-hook	G	1 Bric	160	.050	Ea.	6.75	1.68		8.43	10.20

05 31 Steel Decking

05 31 13 – Steel Floor Decking

05 31 13.50 Floor Decking			Crew	Daily Output	Labor-Hours	Unit	Material	Labor	Equipment	Total	Total Incl O&P
0010	**FLOOR DECKING**										
0015	Made from recycled materials	G									
5100	Non-cellular composite decking, galvanized, 1-1/2" deep, 16 gauge	G	E-4	3500	.009	S.F.	2.39	.34	.03	2.76	3.30
5120	18 gauge	G		3650	.009		1.92	.32	.03	2.27	2.76
5140	20 gauge	G		3800	.008		1.55	.31	.03	1.89	2.32
5200	2" deep, 22 gauge	G		3860	.008		1.35	.31	.03	1.69	2.10
5300	20 gauge	G		3600	.009		1.49	.33	.03	1.85	2.30
5400	18 gauge	G		3380	.009		1.88	.35	.03	2.26	2.78
5500	16 gauge	G		3200	.010		2.37	.37	.03	2.77	3.34
5700	3" deep, 22 gauge	G		3200	.010		1.47	.37	.03	1.87	2.36

05 31 Steel Decking

05 31 13 – Steel Floor Decking

05 31 13.50 Floor Decking			Crew	Daily Output	Labor-Hours	Unit	Material	2011 Bare Costs Labor	Equipment	Total	Total Incl O&P
5800	20 gauge	G	E-4	3000	.011	S.F.	1.64	.39	.04	2.07	2.59
5900	18 gauge	G		2850	.011		2	.41	.04	2.45	3.03
6000	16 gauge	G	↓	2700	.012	↓	2.69	.44	.04	3.17	3.83

05 31 23 – Steel Roof Decking

05 31 23.50 Roof Decking			Crew	Daily Output	Labor-Hours	Unit	Material	Labor	Equipment	Total	Total Incl O&P
0010	**ROOF DECKING**										
0015	Made from recycled materials	G									
2100	Open type, 1-1/2" deep, Type B, wide rib, galv., 22 gauge, under 50 sq.	G	E-4	4500	.007	S.F.	1.42	.26	.02	1.70	2.09
2600	20 gauge, under 50 squares	G		3865	.008		1.67	.31	.03	2.01	2.45
2900	18 gauge, under 50 squares	G		3800	.008		2.15	.31	.03	2.49	2.99
3700	4-1/2" deep, 20 gauge, over 50 squares	G	↓	2700	.012	↓	3.64	.44	.04	4.12	4.87

05 31 33 – Steel Form Decking

05 31 33.50 Form Decking			Crew	Daily Output	Labor-Hours	Unit	Material	Labor	Equipment	Total	Total Incl O&P
0010	**FORM DECKING**										
0015	Made from recycled materials	G									
6100	Slab form, steel, 28 gauge, 9/16" deep, type UFS, uncoated	G	E-4	4000	.008	S.F.	1.07	.30	.03	1.40	1.77
6200	Galvanized	G		4000	.008		.95	.30	.03	1.28	1.64
6220	24 gauge, 1" deep , type UF1X, uncoated	G		3900	.008		1.04	.30	.03	1.37	1.75
6240	Galvanized	G		3900	.008		1.22	.30	.03	1.55	1.95
6300	24 gauge, 1-5/16" deep, type UFX, uncoated	G		3800	.008		1.10	.31	.03	1.44	1.83
6400	Galvanized	G		3800	.008		1.29	.31	.03	1.63	2.04
6500	22 gauge, 1-5/16" deep, uncoated	G		3700	.009		1.38	.32	.03	1.73	2.16
6600	Galvanized	G		3700	.009		1.41	.32	.03	1.76	2.19
6700	22 gauge, 2" deep, uncoated	G		3600	.009		1.82	.33	.03	2.18	2.66
6800	Galvanized	G	↓	3600	.009	↓	1.78	.33	.03	2.14	2.62

05 41 Structural Metal Stud Framing

05 41 13 – Load-Bearing Metal Stud Framing

05 41 13.05 Bracing			Crew	Daily Output	Labor-Hours	Unit	Material	Labor	Equipment	Total	Total Incl O&P
0010	**BRACING**, shear wall X-bracing, per 10' x 10' bay, one face										
0015	Made of recycled materials	G									
0120	Metal strap, 20 ga x 4" wide	G	2 Carp	18	.889	Ea.	16.65	30		46.65	68.50
0130	6" wide	G		18	.889		25	30		55	77.50
0160	18 ga x 4" wide	G		16	1		26	33.50		59.50	84.50
0170	6" wide	G	↓	16	1	↓	38.50	33.50		72	98.50
0410	Continuous strap bracing, per horizontal row on both faces										
0420	Metal strap, 20 ga x 2" wide, studs 12" O.C.	G	1 Carp	7	1.143	C.L.F.	46	38.50		84.50	115
0430	16" O.C.	G		8	1		46	33.50		79.50	107
0440	24" O.C.	G		10	.800		46	27		73	96
0450	18 ga x 2" wide, studs 12" O.C.	G		6	1.333		64	44.50		108.50	145
0460	16" O.C.	G		7	1.143		64	38.50		102.50	134
0470	24" O.C.	G	↓	8	1	↓	64	33.50		97.50	126

05 41 13.10 Bridging			Crew	Daily Output	Labor-Hours	Unit	Material	Labor	Equipment	Total	Total Incl O&P
0010	**BRIDGING**, solid between studs w/1-1/4" leg track, per stud bay										
0015	Made from recycled materials	G									
0200	Studs 12" O.C., 18 ga x 2-1/2" wide	G	1 Carp	125	.064	Ea.	.76	2.15		2.91	4.43
0210	3-5/8" wide	G		120	.067		.91	2.24		3.15	4.75
0220	4" wide	G		120	.067		.98	2.24		3.22	4.82
0230	6" wide	G		115	.070		1.26	2.33		3.59	5.30
0240	8" wide	G	↓	110	.073	↓	1.57	2.44		4.01	5.80

05 41 Structural Metal Stud Framing

05 41 13 – Load-Bearing Metal Stud Framing

05 41 13.10 Bridging			Crew	Daily Output	Labor-Hours	Unit	Material	2011 Bare Costs Labor	Equipment	Total	Total Incl O&P
0300	16 ga x 2-1/2" wide	G	1 Carp	115	.070	Ea.	.97	2.33		3.30	4.97
0310	3-5/8" wide	G		110	.073		1.17	2.44		3.61	5.35
0320	4" wide	G		110	.073		1.25	2.44		3.69	5.45
0330	6" wide	G		105	.076		1.60	2.56		4.16	6.05
0340	8" wide	G		100	.080		2.01	2.68		4.69	6.70
1200	Studs 16" O.C., 18 ga x 2-1/2" wide	G		125	.064		.97	2.15		3.12	4.66
1210	3-5/8" wide	G		120	.067		1.17	2.24		3.41	5.05
1220	4" wide	G		120	.067		1.25	2.24		3.49	5.10
1230	6" wide	G		115	.070		1.62	2.33		3.95	5.70
1240	8" wide	G		110	.073		2.01	2.44		4.45	6.30
1300	16 ga x 2-1/2" wide	G		115	.070		1.24	2.33		3.57	5.25
1310	3-5/8" wide	G		110	.073		1.50	2.44		3.94	5.75
1320	4" wide	G		110	.073		1.60	2.44		4.04	5.85
1330	6" wide	G		105	.076		2.05	2.56		4.61	6.55
1340	8" wide	G		100	.080		2.57	2.68		5.25	7.30
2200	Studs 24" O.C., 18 ga x 2-1/2" wide	G		125	.064		1.41	2.15		3.56	5.15
2210	3-5/8" wide	G		120	.067		1.69	2.24		3.93	5.60
2220	4" wide	G		120	.067		1.81	2.24		4.05	5.75
2230	6" wide	G		115	.070		2.34	2.33		4.67	6.50
2240	8" wide	G		110	.073		2.91	2.44		5.35	7.30
2300	16 ga x 2-1/2" wide	G		115	.070		1.79	2.33		4.12	5.90
2310	3-5/8" wide	G		110	.073		2.17	2.44		4.61	6.45
2320	4" wide	G		110	.073		2.32	2.44		4.76	6.65
2330	6" wide	G		105	.076		2.96	2.56		5.52	7.55
2340	8" wide	G		100	.080		3.72	2.68		6.40	8.60
3000	Continuous bridging, per row										
3100	16 ga x 1-1/2" channel thru studs 12" O.C.	G	1 Carp	6	1.333	C.L.F.	41	44.50		85.50	120
3110	16" O.C.	G		7	1.143		41	38.50		79.50	109
3120	24" O.C.	G		8.80	.909		41	30.50		71.50	96
4100	2" x 2" angle x 18 ga, studs 12" O.C.	G		7	1.143		64	38.50		102.50	134
4110	16" O.C.	G		9	.889		64	30		94	120
4120	24" O.C.	G		12	.667		64	22.50		86.50	108
4200	16 ga, studs 12" O.C.	G		5	1.600		81.50	53.50		135	180
4210	16" O.C.	G		7	1.143		81.50	38.50		120	154
4220	24" O.C.	G		10	.800		81.50	27		108.50	135
05 41 13.25 Framing, Boxed Headers/Beams											
0010	**FRAMING, BOXED HEADERS/BEAMS**										
0015	Made from recycled materials	G									
0200	Double, 18 ga x 6" deep	G	2 Carp	220	.073	L.F.	4.40	2.44		6.84	8.90
0210	8" deep	G		210	.076		4.84	2.56		7.40	9.65
0220	10" deep	G		200	.080		5.90	2.68		8.58	11
0230	12" deep	G		190	.084		6.45	2.83		9.28	11.80
0300	16 ga x 8" deep	G		180	.089		5.60	2.98		8.58	11.15
0310	10" deep	G		170	.094		6.75	3.16		9.91	12.70
0320	12" deep	G		160	.100		7.30	3.36		10.66	13.65
0400	14 ga x 10" deep	G		140	.114		7.75	3.83		11.58	14.95
0410	12" deep	G		130	.123		8.50	4.13		12.63	16.25
1210	Triple, 18 ga x 8" deep	G		170	.094		7	3.16		10.16	13
1220	10" deep	G		165	.097		8.45	3.25		11.70	14.75
1230	12" deep	G		160	.100		9.25	3.36		12.61	15.80
1300	16 ga x 8" deep	G		145	.110		8.10	3.70		11.80	15.10
1310	10" deep	G		140	.114		9.70	3.83		13.53	17.05

05 41 Structural Metal Stud Framing

05 41 13 – Load-Bearing Metal Stud Framing

05 41 13.25 Framing, Boxed Headers/Beams		Crew	Daily Output	Labor-Hours	Unit	Material	Labor	Equipment	Total	Total Incl O&P
						2011 Bare Costs				
1320	12" deep G	2 Carp	135	.119	L.F.	10.55	3.98		14.53	18.30
1400	14 ga x 10" deep G		115	.139		10.60	4.67		15.27	19.45
1410	12" deep G		110	.145		11.70	4.88		16.58	21

05 41 13.30 Framing, Stud Walls		Crew	Daily Output	Labor-Hours	Unit	Material	Labor	Equipment	Total	Total Incl O&P
0010	**FRAMING, STUD WALLS** w/top & bottom track, no openings,									
0020	Headers, beams, bridging or bracing									
0025	Made from recycled materials G									
4100	8' high walls, 18 ga x 2-1/2" wide, studs 12" O.C. G	2 Carp	54	.296	L.F.	7.30	9.95		17.25	24.50
4110	16" O.C. G		77	.208		5.85	6.95		12.80	18.05
4120	24" O.C. G		107	.150		4.37	5		9.37	13.20
4130	3-5/8" wide, studs 12" O.C. G		53	.302		8.60	10.15		18.75	26.50
4140	16" O.C. G		76	.211		6.90	7.05		13.95	19.40
4150	24" O.C. G		105	.152		5.20	5.10		10.30	14.25
4160	4" wide, studs 12" O.C. G		52	.308		9.05	10.30		19.35	27.50
4170	16" O.C. G		74	.216		7.25	7.25		14.50	20
4180	24" O.C. G		103	.155		5.45	5.20		10.65	14.70
4190	6" wide, studs 12" O.C. G		51	.314		11.45	10.55		22	30
4200	16" O.C. G		73	.219		9.20	7.35		16.55	22.50
4210	24" O.C. G		101	.158		6.90	5.30		12.20	16.50
4220	8" wide, studs 12" O.C. G		50	.320		13.85	10.75		24.60	33
4230	16" O.C. G		72	.222		11.15	7.45		18.60	25
4240	24" O.C. G		100	.160		8.45	5.35		13.80	18.25
4300	16 ga x 2-1/2" wide, studs 12" O.C. G		47	.340		8.65	11.40		20.05	28.50
4310	16" O.C. G		68	.235		6.85	7.90		14.75	21
4320	24" O.C. G		94	.170		5.05	5.70		10.75	15.10
4330	3-5/8" wide, studs 12" O.C. G		46	.348		10.20	11.65		21.85	31
4340	16" O.C. G		66	.242		8.10	8.15		16.25	22.50
4350	24" O.C. G		92	.174		6	5.85		11.85	16.35
4360	4" wide, studs 12" O.C. G		45	.356		10.80	11.95		22.75	32
4370	16" O.C. G		65	.246		8.60	8.25		16.85	23.50
4380	24" O.C. G		90	.178		6.35	5.95		12.30	17
4390	6" wide, studs 12" O.C. G		44	.364		13.50	12.20		25.70	35.50
4400	16" O.C. G		64	.250		10.75	8.40		19.15	26
4410	24" O.C. G		88	.182		7.95	6.10		14.05	18.95
4420	8" wide, studs 12" O.C. G		43	.372		16.65	12.50		29.15	39.50
4430	16" O.C. G		63	.254		13.25	8.50		21.75	29
4440	24" O.C. G		86	.186		9.85	6.25		16.10	21.50
5100	10' high walls, 18 ga x 2-1/2" wide, studs 12" O.C. G		54	.296		8.75	9.95		18.70	26.50
5110	16" O.C. G		77	.208		6.90	6.95		13.85	19.25
5120	24" O.C. G		107	.150		5.10	5		10.10	14
5130	3-5/8" wide, studs 12" O.C. G		53	.302		10.35	10.15		20.50	28.50
5140	16" O.C. G		76	.211		8.20	7.05		15.25	21
5150	24" O.C. G		105	.152		6.05	5.10		11.15	15.20
5160	4" wide, studs 12" O.C. G		52	.308		10.85	10.30		21.15	29.50
5170	16" O.C. G		74	.216		8.60	7.25		15.85	21.50
5180	24" O.C. G		103	.155		6.35	5.20		11.55	15.70
5190	6" wide, studs 12" O.C. G		51	.314		13.70	10.55		24.25	32.50
5200	16" O.C. G		73	.219		10.90	7.35		18.25	24.50
5210	24" O.C. G		101	.158		8.05	5.30		13.35	17.75
5220	8" wide, studs 12" O.C. G		50	.320		16.60	10.75		27.35	36
5230	16" O.C. G		72	.222		13.20	7.45		20.65	27
5240	24" O.C. G		100	.160		9.80	5.35		15.15	19.75

05 41 Structural Metal Stud Framing

05 41 13 – Load-Bearing Metal Stud Framing

05 41 13.30 Framing, Stud Walls			Crew	Daily Output	Labor-Hours	Unit	Material	2011 Bare Costs Labor	Equipment	Total	Total Incl O&P
5300	16 ga x 2-1/2" wide, studs 12" O.C.	G	2 Carp	47	.340	L.F.	10.45	11.40		21.85	30.50
5310	16" O.C.	G		68	.235		8.20	7.90		16.10	22
5320	24" O.C.	G		94	.170		5.95	5.70		11.65	16.10
5330	3-5/8" wide, studs 12" O.C.	G		46	.348		12.35	11.65		24	33
5340	16" O.C.	G		66	.242		9.70	8.15		17.85	24.50
5350	24" O.C.	G		92	.174		7.05	5.85		12.90	17.50
5360	4" wide, studs 12" O.C.	G		45	.356		13.05	11.95		25	34.50
5370	16" O.C.	G		65	.246		10.25	8.25		18.50	25
5380	24" O.C.	G		90	.178		7.45	5.95		13.40	18.20
5390	6" wide, studs 12" O.C.	G		44	.364		16.30	12.20		28.50	38.50
5400	16" O.C.	G		64	.250		12.85	8.40		21.25	28
5410	24" O.C.	G		88	.182		9.35	6.10		15.45	20.50
5420	8" wide, studs 12" O.C.	G		43	.372		20	12.50		32.50	43
5430	16" O.C.	G		63	.254		15.80	8.50		24.30	31.50
5440	24" O.C.	G		86	.186		11.55	6.25		17.80	23
6190	12' high walls, 18 ga x 6" wide, studs 12" O.C.	G		41	.390		15.95	13.10		29.05	39.50
6200	16" O.C.	G		58	.276		12.55	9.25		21.80	29.50
6210	24" O.C.	G		81	.198		9.20	6.65		15.85	21
6220	8" wide, studs 12" O.C.	G		40	.400		19.30	13.40		32.70	43.50
6230	16" O.C.	G		57	.281		15.25	9.40		24.65	32.50
6240	24" O.C.	G		80	.200		11.15	6.70		17.85	23.50
6390	16 ga x 6" wide, studs 12" O.C.	G		35	.457		19.10	15.35		34.45	46.50
6400	16" O.C.	G		51	.314		14.90	10.55		25.45	34
6410	24" O.C.	G		70	.229		10.75	7.65		18.40	24.50
6420	8" wide, studs 12" O.C.	G		34	.471		23.50	15.80		39.30	52.50
6430	16" O.C.	G		50	.320		18.40	10.75		29.15	38
6440	24" O.C.	G		69	.232		13.25	7.80		21.05	27.50
6530	14 ga x 3-5/8" wide, studs 12" O.C.	G		34	.471		18.20	15.80		34	46.50
6540	16" O.C.	G		48	.333		14.20	11.20		25.40	34.50
6550	24" O.C.	G		65	.246		10.20	8.25		18.45	25
6560	4" wide, studs 12" O.C.	G		33	.485		19.20	16.25		35.45	48
6570	16" O.C.	G		47	.340		15	11.40		26.40	35.50
6580	24" O.C.	G		64	.250		10.80	8.40		19.20	26
6730	12 ga x 3-5/8" wide, studs 12" O.C.	G		31	.516		25	17.30		42.30	56.50
6740	16" O.C.	G		43	.372		19.40	12.50		31.90	42.50
6750	24" O.C.	G		59	.271		13.70	9.10		22.80	30.50
6760	4" wide, studs 12" O.C.	G		30	.533		27	17.90		44.90	59.50
6770	16" O.C.	G		42	.381		20.50	12.80		33.30	44.50
6780	24" O.C.	G		58	.276		14.60	9.25		23.85	31.50
7390	16' high walls, 16 ga x 6" wide, studs 12" O.C.	G		33	.485		24.50	16.25		40.75	54
7400	16" O.C.	G		48	.333		19.10	11.20		30.30	39.50
7410	24" O.C.	G		67	.239		13.50	8		21.50	28.50
7420	8" wide, studs 12" O.C.	G		32	.500		30.50	16.80		47.30	61.50
7430	16" O.C.	G		47	.340		23.50	11.40		34.90	45
7440	24" O.C.	G		66	.242		16.65	8.15		24.80	32
7560	14 ga x 4" wide, studs 12" O.C.	G		31	.516		25	17.30		42.30	56.50
7570	16" O.C.	G		45	.356		19.20	11.95		31.15	41
7580	24" O.C.	G		61	.262		13.60	8.80		22.40	29.50
7590	6" wide, studs 12" O.C.	G		30	.533		31	17.90		48.90	64.50
7600	16" O.C.	G		44	.364		24	12.20		36.20	47
7610	24" O.C.	G		60	.267		17.10	8.95		26.05	34
7760	12 ga x 4" wide, studs 12" O.C.	G		29	.552		35	18.50		53.50	69.50
7770	16" O.C.	G	↓	40	.400	↓	27	13.40		40.40	52

05 41 Structural Metal Stud Framing

05 41 13 – Load-Bearing Metal Stud Framing

05 41 13.30 Framing, Stud Walls			Crew	Daily Output	Labor-Hours	Unit	2011 Bare Costs Material	Labor	Equipment	Total	Total Incl O&P
7780	24" O.C.	G	2 Carp	55	.291	L.F.	18.70	9.75		28.45	37
7790	6" wide, studs 12" O.C.	G		28	.571		44	19.15		63.15	80.50
7800	16" O.C.	G		39	.410		34	13.75		47.75	60.50
7810	24" O.C.	G		54	.296		23.50	9.95		33.45	42.50
8590	20' high walls, 14 ga x 6" wide, studs 12" O.C.	G		29	.552		38	18.50		56.50	73
8600	16" O.C.	G		42	.381		29.50	12.80		42.30	54
8610	24" O.C.	G		57	.281		20.50	9.40		29.90	38.50
8620	8" wide, studs 12" O.C.	G		28	.571		46.50	19.15		65.65	83.50
8630	16" O.C.	G		41	.390		36	13.10		49.10	61.50
8640	24" O.C.	G		56	.286		25	9.60		34.60	43.50
8790	12 ga x 6" wide, studs 12" O.C.	G		27	.593		54.50	19.90		74.40	93.50
8800	16" O.C.	G		37	.432		41.50	14.50		56	70.50
8810	24" O.C.	G		51	.314		28.50	10.55		39.05	49
8820	8" wide, studs 12" O.C.	G		26	.615		66	20.50		86.50	107
8830	16" O.C.	G		36	.444		50.50	14.90		65.40	80.50
8840	24" O.C.	G		50	.320		35	10.75		45.75	56.50

05 42 Cold-Formed Metal Joist Framing

05 42 13 – Cold-Formed Metal Floor Joist Framing

05 42 13.05 Bracing

			Crew	Daily Output	Labor-Hours	Unit	Material	Labor	Equipment	Total	Total Incl O&P
0010	**BRACING**, continuous, per row, top & bottom										
0015	Made from recycled materials	G									
0120	Flat strap, 20 ga x 2" wide, joists at 12" O.C.	G	1 Carp	4.67	1.713	C.L.F.	48.50	57.50		106	149
0130	16" O.C.	G		5.33	1.501		46.50	50.50		97	136
0140	24" O.C.	G		6.66	1.201		45	40.50		85.50	117
0150	18 ga x 2" wide, joists at 12" O.C.	G		4	2		63	67		130	182
0160	16" O.C.	G		4.67	1.713		62	57.50		119.50	165
0170	24" O.C.	G		5.33	1.501		61	50.50		111.50	152

05 42 13.10 Bridging

			Crew	Daily Output	Labor-Hours	Unit	Material	Labor	Equipment	Total	Total Incl O&P
0010	**BRIDGING**, solid between joists w/1-1/4" leg track, per joist bay										
0015	Made from recycled materials	G									
0230	Joists 12" O.C., 18 ga track x 6" wide	G	1 Carp	80	.100	Ea.	1.26	3.36		4.62	7
0240	8" wide	G		75	.107		1.57	3.58		5.15	7.75
0250	10" wide	G		70	.114		1.96	3.83		5.79	8.55
0260	12" wide	G		65	.123		2.23	4.13		6.36	9.35
0330	16 ga track x 6" wide	G		70	.114		1.60	3.83		5.43	8.15
0340	8" wide	G		65	.123		2.01	4.13		6.14	9.10
0350	10" wide	G		60	.133		2.48	4.47		6.95	10.25
0360	12" wide	G		55	.145		2.86	4.88		7.74	11.30
0440	14 ga track x 8" wide	G		60	.133		2.51	4.47		6.98	10.25
0450	10" wide	G		55	.145		3.11	4.88		7.99	11.60
0460	12" wide	G		50	.160		3.59	5.35		8.94	12.95
0550	12 ga track x 10" wide	G		45	.178		4.54	5.95		10.49	15
0560	12" wide	G		40	.200		4.66	6.70		11.36	16.35
1230	16" O.C., 18 ga track x 6" wide	G		80	.100		1.62	3.36		4.98	7.40
1240	8" wide	G		75	.107		2.01	3.58		5.59	8.20
1250	10" wide	G		70	.114		2.51	3.83		6.34	9.15
1260	12" wide	G		65	.123		2.85	4.13		6.98	10.05
1330	16 ga track x 6" wide	G		70	.114		2.05	3.83		5.88	8.65
1340	8" wide	G		65	.123		2.57	4.13		6.70	9.75
1350	10" wide	G		60	.133		3.18	4.47		7.65	11

05 42 Cold-Formed Metal Joist Framing

05 42 13 – Cold-Formed Metal Floor Joist Framing

05 42 13.10 Bridging			Crew	Daily Output	Labor-Hours	Unit	2011 Bare Costs Material	Labor	Equipment	Total	Total Incl O&P
1360	12″ wide	G	1 Carp	55	.145	Ea.	3.66	4.88		8.54	12.20
1440	14 ga track x 8″ wide	G		60	.133		3.22	4.47		7.69	11.05
1450	10″ wide	G		55	.145		3.99	4.88		8.87	12.55
1460	12″ wide	G		50	.160		4.60	5.35		9.95	14.05
1550	12 ga track x 10″ wide	G		45	.178		5.80	5.95		11.75	16.40
1560	12″ wide	G		40	.200		5.95	6.70		12.65	17.80
2230	24″ O.C., 18 ga track x 6″ wide	G		80	.100		2.34	3.36		5.70	8.15
2240	8″ wide	G		75	.107		2.91	3.58		6.49	9.20
2250	10″ wide	G		70	.114		3.63	3.83		7.46	10.40
2260	12″ wide	G		65	.123		4.13	4.13		8.26	11.45
2330	16 ga track x 6″ wide	G		70	.114		2.96	3.83		6.79	9.65
2340	8″ wide	G		65	.123		3.72	4.13		7.85	11
2350	10″ wide	G		60	.133		4.61	4.47		9.08	12.55
2360	12″ wide	G		55	.145		5.30	4.88		10.18	14
2440	14 ga track x 8″ wide	G		60	.133		4.65	4.47		9.12	12.60
2450	10″ wide	G		55	.145		5.80	4.88		10.68	14.50
2460	12″ wide	G		50	.160		6.65	5.35		12	16.35
2550	12 ga track x 10″ wide	G		45	.178		8.45	5.95		14.40	19.25
2560	12″ wide	G		40	.200		8.65	6.70		15.35	21

05 42 13.25 Framing, Band Joist			Crew	Daily Output	Labor-Hours	Unit	Material	Labor	Equipment	Total	Total Incl O&P
0010	**FRAMING, BAND JOIST** (track) fastened to bearing wall										
0015	Made from recycled materials	G									
0220	18 ga track x 6″ deep	G	2 Carp	1000	.016	L.F.	1.03	.54		1.57	2.03
0230	8″ deep	G		920	.017		1.28	.58		1.86	2.39
0240	10″ deep	G		860	.019		1.60	.62		2.22	2.80
0320	16 ga track x 6″ deep	G		900	.018		1.30	.60		1.90	2.43
0330	8″ deep	G		840	.019		1.64	.64		2.28	2.87
0340	10″ deep	G		780	.021		2.03	.69		2.72	3.38
0350	12″ deep	G		740	.022		2.33	.73		3.06	3.77
0430	14 ga track x 8″ deep	G		750	.021		2.05	.72		2.77	3.45
0440	10″ deep	G		720	.022		2.54	.75		3.29	4.05
0450	12″ deep	G		700	.023		2.93	.77		3.70	4.50
0540	12 ga track x 10″ deep	G		670	.024		3.71	.80		4.51	5.40
0550	12″ deep	G		650	.025		3.80	.83		4.63	5.55

05 42 13.30 Framing, Boxed Headers/Beams			Crew	Daily Output	Labor-Hours	Unit	Material	Labor	Equipment	Total	Total Incl O&P
0010	**FRAMING, BOXED HEADERS/BEAMS**										
0015	Made from recycled materials	G									
0200	Double, 18 ga x 6″ deep	G	2 Carp	220	.073	L.F.	4.40	2.44		6.84	8.90
0210	8″ deep	G		210	.076		4.84	2.56		7.40	9.65
0220	10″ deep	G		200	.080		5.90	2.68		8.58	11
0230	12″ deep	G		190	.084		6.45	2.83		9.28	11.80
0300	16 ga x 8″ deep	G		180	.089		5.60	2.98		8.58	11.15
0310	10″ deep	G		170	.094		6.75	3.16		9.91	12.70
0320	12″ deep	G		160	.100		7.30	3.36		10.66	13.65
0400	14 ga x 10″ deep	G		140	.114		7.75	3.83		11.58	14.95
0410	12″ deep	G		130	.123		8.50	4.13		12.63	16.25
0500	12 ga x 10″ deep	G		110	.145		10.25	4.88		15.13	19.40
0510	12″ deep	G		100	.160		11.35	5.35		16.70	21.50
1210	Triple, 18 ga x 8″ deep	G		170	.094		7	3.16		10.16	13
1220	10″ deep	G		165	.097		8.45	3.25		11.70	14.75
1230	12″ deep	G		160	.100		9.25	3.36		12.61	15.80
1300	16 ga x 8″ deep	G		145	.110		8.10	3.70		11.80	15.10

05 42 Cold-Formed Metal Joist Framing

05 42 13 – Cold-Formed Metal Floor Joist Framing

05 42 13.30 Framing, Boxed Headers/Beams			Crew	Daily Output	Labor-Hours	Unit	Material	2011 Bare Costs Labor	Equipment	Total	Total Incl O&P
1310	10" deep	G	2 Carp	140	.114	L.F.	9.70	3.83		13.53	17.05
1320	12" deep	G		135	.119		10.55	3.98		14.53	18.30
1400	14 ga x 10" deep	G		115	.139		11.25	4.67		15.92	20
1410	12" deep	G		110	.145		12.35	4.88		17.23	22
1500	12 ga x 10" deep	G		90	.178		15	5.95		20.95	26.50
1510	12" deep	G		85	.188		16.60	6.30		22.90	29
05 42 13.40 Framing, Joists											
0010	**FRAMING, JOISTS**, no band joists (track), web stiffeners, headers,										
0020	Beams, bridging or bracing										
0025	Made from recycled materials	G									
0030	Joists (2" flange) and fasteners, materials only										
0220	18 ga x 6" deep	G				L.F.	1.35			1.35	1.49
0230	8" deep	G					1.59			1.59	1.74
0240	10" deep	G					1.87			1.87	2.06
0320	16 ga x 6" deep	G					1.65			1.65	1.81
0330	8" deep	G					1.97			1.97	2.17
0340	10" deep	G					2.31			2.31	2.54
0350	12" deep	G					2.61			2.61	2.88
0430	14 ga x 8" deep	G					2.47			2.47	2.71
0440	10" deep	G					2.86			2.86	3.14
0450	12" deep	G					3.24			3.24	3.57
0540	12 ga x 10" deep	G					4.16			4.16	4.57
0550	12" deep	G					4.73			4.73	5.20
1010	Installation of joists to band joists, beams & headers, labor only										
1220	18 ga x 6" deep		2 Carp	110	.145	Ea.		4.88		4.88	8.15
1230	8" deep			90	.178			5.95		5.95	10
1240	10" deep			80	.200			6.70		6.70	11.25
1320	16 ga x 6" deep			95	.168			5.65		5.65	9.45
1330	8" deep			70	.229			7.65		7.65	12.85
1340	10" deep			60	.267			8.95		8.95	14.95
1350	12" deep			55	.291			9.75		9.75	16.35
1430	14 ga x 8" deep			65	.246			8.25		8.25	13.80
1440	10" deep			45	.356			11.95		11.95	19.95
1450	12" deep			35	.457			15.35		15.35	25.50
1540	12 ga x 10" deep			40	.400			13.40		13.40	22.50
1550	12" deep			30	.533			17.90		17.90	30
05 42 13.45 Framing, Web Stiffeners											
0010	**FRAMING, WEB STIFFENERS** at joist bearing, fabricated from										
0020	Stud piece (1-5/8" flange) to stiffen joist (2" flange)										
0025	Made from recycled materials	G									
2120	For 6" deep joist, with 18 ga x 2-1/2" stud	G	1 Carp	120	.067	Ea.	1.61	2.24		3.85	5.50
2130	3-5/8" stud	G		110	.073		1.76	2.44		4.20	6
2140	4" stud	G		105	.076		1.71	2.56		4.27	6.15
2150	6" stud	G		100	.080		1.86	2.68		4.54	6.55
2160	8" stud	G		95	.084		1.90	2.83		4.73	6.80
2220	8" deep joist, with 2-1/2" stud	G		120	.067		1.76	2.24		4	5.70
2230	3-5/8" stud	G		110	.073		1.90	2.44		4.34	6.15
2240	4" stud	G		105	.076		1.87	2.56		4.43	6.35
2250	6" stud	G		100	.080		2.04	2.68		4.72	6.75
2260	8" stud	G		95	.084		2.19	2.83		5.02	7.15
2320	10" deep joist, with 2-1/2" stud	G		110	.073		2.48	2.44		4.92	6.80
2330	3-5/8" stud	G		100	.080		2.71	2.68		5.39	7.45

05 42 Cold-Formed Metal Joist Framing

05 42 13 – Cold-Formed Metal Floor Joist Framing

05 42 13.45 Framing, Web Stiffeners			Crew	Daily Output	Labor-Hours	Unit	2011 Bare Costs Material	Labor	Equipment	Total	Total Incl O&P
2340	4" stud	G	1 Carp	95	.084	Ea.	2.69	2.83		5.52	7.70
2350	6" stud	G		90	.089		2.91	2.98		5.89	8.20
2360	8" stud	G		85	.094		2.93	3.16		6.09	8.55
2420	12" deep joist, with 2-1/2" stud	G		110	.073		2.63	2.44		5.07	6.95
2430	3-5/8" stud	G		100	.080		2.84	2.68		5.52	7.60
2440	4" stud	G		95	.084		2.79	2.83		5.62	7.80
2450	6" stud	G		90	.089		3.05	2.98		6.03	8.35
2460	8" stud	G		85	.094		3.26	3.16		6.42	8.90
3130	For 6" deep joist, with 16 ga x 3-5/8" stud	G		100	.080		1.86	2.68		4.54	6.55
3140	4" stud	G		95	.084		1.85	2.83		4.68	6.75
3150	6" stud	G		90	.089		2.02	2.98		5	7.20
3160	8" stud	G		85	.094		2.14	3.16		5.30	7.65
3230	8" deep joist, with 3-5/8" stud	G		100	.080		2.06	2.68		4.74	6.75
3240	4" stud	G		95	.084		2.03	2.83		4.86	6.95
3250	6" stud	G		90	.089		2.24	2.98		5.22	7.45
3260	8" stud	G		85	.094		2.41	3.16		5.57	7.95
3330	10" deep joist, with 3-5/8" stud	G		85	.094		2.82	3.16		5.98	8.40
3340	4" stud	G		80	.100		2.88	3.36		6.24	8.75
3350	6" stud	G		75	.107		3.11	3.58		6.69	9.45
3360	8" stud	G		70	.114		3.26	3.83		7.09	10
3430	12" deep joist, with 3-5/8" stud	G		85	.094		3.07	3.16		6.23	8.70
3440	4" stud	G		80	.100		3.02	3.36		6.38	8.95
3450	6" stud	G		75	.107		3.34	3.58		6.92	9.65
3460	8" stud	G		70	.114		3.59	3.83		7.42	10.35
4230	For 8" deep joist, with 14 ga x 3-5/8" stud	G		90	.089		2.67	2.98		5.65	7.95
4240	4" stud	G		85	.094		2.72	3.16		5.88	8.30
4250	6" stud	G		80	.100		2.95	3.36		6.31	8.85
4260	8" stud	G		75	.107		3.15	3.58		6.73	9.45
4330	10" deep joist, with 3-5/8" stud	G		75	.107		3.75	3.58		7.33	10.15
4340	4" stud	G		70	.114		3.72	3.83		7.55	10.50
4350	6" stud	G		65	.123		4.09	4.13		8.22	11.40
4360	8" stud	G		60	.133		4.26	4.47		8.73	12.20
4430	12" deep joist, with 3-5/8" stud	G		75	.107		3.99	3.58		7.57	10.40
4440	4" stud	G		70	.114		4.06	3.83		7.89	10.85
4450	6" stud	G		65	.123		4.40	4.13		8.53	11.75
4460	8" stud	G		60	.133		4.71	4.47		9.18	12.70
5330	For 10" deep joist, with 12 ga x 3-5/8" stud	G		65	.123		3.96	4.13		8.09	11.25
5340	4" stud	G		60	.133		4.06	4.47		8.53	11.95
5350	6" stud	G		55	.145		4.48	4.88		9.36	13.10
5360	8" stud	G		50	.160		4.90	5.35		10.25	14.40
5430	12" deep joist, with 3-5/8" stud	G		65	.123		4.39	4.13		8.52	11.75
5440	4" stud	G		60	.133		4.28	4.47		8.75	12.20
5450	6" stud	G		55	.145		4.88	4.88		9.76	13.50
5460	8" stud	G		50	.160		5.60	5.35		10.95	15.15

05 42 23 – Cold-Formed Metal Roof Joist Framing

05 42 23.05 Framing, Bracing			Crew	Daily Output	Labor-Hours	Unit	Material	Labor	Equipment	Total	Total Incl O&P
0010	**FRAMING, BRACING**										
0015	Made from recycled materials	G									
0020	Continuous bracing, per row										
0100	16 ga x 1-1/2" channel thru rafters/trusses @ 16" O.C.	G	1 Carp	4.50	1.778	C.L.F.	41	59.50		100.50	145
0120	24" O.C.	G		6	1.333		41	44.50		85.50	120
0300	2" x 2" angle x 18 ga, rafters/trusses @ 16" O.C.	G		6	1.333		64	44.50		108.50	145

05 42 Cold-Formed Metal Joist Framing

05 42 23 – Cold-Formed Metal Roof Joist Framing

05 42 23.05 Framing, Bracing			Crew	Daily Output	Labor-Hours	Unit	2011 Bare Costs Material	Labor	Equipment	Total	Total Incl O&P
0320	24" O.C.	G	1 Carp	8	1	C.L.F.	64	33.50		97.50	126
0400	16 ga, rafters/trusses @ 16" O.C.	G		4.50	1.778		81.50	59.50		141	190
0420	24" O.C.	G		6.50	1.231		81.50	41.50		123	159

05 42 23.10 Framing, Bridging			Crew	Daily Output	Labor-Hours	Unit	Material	Labor	Equipment	Total	Total Incl O&P
0010	**FRAMING, BRIDGING**										
0015	Made from recycled materials	G									
0020	Solid, between rafters w/1-1/4" leg track, per rafter bay										
1200	Rafters 16" O.C., 18 ga x 4" deep	G	1 Carp	60	.133	Ea.	1.25	4.47		5.72	8.90
1210	6" deep	G		57	.140		1.62	4.71		6.33	9.70
1220	8" deep	G		55	.145		2.01	4.88		6.89	10.35
1230	10" deep	G		52	.154		2.51	5.15		7.66	11.40
1240	12" deep	G		50	.160		2.85	5.35		8.20	12.15
2200	24" O.C., 18 ga x 4" deep	G		60	.133		1.81	4.47		6.28	9.50
2210	6" deep	G		57	.140		2.34	4.71		7.05	10.45
2220	8" deep	G		55	.145		2.91	4.88		7.79	11.35
2230	10" deep	G		52	.154		3.63	5.15		8.78	12.65
2240	12" deep	G		50	.160		4.13	5.35		9.48	13.55

05 42 23.50 Framing, Parapets			Crew	Daily Output	Labor-Hours	Unit	Material	Labor	Equipment	Total	Total Incl O&P
0010	**FRAMING, PARAPETS**										
0015	Made from recycled materials	G									
0100	3' high installed on 1st story, 18 ga x 4" wide studs, 12" O.C.	G	2 Carp	100	.160	L.F.	4.56	5.35		9.91	14
0110	16" O.C.	G		150	.107		3.89	3.58		7.47	10.30
0120	24" O.C.	G		200	.080		3.21	2.68		5.89	8
0200	6" wide studs, 12" O.C.	G		100	.160		5.80	5.35		11.15	15.35
0210	16" O.C.	G		150	.107		4.94	3.58		8.52	11.45
0220	24" O.C.	G		200	.080		4.10	2.68		6.78	9
1100	Installed on 2nd story, 18 ga x 4" wide studs, 12" O.C.	G		95	.168		4.56	5.65		10.21	14.45
1110	16" O.C.	G		145	.110		3.89	3.70		7.59	10.50
1120	24" O.C.	G		190	.084		3.21	2.83		6.04	8.25
1200	6" wide studs, 12" O.C.	G		95	.168		5.80	5.65		11.45	15.80
1210	16" O.C.	G		145	.110		4.94	3.70		8.64	11.65
1220	24" O.C.	G		190	.084		4.10	2.83		6.93	9.25
2100	Installed on gable, 18 ga x 4" wide studs, 12" O.C.	G		85	.188		4.56	6.30		10.86	15.55
2110	16" O.C.	G		130	.123		3.89	4.13		8.02	11.20
2120	24" O.C.	G		170	.094		3.21	3.16		6.37	8.85
2200	6" wide studs, 12" O.C.	G		85	.188		5.80	6.30		12.10	16.90
2210	16" O.C.	G		130	.123		4.94	4.13		9.07	12.35
2220	24" O.C.	G		170	.094		4.10	3.16		7.26	9.80

05 42 23.60 Framing, Roof Rafters			Crew	Daily Output	Labor-Hours	Unit	Material	Labor	Equipment	Total	Total Incl O&P
0010	**FRAMING, ROOF RAFTERS**										
0015	Made from recycled materials	G									
0100	Boxed ridge beam, double, 18 ga x 6" deep	G	2 Carp	160	.100	L.F.	4.40	3.36		7.76	10.45
0110	8" deep	G		150	.107		4.84	3.58		8.42	11.35
0120	10" deep	G		140	.114		5.90	3.83		9.73	12.90
0130	12" deep	G		130	.123		6.45	4.13		10.58	13.95
0200	16 ga x 6" deep	G		150	.107		4.96	3.58		8.54	11.45
0210	8" deep	G		140	.114		5.60	3.83		9.43	12.55
0220	10" deep	G		130	.123		6.75	4.13		10.88	14.30
0230	12" deep	G		120	.133		7.30	4.47		11.77	15.55
1100	Rafters, 2" flange, material only, 18 ga x 6" deep	G					1.35			1.35	1.49
1110	8" deep	G					1.59			1.59	1.74
1120	10" deep	G					1.87			1.87	2.06

05 42 Cold-Formed Metal Joist Framing

05 42 23 – Cold-Formed Metal Roof Joist Framing

05 42 23.60 Framing, Roof Rafters		Crew	Daily Output	Labor-Hours	Unit	Material	2011 Bare Costs Labor	Equipment	Total	Total Incl O&P
1130	12" deep	G			L.F.	2.15			2.15	2.37
1200	16 ga x 6" deep	G				1.65			1.65	1.81
1210	8" deep	G				1.97			1.97	2.17
1220	10" deep	G				2.31			2.31	2.54
1230	12" deep	G				2.61			2.61	2.88
2100	Installation only, ordinary rafter to 4:12 pitch, 18 ga x 6" deep		2 Carp	35	.457	Ea.		15.35		15.35	25.50
2110	8" deep			30	.533			17.90		17.90	30
2120	10" deep			25	.640			21.50		21.50	36
2130	12" deep			20	.800			27		27	45
2200	16 ga x 6" deep			30	.533			17.90		17.90	30
2210	8" deep			25	.640			21.50		21.50	36
2220	10" deep			20	.800			27		27	45
2230	12" deep			15	1.067			36		36	60
8100	Add to labor, ordinary rafters on steep roofs							25%			
8110	Dormers & complex roofs							50%			
8200	Hip & valley rafters to 4:12 pitch							25%			
8210	Steep roofs							50%			
8220	Dormers & complex roofs							75%			
8300	Hip & valley jack rafters to 4:12 pitch							50%			
8310	Steep roofs							75%			
8320	Dormers & complex roofs							100%			

05 42 23.70 Framing, Soffits and Canopies			Crew	Daily Output	Labor-Hours	Unit	Material	Labor	Equipment	Total	Total Incl O&P
0010	**FRAMING, SOFFITS & CANOPIES**										
0015	Made from recycled materials	G									
0130	Continuous ledger track @ wall, studs @ 16" O.C., 18 ga x 4" wide	G	2 Carp	535	.030	L.F.	.84	1		1.84	2.60
0140	6" wide	G		500	.032		1.08	1.07		2.15	2.99
0150	8" wide	G		465	.034		1.34	1.15		2.49	3.41
0160	10" wide	G		430	.037		1.67	1.25		2.92	3.93
0230	Studs @ 24" O.C., 18 ga x 4" wide	G		800	.020		.80	.67		1.47	2
0240	6" wide	G		750	.021		1.03	.72		1.75	2.33
0250	8" wide	G		700	.023		1.28	.77		2.05	2.69
0260	10" wide	G		650	.025		1.60	.83		2.43	3.14
1000	Horizontal soffit and canopy members, material only										
1030	1-5/8" flange studs, 18 ga x 4" deep	G				L.F.	1.08			1.08	1.19
1040	6" deep	G					1.36			1.36	1.49
1050	8" deep	G					1.63			1.63	1.80
1140	2" flange joists, 18 ga x 6" deep	G					1.55			1.55	1.70
1150	8" deep	G					1.81			1.81	1.99
1160	10" deep	G					2.14			2.14	2.35
4030	Installation only, 18 ga, 1-5/8" flange x 4" deep		2 Carp	130	.123	Ea.		4.13		4.13	6.90
4040	6" deep			110	.145			4.88		4.88	8.15
4050	8" deep			90	.178			5.95		5.95	10
4140	2" flange, 18 ga x 6" deep			110	.145			4.88		4.88	8.15
4150	8" deep			90	.178			5.95		5.95	10
4160	10" deep			80	.200			6.70		6.70	11.25
6010	Clips to attach facia to rafter tails, 2" x 2" x 18 ga angle	G	1 Carp	120	.067		.75	2.24		2.99	4.57
6020	16 ga angle	G	"	100	.080		.96	2.68		3.64	5.55

05 44 Cold-Formed Metal Trusses

05 44 13 – Cold-Formed Metal Roof Trusses

05 44 13.60 Framing, Roof Trusses			Crew	Daily Output	Labor-Hours	Unit	Material	2011 Bare Costs Labor	Equipment	Total	Total Incl O&P
0010	**FRAMING, ROOF TRUSSES**										
0015	Made from recycled materials	G									
0020	Fabrication of trusses on ground, Fink (W) or King Post, to 4:12 pitch										
0120	18 ga x 4" chords, 16' span	G	2 Carp	12	1.333	Ea.	50.50	44.50		95	131
0130	20' span	G		11	1.455		63	49		112	151
0140	24' span	G		11	1.455		75.50	49		124.50	165
0150	28' span	G		10	1.600		88	53.50		141.50	187
0160	32' span	G		10	1.600		101	53.50		154.50	201
0250	6" chords, 28' span	G		9	1.778		111	59.50		170.50	222
0260	32' span	G		9	1.778		127	59.50		186.50	239
0270	36' span	G		8	2		142	67		209	269
0280	40' span	G		8	2		158	67		225	286
1120	5:12 to 8:12 pitch, 18 ga x 4" chords, 16' span	G		10	1.600		57.50	53.50		111	154
1130	20' span	G		9	1.778		72	59.50		131.50	179
1140	24' span	G		9	1.778		86.50	59.50		146	195
1150	28' span	G		8	2		101	67		168	223
1160	32' span	G		8	2		115	67		182	239
1250	6" chords, 28' span	G		7	2.286		127	76.50		203.50	267
1260	32' span	G		7	2.286		145	76.50		221.50	287
1270	36' span	G		6	2.667		163	89.50		252.50	330
1280	40' span	G		6	2.667		181	89.50		270.50	350
2120	9:12 to 12:12 pitch, 18 ga x 4" chords, 16' span	G		8	2		72	67		139	191
2130	20' span	G		7	2.286		90	76.50		166.50	227
2140	24' span	G		7	2.286		108	76.50		184.50	247
2150	28' span	G		6	2.667		126	89.50		215.50	289
2160	32' span	G		6	2.667		144	89.50		233.50	310
2250	6" chords, 28' span	G		5	3.200		158	107		265	355
2260	32' span	G		5	3.200		181	107		288	380
2270	36' span	G		4	4		203	134		337	450
2280	40' span	G		4	4		226	134		360	475
5120	Erection only of roof trusses, to 4:12 pitch, 16' span		F-6	48	.833			26	13.65	39.65	58
5130	20' span			46	.870			27	14.25	41.25	60.50
5140	24' span			44	.909			28.50	14.90	43.40	63.50
5150	28' span			42	.952			29.50	15.60	45.10	66.50
5160	32' span			40	1			31	16.40	47.40	70
5170	36' span			38	1.053			33	17.25	50.25	73.50
5180	40' span			36	1.111			34.50	18.20	52.70	77.50
5220	5:12 to 8:12 pitch, 16' span			42	.952			29.50	15.60	45.10	66.50
5230	20' span			40	1			31	16.40	47.40	70
5240	24' span			38	1.053			33	17.25	50.25	73.50
5250	28' span			36	1.111			34.50	18.20	52.70	77.50
5260	32' span			34	1.176			36.50	19.30	55.80	82
5270	36' span			32	1.250			39	20.50	59.50	87.50
5280	40' span			30	1.333			41.50	22	63.50	93
5320	9:12 to 12:12 pitch, 16' span			36	1.111			34.50	18.20	52.70	77.50
5330	20' span			34	1.176			36.50	19.30	55.80	82
5340	24' span			32	1.250			39	20.50	59.50	87.50
5350	28' span			30	1.333			41.50	22	63.50	93
5360	32' span			28	1.429			44.50	23.50	68	100
5370	36' span			26	1.538			48	25	73	108
5380	40' span			24	1.667			52	27.50	79.50	117

05 51 Metal Stairs

05 51 13 – Metal Pan Stairs

05 51 13.50 Pan Stairs			Crew	Daily Output	Labor-Hours	Unit	Material	2011 Bare Costs Labor	Equipment	Total	Total Incl O&P
0010	**PAN STAIRS**, shop fabricated, steel stringers										
0015	Made from recycled materials	G									
1700	Pre-erected, steel pan tread, 3′-6″ wide, 2 line pipe rail	G	E-2	87	.552	Riser	465	20	18.65	503.65	570
1800	With flat bar picket rail	G	"	87	.552	"	520	20	18.65	558.65	630

05 51 19 – Metal Grating Stairs

05 51 19.50 Grating Stairs			Crew	Daily Output	Labor-Hours	Unit	Material	Labor	Equipment	Total	Total Incl O&P
0010	**GRATING STAIRS**, shop fabricated, steel stringers, safety nosing on treads										
0015	Made from recycled materials	G									
0020	Grating tread and pipe railing, 3′-6″ wide	G	E-4	35	.914	Riser	281	33.50	3.12	317.62	380
0100	4′-0″ wide	G	"	30	1.067	"	365	39.50	3.64	408.14	480

05 51 23 – Metal Fire Escapes

05 51 23.50 Fire Escape Stairs			Crew	Daily Output	Labor-Hours	Unit	Material	Labor	Equipment	Total	Total Incl O&P
0010	**FIRE ESCAPE STAIRS**, shop fabricated										
0020	One story, disappearing, stainless steel	G	2 Sswk	20	.800	V.L.F.	220	29		249	298
0100	Portable ladder					Ea.	120			120	132
1100	Fire escape, galvanized steel, 8′-0″ to 10′-4″ ceiling	G	2 Carp	1	16		1,600	535		2,135	2,650
1110	10′-6″ to 13′-6″ ceiling	G	"	1	16		2,100	535		2,635	3,225

05 51 33 – Metal Ladders

05 51 33.13 Vertical Metal Ladders			Crew	Daily Output	Labor-Hours	Unit	Material	Labor	Equipment	Total	Total Incl O&P
0010	**VERTICAL METAL LADDERS**, shop fabricated										
0015	Made from recycled materials	G									
0020	Steel, 20″ wide, bolted to concrete, with cage	G	E-4	50	.640	V.L.F.	65	23.50	2.18	90.68	119
0100	Without cage	G		85	.376		36	13.90	1.28	51.18	67.50
0300	Aluminum, bolted to concrete, with cage	G		50	.640		111	23.50	2.18	136.68	170
0400	Without cage	G		85	.376		48	13.90	1.28	63.18	81

05 51 33.23 Alternating Tread Ladders			Crew	Daily Output	Labor-Hours	Unit	Material	Labor	Equipment	Total	Total Incl O&P
0010	**ALTERNATING TREAD LADDERS**, shop fabricated										
0015	Made from recycled materials	G									
1350	Alternating tread stair, 56/68°, steel, standard paint color	G	2 Sswk	50	.320	V.L.F.	258	11.65		269.65	305
1360	Non-standard paint color	G		50	.320		292	11.65		303.65	340
1370	Galvanized steel	G		50	.320		300	11.65		311.65	350
1380	Stainless steel	G		50	.320		435	11.65		446.65	495
1390	68°, aluminum	G		50	.320		315	11.65		326.65	370

05 52 Metal Railings

05 52 13 – Pipe and Tube Railings

05 52 13.50 Railings, Pipe			Crew	Daily Output	Labor-Hours	Unit	Material	Labor	Equipment	Total	Total Incl O&P
0010	**RAILINGS, PIPE**, shop fab'd, 3′-6″ high, posts @ 5′ O.C.										
0015	Made from recycled materials	G									
0020	Aluminum, 2 rail, satin finish, 1-1/4″ diameter	G	E-4	160	.200	L.F.	31.50	7.40	.68	39.58	49.50
0030	Clear anodized	G		160	.200		39	7.40	.68	47.08	58
0040	Dark anodized	G		160	.200		44	7.40	.68	52.08	63.50
0080	1-1/2″ diameter, satin finish	G		160	.200		37.50	7.40	.68	45.58	56.50
0090	Clear anodized	G		160	.200		42	7.40	.68	50.08	61
0100	Dark anodized	G		160	.200		46.50	7.40	.68	54.58	66
0140	Aluminum, 3 rail, 1-1/4″ diam., satin finish	G		137	.234		48	8.60	.80	57.40	70.50
0150	Clear anodized	G		137	.234		60	8.60	.80	69.40	83.50
0160	Dark anodized	G		137	.234		66.50	8.60	.80	75.90	90.50
0200	1-1/2″ diameter, satin finish	G		137	.234		57.50	8.60	.80	66.90	81

05 52 Metal Railings

05 52 13 – Pipe and Tube Railings

05 52 13.50 Railings, Pipe			Crew	Daily Output	Labor-Hours	Unit	Material	Labor (2011 Bare Costs)	Equipment (2011 Bare Costs)	Total (2011 Bare Costs)	Total Incl O&P
0210	Clear anodized	G	E-4	137	.234	L.F.	65.50	8.60	.80	74.90	89.50
0220	Dark anodized	G		137	.234		71.50	8.60	.80	80.90	96
0500	Steel, 2 rail, on stairs, primed, 1-1/4" diameter	G		160	.200		21	7.40	.68	29.08	38
0520	1-1/2" diameter	G		160	.200		23	7.40	.68	31.08	40.50
0540	Galvanized, 1-1/4" diameter	G		160	.200		29	7.40	.68	37.08	47
0560	1-1/2" diameter	G		160	.200		32.50	7.40	.68	40.58	51
0580	Steel, 3 rail, primed, 1-1/4" diameter	G		137	.234		31.50	8.60	.80	40.90	52
0600	1-1/2" diameter	G		137	.234		33	8.60	.80	42.40	54
0620	Galvanized, 1-1/4" diameter	G		137	.234		44	8.60	.80	53.40	66
0640	1-1/2" diameter	G		137	.234		51.50	8.60	.80	60.90	74
0700	Stainless steel, 2 rail, 1-1/4" diam. #4 finish	G		137	.234		94	8.60	.80	103.40	120
0720	High polish	G		137	.234		152	8.60	.80	161.40	184
0740	Mirror polish	G		137	.234		190	8.60	.80	199.40	226
0760	Stainless steel, 3 rail, 1-1/2" diam., #4 finish	G		120	.267		142	9.85	.91	152.76	176
0770	High polish	G		120	.267		235	9.85	.91	245.76	279
0780	Mirror finish	G		120	.267		285	9.85	.91	295.76	335
0900	Wall rail, alum. pipe, 1-1/4" diam., satin finish	G		213	.150		17.95	5.55	.51	24.01	31
0905	Clear anodized	G		213	.150		22	5.55	.51	28.06	35
0910	Dark anodized	G		213	.150		26.50	5.55	.51	32.56	40
0915	1-1/2" diameter, satin finish	G		213	.150		20	5.55	.51	26.06	33
0920	Clear anodized	G		213	.150		25	5.55	.51	31.06	38.50
0925	Dark anodized	G		213	.150		31	5.55	.51	37.06	45
0930	Steel pipe, 1-1/4" diameter, primed	G		213	.150		12.70	5.55	.51	18.76	25
0935	Galvanized	G		213	.150		18.45	5.55	.51	24.51	31.50
0940	1-1/2" diameter	G		176	.182		13.10	6.70	.62	20.42	28
0945	Galvanized	G		213	.150		18.50	5.55	.51	24.56	31.50
0955	Stainless steel pipe, 1-1/2" diam., #4 finish	G		107	.299		75	11.05	1.02	87.07	105
0960	High polish	G		107	.299		153	11.05	1.02	165.07	190
0965	Mirror polish	G		107	.299		181	11.05	1.02	193.07	221
2000	2-line pipe rail (1-1/2" T&B) with 1/2" pickets @ 4-1/2" O.C.,										
2005	attached handrail on brackets										
2010	42" high aluminum, satin finish, straight & level	G	E-4	120	.267	L.F.	175	9.85	.91	185.76	213
2050	42" high steel, primed, straight & level	G	"	120	.267		107	9.85	.91	117.76	138
4000	For curved and level rails, add						10%	10%			
4100	For sloped rails for stairs, add						30%	30%			

05 58 Formed Metal Fabrications

05 58 25 – Formed Lamp Posts

05 58 25.40 Lamp Posts			Crew	Daily Output	Labor-Hours	Unit	Material	Labor (2011 Bare Costs)	Equipment (2011 Bare Costs)	Total (2011 Bare Costs)	Total Incl O&P
0010	**LAMP POSTS**										
0020	Aluminum, 7' high, stock units, post only	G	1 Carp	16	.500	Ea.	49	16.80		65.80	82
0100	Mild steel, plain	G	"	16	.500	"	57	16.80		73.80	90.50

05 71 Decorative Metal Stairs

05 71 13 – Fabricated Metal Spiral Stairs

05 71 13.50 Spiral Stairs			Crew	Daily Output	Labor-Hours	Unit	Material	2011 Bare Costs Labor	Equipment	Total	Total Incl O&P
0010	**SPIRAL STAIRS**, shop fabricated										
1810	Spiral aluminum, 5'-0" diameter, stock units	G	E-4	45	.711	Riser	510	26	2.42	538.42	615
1820	Custom units	G		45	.711		965	26	2.42	993.42	1,100
1900	Spiral, cast iron, 4'-0" diameter, ornamental, minimum	G		45	.711		455	26	2.42	483.42	555
1920	Maximum	G	↓	25	1.280	↓	620	47	4.36	671.36	775

05 73 Decorative Metal Railings

05 73 16 – Wire Rope Decorative Metal Railings

05 73 16.10 Cable Railings			Crew	Daily Output	Labor-Hours	Unit	Material	Labor	Equipment	Total	Total Incl O&P
0010	**CABLE RAILINGS**, with 316 stainless steel 1 x 19 cable, 3/16" diameter										
0015	Made from recycled materials	G									
0100	1-3/4" diameter stainless steel posts x 42" high, cables 4" OC	G	2 Sswk	25	.640	L.F.	61.50	23.50		85	113

05 73 23 – Ornamental Railings

05 73 23.50 Railings, Ornamental			Crew	Daily Output	Labor-Hours	Unit	Material	Labor	Equipment	Total	Total Incl O&P
0010	**RAILINGS, ORNAMENTAL**, 3'-6" high, posts @ 6' O.C.										
0020	Bronze or stainless, hand forged, minimum	G	2 Sswk	24	.667	L.F.	112	24.50		136.50	170
0100	Maximum			18	.889		167	32.50		199.50	245
0200	Aluminum, panelized, minimum			24	.667		19.85	24.50		44.35	68.50
0300	Maximum			18	.889		36	32.50		68.50	101
0400	Wrought iron, hand forged, minimum	G		24	.667		74.50	24.50		99	129
0500	Maximum			18	.889		119	32.50		151.50	193
0550	Steel, panelized, minimum			24	.667		18.70	24.50		43.20	67
0560	Maximum			18	.889		24	32.50		56.50	87.50
0600	Composite metal/wood/glass, minimum			18	.889		102	32.50		134.50	174
0700	Maximum		↓	12	1.333	↓	204	48.50		252.50	315

Division Notes

		CREW	DAILY OUTPUT	LABOR-HOURS	UNIT	2011 BARE COSTS				TOTAL INCL O&P
						MAT.	LABOR	EQUIP.	TOTAL	

Estimating Tips

06 05 00 Common Work Results for Wood, Plastics, and Composites

- Common to any wood-framed structure are the accessory connector items such as screws, nails, adhesives, hangers, connector plates, straps, angles, and hold-downs. For typical wood-framed buildings, such as residential projects, the aggregate total for these items can be significant, especially in areas where seismic loading is a concern. For floor and wall framing, the material cost is based on 10 to 25 lbs. per MBF. Hold-downs, hangers, and other connectors should be taken off by the piece.

  Included with material costs are fasteners for a normal installation. RSMeans engineers use manufacturer's recommendations, written specifications, and/or standard construction practice for size and spacing of fasteners. Prices for various fasteners are shown for informational purposes only. Adjustments should be made if unusual fastening conditions exist.

06 10 00 Carpentry

- Lumber is a traded commodity and therefore sensitive to supply and demand in the marketplace. Even in "budgetary" estimating of wood-framed projects, it is advisable to call local suppliers for the latest market pricing.
- Common quantity units for wood-framed projects are "thousand board feet" (MBF). A board foot is a volume of wood, 1″ x 1′ x 1′, or 144 cubic inches. Board-foot quantities are generally calculated using nominal material dimensions—dressed sizes are ignored. Board foot per lineal foot of any stick of lumber can be calculated by dividing the nominal cross-sectional area by 12. As an example, 2,000 lineal feet of 2 x 12 equates to 4 MBF by dividing the nominal area, 2 x 12, by 12, which equals 2, and multiplying by 2,000 to give 4,000 board feet. This simple rule applies to all nominal dimensioned lumber.
- Waste is an issue of concern at the quantity takeoff for any area of construction. Framing lumber is sold in even foot lengths, i.e., 10′, 12′, 14′, 16′, and depending on spans, wall heights, and the grade of lumber, waste is inevitable. A rule of thumb for lumber waste is 5%-10% depending on material quality and the complexity of the framing.
- Wood in various forms and shapes is used in many projects, even where the main structural framing is steel, concrete, or masonry. Plywood as a back-up partition material and 2x boards used as blocking and cant strips around roof edges are two common examples. The estimator should ensure that the costs of all wood materials are included in the final estimate.

06 20 00 Finish Carpentry

- It is necessary to consider the grade of workmanship when estimating labor costs for erecting millwork and interior finish. In practice, there are three grades: premium, custom, and economy. The RSMeans daily output for base and case moldings is in the range of 200 to 250 L.F. per carpenter per day. This is appropriate for most average custom-grade projects. For premium projects, an adjustment to productivity of 25%-50% should be made, depending on the complexity of the job.

Reference Numbers

Reference numbers are shown in shaded boxes at the beginning of some major classifications. These numbers refer to related items in the Reference Section. The reference information may be an estimating procedure, an alternate pricing method, or technical information.

Note: Not all subdivisions listed here necessarily appear in this publication.

06 05 Common Work Results for Wood, Plastics, and Composites

06 05 05 – Selective Wood and Plastics Demolition

06 05 05.10 Selective Demolition Wood Framing		Crew	Daily Output	Labor-Hours	Unit	Material	2011 Bare Costs Labor	Equipment	Total	Total Incl O&P
0010	SELECTIVE DEMOLITION WOOD FRAMING R024119-10									
0100	Timber connector, nailed, small	1 Clab	96	.083	Ea.		2.20		2.20	3.69
0110	Medium		60	.133			3.53		3.53	5.90
0120	Large		48	.167			4.41		4.41	7.40
0130	Bolted, small		48	.167			4.41		4.41	7.40
0140	Medium		32	.250			6.60		6.60	11.10
0150	Large		24	.333			8.80		8.80	14.75
2958	Beams, 2" x 6"	2 Clab	1100	.015	L.F.		.38		.38	.64
2960	2" x 8"		825	.019			.51		.51	.86
2965	2" x 10"		665	.024			.64		.64	1.07
2970	2" x 12"		550	.029			.77		.77	1.29
2972	2" x 14"		470	.034			.90		.90	1.51
2975	4" x 8"	B-1	413	.058			1.58		1.58	2.64
2980	4" x 10"		330	.073			1.97		1.97	3.30
2985	4" x 12"		275	.087			2.37		2.37	3.96
3000	6" x 8"		275	.087			2.37		2.37	3.96
3040	6" x 10"		220	.109			2.96		2.96	4.95
3080	6" x 12"		185	.130			3.52		3.52	5.90
3120	8" x 12"		140	.171			4.65		4.65	7.80
3160	10" x 12"		110	.218			5.90		5.90	9.90
3162	Alternate pricing method		1.10	21.818	M.B.F.		590		590	990
3170	Blocking, in 16" OC wall framing, 2" x 4"	1 Clab	600	.013	L.F.		.35		.35	.59
3172	2" x 6"		400	.020			.53		.53	.89
3174	In 24" OC wall framing, 2" x 4"		600	.013			.35		.35	.59
3176	2" x 6"		400	.020			.53		.53	.89
3178	Alt method, wood blocking removal from wood framing		.40	20	M.B.F.		530		530	885
3179	Wood blocking removal from steel framing		.36	22.222	"		590		590	985
3180	Bracing, let in, 1" x 3", studs 16" OC		1050	.008	L.F.		.20		.20	.34
3181	Studs 24" OC		1080	.007			.20		.20	.33
3182	1" x 4", studs 16" OC		1050	.008			.20		.20	.34
3183	Studs 24" OC		1080	.007			.20		.20	.33
3184	1" x 6", studs 16" OC		1050	.008			.20		.20	.34
3185	Studs 24" OC		1080	.007			.20		.20	.33
3186	2" x 3", studs 16" OC		800	.010			.26		.26	.44
3187	Studs 24" OC		830	.010			.26		.26	.43
3188	2" x 4", studs 16" OC		800	.010			.26		.26	.44
3189	Studs 24" OC		830	.010			.26		.26	.43
3190	2" x 6", studs 16" OC		800	.010			.26		.26	.44
3191	Studs 24" OC		830	.010			.26		.26	.43
3192	2" x 8", studs 16" OC		800	.010			.26		.26	.44
3193	Studs 24" OC		830	.010			.26		.26	.43
3194	"T" shaped metal bracing, studs at 16" OC		1060	.008			.20		.20	.33
3195	Studs at 24" OC		1200	.007			.18		.18	.30
3196	Metal straps, studs at 16" OC		1200	.007			.18		.18	.30
3197	Studs at 24" OC		1240	.006			.17		.17	.29
3200	Columns, round, 8' to 14' tall		40	.200	Ea.		5.30		5.30	8.85
3202	Dimensional lumber sizes	2 Clab	1.10	14.545	M.B.F.		385		385	645
3250	Blocking, between joists	1 Clab	320	.025	Ea.		.66		.66	1.11
3252	Bridging, metal strap, between joists		320	.025	Pr.		.66		.66	1.11
3254	Wood, between joists		320	.025	"		.66		.66	1.11
3260	Door buck, studs, header & access., 8' high 2" x 4" wall, 3' wide		32	.250	Ea.		6.60		6.60	11.10
3261	4' wide		32	.250			6.60		6.60	11.10
3262	5' wide		32	.250			6.60		6.60	11.10

06 05 Common Work Results for Wood, Plastics, and Composites

06 05 05 – Selective Wood and Plastics Demolition

06 05 05.10 Selective Demolition Wood Framing		Crew	Daily Output	Labor-Hours	Unit	Material	2011 Bare Costs Labor	Equipment	Total	Total Incl O&P
3263	6′ wide	1 Clab	32	.250	Ea.		6.60		6.60	11.10
3264	8′ wide		30	.267			7.05		7.05	11.80
3265	10′ wide		30	.267			7.05		7.05	11.80
3266	12′ wide		30	.267			7.05		7.05	11.80
3267	2″ x 6″ wall, 3′ wide		32	.250			6.60		6.60	11.10
3268	4′ wide		32	.250			6.60		6.60	11.10
3269	5′ wide		32	.250			6.60		6.60	11.10
3270	6′ wide		32	.250			6.60		6.60	11.10
3271	8′ wide		30	.267			7.05		7.05	11.80
3272	10′ wide		30	.267			7.05		7.05	11.80
3273	12′ wide		30	.267			7.05		7.05	11.80
3274	Window buck, studs, header & access, 8′ high 2″ x 4″ wall, 2′ wide		24	.333			8.80		8.80	14.75
3275	3′ wide		24	.333			8.80		8.80	14.75
3276	4′ wide		24	.333			8.80		8.80	14.75
3277	5′ wide		24	.333			8.80		8.80	14.75
3278	6′ wide		24	.333			8.80		8.80	14.75
3279	7′ wide		24	.333			8.80		8.80	14.75
3280	8′ wide		22	.364			9.60		9.60	16.10
3281	10′ wide		22	.364			9.60		9.60	16.10
3282	12′ wide		22	.364			9.60		9.60	16.10
3283	2″ x 6″ wall, 2′ wide		24	.333			8.80		8.80	14.75
3284	3′ wide		24	.333			8.80		8.80	14.75
3285	4′ wide		24	.333			8.80		8.80	14.75
3286	5′ wide		24	.333			8.80		8.80	14.75
3287	6′ wide		24	.333			8.80		8.80	14.75
3288	7′ wide		24	.333			8.80		8.80	14.75
3289	8′ wide		22	.364			9.60		9.60	16.10
3290	10′ wide		22	.364			9.60		9.60	16.10
3291	12′ wide		22	.364			9.60		9.60	16.10
3400	Fascia boards, 1″ x 6″		500	.016	L.F.		.42		.42	.71
3440	1″ x 8″		450	.018			.47		.47	.79
3480	1″ x 10″		400	.020			.53		.53	.89
3490	2″ x 6″		450	.018			.47		.47	.79
3500	2″ x 8″		400	.020			.53		.53	.89
3510	2″ x 10″		350	.023			.60		.60	1.01
3610	Furring, on wood walls or ceiling		4000	.002	S.F.		.05		.05	.09
3620	On masonry or concrete walls or ceiling		1200	.007	″		.18		.18	.30
3800	Headers over openings, 2 @ 2″ x 6″		110	.073	L.F.		1.92		1.92	3.22
3840	2 @ 2″ x 8″		100	.080			2.12		2.12	3.54
3880	2 @ 2″ x 10″		90	.089			2.35		2.35	3.94
3885	Alternate pricing method		.26	30.651	M.B.F.		810		810	1,350
3920	Joists, 1″ x 4″		1250	.006	L.F.		.17		.17	.28
3930	1″ x 6″		1135	.007			.19		.19	.31
3950	1″ x 10″		895	.009			.24		.24	.40
3960	1″ x 12″		765	.010			.28		.28	.46
4200	2″ x 4″	2 Clab	1000	.016			.42		.42	.71
4230	2″ x 6″		970	.016			.44		.44	.73
4240	2″ x 8″		940	.017			.45		.45	.75
4250	2″ x 10″		910	.018			.47		.47	.78
4280	2″ x 12″		880	.018			.48		.48	.81
4281	2″ x 14″		850	.019			.50		.50	.83
4282	Composite joists, 9-1/2″		960	.017			.44		.44	.74
4283	11-7/8″		930	.017			.45		.45	.76

06 05 Common Work Results for Wood, Plastics, and Composites

06 05 05 – Selective Wood and Plastics Demolition

06 05 05.10 Selective Demolition Wood Framing		Crew	Daily Output	Labor-Hours	Unit	Material	2011 Bare Costs Labor	Equipment	Total	Total Incl O&P
4284	14"	2 Clab	897	.018	L.F.		.47		.47	.79
4285	16"		865	.019			.49		.49	.82
4290	Wood joists, alternate pricing method		1.50	10.667	M.B.F.		282		282	475
4500	Open web joist, 12" deep		500	.032	L.F.		.85		.85	1.42
4505	14" deep		475	.034			.89		.89	1.49
4510	16" deep		450	.036			.94		.94	1.58
4530	24" deep		400	.040			1.06		1.06	1.77
4550	Ledger strips, 1" x 2"	1 Clab	1200	.007			.18		.18	.30
4560	1" x 3"		1200	.007			.18		.18	.30
4570	1" x 4"		1200	.007			.18		.18	.30
4580	2" x 2"		1100	.007			.19		.19	.32
4590	2" x 4"		1000	.008			.21		.21	.35
4600	2" x 6"		1000	.008			.21		.21	.35
4601	2" x 8 or 2" x 10"		800	.010			.26		.26	.44
4602	4" x 6"		600	.013			.35		.35	.59
4604	4" x 8"		450	.018			.47		.47	.79
5400	Posts, 4" x 4"	2 Clab	800	.020			.53		.53	.89
5405	4" x 6"		550	.029			.77		.77	1.29
5410	4" x 8"		440	.036			.96		.96	1.61
5425	4" x 10"		390	.041			1.09		1.09	1.82
5430	4" x 12"		350	.046			1.21		1.21	2.03
5440	6" x 6"		400	.040			1.06		1.06	1.77
5445	6" x 8"		350	.046			1.21		1.21	2.03
5450	6" x 10"		320	.050			1.32		1.32	2.22
5455	6" x 12"		290	.055			1.46		1.46	2.44
5480	8" x 8"		300	.053			1.41		1.41	2.36
5500	10" x 10"		240	.067			1.76		1.76	2.95
5660	Tongue and groove floor planks		2	8	M.B.F.		212		212	355
5682	16" OC, 2" x 4"		880	.018	S.F.		.48		.48	.81
5683	2" x 6"		840	.019			.50		.50	.84
5684	2" x 8"		820	.020			.52		.52	.86
5685	2" x 10"		820	.020			.52		.52	.86
5686	2" x 12"		810	.020			.52		.52	.87
5687	24" OC, 2" x 4"		1170	.014			.36		.36	.61
5688	2" x 6"		1117	.014			.38		.38	.63
5689	2" x 8"		1091	.015			.39		.39	.65
5690	2" x 10"		1091	.015			.39		.39	.65
5691	2" x 12"		1077	.015			.39		.39	.66
5795	Rafters, ordinary, 2" x 4" (alternate method)		862	.019	L.F.		.49		.49	.82
5800	2" x 6" (alternate method)		850	.019			.50		.50	.83
5840	2" x 8" (alternate method)		837	.019			.51		.51	.85
5855	2" x 10" (alternate method)		825	.019			.51		.51	.86
5865	2" x 12" (alternate method)		812	.020			.52		.52	.87
5870	Sill plate, 2" x 4"	1 Clab	1170	.007			.18		.18	.30
5871	2" x 6"		780	.010			.27		.27	.45
5872	2" x 8"		586	.014			.36		.36	.60
5873	Alternate pricing method		.78	10.256	M.B.F.		271		271	455
5885	Ridge board, 1" x 4"	2 Clab	900	.018	L.F.		.47		.47	.79
5886	1" x 6"		875	.018			.48		.48	.81
5887	1" x 8"		850	.019			.50		.50	.83
5888	1" x 10"		825	.019			.51		.51	.86
5889	1" x 12"		800	.020			.53		.53	.89
5890	2" x 4"		900	.018			.47		.47	.79

06 05 Common Work Results for Wood, Plastics, and Composites

06 05 05 – Selective Wood and Plastics Demolition

	06 05 05.10 Selective Demolition Wood Framing	Crew	Daily Output	Labor-Hours	Unit	2011 Bare Costs Material	Labor	Equipment	Total	Total Incl O&P
5892	2" x 6"	2 Clab	875	.018	L.F.		.48		.48	.81
5894	2" x 8"		850	.019			.50		.50	.83
5896	2" x 10"		825	.019			.51		.51	.86
5898	2" x 12"		800	.020			.53		.53	.89
6050	Rafter tie, 1" x 4"		1250	.013			.34		.34	.57
6052	1" x 6"		1135	.014			.37		.37	.62
6054	2" x 4"		1000	.016			.42		.42	.71
6056	2" x 6"		970	.016			.44		.44	.73
6070	Sleepers, on concrete, 1" x 2"	1 Clab	4700	.002			.05		.05	.08
6075	1" x 3"		4000	.002			.05		.05	.09
6080	2" x 4"		3000	.003			.07		.07	.12
6085	2" x 6"		2600	.003			.08		.08	.14
6086	Sheathing from roof, 5/16"	2 Clab	1600	.010	S.F.		.26		.26	.44
6088	3/8"		1525	.010			.28		.28	.46
6090	1/2"		1400	.011			.30		.30	.51
6092	5/8"		1300	.012			.33		.33	.55
6094	3/4"		1200	.013			.35		.35	.59
6096	Board sheathing from roof		1400	.011			.30		.30	.51
6100	Sheathing, from walls, 1/4"		1200	.013			.35		.35	.59
6110	5/16"		1175	.014			.36		.36	.60
6120	3/8"		1150	.014			.37		.37	.62
6130	1/2"		1125	.014			.38		.38	.63
6140	5/8"		1100	.015			.38		.38	.64
6150	3/4"		1075	.015			.39		.39	.66
6152	Board sheathing from walls		1500	.011			.28		.28	.47
6158	Subfloor, with boards		1050	.015			.40		.40	.68
6160	Plywood, 1/2" thick		768	.021			.55		.55	.92
6162	5/8" thick		760	.021			.56		.56	.93
6164	3/4" thick		750	.021			.56		.56	.94
6165	1-1/8" thick		720	.022			.59		.59	.98
6166	Underlayment, particle board, 3/8" thick	1 Clab	780	.010			.27		.27	.45
6168	1/2" thick		768	.010			.28		.28	.46
6170	5/8" thick		760	.011			.28		.28	.47
6172	3/4" thick		750	.011			.28		.28	.47
6200	Stairs and stringers, minimum	2 Clab	40	.400	Riser		10.60		10.60	17.70
6240	Maximum	"	26	.615	"		16.30		16.30	27.50
6300	Components, tread	1 Clab	110	.073	Ea.		1.92		1.92	3.22
6320	Riser		80	.100	"		2.65		2.65	4.43
6390	Stringer, 2" x 10"		260	.031	L.F.		.81		.81	1.36
6400	2" x 12"		260	.031			.81		.81	1.36
6410	3" x 10"		250	.032			.85		.85	1.42
6420	3" x 12"		250	.032			.85		.85	1.42
6590	Wood studs, 2" x 3"	2 Clab	3076	.005			.14		.14	.23
6600	2" x 4"		2000	.008			.21		.21	.35
6640	2" x 6"		1600	.010			.26		.26	.44
6720	Wall framing, including studs, plates and blocking, 2" x 4"	1 Clab	600	.013	S.F.		.35		.35	.59
6740	2" x 6"		480	.017	"		.44		.44	.74
6750	Headers, 2" x 4"		1125	.007	L.F.		.19		.19	.32
6755	2" x 6"		1125	.007			.19		.19	.32
6760	2" x 8"		1050	.008			.20		.20	.34
6765	2" x 10"		1050	.008			.20		.20	.34
6770	2" x 12"		1000	.008			.21		.21	.35
6780	4" x 10"		525	.015			.40		.40	.68

06 05 Common Work Results for Wood, Plastics, and Composites

06 05 05 – Selective Wood and Plastics Demolition

06 05 05.10 Selective Demolition Wood Framing		Crew	Daily Output	Labor-Hours	Unit	2011 Bare Costs Material	Labor	Equipment	Total	Total Incl O&P
6785	4" x 12"	1 Clab	500	.016	L.F.		.42		.42	.71
6790	6" x 8"		560	.014			.38		.38	.63
6795	6" x 10"		525	.015			.40		.40	.68
6797	6" x 12"		500	.016			.42		.42	.71
7000	Trusses									
7050	12' span	2 Clab	74	.216	Ea.		5.70		5.70	9.60
7150	24' span	F-3	66	.606			18.60	9.95	28.55	42
7200	26' span		64	.625			19.20	10.25	29.45	43.50
7250	28' span		62	.645			19.80	10.55	30.35	44.50
7300	30' span		58	.690			21	11.30	32.30	47.50
7350	32' span		56	.714			22	11.70	33.70	49.50
7400	34' span		54	.741			23	12.15	35.15	51.50
7450	36' span		52	.769			23.50	12.60	36.10	53.50
8000	Soffit, T & G wood	1 Clab	520	.015	S.F.		.41		.41	.68
8010	Hardboard, vinyl or aluminum	"	640	.013			.33		.33	.55
8030	Plywood	2 Carp	315	.051			1.70		1.70	2.85
9500	See Section 02 41 19.23 for rubbish handling									

06 05 05.20 Selective Demolition Millwork and Trim		Crew	Daily Output	Labor-Hours	Unit	Material	Labor	Equipment	Total	Total Incl O&P
0010	SELECTIVE DEMOLITION MILLWORK AND TRIM R024119-10									
1000	Cabinets, wood, base cabinets, per L.F.	2 Clab	80	.200	L.F.		5.30		5.30	8.85
1020	Wall cabinets, per L.F.		80	.200			5.30		5.30	8.85
1100	Steel, painted, base cabinets		60	.267			7.05		7.05	11.80
1500	Counter top, minimum		200	.080			2.12		2.12	3.54
1510	Maximum		120	.133			3.53		3.53	5.90
2000	Paneling, 4' x 8' sheets		2000	.008	S.F.		.21		.21	.35
2100	Boards, 1" x 4"		700	.023			.60		.60	1.01
2120	1" x 6"		750	.021			.56		.56	.94
2140	1" x 8"		800	.020			.53		.53	.89
3000	Trim, baseboard, to 6" wide		1200	.013	L.F.		.35		.35	.59
3040	Greater than 6" and up to 12" wide		1000	.016			.42		.42	.71
3100	Ceiling trim		1000	.016			.42		.42	.71
3120	Chair rail		1200	.013			.35		.35	.59
3140	Railings with balusters		240	.067			1.76		1.76	2.95
3160	Wainscoting		700	.023	S.F.		.60		.60	1.01
4000	Curtain rod	1 Clab	80	.100	L.F.		2.65		2.65	4.43

06 05 23 – Wood, Plastic, and Composite Fastenings

06 05 23.10 Nails		Crew	Daily Output	Labor-Hours	Unit	Material	Labor	Equipment	Total	Total Incl O&P
0010	**NAILS**, material only, based upon 50# box purchase									
0020	Copper nails, plain				Lb.	10			10	11
0400	Stainless steel, plain					7.80			7.80	8.60
0500	Box, 3d to 20d, bright					1.25			1.25	1.38
0520	Galvanized					1.77			1.77	1.95
0600	Common, 3d to 60d, plain					1.30			1.30	1.43
0700	Galvanized					1.56			1.56	1.72
0800	Aluminum					5.80			5.80	6.40
1000	Annular or spiral thread, 4d to 60d, plain					1.80			1.80	1.98
1200	Galvanized					3			3	3.30
1400	Drywall nails, plain					3.59			3.59	3.95
1600	Galvanized					2.19			2.19	2.41
1800	Finish nails, 4d to 10d, plain					1.36			1.36	1.50
2000	Galvanized					1.79			1.79	1.97
2100	Aluminum					5.80			5.80	6.40

06 05 Common Work Results for Wood, Plastics, and Composites

06 05 23 – Wood, Plastic, and Composite Fastenings

06 05 23.10 Nails		Crew	Daily Output	Labor-Hours	Unit	Material	2011 Bare Costs Labor	Equipment	Total	Total Ind O&P
2300	Flooring nails, hardened steel, 2d to 10d, plain				Lb.	3.10			3.10	3.41
2400	Galvanized					4			4	4.40
2500	Gypsum lath nails, 1-1/8", 13 ga. flathead, blued					3.59			3.59	3.95
2600	Masonry nails, hardened steel, 3/4" to 3" long, plain					2.20			2.20	2.42
2700	Galvanized					3.75			3.75	4.13
2900	Roofing nails, threaded, galvanized					1.63			1.63	1.79
3100	Aluminum					4.80			4.80	5.30
3300	Compressed lead head, threaded, galvanized					2.47			2.47	2.72
3600	Siding nails, plain shank, galvanized					2			2	2.20
3800	Aluminum					5.80			5.80	6.40
5000	Add to prices above for cement coating					.11			.11	.12
5200	Zinc or tin plating					.14			.14	.15
5500	Vinyl coated sinkers, 8d to 16d				↓	.65			.65	.72
06 05 23.40 Sheet Metal Screws										
0010	**SHEET METAL SCREWS**									
0020	Steel, standard, #8 x 3/4", plain				C	2.92			2.92	3.21
0100	Galvanized					2.92			2.92	3.21
0300	#10 x 1", plain					4.03			4.03	4.43
0400	Galvanized					4.03			4.03	4.43
1500	Self-drilling, with washers, (pinch point) #8 x 3/4", plain					7.80			7.80	8.55
1600	Galvanized					7.80			7.80	8.55
1800	#10 x 3/4", plain					8.90			8.90	9.80
1900	Galvanized					8.90			8.90	9.80
3000	Stainless steel w/aluminum or neoprene washers, #14 x 1", plain					30			30	33
3100	#14 x 2", plain				↓	40			40	44
06 05 23.50 Wood Screws										
0010	**WOOD SCREWS**									
0020	#8 x 1" long, steel				C	3.41			3.41	3.75
0100	Brass					10.50			10.50	11.55
0200	#8, 2" long, steel					5.25			5.25	5.80
0300	Brass					19.95			19.95	22
0400	#10, 1" long, steel					3.65			3.65	4.02
0500	Brass					14.15			14.15	15.60
0600	#10, 2" long, steel					5.85			5.85	6.45
0700	Brass					24.50			24.50	27
0800	#10, 3" long, steel					9.80			9.80	10.75
1000	#12, 2" long, steel					8.05			8.05	8.85
1100	Brass					29.50			29.50	32
1500	#12, 3" long, steel					11.65			11.65	12.80
2000	#12, 4" long, steel				↓	21.50			21.50	23.50
06 05 23.60 Timber Connectors										
0010	**TIMBER CONNECTORS**									
0020	Add up cost of each part for total cost of connection									
0100	Connector plates, steel, with bolts, straight	2 Carp	75	.213	Ea.	28	7.15		35.15	42.50
0110	Tee, 7 gauge		50	.320		32	10.75		42.75	53
0120	T- Strap, 14 gauge, 12" x 8" x 2"		50	.320		32	10.75		42.75	53
0150	Anchor plates, 7 ga, 9" x 7"	↓	75	.213		28	7.15		35.15	42.50
0200	Bolts, machine, sq. hd. with nut & washer, 1/2" diameter, 4" long	1 Carp	140	.057		.78	1.92		2.70	4.07
0300	7-1/2" long		130	.062		1.41	2.06		3.47	5
0500	3/4" diameter, 7-1/2" long		130	.062		3.35	2.06		5.41	7.15
0610	Machine bolts, w/nut, washer, 3/4" diam., 15" L, HD's & beam hangers		95	.084	↓	6.25	2.83		9.08	11.60
0720	Machine bolts, sq. hd. w/nut & wash	↓	150	.053	Lb.	3.17	1.79		4.96	6.50

06 05 Common Work Results for Wood, Plastics, and Composites

06 05 23 – Wood, Plastic, and Composite Fastenings

06 05 23.60 Timber Connectors		Crew	Daily Output	Labor-Hours	Unit	2011 Bare Costs Material	Labor	Equipment	Total	Total Incl O&P
0800	Drilling bolt holes in timber, 1/2" diameter	1 Carp	450	.018	Inch		.60		.60	1
0900	1" diameter		350	.023	"		.77		.77	1.28
1100	Framing anchor, angle, 3" x 3" x 1-1/2", 12 ga		175	.046	Ea.	2.40	1.53		3.93	5.20
1150	Framing anchors, 18 gauge, 4-1/2" x 2-3/4"		175	.046		2.40	1.53		3.93	5.20
1160	Framing anchors, 18 gauge, 4-1/2" x 3"		175	.046		2.40	1.53		3.93	5.20
1170	Clip anchors plates, 18 gauge, 12" x 1-1/8"		175	.046		2.40	1.53		3.93	5.20
1250	Holdowns, 3 gauge base, 10 gauge body		8	1		23	33.50		56.50	81
1260	Holdowns, 7 gauge 11-1/16" x 3-1/4"		8	1		23	33.50		56.50	81
1270	Holdowns, 7 gauge 14-3/8" x 3-1/8"		8	1		23	33.50		56.50	81
1275	Holdowns, 12 gauge 8" x 2-1/2"		8	1		23	33.50		56.50	81
1300	Joist and beam hangers, 18 ga. galv., for 2" x 4" joist		175	.046		.67	1.53		2.20	3.31
1400	2" x 6" to 2" x 10" joist		165	.048		1.28	1.63		2.91	4.13
1600	16 ga. galv., 3" x 6" to 3" x 10" joist		160	.050		2.72	1.68		4.40	5.80
1700	3" x 10" to 3" x 14" joist		160	.050		4.52	1.68		6.20	7.80
1800	4" x 6" to 4" x 10" joist		155	.052		2.82	1.73		4.55	6
1900	4" x 10" to 4" x 14" joist		155	.052		4.62	1.73		6.35	8
2000	Two-2" x 6" to two-2" x 10" joists		150	.053		3.85	1.79		5.64	7.25
2100	Two-2" x 10" to two-2" x 14" joists		150	.053		4.30	1.79		6.09	7.70
2300	3/16" thick, 6" x 8" joist		145	.055		59.50	1.85		61.35	68.50
2400	6" x 10" joist		140	.057		62	1.92		63.92	71.50
2500	6" x 12" joist		135	.059		64.50	1.99		66.49	74.50
2700	1/4" thick, 6" x 14" joist		130	.062		67.50	2.06		69.56	77.50
2900	Plywood clips, extruded aluminum H clip, for 3/4" panels					.22			.22	.24
3000	Galvanized 18 ga. back-up clip					.17			.17	.19
3200	Post framing, 16 ga. galv. for 4" x 4" base, 2 piece	1 Carp	130	.062		15.05	2.06		17.11	20
3300	Cap		130	.062		21	2.06		23.06	26.50
3500	Rafter anchors, 18 ga. galv., 1-1/2" wide, 5-1/4" long		145	.055		.45	1.85		2.30	3.60
3600	10-3/4" long		145	.055		1.34	1.85		3.19	4.57
3800	Shear plates, 2-5/8" diameter		120	.067		2.19	2.24		4.43	6.15
3900	4" diameter		115	.070		5.20	2.33		7.53	9.60
4000	Sill anchors, embedded in concrete or block, 25-1/2" long		115	.070		11.65	2.33		13.98	16.75
4100	Spike grids, 3" x 6"		120	.067		.88	2.24		3.12	4.71
4400	Split rings, 2-1/2" diameter		120	.067		1.80	2.24		4.04	5.70
4500	4" diameter		110	.073		2.73	2.44		5.17	7.10
4550	Tie plate, 20 gauge, 7" x 3 1/8"		110	.073		2.73	2.44		5.17	7.10
4560	Tie plate, 20 gauge, 5" x 4 1/8"		110	.073		2.73	2.44		5.17	7.10
4575	Twist straps, 18 gauge, 12" x 1 1/4"		110	.073		2.73	2.44		5.17	7.10
4580	Twist straps, 18 gauge, 16" x 1 1/4"		110	.073		2.73	2.44		5.17	7.10
4600	Strap ties, 20 ga., 2 -1/16" wide, 12 13/16" long		180	.044		.87	1.49		2.36	3.46
4700	Strap ties, 16 ga., 1-3/8" wide, 12" long		180	.044		.87	1.49		2.36	3.46
4800	21-5/8" x 1-1/4"		160	.050		2.72	1.68		4.40	5.80
5000	Toothed rings, 2-5/8" or 4" diameter		90	.089		1.63	2.98		4.61	6.80
5200	Truss plates, nailed, 20 gauge, up to 32' span		17	.471	Truss	11.85	15.80		27.65	39.50
5400	Washers, 2" x 2" x 1/8"				Ea.	.37			.37	.41
5500	3" x 3" x 3/16"				"	.98			.98	1.08
06 05 23.70 Rough Hardware										
0010	**ROUGH HARDWARE**, average percent of carpentry material									
0020	Minimum				Job	.50%				
0200	Maximum					1.50%				
0210	In seismic or hurricane areas, up to					10%				

06 05 Common Work Results for Wood, Plastics, and Composites

06 05 23 – Wood, Plastic, and Composite Fastenings

06 05 23.80 Metal Bracing		Crew	Daily Output	Labor-Hours	Unit	Material	2011 Bare Costs Labor	Equipment	Total	Total Incl O&P
0010	METAL BRACING R051223-20									
0302	Let-in, "T" shaped, 22 ga. galv. steel, studs at 16" O.C.	1 Carp	580	.014	L.F.	.75	.46		1.21	1.60
0402	Studs at 24" O.C.		600	.013		.75	.45		1.20	1.58
0502	16 ga. galv. steel straps, studs at 16" O.C.		600	.013		.97	.45		1.42	1.82
0602	Studs at 24" O.C.	↓	620	.013	↓	.97	.43		1.40	1.79

06 05 23.90 Recycled Plastic Shims		Crew	Daily Output	Labor-Hours	Unit	Material	Labor	Equipment	Total	Total Incl O&P
0010	RECYCLED PLASTIC SHIMS									
5000	Made from recycled plastic, 8"L x 1-1/4"W G				Ea.	.15			.15	.17
5010	Case of 360 G				"	55.50			55.50	61

06 11 Wood Framing

06 11 10 – Framing with Dimensional, Engineered or Composite Lumber

06 11 10.01 Forest Stewardship Council Certification		Crew	Daily Output	Labor-Hours	Unit	Material	Labor	Equipment	Total	Total Incl O&P
0010	FOREST STEWARDSHIP COUNCIL CERTIFICATION									
0020	For Forest Stewardship Council (FSC) cert dimension lumber, add G					65%				

06 11 10.02 Blocking		Crew	Daily Output	Labor-Hours	Unit	Material	Labor	Equipment	Total	Total Incl O&P
0010	BLOCKING									
1950	Miscellaneous, to wood construction									
2000	2" x 4"	1 Carp	250	.032	L.F.	.33	1.07		1.40	2.16
2005	Pneumatic nailed		305	.026		.33	.88		1.21	1.83
2050	2" x 6"		222	.036		.53	1.21		1.74	2.60
2055	Pneumatic nailed		271	.030		.53	.99		1.52	2.24
2100	2" x 8"		200	.040		.73	1.34		2.07	3.05
2105	Pneumatic nailed		244	.033		.73	1.10		1.83	2.64
2150	2" x 10"		178	.045		1.11	1.51		2.62	3.74
2155	Pneumatic nailed		217	.037		1.11	1.24		2.35	3.29
2200	2" x 12"		151	.053		1.36	1.78		3.14	4.47
2205	Pneumatic nailed	↓	185	.043	↓	1.36	1.45		2.81	3.93
2300	To steel construction									
2320	2" x 4"	1 Carp	208	.038	L.F.	.33	1.29		1.62	2.52
2340	2" x 6"		180	.044		.53	1.49		2.02	3.08
2360	2" x 8"		158	.051		.73	1.70		2.43	3.64
2380	2" x 10"		136	.059		1.11	1.97		3.08	4.52
2400	2" x 12"	↓	109	.073	↓	1.36	2.46		3.82	5.60

06 11 10.04 Wood Bracing		Crew	Daily Output	Labor-Hours	Unit	Material	Labor	Equipment	Total	Total Incl O&P
0010	WOOD BRACING									
0011	Let-in, with 1" x 6" boards, studs 16" O.C.	1 Carp	1.50	5.333	C.L.F.	67.50	179		246.50	375
0200	Studs @ 24" O.C.	"	2.30	3.478	"	67.50	117		184.50	269

06 11 10.06 Bridging		Crew	Daily Output	Labor-Hours	Unit	Material	Labor	Equipment	Total	Total Incl O&P
0010	BRIDGING									
0011	Wood, for joists 16" O.C., 1" x 3"	1 Carp	1.30	6.154	C.Pr.	51	206		257	400
0015	Pneumatic nailed		1.70	4.706		51	158		209	320
0100	2" x 3" bridging		1.30	6.154		49.50	206		255.50	400
0105	Pneumatic nailed		1.70	4.706		49.50	158		207.50	320
0300	Steel, galvanized, 18 ga., for 2" x 10" joists at 12" O.C.		1.30	6.154		172	206		378	535
0400	24" O.C.		1.40	5.714		185	192		377	525
0600	For 2" x 14" joists at 16" O.C.		1.30	6.154		185	206		391	550
0900	Compression type, 16" O.C., 2" x 8" joists		2	4		178	134		312	420
1000	2" x 12" joists	↓	2	4	↓	178	134		312	420

06 11 Wood Framing

06 11 10 – Framing with Dimensional, Engineered or Composite Lumber

06 11 10.10 Beam and Girder Framing			Crew	Daily Output	Labor-Hours	Unit	2011 Bare Costs Material	Labor	Equipment	Total	Total Incl O&P
0010	**BEAM AND GIRDER FRAMING**	R061110-30									
1000	Single, 2″ x 6″		2 Carp	700	.023	L.F.	.53	.77		1.30	1.86
1005	Pneumatic nailed			812	.020		.53	.66		1.19	1.69
1020	2″ x 8″			650	.025		.73	.83		1.56	2.18
1025	Pneumatic nailed			754	.021		.73	.71		1.44	1.99
1040	2″ x 10″			600	.027		1.11	.89		2	2.72
1045	Pneumatic nailed			696	.023		1.11	.77		1.88	2.51
1060	2″ x 12″			550	.029		1.36	.98		2.34	3.13
1065	Pneumatic nailed			638	.025		1.36	.84		2.20	2.91
1080	2″ x 14″			500	.032		1.70	1.07		2.77	3.67
1085	Pneumatic nailed			580	.028		1.70	.93		2.63	3.42
1100	3″ x 8″			550	.029		2.12	.98		3.10	3.96
1120	3″ x 10″			500	.032		2.68	1.07		3.75	4.75
1140	3″ x 12″			450	.036		3.27	1.19		4.46	5.60
1160	3″ x 14″			400	.040		3.85	1.34		5.19	6.50
1180	4″ x 8″		F-3	1000	.040		2.98	1.23	.66	4.87	6.05
1200	4″ x 10″			950	.042		3.79	1.29	.69	5.77	7.10
1220	4″ x 12″			900	.044		5.50	1.37	.73	7.60	9.10
1240	4″ x 14″			850	.047		7.15	1.45	.77	9.37	11.10
1250	6″ x 8″	G		525	.076		4.82	2.34	1.25	8.41	10.55
1290	8″ x 12″	G		300	.133		11.95	4.10	2.19	18.24	22.50
1300	Treated, single, 2 x 4	G	2 Carp	700	.023		.36	.77		1.13	1.68
1320	2 x 6	G		700	.023		.62	.77		1.39	1.96
1340	2 x 8	G		650	.025		.95	.83		1.78	2.43
1360	2 x 10	G		600	.027		1.29	.89		2.18	2.92
1380	2 x 12	G		550	.029		1.70	.98		2.68	3.50
1400	2 x 14	G		500	.032		2.37	1.07		3.44	4.41
1420	3 x 8	G		550	.029		2.93	.98		3.91	4.85
1440	3 x 10	G		500	.032		3.68	1.07		4.75	5.85
1460	3 x 12	G		450	.036		4.43	1.19		5.62	6.85
1480	3 x 14	G		400	.040		6.15	1.34		7.49	9
1500	4 x 8	G	F-3	1000	.040		4.68	1.23	.66	6.57	7.90
1520	4 x 10	G		950	.042		5.85	1.29	.69	7.83	9.35
1540	4 x 12	G		900	.044		7	1.37	.73	9.10	10.75
1560	4 x 14	G		850	.047		8.25	1.45	.77	10.47	12.30
2000	Double, 2″ x 6″		2 Carp	625	.026		1.05	.86		1.91	2.60
2005	Pneumatic nailed			725	.022		1.05	.74		1.79	2.40
2020	2″ x 8″			575	.028		1.45	.93		2.38	3.16
2025	Pneumatic nailed			667	.024		1.45	.80		2.25	2.95
2040	2″ x 10″			550	.029		2.23	.98		3.21	4.08
2045	Pneumatic nailed			638	.025		2.23	.84		3.07	3.86
2060	2″ x 12″			525	.030		2.72	1.02		3.74	4.70
2065	Pneumatic nailed			610	.026		2.72	.88		3.60	4.46
2080	2″ x 14″			475	.034		3.39	1.13		4.52	5.60
2085	Pneumatic nailed			551	.029		3.39	.97		4.36	5.35
3000	Triple, 2″ x 6″			550	.029		1.58	.98		2.56	3.37
3005	Pneumatic nailed			638	.025		1.58	.84		2.42	3.15
3020	2″ x 8″			525	.030		2.18	1.02		3.20	4.10
3025	Pneumatic nailed			609	.026		2.18	.88		3.06	3.87
3040	2″ x 10″			500	.032		3.34	1.07		4.41	5.45
3045	Pneumatic nailed			580	.028		3.34	.93		4.27	5.20
3060	2″ x 12″			475	.034		4.08	1.13		5.21	6.40

06 11 Wood Framing

06 11 10 – Framing with Dimensional, Engineered or Composite Lumber

06 11 10.10 Beam and Girder Framing		Crew	Daily Output	Labor-Hours	Unit	2011 Bare Costs Material	Labor	Equipment	Total	Total Incl O&P
3065	Pneumatic nailed	2 Carp	551	.029	L.F.	4.08	.97		5.05	6.10
3080	2" x 14"		450	.036		5.10	1.19		6.29	7.60
3085	Pneumatic nailed		522	.031		5.10	1.03		6.13	7.30

06 11 10.12 Ceiling Framing

		Crew	Daily Output	Labor-Hours	Unit	Material	Labor	Equipment	Total	Total Incl O&P
0010	**CEILING FRAMING**									
6000	Suspended, 2" x 3"	2 Carp	1000	.016	L.F.	.33	.54		.87	1.26
6050	2" x 4"		900	.018		.33	.60		.93	1.36
6100	2" x 6"		800	.020		.53	.67		1.20	1.70
6150	2" x 8"		650	.025		.73	.83		1.56	2.18

06 11 10.14 Posts and Columns

		Crew	Daily Output	Labor-Hours	Unit	Material	Labor	Equipment	Total	Total Incl O&P
0010	**POSTS AND COLUMNS**									
0100	4" x 4"	2 Carp	390	.041	L.F.	1.47	1.38		2.85	3.92
0150	4" x 6"		275	.058		2.35	1.95		4.30	5.85
0200	4" x 8"		220	.073		2.98	2.44		5.42	7.35
0250	6" x 6"		215	.074		3.62	2.50		6.12	8.15
0300	6" x 8"		175	.091		4.82	3.07		7.89	10.45
0350	6" x 10"		150	.107		6.50	3.58		10.08	13.15

06 11 10.18 Joist Framing

		Crew	Daily Output	Labor-Hours	Unit	Material	Labor	Equipment	Total	Total Incl O&P
0010	**JOIST FRAMING**									
2000	Joists, 2" x 4"	2 Carp	1250	.013	L.F.	.33	.43		.76	1.08
2005	Pneumatic nailed		1438	.011		.33	.37		.70	.98
2100	2" x 6"		1250	.013		.53	.43		.96	1.30
2105	Pneumatic nailed		1438	.011		.53	.37		.90	1.20
2150	2" x 8"		1100	.015		.73	.49		1.22	1.62
2155	Pneumatic nailed		1265	.013		.73	.42		1.15	1.51
2200	2" x 10"		900	.018		1.11	.60		1.71	2.22
2205	Pneumatic nailed		1035	.015		1.11	.52		1.63	2.09
2250	2" x 12"		875	.018		1.36	.61		1.97	2.53
2255	Pneumatic nailed		1006	.016		1.36	.53		1.89	2.39
2300	2" x 14"		770	.021		1.70	.70		2.40	3.04
2305	Pneumatic nailed		886	.018		1.70	.61		2.31	2.88
2350	3" x 6"		925	.017		1.57	.58		2.15	2.70
2400	3" x 10"		780	.021		2.68	.69		3.37	4.10
2450	3" x 12"		600	.027		3.27	.89		4.16	5.10
2500	4" x 6"		800	.020		2.35	.67		3.02	3.71
2550	4" x 10"		600	.027		3.79	.89		4.68	5.65
2600	4" x 12"		450	.036		5.50	1.19		6.69	8.05
2605	Sister joist, 2" x 6"		800	.020		.53	.67		1.20	1.70
2606	Pneumatic nailed		960	.017		.53	.56		1.09	1.52
3000	Composite wood joist 9-1/2" deep		.90	17.778	M.L.F.	1,550	595		2,145	2,700
3010	11-1/2" deep		.88	18.182		1,725	610		2,335	2,925
3020	14" deep		.82	19.512		2,400	655		3,055	3,725
3030	16" deep		.78	20.513		2,975	690		3,665	4,400
4000	Open web joist 12" deep		.88	18.182		2,775	610		3,385	4,075
4010	14" deep		.82	19.512		2,975	655		3,630	4,375
4020	16" deep		.78	20.513		2,950	690		3,640	4,400
4030	18" deep		.74	21.622		3,175	725		3,900	4,725

06 11 10.24 Miscellaneous Framing

		Crew	Daily Output	Labor-Hours	Unit	Material	Labor	Equipment	Total	Total Incl O&P
0010	**MISCELLANEOUS FRAMING**									
2000	Firestops, 2" x 4"	2 Carp	780	.021	L.F.	.33	.69		1.02	1.51
2005	Pneumatic nailed R061110-30		952	.017		.33	.56		.89	1.30
2100	2" x 6"		600	.027		.53	.89		1.42	2.08

06 11 Wood Framing

06 11 10 – Framing with Dimensional, Engineered or Composite Lumber

06 11 10.24 Miscellaneous Framing		Crew	Daily Output	Labor-Hours	Unit	Material	2011 Bare Costs Labor	Equipment	Total	Total Incl O&P
2105	Pneumatic nailed	2 Carp	732	.022	L.F.	.53	.73		1.26	1.81
5000	Nailers, treated, wood construction, 2" x 4"		800	.020		.36	.67		1.03	1.52
5005	Pneumatic nailed		960	.017		.36	.56		.92	1.34
5100	2" x 6"		750	.021		.62	.72		1.34	1.88
5105	Pneumatic nailed		900	.018		.62	.60		1.22	1.68
5120	2" x 8"		700	.023		.95	.77		1.72	2.33
5125	Pneumatic nailed		840	.019		.95	.64		1.59	2.12
5200	Steel construction, 2" x 4"		750	.021		.36	.72		1.08	1.60
5220	2" x 6"		700	.023		.62	.77		1.39	1.96
5240	2" x 8"		650	.025		.95	.83		1.78	2.43
7000	Rough bucks, treated, for doors or windows, 2" x 6"		400	.040		.62	1.34		1.96	2.93
7005	Pneumatic nailed		480	.033		.62	1.12		1.74	2.55
7100	2" x 8"		380	.042		.95	1.41		2.36	3.41
7105	Pneumatic nailed		456	.035		.95	1.18		2.13	3.02
8000	Stair stringers, 2" x 10"		130	.123		1.11	4.13		5.24	8.10
8100	2" x 12"		130	.123		1.36	4.13		5.49	8.40
8150	3" x 10"		125	.128		2.68	4.29		6.97	10.15
8200	3" x 12"	↓	125	.128	↓	3.27	4.29		7.56	10.80

06 11 10.26 Partitions		Crew	Daily Output	Labor-Hours	Unit	Material	Labor	Equipment	Total	Total Incl O&P
0010	**PARTITIONS**									
0020	Single bottom and double top plate, no waste, std. & better lumber									
0180	2" x 4" studs, 8' high, studs 12" O.C.	2 Carp	80	.200	L.F.	3.64	6.70		10.34	15.25
0185	12" O.C., pneumatic nailed		96	.167		3.64	5.60		9.24	13.35
0200	16" O.C.		100	.160		2.98	5.35		8.33	12.25
0205	16" O.C., pneumatic nailed		120	.133		2.98	4.47		7.45	10.75
0300	24" O.C.		125	.128		2.31	4.29		6.60	9.75
0305	24" O.C., pneumatic nailed		150	.107		2.31	3.58		5.89	8.55
0380	10' high, studs 12" O.C.		80	.200		4.30	6.70		11	16
0385	12" O.C., pneumatic nailed		96	.167		4.30	5.60		9.90	14.10
0400	16" O.C.		100	.160		3.47	5.35		8.82	12.80
0405	16" O.C., pneumatic nailed		120	.133		3.47	4.47		7.94	11.30
0500	24" O.C.		125	.128		2.65	4.29		6.94	10.10
0505	24" O.C., pneumatic nailed		150	.107		2.65	3.58		6.23	8.90
0580	12' high, studs 12" O.C.		65	.246		4.96	8.25		13.21	19.25
0585	12" O.C., pneumatic nailed		78	.205		4.96	6.90		11.86	16.95
0600	16" O.C.		80	.200		3.97	6.70		10.67	15.60
0605	16" O.C., pneumatic nailed		96	.167		3.97	5.60		9.57	13.70
0700	24" O.C.		100	.160		2.98	5.35		8.33	12.25
0705	24" O.C., pneumatic nailed		120	.133		2.98	4.47		7.45	10.75
0780	2" x 6" studs, 8' high, studs 12" O.C.		70	.229		5.80	7.65		13.45	19.25
0785	12" O.C., pneumatic nailed		84	.190		5.80	6.40		12.20	17.10
0800	16" O.C.		90	.178		4.74	5.95		10.69	15.20
0805	16" O.C., pneumatic nailed		108	.148		4.74	4.97		9.71	13.50
0900	24" O.C.		115	.139		3.69	4.67		8.36	11.85
0905	24" O.C., pneumatic nailed		138	.116		3.69	3.89		7.58	10.55
0980	10' high, studs 12" O.C.		70	.229		6.85	7.65		14.50	20.50
0985	12" O.C., pneumatic nailed		84	.190		6.85	6.40		13.25	18.25
1000	16" O.C.		90	.178		5.55	5.95		11.50	16.10
1005	16" O.C., pneumatic nailed		108	.148		5.55	4.97		10.52	14.40
1100	24" O.C.		115	.139		4.22	4.67		8.89	12.45
1105	24" O.C., pneumatic nailed		138	.116		4.22	3.89		8.11	11.15
1180	12' high, studs 12" O.C.	↓	55	.291	↓	7.90	9.75		17.65	25

06 11 Wood Framing

06 11 10 – Framing with Dimensional, Engineered or Composite Lumber

06 11 10.26 Partitions

	06 11 10.26 Partitions	Crew	Daily Output	Labor-Hours	Unit	2011 Bare Costs Material	Labor	Equipment	Total	Total Incl O&P
1185	12" O.C., pneumatic nailed	2 Carp	66	.242	L.F.	7.90	8.15		16.05	22.50
1200	16" O.C.		70	.229		6.30	7.65		13.95	19.80
1205	16" O.C., pneumatic nailed		84	.190		6.30	6.40		12.70	17.65
1300	24" O.C.		90	.178		4.74	5.95		10.69	15.20
1305	24" O.C., pneumatic nailed		108	.148		4.74	4.97		9.71	13.50
1400	For horizontal blocking, 2" x 4", add		600	.027		.33	.89		1.22	1.86
1500	2" x 6", add		600	.027		.53	.89		1.42	2.08
1600	For openings, add		250	.064			2.15		2.15	3.59
1700	Headers for above openings, material only, add				M.B.F.	545			545	600

06 11 10.28 Porch or Deck Framing

	06 11 10.28 Porch or Deck Framing	Crew	Daily Output	Labor-Hours	Unit	Material	Labor	Equipment	Total	Total Incl O&P
0010	**PORCH OR DECK FRAMING**									
0100	Treated lumber, posts or columns, 4" x 4"	2 Carp	390	.041	L.F.	1.37	1.38		2.75	3.80
0110	4" x 6"		275	.058		2.46	1.95		4.41	6
0120	4" x 8"		220	.073		4.74	2.44		7.18	9.30
0130	Girder, single, 4" x 4"		675	.024		1.37	.80		2.17	2.83
0140	4" x 6"		600	.027		2.46	.89		3.35	4.21
0150	4" x 8"		525	.030		4.74	1.02		5.76	6.90
0160	Double, 2" x 4"		625	.026		.75	.86		1.61	2.26
0170	2" x 6"		600	.027		1.28	.89		2.17	2.90
0180	2" x 8"		575	.028		1.97	.93		2.90	3.73
0190	2" x 10"		550	.029		2.65	.98		3.63	4.55
0200	2" x 12"		525	.030		3.49	1.02		4.51	5.55
0210	Triple, 2" x 4"		575	.028		1.12	.93		2.05	2.80
0220	2" x 6"		550	.029		1.92	.98		2.90	3.74
0230	2" x 8"		525	.030		2.96	1.02		3.98	4.96
0240	2" x 10"		500	.032		3.98	1.07		5.05	6.20
0250	2" x 12"		475	.034		5.25	1.13		6.38	7.65
0260	Ledger, bolted 4' O.C., 2" x 4"		400	.040		.48	1.34		1.82	2.78
0270	2" x 6"		395	.041		.74	1.36		2.10	3.08
0280	2" x 8"		390	.041		1.07	1.38		2.45	3.48
0300	2" x 12"		380	.042		1.82	1.41		3.23	4.36
0310	Joists, 2" x 4"		1250	.013		.38	.43		.81	1.13
0320	2" x 6"		1250	.013		.64	.43		1.07	1.43
0330	2" x 8"		1100	.015		.99	.49		1.48	1.91
0340	2" x 10"		900	.018		1.33	.60		1.93	2.47
0350	2" x 12"		875	.018		1.60	.61		2.21	2.79
0440	Balusters, square, 2" x 2"		660	.024		.36	.81		1.17	1.75
0450	Turned, 2" x 2"		420	.038		.48	1.28		1.76	2.66
0460	Stair stringer, 2" x 10"		130	.123		1.33	4.13		5.46	8.35
0470	2" x 12"		130	.123		1.60	4.13		5.73	8.65
0480	Stair treads, 1" x 4"		140	.114		1.56	3.83		5.39	8.10
0490	2" x 4"		140	.114		.38	3.83		4.21	6.80
0500	2" x 6"		160	.100		.57	3.36		3.93	6.20
0510	5/4" x 6"		160	.100		.98	3.36		4.34	6.65
0520	Turned handrail post, 4" x 4"		64	.250	Ea.	32	8.40		40.40	49
0530	Lattice panel, 4' x 8', 1/2"		1600	.010	S.F.	.68	.34		1.02	1.31
0535	3/4"		1600	.010	"	1.02	.34		1.36	1.68
0540	Cedar, posts or columns, 4" x 4"		390	.041	L.F.	3.01	1.38		4.39	5.60
0550	4" x 6"		275	.058		5.75	1.95		7.70	9.55
0560	4" x 8"		220	.073		9.70	2.44		12.14	14.80
0800	Decking, 1" x 4"		550	.029		1.49	.98		2.47	3.27
0810	2" x 4"		600	.027		2.98	.89		3.87	4.78

06 11 Wood Framing

06 11 10 – Framing with Dimensional, Engineered or Composite Lumber

06 11 10.28 Porch or Deck Framing		Crew	Daily Output	Labor-Hours	Unit	2011 Bare Costs Material	Labor	Equipment	Total	Total Incl O&P
0820	2" x 6"	2 Carp	640	.025	L.F.	5.25	.84		6.09	7.20
0830	5/4" x 6"		640	.025		3.50	.84		4.34	5.25
0840	Railings and trim, 1" x 4"		600	.027		1.49	.89		2.38	3.14
0860	2" x 4"		600	.027		2.98	.89		3.87	4.78
0870	2" x 6"		600	.027		5.25	.89		6.14	7.30
0920	Stair treads, 1" x 4"		140	.114		1.49	3.83		5.32	8.05
0930	2" x 4"		140	.114		2.98	3.83		6.81	9.70
0940	2" x 6"		160	.100		5.25	3.36		8.61	11.40
0950	5/4" x 6"		160	.100		3.50	3.36		6.86	9.45
0980	Redwood, posts or columns, 4" x 4"		390	.041		6.40	1.38		7.78	9.35
0990	4" x 6"		275	.058		11.20	1.95		13.15	15.55
1000	4" x 8"		220	.073		20.50	2.44		22.94	26.50
1280	Railings and trim, 1" x 4"		600	.027		1.17	.89		2.06	2.78
1310	2" x 6"		600	.027		7.50	.89		8.39	9.75
1420	Alternative decking, wood/plastic composite, 5/4" x 6" G		640	.025		3	.84		3.84	4.70
1430	Vinyl, 1-1/2" x 5-1/2"		640	.025		4.15	.84		4.99	5.95
1440	1" x 4" square edge fir		550	.029		1.57	.98		2.55	3.35
1450	1" x 4" tongue and groove fir		450	.036		1.41	1.19		2.60	3.55
1460	1" x 4" mahogany		550	.029		2.51	.98		3.49	4.39
1462	5/4" x 6" PVC		550	.029		2.72	.98		3.70	4.62
1465	Framing, porch or deck, alt deck fastening, screws, add	1 Carp	240	.033	S.F.		1.12		1.12	1.87
1470	Accessories, joist hangers, 2" x 4"		160	.050	Ea.	.67	1.68		2.35	3.55
1480	2" x 6" through 2" x 12"		150	.053		1.28	1.79		3.07	4.40
1530	Post footing, incl excav, backfill, tube form & concrete, 4' deep, 8" dia	F-7	12	2.667		10.65	80		90.65	146
1540	10" diameter		11	2.909		15.80	87.50		103.30	163
1550	12" diameter		10	3.200		21	96		117	184

06 11 10.30 Roof Framing		Crew	Daily Output	Labor-Hours	Unit	Material	Labor	Equipment	Total	Total Incl O&P
0010	**ROOF FRAMING**									
2000	Fascia boards, 2" x 8"	2 Carp	225	.071	L.F.	.73	2.39		3.12	4.79
2100	2" x 10"		180	.089		1.11	2.98		4.09	6.20
5000	Rafters, to 4 in 12 pitch, 2" x 6", ordinary		1000	.016		.53	.54		1.07	1.48
5020	On steep roofs		800	.020		.53	.67		1.20	1.70
5040	On dormers or complex roofs		590	.027		.53	.91		1.44	2.10
5060	2" x 8", ordinary		950	.017		.73	.57		1.30	1.75
5080	On steep roofs		750	.021		.73	.72		1.45	2
5100	On dormers or complex roofs		540	.030		.73	.99		1.72	2.46
5120	2" x 10", ordinary		630	.025		1.11	.85		1.96	2.65
5140	On steep roofs		495	.032		1.11	1.08		2.19	3.03
5160	On dormers or complex roofs		425	.038		1.11	1.26		2.37	3.33
5180	2" x 12", ordinary		575	.028		1.36	.93		2.29	3.06
5200	On steep roofs		455	.035		1.36	1.18		2.54	3.47
5220	On dormers or complex roofs		395	.041		1.36	1.36		2.72	3.77
5300	Hip and valley rafters, 2" x 6", ordinary		760	.021		.53	.71		1.24	1.76
5320	On steep roofs		585	.027		.53	.92		1.45	2.12
5340	On dormers or complex roofs		510	.031		.53	1.05		1.58	2.34
5360	2" x 8", ordinary		720	.022		.73	.75		1.48	2.05
5380	On steep roofs		545	.029		.73	.99		1.72	2.45
5400	On dormers or complex roofs		470	.034		.73	1.14		1.87	2.71
5420	2" x 10", ordinary		570	.028		1.11	.94		2.05	2.80
5440	On steep roofs		440	.036		1.11	1.22		2.33	3.26
5460	On dormers or complex roofs		380	.042		1.11	1.41		2.52	3.58
5480	2" x 12", ordinary		525	.030		1.36	1.02		2.38	3.21

06 11 Wood Framing

06 11 10 – Framing with Dimensional, Engineered or Composite Lumber

06 11 10.30 Roof Framing		Crew	Daily Output	Labor-Hours	Unit	Material	2011 Bare Costs Labor	Equipment	Total	Total Incl O&P
5500	On steep roofs	2 Carp	410	.039	L.F.	1.36	1.31		2.67	3.69
5520	On dormers or complex roofs		355	.045		1.36	1.51		2.87	4.03
5540	Hip and valley jacks, 2" x 6", ordinary		600	.027		.53	.89		1.42	2.08
5560	On steep roofs		475	.034		.53	1.13		1.66	2.47
5580	On dormers or complex roofs		410	.039		.53	1.31		1.84	2.77
5600	2" x 8", ordinary		490	.033		.73	1.10		1.83	2.63
5620	On steep roofs		385	.042		.73	1.39		2.12	3.13
5640	On dormers or complex roofs		335	.048		.73	1.60		2.33	3.48
5660	2" x 10", ordinary		450	.036		1.11	1.19		2.30	3.22
5680	On steep roofs		350	.046		1.11	1.53		2.64	3.79
5700	On dormers or complex roofs		305	.052		1.11	1.76		2.87	4.17
5720	2" x 12", ordinary		375	.043		1.36	1.43		2.79	3.90
5740	On steep roofs		295	.054		1.36	1.82		3.18	4.55
5760	On dormers or complex roofs		255	.063		1.36	2.11		3.47	5
5780	Rafter tie, 1" x 4", #3		800	.020		.53	.67		1.20	1.70
5790	2" x 4", #3		800	.020		.33	.67		1	1.48
5800	Ridge board, #2 or better, 1" x 6"		600	.027		.67	.89		1.56	2.24
5820	1" x 8"		550	.029		.97	.98		1.95	2.69
5840	1" x 10"		500	.032		1.21	1.07		2.28	3.13
5860	2" x 6"		500	.032		.53	1.07		1.60	2.38
5880	2" x 8"		450	.036		.73	1.19		1.92	2.80
5900	2" x 10"		400	.040		1.11	1.34		2.45	3.47
5920	Roof cants, split, 4" x 4"		650	.025		1.47	.83		2.30	3
5940	6" x 6"		600	.027		3.62	.89		4.51	5.50
5960	Roof curbs, untreated, 2" x 6"		520	.031		.53	1.03		1.56	2.31
5980	2" x 12"		400	.040		1.36	1.34		2.70	3.75
6000	Sister rafters, 2" x 6"		800	.020		.53	.67		1.20	1.70
6020	2" x 8"		640	.025		.73	.84		1.57	2.20
6040	2" x 10"		535	.030		1.11	1		2.11	2.90
6060	2" x 12"	↓	455	.035	↓	1.36	1.18		2.54	3.47
06 11 10.32 Sill and Ledger Framing										
0010	**SILL AND LEDGER FRAMING**									
2000	Ledgers, nailed, 2" x 4"	2 Carp	755	.021	L.F.	.33	.71		1.04	1.55
2050	2" x 6"		600	.027		.53	.89		1.42	2.08
2100	Bolted, not including bolts, 3" x 6"		325	.049		1.57	1.65		3.22	4.49
2150	3" x 12"		233	.069		3.27	2.30		5.57	7.45
2600	Mud sills, redwood, construction grade, 2" x 4"		895	.018		2.26	.60		2.86	3.49
2620	2" x 6"		780	.021		3.40	.69		4.09	4.89
4000	Sills, 2" x 4"		600	.027		.33	.89		1.22	1.86
4050	2" x 6"		550	.029		.53	.98		1.51	2.21
4080	2" x 8"		500	.032		.73	1.07		1.80	2.60
4100	2" x 10"		450	.036		1.11	1.19		2.30	3.22
4120	2" x 12"		400	.040		1.36	1.34		2.70	3.75
4200	Treated, 2" x 4"		550	.029		.36	.98		1.34	2.03
4220	2" x 6"		500	.032		.62	1.07		1.69	2.48
4240	2" x 8"		450	.036		.95	1.19		2.14	3.05
4260	2" x 10"		400	.040		1.29	1.34		2.63	3.67
4280	2" x 12"		350	.046		1.70	1.53		3.23	4.44
4400	4" x 4"		450	.036		1.33	1.19		2.52	3.47
4420	4" x 6"		350	.046		2.42	1.53		3.95	5.25
4460	4" x 8"		300	.053		4.68	1.79		6.47	8.15
4480	4" x 10"	↓	260	.062	↓	5.85	2.06		7.91	9.90

06 11 Wood Framing

06 11 10 – Framing with Dimensional, Engineered or Composite Lumber

06 11 10.34 Sleepers

	06 11 10.34 Sleepers	Crew	Daily Output	Labor-Hours	Unit	Material	Labor (2011 Bare Costs)	Equipment	Total	Total Incl O&P
0010	**SLEEPERS**									
0100	On concrete, treated, 1" x 2"	2 Carp	2350	.007	L.F.	.29	.23		.52	.69
0150	1" x 3"		2000	.008		.53	.27		.80	1.03
0200	2" x 4"		1500	.011		.36	.36		.72	1
0250	2" x 6"		1300	.012		.62	.41		1.03	1.37

06 11 10.36 Soffit and Canopy Framing

	06 11 10.36 Soffit and Canopy Framing	Crew	Daily Output	Labor-Hours	Unit	Material	Labor	Equipment	Total	Total Incl O&P
0010	**SOFFIT AND CANOPY FRAMING**									
1000	Canopy or soffit framing , 1" x 4"	2 Carp	900	.018	L.F.	.53	.60		1.13	1.58
1020	1" x 6"		850	.019		.67	.63		1.30	1.80
1040	1" x 8"		750	.021		.97	.72		1.69	2.26
1100	2" x 4"		620	.026		.33	.87		1.20	1.81
1120	2" x 6"		560	.029		.53	.96		1.49	2.18
1140	2" x 8"		500	.032		.73	1.07		1.80	2.60
1200	3" x 4"		500	.032		.85	1.07		1.92	2.74
1220	3" x 6"		400	.040		1.57	1.34		2.91	3.98
1240	3" x 10"		300	.053		2.68	1.79		4.47	5.95

06 11 10.38 Treated Lumber Framing Material

	06 11 10.38 Treated Lumber Framing Material	Crew	Daily Output	Labor-Hours	Unit	Material	Labor	Equipment	Total	Total Incl O&P
0010	**TREATED LUMBER FRAMING MATERIAL**									
0100	2" x 4"				M.B.F.	540			540	595
0110	2" x 6"					615			615	675
0120	2" x 8"					715			715	790
0130	2" x 10"					775			775	850
0140	2" x 12"					850			850	935
0200	4" x 4"					1,000			1,000	1,100
0210	4" x 6"					1,200			1,200	1,325
0220	4" x 8"					1,750			1,750	1,925

06 11 10.40 Wall Framing

	06 11 10.40 Wall Framing	Crew	Daily Output	Labor-Hours	Unit	Material	Labor	Equipment	Total	Total Incl O&P
0010	**WALL FRAMING**									
2000	Headers over openings, 2" x 6"	2 Carp	360	.044	L.F.	.53	1.49		2.02	3.08
2005	2" x 6", pneumatic nailed R061110-30		432	.037		.53	1.24		1.77	2.66
2050	2" x 8"		340	.047		.73	1.58		2.31	3.44
2055	2" x 8", pneumatic nailed		408	.039		.73	1.32		2.05	3
2100	2" x 10"		320	.050		1.11	1.68		2.79	4.03
2105	2" x 10", pneumatic nailed		384	.042		1.11	1.40		2.51	3.56
2150	2" x 12"		300	.053		1.36	1.79		3.15	4.49
2155	2" x 12", pneumatic nailed		360	.044		1.36	1.49		2.85	4
2190	4" x 10"		240	.067		3.79	2.24		6.03	7.90
2195	4" x 10", pneumatic nailed		288	.056		3.79	1.86		5.65	7.30
2200	4" x 12"		190	.084		5.50	2.83		8.33	10.80
2205	4" x 12", pneumatic nailed		228	.070		5.50	2.35		7.85	10
2240	6" x 10"		165	.097		6.50	3.25		9.75	12.60
2245	6" x 10", pneumatic nailed		198	.081		6.50	2.71		9.21	11.70
2250	6" x 12"		140	.114		10.15	3.83		13.98	17.60
2255	6" x 12", pneumatic nailed		168	.095		10.15	3.20		13.35	16.55
5000	Plates, untreated, 2" x 3"		850	.019		.33	.63		.96	1.42
5005	2" x 3", pneumatic nailed		1020	.016		.33	.53		.86	1.24
5020	2" x 4"		800	.020		.33	.67		1	1.48
5025	2" x 4", pneumatic nailed		960	.017		.33	.56		.89	1.30
5040	2" x 6"		750	.021		.53	.72		1.25	1.78
5045	2" x 6", pneumatic nailed		900	.018		.53	.60		1.13	1.58
5060	Treated, 2" x 3"		850	.019		.55	.63		1.18	1.66

06 11 Wood Framing

06 11 10 – Framing with Dimensional, Engineered or Composite Lumber

06 11 10.40 Wall Framing			Crew	Daily Output	Labor-Hours	Unit	2011 Bare Costs Material	Labor	Equipment	Total	Total Incl O&P
5065	2" x 3", treated, pneumatic nailed		2 Carp	1020	.016	L.F.	.55	.53		1.08	1.48
5080	2" x 4"			800	.020		.36	.67		1.03	1.52
5085	2" x 4", treated, pneumatic nailed			960	.017		.36	.56		.92	1.34
5100	2" x 6"			750	.021		.62	.72		1.34	1.88
5102	2 x 8	G		725	.022		.95	.74		1.69	2.29
5103	2" x 6", treated, pneumatic nailed			900	.018		.62	.60		1.22	1.68
5104	2 x 10	G		700	.023		1.29	.77		2.06	2.70
5105	2" x 12"	G		670	.024		1.70	.80		2.50	3.21
5107	2" x 8", pneumatic nailed	G		870	.018		.95	.62		1.57	2.08
5108	2" x 10", pneumatic nailed	G		840	.019		1.29	.64		1.93	2.49
5109	2" x 12", pneumatic nailed	G		804	.020		1.70	.67		2.37	2.99
5120	Studs, 8' high wall, 2" x 3"			1200	.013		.33	.45		.78	1.11
5125	2" x 3", pneumatic nailed			1440	.011		.33	.37		.70	.98
5140	2" x 4"			1100	.015		.33	.49		.82	1.18
5146	2" x 4", pneumatic nailed			1320	.012		.33	.41		.74	1.04
5160	2" x 6"			1000	.016		.53	.54		1.07	1.48
5166	2" x 6", pneumatic nailed			1200	.013		.53	.45		.98	1.33
5180	3" x 4"			800	.020		.85	.67		1.52	2.06
5185	3" x 4", pneumatic nailed			960	.017		.85	.56		1.41	1.88
5200	Installed on second story, 2" x 3"			1170	.014		.33	.46		.79	1.13
5205	2" x 3", pneumatic nailed			1404	.011		.33	.38		.71	1
5220	2" x 4"			1015	.016		.33	.53		.86	1.24
5225	2" x 4", pneumatic nailed			1218	.013		.33	.44		.77	1.10
5240	2" x 6"			890	.018		.53	.60		1.13	1.59
5245	2" x 6", pneumatic nailed			1080	.015		.53	.50		1.03	1.41
5260	3" x 4"			800	.020		.85	.67		1.52	2.06
5265	3" x 4", pneumatic nailed			960	.017		.85	.56		1.41	1.88
5280	Installed on dormer or gable, 2" x 3"			1045	.015		.33	.51		.84	1.22
5285	2" x 3", pneumatic nailed			1254	.013		.33	.43		.76	1.08
5300	2" x 4"			905	.018		.33	.59		.92	1.35
5305	2" x 4", pneumatic nailed			1086	.015		.33	.49		.82	1.19
5320	2" x 6"			800	.020		.53	.67		1.20	1.70
5325	2" x 6", pneumatic nailed			960	.017		.53	.56		1.09	1.52
5340	3" x 4"			700	.023		.85	.77		1.62	2.22
5345	3" x 4", pneumatic nailed			840	.019		.85	.64		1.49	2.01
5360	6' high wall, 2" x 3"			970	.016		.33	.55		.88	1.29
5365	2" x 3", pneumatic nailed			1164	.014		.33	.46		.79	1.13
5380	2" x 4"			850	.019		.33	.63		.96	1.42
5385	2" x 4", pneumatic nailed			1020	.016		.33	.53		.86	1.24
5400	2" x 6"			740	.022		.53	.73		1.26	1.79
5405	2" x 6", pneumatic nailed			888	.018		.53	.60		1.13	1.59
5420	3" x 4"			600	.027		.85	.89		1.74	2.44
5425	3" x 4", pneumatic nailed			720	.022		.85	.75		1.60	2.19
5440	Installed on second story, 2" x 3"			950	.017		.33	.57		.90	1.31
5445	2" x 3", pneumatic nailed			1140	.014		.33	.47		.80	1.15
5460	2" x 4"			810	.020		.33	.66		.99	1.47
5465	2" x 4", pneumatic nailed			972	.016		.33	.55		.88	1.28
5480	2" x 6"			700	.023		.53	.77		1.30	1.86
5485	2" x 6", pneumatic nailed			840	.019		.53	.64		1.17	1.65
5500	3" x 4"			550	.029		.85	.98		1.83	2.57
5505	3" x 4", pneumatic nailed			660	.024		.85	.81		1.66	2.30
5520	Installed on dormer or gable, 2" x 3"			850	.019		.33	.63		.96	1.42
5525	2" x 3", pneumatic nailed			1020	.016		.33	.53		.86	1.24

06 11 Wood Framing

06 11 10 – Framing with Dimensional, Engineered or Composite Lumber

06 11 10.40 Wall Framing		Crew	Daily Output	Labor-Hours	Unit	2011 Bare Costs Material	Labor	Equipment	Total	Total Incl O&P
5540	2" x 4"	2 Carp	720	.022	L.F.	.33	.75		1.08	1.61
5545	2" x 4", pneumatic nailed		864	.019		.33	.62		.95	1.40
5560	2" x 6"		620	.026		.53	.87		1.40	2.03
5565	2" x 6", pneumatic nailed		744	.022		.53	.72		1.25	1.79
5580	3" x 4"		480	.033		.85	1.12		1.97	2.81
5585	3" x 4", pneumatic nailed		576	.028		.85	.93		1.78	2.50
5600	3' high wall, 2" x 3"		740	.022		.33	.73		1.06	1.57
5605	2" x 3", pneumatic nailed		888	.018		.33	.60		.93	1.37
5620	2" x 4"		640	.025		.33	.84		1.17	1.76
5625	2" x 4", pneumatic nailed		768	.021		.33	.70		1.03	1.53
5640	2" x 6"		550	.029		.53	.98		1.51	2.21
5645	2" x 6", pneumatic nailed		660	.024		.53	.81		1.34	1.94
5660	3" x 4"		440	.036		.85	1.22		2.07	2.98
5665	3" x 4", pneumatic nailed		528	.030		.85	1.02		1.87	2.64
5680	Installed on second story, 2" x 3"		700	.023		.33	.77		1.10	1.64
5685	2" x 3", pneumatic nailed		840	.019		.33	.64		.97	1.43
5700	2" x 4"		610	.026		.33	.88		1.21	1.83
5705	2" x 4", pneumatic nailed		732	.022		.33	.73		1.06	1.59
5720	2" x 6"		520	.031		.53	1.03		1.56	2.31
5725	2" x 6", pneumatic nailed		624	.026		.53	.86		1.39	2.02
5740	3" x 4"		430	.037		.85	1.25		2.10	3.03
5745	3" x 4", pneumatic nailed		516	.031		.85	1.04		1.89	2.68
5760	Installed on dormer or gable, 2" x 3"		625	.026		.33	.86		1.19	1.80
5765	2" x 3", pneumatic nailed		750	.021		.33	.72		1.05	1.56
5780	2" x 4"		545	.029		.33	.99		1.32	2.01
5785	2" x 4", pneumatic nailed		654	.024		.33	.82		1.15	1.73
5800	2" x 6"		465	.034		.53	1.15		1.68	2.51
5805	2" x 6", pneumatic nailed		558	.029		.53	.96		1.49	2.19
5820	3" x 4"		380	.042		.85	1.41		2.26	3.30
5825	3" x 4", pneumatic nailed		456	.035		.85	1.18		2.03	2.91
8250	For second story & above, add						5%			
8300	For dormer & gable, add						15%			

06 11 10.42 Furring		Crew	Daily Output	Labor-Hours	Unit	Material	Labor	Equipment	Total	Total Incl O&P
0010	**FURRING**									
0012	Wood strips, 1" x 2", on walls, on wood	1 Carp	550	.015	L.F.	.25	.49		.74	1.09
0015	On wood, pneumatic nailed		710	.011		.25	.38		.63	.90
0300	On masonry		495	.016		.25	.54		.79	1.18
0400	On concrete		260	.031		.25	1.03		1.28	2
0600	1" x 3", on walls, on wood		550	.015		.34	.49		.83	1.20
0605	On wood, pneumatic nailed		710	.011		.34	.38		.72	1.01
0700	On masonry		495	.016		.34	.54		.88	1.29
0800	On concrete		260	.031		.34	1.03		1.37	2.11
0850	On ceilings, on wood		350	.023		.34	.77		1.11	1.66
0855	On wood, pneumatic nailed		450	.018		.34	.60		.94	1.38
0900	On masonry		320	.025		.34	.84		1.18	1.78
0950	On concrete		210	.038		.34	1.28		1.62	2.52

06 11 10.44 Grounds		Crew	Daily Output	Labor-Hours	Unit	Material	Labor	Equipment	Total	Total Incl O&P
0010	**GROUNDS**									
0020	For casework, 1" x 2" wood strips, on wood	1 Carp	330	.024	L.F.	.25	.81		1.06	1.63
0100	On masonry		285	.028		.25	.94		1.19	1.85
0200	On concrete		250	.032		.25	1.07		1.32	2.07
0400	For plaster, 3/4" deep, on wood		450	.018		.25	.60		.85	1.27

06 11 Wood Framing

06 11 10 – Framing with Dimensional, Engineered or Composite Lumber

06 11 10.44 Grounds		Crew	Daily Output	Labor-Hours	Unit	2011 Bare Costs Material	Labor	Equipment	Total	Total Incl O&P
0500	On masonry	1 Carp	225	.036	L.F.	.25	1.19		1.44	2.27
0600	On concrete		175	.046		.25	1.53		1.78	2.84
0700	On metal lath		200	.040		.25	1.34		1.59	2.52

06 12 Structural Panels

06 12 10 – Structural Insulated Panels

06 12 10.10 OSB Faced Panels

Code	Description		Crew	Daily Output	Labor-Hours	Unit	Material	Labor	Equipment	Total	Total Incl O&P
0010	**OSB FACED PANELS**										
0100	Structural insul. panels, 7/16" OSB both faces, EPS insul, 3-5/8" T	G	F-3	2075	.019	S.F.	2.72	.59	.32	3.63	4.32
0110	5-5/8" thick	G		1725	.023		3.05	.71	.38	4.14	4.96
0120	7-3/8" thick	G		1425	.028		3.33	.86	.46	4.65	5.60
0130	9-3/8" thick	G		1125	.036		3.64	1.09	.58	5.31	6.45
0140	7/16" OSB one face, EPS insul, 3-5/8" thick	G		2175	.018		2.90	.56	.30	3.76	4.46
0150	5-5/8" thick	G		1825	.022		3.32	.67	.36	4.35	5.15
0160	7-3/8" thick	G		1525	.026		3.70	.81	.43	4.94	5.90
0170	9-3/8" thick	G		1225	.033		4.12	1	.54	5.66	6.80
0190	7/16" OSB - 1/2" GWB faces , EPS insul, 3-5/8" T	G		2075	.019		2.87	.59	.32	3.78	4.49
0200	5-5/8" thick	G		1725	.023		3.29	.71	.38	4.38	5.20
0210	7-3/8" thick	G		1425	.028		3.66	.86	.46	4.98	5.95
0220	9-3/8" thick	G		1125	.036		4.29	1.09	.58	5.96	7.20
0240	7/16" OSB - 1/2" MRGWB faces , EPS insul, 3-5/8" T	G		2075	.019		2.87	.59	.32	3.78	4.49
0250	5-5/8" thick	G		1725	.023		3.29	.71	.38	4.38	5.20
0260	7-3/8" thick	G		1425	.028		3.66	.86	.46	4.98	5.95
0270	9-3/8" thick	G		1125	.036		4.29	1.09	.58	5.96	7.20
0300	For 1/2" GWB added to OSB skin, add	G					.60			.60	.66
0310	For 1/2" MRGWB added to OSB skin, add	G					.60			.60	.66
0320	For one T1-11 skin, add to OSB-OSB	G					1.39			1.39	1.53
0330	For one 19/32" CDX skin, add to OSB-OSB	G					.70			.70	.77
0500	Structural insulated panel, 7/16" OSB both sides, straw core										
0510	4-3/8" T, walls (w/sill, splines, plates)	G	F-6	2400	.017	S.F.	6.60	.52	.27	7.39	8.40
0520	Floors (w/splines)	G		2400	.017		6.60	.52	.27	7.39	8.40
0530	Roof (w/splines)	G		2400	.017		6.60	.52	.27	7.39	8.40
0550	7-7/8" T, walls (w/sill, splines, plates)	G		2400	.017		9.95	.52	.27	10.74	12.10
0560	Floors (w/splines)	G		2400	.017		9.95	.52	.27	10.74	12.10
0570	Roof (w/splines)	G		2400	.017		9.95	.52	.27	10.74	12.10

06 12 19 – Composite Shearwall Panels

06 12 19.10 Steel and Wood Composite Shearwall Panels

Code	Description	Crew	Daily Output	Labor-Hours	Unit	Material	Labor	Equipment	Total	Total Incl O&P
0010	**STEEL & WOOD COMPOSITE SHEARWALL PANELS**									
0020	Anchor bolts, 36" long (must be placed in wet concrete)	1 Carp	150	.053	Ea.	32.50	1.79		34.29	38.50
0030	On concrete, 2 x 4 & 2 x 6 walls, 7' - 10' high, 360 lb. shear, 12" wide	2 Carp	8	2		278	67		345	415
0040	715 lb. shear, 15" wide		8	2		320	67		387	460
0050	1860 lb. shear, 18" wide		8	2		355	67		422	500
0060	2780 lb. shear, 21" wide		8	2		365	67		432	510
0070	3790 lb. shear, 24" wide		8	2		460	67		527	615
0080	2 x 6 walls, 11' to 13' high, 1180 lb. shear, 18" wide		6	2.667		350	89.50		439.50	535
0090	1555 lb. shear, 21" wide		6	2.667		510	89.50		599.50	710
0100	2280 lb. shear, 24" wide		6	2.667		620	89.50		709.50	830
0110	For installing above on wood floor frame, add									
0120	Coupler nuts, threaded rods, bolts, shear transfer plate kit	1 Carp	16	.500	Ea.	48	16.80		64.80	81
0130	Framing anchors, angle (2 required)	"	96	.083	"	2.40	2.80		5.20	7.30

06 12 Structural Panels

06 12 19 – Composite Shearwall Panels

06 12 19.10 Steel and Wood Composite Shearwall Panels		Crew	Daily Output	Labor-Hours	Unit	2011 Bare Costs Material	Labor	Equipment	Total	Total Incl O&P
0140	For blocking see Section 06 11 10.02									
0150	For installing above, first floor to second floor, wood floor frame, add									
0160	Add stack option to first floor wall panel				Ea.	53			53	58.50
0170	Threaded rods, bolts, shear transfer plate kit	1 Carp	16	.500		57.50	16.80		74.30	91.50
0180	Framing anchors, angle (2 required)	"	96	.083		2.40	2.80		5.20	7.30
0200	For installing stacked panels, balloon framing									
0210	Add stack option to first floor wall panel				Ea.	53			53	58.50
0220	Threaded rods, bolts kit	1 Carp	16	.500	"	31	16.80		47.80	62

06 13 Heavy Timber Construction

06 13 23 – Heavy Timber Framing

06 13 23.10 Heavy Framing

Line	Description	Crew	Daily Output	Labor-Hours	Unit	Material	Labor	Equipment	Total	Total Incl O&P
0010	**HEAVY FRAMING**									
0020	Beams, single 6″ x 10″	2 Carp	1.10	14.545	M.B.F.	1,400	490		1,890	2,375
0100	Single 8″ x 16″		1.20	13.333		1,975	445		2,420	2,925
0200	Built from 2″ lumber, multiple 2″ x 14″		.90	17.778		725	595		1,320	1,800
0210	Built from 3″ lumber, multiple 3″ x 6″		.70	22.857		1,050	765		1,815	2,425
0220	Multiple 3″ x 8″		.80	20		1,050	670		1,720	2,300
0230	Multiple 3″ x 10″		.90	17.778		1,075	595		1,670	2,175
0240	Multiple 3″ x 12″		1	16		1,100	535		1,635	2,100
0250	Built from 4″ lumber, multiple 4″ x 6″		.80	20		1,175	670		1,845	2,425
0260	Multiple 4″ x 8″		.90	17.778		1,125	595		1,720	2,225
0270	Multiple 4″ x 10″		1	16		1,150	535		1,685	2,150
0280	Multiple 4″ x 12″		1.10	14.545		1,375	490		1,865	2,350
0290	Columns, structural grade, 1500f, 4″ x 4″		.60	26.667		1,275	895		2,170	2,900
0300	6″ x 6″		.65	24.615		1,300	825		2,125	2,800
0400	8″ x 8″		.70	22.857		1,350	765		2,115	2,750
0500	10″ x 10″		.75	21.333		1,450	715		2,165	2,800
0600	12″ x 12″		.80	20		1,525	670		2,195	2,800
0800	Floor planks, 2″ thick, T & G, 2″ x 6″		1.05	15.238		1,050	510		1,560	2,000
0900	2″ x 10″		1.10	14.545		1,050	490		1,540	1,975
1100	3″ thick, 3″ x 6″		1.05	15.238		1,250	510		1,760	2,225
1200	3″ x 10″		1.10	14.545		1,250	490		1,740	2,200
1400	Girders, structural grade, 12″ x 12″		.80	20		1,525	670		2,195	2,800
1500	10″ x 16″		1	16		2,425	535		2,960	3,575
2300	Roof purlins, 4″ thick, structural grade	↓	1.05	15.238	↓	1,125	510		1,635	2,075

06 15 Wood Decking

06 15 16 – Wood Roof Decking

06 15 16.10 Solid Wood Roof Decking

Line	Description	Crew	Daily Output	Labor-Hours	Unit	Material	Labor	Equipment	Total	Total Incl O&P
0010	**SOLID WOOD ROOF DECKING**									
0400	Cedar planks, 3″ thick	2 Carp	320	.050	S.F.	7.70	1.68		9.38	11.30
0500	4″ thick		250	.064		10.30	2.15		12.45	14.95
0700	Douglas fir, 3″ thick		320	.050		3.60	1.68		5.28	6.75
0800	4″ thick		250	.064		4.80	2.15		6.95	8.90
1000	Hemlock, 3″ thick		320	.050		3.28	1.68		4.96	6.40
1100	4″ thick		250	.064		4.36	2.15		6.51	8.40
1300	Western white spruce, 3″ thick		320	.050		3.28	1.68		4.96	6.40
1400	4″ thick	↓	250	.064	↓	4.36	2.15		6.51	8.40

06 15 Wood Decking

06 15 23 – Laminated Wood Decking

06 15 23.10 Laminated Roof Deck		Crew	Daily Output	Labor-Hours	Unit	Material	2011 Bare Costs Labor	Equipment	Total	Total Incl O&P
0010	**LAMINATED ROOF DECK**									
0020	Pine or hemlock, 3" thick	2 Carp	425	.038	S.F.	3.75	1.26		5.01	6.25
0100	4" thick		325	.049		5	1.65		6.65	8.25
0300	Cedar, 3" thick		425	.038		4.11	1.26		5.37	6.65
0400	4" thick		325	.049		5.25	1.65		6.90	8.50
0600	Fir, 3" thick		425	.038		3.80	1.26		5.06	6.30
0700	4" thick	↓	325	.049	↓	5.05	1.65		6.70	8.35

06 16 Sheathing

06 16 23 – Subflooring

06 16 23.10 Subfloor		Crew	Daily Output	Labor-Hours	Unit	Material	Labor	Equipment	Total	Total Incl O&P
0010	**SUBFLOOR** R061636-20									
0011	Plywood, CDX, 1/2" thick	2 Carp	1500	.011	SF Flr.	.58	.36		.94	1.24
0015	Pneumatic nailed		1860	.009		.58	.29		.87	1.12
0100	5/8" thick		1350	.012		.70	.40		1.10	1.44
0105	Pneumatic nailed		1674	.010		.70	.32		1.02	1.31
0200	3/4" thick		1250	.013		.82	.43		1.25	1.62
0205	Pneumatic nailed		1550	.010		.82	.35		1.17	1.48
0300	1-1/8" thick, 2-4-1 including underlayment		1050	.015		1.53	.51		2.04	2.54
0440	With boards, 1" x 6", S4S, laid regular		900	.018		1.47	.60		2.07	2.61
0450	1" x 8", laid regular		1000	.016		1.55	.54		2.09	2.60
0460	Laid diagonal		850	.019		1.55	.63		2.18	2.76
0500	1" x 10", laid regular		1100	.015		1.52	.49		2.01	2.49
0600	Laid diagonal		900	.018	↓	1.52	.60		2.12	2.67
1500	OSB, 5/8" thick		1330	.012	S.F.	.46	.40		.86	1.19
1600	3/4" thick	↓	1230	.013	"	.65	.44		1.09	1.45
8990	Subfloor adhesive, 3/8" bead	1 Carp	2300	.003	L.F.	.09	.12		.21	.30

06 16 26 – Underlayment

06 16 26.10 Wood Product Underlayment		Crew	Daily Output	Labor-Hours	Unit	Material	Labor	Equipment	Total	Total Incl O&P
0010	**WOOD PRODUCT UNDERLAYMENT** R061636-20									
0030	Plywood, underlayment grade, 3/8" thick	2 Carp	1500	.011	SF Flr.	.77	.36		1.13	1.45
0070	Pneumatic nailed		1860	.009		.77	.29		1.06	1.33
0100	1/2" thick		1450	.011		.80	.37		1.17	1.50
0105	Pneumatic nailed		1798	.009		.80	.30		1.10	1.38
0200	5/8" thick		1400	.011		1.01	.38		1.39	1.75
0205	Pneumatic nailed		1736	.009		1.01	.31		1.32	1.63
0300	3/4" thick		1300	.012		1.23	.41		1.64	2.04
0305	Pneumatic nailed		1612	.010		1.23	.33		1.56	1.91
0500	Particle board, 3/8" thick G		1500	.011		.33	.36		.69	.96
0505	Pneumatic nailed G		1860	.009		.33	.29		.62	.84
0600	1/2" thick G		1450	.011		.36	.37		.73	1.02
0605	Pneumatic nailed G		1798	.009		.36	.30		.66	.90
0800	5/8" thick G		1400	.011		.46	.38		.84	1.15
0805	Pneumatic nailed G		1736	.009		.46	.31		.77	1.03
0900	3/4" thick G		1300	.012		.65	.41		1.06	1.41
0905	Pneumatic nailed G		1612	.010		.65	.33		.98	1.28
1100	Hardboard, underlayment grade, 4' x 4', .215" thick G	↓	1500	.011	↓	.57	.36		.93	1.23

06 16 Sheathing

06 16 36 – Wood Panel Product Sheathing

06 16 36.10 Sheathing			Crew	Daily Output	Labor-Hours	Unit	Material	2011 Bare Costs Labor	Equipment	Total	Total Incl O&P
0010	**SHEATHING**	R061636-20									
0012	Plywood on roofs, CDX										
0030	5/16" thick		2 Carp	1600	.010	S.F.	.55	.34		.89	1.16
0035	Pneumatic nailed			1952	.008		.55	.28		.83	1.06
0050	3/8" thick			1525	.010		.47	.35		.82	1.11
0055	Pneumatic nailed			1860	.009		.47	.29		.76	1
0100	1/2" thick			1400	.011		.58	.38		.96	1.28
0105	Pneumatic nailed			1708	.009		.58	.31		.89	1.17
0200	5/8" thick			1300	.012		.70	.41		1.11	1.46
0205	Pneumatic nailed			1586	.010		.70	.34		1.04	1.34
0300	3/4" thick			1200	.013		.82	.45		1.27	1.65
0305	Pneumatic nailed			1464	.011		.82	.37		1.19	1.51
0500	Plywood on walls with exterior CDX, 3/8" thick			1200	.013		.47	.45		.92	1.27
0505	Pneumatic nailed			1488	.011		.47	.36		.83	1.12
0600	1/2" thick			1125	.014		.58	.48		1.06	1.44
0605	Pneumatic nailed			1395	.011		.58	.38		.96	1.28
0700	5/8" thick			1050	.015		.70	.51		1.21	1.63
0705	Pneumatic nailed			1302	.012		.70	.41		1.11	1.46
0800	3/4" thick			975	.016		.82	.55		1.37	1.82
0805	Pneumatic nailed			1209	.013		.82	.44		1.26	1.64
0840	Oriented strand board, 7/16" thick	G		1400	.011		.33	.38		.71	1
0845	Pneumatic nailed	G		1736	.009		.33	.31		.64	.88
0846	1/2" thick	G		1325	.012		.33	.41		.74	1.04
0847	Pneumatic nailed	G		1643	.010		.33	.33		.66	.91
0852	5/8" thick	G		1250	.013		.79	.43		1.22	1.59
0857	Pneumatic nailed	G		1550	.010		.79	.35		1.14	1.45
0900	For exterior C-C grade plywood, add	G					15%				
0920	For application to metal studs, joists, rafters, add	G						20%			
1000	For shear wall construction, add							20%			
1200	For structural 1 exterior plywood, add					S.F.	10%				
1400	With boards, on roof 1" x 6" boards, laid horizontal		2 Carp	725	.022		1.47	.74		2.21	2.85
1500	Laid diagonal			650	.025		1.47	.83		2.30	2.99
1700	1" x 8" boards, laid horizontal			875	.018		1.55	.61		2.16	2.73
1800	Laid diagonal			725	.022		1.55	.74		2.29	2.94
2000	For steep roofs, add							40%			
2200	For dormers, hips and valleys, add						5%	50%			
2400	Boards on walls, 1" x 6" boards, laid regular		2 Carp	650	.025		1.47	.83		2.30	2.99
2500	Laid diagonal			585	.027		1.47	.92		2.39	3.15
2700	1" x 8" boards, laid regular			765	.021		1.55	.70		2.25	2.87
2800	Laid diagonal			650	.025		1.55	.83		2.38	3.08
2850	Gypsum, weatherproof, 1/2" thick			1125	.014		.38	.48		.86	1.22
2900	With embedded glass mats			1100	.015		.75	.49		1.24	1.65
3000	Wood fiber, regular, no vapor barrier, 1/2" thick			1200	.013		.56	.45		1.01	1.37
3100	5/8" thick			1200	.013		.71	.45		1.16	1.53
3300	No vapor barrier, in colors, 1/2" thick			1200	.013		.68	.45		1.13	1.50
3400	5/8" thick			1200	.013		.72	.45		1.17	1.54
3600	With vapor barrier one side, white, 1/2" thick			1200	.013		.55	.45		1	1.36
3700	Vapor barrier 2 sides, 1/2" thick			1200	.013		.77	.45		1.22	1.60
3800	Asphalt impregnated, 25/32" thick			1200	.013		.28	.45		.73	1.06
3850	Intermediate, 1/2" thick			1200	.013		.20	.45		.65	.97
4000	OSB on roof, 1/2" thick	G		1455	.011		.33	.37		.70	.98
4100	5/8" thick	G		1330	.012		.79	.40		1.19	1.55

06 17 Shop-Fabricated Structural Wood

06 17 33 – Wood I-Joists

06 17 33.10 Wood and Composite I-Joists		Crew	Daily Output	Labor-Hours	Unit	Material	2011 Bare Costs Labor	Equipment	Total	Total Incl O&P
0010	**WOOD AND COMPOSITE I-JOISTS**									
0100	Plywood webs, incl. bridging & blocking, panels 24" O.C.									
1200	15' to 24' span, 50 psf live load	F-5	2400	.013	SF Flr.	1.76	.39		2.15	2.59
1300	55 psf live load		2250	.014		1.98	.42		2.40	2.88
1400	24' to 30' span, 45 psf live load		2600	.012		2.73	.36		3.09	3.61
1500	55 psf live load	↓	2400	.013	↓	3.38	.39		3.77	4.38

06 17 53 – Shop-Fabricated Wood Trusses

06 17 53.10 Roof Trusses		Crew	Daily Output	Labor-Hours	Unit	Material	Labor	Equipment	Total	Total Incl O&P
0010	**ROOF TRUSSES**									
5000	Common wood, 2" x 4" metal plate connected, 24" O.C., 4/12 slope									
5010	1' overhang, 12' span	F-5	55	.582	Ea.	43	17.15		60.15	76
5050	20' span	F-6	62	.645		68.50	20	10.55	99.05	120
5100	24' span		60	.667		77	21	10.95	108.95	131
5150	26' span		57	.702		80	22	11.50	113.50	137
5200	28' span		53	.755		92.50	23.50	12.35	128.35	155
5240	30' span		51	.784		88.50	24.50	12.85	125.85	152
5250	32' span		50	.800		109	25	13.10	147.10	175
5280	34' span		48	.833		117	26	13.65	156.65	186
5350	8/12 pitch, 1' overhang, 20' span		57	.702		76.50	22	11.50	110	133
5400	24' span		55	.727		97	22.50	11.90	131.40	158
5450	26' span		52	.769		101	24	12.60	137.60	165
5500	28' span		49	.816		111	25.50	13.40	149.90	180
5550	32' span		45	.889		134	27.50	14.55	176.05	209
5600	36' span		41	.976		167	30.50	16	213.50	251
5650	38' span		40	1		182	31	16.40	229.40	270
5700	40' span	↓	40	1	↓	188	31	16.40	235.40	276

06 18 Glued-Laminated Construction

06 18 13 – Glued-Laminated Beams

06 18 13.10 Laminated Beams		Crew	Daily Output	Labor-Hours	Unit	Material	Labor	Equipment	Total	Total Incl O&P
0010	**LAMINATED BEAMS**									
0050	3-1/2" x 18"	F-3	480	.083	L.F.	30.50	2.56	1.37	34.43	40
0100	5-1/4" x 11-7/8"		450	.089		29.50	2.73	1.46	33.69	38.50
0150	5-1/4" x 16"		360	.111		40	3.41	1.82	45.23	51.50
0200	5-1/4" x 18"		290	.138		45	4.24	2.26	51.50	59
0250	5-1/4" x 24"		220	.182		60	5.60	2.98	68.58	78.50
0300	7" x 11-7/8"		320	.125		39.50	3.84	2.05	45.39	52
0350	7" x 16"		260	.154		53.50	4.73	2.52	60.75	69.50
0400	7" x 18"	↓	210	.190		60	5.85	3.12	68.97	79
0500	For premium appearance, add to S.F. prices					5%				
0550	For industrial type, deduct					15%				
0600	For stain and varnish, add					5%				
0650	For 3/4" laminations, add				↓	25%				

06 18 13.20 Laminated Framing		Crew	Daily Output	Labor-Hours	Unit	Material	Labor	Equipment	Total	Total Incl O&P
0010	**LAMINATED FRAMING**									
0020	30 lb., short term live load, 15 lb. dead load									
0200	Straight roof beams, 20' clear span, beams 8' O.C.	F-3	2560	.016	SF Flr.	1.99	.48	.26	2.73	3.27
0300	Beams 16' O.C.		3200	.013		1.44	.38	.20	2.02	2.45
0500	40' clear span, beams 8' O.C.		3200	.013		3.80	.38	.20	4.38	5.05
0600	Beams 16' O.C.	↓	3840	.010	↓	3.11	.32	.17	3.60	4.14

06 18 Glued-Laminated Construction

06 18 13 – Glued-Laminated Beams

06 18 13.20 Laminated Framing		Crew	Daily Output	Labor-Hours	Unit	2011 Bare Costs Material	Labor	Equipment	Total	Total Incl O&P
0800	60' clear span, beams 8' O.C.	F-4	2880	.014	SF Flr.	6.55	.43	.42	7.40	8.35
0900	Beams 16' O.C.	"	3840	.010		4.87	.32	.31	5.50	6.25
1100	Tudor arches, 30' to 40' clear span, frames 8' O.C.	F-3	1680	.024		8.50	.73	.39	9.62	11
1200	Frames 16' O.C.	"	2240	.018		6.70	.55	.29	7.54	8.60
1400	50' to 60' clear span, frames 8' O.C.	F-4	2200	.018		9.20	.56	.55	10.31	11.65
1500	Frames 16' O.C.		2640	.015		7.80	.47	.46	8.73	9.85
1700	Radial arches, 60' clear span, frames 8' O.C.		1920	.021		8.60	.64	.63	9.87	11.20
1800	Frames 16' O.C.		2880	.014		6.60	.43	.42	7.45	8.40
2000	100' clear span, frames 8' O.C.		1600	.025		8.90	.77	.76	10.43	11.90
2100	Frames 16' O.C.		2400	.017		7.80	.51	.50	8.81	10
2300	120' clear span, frames 8' O.C.		1440	.028		11.85	.85	.84	13.54	15.35
2400	Frames 16' O.C.		1920	.021		10.80	.64	.63	12.07	13.65
2600	Bowstring trusses, 20' O.C., 40' clear span	F-3	2400	.017		5.35	.51	.27	6.13	7
2700	60' clear span	F-4	3600	.011		4.79	.34	.34	5.47	6.20
2800	100' clear span		4000	.010		6.80	.31	.30	7.41	8.30
2900	120' clear span		3600	.011		7.25	.34	.34	7.93	8.95
3100	For premium appearance, add to S.F. prices					5%				
3300	For industrial type, deduct					15%				
3500	For stain and varnish, add					5%				
3900	For 3/4" laminations, add to straight					25%				
4100	Add to curved					15%				
4300	Alternate pricing method: (use nominal footage of									
4310	components). Straight beams, camber less than 6"	F-3	3.50	11.429	M.B.F.	2,950	350	187	3,487	4,050
4400	Columns, including hardware		2	20		3,175	615	330	4,120	4,875
4600	Curved members, radius over 32'		2.50	16		3,250	490	262	4,002	4,675
4700	Radius 10' to 32'		3	13.333		3,225	410	219	3,854	4,450
4900	For complicated shapes, add maximum					100%				
5100	For pressure treating, add to straight					35%				
5200	Add to curved					45%				
6000	Laminated veneer members, southern pine or western species									
6050	1-3/4" wide x 5-1/2" deep	2 Carp	480	.033	L.F.	3.19	1.12		4.31	5.40
6100	9-1/2" deep		480	.033		4.56	1.12		5.68	6.85
6150	14" deep		450	.036		7.30	1.19		8.49	10.05
6200	18" deep		450	.036		9.90	1.19		11.09	12.85
6300	Parallel strand members, southern pine or western species									
6350	1-3/4" wide x 9-1/4" deep	2 Carp	480	.033	L.F.	4.56	1.12		5.68	6.85
6400	11-1/4" deep		450	.036		5.70	1.19		6.89	8.25
6450	14" deep		400	.040		7.30	1.34		8.64	10.30
6500	3-1/2" wide x 9-1/4" deep		480	.033		15.85	1.12		16.97	19.30
6550	11-1/4" deep		450	.036		19.15	1.19		20.34	23
6600	14" deep		400	.040		23.50	1.34		24.84	28
6650	7" wide x 9-1/4" deep		450	.036		31.50	1.19		32.69	37
6700	11-1/4" deep		420	.038		39.50	1.28		40.78	45.50
6750	14" deep		400	.040		46.50	1.34		47.84	54

06 22 Millwork

06 22 13 – Standard Pattern Wood Trim

06 22 13.10 Millwork

Code	Description	Crew	Daily Output	Labor-Hours	Unit	Material	2011 Bare Costs Labor	Equipment	Total	Total Incl O&P
0010	**MILLWORK** R061110-30									
1020	1" x 12", custom birch				L.F.	5.60			5.60	6.15
1040	Cedar					5.80			5.80	6.35
1060	Oak					3.91			3.91	4.30
1080	Redwood					4.68			4.68	5.15
1100	Southern yellow pine					3.35			3.35	3.69
1120	Sugar pine					3.50			3.50	3.85
1140	Teak					29.50			29.50	32.50
1160	Walnut					6.15			6.15	6.75
1180	White pine					4.24			4.24	4.66

06 22 13.15 Moldings, Base

Code	Description	Crew	Daily Output	Labor-Hours	Unit	Material	Labor	Equipment	Total	Total Incl O&P
0010	**MOLDINGS, BASE**									
0500	Base, stock pine, 9/16" x 3-1/2"	1 Carp	240	.033	L.F.	2.22	1.12		3.34	4.31
0501	Oak or birch, 9/16" x 3-1/2"		240	.033		3.38	1.12		4.50	5.60
0550	9/16" x 4-1/2"		200	.040		2.46	1.34		3.80	4.96
0561	Base shoe, oak, 3/4" x 1"		240	.033		1.13	1.12		2.25	3.11
0570	Base, prefinished, 2-1/2" x 9/16"		242	.033		.95	1.11		2.06	2.91
0580	Shoe, prefinished, 3/8" x 5/8"		266	.030		.45	1.01		1.46	2.19
0585	Flooring cant strip, 3/4" x 1/2"		500	.016		.52	.54		1.06	1.47

06 22 13.30 Moldings, Casings

Code	Description	Crew	Daily Output	Labor-Hours	Unit	Material	Labor	Equipment	Total	Total Incl O&P
0010	**MOLDINGS, CASINGS**									
0090	Apron, stock pine, 5/8" x 2"	1 Carp	250	.032	L.F.	1.22	1.07		2.29	3.14
0110	5/8" x 3-1/2"		220	.036		1.62	1.22		2.84	3.82
0300	Band, stock pine, 11/16" x 1-1/8"		270	.030		.82	.99		1.81	2.56
0350	11/16" x 1-3/4"		250	.032		1.06	1.07		2.13	2.97
0700	Casing, stock pine, 11/16" x 2-1/2"		240	.033		1.33	1.12		2.45	3.33
0701	Oak or birch		240	.033		2.58	1.12		3.70	4.71
0750	11/16" x 3-1/2"		215	.037		1.62	1.25		2.87	3.87
0760	Door & window casing, exterior, 1-1/4" x 2"		200	.040		2.14	1.34		3.48	4.60
0770	Finger jointed, 1-1/4" x 2"		200	.040		.99	1.34		2.33	3.34
4600	Mullion casing, stock pine, 5/16" x 2"		200	.040		.84	1.34		2.18	3.17
4601	Oak or birch, 9/16" x 2-1/2"		200	.040		2.79	1.34		4.13	5.30
4700	Teak, custom, nominal 1" x 1"		215	.037		2	1.25		3.25	4.29
4800	Nominal 1" x 3"		200	.040		5.70	1.34		7.04	8.50

06 22 13.35 Moldings, Ceilings

Code	Description	Crew	Daily Output	Labor-Hours	Unit	Material	Labor	Equipment	Total	Total Incl O&P
0010	**MOLDINGS, CEILINGS**									
0600	Bed, stock pine, 9/16" x 1-3/4"	1 Carp	270	.030	L.F.	1.02	.99		2.01	2.78
0650	9/16" x 2"		240	.033		1.22	1.12		2.34	3.21
1200	Cornice molding, stock pine, 9/16" x 1-3/4"		330	.024		1.02	.81		1.83	2.48
1300	9/16" x 2-1/4"		300	.027		1.24	.89		2.13	2.86
2400	Cove scotia, stock pine, 9/16" x 1-3/4"		270	.030		.67	.99		1.66	2.40
2401	Oak or birch, 9/16" x 1-3/4"		270	.030		.89	.99		1.88	2.64
2500	11/16" x 2-3/4"		255	.031		1.86	1.05		2.91	3.81
2600	Crown, stock pine, 9/16" x 3-5/8"		250	.032		2.30	1.07		3.37	4.33
2700	11/16" x 4-5/8"		220	.036		3.87	1.22		5.09	6.30

06 22 13.40 Moldings, Exterior

Code	Description	Crew	Daily Output	Labor-Hours	Unit	Material	Labor	Equipment	Total	Total Incl O&P
0010	**MOLDINGS, EXTERIOR**									
1500	Cornice, boards, pine, 1" x 2"	1 Carp	330	.024	L.F.	.25	.81		1.06	1.63
1600	1" x 4"		250	.032		.53	1.07		1.60	2.38
1700	1" x 6"		250	.032		.67	1.07		1.74	2.54
1800	1" x 8"		200	.040		.97	1.34		2.31	3.31
1900	1" x 10"		180	.044		1.21	1.49		2.70	3.83

06 22 Millwork

06 22 13 – Standard Pattern Wood Trim

06 22 13.40 Moldings, Exterior		Crew	Daily Output	Labor-Hours	Unit	Material	2011 Bare Costs Labor	Equipment	Total	Total Incl O&P
2000	1" x 12"	1 Carp	180	.044	L.F.	1.46	1.49		2.95	4.11
2200	Three piece, built-up, pine, minimum		80	.100		1.45	3.36		4.81	7.20
2300	Maximum		65	.123		3.64	4.13		7.77	10.90
3000	Corner board, sterling pine, 1" x 4"		200	.040		.79	1.34		2.13	3.12
3100	1" x 6"		200	.040		1.05	1.34		2.39	3.41
3350	Fascia, sterling pine, 1" x 6"		250	.032		1.05	1.07		2.12	2.96
3370	1" x 8"		225	.036		1.43	1.19		2.62	3.57
3376	1" x 10"		180	.044		1.99	1.49		3.48	4.69
3395	Grounds, 1" x 1" redwood		300	.027		.25	.89		1.14	1.77
3400	Trim, back band, 11/16" x 1-1/16"		250	.032		.83	1.07		1.90	2.71
3500	Casing, 11/16" x 4-1/4"		250	.032		2.74	1.07		3.81	4.81
3600	Crown, 11/16" x 4-1/4"		250	.032		3.85	1.07		4.92	6.05
3700	1" x 4" and 1" x 6" railing w/balusters 4" O.C.		22	.364		22.50	12.20		34.70	45
3800	Insect screen framing stock, 1-1/16" x 1-3/4"		395	.020		2.03	.68		2.71	3.37
4100	Verge board, sterling pine, 1" x 4"		200	.040		.79	1.34		2.13	3.12
4200	1" x 6"		200	.040		1.05	1.34		2.39	3.41
4300	2" x 6"		165	.048		2.10	1.63		3.73	5.05
4400	2" x 8"	↓	165	.048	↓	2.86	1.63		4.49	5.85
4700	For redwood trim, add					200%				
5000	Casing/fascia, rough-sawn cedar									
5100	1" x 2"	1 Carp	275	.029	L.F.	.42	.98		1.40	2.09
5200	1" x 6"		250	.032		1.27	1.07		2.34	3.20
5300	1" x 8"		230	.035		2	1.17		3.17	4.15
5400	2" x 4"		220	.036		1.57	1.22		2.79	3.77
5500	2" x 6"		220	.036		2.56	1.22		3.78	4.86
5600	2" x 8"		200	.040		3.89	1.34		5.23	6.55
5700	2" x 10"		180	.044		4.69	1.49		6.18	7.65
5800	2" x 12"	↓	170	.047	↓	5.85	1.58		7.43	9.10

06 22 13.45 Moldings, Trim		Crew	Daily Output	Labor-Hours	Unit	Material	Labor	Equipment	Total	Total Incl O&P
0010	**MOLDINGS, TRIM**									
0200	Astragal, stock pine, 11/16" x 1-3/4"	1 Carp	255	.031	L.F.	1.22	1.05		2.27	3.10
0250	1-5/16" x 2-3/16"		240	.033		2.57	1.12		3.69	4.70
0800	Chair rail, stock pine, 5/8" x 2-1/2"		270	.030		1.43	.99		2.42	3.23
0900	5/8" x 3-1/2"		240	.033		2.26	1.12		3.38	4.36
1000	Closet pole, stock pine, 1-1/8" diameter		200	.040		1.05	1.34		2.39	3.41
1100	Fir, 1-5/8" diameter		200	.040		1.93	1.34		3.27	4.37
1150	Corner, inside, 5/16" x 1"		225	.036		.34	1.19		1.53	2.37
1160	Outside, 1-1/16" x 1-1/16"		240	.033		1.32	1.12		2.44	3.32
1161	1-5/16" x 1-5/16"		240	.033		1.73	1.12		2.85	3.77
3300	Half round, stock pine, 1/4" x 1/2"		270	.030		.24	.99		1.23	1.92
3350	1/2" x 1"	↓	255	.031	↓	.62	1.05		1.67	2.44
3400	Handrail, fir, single piece, stock, hardware not included									
3450	1-1/2" x 1-3/4"	1 Carp	80	.100	L.F.	2.03	3.36		5.39	7.85
3470	Pine, 1-1/2" x 1-3/4"		80	.100		2.03	3.36		5.39	7.85
3500	1-1/2" x 2-1/2"		76	.105		3.04	3.53		6.57	9.25
3600	Lattice, stock pine, 1/4" x 1-1/8"		270	.030		.42	.99		1.41	2.12
3700	1/4" x 1-3/4"		250	.032		.65	1.07		1.72	2.52
3800	Miscellaneous, custom, pine, 1" x 1"		270	.030		.35	.99		1.34	2.05
3900	1" x 3"		240	.033		1.06	1.12		2.18	3.04
4100	Birch or oak, nominal 1" x 1"		240	.033		.47	1.12		1.59	2.38
4200	Nominal 1" x 3"		215	.037		1.40	1.25		2.65	3.63
4400	Walnut, nominal 1" x 1"	↓	215	.037	↓	.51	1.25		1.76	2.65

06 22 Millwork

06 22 13 – Standard Pattern Wood Trim

06 22 13.45 Moldings, Trim		Crew	Daily Output	Labor-Hours	Unit	Material	2011 Bare Costs Labor	Equipment	Total	Total Incl O&P
4500	Nominal 1" x 3"	1 Carp	200	.040	L.F.	1.53	1.34		2.87	3.94
4700	Teak, nominal 1" x 1"		215	.037		2.47	1.25		3.72	4.81
4800	Nominal 1" x 3"		200	.040		7.40	1.34		8.74	10.40
4900	Quarter round, stock pine, 1/4" x 1/4"		275	.029		.22	.98		1.20	1.87
4950	3/4" x 3/4"		255	.031		.60	1.05		1.65	2.42
5600	Wainscot moldings, 1-1/8" x 9/16", 2' high, minimum		76	.105	S.F.	11.15	3.53		14.68	18.20
5700	Maximum		65	.123	"	20	4.13		24.13	29.50
06 22 13.50 Moldings, Window and Door										
0010	**MOLDINGS, WINDOW AND DOOR**									
2800	Door moldings, stock, decorative, 1-1/8" wide, plain	1 Carp	17	.471	Set	45	15.80		60.80	76
2900	Detailed		17	.471	"	87.50	15.80		103.30	123
2960	Clear pine door jamb, no stops, 11/16" x 4-9/16"		240	.033	L.F.	5.15	1.12		6.27	7.50
3150	Door trim set, 1 head and 2 sides, pine, 2-1/2 wide		12	.667	Opng.	22.50	22.50		45	62.50
3170	3-1/2" wide		11	.727	"	27.50	24.50		52	71.50
3250	Glass beads, stock pine, 3/8" x 1/2"		275	.029	L.F.	.31	.98		1.29	1.97
3270	3/8" x 7/8"		270	.030		.41	.99		1.40	2.11
4850	Parting bead, stock pine, 3/8" x 3/4"		275	.029		.36	.98		1.34	2.03
4870	1/2" x 3/4"		255	.031		.45	1.05		1.50	2.26
5000	Stool caps, stock pine, 11/16" x 3-1/2"		200	.040		2.27	1.34		3.61	4.75
5100	1-1/16" x 3-1/4"		150	.053		3.56	1.79		5.35	6.90
5300	Threshold, oak, 3' long, inside, 5/8" x 3-5/8"		32	.250	Ea.	9	8.40		17.40	24
5400	Outside, 1-1/2" x 7-5/8"		16	.500	"	38	16.80		54.80	70
5900	Window trim sets, including casings, header, stops,									
5910	stool and apron, 2-1/2" wide, minimum	1 Carp	13	.615	Opng.	32.50	20.50		53	70
5950	Average		10	.800		38	27		65	87
6000	Maximum		6	1.333		62.50	44.50		107	144
06 22 13.60 Moldings, Soffits										
0010	**MOLDINGS, SOFFITS**									
0200	Soffits, pine, 1" x 4"	2 Carp	420	.038	L.F.	.53	1.28		1.81	2.72
0210	1" x 6"		420	.038		.67	1.28		1.95	2.88
0220	1" x 8"		420	.038		.97	1.28		2.25	3.20
0230	1" x 10"		400	.040		1.21	1.34		2.55	3.58
0240	1" x 12"		400	.040		1.46	1.34		2.80	3.86
0250	STK cedar, 1" x 4"		420	.038		.43	1.28		1.71	2.61
0260	1" x 6"		420	.038		.76	1.28		2.04	2.98
0270	1" x 8"		420	.038		1.18	1.28		2.46	3.44
0280	1" x 10"		400	.040		1.68	1.34		3.02	4.10
0290	1" x 12"		400	.040		2.49	1.34		3.83	4.99
1000	Exterior AC plywood, 1/4" thick		420	.038	S.F.	.62	1.28		1.90	2.82
1100	1/2" thick		420	.038		.80	1.28		2.08	3.02
1150	Polyvinyl chloride, white, solid	1 Carp	230	.035		1.96	1.17		3.13	4.11
1160	Perforated	"	230	.035		1.96	1.17		3.13	4.11
1170	Accessories, "J" channel 5/8"	2 Carp	700	.023	L.F.	.36	.77		1.13	1.68

06 25 Prefinished Paneling

06 25 13 – Prefinished Hardboard Paneling

06 25 13.10 Paneling, Hardboard			Crew	Daily Output	Labor-Hours	Unit	2011 Bare Costs Material	Labor	Equipment	Total	Total Incl O&P
0010	**PANELING, HARDBOARD**										
0050	Not incl. furring or trim, hardboard, tempered, 1/8" thick	G	2 Carp	500	.032	S.F.	.38	1.07		1.45	2.22
0100	1/4" thick	G		500	.032		.55	1.07		1.62	2.41
0300	Tempered pegboard, 1/8" thick	G		500	.032		.45	1.07		1.52	2.30
0400	1/4" thick	G		500	.032		.64	1.07		1.71	2.50
0600	Untempered hardboard, natural finish, 1/8" thick	G		500	.032		.38	1.07		1.45	2.22
0700	1/4" thick	G		500	.032		.46	1.07		1.53	2.31
0900	Untempered pegboard, 1/8" thick	G		500	.032		.39	1.07		1.46	2.23
1000	1/4" thick	G		500	.032		.42	1.07		1.49	2.26
1200	Plastic faced hardboard, 1/8" thick	G		500	.032		.68	1.07		1.75	2.55
1300	1/4" thick	G		500	.032		.84	1.07		1.91	2.72
1500	Plastic faced pegboard, 1/8" thick	G		500	.032		.62	1.07		1.69	2.48
1600	1/4" thick	G		500	.032		.74	1.07		1.81	2.61
1800	Wood grained, plain or grooved, 1/4" thick, minimum	G		500	.032		.58	1.07		1.65	2.44
1900	Maximum	G		425	.038		1.12	1.26		2.38	3.34
2100	Moldings for hardboard, wood or aluminum, minimum			500	.032	L.F.	.39	1.07		1.46	2.23
2200	Maximum			425	.038	"	1.09	1.26		2.35	3.31

06 25 16 – Prefinished Plywood Paneling

06 25 16.10 Paneling, Plywood		Crew	Daily Output	Labor-Hours	Unit	Material	Labor	Equipment	Total	Total Incl O&P
0010	**PANELING, PLYWOOD**									
2400	Plywood, prefinished, 1/4" thick, 4' x 8' sheets									
2410	with vertical grooves. Birch faced, minimum	2 Carp	500	.032	S.F.	.89	1.07		1.96	2.78
2420	Average		420	.038		1.35	1.28		2.63	3.63
2430	Maximum		350	.046		1.97	1.53		3.50	4.74
2600	Mahogany, African		400	.040		2.51	1.34		3.85	5
2700	Philippine (Lauan)		500	.032		1.08	1.07		2.15	2.99
2900	Oak or Cherry, minimum		500	.032		2.10	1.07		3.17	4.11
3000	Maximum		400	.040		3.22	1.34		4.56	5.80
3200	Rosewood		320	.050		4.58	1.68		6.26	7.85
3400	Teak		400	.040		3.22	1.34		4.56	5.80
3600	Chestnut		375	.043		4.77	1.43		6.20	7.65
3800	Pecan		400	.040		2.06	1.34		3.40	4.52
3900	Walnut, minimum		500	.032		2.75	1.07		3.82	4.83
3950	Maximum		400	.040		5.20	1.34		6.54	8
4000	Plywood, prefinished, 3/4" thick, stock grades, minimum		320	.050		1.24	1.68		2.92	4.17
4100	Maximum		224	.071		5.40	2.40		7.80	9.90
4300	Architectural grade, minimum		224	.071		3.98	2.40		6.38	8.40
4400	Maximum		160	.100		6.10	3.36		9.46	12.30
4600	Plywood, "A" face, birch, V.C., 1/2" thick, natural		450	.036		1.89	1.19		3.08	4.08
4700	Select		450	.036		2.06	1.19		3.25	4.27
4900	Veneer core, 3/4" thick, natural		320	.050		1.99	1.68		3.67	5
5000	Select		320	.050		2.23	1.68		3.91	5.25
5200	Lumber core, 3/4" thick, natural		320	.050		2.99	1.68		4.67	6.10
5500	Plywood, knotty pine, 1/4" thick, A2 grade		450	.036		1.63	1.19		2.82	3.79
5600	A3 grade		450	.036		2.06	1.19		3.25	4.27
5800	3/4" thick, veneer core, A2 grade		320	.050		2.11	1.68		3.79	5.15
5900	A3 grade		320	.050		2.38	1.68		4.06	5.45
6100	Aromatic cedar, 1/4" thick, plywood		400	.040		2.08	1.34		3.42	4.54
6200	1/4" thick, particle board		400	.040		1.01	1.34		2.35	3.36

06 25 Prefinished Paneling

06 25 26 – Panel System

06 25 26.10 Panel Systems		Crew	Daily Output	Labor-Hours	Unit	Material	2011 Bare Costs Labor	Equipment	Total	Total Incl O&P
0010	**PANEL SYSTEMS**									
0100	Raised panel, eng. wood core w/wood veneer, std., paint grade	2 Carp	300	.053	S.F.	12.20	1.79		13.99	16.40
0110	Oak veneer		300	.053		20	1.79		21.79	25
0120	Maple veneer		300	.053		24.50	1.79		26.29	29.50
0130	Cherry veneer		300	.053		29	1.79		30.79	35
0300	Class I fire rated, paint grade		300	.053		14.25	1.79		16.04	18.65
0310	Oak veneer		300	.053		23.50	1.79		25.29	29
0320	Maple veneer		300	.053		30	1.79		31.79	36
0330	Cherry veneer		300	.053		38.50	1.79		40.29	45
0510	Beadboard, 5/8" MDF, standard, primed		300	.053		6.60	1.79		8.39	10.25
0520	Oak veneer, unfinished		300	.053		10.05	1.79		11.84	14.05
0530	Maple veneer, unfinished		300	.053		11.20	1.79		12.99	15.30
0610	Rustic paneling, 5/8" MDF, standard, maple veneer, unfinished	↓	300	.053	↓	12.70	1.79		14.49	16.95

06 26 Board Paneling

06 26 13 – Profile Board Paneling

06 26 13.10 Paneling, Boards		Crew	Daily Output	Labor-Hours	Unit	Material	Labor	Equipment	Total	Total Incl O&P
0010	**PANELING, BOARDS**									
6400	Wood board paneling, 3/4" thick, knotty pine	2 Carp	300	.053	S.F.	1.89	1.79		3.68	5.05
6500	Rough sawn cedar		300	.053		2.86	1.79		4.65	6.15
6700	Redwood, clear, 1" x 4" boards		300	.053		4.96	1.79		6.75	8.45
6900	Aromatic cedar, closet lining, boards	↓	275	.058	↓	2	1.95		3.95	5.45

06 43 Wood Stairs and Railings

06 43 13 – Wood Stairs

06 43 13.20 Prefabricated Wood Stairs		Crew	Daily Output	Labor-Hours	Unit	Material	Labor	Equipment	Total	Total Incl O&P
0010	**PREFABRICATED WOOD STAIRS**									
0100	Box stairs, prefabricated, 3'-0" wide									
0110	Oak treads, up to 14 risers	2 Carp	39	.410	Riser	85	13.75		98.75	117
0600	With pine treads for carpet, up to 14 risers	"	39	.410	"	55	13.75		68.75	83.50
1100	For 4' wide stairs, add				Flight	25%				
1550	Stairs, prefabricated stair handrail with balusters	1 Carp	30	.267	L.F.	75	8.95		83.95	97
1700	Basement stairs, prefabricated, pine treads									
1710	Pine risers, 3' wide, up to 14 risers	2 Carp	52	.308	Riser	55	10.30		65.30	78
4000	Residential, wood, oak treads, prefabricated		1.50	10.667	Flight	1,100	360		1,460	1,800
4200	Built in place	↓	.44	36.364	"	1,750	1,225		2,975	3,975
4400	Spiral, oak, 4'-6" diameter, unfinished, prefabricated,									
4500	incl. railing, 9' high	2 Carp	1.50	10.667	Flight	3,625	360		3,985	4,575

06 43 13.40 Wood Stair Parts		Crew	Daily Output	Labor-Hours	Unit	Material	Labor	Equipment	Total	Total Incl O&P
0010	**WOOD STAIR PARTS**									
0020	Pin top balusters, 1-1/4", oak, 34"	1 Carp	96	.083	Ea.	7.80	2.80		10.60	13.30
0030	38"		96	.083		9.85	2.80		12.65	15.55
0040	42"		96	.083		12.10	2.80		14.90	18
0050	Poplar, 34"		96	.083		3.05	2.80		5.85	8.05
0060	38"		96	.083		3.62	2.80		6.42	8.65
0070	42"		96	.083		8.55	2.80		11.35	14.15
0080	Maple, 34"		96	.083		4.90	2.80		7.70	10.10
0090	38"	↓	96	.083	↓	6.20	2.80		9	11.50

06 43 Wood Stairs and Railings

06 43 13 – Wood Stairs

06 43 13.40 Wood Stair Parts		Crew	Daily Output	Labor-Hours	Unit	Material	2011 Bare Costs Labor	Equipment	Total	Total Incl O&P
0100	42"	1 Carp	96	.083	Ea.	7.60	2.80		10.40	13.05
0130	Primed, 34"		96	.083		2.85	2.80		5.65	7.80
0140	38"		96	.083		3.62	2.80		6.42	8.65
0150	42"		96	.083		4.42	2.80		7.22	9.55
0180	Box top balusters, 1-1/4", oak, 34"		60	.133		10.85	4.47		15.32	19.45
0190	38"		60	.133		12.50	4.47		16.97	21.50
0200	42"		60	.133		12.95	4.47		17.42	22
0210	Poplar, 34"		60	.133		8.65	4.47		13.12	17
0220	38"		60	.133		9.50	4.47		13.97	17.95
0230	42"		60	.133		10	4.47		14.47	18.50
0240	Maple, 34"		60	.133		11.90	4.47		16.37	20.50
0250	38"		60	.133		13.80	4.47		18.27	22.50
0260	42"		60	.133		14.20	4.47		18.67	23
0290	Primed, 34"		60	.133		6.95	4.47		11.42	15.15
0300	38"		60	.133		8	4.47		12.47	16.30
0310	42"		60	.133		8.35	4.47		12.82	16.70
0340	Square balusters, cut from lineal stock, pine, 1-1/16" x 1-1/16"		180	.044	L.F.	1.32	1.49		2.81	3.95
0350	1-5/16" x 1-5/16"		180	.044		1.70	1.49		3.19	4.37
0360	1-5/8" x 1-5/8"		180	.044		2.82	1.49		4.31	5.60
0370	Turned newel, oak, 3-1/2" square, 48" high		8	1	Ea.	94.50	33.50		128	160
0380	62" high		8	1		138	33.50		171.50	208
0390	Poplar, 3-1/2" square, 48" high		8	1		52	33.50		85.50	113
0400	62" high		8	1		75	33.50		108.50	139
0410	Maple, 3-1/2" square, 48" high		8	1		66	33.50		99.50	129
0420	62" high		8	1		93	33.50		126.50	158
0430	Square newel, oak, 3-1/2" square, 48" high		8	1		50	33.50		83.50	111
0440	58" high		8	1		70	33.50		103.50	133
0450	Poplar, 3-1/2" square, 48" high		8	1		49	33.50		82.50	110
0460	58" high		8	1		54	33.50		87.50	116
0470	Maple, 3" square, 48" high		8	1		50	33.50		83.50	111
0480	58" high		8	1		62	33.50		95.50	124
0490	Railings, oak, minimum		96	.083	L.F.	10.40	2.80		13.20	16.15
0500	Average		96	.083		12.50	2.80		15.30	18.45
0510	Maximum		96	.083		15.50	2.80		18.30	21.50
0520	Maple, minimum		96	.083		15	2.80		17.80	21
0530	Average		96	.083		15.25	2.80		18.05	21.50
0540	Maximum		96	.083		15.50	2.80		18.30	21.50
0550	Oak, for bending rail, minimum		48	.167		21	5.60		26.60	32.50
0560	Average		48	.167		22	5.60		27.60	33.50
0570	Maximum		48	.167		23	5.60		28.60	35
0580	Maple, for bending rail, minimum		48	.167		23	5.60		28.60	34.50
0590	Average		48	.167		25	5.60		30.60	36.50
0600	Maximum		48	.167		27	5.60		32.60	39
0610	Risers, oak, 3/4" x 8", 36" long		80	.100	Ea.	18	3.36		21.36	25.50
0620	42" long		80	.100		21	3.36		24.36	28.50
0630	48" long		80	.100		24	3.36		27.36	32
0640	54" long		80	.100		27	3.36		30.36	35
0650	60" long		80	.100		30	3.36		33.36	38.50
0660	72" long		80	.100		36	3.36		39.36	45
0670	Poplar, 3/4" x 8", 36" long		80	.100		11.10	3.36		14.46	17.80
0680	42" long		80	.100		12.95	3.36		16.31	19.85
0690	48" long		80	.100		14.80	3.36		18.16	22
0700	54" long		80	.100		16.65	3.36		20.01	24

06 43 Wood Stairs and Railings

06 43 13 – Wood Stairs

06 43 13.40 Wood Stair Parts		Crew	Daily Output	Labor-Hours	Unit	Material	2011 Bare Costs Labor	Equipment	Total	Total Incl O&P
0710	60" long	1 Carp	80	.100	Ea.	18.50	3.36		21.86	26
0720	72" long		80	.100		22	3.36		25.36	30
0730	Pine, 1" x 8", 36" long		80	.100		2.90	3.36		6.26	8.80
0740	42" long		80	.100		3.38	3.36		6.74	9.30
0750	48" long		80	.100		3.87	3.36		7.23	9.85
0760	54" long		80	.100		4.35	3.36		7.71	10.40
0770	60" long		80	.100		4.83	3.36		8.19	10.90
0780	72" long		80	.100		5.80	3.36		9.16	12
0790	Treads, oak, no returns, 1-1/32" x 11-1/2" x 36" long		32	.250		32	8.40		40.40	49
0800	42" long		32	.250		37.50	8.40		45.90	55
0810	48" long		32	.250		42.50	8.40		50.90	61
0820	54" long		32	.250		48	8.40		56.40	67
0830	60" long		32	.250		53.50	8.40		61.90	72.50
0840	72" long		32	.250		64	8.40		72.40	84.50
0850	Mitred return one end, 1-1/32" x 11-1/2" x 36" long		24	.333		46	11.20		57.20	69
0860	42" long		24	.333		53.50	11.20		64.70	77.50
0870	48" long		24	.333		61.50	11.20		72.70	86
0880	54" long		24	.333		69	11.20		80.20	94.50
0890	60" long		24	.333		76.50	11.20		87.70	103
0900	72" long		24	.333		92	11.20		103.20	120
0910	Mitred return two ends, 1-1/32" x 11-1/2" x 36" long		12	.667		49	22.50		71.50	91.50
0920	42" long		12	.667		57	22.50		79.50	101
0930	48" long		12	.667		65.50	22.50		88	110
0940	54" long		12	.667		73.50	22.50		96	119
0950	60" long		12	.667		81.50	22.50		104	128
0960	72" long		12	.667		98	22.50		120.50	146
0970	Starting step, oak, 48", bullnose		8	1		192	33.50		225.50	267
0980	Double end bullnose		8	1		345	33.50		378.50	435
1030	Skirt board, pine, 1" x 10"		55	.145	L.F.	1.21	4.88		6.09	9.50
1040	1" x 12"		52	.154	"	1.46	5.15		6.61	10.25
1050	Oak landing tread, 1-1/16" thick		54	.148	S.F.	8.25	4.97		13.22	17.40
1060	Oak cove molding		96	.083	L.F.	1.45	2.80		4.25	6.30
1070	Oak stringer molding		96	.083	"	2.75	2.80		5.55	7.70
1090	Rail bolt, 5/16" x 3-1/2"		48	.167	Ea.	2.75	5.60		8.35	12.40
1100	5/16" x 4-1/2"		48	.167		2.75	5.60		8.35	12.40
1120	Newel post anchor		16	.500		13.50	16.80		30.30	43
1130	Tapered plug, 1/2"		240	.033		.30	1.12		1.42	2.20
1140	1"		240	.033		.40	1.12		1.52	2.31

06 43 16 – Wood Railings

06 43 16.10 Wood Handrails and Railings		Crew	Daily Output	Labor-Hours	Unit	Material	Labor	Equipment	Total	Total Incl O&P
0010	**WOOD HANDRAILS AND RAILINGS**									
0020	Custom design, architectural grade, hardwood, minimum	1 Carp	38	.211	L.F.	9.50	7.05		16.55	22.50
0100	Maximum		30	.267		50	8.95		58.95	70
0300	Stock interior railing with spindles 4" O.C., 4' long		40	.200		37	6.70		43.70	52
0400	8' long		48	.167		37	5.60		42.60	50

06 44 Ornamental Woodwork

06 44 19 – Wood Grilles

06 44 19.10 Grilles		Crew	Daily Output	Labor-Hours	Unit	2011 Bare Costs Material	Labor	Equipment	Total	Total Incl O&P
0010	**GRILLES** and panels, hardwood, sanded									
0020	2′ x 4′ to 4′ x 8′, custom designs, unfinished, minimum	1 Carp	38	.211	S.F.	54	7.05		61.05	71.50
0050	Average		30	.267		62.50	8.95		71.45	84
0100	Maximum		19	.421		71	14.15		85.15	102

06 44 33 – Wood Mantels

06 44 33.10 Fireplace Mantels		Crew	Daily Output	Labor-Hours	Unit	2011 Bare Costs Material	Labor	Equipment	Total	Total Incl O&P
0010	**FIREPLACE MANTELS**									
0015	6″ molding, 6′ x 3′-6″ opening, minimum	1 Carp	5	1.600	Opng.	228	53.50		281.50	340
0100	Maximum		5	1.600		410	53.50		463.50	545
0300	Prefabricated pine, colonial type, stock, deluxe		2	4		1,250	134		1,384	1,625
0400	Economy		3	2.667		530	89.50		619.50	735

06 44 33.20 Fireplace Mantel Beam		Crew	Daily Output	Labor-Hours	Unit	2011 Bare Costs Material	Labor	Equipment	Total	Total Incl O&P
0010	**FIREPLACE MANTEL BEAM**									
0020	Rough texture wood, 4″ x 8″	1 Carp	36	.222	L.F.	8.30	7.45		15.75	21.50
0100	4″ x 10″		35	.229	″	10.90	7.65		18.55	25
0300	Laminated hardwood, 2-1/4″ x 10-1/2″ wide, 6′ long		5	1.600	Ea.	110	53.50		163.50	211
0400	8′ long		5	1.600	″	150	53.50		203.50	255
0600	Brackets for above, rough sawn		12	.667	Pr.	10	22.50		32.50	48.50
0700	Laminated		12	.667	″	15	22.50		37.50	54

06 44 39 – Wood Posts and Columns

06 44 39.10 Decorative Beams		Crew	Daily Output	Labor-Hours	Unit	2011 Bare Costs Material	Labor	Equipment	Total	Total Incl O&P
0010	**DECORATIVE BEAMS**									
0020	Rough sawn cedar, non-load bearing, 4″ x 4″	2 Carp	180	.089	L.F.	1.31	2.98		4.29	6.45
0100	4″ x 6″		170	.094		1.97	3.16		5.13	7.45
0200	4″ x 8″		160	.100		2.63	3.36		5.99	8.50
0300	4″ x 10″		150	.107		3.99	3.58		7.57	10.40
0400	4″ x 12″		140	.114		5.05	3.83		8.88	12
0500	8″ x 8″		130	.123		5.25	4.13		9.38	12.70
0600	Plastic beam, "hewn finish", 6″ x 2″		240	.067		3.10	2.24		5.34	7.15
0601	6″ x 4″		220	.073		3.50	2.44		5.94	7.95

06 44 39.20 Columns		Crew	Daily Output	Labor-Hours	Unit	2011 Bare Costs Material	Labor	Equipment	Total	Total Incl O&P
0010	**COLUMNS**									
0050	Aluminum, round colonial, 6″ diameter	2 Carp	80	.200	V.L.F.	16	6.70		22.70	29
0100	8″ diameter		62.25	.257		19.30	8.60		27.90	35.50
0200	10″ diameter		55	.291		23.50	9.75		33.25	42.50
0250	Fir, stock units, hollow round, 6″ diameter		80	.200		26.50	6.70		33.20	40.50
0300	8″ diameter		80	.200		31	6.70		37.70	46
0350	10″ diameter		70	.229		39.50	7.65		47.15	56.50
0400	Solid turned, to 8′ high, 3-1/2″ diameter		80	.200		9	6.70		15.70	21
0500	4-1/2″ diameter		75	.213		14	7.15		21.15	27.50
0600	5-1/2″ diameter		70	.229		18.40	7.65		26.05	33
0800	Square columns, built-up, 5″ x 5″		65	.246		15.50	8.25		23.75	31
0900	Solid, 3-1/2″ x 3-1/2″		130	.123		9	4.13		13.13	16.80
1600	Hemlock, tapered, T & G, 12″ diam., 10′ high		100	.160		53	5.35		58.35	67
1700	16′ high		65	.246		79.50	8.25		87.75	101
1900	10′ high, 14″ diameter		100	.160		121	5.35		126.35	142
2000	18′ high		65	.246		107	8.25		115.25	132
2200	18″ diameter, 12′ high		65	.246		153	8.25		161.25	183
2300	20′ high		50	.320		148	10.75		158.75	181
2500	20″ diameter, 14′ high		40	.400		159	13.40		172.40	198
2600	20′ high		35	.457		149	15.35		164.35	189

06 44 Ornamental Woodwork

06 44 39 – Wood Posts and Columns

06 44 39.20 Columns		Crew	Daily Output	Labor-Hours	Unit	Material	2011 Bare Costs Labor	Equipment	Total	Total Incl O&P
2800	For flat pilasters, deduct				V.L.F.	33%				
3000	For splitting into halves, add				Ea.	116			116	128
4000	Rough sawn cedar posts, 4" x 4"	2 Carp	250	.064	V.L.F.	2.98	2.15		5.13	6.85
4100	4" x 6"		235	.068		5.70	2.28		7.98	10.05
4200	6" x 6"		220	.073		10.30	2.44		12.74	15.45
4300	8" x 8"		200	.080		19.30	2.68		21.98	26

06 48 Wood Frames

06 48 13 – Exterior Wood Door Frames

06 48 13.10 Exterior Wood Door Frames and Accessories		Crew	Daily Output	Labor-Hours	Unit	Material	Labor	Equipment	Total	Total Incl O&P
0010	**EXTERIOR WOOD DOOR FRAMES AND ACCESSORIES**									
0400	Exterior frame, incl. ext. trim, pine, 5/4 x 4-9/16" deep	2 Carp	375	.043	L.F.	5.55	1.43		6.98	8.50
0420	5-3/16" deep		375	.043		11.35	1.43		12.78	14.90
0440	6-9/16" deep		375	.043		8.50	1.43		9.93	11.75
0600	Oak, 5/4 x 4-9/16" deep		350	.046		10.70	1.53		12.23	14.35
0620	5-3/16" deep		350	.046		11.80	1.53		13.33	15.50
0640	6-9/16" deep		350	.046		14.50	1.53		16.03	18.50
1000	Sills, 8/4 x 8" deep, oak, no horns		100	.160		17	5.35		22.35	27.50
1020	2" horns		100	.160		17.10	5.35		22.45	28
1040	3" horns		100	.160		17.10	5.35		22.45	28
1100	8/4 x 10" deep, oak, no horns		90	.178		20.50	5.95		26.45	32.50
1120	2" horns		90	.178		21	5.95		26.95	33.50
1140	3" horns		90	.178		21	5.95		26.95	33.50
2000	Exterior, colonial, frame & trim, 3' opng., in-swing, minimum		22	.727	Ea.	315	24.50		339.50	390
2010	Average		21	.762		470	25.50		495.50	560
2020	Maximum		20	.800		1,075	27		1,102	1,225
2100	5'-4" opening, in-swing, minimum		17	.941		360	31.50		391.50	450
2120	Maximum		15	1.067		1,075	36		1,111	1,225
2140	Out-swing, minimum		17	.941		370	31.50		401.50	460
2160	Maximum		15	1.067		1,100	36		1,136	1,275
2400	6'-0" opening, in-swing, minimum		16	1		345	33.50		378.50	435
2420	Maximum		10	1.600		1,100	53.50		1,153.50	1,325
2460	Out-swing, minimum		16	1		375	33.50		408.50	470
2480	Maximum		10	1.600		1,300	53.50		1,353.50	1,525
2600	For two sidelights, add, minimum		30	.533	Opng.	355	17.90		372.90	420
2620	Maximum		20	.800	"	1,125	27		1,152	1,300
2700	Custom birch frame, 3'-0" opening		16	1	Ea.	205	33.50		238.50	282
2750	6'-0" opening		16	1		305	33.50		338.50	390
2900	Exterior, modern, plain trim, 3' opng., in-swing, minimum		26	.615		35.50	20.50		56	73.50
2920	Average		24	.667		42	22.50		64.50	83.50
2940	Maximum		22	.727		50.50	24.50		75	96.50

06 48 16 – Interior Wood Door Frames

06 48 16.10 Interior Wood Door Jamb and Frames		Crew	Daily Output	Labor-Hours	Unit	Material	Labor	Equipment	Total	Total Incl O&P
0010	**INTERIOR WOOD DOOR JAMB AND FRAMES**									
3000	Interior frame, pine, 11/16" x 3-5/8" deep	2 Carp	375	.043	L.F.	5.55	1.43		6.98	8.50
3020	4-9/16" deep		375	.043		6.50	1.43		7.93	9.55
3200	Oak, 11/16" x 3-5/8" deep		350	.046		4.35	1.53		5.88	7.35
3220	4-9/16" deep		350	.046		4.50	1.53		6.03	7.50
3240	5-3/16" deep		350	.046		4.58	1.53		6.11	7.60
3400	Walnut, 11/16" x 3-5/8" deep		350	.046		7.30	1.53		8.83	10.60

06 48 Wood Frames

06 48 16 – Interior Wood Door Frames

06 48 16.10 Interior Wood Door Jamb and Frames		Crew	Daily Output	Labor-Hours	Unit	2011 Bare Costs Material	Labor	Equipment	Total	Total Incl O&P
3420	4-9/16" deep	2 Carp	350	.046	L.F.	7.50	1.53		9.03	10.80
3440	5-3/16" deep		350	.046		7.90	1.53		9.43	11.20
3800	Threshold, oak, 5/8" x 3-5/8" deep		200	.080		3.83	2.68		6.51	8.70
3820	4-5/8" deep		190	.084		4.58	2.83		7.41	9.80
3840	5-5/8" deep		180	.089		7.15	2.98		10.13	12.85

06 49 Wood Screens and Exterior Wood Shutters

06 49 19 – Exterior Wood Shutters

06 49 19.10 Shutters, Exterior		Crew	Daily Output	Labor-Hours	Unit	2011 Bare Costs Material	Labor	Equipment	Total	Total Incl O&P
0010	**SHUTTERS, EXTERIOR**									
0012	Aluminum, louvered, 1'-4" wide, 3'-0" long	1 Carp	10	.800	Pr.	61	27		88	113
0200	4'-0" long		10	.800		73	27		100	126
0300	5'-4" long		10	.800		96.50	27		123.50	151
0400	6'-8" long		9	.889		123	30		153	185
1000	Pine, louvered, primed, each 1'-2" wide, 3'-3" long		10	.800		106	27		133	162
1100	4'-7" long		10	.800		144	27		171	203
1250	Each 1'-4" wide, 3'-0" long		10	.800		108	27		135	164
1350	5'-3" long		10	.800		163	27		190	224
1500	Each 1'-6" wide, 3'-3" long		10	.800		113	27		140	169
1600	4'-7" long		10	.800		158	27		185	219
1620	Hemlock, louvered, 1'-2" wide, 5'-7" long		10	.800		174	27		201	236
1630	Each 1'-4" wide, 2'-2" long		10	.800		108	27		135	164
1670	4'-3" long		10	.800		130	27		157	188
1680	5'-3" long		10	.800		162	27		189	223
1690	5'-11" long		10	.800		183	27		210	246
1700	Door blinds, 6'-9" long, each 1'-3" wide		9	.889		184	30		214	252
1710	1'-6" wide		9	.889		198	30		228	268
2500	Polystyrene, solid raised panel, each 1'-4" wide, 3'-3" long		10	.800		41	27		68	90
2600	3'-11" long		10	.800		47	27		74	97
2700	4'-7" long		10	.800		51.50	27		78.50	102
2800	5'-3" long		10	.800		55.50	27		82.50	107
2900	6'-8" long		9	.889		66.50	30		96.50	123
4500	Polystyrene, louvered, each 1'-2" wide, 3'-3" long		10	.800		28	27		55	75.50
4600	4'-7" long		10	.800		34.50	27		61.50	83
4750	5'-3" long		10	.800		39.50	27		66.50	88.50
4850	6'-8" long		9	.889		47	30		77	102
6000	Vinyl, louvered, each 1'-2" x 4'-7" long		10	.800		34.50	27		61.50	82.50
6200	Each 1'-4" x 6'-8" long		9	.889		48	30		78	103
8000	PVC exterior rolling shutters									
8100	including crank control	1 Carp	8	1	Ea.	480	33.50		513.50	585
8500	Insulative - 6' x 6'8" stock unit	"	8	1	"	685	33.50		718.50	810

06 63 Plastic Railings

06 63 10 – Plastic (PVC) Railings

06 63 10.10 Plastic Railings		Crew	Daily Output	Labor-Hours	Unit	2011 Bare Costs Material	Labor	Equipment	Total	Total Incl O&P
0010	**PLASTIC RAILINGS**									
0100	Horizontal PVC handrail with balusters, 3-1/2" wide, 36" high	1 Carp	96	.083	L.F.	24	2.80		26.80	31
0150	42" high		96	.083		27	2.80		29.80	34
0200	Angled PVC handrail with balusters, 3-1/2" wide, 36" high		72	.111		28	3.73		31.73	37
0250	42" high		72	.111		30.50	3.73		34.23	40
0300	Post sleeve for 4 x 4 post		96	.083		12.55	2.80		15.35	18.50
0400	Post cap for 4 x 4 post, flat profile		48	.167	Ea.	11.40	5.60		17	22
0450	Newel post style profile		48	.167		21	5.60		26.60	32.50
0500	Raised corbeled profile		48	.167		23	5.60		28.60	34.50
0550	Post base trim for 4 x 4 post		96	.083		12.80	2.80		15.60	18.80

06 65 Plastic Simulated Wood Trim

06 65 10 – PVC Trim

06 65 10.10 PVC Trim, Exterior		Crew	Daily Output	Labor-Hours	Unit	Material	Labor	Equipment	Total	Total Incl O&P
0010	**PVC TRIM, EXTERIOR**									
0100	Cornerboards, 5/4" x 6" x 6"	1 Carp	240	.033	L.F.	7.30	1.12		8.42	9.90
0110	Door/window casing, 1" x 4"		200	.040		1.42	1.34		2.76	3.81
0120	1" x 6"		200	.040		2.24	1.34		3.58	4.71
0130	1" x 8"		195	.041		2.96	1.38		4.34	5.55
0140	1" x 10"		195	.041		3.74	1.38		5.12	6.40
0150	1" x 12"		190	.042		4.57	1.41		5.98	7.40
0160	5/4" x 4"		195	.041		2.05	1.38		3.43	4.56
0170	5/4" x 6"		195	.041		3.24	1.38		4.62	5.85
0180	5/4" x 8"		190	.042		4.24	1.41		5.65	7
0190	5/4" x 10"		190	.042		5.40	1.41		6.81	8.30
0200	5/4" x 12"		185	.043		6.60	1.45		8.05	9.70
0210	Fascia, 1" x 4"		250	.032		1.42	1.07		2.49	3.36
0220	1" x 6"		250	.032		2.24	1.07		3.31	4.26
0230	1" x 8"		225	.036		2.96	1.19		4.15	5.25
0240	1" x 10"		225	.036		3.74	1.19		4.93	6.10
0250	1" x 12"		200	.040		4.57	1.34		5.91	7.30
0260	5/4" x 4"		240	.033		2.05	1.12		3.17	4.13
0270	5/4" x 6"		240	.033		3.24	1.12		4.36	5.45
0280	5/4" x 8"		215	.037		4.24	1.25		5.49	6.75
0290	5/4" x 10"		215	.037		5.40	1.25		6.65	8.05
0300	5/4" x 12"		190	.042		6.60	1.41		8.01	9.60
0310	Frieze, 1" x 4"		250	.032		1.42	1.07		2.49	3.36
0320	1" x 6"		250	.032		2.24	1.07		3.31	4.26
0330	1" x 8"		225	.036		2.96	1.19		4.15	5.25
0340	1" x 10"		225	.036		3.74	1.19		4.93	6.10
0350	1" x 12"		200	.040		4.57	1.34		5.91	7.30
0360	5/4" x 4"		240	.033		2.05	1.12		3.17	4.13
0370	5/4" x 6"		240	.033		3.24	1.12		4.36	5.45
0380	5/4" x 8"		215	.037		4.24	1.25		5.49	6.75
0390	5/4" x 10"		215	.037		5.40	1.25		6.65	8.05
0400	5/4" x 12"		190	.042		6.60	1.41		8.01	9.60
0410	Rake, 1" x 4"		200	.040		1.42	1.34		2.76	3.81
0420	1" x 6"		200	.040		2.24	1.34		3.58	4.71
0430	1" x 8"		190	.042		2.96	1.41		4.37	5.60
0440	1" x 10"		190	.042		3.74	1.41		5.15	6.45
0450	1" x 12"		180	.044		4.57	1.49		6.06	7.55

06 65 Plastic Simulated Wood Trim

06 65 10 – PVC Trim

06 65 10.10 PVC Trim, Exterior		Crew	Daily Output	Labor-Hours	Unit	2011 Bare Costs Material	Labor	Equipment	Total	Total Incl O&P
0460	5/4" x 4"	1 Carp	195	.041	L.F.	2.05	1.38		3.43	4.56
0470	5/4" x 6"		195	.041		3.24	1.38		4.62	5.85
0480	5/4" x 8"		185	.043		4.24	1.45		5.69	7.10
0490	5/4" x 10"		185	.043		5.40	1.45		6.85	8.40
0500	5/4" x 12"		175	.046		6.60	1.53		8.13	9.80
0510	Rake trim, 1" x 4"		225	.036		1.42	1.19		2.61	3.56
0520	1" x 6"		225	.036		2.24	1.19		3.43	4.46
0560	5/4" x 4"		220	.036		2.05	1.22		3.27	4.30
0570	5/4" x 6"		220	.036		3.24	1.22		4.46	5.60
0610	Soffit, 1" x 4"	2 Carp	420	.038		1.42	1.28		2.70	3.70
0620	1" x 6"		420	.038		2.24	1.28		3.52	4.60
0630	1" x 8"		420	.038		2.96	1.28		4.24	5.40
0640	1" x 10"		400	.040		3.74	1.34		5.08	6.35
0650	1" x 12"		400	.040		4.57	1.34		5.91	7.30
0660	5/4" x 4"		410	.039		2.05	1.31		3.36	4.45
0670	5/4" x 6"		410	.039		3.24	1.31		4.55	5.75
0680	5/4" x 8"		410	.039		4.24	1.31		5.55	6.85
0690	5/4" x 10"		390	.041		5.40	1.38		6.78	8.25
0700	5/4" x 12"		390	.041		6.60	1.38		7.98	9.55

Estimating Tips

07 10 00 Dampproofing and Waterproofing

- Be sure of the job specifications before pricing this subdivision. The difference in cost between waterproofing and dampproofing can be great. Waterproofing will hold back standing water. Dampproofing prevents the transmission of water vapor. Also included in this section are vapor retarding membranes.

07 20 00 Thermal Protection

- Insulation and fireproofing products are measured by area, thickness, volume or R-value. Specifications may give only what the specific R-value should be in a certain situation. The estimator may need to choose the type of insulation to meet that R-value.

07 30 00 Steep Slope Roofing
07 40 00 Roofing and Siding Panels

- Many roofing and siding products are bought and sold by the square. One square is equal to an area that measures 100 square feet.

This simple change in unit of measure could create a large error if the estimator is not observant. Accessories necessary for a complete installation must be figured into any calculations for both material and labor.

07 50 00 Membrane Roofing
07 60 00 Flashing and Sheet Metal
07 70 00 Roofing and Wall Specialties and Accessories

- The items in these subdivisions compose a roofing system. No one component completes the installation, and all must be estimated. Built-up or single-ply membrane roofing systems are made up of many products and installation trades. Wood blocking at roof perimeters or penetrations, parapet coverings, reglets, roof drains, gutters, downspouts, sheet metal flashing, skylights, smoke vents, and roof hatches all need to be considered along with the roofing material. Several different installation trades will need to work together on the roofing system. Inherent difficulties in the scheduling and coordination of various trades must be accounted for when estimating labor costs.

07 90 00 Joint Protection

- To complete the weather-tight shell, the sealants and caulkings must be estimated. Where different materials meet—at expansion joints, at flashing penetrations, and at hundreds of other locations throughout a construction project—they provide another line of defense against water penetration. Often, an entire system is based on the proper location and placement of caulking or sealants. The detailed drawings that are included as part of a set of architectural plans show typical locations for these materials. When caulking or sealants are shown at typical locations, this means the estimator must include them for all the locations where this detail is applicable. Be careful to keep different types of sealants separate, and remember to consider backer rods and primers if necessary.

Reference Numbers

Reference numbers are shown in shaded boxes at the beginning of some major classifications. These numbers refer to related items in the Reference Section. The reference information may be an estimating procedure, an alternate pricing method, or technical information.

Note: Not all subdivisions listed here necessarily appear in this publication.

07 01 Operation and Maint. of Thermal and Moisture Protection

07 01 50 – Maintenance of Membrane Roofing

07 01 50.10 Roof Coatings		Crew	Daily Output	Labor-Hours	Unit	2011 Bare Costs Material	Labor	Equipment	Total	Total Incl O&P
0010	**ROOF COATINGS**									
0012	Asphalt, brush grade, material only				Gal.	7.90			7.90	8.65
0200	Asphalt base, fibered aluminum coating G				"	10.75			10.75	11.85
0210	Asphalt aluminum coating G				S.F.	.33			.33	.36
0300	Asphalt primer, 5 gallon				Gal.	6.40			6.40	7.05
0310	Primer, asphalt				S.F.	.06			.06	.07
0600	Coal tar pitch, 200 lb. barrels				Ton	1,100			1,100	1,200
0610	Coal tar pitch, 200# barrels				S.F.	1.07			1.07	1.18
0700	Tar roof cement, 5 gal. lots				Gal.	14			14	15.40
0710	Tar roof cement				S.F.	1.17			1.17	1.28
0800	Glass fibered roof & patching cement, 5 gallon				Gal.	7.40			7.40	8.15
0810	Roof & patching cement, glass fiber				S.F.	.62			.62	.68
0900	Reinforcing glass membrane, 450 S.F./roll				Ea.	111			111	122
1000	Neoprene roof coating, 5 gal., 2 gal./sq.				Gal.	27.50			27.50	30.50
1010	Neoprene roof coating				S.F.	1.38			1.38	1.52
1100	Roof patch & flashing cement, 5 gallon				Gal.	8.90			8.90	9.75
1110	Roof patch & flashing cement				S.F.	.74			.74	.81
1200	Roof resaturant, glass fibered, 3 gal./sq.				Gal.	5.35			5.35	5.85
1600	Reflective roof coating, white, elastomeric, approx. 50 S.F. per gal. G				"	16.20			16.20	17.85

07 01 90 – Maintenance of Joint Protection

07 01 90.81 Joint Sealant Replacement		Crew	Daily Output	Labor-Hours	Unit	Material	Labor	Equipment	Total	Total Incl O&P
0010	**JOINT SEALANT REPLACEMENT**									
0050	Control joints in concrete floors/slabs									
0100	Option 1 for joints with hard dry sealant									
0110	Step 1: Sawcut to remove 95% of old sealant									
0112	1/4" wide x 1/2" deep, with single saw blade	C-27	4800	.003	L.F.	.02	.11	.03	.16	.22
0114	3/8" wide x 3/4" deep, with single saw blade		4000	.004		.04	.13	.03	.20	.29
0116	1/2" wide x 1" deep, with double saw blades		3600	.004		.08	.14	.04	.26	.36
0118	3/4" wide x 1-1/2" deep, with double saw blades		3200	.005		.17	.16	.04	.37	.48
0120	Step 2: Water blast joint faces and edges	C-29	2500	.003			.08	.02	.10	.17
0130	Step 3: Air blast joint faces and edges	C-28	2000	.004			.13	.01	.14	.21
0140	Step 4: Sand blast joint faces and edges	E-11	2000	.016			.48	.10	.58	.98
0150	Step 5: Air blast joint faces and edges	C-28	2000	.004			.13	.01	.14	.21
0200	Option 2 for joints with soft pliable sealant									
0210	Step 1: Plow joint with rectangular blade	B-62	2600	.009	L.F.		.27	.06	.33	.50
0220	Step 2: Sawcut to re-face joint faces									
0222	1/4" wide x 1/2" deep, with single saw blade	C-27	2400	.007	L.F.	.03	.21	.06	.30	.43
0224	3/8" wide x 3/4" deep, with single saw blade		2000	.008		.06	.25	.07	.38	.54
0226	1/2" wide x 1" deep, with double saw blades		1800	.009		.11	.28	.07	.46	.65
0228	3/4" wide x 1-1/2" deep, with double saw blades		1600	.010		.22	.32	.08	.62	.84
0230	Step 3: Water blast joint faces and edges	C-29	2500	.003			.08	.02	.10	.17
0240	Step 4: Air blast joint faces and edges	C-28	2000	.004			.13	.01	.14	.21
0250	Step 5: Sand blast joint faces and edges	E-11	2000	.016			.48	.10	.58	.98
0260	Step 6: Air blast joint faces and edges	C-28	2000	.004			.13	.01	.14	.21
0290	For saw cutting new control joints, see Section 03 35 29.35									
8910	For backer rod, see Section 07 91 23.10									
8920	For joint sealant, see Section 07 92 13.20, if available									

07 05 Common Work Results for Thermal and Moisture Protection

07 05 05 – Selective Demolition

07 05 05.10 Selective Demo., Thermal and Moist. Protection		Crew	Daily Output	Labor-Hours	Unit	Material	2011 Bare Costs Labor	Equipment	Total	Total Ind O&P
0010	**SELECTIVE DEMO., THERMAL AND MOISTURE PROTECTION**									
0020	Caulking/sealant, to 1" x 1" joint R024119-10	1 Clab	600	.013	L.F.		.35		.35	.59
0120	Downspouts, including hangers		350	.023	"		.60		.60	1.01
0220	Flashing, sheet metal		290	.028	S.F.		.73		.73	1.22
0420	Gutters, aluminum or wood, edge hung		240	.033	L.F.		.88		.88	1.48
0520	Built-in		100	.080	"		2.12		2.12	3.54
0620	Insulation, air/vapor barrier		3500	.002	S.F.		.06		.06	.10
0670	Batts or blankets		1400	.006	C.F.		.15		.15	.25
0720	Foamed or sprayed in place	2 Clab	1000	.016	B.F.		.42		.42	.71
0770	Loose fitting	1 Clab	3000	.003	C.F.		.07		.07	.12
0870	Rigid board		3450	.002	B.F.		.06		.06	.10
1120	Roll roofing, cold adhesive		12	.667	Sq.		17.65		17.65	29.50
1170	Roof accessories, adjustable metal chimney flashing		9	.889	Ea.		23.50		23.50	39.50
1325	Plumbing vent flashing		32	.250	"		6.60		6.60	11.10
1375	Ridge vent strip, aluminum		310	.026	L.F.		.68		.68	1.14
1620	Skylight to 10 S.F.		8	1	Ea.		26.50		26.50	44.50
2120	Roof edge, aluminum soffit and fascia		570	.014	L.F.		.37		.37	.62
2170	Concrete coping, up to 12" wide	2 Clab	160	.100			2.65		2.65	4.43
2220	Drip edge	1 Clab	1650	.005			.13		.13	.21
2270	Gravel stop		1650	.005			.13		.13	.21
2370	Sheet metal coping, up to 12" wide		240	.033			.88		.88	1.48
2470	Roof insulation board, over 2" thick	B-2	7800	.005	B.F.		.14		.14	.23
2520	Up to 2" thick	"	3900	.010	S.F.		.28		.28	.46
2620	Roof ventilation, louvered gable vent	1 Clab	16	.500	Ea.		13.25		13.25	22
2670	Remove, roof hatch	G-3	15	2.133			69		69	115
2675	Rafter vents	1 Clab	960	.008			.22		.22	.37
2720	Soffit vent and/or fascia vent		575	.014	L.F.		.37		.37	.62
2775	Soffit vent strip, aluminum, 3" to 4" wide		160	.050			1.32		1.32	2.22
2820	Roofing accessories, shingle moulding, to 1" x 4"		1600	.005			.13		.13	.22
2870	Cant strip	B-2	2000	.020			.54		.54	.90
2920	Concrete block walkway	1 Clab	230	.035			.92		.92	1.54
3070	Roofing, felt paper, 15#		70	.114	Sq.		3.02		3.02	5.05
3125	#30 felt		30	.267	"		7.05		7.05	11.80
3170	Asphalt shingles, 1 layer	B-2	3500	.011	S.F.		.31		.31	.51
3180	2 layers		1750	.023	"		.61		.61	1.03
3370	Modified bitumen		26	1.538	Sq.		41.50		41.50	69
3420	Built-up, no gravel, 3 ply		25	1.600			43		43	72
3470	4 ply		21	1.905			51		51	85.50
3620	5 ply		1600	.025	S.F.		.67		.67	1.12
3725	Gravel removal, minimum		5000	.008			.21		.21	.36
3730	Maximum		2000	.020			.54		.54	.90
3870	Fiberglass sheet		1200	.033			.89		.89	1.50
4120	Slate shingles		1900	.021			.57		.57	.95
4170	Ridge shingles, clay or slate		2000	.020	L.F.		.54		.54	.90
4320	Single ply membrane, attached at seams		52	.769	Sq.		20.50		20.50	34.50
4370	Ballasted		75	.533			14.30		14.30	24
4420	Fully adhered		39	1.026			27.50		27.50	46
4550	Roof hatch, 2'-6" x 3'-0"	1 Clab	10	.800	Ea.		21		21	35.50
4670	Wood shingles	B-2	2200	.018	S.F.		.49		.49	.82
4820	Sheet metal roofing	"	2150	.019			.50		.50	.84
4970	Siding, horizontal wood clapboards	1 Clab	380	.021			.56		.56	.93
5025	Exterior insulation finish system	"	120	.067			1.76		1.76	2.95
5070	Tempered hardboard, remove and reset	1 Carp	380	.021			.71		.71	1.18

07 05 Common Work Results for Thermal and Moisture Protection

07 05 05 – Selective Demolition

07 05 05.10 Selective Demo., Thermal and Moist. Protection		Crew	Daily Output	Labor-Hours	Unit	2011 Bare Costs Material	Labor	Equipment	Total	Total Incl O&P
5120	Tempered hardboard sheet siding	1 Carp	375	.021	S.F.		.72		.72	1.20
5170	Metal, corner strips	1 Clab	850	.009	L.F.		.25		.25	.42
5225	Horizontal strips		444	.018	S.F.		.48		.48	.80
5320	Vertical strips		400	.020			.53		.53	.89
5520	Wood shingles		350	.023			.60		.60	1.01
5620	Stucco siding		360	.022			.59		.59	.98
5670	Textured plywood		725	.011			.29		.29	.49
5720	Vinyl siding		510	.016			.42		.42	.70
5770	Corner strips		900	.009	L.F.		.24		.24	.39
5870	Wood, boards, vertical		400	.020	S.F.		.53		.53	.89
5920	Waterproofing, protection/drain board	2 Clab	3900	.004	B.F.		.11		.11	.18
5970	Over 1/2" thick		1750	.009	S.F.		.24		.24	.40
6020	To 1/2" thick		2000	.008	"		.21		.21	.35

07 11 Dampproofing

07 11 13 – Bituminous Dampproofing

07 11 13.10 Bituminous Asphalt Coating

Code	Description	Crew	Daily Output	Labor-Hours	Unit	Material	Labor	Equipment	Total	Total Incl O&P
0010	**BITUMINOUS ASPHALT COATING**									
0030	Brushed on, below grade, 1 coat	1 Rofc	665	.012	S.F.	.20	.34		.54	.83
0100	2 coat		500	.016		.39	.45		.84	1.24
0300	Sprayed on, below grade, 1 coat, 25.6 S.F./gal.		830	.010		.20	.27		.47	.71
0400	2 coat, 20.5 S.F./gal.		500	.016		.38	.45		.83	1.23
0600	Troweled on, asphalt with fibers, 1/16" thick		500	.016		.47	.45		.92	1.32
0700	1/8" thick		400	.020		.83	.56		1.39	1.93
1000	1/2" thick		350	.023		2.69	.64		3.33	4.12

07 11 16 – Cementitious Dampproofing

07 11 16.20 Cementitious Parging

Code	Description	Crew	Daily Output	Labor-Hours	Unit	Material	Labor	Equipment	Total	Total Incl O&P
0010	**CEMENTITIOUS PARGING**									
0020	Portland cement, 2 coats, 1/2" thick	D-1	250	.064	S.F.	.29	1.95		2.24	3.51
0100	Waterproofed Portland cement, 1/2" thick, 2 coats	"	250	.064	"	3.01	1.95		4.96	6.50

07 19 Water Repellents

07 19 19 – Silicone Water Repellents

07 19 19.10 Silicone Based Water Repellents

Code	Description	Crew	Daily Output	Labor-Hours	Unit	Material	Labor	Equipment	Total	Total Incl O&P
0010	**SILICONE BASED WATER REPELLENTS**									
0020	Water base liquid, roller applied	2 Rofc	7000	.002	S.F.	.86	.06		.92	1.07
0200	Silicone or stearate, sprayed on CMU, 1 coat	1 Rofc	4000	.002		.36	.06		.42	.50
0300	2 coats	"	3000	.003		.73	.08		.81	.94

07 21 Thermal Insulation

07 21 13 – Board Insulation

07 21 13.10 Rigid Insulation		Crew	Daily Output	Labor-Hours	Unit	Material	2011 Bare Costs Labor	Equipment	Total	Total Incl O&P
0010	RIGID INSULATION, for walls									
0020	Fiberboard, 3/4" thick, R2.08 G	1 Carp	1100	.007	S.F.	.36	.24		.60	.81
0025	1" thick, R2.78 G		800	.010		.47	.34		.81	1.08
0030	2" thick, R5.26 G		730	.011		.94	.37		1.31	1.65
0040	Fiberglass, 1.5#/C.F., unfaced, 1" thick, R4.1 G		1000	.008		.40	.27		.67	.89
0060	1-1/2" thick, R6.2 G		1000	.008		.58	.27		.85	1.09
0080	2" thick, R8.3 G		1000	.008		.64	.27		.91	1.15
0120	3" thick, R12.4 G		800	.010		.76	.34		1.10	1.40
0370	3#/C.F., unfaced, 1" thick, R4.3 G		1000	.008		.46	.27		.73	.96
0390	1-1/2" thick, R6.5 G		1000	.008		.86	.27		1.13	1.40
0400	2" thick, R8.7 G		890	.009		.91	.30		1.21	1.50
0420	2-1/2" thick, R10.9 G		800	.010		.96	.34		1.30	1.62
0440	3" thick, R13 G		800	.010		1.39	.34		1.73	2.09
0520	Foil faced, 1" thick, R4.3 G		1000	.008		.83	.27		1.10	1.36
0540	1-1/2" thick, R6.5 G		1000	.008		1.24	.27		1.51	1.81
0560	2" thick, R8.7 G		890	.009		1.55	.30		1.85	2.21
0580	2-1/2" thick, R10.9 G		800	.010		1.82	.34		2.16	2.56
0600	3" thick, R13 G	↓	800	.010	↓	2	.34		2.34	2.76
1600	Isocyanurate, 4' x 8' sheet, foil faced, both sides									
1610	1/2" thick G	1 Carp	800	.010	S.F.	.29	.34		.63	.88
1620	5/8" thick G		800	.010		.32	.34		.66	.91
1630	3/4" thick G		800	.010		.33	.34		.67	.92
1640	1" thick G		800	.010		.51	.34		.85	1.12
1650	1-1/2" thick G		730	.011		.64	.37		1.01	1.32
1660	2" thick G		730	.011		.79	.37		1.16	1.49
1670	3" thick G		730	.011		1.78	.37		2.15	2.58
1680	4" thick G		730	.011		2	.37		2.37	2.82
1700	Perlite, 1" thick, R2.77 G		800	.010		.30	.34		.64	.89
1750	2" thick, R5.55 G		730	.011		.60	.37		.97	1.28
1900	Extruded polystyrene, 25 PSI compressive strength, 1" thick, R5 G		800	.010		.50	.34		.84	1.11
1940	2" thick R10 G		730	.011		1	.37		1.37	1.72
1960	3" thick, R15 G		730	.011		1.44	.37		1.81	2.20
2100	Expanded polystyrene, 1" thick, R3.85 G		800	.010		.29	.34		.63	.88
2120	2" thick, R7.69 G		730	.011		.58	.37		.95	1.26
2140	3" thick, R11.49 G		730	.011		.87	.37		1.24	1.58
2200	Sound board, 1/2" thick G		625	.013		.35	.43		.78	1.11
2360	Fiberboard, low density, 1/2" thick, R1.39 G		750	.011		.26	.36		.62	.89
2400	Wood fiber, 1" thick, R3.85 G		1000	.008		.35	.27		.62	.84
2410	2" thick, R7.7 G		1000	.008		.70	.27		.97	1.22
2680	Mineral fiberboard, rigid, 1" thick, R4.2 G		800	.010		.34	.34		.68	.93
2700	2" thick, R8.4 G	↓	730	.011	↓	.66	.37		1.03	1.35
07 21 13.13 Foam Board Insulation										
0010	FOAM BOARD INSULATION									
0600	Polystyrene, expanded, 1" thick, R4 G	1 Carp	680	.012	S.F.	.29	.39		.68	.98
0700	2" thick, R8 G	"	675	.012	"	.58	.40		.98	1.31

07 21 16 – Blanket Insulation

07 21 16.10 Blanket Insulation for Floors/Ceilings		Crew	Daily Output	Labor-Hours	Unit	Material	Labor	Equipment	Total	Total Incl O&P
0010	BLANKET INSULATION FOR FLOORS/CEILINGS									
0020	Including spring type wire fasteners									
2000	Fiberglass, blankets or batts, paper or foil backing									
2100	3-1/2" thick, R13 G	1 Carp	700	.011	S.F.	.32	.38		.70	.99
2150	6-1/4" thick, R19 G	↓	600	.013	↓	.42	.45		.87	1.21

07 21 Thermal Insulation

07 21 16 – Blanket Insulation

07 21 16.10 Blanket Insulation for Floors/Ceilings			Crew	Daily Output	Labor-Hours	Unit	2011 Bare Costs Material	Labor	Equipment	Total	Total Incl O&P
2210	9-1/2" thick, R30	G	1 Carp	500	.016	S.F.	.70	.54		1.24	1.67
2220	12" thick, R38	G		475	.017		.85	.57		1.42	1.89
3000	Unfaced, 3-1/2" thick, R13	G		600	.013		.28	.45		.73	1.06
3010	6-1/4" thick, R19	G		500	.016		.44	.54		.98	1.38
3020	9-1/2" thick, R30	G		450	.018		.70	.60		1.30	1.77
3030	12" thick, R38	G		425	.019		.73	.63		1.36	1.86

07 21 16.20 Blanket Insulation for Walls			Crew	Daily Output	Labor-Hours	Unit	Material	Labor	Equipment	Total	Total Incl O&P
0010	**BLANKET INSULATION FOR WALLS**										
0020	Kraft faced fiberglass, 3-1/2" thick, R11, 15" wide	G	1 Carp	1350	.006	S.F.	.28	.20		.48	.64
0030	23" wide	G		1600	.005		.28	.17		.45	.59
0060	R13, 11" wide	G		1150	.007		.27	.23		.50	.69
0110	R-15, 11" wide	G		1150	.007		.41	.23		.64	.84
0140	6" thick, R19, 11" wide	G		1150	.007		.37	.23		.60	.80
0182	R21, 11" wide	G		1150	.007		.51	.23		.74	.95
0184	15" wide	G		1350	.006		.51	.20		.71	.89
0186	23" wide	G		1600	.005		.51	.17		.68	.84
0188	9" thick, R-30, 11" wide	G		985	.008		.71	.27		.98	1.24
0200	15" wide	G		1150	.007		.71	.23		.94	1.17
0230	12" thick, R38, 11" wide	G		985	.008		.95	.27		1.22	1.51
0240	15" wide	G		1150	.007		.95	.23		1.18	1.44
0410	Foil faced fiberglass, 3-1/2" thick, R13, 11" wide	G		1150	.007		.54	.23		.77	.98
0420	15" wide	G		1350	.006		.54	.20		.74	.92
0442	R15, 11" wide	G		1150	.007		.56	.23		.79	1.01
0444	15" wide	G		1350	.006		.56	.20		.76	.95
0448	6" thick, R19, 11" wide	G		1150	.007		.77	.23		1	1.24
0460	15" wide	G		1350	.006		.77	.20		.97	1.18
0482	R-21, 11" wide	G		1150	.007		.82	.23		1.05	1.29
0484	15" wide	G		1350	.006		.82	.20		1.02	1.23
0488	9" thick, R-30, 11" wide	G		985	.008		.89	.27		1.16	1.44
0500	9" thick, R30, 15" wide	G		1150	.007		.89	.23		1.12	1.37
0560	12" thick, R-38, 11" wide	G		985	.008		.90	.27		1.17	1.45
0570	15" wide	G		1150	.007		.90	.23		1.13	1.38
0620	Unfaced fiberglass, 3-1/2" thick, R-13, 11" wide	G		1150	.007		.28	.23		.51	.70
0820	15" wide	G		1350	.006		.28	.20		.48	.64
0832	R15, 11" wide	G		1150	.007		.40	.23		.63	.83
0838	6" thick, R19, 11" wide	G		1150	.007		.44	.23		.67	.87
0860	15" wide	G		1150	.007		.44	.23		.67	.87
0882	R-21, 11" wide	G		1150	.007		.46	.23		.69	.90
0886	15" wide	G		1350	.006		.46	.20		.66	.84
0890	9" thick, R30, 11" wide	G		985	.008		.70	.27		.97	1.23
0900	15" wide	G		1150	.007		.70	.23		.93	1.16
0930	12" thick, R38, 11" wide	G		985	.008		.73	.27		1	1.26
0940	15" wide	G		1000	.008		.73	.27		1	1.25
1300	Mineral fiber batts, kraft faced										
1320	3-1/2" thick, R12	G	1 Carp	1600	.005	S.F.	.38	.17		.55	.70
1340	6" thick, R19	G		1600	.005		.42	.17		.59	.74
1380	10" thick, R30	G		1350	.006		.67	.20		.87	1.07
1700	Non-rigid insul, recycled blue cotton fiber, unfaced batts, R-13, 16" wide	G		1600	.005		.71	.17		.88	1.06
1710	R-19, 16" wide	G		1600	.005		1.04	.17		1.21	1.42
1850	Friction fit wire insulation supports, 16" O.C.			960	.008	Ea.	.05	.28		.33	.53

07 21 Thermal Insulation

07 21 23 – Loose-Fill Insulation

07 21 23.10 Poured Loose-Fill Insulation		Crew	Daily Output	Labor-Hours	Unit	Material	2011 Bare Costs Labor	Equipment	Total	Total Incl O&P
0010	**POURED LOOSE-FILL INSULATION**									
0020	Cellulose fiber, R3.8 per inch G	1 Carp	200	.040	C.F.	.56	1.34		1.90	2.87
0080	Fiberglass wool, R4 per inch G		200	.040		.42	1.34		1.76	2.71
0100	Mineral wool, R3 per inch G		200	.040		.39	1.34		1.73	2.68
0300	Polystyrene, R4 per inch G		200	.040		3.09	1.34		4.43	5.65
0400	Perlite, R2.7 per inch G		200	.040		1.95	1.34		3.29	4.40

07 21 23.20 Masonry Loose-Fill Insulation		Crew	Daily Output	Labor-Hours	Unit	Material	Labor	Equipment	Total	Total Incl O&P
0010	**MASONRY LOOSE-FILL INSULATION,** vermiculite or perlite									
0100	In cores of concrete block, 4" thick wall, .115 C.F./S.F. G	D-1	4800	.003	S.F.	.22	.10		.32	.42
0700	Foamed in place, urethane in 2-5/8" cavity G	G-2A	1035	.023		.49	.58	.57	1.64	2.20
0800	For each 1" added thickness, add G	"	2372	.010		.12	.25	.25	.62	.86

07 21 26 – Blown Insulation

07 21 26.10 Blown Insulation		Crew	Daily Output	Labor-Hours	Unit	Material	Labor	Equipment	Total	Total Incl O&P
0010	**BLOWN INSULATION** Ceilings, with open access									
0020	Cellulose, 3-1/2" thick, R13 G	G-4	5000	.005	S.F.	.17	.13	.06	.36	.48
0030	5-3/16" thick, R19 G		3800	.006		.25	.17	.08	.50	.66
0050	6-1/2" thick, R22 G		3000	.008		.32	.22	.10	.64	.83
1000	Fiberglass, 5.5" thick, R11 G		3800	.006		.15	.17	.08	.40	.55
1050	6" thick, R12 G		3000	.008		.19	.22	.10	.51	.67
1100	8.8" thick, R19 G		2200	.011		.26	.30	.13	.69	.94
1300	11.5" thick, R26 G		1500	.016		.37	.43	.20	1	1.36
1350	13" thick, R30 G		1400	.017		.40	.46	.21	1.07	1.45
1450	16" thick, R38 G		1145	.021		.49	.57	.26	1.32	1.77
1500	20" thick, R49 G		920	.026		.62	.71	.32	1.65	2.22

07 21 27 – Reflective Insulation

07 21 27.10 Reflective Insulation Options		Crew	Daily Output	Labor-Hours	Unit	Material	Labor	Equipment	Total	Total Incl O&P
0010	**REFLECTIVE INSULATION OPTIONS**									
0020	Aluminum foil on reinforced scrim G	1 Carp	19	.421	C.S.F.	14.20	14.15		28.35	39
0100	Reinforced with woven polyolefin G		19	.421		18	14.15		32.15	43.50
0500	With single bubble air space, R8.8 G		15	.533		28	17.90		45.90	61
0600	With double bubble air space, R9.8 G		15	.533		29	17.90		46.90	61.50

07 21 29 – Sprayed Insulation

07 21 29.10 Sprayed-On Insulation		Crew	Daily Output	Labor-Hours	Unit	Material	Labor	Equipment	Total	Total Incl O&P
0010	**SPRAYED-ON INSULATION**									
0300	Closed cell, spray polyurethane foam, 2 pounds per cubic foot density									
0310	1" thick G	G-2A	6000	.004	S.F.	.41	.10	.10	.61	.74
0320	2" thick G		3000	.008		.82	.20	.20	1.22	1.47
0330	3" thick G		2000	.012		1.23	.30	.30	1.83	2.22
0335	3-1/2" thick G		1715	.014		1.44	.35	.35	2.14	2.58
0340	4" thick G		1500	.016		1.64	.40	.40	2.44	2.95
0350	5" thick G		1200	.020		2.05	.50	.49	3.04	3.69
0355	5-1/2" thick G		1090	.022		2.26	.55	.54	3.35	4.05
0360	6" thick G		1000	.024		2.46	.60	.59	3.65	4.42

07 22 Roof and Deck Insulation

07 22 16 – Roof Board Insulation

07 22 16.10 Roof Deck Insulation			Crew	Daily Output	Labor-Hours	Unit	2011 Bare Costs Material	Labor	Equipment	Total	Total Incl O&P
0010	**ROOF DECK INSULATION**										
0020	Fiberboard low density, 1/2" thick R1.39	G	1 Rofc	1000	.008	S.F.	.26	.23		.49	.70
0030	1" thick R2.78	G		800	.010		.46	.28		.74	1.02
0080	1 1/2" thick R4.17	G		800	.010		.70	.28		.98	1.28
0100	2" thick R5.56	G		800	.010		.93	.28		1.21	1.53
0110	Fiberboard high density, 1/2" thick R1.3	G		1000	.008		.24	.23		.47	.67
0120	1" thick R2.5	G		800	.010		.48	.28		.76	1.04
0130	1-1/2" thick R3.8	G		800	.010		.72	.28		1	1.30
0200	Fiberglass, 3/4" thick R2.78	G		1000	.008		.53	.23		.76	.99
0400	15/16" thick R3.70	G		1000	.008		.71	.23		.94	1.19
0460	1-1/16" thick R4.17	G		1000	.008		.89	.23		1.12	1.39
0600	1-5/16" thick R5.26	G		1000	.008		1.20	.23		1.43	1.73
0650	2-1/16" thick R8.33	G		800	.010		1.27	.28		1.55	1.91
0700	2-7/16" thick R10	G		800	.010		1.47	.28		1.75	2.13
1650	Perlite, 1/2" thick R1.32	G		1050	.008		.34	.21		.55	.76
1655	3/4" thick R2.08	G		800	.010		.37	.28		.65	.92
1660	1" thick R2.78	G		800	.010		.46	.28		.74	1.02
1670	1-1/2" thick R4.17	G		800	.010		.48	.28		.76	1.04
1680	2" thick R5.56	G		700	.011		.80	.32		1.12	1.46
1685	2-1/2" thick R6.67	G		700	.011		.95	.32		1.27	1.63
1690	Tapered for drainage	G		800	.010	B.F.	.73	.28		1.01	1.31
1700	Polyisocyanurate, 2#/C.F. density, 3/4" thick	G		1500	.005	S.F.	.44	.15		.59	.75
1705	1" thick	G		1400	.006		.51	.16		.67	.85
1715	1-1/2" thick	G		1250	.006		.66	.18		.84	1.06
1725	2" thick	G		1100	.007		.78	.20		.98	1.23
1735	2-1/2" thick	G		1050	.008		1.01	.21		1.22	1.50
1745	3" thick	G		1000	.008		1.22	.23		1.45	1.75
1755	3-1/2" thick	G		1000	.008		1.95	.23		2.18	2.56
1765	Tapered for drainage	G		1400	.006	B.F.	1.95	.16		2.11	2.44
1900	Extruded Polystyrene										
1910	15 PSI compressive strength, 1" thick, R5	G	1 Rofc	1500	.005	S.F.	.44	.15		.59	.75
1920	2" thick, R10	G		1250	.006		.56	.18		.74	.95
1930	3" thick R15	G		1000	.008		1.15	.23		1.38	1.68
1932	4" thick R20	G		1000	.008		1.52	.23		1.75	2.08
1934	Tapered for drainage	G		1500	.005	B.F.	.51	.15		.66	.83
1940	25 PSI compressive strength, 1" thick R5	G		1500	.005	S.F.	.61	.15		.76	.94
1942	2" thick R10	G		1250	.006		1.15	.18		1.33	1.60
1944	3" thick R15	G		1000	.008		1.75	.23		1.98	2.34
1946	4" thick R20	G		1000	.008		2.50	.23		2.73	3.16
1948	Tapered for drainage	G		1500	.005	B.F.	.53	.15		.68	.85
1950	40 psi compressive strength, 1" thick R5	G		1500	.005	S.F.	.47	.15		.62	.79
1952	2" thick R10	G		1250	.006		.88	.18		1.06	1.30
1954	3" thick R15	G		1000	.008		1.28	.23		1.51	1.82
1956	4" thick R20	G		1000	.008		1.70	.23		1.93	2.28
1958	Tapered for drainage	G		1400	.006	B.F.	.67	.16		.83	1.03
1960	60 PSI compressive strength, 1" thick R5	G		1450	.006	S.F.	.65	.16		.81	1
1962	2" thick R10	G		1200	.007		1.25	.19		1.44	1.72
1964	3" thick R15	G		975	.008		2	.23		2.23	2.62
1966	4" thick R20	G		950	.008		2.50	.24		2.74	3.18
1968	Tapered for drainage	G		1400	.006	B.F.	.85	.16		1.01	1.23
2010	Expanded polystyrene, 1#/C.F. density, 3/4" thick R2.89	G		1500	.005	S.F.	.22	.15		.37	.51
2020	1" thick R3.85	G		1500	.005		.29	.15		.44	.59
2100	2" thick R7.69	G		1250	.006		.58	.18		.76	.97

07 22 Roof and Deck Insulation

07 22 16 – Roof Board Insulation

07 22 16.10 Roof Deck Insulation			Crew	Daily Output	Labor-Hours	Unit	Material	2011 Bare Costs Labor	Equipment	Total	Total Incl O&P
2110	3" thick R11.49	G	1 Rofc	1250	.006	S.F.	.87	.18		1.05	1.29
2120	4" thick R15.38	G		1200	.007		1.16	.19		1.35	1.62
2130	5" thick R19.23	G		1150	.007		1.45	.20		1.65	1.95
2140	6" thick R23.26	G		1150	.007		1.74	.20		1.94	2.26
2150	Tapered for drainage	G		1500	.005	B.F.	.54	.15		.69	.86
2400	Composites with 2" EPS										
2410	1" fiberboard	G	1 Rofc	950	.008	S.F.	1.28	.24		1.52	1.84
2420	7/16" oriented strand board	G		800	.010		1	.28		1.28	1.61
2430	1/2" plywood	G		800	.010		1.20	.28		1.48	1.83
2440	1" perlite	G		800	.010		1.18	.28		1.46	1.81
2450	Composites with 1-1/2" polyisocyanurate										
2460	1" fiberboard	G	1 Rofc	800	.010	S.F.	1.20	.28		1.48	1.83
2470	1" perlite	G		850	.009		1.08	.26		1.34	1.67
2480	7/16" oriented strand board	G		800	.010		.84	.28		1.12	1.43

07 24 Exterior Insulation and Finish Systems

07 24 13 – Polymer-Based Exterior Insulation and Finish System

07 24 13.10 Exterior Insulation and Finish Systems			Crew	Daily Output	Labor-Hours	Unit	Material	Labor	Equipment	Total	Total Incl O&P
0010	**EXTERIOR INSULATION AND FINISH SYSTEMS**										
0095	Field applied, 1" EPS insulation	G	J-1	390	.103	S.F.	2.09	3.01	.32	5.42	7.55
0105	2" EPS insulation	G		390	.103		2.38	3.01	.32	5.71	7.85
0115	3" EPS insulation	G		390	.103		2.67	3.01	.32	6	8.20
0125	4" EPS insulation	G		390	.103		2.96	3.01	.32	6.29	8.50
0140	Premium finish add			1265	.032		.31	.93	.10	1.34	1.96
0440	For higher than one story, add							25%			

07 25 Weather Barriers

07 25 10 – Weather Barriers or Wraps

07 25 10.10 Weather Barriers		Crew	Daily Output	Labor-Hours	Unit	Material	Labor	Equipment	Total	Total Incl O&P
0010	**WEATHER BARRIERS**									
0400	Asphalt felt paper, 15#	1 Carp	37	.216	Sq.	5.65	7.25		12.90	18.35
0401	Per square foot	"	3700	.002	S.F.	.06	.07		.13	.18
0450	Housewrap, exterior, spun bonded polypropylene									
0470	Small roll	1 Carp	3800	.002	S.F.	.21	.07		.28	.36
0480	Large roll	"	4000	.002	"	.13	.07		.20	.25
2100	Asphalt felt roof deck vapor barrier, class 1 metal decks	1 Rofc	37	.216	Sq.	19.70	6.10		25.80	32.50
2200	For all other decks	"	37	.216		14.05	6.10		20.15	26.50
2800	Asphalt felt, 50% recycled content, 15 lb, 4 sq per roll	1 Carp	36	.222		5.65	7.45		13.10	18.70
2810	30 lb, 2 sq per roll	"	36	.222		9.45	7.45		16.90	23
3000	Building wrap, spunbonded polyethylene	2 Carp	8000	.002	S.F.	.14	.07		.21	.26

07 26 Vapor Retarders

07 26 10 – Above-Grade Vapor Retarders

07 26 10.10 Vapor Retarders			Crew	Daily Output	Labor-Hours	Unit	Material	Labor (2011 Bare Costs)	Equipment (2011 Bare Costs)	Total (2011 Bare Costs)	Total Incl O&P
0010	VAPOR RETARDERS										
0020	Aluminum and kraft laminated, foil 1 side	G	1 Carp	37	.216	Sq.	9.35	7.25		16.60	22.50
0100	Foil 2 sides	G		37	.216		9.75	7.25		17	23
0600	Polyethylene vapor barrier, standard, .002" thick	G		37	.216		1.19	7.25		8.44	13.45
0700	.004" thick	G		37	.216		3.34	7.25		10.59	15.80
0900	.006" thick	G		37	.216		5.25	7.25		12.50	17.90
1200	.010" thick	G		37	.216		6.70	7.25		13.95	19.50
1800	Reinf. waterproof, .002" polyethylene backing, 1 side			37	.216		5.65	7.25		12.90	18.35
1900	2 sides			37	.216		7.45	7.25		14.70	20.50

07 31 Shingles and Shakes

07 31 13 – Asphalt Shingles

07 31 13.10 Asphalt Roof Shingles

Line	Description	Crew	Daily Output	Labor-Hours	Unit	Material	Labor	Equipment	Total	Total Incl O&P
0010	ASPHALT ROOF SHINGLES									
0100	Standard strip shingles									
0150	Inorganic, class A, 210-235 lb./sq.	1 Rofc	5.50	1.455	Sq.	74	41		115	156
0155	Pneumatic nailed		7	1.143		74	32		106	140
0200	Organic, class C, 235-240 lb./sq.		5	1.600		72.50	45		117.50	162
0205	Pneumatic nailed		6.25	1.280		72.50	36		108.50	145
0250	Standard, laminated multi-layered shingles									
0300	Class A, 240-260 lb./sq.	1 Rofc	4.50	1.778	Sq.	97.50	50		147.50	198
0305	Pneumatic nailed		5.63	1.422		97.50	40		137.50	180
0350	Class C, 260-300 lb/square, 4 bundles/square		4	2		120	56.50		176.50	234
0355	Pneumatic nailed		5	1.600		120	45		165	214
0400	Premium, laminated multi-layered shingles									
0450	Class A, 260-300 lb, 4 bundles/sq	1 Rofc	3.50	2.286	Sq.	157	64.50		221.50	289
0455	Pneumatic nailed		4.37	1.831		157	51.50		208.50	266
0500	Class C, 300-385 lb/square, 5 bundles/square		3	2.667		150	75		225	300
0505	Pneumatic nailed		3.75	2.133		150	60		210	273
0800	#15 felt underlayment		64	.125		5.65	3.52		9.17	12.55
0825	#30 felt underlayment		58	.138		9.45	3.88		13.33	17.35
0850	Self adhering polyethylene and rubberized asphalt underlayment		22	.364		70	10.25		80.25	95.50
0900	Ridge shingles		330	.024	L.F.	1.90	.68		2.58	3.32
0905	Pneumatic nailed		412.50	.019	"	1.90	.55		2.45	3.08
1000	For steep roofs (7 to 12 pitch or greater), add						50%			

07 31 16 – Metal Shingles

07 31 16.10 Aluminum Shingles

Line	Description	Crew	Daily Output	Labor-Hours	Unit	Material	Labor	Equipment	Total	Total Incl O&P
0010	ALUMINUM SHINGLES									
0020	Mill finish, .019 thick	1 Carp	5	1.600	Sq.	191	53.50		244.50	300
0100	.020" thick	"	5	1.600		214	53.50		267.50	325
0300	For colors, add					20			20	22
0600	Ridge cap, .024" thick	1 Carp	170	.047	L.F.	2.18	1.58		3.76	5.05
0700	End wall flashing, .024" thick		170	.047		1.78	1.58		3.36	4.60
0900	Valley section, .024" thick		170	.047		2.90	1.58		4.48	5.85
1000	Starter strip, .024" thick		400	.020		1.41	.67		2.08	2.67
1200	Side wall flashing, .024" thick		170	.047		1.78	1.58		3.36	4.60

07 31 26 – Slate Shingles

07 31 26.10 Slate Roof Shingles

Line	Description		Crew	Daily Output	Labor-Hours	Unit	Material	Labor	Equipment	Total	Total Incl O&P
0010	SLATE ROOF SHINGLES	R073126-20									
0100	Buckingham Virginia black, 3/16" - 1/4" thick	G	1 Rots	1.75	4.571	Sq.	475	129		604	760

07 31 Shingles and Shakes

07 31 26 – Slate Shingles

07 31 26.10 Slate Roof Shingles		Crew	Daily Output	Labor-Hours	Unit	Material	2011 Bare Costs Labor	Equipment	Total	Total Incl O&P
0200	1/4" thick G	1 Rots	1.75	4.571	Sq.	475	129		604	760
0900	Pennsylvania black, Bangor, #1 clear G		1.75	4.571		490	129		619	775
1200	Vermont, unfading, green, mottled green G		1.75	4.571		480	129		609	760
1300	Semi-weathering green & gray G		1.75	4.571		350	129		479	615
1400	Purple G		1.75	4.571		425	129		554	700
1500	Black or gray G		1.75	4.571		460	129		589	740
2700	Ridge shingles, slate		200	.040	L.F.	9.30	1.13		10.43	12.25

07 31 29 – Wood Shingles and Shakes

07 31 29.13 Wood Shingles

		Crew	Daily Output	Labor-Hours	Unit	Material	Labor	Equipment	Total	Total Incl O&P
0010	**WOOD SHINGLES** R061110-30									
0012	16" No. 1 red cedar shingles, 5" exposure, on roof	1 Carp	2.50	3.200	Sq.	215	107		322	415
0015	Pneumatic nailed		3.25	2.462		215	82.50		297.50	375
0200	7-1/2" exposure, on walls		2.05	3.902		143	131		274	375
0205	Pneumatic nailed		2.67	2.996		143	101		244	325
0300	18" No. 1 red cedar perfections, 5-1/2" exposure, on roof		2.75	2.909		220	97.50		317.50	405
0305	Pneumatic nailed		3.57	2.241		220	75		295	370
0500	7-1/2" exposure, on walls		2.25	3.556		162	119		281	380
0505	Pneumatic nailed		2.92	2.740		162	92		254	330
0600	Resquared, and rebutted, 5-1/2" exposure, on roof		3	2.667		238	89.50		327.50	410
0605	Pneumatic nailed		3.90	2.051		238	69		307	375
0900	7-1/2" exposure, on walls		2.45	3.265		175	110		285	375
0905	Pneumatic nailed		3.18	2.516		175	84.50		259.50	335
1000	Add to above for fire retardant shingles					55			55	60.50
1060	Preformed ridge shingles	1 Carp	400	.020	L.F.	3.60	.67		4.27	5.10
2000	White cedar shingles, 16" long, extras, 5" exposure, on roof		2.40	3.333	Sq.	149	112		261	350
2005	Pneumatic nailed		3.12	2.564		149	86		235	310
2050	5" exposure on walls		2	4		149	134		283	390
2055	Pneumatic nailed		2.60	3.077		149	103		252	335
2100	7-1/2" exposure, on walls		2	4		106	134		240	340
2105	Pneumatic nailed		2.60	3.077		106	103		209	290
2150	"B" grade, 5" exposure on walls		2	4		136	134		270	375
2155	Pneumatic nailed		2.60	3.077		136	103		239	320
2300	For 15# organic felt underlayment on roof, 1 layer, add		64	.125		5.65	4.19		9.84	13.20
2400	2 layers, add		32	.250		11.25	8.40		19.65	26.50
2600	For steep roofs (7/12 pitch or greater), add to above						50%			
3000	Ridge shakes or shingle wood	1 Carp	280	.029	L.F.	3.60	.96		4.56	5.55

07 31 29.16 Wood Shakes

		Crew	Daily Output	Labor-Hours	Unit	Material	Labor	Equipment	Total	Total Incl O&P
0010	**WOOD SHAKES**									
1100	Hand-split red cedar shakes, 1/2" thick x 24" long, 10" exp. on roof	1 Carp	2.50	3.200	Sq.	233	107		340	435
1105	Pneumatic nailed		3.25	2.462		233	82.50		315.50	395
1110	3/4" thick x 24" long, 10" exp. on roof		2.25	3.556		233	119		352	455
1115	Pneumatic nailed		2.92	2.740		233	92		325	410
1200	1/2" thick, 18" long, 8-1/2" exp. on roof		2	4		159	134		293	400
1205	Pneumatic nailed		2.60	3.077		159	103		262	350
1210	3/4" thick x 18" long, 8 1/2" exp. on roof		1.80	4.444		159	149		308	425
1215	Pneumatic nailed		2.34	3.419		159	115		274	365
1255	10" exp. on walls		2	4		154	134		288	395
1260	10" exposure on walls, pneumatic nailed		2.60	3.077		154	103		257	340
1700	Add to above for fire retardant shakes, 24" long					55			55	60.50
1800	18" long					55			55	60.50
1810	Ridge shakes	1 Carp	350	.023	L.F.	3.60	.77		4.37	5.25

07 32 Roof Tiles

07 32 13 – Clay Roof Tiles

07 32 13.10 Clay Tiles			Crew	Daily Output	Labor-Hours	Unit	Material	2011 Bare Costs Labor	Equipment	Total	Total Incl O&P
0010	**CLAY TILES**										
0200	Lanai tile or Classic tile, 158 pc per sq	G	1 Rots	1.65	4.848	Sq.	200	137		337	470
0300	Americana, 158 pc per sq, most colors	G		1.65	4.848		440	137		577	735
0350	Green, gray or brown	G		1.65	4.848		440	137		577	735
0400	Blue	G		1.65	4.848		440	137		577	735
0600	Spanish tile, 171 pc per sq, red	G		1.80	4.444		325	126		451	585
0800	Buff, green, gray, brown	G		1.80	4.444		550	126		676	830
0900	Glazed white	G		1.80	4.444		630	126		756	920
1100	Mission tile, 192 pc per sq, machine scored finish, red	G		1.15	6.957		740	197		937	1,175
1700	French tile, 133 pc per sq, smooth finish, red	G		1.35	5.926		675	168		843	1,050
1750	Blue or green	G		1.35	5.926		870	168		1,038	1,250
1800	Norman black 317 pc per sq	G		1	8		1,050	226		1,276	1,550
2200	Williamsburg tile, 158 pc per sq, aged cedar	G		1.35	5.926		660	168		828	1,025
2250	Gray or green	G		1.35	5.926		595	168		763	960
2350	Ridge shingles, clay tile	G		200	.040	L.F.	10.70	1.13		11.83	13.85
3000	For steep roofs (7/12 pitch or greater), add to above					Sq.		50%			
3010	Clay tile, #15 felt underlayment		1 Rofc	64	.125		5.65	3.52		9.17	12.55
3020	Clay tile, #30 felt underlayment			58	.138		9.45	3.88		13.33	17.35
3040	Clay tile, polyethylene and rubberized asph. underlayment			22	.364		70	10.25		80.25	95.50

07 32 16 – Concrete Roof Tiles

07 32 16.10 Concrete Tiles			Crew	Daily Output	Labor-Hours	Unit	Material	Labor	Equipment	Total	Total Incl O&P
0010	**CONCRETE TILES**										
0020	Corrugated, 13" x 16-1/2", 90 per sq, 950 lb per sq										
0050	Earthtone colors, nailed to wood deck		1 Rots	1.35	5.926	Sq.	97.50	168		265.50	410
0150	Blues			1.35	5.926		101	168		269	415
0200	Greens			1.35	5.926		101	168		269	415
0250	Premium colors			1.35	5.926		101	168		269	415
0500	Shakes, 13" x 16-1/2", 90 per sq, 950 lb per sq										
0600	All colors, nailed to wood deck		1 Rots	1.50	5.333	Sq.	260	151		411	560
1500	Accessory pieces, ridge & hip, 10" x 16-1/2", 8 lb. each		"	120	.067	Ea.	3.64	1.89		5.53	7.40
1700	Rake, 6-1/2" x 16-3/4", 9 lb. each						3.64			3.64	4
1800	Mansard hip, 10" x 16-1/2", 9.2 lb. each						3.64			3.64	4
1900	Hip starter, 10" x 16-1/2", 10.5 lb. each						11.45			11.45	12.60
2000	3 or 4 way apex, 10" each side, 11.5 lb. each						13.25			13.25	14.55

07 32 19 – Metal Roof Tiles

07 32 19.10 Metal Roof Tiles			Crew	Daily Output	Labor-Hours	Unit	Material	Labor	Equipment	Total	Total Incl O&P
0010	**METAL ROOF TILES**										
0020	Accessories included, .032" thick aluminum, mission tile		1 Carp	2.50	3.200	Sq.	815	107		922	1,075
0200	Spanish tiles		"	3	2.667	"	560	89.50		649.50	765

07 41 Roof Panels

07 41 13 – Metal Roof Panels

07 41 13.10 Aluminum Roof Panels			Crew	Daily Output	Labor-Hours	Unit	Material	2011 Bare Costs Labor	Equipment	Total	Total Incl O&P
0010	**ALUMINUM ROOF PANELS**										
0020	Corrugated or ribbed, .0155" thick, natural		G-3	1200	.027	S.F.	.96	.86		1.82	2.49
0300	Painted		"	1200	.027	"	1.40	.86		2.26	2.97

07 41 13.20 Steel Roofing Panels

Code	Description		Crew	Daily Output	Labor-Hours	Unit	Material	Labor	Equipment	Total	Total Incl O&P
0010	**STEEL ROOFING PANELS**										
0012	Corrugated or ribbed, on steel framing, 30 ga galv		G-3	1100	.029	S.F.	1.68	.94		2.62	3.41
0100	28 ga			1050	.030		1.79	.99		2.78	3.61
0300	26 ga			1000	.032		1.85	1.03		2.88	3.76
0400	24 ga			950	.034		2.91	1.09		4	5
0510	Painted, including fasteners, 18 ga	G		850	.038		3.57	1.22		4.79	5.95
0520	20 ga	G		875	.037		3.07	1.18		4.25	5.35
0530	22 ga	G		900	.036		2.96	1.15		4.11	5.15
0600	Colored, 28 ga			1050	.030		1.68	.99		2.67	3.49
0700	26 ga			1000	.032		1.99	1.03		3.02	3.91
0710	Flat profile, 1-3/4" standing seams, 10" wide, standard finish, 26 ga			1000	.032		3.89	1.03		4.92	6
0715	24 ga			950	.034		4.51	1.09		5.60	6.75
0720	22 ga			900	.036		5.55	1.15		6.70	8.05
0725	Zinc aluminum alloy finish, 26 ga			1000	.032		3.05	1.03		4.08	5.10
0730	24 ga			950	.034		3.64	1.09		4.73	5.80
0735	22 ga			900	.036		4.17	1.15		5.32	6.50
0740	12" wide, standard finish, 26 ga			1000	.032		3.88	1.03		4.91	6
0745	24 ga			950	.034		5.10	1.09		6.19	7.40
0750	Zinc aluminum alloy finish, 26 ga			1000	.032		4.41	1.03		5.44	6.55
0755	24 ga			950	.034		3.63	1.09		4.72	5.80
0840	Flat profile, 1" x 3/8" batten, 12" wide, standard finish, 26 ga			1000	.032		3.42	1.03		4.45	5.50
0845	24 ga			950	.034		4.01	1.09		5.10	6.20
0850	22 ga			900	.036		4.80	1.15		5.95	7.20
0855	Zinc aluminum alloy finish, 26 ga			1000	.032		3.28	1.03		4.31	5.35
0860	24 ga			950	.034		3.66	1.09		4.75	5.85
0865	22 ga			900	.036		4.23	1.15		5.38	6.55
0870	16-1/2" wide, standard finish, 24 ga			950	.034		3.95	1.09		5.04	6.15
0875	22 ga			900	.036		4.42	1.15		5.57	6.75
0880	Zinc aluminum alloy finish, 24 ga			950	.034		3.45	1.09		4.54	5.60
0885	22 ga			900	.036		3.85	1.15		5	6.15
0890	Flat profile, 2" x 2" batten, 12" wide, standard finish, 26 ga			1000	.032		3.93	1.03		4.96	6.05
0895	24 ga			950	.034		4.69	1.09		5.78	6.95
0900	22 ga			900	.036		5.75	1.15		6.90	8.25
0905	Zinc aluminum alloy finish, 26 ga			1000	.032		3.66	1.03		4.69	5.75
0910	24 ga			950	.034		4.18	1.09		5.27	6.40
0915	22 ga			900	.036		4.87	1.15		6.02	7.25
0920	16-1/2" wide, standard finish, 24 ga			950	.034		4.32	1.09		5.41	6.55
0925	22 ga			900	.036		5.05	1.15		6.20	7.45
0930	Zinc aluminum alloy finish, 24 ga			950	.034		3.91	1.09		5	6.10
0935	22 ga			900	.036		4.45	1.15		5.60	6.80
0950	Box rib roof panels, painted, including fasteners, 18 ga	G		850	.038		3.48	1.22		4.70	5.85
0960	20 ga	G		875	.037		3.16	1.18		4.34	5.45
0970	22 ga	G		900	.036		2.74	1.15		3.89	4.92
1000	4" rib panel, painted, including fasteners, 22 ga	G		900	.036		2.73	1.15		3.88	4.91
1010	20 ga	G		875	.037		3.04	1.18		4.22	5.30
1020	18 ga	G		850	.038		6.20	1.22		7.42	8.80
1050	On substrate, 2" standing seam panel, painted, 22 ga	G		900	.036		4.03	1.15		5.18	6.35
1060	24 ga	G		950	.034		3.63	1.09		4.72	5.80

07 41 Roof Panels

07 41 13 – Metal Roof Panels

07 41 13.20 Steel Roofing Panels			Crew	Daily Output	Labor-Hours	Unit	Material	2011 Bare Costs Labor	Equipment	Total	Total Incl O&P
1070	26 ga	G	G-3	1000	.032	S.F.	2.80	1.03		3.83	4.80
1203	14" wide	G	2 Shee	316	.051	L.F.	3.82	1.93		5.75	7.40

07 41 33 – Plastic Roof Panels

07 41 33.10 Fiberglass Panels		Crew	Daily Output	Labor-Hours	Unit	Material	Labor	Equipment	Total	Total Incl O&P
0010	**FIBERGLASS PANELS**									
0012	Corrugated panels, roofing, 8 oz per S.F.	G-3	1000	.032	S.F.	1.74	1.03		2.77	3.63
0300	Corrugated siding, 6 oz per S.F.		880	.036		1.53	1.18		2.71	3.63
0400	8 oz per S.F.		880	.036		1.74	1.18		2.92	3.86
0600	12 oz. siding, textured		880	.036		3.31	1.18		4.49	5.60
0900	Flat panels, 6 oz per S.F., clear or colors		880	.036		1.95	1.18		3.13	4.10
1300	8 oz per S.F., clear or colors		880	.036		2.15	1.18		3.33	4.32

07 42 Wall Panels

07 42 13 – Metal Wall Panels

07 42 13.20 Aluminum Siding		Crew	Daily Output	Labor-Hours	Unit	Material	Labor	Equipment	Total	Total Incl O&P
0011	**ALUMINUM SIDING**									
6040	.024 thick smooth white single 8" wide	2 Carp	515	.031	S.F.	2.17	1.04		3.21	4.13
6060	Double 4" pattern		515	.031		2.14	1.04		3.18	4.09
6080	Double 5" pattern		550	.029		2.21	.98		3.19	4.06
6120	Embossed white, 8" wide		515	.031		1.82	1.04		2.86	3.74
6140	Double 4" pattern		515	.031		2.26	1.04		3.30	4.23
6160	Double 5" pattern		550	.029		2.24	.98		3.22	4.09
6170	Vertical, embossed white, 12" wide		590	.027		2.24	.91		3.15	3.98
6320	.019 thick, insulated, smooth white, 8" wide		515	.031		1.98	1.04		3.02	3.92
6340	Double 4" pattern		515	.031		1.96	1.04		3	3.90
6360	Double 5" pattern		550	.029		1.96	.98		2.94	3.79
6400	Embossed white, 8" wide		515	.031		2.29	1.04		3.33	4.26
6420	Double 4" pattern		515	.031		2.32	1.04		3.36	4.29
6440	Double 5" pattern		550	.029		2.32	.98		3.30	4.18
6500	Shake finish 10" wide white		550	.029		2.47	.98		3.45	4.35
6600	Vertical pattern, 12" wide, white		590	.027		2.06	.91		2.97	3.79
6640	For colors add					.13			.13	.14
6700	Accessories, white									
6720	Starter strip 2-1/8"	2 Carp	610	.026	L.F.	.43	.88		1.31	1.94
6740	Sill trim		450	.036		.58	1.19		1.77	2.64
6760	Inside corner		610	.026		1.50	.88		2.38	3.12
6780	Outside corner post		610	.026		3.19	.88		4.07	4.98
6800	Door & window trim		440	.036		.55	1.22		1.77	2.65
6820	For colors add					.12			.12	.13
6900	Soffit & fascia 1' overhang solid	2 Carp	110	.145		4.10	4.88		8.98	12.65
6920	Vented		110	.145		4.10	4.88		8.98	12.65
6940	2' overhang solid		100	.160		6.05	5.35		11.40	15.65
6960	Vented		100	.160		6.05	5.35		11.40	15.65

07 42 13.30 Steel Siding		Crew	Daily Output	Labor-Hours	Unit	Material	Labor	Equipment	Total	Total Incl O&P
0010	**STEEL SIDING**									
0020	Beveled, vinyl coated, 8" wide	1 Carp	265	.030	S.F.	1.97	1.01		2.98	3.87
0050	10" wide	"	275	.029		2.11	.98		3.09	3.95
0080	Galv, corrugated or ribbed, on steel frame, 30 gauge	G-3	800	.040		1.32	1.29		2.61	3.60
0100	28 gauge		795	.040		1.38	1.30		2.68	3.68
0300	26 gauge		790	.041		1.94	1.31		3.25	4.31

07 42 Wall Panels

07 42 13 – Metal Wall Panels

07 42 13.30 Steel Siding		Crew	Daily Output	Labor-Hours	Unit	2011 Bare Costs Material	Labor	Equipment	Total	Total Incl O&P
0400	24 gauge	G-3	785	.041	S.F.	1.95	1.32		3.27	4.34
0600	22 gauge		770	.042		2.25	1.34		3.59	4.71
0700	Colored, corrugated/ribbed, on steel frame, 10 yr. finish, 28 ga.		800	.040		2.05	1.29		3.34	4.41
0900	26 gauge		795	.040		2.14	1.30		3.44	4.51
1000	24 gauge		790	.041		2.49	1.31		3.80	4.92
1020	20 gauge		785	.041		3.16	1.32		4.48	5.65

07 46 Siding

07 46 23 – Wood Siding

07 46 23.10 Wood Board Siding		Crew	Daily Output	Labor-Hours	Unit	Material	Labor	Equipment	Total	Total Incl O&P
0010	**WOOD BOARD SIDING**									
2000	Board & batten, cedar, "B" grade, 1" x 10"	1 Carp	375	.021	S.F.	2.44	.72		3.16	3.88
2200	Redwood, clear, vertical grain, 1" x 10"		375	.021		5.10	.72		5.82	6.85
2400	White pine, #2 & better, 1" x 10"		375	.021		2.97	.72		3.69	4.47
2410	Board & batten siding, white pine #2, 1" x 12"		420	.019		2.97	.64		3.61	4.34
3200	Wood, cedar bevel, A grade, 1/2" x 6"		295	.027		4.39	.91		5.30	6.35
3300	1/2" x 8"		330	.024		5.85	.81		6.66	7.75
3500	3/4" x 10", clear grade		375	.021		6.55	.72		7.27	8.40
3600	"B" grade		375	.021		3.79	.72		4.51	5.35
3800	Cedar, rough sawn, 1" x 4", A grade, natural		220	.036		3.88	1.22		5.10	6.30
3900	Stained		220	.036		4.25	1.22		5.47	6.70
4100	1" x 12", board & batten, #3 & Btr., natural		420	.019		4.02	.64		4.66	5.50
4200	Stained		420	.019		4.25	.64		4.89	5.75
4400	1" x 8" channel siding, #3 & Btr., natural		330	.024		2.92	.81		3.73	4.57
4500	Stained		330	.024		2.67	.81		3.48	4.30
4700	Redwood, clear, beveled, vertical grain, 1/2" x 4"		220	.036		3.85	1.22		5.07	6.30
4750	1/2" x 6"		295	.027		3.85	.91		4.76	5.75
4800	1/2" x 8"		330	.024		4.10	.81		4.91	5.85
5000	3/4" x 10"		375	.021		4.12	.72		4.84	5.75
5200	Channel siding, 1" x 10", B grade		375	.021		4.12	.72		4.84	5.75
5250	Redwood, T&G boards, B grade, 1" x 4"		220	.036		2.54	1.22		3.76	4.83
5270	1" x 8"		330	.024		4.10	.81		4.91	5.85
5400	White pine, rough sawn, 1" x 8", natural		330	.024		2.20	.81		3.01	3.78
5500	Stained		330	.024		2.60	.81		3.41	4.22
5600	Tongue and groove, 1" x 8"		330	.024		2.20	.81		3.01	3.78

07 46 29 – Plywood Siding

07 46 29.10 Plywood Siding Options		Crew	Daily Output	Labor-Hours	Unit	Material	Labor	Equipment	Total	Total Incl O&P
0010	**PLYWOOD SIDING OPTIONS**									
0900	Plywood, medium density overlaid, 3/8" thick	2 Carp	750	.021	S.F.	1.25	.72		1.97	2.58
1000	1/2" thick		700	.023		1.22	.77		1.99	2.62
1100	3/4" thick		650	.025		1.55	.83		2.38	3.09
1600	Texture 1-11, cedar, 5/8" thick, natural		675	.024		2.43	.80		3.23	4
1700	Factory stained		675	.024		1.95	.80		2.75	3.48
1900	Texture 1-11, fir, 5/8" thick, natural		675	.024		1.37	.80		2.17	2.84
2000	Factory stained		675	.024		1.73	.80		2.53	3.23
2050	Texture 1-11, S.Y.P., 5/8" thick, natural		675	.024		1.19	.80		1.99	2.64
2100	Factory stained		675	.024		1.24	.80		2.04	2.69
2200	Rough sawn cedar, 3/8" thick, natural		675	.024		1.20	.80		2	2.65
2300	Factory stained		675	.024		.96	.80		1.76	2.39
2500	Rough sawn fir, 3/8" thick, natural		675	.024		.76	.80		1.56	2.17

07 46 Siding

07 46 29 – Plywood Siding

07 46 29.10 Plywood Siding Options		Crew	Daily Output	Labor-Hours	Unit	2011 Bare Costs Material	Labor	Equipment	Total	Total Incl O&P
2600	Factory stained	2 Carp	675	.024	S.F.	.96	.80		1.76	2.39
2800	Redwood, textured siding, 5/8" thick	↓	675	.024	↓	1.92	.80		2.72	3.44

07 46 33 – Plastic Siding

07 46 33.10 Vinyl Siding		Crew	Daily Output	Labor-Hours	Unit	2011 Bare Costs Material	Labor	Equipment	Total	Total Incl O&P
0010	**VINYL SIDING**									
3995	Clapboard profile, woodgrain texture, .048 thick, double 4	2 Carp	495	.032	S.F.	.87	1.08		1.95	2.77
4000	Double 5		550	.029		.87	.98		1.85	2.59
4005	Single 8		495	.032		.94	1.08		2.02	2.85
4010	Single 10		550	.029		1.13	.98		2.11	2.87
4015	.044 thick, double 4		495	.032		.82	1.08		1.90	2.72
4020	Double 5		550	.029		.82	.98		1.80	2.54
4025	.042 thick, double 4		495	.032		.82	1.08		1.90	2.72
4030	Double 5		550	.029		.82	.98		1.80	2.54
4035	Cross sawn texture, .040 thick, double 4		495	.032		.64	1.08		1.72	2.52
4040	Double 5		550	.029		.64	.98		1.62	2.34
4045	Smooth texture, .042 thick, double 4		495	.032		.68	1.08		1.76	2.56
4050	Double 5		550	.029		.73	.98		1.71	2.44
4055	Single 8		495	.032		.94	1.08		2.02	2.85
4060	Cedar texture, .044 thick, double 4		495	.032		.85	1.08		1.93	2.75
4065	Double 6		600	.027		.94	.89		1.83	2.54
4070	Dutch lap profile, woodgrain texture, .048 thick, double 5		550	.029		.91	.98		1.89	2.64
4075	.044 thick, double 4.5		525	.030		.87	1.02		1.89	2.67
4080	.042 thick, double 4.5		525	.030		.72	1.02		1.74	2.51
4085	.040 thick, double 4.5		525	.030		.65	1.02		1.67	2.43
4090	Shingle profile, random grooves, double 7		400	.040		2.71	1.34		4.05	5.25
4095	Triple 5		400	.040		2.71	1.34		4.05	5.25
4100	Shake profile, 10" wide		400	.040		2.14	1.34		3.48	4.61
4105	Vertical pattern, .046 thick, double 5		550	.029		1.27	.98		2.25	3.03
4110	.044 thick, triple 3		550	.029		1.31	.98		2.29	3.08
4115	.040 thick, triple 4		550	.029		1.25	.98		2.23	3.01
4120	.040 thick, triple 2.66		550	.029		1.41	.98		2.39	3.19
4125	Insulation, fan folded extruded polystyrene, 1/4"		2000	.008		.23	.27		.50	.71
4130	3/8"		2000	.008	↓	.26	.27		.53	.73
4135	Accessories, J channel, 5/8" pocket		700	.023	L.F.	.37	.77		1.14	1.68
4140	3/4" pocket		695	.023		.38	.77		1.15	1.70
4145	1-1/4" pocket		680	.024		.59	.79		1.38	1.97
4150	Flexible, 3/4" pocket		600	.027		1.92	.89		2.81	3.61
4155	Under sill finish trim		500	.032		.39	1.07		1.46	2.23
4160	Vinyl starter strip		700	.023		.39	.77		1.16	1.71
4165	Aluminum starter strip		700	.023		.26	.77		1.03	1.56
4170	Window casing, 2-1/2" wide, 3/4" pocket		510	.031		1.22	1.05		2.27	3.10
4175	Outside corner, woodgrain finish, 4" face, 3/4" pocket		700	.023		1.58	.77		2.35	3.02
4180	5/8" pocket		700	.023		1.66	.77		2.43	3.11
4185	Smooth finish, 4" face, 3/4" pocket		700	.023		1.58	.77		2.35	3.02
4190	7/8" pocket		690	.023		1.46	.78		2.24	2.91
4195	1-1/4" pocket		700	.023		1.05	.77		1.82	2.44
4200	Soffit and fascia, 1' overhang, solid		120	.133		3.38	4.47		7.85	11.20
4205	Vented		120	.133		3.38	4.47		7.85	11.20
4207	18" overhang, solid		110	.145		4.02	4.88		8.90	12.55
4208	Vented		110	.145		4.02	4.88		8.90	12.55
4210	2' overhang, solid		100	.160		4.65	5.35		10	14.10
4215	Vented	↓	100	.160	↓	4.65	5.35		10	14.10

07 46 Siding

07 46 33 – Plastic Siding

07 46 33.10 Vinyl Siding		Crew	Daily Output	Labor-Hours	Unit	Material	2011 Bare Costs Labor	Equipment	Total	Total Incl O&P
4217	3′ overhang, solid	2 Carp	100	.160	L.F.	5.90	5.35		11.25	15.50
4218	Vented		100	.160		5.90	5.35		11.25	15.50
4220	Colors for siding and soffits, add				S.F.	.16			.16	.18
4225	Colors for accessories and trim, add				L.F.	.32			.32	.35

07 46 46 – Fiber Cement Siding

07 46 46.10 Fiber Cement Siding		Crew	Daily Output	Labor-Hours	Unit	Material	Labor	Equipment	Total	Total Incl O&P
0010	**FIBER CEMENT SIDING**									
0020	Lap siding, 5/16″ thick, 6″ wide, 4-3/4″ exposure, smooth texture	2 Carp	415	.039	S.F.	1.16	1.29		2.45	3.44
0025	Woodgrain texture		415	.039		1.16	1.29		2.45	3.44
0030	7-1/2″ wide, 6-1/4″ exposure, smooth texture		425	.038		1.04	1.26		2.30	3.25
0035	Woodgrain texture		425	.038		1.04	1.26		2.30	3.25
0040	8″ wide, 6-3/4″ exposure, smooth texture		425	.038		1.12	1.26		2.38	3.34
0045	Roughsawn texture		425	.038		1.12	1.26		2.38	3.34
0050	9-1/2″ wide, 8-1/4″ exposure, smooth texture		440	.036		1.16	1.22		2.38	3.32
0055	Woodgrain texture		440	.036		1.16	1.22		2.38	3.32
0060	12″ wide, 10-3/8″ exposure, smooth texture		455	.035		1.28	1.18		2.46	3.38
0065	Woodgrain texture		455	.035		1.28	1.18		2.46	3.38
0070	Panel siding, 5/16″ thick, smooth texture		750	.021		1.05	.72		1.77	2.36
0075	Stucco texture		750	.021		1.05	.72		1.77	2.36
0080	Grooved woodgrain texture		750	.021		1.05	.72		1.77	2.36
0085	V - grooved woodgrain texture		750	.021		1.05	.72		1.77	2.36
0090	Wood starter strip		400	.040	L.F.	.53	1.34		1.87	2.83

07 51 Built-Up Bituminous Roofing

07 51 13 – Built-Up Asphalt Roofing

07 51 13.10 Built-Up Roofing Components		Crew	Daily Output	Labor-Hours	Unit	Material	Labor	Equipment	Total	Total Incl O&P
0010	**BUILT-UP ROOFING COMPONENTS**									
0012	Asphalt saturated felt, #30, 2 square per roll	1 Rofc	58	.138	Sq.	9.45	3.88		13.33	17.35
0200	#15, 4 sq per roll, plain or perforated, not mopped		58	.138		5.65	3.88		9.53	13.20
0250	Perforated		58	.138		5.65	3.88		9.53	13.20
0300	Roll roofing, smooth, #65		15	.533		9.65	15		24.65	37.50
0500	#90		12	.667		34.50	18.75		53.25	72
0520	Mineralized		12	.667		34.50	18.75		53.25	72
0540	D.C. (Double coverage), 19″ selvage edge		10	.800		44	22.50		66.50	88.50
0580	Adhesive (lap cement)				Gal.	4.79			4.79	5.25

07 51 13.13 Cold-Applied Built-Up Asphalt Roofing		Crew	Daily Output	Labor-Hours	Unit	Material	Labor	Equipment	Total	Total Incl O&P
0010	**COLD-APPLIED BUILT-UP ASPHALT ROOFING**									
0020	3 ply system, installation only (components listed below)	G-5	50	.800	Sq.		20.50	3.37	23.87	40.50
0100	Spunbond poly. fabric, 1.35 oz./S.Y., 36″W, 10.8 sq./roll				Ea.	175			175	193
0200	49″ wide, 14.6 sq./roll					244			244	268
0300	2.10 oz./S.Y., 36″ wide, 10.8 sq./roll					266			266	293
0400	49″ wide, 14.6 sq./roll					360			360	395
0500	Base & finish coat, 3 gal./sq., 5 gal./can				Gal.	5			5	5.50
0600	Coating, ceramic granules, 1/2 sq./bag				Ea.	16.75			16.75	18.45
0700	Aluminum, 2 gal./sq.				Gal.	13.50			13.50	14.85
0800	Emulsion, fibered or non-fibered, 4 gal./sq.				"	4.70			4.70	5.15

07 51 13.20 Built-Up Roofing Systems		Crew	Daily Output	Labor-Hours	Unit	Material	Labor	Equipment	Total	Total Incl O&P
0010	**BUILT-UP ROOFING SYSTEMS**									
0120	Asphalt flood coat with gravel/slag surfacing, not including									
0140	Insulation, flashing or wood nailers									

07 51 Built-Up Bituminous Roofing

07 51 13 – Built-Up Asphalt Roofing

07 51 13.20 Built-Up Roofing Systems		Crew	Daily Output	Labor-Hours	Unit	2011 Bare Costs Material	Labor	Equipment	Total	Total Incl O&P
0200	Asphalt base sheet, 3 plies #15 asphalt felt, mopped	G-1	22	2.545	Sq.	82.50	67	18.60	168.10	233
0350	On nailable decks		21	2.667		87.50	70.50	19.50	177.50	245
0500	4 plies #15 asphalt felt, mopped		20	2.800		115	74	20.50	209.50	282
0550	On nailable decks		19	2.947		103	77.50	21.50	202	277
0700	Coated glass base sheet, 2 plies glass (type IV), mopped		22	2.545		86	67	18.60	171.60	236
0850	3 plies glass, mopped		20	2.800		103	74	20.50	197.50	269
0950	On nailable decks		19	2.947		97	77.50	21.50	196	270
1100	4 plies glass fiber felt (type IV), mopped		20	2.800		125	74	20.50	219.50	294
1150	On nailable decks		19	2.947		113	77.50	21.50	212	289
1200	Coated & saturated base sheet, 3 plies #15 asph. felt, mopped		20	2.800		92.50	74	20.50	187	258
1250	On nailable decks		19	2.947		86.50	77.50	21.50	185.50	259
1300	4 plies #15 asphalt felt, mopped		22	2.545		108	67	18.60	193.60	261
2000	Asphalt flood coat, smooth surface									
2200	Asphalt base sheet & 3 plies #15 asphalt felt, mopped	G-1	24	2.333	Sq.	87.50	61.50	17.05	166.05	226
2400	On nailable decks		23	2.435		81.50	64	17.80	163.30	226
2600	4 plies #15 asphalt felt, mopped		24	2.333		103	61.50	17.05	181.55	243
2700	On nailable decks		23	2.435		97	64	17.80	178.80	243
2900	Coated glass fiber base sheet, mopped, and 2 plies of									
2910	glass fiber felt (type IV)	G-1	25	2.240	Sq.	80.50	59	16.35	155.85	214
3100	On nailable decks		24	2.333		76	61.50	17.05	154.55	213
3200	3 plies, mopped		23	2.435		97	64	17.80	178.80	243
3300	On nailable decks		22	2.545		91	67	18.60	176.60	242
3800	4 plies glass fiber felt (type IV), mopped		23	2.435		114	64	17.80	195.80	261
3900	On nailable decks		22	2.545		108	67	18.60	193.60	260
4000	Coated & saturated base sheet, 3 plies #15 asph. felt, mopped		24	2.333		86.50	61.50	17.05	165.05	225
4200	On nailable decks		23	2.435		81	64	17.80	162.80	225
4300	4 plies #15 organic felt, mopped		22	2.545		102	67	18.60	187.60	254
4500	Coal tar pitch with gravel/slag surfacing									
4600	4 plies #15 tarred felt, mopped	G-1	21	2.667	Sq.	176	70.50	19.50	266	345
4800	3 plies glass fiber felt (type IV), mopped	"	19	2.947	"	146	77.50	21.50	245	325
5000	Coated glass fiber base sheet, and 2 plies of									
5010	glass fiber felt, (type IV), mopped	G-1	19	2.947	Sq.	150	77.50	21.50	249	330
5300	On nailable decks		18	3.111		131	82	22.50	235.50	320
5600	4 plies glass fiber felt (type IV), mopped		21	2.667		202	70.50	19.50	292	370
5800	On nailable decks		20	2.800		183	74	20.50	277.50	360

07 51 13.30 Cants		Crew	Daily Output	Labor-Hours	Unit	Material	Labor	Equipment	Total	Total Incl O&P
0010	**CANTS**									
0012	Lumber, treated, 4" x 4" cut diagonally	1 Rofc	325	.025	L.F.	2	.69		2.69	3.45
0100	Foamglass		325	.025		2.23	.69		2.92	3.70
0300	Mineral or fiber, trapezoidal, 1" x 4" x 48"		325	.025		.19	.69		.88	1.46
0400	1-1/2" x 5-5/8" x 48"		325	.025		.31	.69		1	1.59

07 51 13.40 Felts		Crew	Daily Output	Labor-Hours	Unit	Material	Labor	Equipment	Total	Total Incl O&P
0010	**FELTS**									
0012	Glass fibered roofing felt, #15, not mopped	1 Rofc	58	.138	Sq.	6.90	3.88		10.78	14.55
0300	Base sheet, #45, channel vented		58	.138		32	3.88		35.88	42.50
0400	#50, coated		58	.138		16.45	3.88		20.33	25
0500	Cap, mineral surfaced		58	.138		45	3.88		48.88	56.50
0600	Flashing membrane, #65		16	.500		9.65	14.10		23.75	36
0800	Coal tar fibered, #15, no mopping		58	.138		14	3.88		17.88	22.50
0900	Asphalt felt, #15, 4 sq per roll, no mopping		58	.138		5.65	3.88		9.53	13.20
1100	#30, 2 sq per roll		58	.138		9.45	3.88		13.33	17.35
1200	Double coated, #33		58	.138		9.90	3.88		13.78	17.90

07 51 Built-Up Bituminous Roofing

07 51 13 – Built-Up Asphalt Roofing

07 51 13.40 Felts		Crew	Daily Output	Labor-Hours	Unit	2011 Bare Costs Material	Labor	Equipment	Total	Total Incl O&P
1400	#40, base sheet	1 Rofc	58	.138	Sq.	10.90	3.88		14.78	19
1450	Coated and saturated		58	.138		9.95	3.88		13.83	17.95
1500	Tarred felt, organic, #15, 4 sq rolls		58	.138		11.50	3.88		15.38	19.65
1550	#30, 2 sq roll	↓	58	.138		23	3.88		26.88	32.50
1700	Add for mopping above felts, per ply, asphalt, 24 lb per sq	G-1	192	.292		8.40	7.70	2.13	18.23	25.50
1800	Coal tar mopping, 30 lb per sq		186	.301		16.50	7.95	2.20	26.65	35
1900	Flood coat, with asphalt, 60 lb per sq		60	.933		21	24.50	6.80	52.30	75
2000	With coal tar, 75 lb per sq	↓	56	1	↓	41.50	26.50	7.30	75.30	101

07 52 Modified Bituminous Membrane Roofing

07 52 13 – Atactic-Polypropylene-Modified Bituminous Membrane Roofing

07 52 13.10 APP Modified Bituminous Membrane

		Crew	Daily Output	Labor-Hours	Unit	Material	Labor	Equipment	Total	Total Incl O&P
0010	**APP MODIFIED BITUMINOUS MEMBRANE** R075213-30									
0020	Base sheet, #15 glass fiber felt, nailed to deck	1 Rofc	58	.138	Sq.	7.80	3.88		11.68	15.60
0030	Spot mopped to deck	G-1	295	.190		11.10	5	1.39	17.49	23
0040	Fully mopped to deck	"	192	.292		15.30	7.70	2.13	25.13	33
0050	#15 organic felt, nailed to deck	1 Rofc	58	.138		6.55	3.88		10.43	14.20
0060	Spot mopped to deck	G-1	295	.190		9.85	5	1.39	16.24	21.50
0070	Fully mopped to deck	"	192	.292	↓	14.05	7.70	2.13	23.88	31.50
2100	APP mod., smooth surf. cap sheet, poly. reinf., torched, 160 mils	G-5	2100	.019	S.F.	.48	.49	.08	1.05	1.50
2150	170 mils		2100	.019		.55	.49	.08	1.12	1.58
2200	Granule surface cap sheet, poly. reinf., torched, 180 mils		2000	.020		.59	.51	.08	1.18	1.67
2250	Smooth surface flashing, torched, 160 mils		1260	.032		.48	.81	.13	1.42	2.15
2300	170 mils		1260	.032		.55	.81	.13	1.49	2.23
2350	Granule surface flashing, torched, 180 mils	↓	1260	.032		.59	.81	.13	1.53	2.27
2400	Fibrated aluminum coating	1 Rofc	3800	.002	↓	.11	.06		.17	.23

07 52 16 – Styrene-Butadiene-Styrene Modified Bituminous Membrane Roofing

07 52 16.10 SBS Modified Bituminous Membrane

		Crew	Daily Output	Labor-Hours	Unit	Material	Labor	Equipment	Total	Total Incl O&P
0010	**SBS MODIFIED BITUMINOUS MEMBRANE**									
0080	SBS modified, granule surf cap sheet, polyester rein., mopped									
0600	150 mils	G-1	2000	.028	S.F.	.58	.74	.20	1.52	2.20
1100	160 mils		2000	.028		.77	.74	.20	1.71	2.41
1500	Glass fiber reinforced, mopped, 160 mils		2000	.028		.49	.74	.20	1.43	2.10
1600	Smooth surface cap sheet, mopped, 145 mils		2100	.027		.49	.70	.20	1.39	2.02
1700	Smooth surface flashing, 145 mils		1260	.044		.49	1.17	.32	1.98	3.01
1800	150 mils		1260	.044		.48	1.17	.32	1.97	3
1900	Granular surface flashing, 150 mils		1260	.044		.58	1.17	.32	2.07	3.11
2000	160 mils	↓	1260	.044	↓	.77	1.17	.32	2.26	3.32

07 53 Elastomeric Membrane Roofing

07 53 16 – Chlorosulfonate-Polyethylene Roofing

07 53 16.10 Chlorosulfonated Polyethylene Roofing		Crew	Daily Output	Labor-Hours	Unit	Material	2011 Bare Costs Labor	Equipment	Total	Total Incl O&P
0010	**CHLOROSULFONATED POLYETHYLENE ROOFING**									
0800	Chlorosulfonated polyethylene-hypalon (CSPE), 45 mils,									
0900	0.29 P.S.F., fully adhered	G-5	26	1.538	Sq.	185	39.50	6.50	231	282
1100	Loose-laid & ballasted with stone (10 P.S.F.)		51	.784		193	20	3.30	216.30	252
1200	Mechanically attached		35	1.143		186	29.50	4.81	220.31	263
1300	Plates with adhesive attachment		35	1.143		183	29.50	4.81	217.31	259

07 53 23 – Ethylene-Propylene-Diene-Monomer Roofing

07 53 23.20 Ethylene-Propylene-Diene-Monomer Roofing		Crew	Daily Output	Labor-Hours	Unit	Material	Labor	Equipment	Total	Total Incl O&P
0010	**ETHYLENE-PROPYLENE-DIENE-MONOMER ROOFING (E.P.D.M.)**									
3500	Ethylene-propylene-diene-monomer (EPDM), 45 mils, 0.28 P.S.F.									
3600	Loose-laid & ballasted with stone (10 P.S.F.)	G-5	51	.784	Sq.	73.50	20	3.30	96.80	121
3700	Mechanically attached		35	1.143		65.50	29.50	4.81	99.81	130
3800	Fully adhered with adhesive		26	1.538		101	39.50	6.50	147	189
4500	60 mils, 0.40 P.S.F.									
4600	Loose-laid & ballasted with stone (10 P.S.F.)	G-5	51	.784	Sq.	87	20	3.30	110.30	136
4700	Mechanically attached		35	1.143		77.50	29.50	4.81	111.81	143
4800	Fully adhered with adhesive		26	1.538		113	39.50	6.50	159	202
4810	45 mil, .28 PSF, membrane only					39.50			39.50	43.50
4820	60 mil, .40 PSF, membrane only					50.50			50.50	55.50
4850	Seam tape for membrane, 3" x 100' roll				Ea.	48			48	53
4900	Batten strips, 10' sections					3.05			3.05	3.36
4910	Cover tape for batten strips, 6" x 100' roll					110			110	121
4930	Plate anchors				M	105			105	116
4970	Adhesive for fully adhered systems, 60 S.F./gal.				Gal.	21			21	23

07 53 29 – Polyisobutylene Roofing

07 53 29.10 Polyisobutylene Roofing		Crew	Daily Output	Labor-Hours	Unit	Material	Labor	Equipment	Total	Total Incl O&P
0010	**POLYISOBUTYLENE ROOFING**									
7500	Polyisobutylene (PIB), 100 mils, 0.57 P.S.F.									
7600	Loose-laid & ballasted with stone/gravel (10 P.S.F.)	G-5	51	.784	Sq.	181	20	3.30	204.30	239
7700	Partially adhered with adhesive		35	1.143		227	29.50	4.81	261.31	310
7800	Hot asphalt attachment		35	1.143		217	29.50	4.81	251.31	296
7900	Fully adhered with contact cement		26	1.538		234	39.50	6.50	280	335

07 54 Thermoplastic Membrane Roofing

07 54 19 – Polyvinyl-Chloride Roofing

07 54 19.10 Polyvinyl-Chloride Roofing (P.V.C.)		Crew	Daily Output	Labor-Hours	Unit	Material	Labor	Equipment	Total	Total Incl O&P
0010	**POLYVINYL-CHLORIDE ROOFING (P.V.C.)**									
8200	Heat welded seams									
8700	Reinforced, 48 mils, 0.33 P.S.F.									
8750	Loose-laid & ballasted with stone/gravel (12 P.S.F.)	G-5	51	.784	Sq.	102	20	3.30	125.30	152
8800	Mechanically attached		35	1.143		94	29.50	4.81	128.31	161
8850	Fully adhered with adhesive		26	1.538		127	39.50	6.50	173	217
8860	Reinforced, 60 mils, .40 P.S.F.									
8870	Loose-laid & ballasted with stone/gravel (12 P.S.F.)	G-5	51	.784	Sq.	103	20	3.30	126.30	154
8880	Mechanically attached		35	1.143		95.50	29.50	4.81	129.81	163
8890	Fully adhered with adhesive		26	1.538		128	39.50	6.50	174	219

07 54 Thermoplastic Membrane Roofing

07 54 23 – Thermoplastic-Polyolefin Roofing

07 54 23.10 Thermoplastic Polyolefin Roofing (T.P.O)		Crew	Daily Output	Labor-Hours	Unit	Material	Labor	Equipment	Total	Total Incl O&P
						2011 Bare Costs				
0010	**THERMOPLASTIC POLYOLEFIN ROOFING (T.P.O.)**									
0100	45 mils, loose laid & ballasted with stone(1/2 ton/sq.)	G-5	51	.784	Sq.	83.50	20	3.30	106.80	132
0120	Fully adhered		25	1.600		72.50	41	6.75	120.25	161
0140	Mechanically attached		34	1.176		72	30	4.95	106.95	139
0160	Self adhered		35	1.143		88	29.50	4.81	122.31	155
0180	60 mil membrane, heat welded seams, ballasted		50	.800		89	20.50	3.37	112.87	139
0200	Fully adhered		25	1.600		77	41	6.75	124.75	166
0220	Mechanically attached		34	1.176		80.50	30	4.95	115.45	148
0240	Self adhered	↓	35	1.143	↓	109	29.50	4.81	143.31	178

07 57 Coated Foamed Roofing

07 57 13 – Sprayed Polyurethane Foam Roofing

07 57 13.10 Sprayed Polyurethane Foam Roofing (S.P.F.)		Crew	Daily Output	Labor-Hours	Unit	Material	Labor	Equipment	Total	Total Incl O&P
0010	**SPRAYED POLYURETHANE FOAM ROOFING (S.P.F.)**									
0100	Primer for metal substrate (when required)	G-2A	3000	.008	S.F.	.42	.20	.20	.82	1.03
0200	Primer for non-metal substrate (when required)		3000	.008		.16	.20	.20	.56	.75
0300	Closed cell spray, polyurethane foam, 3 lb. per C.F. density, 1", R6.7		15000	.002		.55	.04	.04	.63	.72
0400	2", R13.4		13125	.002		1.10	.05	.05	1.20	1.34
0500	3", R18.6		11485	.002		1.66	.05	.05	1.76	1.97
0550	4", R24.8		10080	.002		2.21	.06	.06	2.33	2.60
0700	Spray-on silicone coating	↓	2500	.010		.92	.24	.24	1.40	1.70
0800	Warranty 5-20 year manufacturer's									.15
0900	Warranty 20 year, no dollar limit				↓					.20

07 58 Roll Roofing

07 58 10 – Asphalt Roll Roofing

07 58 10.10 Roll Roofing		Crew	Daily Output	Labor-Hours	Unit	Material	Labor	Equipment	Total	Total Incl O&P
0010	**ROLL ROOFING**									
0100	Asphalt, mineral surface									
0200	1 ply #15 organic felt, 1 ply mineral surfaced									
0300	Selvage roofing, lap 19", nailed & mopped	G-1	27	2.074	Sq.	60.50	54.50	15.15	130.15	182
0400	3 plies glass fiber felt (type IV), 1 ply mineral surfaced									
0500	Selvage roofing, lapped 19", mopped	G-1	25	2.240	Sq.	98	59	16.35	173.35	233
0600	Coated glass fiber base sheet, 2 plies of glass fiber									
0700	Felt (type IV), 1 ply mineral surfaced selvage									
0800	Roofing, lapped 19", mopped	G-1	25	2.240	Sq.	108	59	16.35	183.35	244
0900	On nailable decks	"	24	2.333	"	99.50	61.50	17.05	178.05	239
1000	3 plies glass fiber felt (type III), 1 ply mineral surfaced									
1100	Selvage roofing, lapped 19", mopped	G-1	25	2.240	Sq.	98	59	16.35	173.35	233

07 61 Sheet Metal Roofing

07 61 13 – Standing Seam Sheet Metal Roofing

07 61 13.10 Standing Seam Sheet Metal Roofing, Field Fab.		Crew	Daily Output	Labor-Hours	Unit	Material	2011 Bare Costs Labor	Equipment	Total	Total Incl O&P
0010	**STANDING SEAM SHEET METAL ROOFING, FIELD FABRICATED**									
0400	Copper standing seam roofing, over 10 squares, 16 oz, 125 lb per sq	1 Shee	1.30	6.154	Sq.	735	235		970	1,200
0600	18 oz, 140 lb per sq	"	1.20	6.667		825	255		1,080	1,325
1200	For abnormal conditions or small areas, add					25%	100%			
1300	For lead-coated copper, add					25%				

07 61 16 – Batten Seam Sheet Metal Roofing

07 61 16.10 Batten Seam Sheet Metal Roofing, Field Fabricated		Crew	Daily Output	Labor-Hours	Unit	Material	Labor	Equipment	Total	Total Incl O&P
0010	**BATTEN SEAM SHEET METAL ROOFING, FIELD FABRICATED**									
0012	Copper batten seam roofing, over 10 sq, 16 oz, 130 lb per sq	1 Shee	1.10	7.273	Sq.	935	278		1,213	1,475
0020	Lead batten seam roofing, 5 lb. per S.F.		1.20	6.667		600	255		855	1,075
0100	Zinc/copper alloy batten seam roofing, .020 thick		1.20	6.667		1,100	255		1,355	1,650
0200	Copper roofing, batten seam, over 10 sq, 18 oz, 145 lb per sq		1	8		1,050	305		1,355	1,650
0800	Zinc, copper alloy roofing, batten seam, .027" thick		1.15	6.957		1,550	266		1,816	2,175
0900	.032" thick		1.10	7.273		1,875	278		2,153	2,525
1000	.040" thick		1.05	7.619		2,100	291		2,391	2,775

07 61 19 – Flat Seam Sheet Metal Roofing

07 61 19.10 Flat Seam Sheet Metal Roofing, Field Fabricated		Crew	Daily Output	Labor-Hours	Unit	Material	Labor	Equipment	Total	Total Incl O&P
0010	**FLAT SEAM SHEET METAL ROOFING, FIELD FABRICATED**									
0900	Copper flat seam roofing, over 10 squares, 16 oz, 115 lb./sq.	1 Shee	1.20	6.667	Sq.	685	255		940	1,175
1100	Lead flat seam roofing, 5 lb. per S.F.	"	1.30	6.154	"	600	235		835	1,050

07 65 Flexible Flashing

07 65 10 – Sheet Metal Flashing

07 65 10.10 Sheet Metal Flashing and Counter Flashing		Crew	Daily Output	Labor-Hours	Unit	Material	Labor	Equipment	Total	Total Incl O&P
0010	**SHEET METAL FLASHING AND COUNTER FLASHING**									
0011	Including up to 4 bends									
0020	Aluminum, mill finish, .013" thick	1 Rofc	145	.055	S.F.	.73	1.55		2.28	3.60
0030	.016" thick		145	.055		.84	1.55		2.39	3.72
0060	.019" thick		145	.055		1.07	1.55		2.62	3.98
0100	.032" thick		145	.055		1.25	1.55		2.80	4.18
0200	.040" thick		145	.055		2.14	1.55		3.69	5.15
0300	.050" thick		145	.055		2.25	1.55		3.80	5.30
0325	Mill finish 5" x 7" step flashing, .016" thick		1920	.004	Ea.	.18	.12		.30	.41
0350	Mill finish 12" x 12" step flashing, .016" thick		1600	.005	"	.62	.14		.76	.93
0400	Painted finish, add				S.F.	.34			.34	.37
1000	Mastic-coated 2 sides, .005" thick	1 Rofc	330	.024		1.67	.68		2.35	3.07
1100	.016" thick		330	.024		1.84	.68		2.52	3.25
1600	Copper, 16 oz, sheets, under 1000 lb.		115	.070		5.90	1.96		7.86	10.05
1700	Over 4000 lb.		155	.052		5.55	1.45		7	8.70
1900	20 oz sheets, under 1000 lb.		110	.073		7.35	2.05		9.40	11.80
2000	Over 4000 lb.		145	.055		7	1.55		8.55	10.50
2200	24 oz sheets, under 1000 lb.		105	.076		9.25	2.14		11.39	14.05
2500	32 oz sheets, under 1000 lb.		100	.080		11.10	2.25		13.35	16.25
2700	W shape for valleys, 16 oz, 24" wide		100	.080	L.F.	12	2.25		14.25	17.25
5800	Lead, 2.5 lb. per S.F., up to 12" wide		135	.059	S.F.	5.40	1.67		7.07	8.95
5900	Over 12" wide		135	.059		4.43	1.67		6.10	7.90
8900	Stainless steel sheets, 32 ga, .010" thick		155	.052		4	1.45		5.45	7
9000	28 ga, .015" thick		155	.052		5.65	1.45		7.10	8.80
9100	26 ga, .018" thick		155	.052		7.25	1.45		8.70	10.60
9200	24 ga, .025" thick		155	.052		8.90	1.45		10.35	12.40

07 65 Flexible Flashing

07 65 10 – Sheet Metal Flashing

07 65 10.10 Sheet Metal Flashing and Counter Flashing		Crew	Daily Output	Labor-Hours	Unit	2011 Bare Costs Material	Labor	Equipment	Total	Total Incl O&P
9290	For mechanically keyed flashing, add					40%				
9320	Steel sheets, galvanized, 20 gauge	1 Rofc	130	.062	S.F.	1.57	1.73		3.30	4.86
9340	30 gauge		160	.050		1.46	1.41		2.87	4.15
9400	Terne coated stainless steel, .015" thick, 28 ga		155	.052		7.35	1.45		8.80	10.70
9500	.018" thick, 26 ga		155	.052		8.30	1.45		9.75	11.70
9600	Zinc and copper alloy (brass), .020" thick		155	.052		11.10	1.45		12.55	14.80
9700	.027" thick		155	.052		15.60	1.45		17.05	19.75
9800	.032" thick		155	.052		18.80	1.45		20.25	23
9900	.040" thick		155	.052		21	1.45		22.45	25.50

07 65 12 – Fabric and Mastic Flashings

07 65 12.10 Fabric and Mastic Flashing and Counter Flashing		Crew	Daily Output	Labor-Hours	Unit	Material	Labor	Equipment	Total	Total Incl O&P
0010	**FABRIC AND MASTIC FLASHING AND COUNTER FLASHING**									
1300	Asphalt flashing cement, 5 gallon				Gal.	10.80			10.80	11.85
4900	Fabric, asphalt-saturated cotton, specification grade	1 Rofc	35	.229	S.Y.	2.36	6.45		8.81	14.20
5000	Utility grade		35	.229		1.26	6.45		7.71	13
5300	Close-mesh fabric, saturated, 17 oz per S.Y.		35	.229		2.36	6.45		8.81	14.20
5500	Fiberglass, resin-coated		35	.229		1.23	6.45		7.68	12.95
8500	Shower pan, bituminous membrane, 7 oz		155	.052	S.F.	2	1.45		3.45	4.82

07 65 13 – Laminated Sheet Flashing

07 65 13.10 Laminated Sheet Flashing		Crew	Daily Output	Labor-Hours	Unit	Material	Labor	Equipment	Total	Total Incl O&P
0010	**LAMINATED SHEET FLASHING**, Including up to 4 bends									
0500	Fabric-backed 2 sides, .004" thick	1 Rofc	330	.024	S.F.	1.41	.68		2.09	2.78
0700	.005" thick		330	.024		1.67	.68		2.35	3.07
0750	Mastic-backed, self adhesive		460	.017		3.46	.49		3.95	4.69
0800	Mastic-coated 2 sides, .004" thick		330	.024		1.41	.68		2.09	2.78
2800	Copper, paperbacked 1 side, 2 oz		330	.024		1.25	.68		1.93	2.61
2900	3 oz		330	.024		1.63	.68		2.31	3.02
3100	Paperbacked 2 sides, 2 oz		330	.024		1.26	.68		1.94	2.62
3150	3 oz		330	.024		1.62	.68		2.30	3.01
3200	5 oz		330	.024		2.43	.68		3.11	3.90
3400	Mastic-backed 2 sides, copper, 2 oz		330	.024		1.49	.68		2.17	2.87
3500	3 oz		330	.024		1.83	.68		2.51	3.24
3700	5 oz		330	.024		2.68	.68		3.36	4.18
3800	Fabric-backed 2 sides, copper, 2 oz		330	.024		1.57	.68		2.25	2.96
4000	3 oz		330	.024		2.04	.68		2.72	3.47
4100	5 oz		330	.024		2.76	.68		3.44	4.27
4300	Copper-clad stainless steel, .015" thick, under 500 lb.		115	.070		5.80	1.96		7.76	9.95
4600	.018" thick, under 500 lb.		100	.080		6.55	2.25		8.80	11.25
6100	Lead-coated copper, fabric-backed, 2 oz		330	.024		2.42	.68		3.10	3.89
6200	5 oz		330	.024		2.78	.68		3.46	4.29
6400	Mastic-backed 2 sides, 2 oz		330	.024		1.89	.68		2.57	3.31
6500	5 oz		330	.024		2.34	.68		3.02	3.80
6700	Paperbacked 1 side, 2 oz		330	.024		1.63	.68		2.31	3.02
6800	3 oz		330	.024		1.92	.68		2.60	3.34
7000	Paperbacked 2 sides, 2 oz		330	.024		1.69	.68		2.37	3.09
7100	5 oz		330	.024		2.75	.68		3.43	4.26
8550	3 ply copper and fabric, 3 oz		155	.052		2.32	1.45		3.77	5.15
8600	7 oz		155	.052		3.25	1.45		4.70	6.20
8700	Lead on copper and fabric, 5 oz		155	.052		2.78	1.45		4.23	5.70
8800	7 oz		155	.052		5.25	1.45		6.70	8.40
9300	Stainless steel, paperbacked 2 sides, .005" thick		330	.024		3.62	.68		4.30	5.20

07 65 Flexible Flashing

07 65 19 – Plastic Sheet Flashing

07 65 19.10 Plastic Sheet Flashing and Counter Flashing		Crew	Daily Output	Labor-Hours	Unit	2011 Bare Costs Material	Labor	Equipment	Total	Total Incl O&P
0010	PLASTIC SHEET FLASHING AND COUNTER FLASHING									
7300	Polyvinyl chloride, black, .010" thick	1 Rofc	285	.028	S.F.	.21	.79		1	1.66
7400	.020" thick		285	.028		.30	.79		1.09	1.76
7600	.030" thick		285	.028		.39	.79		1.18	1.86
7700	.056" thick		285	.028		.94	.79		1.73	2.46
7900	Black or white for exposed roofs, .060" thick		285	.028		2	.79		2.79	3.63
8060	PVC tape, 5" x 45 mils, for joint covers, 100 L.F./roll				Ea.	128			128	141

07 65 23 – Rubber Sheet Flashing

07 65 23.10 Rubber Sheet Flashing and Counterflashing		Crew	Daily Output	Labor-Hours	Unit	Material	Labor	Equipment	Total	Total Incl O&P
0010	RUBBER SHEET FLASHING AND COUNTERFLASHING									
4750	EPDM Cured	1 Rofc	285	.028	S.F.	1.01	.79		1.80	2.54
4800	Uncured		285	.028		1.10	.79		1.89	2.64
7150	Neoprene, 60 mil		285	.028		1.92	.79		2.71	3.54
7160	Self-curing		285	.028		2.26	.79		3.05	3.92
7170	Uncured		285	.028		1.92	.79		2.71	3.54
8100	Rubber, butyl, 1/32" thick		285	.028		1.75	.79		2.54	3.36
8200	1/16" thick		285	.028		2.38	.79		3.17	4.05
8300	Neoprene, cured, 1/16" thick		285	.028		1.92	.79		2.71	3.54
8400	1/8" thick		285	.028		4	.79		4.79	5.85

07 71 Roof Specialties

07 71 16 – Manufactured Counterflashing Systems

07 71 16.10 Roof Drain Boot		Crew	Daily Output	Labor-Hours	Unit	Material	Labor	Equipment	Total	Total Incl O&P
0010	ROOF DRAIN BOOT									
0100	Cast iron, 2" x 3"	1 Shee	125	.064	L.F.	114	2.44		116.44	129
0200	3" x 4"		125	.064		227	2.44		229.44	254
0300	4" x 5"		125	.064		238	2.44		240.44	266
0400	4" x 6"		125	.064		286	2.44		288.44	320

07 71 19 – Manufactured Gravel Stops and Fascias

07 71 19.10 Gravel Stop		Crew	Daily Output	Labor-Hours	Unit	Material	Labor	Equipment	Total	Total Incl O&P
0010	GRAVEL STOP									
0020	Aluminum, .050" thick, 4" face height, mill finish	1 Shee	145	.055	L.F.	6.10	2.11		8.21	10.25
0080	Duranodic finish		145	.055		5.90	2.11		8.01	10
0100	Painted		145	.055		6.85	2.11		8.96	11
1350	Galv steel, 24 ga., 4" leg, plain, with continuous cleat, 4" face		145	.055		2.65	2.11		4.76	6.40
1500	Polyvinyl chloride, 6" face height		135	.059		4.90	2.26		7.16	9.15
1800	Stainless steel, 24 ga., 6" face height		135	.059		12.60	2.26		14.86	17.60

07 71 19.30 Fascia		Crew	Daily Output	Labor-Hours	Unit	Material	Labor	Equipment	Total	Total Incl O&P
0010	FASCIA									
0100	Aluminum, reverse board and batten, .032" thick, colored, no furring incl	1 Shee	145	.055	S.F.	5.95	2.11		8.06	10.05
0200	Residential type, aluminum	1 Carp	200	.040	L.F.	1.29	1.34		2.63	3.67
0300	Steel, galv and enameled, stock, no furring, long panels	1 Shee	145	.055	S.F.	4	2.11		6.11	7.90
0600	Short panels	"	115	.070	"	5.45	2.66		8.11	10.40

07 71 23 – Manufactured Gutters and Downspouts

07 71 23.10 Downspouts		Crew	Daily Output	Labor-Hours	Unit	Material	Labor	Equipment	Total	Total Incl O&P
0010	DOWNSPOUTS									
0020	Aluminum 2" x 3", .020" thick, embossed	1 Shee	190	.042	L.F.	1.03	1.61		2.64	3.79
0100	Enameled		190	.042		1.26	1.61		2.87	4.05

07 71 Roof Specialties

07 71 23 – Manufactured Gutters and Downspouts

07 71 23.10 Downspouts		Crew	Daily Output	Labor-Hours	Unit	Material	2011 Bare Costs Labor	Equipment	Total	Total Incl O&P
0300	Enameled, .024" thick, 2" x 3"	1 Shee	180	.044	L.F.	1.89	1.70		3.59	4.88
0400	3" x 4"		140	.057		2.75	2.18		4.93	6.65
0600	Round, corrugated aluminum, 3" diameter, .020" thick		190	.042		1.60	1.61		3.21	4.42
0700	4" diameter, .025" thick		140	.057		2.47	2.18		4.65	6.35
0900	Wire strainer, round, 2" diameter		155	.052	Ea.	3.42	1.97		5.39	7
1000	4" diameter		155	.052		6.30	1.97		8.27	10.20
1200	Rectangular, perforated, 2" x 3"		145	.055		2.30	2.11		4.41	6
1300	3" x 4"		145	.055		3.32	2.11		5.43	7.15
1500	Copper, round, 16 oz., stock, 2" diameter		190	.042	L.F.	6.90	1.61		8.51	10.25
1600	3" diameter		190	.042		6.90	1.61		8.51	10.25
1800	4" diameter		145	.055		10.20	2.11		12.31	14.70
1900	5" diameter		130	.062		14.55	2.35		16.90	19.95
2100	Rectangular, corrugated copper, stock, 2" x 3"		190	.042		7.55	1.61		9.16	10.95
2200	3" x 4"		145	.055		9.95	2.11		12.06	14.45
2400	Rectangular, plain copper, stock, 2" x 3"		190	.042		6.70	1.61		8.31	10
2500	3" x 4"		145	.055		9.05	2.11		11.16	13.50
2700	Wire strainers, rectangular, 2" x 3"		145	.055	Ea.	5.50	2.11		7.61	9.55
2800	3" x 4"		145	.055		6.60	2.11		8.71	10.75
3000	Round, 2" diameter		145	.055		6.75	2.11		8.86	10.90
3100	3" diameter		145	.055		6.75	2.11		8.86	10.90
3300	4" diameter		145	.055		12.20	2.11		14.31	16.90
3400	5" diameter		115	.070		16.90	2.66		19.56	23
3600	Lead-coated copper, round, stock, 2" diameter		190	.042	L.F.	34	1.61		35.61	40
3700	3" diameter		190	.042		30	1.61		31.61	35.50
3900	4" diameter		145	.055		31	2.11		33.11	38
4300	Rectangular, corrugated, stock, 2" x 3"		190	.042		12.70	1.61		14.31	16.60
4500	Plain, stock, 2" x 3"		190	.042		16.35	1.61		17.96	20.50
4600	3" x 4"		145	.055		18.20	2.11		20.31	23.50
4800	Steel, galvanized, round, corrugated, 2" or 3" diameter, 28 gauge		190	.042		1.72	1.61		3.33	4.55
4900	4" diameter, 28 gauge		145	.055		2.35	2.11		4.46	6.05
5700	Rectangular, corrugated, 28 gauge, 2" x 3"		190	.042		1.28	1.61		2.89	4.07
5800	3" x 4"		145	.055		1.64	2.11		3.75	5.30
6000	Rectangular, plain, 28 gauge, galvanized, 2" x 3"		190	.042		1.28	1.61		2.89	4.07
6100	3" x 4"		145	.055		2.33	2.11		4.44	6.05
6300	Epoxy painted, 24 gauge, corrugated, 2" x 3"		190	.042		2	1.61		3.61	4.86
6400	3" x 4"		145	.055		2.50	2.11		4.61	6.25
6600	Wire strainers, rectangular, 2" x 3"		145	.055	Ea.	12	2.11		14.11	16.70
6700	3" x 4"		145	.055		14	2.11		16.11	18.90
6900	Round strainers, 2" or 3" diameter		145	.055		2.80	2.11		4.91	6.55
7000	4" diameter		145	.055		5.20	2.11		7.31	9.20
8200	Vinyl, rectangular, 2" x 3"		210	.038	L.F.	1.94	1.46		3.40	4.53
8300	Round, 2-1/2"		220	.036	"	1.28	1.39		2.67	3.70

07 71 23.20 Downspout Elbows		Crew	Daily Output	Labor-Hours	Unit	Material	Labor	Equipment	Total	Total Incl O&P
0010	**DOWNSPOUT ELBOWS**									
0020	Aluminum, 2" x 3", embossed	1 Shee	100	.080	Ea.	.90	3.06		3.96	6.05
0100	Enameled		100	.080		.96	3.06		4.02	6.10
0200	3" x 4", .025" thick, embossed		100	.080		3.50	3.06		6.56	8.90
0300	Enameled		100	.080		1.75	3.06		4.81	7
0400	Round corrugated, 3", embossed, .020" thick		100	.080		2.27	3.06		5.33	7.55
0500	4", .025" thick		100	.080		4.74	3.06		7.80	10.25
0600	Copper, 16 oz. round, 2" diameter		100	.080		8	3.06		11.06	13.85
0700	3" diameter		100	.080		10.55	3.06		13.61	16.65

07 71 Roof Specialties

07 71 23 – Manufactured Gutters and Downspouts

07 71 23.20 Downspout Elbows		Crew	Daily Output	Labor-Hours	Unit	2011 Bare Costs Material	Labor	Equipment	Total	Total Incl O&P
0800	4" diameter	1 Shee	100	.080	Ea.	12.40	3.06		15.46	18.70
1000	2" x 3" corrugated		100	.080		7.25	3.06		10.31	13.05
1100	3" x 4" corrugated		100	.080		12.15	3.06		15.21	18.40
1300	Vinyl, 2-1/2" diameter, 45° or 75°		100	.080		3.98	3.06		7.04	9.45
1400	Tee Y junction		75	.107		12.90	4.07		16.97	21

07 71 23.30 Gutters		Crew	Daily Output	Labor-Hours	Unit	Material	Labor	Equipment	Total	Total Incl O&P
0010	**GUTTERS**									
0012	Aluminum, stock units, 5" K type, .027" thick, plain	1 Shee	120	.067	L.F.	2.58	2.55		5.13	7.05
0020	Inside corner		25	.320	Ea.	8.65	12.20		20.85	29.50
0030	Outside corner		25	.320	"	8.65	12.20		20.85	29.50
0100	Enameled		120	.067	L.F.	2.51	2.55		5.06	6.95
0110	Inside corner		25	.320	Ea.	8.65	12.20		20.85	29.50
0120	Outside corner		25	.320	"	8.65	12.20		20.85	29.50
3000	Vinyl, O.G., 4" wide	1 Carp	110	.073	L.F.	1.08	2.44		3.52	5.25
3100	5" wide		110	.073		1.38	2.44		3.82	5.60
3200	4" half round, stock units		110	.073		1.08	2.44		3.52	5.25
3250	Joint connectors				Ea.	2.88			2.88	3.17
3300	Wood, clear treated cedar, fir or hemlock, 3" x 4"	1 Carp	100	.080	L.F.	7.95	2.68		10.63	13.25
3400	4" x 5"	"	100	.080	"	11.55	2.68		14.23	17.20

07 71 23.35 Gutter Guard		Crew	Daily Output	Labor-Hours	Unit	Material	Labor	Equipment	Total	Total Incl O&P
0010	**GUTTER GUARD**									
0020	6" wide strip, aluminum mesh	1 Carp	500	.016	L.F.	2.18	.54		2.72	3.30
0100	Vinyl mesh	"	500	.016	"	2.58	.54		3.12	3.74

07 71 26 – Reglets

07 71 26.10 Reglets and Accessories		Crew	Daily Output	Labor-Hours	Unit	Material	Labor	Equipment	Total	Total Incl O&P
0010	**REGLETS AND ACCESSORIES**									
0020	Aluminum, .025" thick, in concrete parapet	1 Carp	225	.036	L.F.	1.62	1.19		2.81	3.78
0100	Copper, 10 oz.		225	.036		2.62	1.19		3.81	4.88
0300	16 oz.		225	.036		5.10	1.19		6.29	7.65
0400	Galvanized steel, 24 gauge		225	.036		.94	1.19		2.13	3.03
0600	Stainless steel, .020" thick		225	.036		3.15	1.19		4.34	5.45
0700	Zinc and copper alloy, 20 oz.		225	.036		2.62	1.19		3.81	4.88
0900	Counter flashing for above, 12" wide, .032" aluminum	1 Shee	150	.053		1.63	2.04		3.67	5.15
1000	Copper, 10 oz.		150	.053		4.95	2.04		6.99	8.80
1200	16 oz.		150	.053		4.80	2.04		6.84	8.65
1300	Galvanized steel, .020" thick		150	.053		1.17	2.04		3.21	4.66
1500	Stainless steel, .020" thick		150	.053		4.69	2.04		6.73	8.50
1600	Zinc and copper alloy, 20 oz.		150	.053		4.78	2.04		6.82	8.60

07 71 43 – Drip Edge

07 71 43.10 Drip Edge, Rake Edge, Ice Belts		Crew	Daily Output	Labor-Hours	Unit	Material	Labor	Equipment	Total	Total Incl O&P
0010	**DRIP EDGE, RAKE EDGE, ICE BELTS**									
0020	Aluminum, .016" thick, 5" wide, mill finish	1 Carp	400	.020	L.F.	.38	.67		1.05	1.54
0100	White finish		400	.020		.38	.67		1.05	1.54
0200	8" wide, mill finish		400	.020		.46	.67		1.13	1.63
0300	Ice belt, 28" wide, mill finish		100	.080		3.85	2.68		6.53	8.75
0310	Vented, mill finish		400	.020		2.76	.67		3.43	4.16
0320	Painted finish		400	.020		2.86	.67		3.53	4.27
0400	Galvanized, 5" wide		400	.020		.36	.67		1.03	1.52
0500	8" wide, mill finish		400	.020		.69	.67		1.36	1.88
0510	Rake edge, aluminum, 1-1/2" x 1-1/2"		400	.020		.28	.67		.95	1.43
0520	3-1/2" x 1-1/2"		400	.020		.38	.67		1.05	1.54

07 72 Roof Accessories

07 72 26 – Ridge Vents

07 72 26.10 Ridge Vents and Accessories		Crew	Daily Output	Labor-Hours	Unit	Material	2011 Bare Costs Labor	Equipment	Total	Total Incl O&P
0010	RIDGE VENTS AND ACCESSORIES									
2300	Ridge vent strip, mill finish	1 Shee	155	.052	L.F.	2.73	1.97		4.70	6.25

07 72 33 – Roof Hatches

07 72 33.10 Roof Hatch Options

		Crew	Daily Output	Labor-Hours	Unit	Material	Labor	Equipment	Total	Total Incl O&P
0010	ROOF HATCH OPTIONS									
0500	2′-6″ x 3′, aluminum curb and cover	G-3	10	3.200	Ea.	985	103		1,088	1,250
0520	Galvanized steel curb and aluminum cover		10	3.200		595	103		698	825
0540	Galvanized steel curb and cover		10	3.200		550	103		653	775
0600	2′-6″ x 4′-6″, aluminum curb and cover		9	3.556		850	115		965	1,125
0800	Galvanized steel curb and aluminum cover		9	3.556		600	115		715	850
0900	Galvanized steel curb and cover		9	3.556		825	115		940	1,100
1100	4′ x 4′ aluminum curb and cover		8	4		1,675	129		1,804	2,050
1120	Galvanized steel curb and aluminum cover		8	4		1,675	129		1,804	2,050
1140	Galvanized steel curb and cover		8	4		1,500	129		1,629	1,875
1200	2′-6″ x 8′-0″, aluminum curb and cover		6.60	4.848		1,675	157		1,832	2,075
1400	Galvanized steel curb and aluminum cover		6.60	4.848		1,600	157		1,757	2,000
1500	Galvanized steel curb and cover		6.60	4.848		1,200	157		1,357	1,575
1800	For plexiglass panels, 2′-6″ x 3′-0″, add to above					400			400	440

07 72 53 – Snow Guards

07 72 53.10 Snow Guard Options

		Crew	Daily Output	Labor-Hours	Unit	Material	Labor	Equipment	Total	Total Incl O&P
0010	SNOW GUARD OPTIONS									
0200	Standing seam metal roofs, fastened with set screws	1 Rofc	48	.167	Ea.	14.60	4.69		19.29	24.50
0300	Surface mount for metal roofs, fastened with solder		48	.167	″	6.60	4.69		11.29	15.70
0400	Double rail pipe type, including pipe		130	.062	L.F.	22.50	1.73		24.23	28

07 72 73 – Pitch Pockets

07 72 73.10 Pitch Pockets, Variable Sizes

		Crew	Daily Output	Labor-Hours	Unit	Material	Labor	Equipment	Total	Total Incl O&P
0010	PITCH POCKETS, VARIABLE SIZES									
0100	Adjustable, 4″ to 7″, welded corners, 4″ deep	1 Rofc	48	.167	Ea.	12.90	4.69		17.59	22.50
0200	Side extenders, 6″	″	240	.033	″	1.97	.94		2.91	3.86

07 72 80 – Vents

07 72 80.30 Vent Options

		Crew	Daily Output	Labor-Hours	Unit	Material	Labor	Equipment	Total	Total Incl O&P
0010	VENT OPTIONS									
0020	Plastic, for insulated decks, 1 per M.S.F., minimum	1 Rofc	40	.200	Ea.	19	5.65		24.65	31
0100	Maximum		20	.400		38	11.25		49.25	62.50
0300	Aluminum		30	.267		24	7.50		31.50	40
0800	Polystyrene baffles, 12″ wide for 16″ O.C. rafter spacing	1 Carp	90	.089		.40	2.98		3.38	5.45
0900	For 24″ O.C. rafter spacing	″	110	.073		.75	2.44		3.19	4.91

07 81 Applied Fireproofing

07 81 16 – Cementitious Fireproofing

07 81 16.10 Sprayed Cementitious Fireproofing

		Crew	Daily Output	Labor-Hours	Unit	Material	Labor	Equipment	Total	Total Incl O&P
0010	SPRAYED CEMENTITIOUS FIREPROOFING									
0050	Not incl tamping or canvas protection									
0100	1″ thick, on flat plate steel	G-2	3000	.008	S.F.	.53	.23	.04	.80	1
0200	Flat decking		2400	.010		.53	.28	.05	.86	1.10
0400	Beams		1500	.016		.53	.45	.08	1.06	1.41
0500	Corrugated or fluted decks		1250	.019		.79	.54	.10	1.43	1.87
0700	Columns, 1-1/8″ thick		1100	.022		.59	.61	.11	1.31	1.79
0800	2-3/16″ thick		700	.034		1.13	.97	.18	2.28	3.03

07 81 Applied Fireproofing

07 81 16 – Cementitious Fireproofing

07 81 16.10 Sprayed Cementitious Fireproofing		Crew	Daily Output	Labor-Hours	Unit	Material (2011 Bare Costs)	Labor (2011 Bare Costs)	Equipment (2011 Bare Costs)	Total (2011 Bare Costs)	Total Incl O&P
0850	For tamping, add						10%			
0900	For canvas protection, add	G-2	5000	.005	S.F.	.07	.14	.03	.24	.33

07 84 Firestopping

07 84 13 – Penetration Firestopping

07 84 13.10 Firestopping

Line	Description	Crew	Daily Output	Labor-Hours	Unit	Material	Labor	Equipment	Total	Total Incl O&P
0010	**FIRESTOPPING** R078413-30									
0100	Metallic piping, non insulated									
0110	Through walls, 2" diameter	1 Carp	16	.500	Ea.	13.75	16.80		30.55	43
0120	4" diameter		14	.571		21	19.15		40.15	55
0130	6" diameter		12	.667		28.50	22.50		51	68.50
0140	12" diameter		10	.800		50	27		77	100
0150	Through floors, 2" diameter		32	.250		8.35	8.40		16.75	23
0160	4" diameter		28	.286		12	9.60		21.60	29.50
0170	6" diameter		24	.333		15.75	11.20		26.95	36
0180	12" diameter		20	.400		26.50	13.40		39.90	52
0190	Metallic piping, insulated									
0200	Through walls, 2" diameter	1 Carp	16	.500	Ea.	19.50	16.80		36.30	49.50
0210	4" diameter		14	.571		26.50	19.15		45.65	61.50
0220	6" diameter		12	.667		34	22.50		56.50	75
0230	12" diameter		10	.800		55.50	27		82.50	106
0240	Through floors, 2" diameter		32	.250		14.10	8.40		22.50	29.50
0250	4" diameter		28	.286		17.70	9.60		27.30	35.50
0260	6" diameter		24	.333		21.50	11.20		32.70	42
0270	12" diameter		20	.400		26.50	13.40		39.90	52
0280	Non metallic piping, non insulated									
0290	Through walls, 2" diameter	1 Carp	12	.667	Ea.	56.50	22.50		79	100
0300	4" diameter		10	.800		71	27		98	123
0310	6" diameter		8	1		99	33.50		132.50	165
0330	Through floors, 2" diameter		16	.500		44.50	16.80		61.30	77
0340	4" diameter		6	1.333		55	44.50		99.50	136
0350	6" diameter		6	1.333		66	44.50		110.50	148
0370	Ductwork, insulated & non insulated, round									
0380	Through walls, 6" diameter	1 Carp	12	.667	Ea.	29	22.50		51.50	69.50
0390	12" diameter		10	.800		57	27		84	108
0400	18" diameter		8	1		93	33.50		126.50	159
0410	Through floors, 6" diameter		16	.500		15.75	16.80		32.55	45.50
0420	12" diameter		14	.571		29	19.15		48.15	64
0430	18" diameter		12	.667		50	22.50		72.50	92.50
0440	Ductwork, insulated & non insulated, rectangular									
0450	With stiffener/closure angle, through walls, 6" x 12"	1 Carp	8	1	Ea.	24	33.50		57.50	82
0460	12" x 24"		6	1.333		31.50	44.50		76	110
0470	24" x 48"		4	2		90.50	67		157.50	212
0480	With stiffener/closure angle, through floors, 6" x 12"		10	.800		12.95	27		39.95	59
0490	12" x 24"		8	1		23.50	33.50		57	81.50
0500	24" x 48"		6	1.333		45.50	44.50		90	125
0510	Multi trade openings									
0520	Through walls, 6" x 12"	1 Carp	2	4	Ea.	50	134		184	280
0530	12" x 24"	"	1	8		202	268		470	670
0540	24" x 48"	2 Carp	1	16		810	535		1,345	1,800
0550	48" x 96"	"	.75	21.333		3,250	715		3,965	4,775

07 84 Firestopping

07 84 13 – Penetration Firestopping

07 84 13.10 Firestopping		Crew	Daily Output	Labor-Hours	Unit	Material	2011 Bare Costs Labor	Equipment	Total	Total Incl O&P
0560	Through floors, 6" x 12"	1 Carp	2	4	Ea.	50	134		184	280
0570	12" x 24"	"	1	8		202	268		470	670
0580	24" x 48"	2 Carp	.75	21.333		810	715		1,525	2,100
0590	48" x 96"	"	.50	32		3,250	1,075		4,325	5,375
0600	Structural penetrations, through walls									
0610	Steel beams, W8 x 10	1 Carp	8	1	Ea.	31.50	33.50		65	91
0620	W12 x 14		6	1.333		50	44.50		94.50	130
0630	W21 x 44		5	1.600		100	53.50		153.50	201
0640	W36 x 135		3	2.667		244	89.50		333.50	420
0650	Bar joists, 18" deep		6	1.333		46	44.50		90.50	126
0660	24" deep		6	1.333		57	44.50		101.50	138
0670	36" deep		5	1.600		86	53.50		139.50	185
0680	48" deep		4	2		100	67		167	223
0690	Construction joints, floor slab at exterior wall									
0700	Precast, brick, block or drywall exterior									
0710	2" wide joint	1 Carp	125	.064	L.F.	7.15	2.15		9.30	11.50
0720	4" wide joint	"	75	.107	"	14.30	3.58		17.88	22
0730	Metal panel, glass or curtain wall exterior									
0740	2" wide joint	1 Carp	40	.200	L.F.	17	6.70		23.70	30
0750	4" wide joint	"	25	.320	"	23.50	10.75		34.25	43.50
0760	Floor slab to drywall partition									
0770	Flat joint	1 Carp	100	.080	L.F.	7.05	2.68		9.73	12.25
0780	Fluted joint		50	.160		14.30	5.35		19.65	25
0790	Etched fluted joint		75	.107		9.30	3.58		12.88	16.25
0800	Floor slab to concrete/masonry partition									
0810	Flat joint	1 Carp	75	.107	L.F.	15.75	3.58		19.33	23.50
0820	Fluted joint	"	50	.160	"	18.65	5.35		24	29.50
0830	Concrete/CMU wall joints									
0840	1" wide	1 Carp	100	.080	L.F.	8.60	2.68		11.28	13.95
0850	2" wide		75	.107		15.75	3.58		19.33	23.50
0860	4" wide		50	.160		30	5.35		35.35	42
0870	Concrete/CMU floor joints									
0880	1" wide	1 Carp	200	.040	L.F.	4.31	1.34		5.65	7
0890	2" wide		150	.053		7.90	1.79		9.69	11.65
0900	4" wide		100	.080		15.05	2.68		17.73	21

07 91 Preformed Joint Seals

07 91 13 – Compression Seals

07 91 13.10 Compression Seals		Crew	Daily Output	Labor-Hours	Unit	Material	Labor	Equipment	Total	Total Incl O&P
0010	**COMPRESSION SEALS**									
4900	O-ring type cord, 1/4"	1 Bric	472	.017	L.F.	.47	.57		1.04	1.46
4910	1/2"		440	.018		1.78	.61		2.39	2.96
4920	3/4"		424	.019		3.61	.64		4.25	5
4930	1"		408	.020		6.95	.66		7.61	8.75
4940	1-1/4"		384	.021		13.60	.70		14.30	16.10
4950	1-1/2"		368	.022		17.20	.73		17.93	20
4960	1-3/4"		352	.023		36	.76		36.76	41
4970	2"		344	.023		63.50	.78		64.28	71.50

07 91 Preformed Joint Seals

07 91 16 – Joint Gaskets

07 91 16.10 Joint Gaskets		Crew	Daily Output	Labor-Hours	Unit	2011 Bare Costs Material	Labor	Equipment	Total	Total Incl O&P
0010	**JOINT GASKETS**									
4400	Joint gaskets, neoprene, closed cell w/adh, 1/8" x 3/8"	1 Bric	240	.033	L.F.	.26	1.12		1.38	2.13
4500	1/4" x 3/4"		215	.037		.58	1.25		1.83	2.69
4700	1/2" x 1"		200	.040		1.30	1.35		2.65	3.64
4800	3/4" x 1-1/2"	↓	165	.048	↓	1.44	1.63		3.07	4.26

07 91 23 – Backer Rods

07 91 23.10 Backer Rods		Crew	Daily Output	Labor-Hours	Unit	Material	Labor	Equipment	Total	Total Incl O&P
0010	**BACKER RODS**									
0030	Backer rod, polyethylene, 1/4" diameter	1 Bric	4.60	1.739	C.L.F.	2.83	58.50		61.33	99
0050	1/2" diameter		4.60	1.739		6.60	58.50		65.10	103
0070	3/4" diameter		4.60	1.739		10.10	58.50		68.60	107
0090	1" diameter	↓	4.60	1.739	↓	18.70	58.50		77.20	117

07 91 26 – Joint Fillers

07 91 26.10 Joint Fillers		Crew	Daily Output	Labor-Hours	Unit	Material	Labor	Equipment	Total	Total Incl O&P
0010	**JOINT FILLERS**									
4360	Butyl rubber filler, 1/4" x 1/4"	1 Bric	290	.028	L.F.	.14	.93		1.07	1.68
4365	1/2" x 1/2"		250	.032		.57	1.08		1.65	2.39
4370	1/2" x 3/4"		210	.038		.85	1.28		2.13	3.04
4375	3/4" x 3/4"		230	.035		1.28	1.17		2.45	3.33
4380	1" x 1"	↓	180	.044	↓	1.71	1.50		3.21	4.33
4390	For coloring, add					12%				
4980	Polyethylene joint backing, 1/4" x 2"	1 Bric	2.08	3.846	C.L.F.	12	129		141	225
4990	1/4" x 6"		1.28	6.250	"	36	210		246	385
5600	Silicone, room temp vulcanizing foam seal, 1/4" x 1/2"		1312	.006	L.F.	.29	.21		.50	.66
5610	1/2" x 1/2"		656	.012		.57	.41		.98	1.30
5620	1/2" x 3/4"		442	.018		.86	.61		1.47	1.95
5630	3/4" x 3/4"		328	.024		1.29	.82		2.11	2.77
5640	1/8" x 1"		1312	.006		.29	.21		.50	.66
5650	1/8" x 3"		442	.018		.86	.61		1.47	1.95
5670	1/4" x 3"		295	.027		1.72	.91		2.63	3.39
5680	1/4" x 6"		148	.054		3.44	1.82		5.26	6.75
5690	1/2" x 6"		82	.098		6.90	3.28		10.18	12.95
5700	1/2" x 9"		52.50	.152		10.30	5.15		15.45	19.75
5710	1/2" x 12"	↓	33	.242	↓	13.75	8.15		21.90	28.50

07 92 Joint Sealants

07 92 13 – Elastomeric Joint Sealants

07 92 13.20 Caulking and Sealant Options		Crew	Daily Output	Labor-Hours	Unit	Material	Labor	Equipment	Total	Total Incl O&P
0010	**CAULKING AND SEALANT OPTIONS**									
0050	Latex acrylic based, bulk				Gal.	27			27	29.50
0055	Bulk in place 1/4" x 1/4" bead	1 Bric	300	.027	L.F.	.09	.90		.99	1.56
0060	1/4" x 3/8"		294	.027		.14	.92		1.06	1.66
0065	1/4" x 1/2"		288	.028		.19	.93		1.12	1.74
0075	3/8" x 3/8"		284	.028		.21	.95		1.16	1.79
0080	3/8" x 1/2"		280	.029		.28	.96		1.24	1.89
0085	3/8" x 5/8"		276	.029		.35	.98		1.33	1.99
0095	3/8" x 3/4"		272	.029		.42	.99		1.41	2.09
0100	1/2" x 1/2"		275	.029		.38	.98		1.36	2.03
0105	1/2" x 5/8"	↓	269	.030	↓	.47	1		1.47	2.16

07 92 Joint Sealants

07 92 13 – Elastomeric Joint Sealants

07 92 13.20 Caulking and Sealant Options		Crew	Daily Output	Labor-Hours	Unit	Material	2011 Bare Costs Labor	Equipment	Total	Total Incl O&P
0110	1/2" x 3/4"	1 Bric	263	.030	L.F.	.57	1.02		1.59	2.31
0115	1/2" x 7/8"		256	.031		.66	1.05		1.71	2.46
0120	1/2" x 1"		250	.032		.76	1.08		1.84	2.60
0125	3/4" x 3/4"		244	.033		.85	1.10		1.95	2.75
0130	3/4" x 1"		225	.036		1.14	1.20		2.34	3.21
0135	1" x 1"	↓	200	.040	↓	1.52	1.35		2.87	3.88
0190	Cartridges				Gal.	22.50			22.50	24.50
0200	11 fl. oz cartridge				Ea.	1.93			1.93	2.12
0500	1/4" x 1/2"	1 Bric	288	.028	L.F.	.16	.93		1.09	1.70
0600	1/2" x 1/2"		275	.029		.32	.98		1.30	1.96
0800	3/4" x 3/4"		244	.033		.71	1.10		1.81	2.59
0900	3/4" x 1"		225	.036		.95	1.20		2.15	3
1000	1" x 1"	↓	200	.040	↓	1.18	1.35		2.53	3.51
1400	Butyl based, bulk				Gal.	26			26	28.50
1500	Cartridges				"	32			32	35
1700	1/4" x 1/2", 154 L.F./gal.	1 Bric	288	.028	L.F.	.17	.93		1.10	1.72
1800	1/2" x 1/2", 77 L.F./gal.	"	275	.029	"	.34	.98		1.32	1.98
2300	Polysulfide compounds, 1 component, bulk				Gal.	56			56	61.50
2400	Cartridges				"	54.50			54.50	60
2600	1 or 2 component, in place, 1/4" x 1/4", 308 L.F./gal.	1 Bric	300	.027	L.F.	.18	.90		1.08	1.67
2700	1/2" x 1/4", 154 L.F./gal.		288	.028		.36	.93		1.29	1.93
2900	3/4" x 3/8", 68 L.F./gal.		272	.029		.82	.99		1.81	2.53
3000	1" x 1/2", 38 L.F./gal.	↓	250	.032	↓	1.47	1.08		2.55	3.39
3200	Polyurethane, 1 or 2 component				Gal.	50			50	55
3300	Cartridges				"	70			70	77
3500	Bulk, in place, 1/4" x 1/4"	1 Bric	300	.027	L.F.	.16	.90		1.06	1.65
3655	1/2" x 1/4"		288	.028		.32	.93		1.25	1.89
3800	3/4" x 3/8"		272	.029		.74	.99		1.73	2.43
3900	1" x 1/2"	↓	250	.032	↓	1.30	1.08		2.38	3.20
4100	Silicone rubber, bulk				Gal.	41			41	45
4200	Cartridges				"	41			41	45

07 92 19 – Acoustical Joint Sealants

07 92 19.10 Acoustical Sealant		Crew	Daily Output	Labor-Hours	Unit	Material	Labor	Equipment	Total	Total Incl O&P
0010	**ACOUSTICAL SEALANT**									
0020	Acoustical sealant, elastomeric, cartridges				Ea.	2.30			2.30	2.53
0025	In place, 1/4" x 1/4"	1 Bric	300	.027	L.F.	.09	.90		.99	1.57
0030	1/4" x 1/2"		288	.028		.19	.93		1.12	1.74
0035	1/2" x 1/2"		275	.029		.38	.98		1.36	2.02
0040	1/2" x 3/4"		263	.030		.56	1.02		1.58	2.30
0045	3/4" x 3/4"		244	.033		.85	1.10		1.95	2.74
0050	1" x 1"	↓	200	.040	↓	1.50	1.35		2.85	3.86

Division Notes

		CREW	DAILY OUTPUT	LABOR-HOURS	UNIT	2011 BARE COSTS MAT.	LABOR	EQUIP.	TOTAL	TOTAL INCL O&P

Estimating Tips

08 10 00 Doors and Frames

All exterior doors should be addressed for their energy conservation (insulation and seals).

- Most metal doors and frames look alike, but there may be significant differences among them. When estimating these items, be sure to choose the line item that most closely compares to the specification or door schedule requirements regarding:
  - — type of metal
  - — metal gauge
  - — door core material
  - — fire rating
  - — finish
- Wood and plastic doors vary considerably in price. The primary determinant is the veneer material. Lauan, birch, and oak are the most common veneers. Other variables include the following:
  - — hollow or solid core
  - — fire rating
  - — flush or raised panel
  - — finish

08 30 00 Specialty Doors and Frames

- There are many varieties of special doors, and they are usually priced per each. Add frames, hardware, or operators required for a complete installation.

08 40 00 Entrances, Storefronts, and Curtain Walls

- Glazed curtain walls consist of the metal tube framing and the glazing material. The cost data in this subdivision is presented for the metal tube framing alone or the composite wall. If your estimate requires a detailed takeoff of the framing, be sure to add the glazing cost.

08 50 00 Windows

- Most metal windows are delivered preglazed. However, some metal windows are priced without glass. Refer to 08 80 00 Glazing for glass pricing. The grade C indicates commercial grade windows, usually ASTM C-35.
- All wood windows and vinyl are priced preglazed. The glazing is insulating glass. Some wood windows may have single pane float glass. Add the cost of screens and grills if required, and not already included.

08 70 00 Hardware

- Hardware costs add considerably to the cost of a door. The most efficient method to determine the hardware requirements for a project is to review the door schedule.
- Door hinges are priced by the pair, with most doors requiring 1-1/2 pairs per door. The hinge prices do not include installation labor because it is included in door installation. Hinges are classified according to the frequency of use.

08 80 00 Glazing

- Different openings require different types of glass. The most common types are:
  - — float
  - — tempered
  - — insulating
  - — impact-resistant
  - — ballistic-resistant
- Most exterior windows are glazed with insulating glass. Entrance doors and window walls, where the glass is less than 18″ from the floor, are generally glazed with tempered glass. Interior windows and some residential windows are glazed with float glass.
- Coastal communities require the use of impact-resistant glass.
- The insulation or 'u' value is a strong consideration, along with solar heat gain, to determine total energy efficiency.

Reference Numbers

Reference numbers are shown in shaded boxes at the beginning of some major classifications. These numbers refer to related items in the Reference Section. The reference information may be an estimating procedure, an alternate pricing method, or technical information.

Note: Not all subdivisions listed here necessarily appear in this publication.

08 01 Operation and Maintenance of Openings

08 01 53 – Operation and Maintenance of Plastic Windows

08 01 53.81 Solid Vinyl Replacement Windows			Crew	Daily Output	Labor-Hours	Unit	Material	2011 Bare Costs Labor	Equipment	Total	Total Incl O&P
0010	SOLID VINYL REPLACEMENT WINDOWS	R085313-20									
0020	Double hung, insulated glass, up to 83 united inches	G	2 Carp	8	2	Ea.	330	67		397	470
0040	84 to 93	G		8	2		360	67		427	505
0060	94 to 101	G		6	2.667		360	89.50		449.50	545
0080	102 to 111	G		6	2.667		380	89.50		469.50	565
0100	112 to 120	G		6	2.667		410	89.50		499.50	605
0120	For each united inch over 120 , add	G		800	.020	Inch	5.20	.67		5.87	6.80
0140	Casement windows, one operating sash , 42 to 60 united inches	G		8	2	Ea.	212	67		279	345
0160	61 to 70	G		8	2		241	67		308	375
0180	71 to 80	G		8	2		261	67		328	400
0200	81 to 96	G		8	2		277	67		344	415
0220	Two operating sash, 58 to 78 united inches	G		8	2		425	67		492	575
0240	79 to 88	G		8	2		450	67		517	605
0260	89 to 98	G		8	2		490	67		557	650
0280	99 to 108	G		6	2.667		515	89.50		604.50	715
0300	109 to 121	G		6	2.667		555	89.50		644.50	760
0320	Two operating.one fixed sash, 73 to 108 united inches	G		8	2		670	67		737	845
0340	109 to 118	G		8	2		705	67		772	885
0360	119 to 128	G		6	2.667		725	89.50		814.50	945
0380	129 to 138	G		6	2.667		775	89.50		864.50	1,000
0400	139 to 156	G		6	2.667		815	89.50		904.50	1,050
0420	Four operating sash, 98 to 118 united inches	G		8	2		960	67		1,027	1,150
0440	119 to 128	G		8	2		1,025	67		1,092	1,225
0460	129 to 138	G		6	2.667		1,100	89.50		1,189.50	1,350
0480	139 to 148	G		6	2.667		1,150	89.50		1,239.50	1,400
0500	149 to 168	G		6	2.667		1,225	89.50		1,314.50	1,500
0520	169 to 178	G		6	2.667		1,325	89.50		1,414.50	1,600
0560	Fixed picture window, up to 63 united inches	G		8	2		144	67		211	270
0580	64 to 83	G		8	2		170	67		237	299
0600	84 to 101	G		8	2		220	67		287	355
0620	For each united inch over 101, add	G		900	.018	Inch	2.50	.60		3.10	3.75
0760	Muntins, between glazing, square, per lite					Ea.	6			6	6.60
0780	Diamond shape, per full or partial diamond					″	8			8	8.80
0800	Cellulose fiber insulation, poured into sash balance cavity	G	1 Carp	36	.222	C.F.	.56	7.45		8.01	13.10
0820	Silicone caulking at perimeter	G	″	800	.010	L.F.	.13	.34		.47	.71

08 05 Common Work Results for Openings

08 05 05 – Selective Windows and Doors Demolition

08 05 05.10 Selective Demolition Doors			Crew	Daily Output	Labor-Hours	Unit	Material	2011 Bare Costs Labor	Equipment	Total	Total Incl O&P
0010	SELECTIVE DEMOLITION DOORS	R024119-10									
0200	Doors, exterior, 1-3/4″ thick, single, 3′ x 7′ high		1 Clab	16	.500	Ea.		13.25		13.25	22
0210	Door demo, doors, exterior, 1-3/4″ thick, single, 3′-0″ x 8′-0″			10	.800			21		21	35.50
0215	Door demo, doors, exterior, double 1-3/4″ thick, 3′-0″ x 8′-0″			6	1.333			35.50		35.50	59
0220	Double, 6′ x 7′ high			12	.667			17.65		17.65	29.50
0500	Interior, 1-3/8″ thick, single, 3′ x 7′ high			20	.400			10.60		10.60	17.70
0520	Double, 6′ x 7′ high			16	.500			13.25		13.25	22
0700	Bi-folding, 3′ x 6′-8″ high			20	.400			10.60		10.60	17.70
0720	6′ x 6′-8″ high			18	.444			11.75		11.75	19.70
0900	Bi-passing, 3′ x 6′-8″ high			16	.500			13.25		13.25	22
0940	6′ x 6′-8″ high			14	.571			15.10		15.10	25.50
1500	Remove and reset, minimum		1 Carp	8	1			33.50		33.50	56

08 05 Common Work Results for Openings

08 05 05 – Selective Windows and Doors Demolition

08 05 05.10 Selective Demolition Doors

	08 05 05.10 Selective Demolition Doors	Crew	Daily Output	Labor-Hours	Unit	Material	2011 Bare Costs Labor	Equipment	Total	Total Incl O&P
1520	Maximum	1 Carp	6	1.333	Ea.		44.50		44.50	75
2000	Frames, including trim, metal		8	1			33.50		33.50	56
2200	Wood	2 Carp	32	.500			16.80		16.80	28
2201	Alternate pricing method	1 Carp	200	.040	L.F.		1.34		1.34	2.25
3000	Special doors, counter doors	2 Carp	6	2.667	Ea.		89.50		89.50	150
3300	Glass, sliding, including frames		12	1.333			44.50		44.50	75
3400	Overhead, commercial, 12′ x 12′ high		4	4			134		134	225
3500	Residential, 9′ x 7′ high		8	2			67		67	112
3540	16′ x 7′ high		7	2.286			76.50		76.50	128
3600	Remove and reset, minimum		4	4			134		134	225
3620	Maximum		2.50	6.400			215		215	360
3660	Remove and reset elec. garage door opener	1 Carp	8	1			33.50		33.50	56
3902	Cafe/bar swing door	2 Clab	8	2			53		53	88.50
4000	Residential lockset, exterior	1 Carp	28	.286			9.60		9.60	16.05
4010	Residential lockset, exterior w/deadbolt		26	.308			10.30		10.30	17.30
4020	Residential lockset, interior		30	.267			8.95		8.95	14.95
4200	Deadbolt lock		32	.250			8.40		8.40	14.05
7100	Remove double swing pneumatic doors, openers and sensors	2 Skwk	.50	32	Opng.		1,100		1,100	1,825
7570	Remove shock absorbing door	2 Sswk	1.90	8.421	"		305		305	585

08 05 05.20 Selective Demolition of Windows

	08 05 05.20 Selective Demolition of Windows	Crew	Daily Output	Labor-Hours	Unit	Material	Labor	Equipment	Total	Total Incl O&P
0010	**SELECTIVE DEMOLITION OF WINDOWS** R024119-10									
0200	Aluminum, including trim, to 12 S.F.	1 Clab	16	.500	Ea.		13.25		13.25	22
0240	To 25 S.F.		11	.727			19.25		19.25	32
0280	To 50 S.F.		5	1.600			42.50		42.50	71
0320	Storm windows/screens, to 12 S.F.		27	.296			7.85		7.85	13.15
0360	To 25 S.F.		21	.381			10.10		10.10	16.90
0400	To 50 S.F.		16	.500			13.25		13.25	22
0600	Glass, minimum		200	.040	S.F.		1.06		1.06	1.77
0620	Maximum		150	.053	"		1.41		1.41	2.36
2000	Wood, including trim, to 12 S.F.		22	.364	Ea.		9.60		9.60	16.10
2020	To 25 S.F.		18	.444			11.75		11.75	19.70
2060	To 50 S.F.		13	.615			16.30		16.30	27.50
2065	To 180 S.F.		8	1			26.50		26.50	44.50
4300	Remove bay/bow window	2 Carp	6	2.667			89.50		89.50	150
4410	Remove skylight, plstc domes, flush/curb mtd	G-3	395	.081	S.F.		2.62		2.62	4.35
5020	Remove and reset window, minimum	1 Carp	6	1.333	Ea.		44.50		44.50	75
5040	Average		4	2			67		67	112
5080	Maximum		2	4			134		134	225
6000	Screening only	1 Clab	4000	.002	S.F.		.05		.05	.09

08 11 Metal Doors and Frames

08 11 16 – Aluminum Doors and Frames

08 11 16.10 Entrance Doors

	08 11 16.10 Entrance Doors	Crew	Daily Output	Labor-Hours	Unit	Material	Labor	Equipment	Total	Total Incl O&P
0010	**ENTRANCE DOORS** Aluminum, narrow stile									
0011	Including standard hardware, clear finish, no glass									
0020	3′-0″ x 7′-0″ opening	2 Sswk	2	8	Ea.	760	291		1,051	1,400
0030	3′-6″ x 7′-0″ opening		2	8		740	291		1,031	1,375
0100	3′-0″ x 10′-0″ opening, 3′ high transom		1.80	8.889		1,200	325		1,525	1,925
0200	3′-6″ x 10′-0″ opening, 3′ high transom		1.80	8.889		1,250	325		1,575	2,000
0280	5′-0″ x 7′-0″ opening		2	8		1,350	291		1,641	2,025
0300	6′-0″ x 7′-0″ opening		1.30	12.308		1,150	450		1,600	2,100

08 11 Metal Doors and Frames

08 11 16 – Aluminum Doors and Frames

08 11 16.10 Entrance Doors		Crew	Daily Output	Labor-Hours	Unit	2011 Bare Costs Material	Labor	Equipment	Total	Total Incl O&P
0400	6'-0" x 10'-0" opening, 3' high transom	2 Sswk	1.10	14.545	Pr.	1,525	530		2,055	2,675
0420	7'-0" x 7'-0" opening		1	16	"	1,275	580		1,855	2,500
0520	3'-0" x 7'-0" opening, wide stile		2	8	Ea.	910	291		1,201	1,550
0540	3'-6" x 7'-0" opening		2	8		1,125	291		1,416	1,800
0560	5'-0" x 7'-0" opening		2	8		1,550	291		1,841	2,250
0580	6'-0" x 7'-0" opening		1.30	12.308	Pr.	1,475	450		1,925	2,475
0600	7'-0" x 7'-0" opening		1	16	"	1,625	580		2,205	2,875
1100	For full vision doors, with 1/2" glass, add				Leaf	55%				
1200	For non-standard size, add					67%				
1300	Light bronze finish, add					36%				
1400	Dark bronze finish, add					18%				
1500	For black finish, add					36%				
1600	Concealed panic device, add					910			910	1,000
1700	Electric striker release, add				Opng.	252			252	277
1800	Floor check, add				Leaf	630			630	695
1900	Concealed closer, add				"	505			505	555
2000	Flush 3' x 7' Insulated, 12" x 12" lite, clear finish	2 Sswk	2	8	Ea.	1,300	291		1,591	1,975

08 11 63 – Metal Screen and Storm Doors and Frames

08 11 63.23 Aluminum Screen and Storm Doors and Frames		Crew	Daily Output	Labor-Hours	Unit	Material	Labor	Equipment	Total	Total Incl O&P
0010	**ALUMINUM SCREEN AND STORM DOORS AND FRAMES**									
0020	Combination storm and screen									
0420	Clear anodic coating, 2'-8" wide	2 Carp	14	1.143	Ea.	158	38.50		196.50	238
0440	3'-0" wide	"	14	1.143	"	158	38.50		196.50	238
0500	For 7' door height, add					5%				
1020	Mill finish, 2'-8" wide	2 Carp	14	1.143	Ea.	225	38.50		263.50	310
1040	3'-0" wide	"	14	1.143		246	38.50		284.50	335
1100	For 7'-0" door, add					5%				
1520	White painted, 2'-8" wide	2 Carp	14	1.143		281	38.50		319.50	375
1540	3'-0" wide		14	1.143		298	38.50		336.50	390
1541	Storm door, painted, alum., insul., 6'-8" x 2'-6" wide		14	1.143		262	38.50		300.50	350
1545	2'-8" wide		14	1.143		262	38.50		300.50	350
1600	For 7'-0" door, add					5%				
1800	Aluminum screen door, minimum, 6'-8" x 2'-8" wide	2 Carp	14	1.143		114	38.50		152.50	189
1810	3'-0" wide		14	1.143		240	38.50		278.50	330
1820	Average, 6'-8" x 2'-8" wide		14	1.143		160	38.50		198.50	240
1830	3'-0" wide		14	1.143		240	38.50		278.50	330
1840	Maximum, 6'-8" x 2'-8" wide		14	1.143		325	38.50		363.50	425
1850	3'-0" wide		14	1.143		262	38.50		300.50	350

08 12 Metal Frames

08 12 13 – Hollow Metal Frames

08 12 13.13 Standard Hollow Metal Frames		Crew	Daily Output	Labor-Hours	Unit	Material	Labor	Equipment	Total	Total Incl O&P
0010	**STANDARD HOLLOW METAL FRAMES**									
0020	16 ga., up to 5-3/4" jamb depth									
0025	6'-8" high, 3'-0" wide, single G	2 Carp	16	1	Ea.	138	33.50		171.50	208
0028	3'-6" wide, single G		16	1		146	33.50		179.50	217
0030	4'-0" wide, single G		16	1		146	33.50		179.50	217
0040	6'-0" wide, double G		14	1.143		195	38.50		233.50	278
0045	8'-0" wide, double G		14	1.143		203	38.50		241.50	287
0100	7'-0" high, 3'-0" wide, single G		16	1		143	33.50		176.50	213

08 12 Metal Frames

08 12 13 – Hollow Metal Frames

08 12 13.13 Standard Hollow Metal Frames			Crew	Daily Output	Labor-Hours	Unit	Material	2011 Bare Costs Labor	Equipment	Total	Total Incl O&P
0110	3'-6" wide, single	G	2 Carp	16	1	Ea.	151	33.50		184.50	222
0112	4'-0" wide, single	G		16	1		151	33.50		184.50	222
0140	6'-0" wide, double	G		14	1.143		185	38.50		223.50	268
0145	8'-0" wide, double	G		14	1.143		219	38.50		257.50	305
1000	16 ga., up to 4-7/8" deep, 7'-0" H, 3'-0" W, single	G		16	1		157	33.50		190.50	228
1140	6'-0" wide, double	G		14	1.143		177	38.50		215.50	259
1200	16 ga., 8-3/4" deep, 7'-0" H, 3'-0" W, single	G		16	1		190	33.50		223.50	265
1240	6'-0" wide, double	G		14	1.143		223	38.50		261.50	310
2800	14 ga., up to 3-7/8" deep, 7'-0" high, 3'-0" wide, single	G		16	1		171	33.50		204.50	244
2840	6'-0" wide, double	G		14	1.143		205	38.50		243.50	289
3000	14 ga., up to 5-3/4" deep, 6'-8" high, 3'-0" wide, single	G		16	1		145	33.50		178.50	215
3002	3'-6" wide, single	G		16	1		188	33.50		221.50	263
3005	4'-0" wide, single	G		16	1		188	33.50		221.50	263
3600	up to 5-3/4" jamb depth, 7'-0" high, 4'-0" wide, single	G		15	1.067		176	36		212	254
3620	6'-0" wide, double	G		12	1.333		224	44.50		268.50	320
3640	8'-0" wide, double	G		12	1.333		235	44.50		279.50	335
3700	8'-0" high, 4'-0" wide, single	G		15	1.067		224	36		260	305
3740	8'-0" wide, double	G		12	1.333		276	44.50		320.50	380
4000	6-3/4" deep, 7'-0" high, 4'-0" wide, single	G		15	1.067		206	36		242	287
4020	6'-0" wide, double	G		12	1.333		257	44.50		301.50	360
4040	8'-0", wide double	G		12	1.333		264	44.50		308.50	365
4100	8'-0" high, 4'-0" wide, single	G		15	1.067		263	36		299	350
4140	8'-0" wide, double	G		12	1.333		305	44.50		349.50	415
4400	8-3/4" deep, 7'-0" high, 4'-0" wide, single	G		15	1.067		241	36		277	325
4440	8'-0" wide, double	G		12	1.333		299	44.50		343.50	405
4500	8'-0" high, 4'-0" wide, single	G		15	1.067		270	36		306	355
4540	8'-0" wide, double	G		12	1.333		335	44.50		379.50	445
4900	For welded frames, add						60			60	66
5380	Steel frames, KD, 14 ga, "B" label, to 5-3/4" throat, to 3'-0" x 7'-0"		2 Carp	15	1.067		204	36		240	285
5400	14 ga., "B" label, up to 5-3/4" deep, 7'-0" high, 4'-0" wide, single	G		15	1.067		197	36		233	277
5440	8'-0" wide, double	G		12	1.333		258	44.50		302.50	360
5800	6-3/4" deep, 7'-0" high, 4'-0" wide, single	G		15	1.067		208	36		244	289
5840	8'-0" wide, double	G		12	1.333		365	44.50		409.50	475
6200	8-3/4" deep, 7'-0" high, 4'-0" wide, single	G		15	1.067		296	36		332	385
6240	8'-0" wide, double	G		12	1.333		340	44.50		384.50	450
6300	For "A" label use same price as "B" label										
6400	For baked enamel finish, add						30%	15%			
6500	For galvanizing, add						15%				
6600	For hospital stop, add					Ea.	282			282	310
6620	For hospital stop, stainless steel add					"	365			365	400
7900	Transom lite frames, fixed, add		2 Carp	155	.103	S.F.	52	3.46		55.46	63
8000	Movable, add		"	130	.123	"	65	4.13		69.13	78.50

08 12 13.25 Channel Metal Frames			Crew	Daily Output	Labor-Hours	Unit	Material	Labor	Equipment	Total	Total Incl O&P
0010	**CHANNEL METAL FRAMES**										
0020	Steel channels with anchors and bar stops										
0100	6" channel @ 8.2#/L.F., 3' x 7' door, weighs 150#	G	E-4	13	2.462	Ea.	203	91	8.40	302.40	405
0200	8" channel @ 11.5#/L.F., 6' x 8' door, weighs 275#	G		9	3.556		370	131	12.10	513.10	675
0300	8' x 12' door, weighs 400#	G		6.50	4.923		540	182	16.80	738.80	960
0400	10" channel @ 15.3#/L.F., 10' x 10' door, weighs 500#	G		6	5.333		675	197	18.20	890.20	1,150
0500	12' x 12' door, weighs 600#	G		5.50	5.818		810	215	19.85	1,044.85	1,325
0600	12" channel @ 20.7#/L.F., 12' x 12' door, weighs 825#	G		4.50	7.111		1,125	262	24.50	1,411.50	1,750
0700	12' x 16' door, weighs 1000#	G		4	8		1,350	295	27.50	1,672.50	2,075

08 12 Metal Frames

08 12 13 – Hollow Metal Frames

08 12 13.25 Channel Metal Frames			Crew	Daily Output	Labor-Hours	Unit	Material	2011 Bare Costs Labor	Equipment	Total	Total Incl O&P
0800	For frames without bar stops, light sections, deduct					Ea.	15%				
0900	Heavy sections, deduct					↓	10%				

08 13 Metal Doors

08 13 13 – Hollow Metal Doors

08 13 13.13 Standard Hollow Metal Doors			Crew	Daily Output	Labor-Hours	Unit	Material	2011 Bare Costs Labor	Equipment	Total	Total Incl O&P
0010	**STANDARD HOLLOW METAL DOORS** R081313-20										
0015	Flush, full panel, hollow core										
0017	When noted doors are prepared but do not include glass or louvers										
0020	1-3/8" thick, 20 ga., 2'-0" x 6'-8"	G	2 Carp	20	.800	Ea.	320	27		347	395
0040	2'-8" x 6'-8"	G		18	.889		335	30		365	420
0060	3'-0" x 6'-8"	G		17	.941		335	31.50		366.50	425
0100	3'-0" x 7'-0"	G	↓	17	.941		345	31.50		376.50	435
0120	For vision lite, add						90			90	99
0140	For narrow lite, add						97.50			97.50	107
0320	Half glass, 20 ga., 2'-0" x 6'-8"	G	2 Carp	20	.800		470	27		497	560
0340	2'-8" x 6'-8"	G		18	.889		490	30		520	590
0360	3'-0" x 6'-8"	G		17	.941		485	31.50		516.50	590
0400	3'-0" x 7'-0"	G		17	.941		595	31.50		626.50	710
0410	1-3/8" thick, 18 ga., 2'-0" x 6'-8"	G		20	.800		380	27		407	465
0420	3'-0" x 6'-8"	G		17	.941		385	31.50		416.50	480
0425	3'-0" x 7'-0"	G	↓	17	.941		400	31.50		431.50	495
0450	For vision lite, add						90			90	99
0452	For narrow lite, add						97.50			97.50	107
0460	Half glass, 18 ga., 2'-0" x 6'-8"	G	2 Carp	20	.800		530	27		557	630
0465	2'-8" x 6'-8"	G		18	.889		550	30		580	655
0470	3'-0" x 6'-8"	G		17	.941		535	31.50		566.50	645
0475	3'-0" x 7'-0"	G		17	.941		550	31.50		581.50	660
0500	Hollow core, 1-3/4" thick, full panel, 20 ga., 2'-8" x 6'-8"	G		18	.889		400	30		430	490
0520	3'-0" x 6'-8"	G		17	.941		400	31.50		431.50	495
0640	3'-0" x 7'-0"	G		17	.941		420	31.50		451.50	515
0680	4'-0" x 7'-0"	G		15	1.067		605	36		641	725
0700	4'-0" x 8'-0"	G		13	1.231		705	41.50		746.50	845
1000	18 ga., 2'-8" x 6'-8"	G		17	.941		465	31.50		496.50	570
1020	3'-0" x 6'-8"	G		16	1		455	33.50		488.50	555
1120	3'-0" x 7'-0"	G		17	.941		475	31.50		506.50	580
1180	4'-0" x 7'-0"	G		14	1.143		605	38.50		643.50	730
1200	4'-0" x 8'-0"	G	↓	17	.941		705	31.50		736.50	830
1212	For vision lite, add						90			90	99
1214	For narrow lite, add						97.50			97.50	107
1230	Half glass, 20 ga., 2'-8" x 6'-8"	G	2 Carp	20	.800		550	27		577	650
1240	3'-0" x 6'-8"	G		18	.889		550	30		580	655
1260	3'-0" x 7'-0"	G		18	.889		565	30		595	670
1320	18 ga., 2'-8" x 6'-8"	G		18	.889		610	30		640	720
1340	3'-0" x 6'-8"	G		17	.941		605	31.50		636.50	720
1360	3'-0" x 7'-0"	G		17	.941		615	31.50		646.50	730
1380	4'-0" x 7'-0"	G		15	1.067		750	36		786	885
1400	4'-0" x 8'-0"	G		14	1.143		850	38.50		888.50	1,000
1720	Insulated, 1-3/4" thick, full panel, 18 ga., 3'-0" x 6'-8"	G		15	1.067		455	36		491	560
1740	2'-8" x 7'-0"	G		16	1		480	33.50		513.50	585
1760	3'-0" x 7'-0"	G	↓	15	1.067	↓	465	36		501	575

08 13 Metal Doors

08 13 13 – Hollow Metal Doors

08 13 13.13 Standard Hollow Metal Doors			Crew	Daily Output	Labor-Hours	Unit	2011 Bare Costs Material	Labor	Equipment	Total	Total Incl O&P
1800	4'-0" x 8'-0"	G	2 Carp	13	1.231	Ea.	705	41.50		746.50	845
1805	For vision lite, add						90			90	99
1810	For narrow lite, add						97.50			97.50	107
1820	Half glass, 18 ga., 3'-0" x 6'-8"	G	2 Carp	16	1		605	33.50		638.50	720
1840	2'-8" x 7'-0"	G		17	.941		630	31.50		661.50	745
1860	3'-0" x 7'-0"	G		16	1		660	33.50		693.50	780
1900	4'-0" x 8'-0"	G		14	1.143		705	38.50		743.50	840
2000	For bottom louver, add						264			264	291
2020	For baked enamel finish, add						30%	15%			
2040	For galvanizing, add						15%				
2540	For soundproofing STC 40, add					Ea.	1,725			1,725	1,900
2560	For 3 hour door, add						370			370	405
2640	For dutch door with shelf, add to standard door						300			300	330

08 13 13.15 Metal Fire Doors			Crew	Daily Output	Labor-Hours	Unit	Material	Labor	Equipment	Total	Total Incl O&P
0010	**METAL FIRE DOORS**	R081313-20									
0015	Steel, flush, "B" label, 90 minute										
0020	Full panel, 20 ga., 2'-0" x 6'-8"		2 Carp	20	.800	Ea.	400	27		427	485
0040	2'-8" x 6'-8"			18	.889		415	30		445	510
0060	3'-0" x 6'-8"			17	.941		415	31.50		446.50	515
0080	3'-0" x 7'-0"			17	.941		430	31.50		461.50	530
0140	18 ga., 3'-0" x 6'-8"			16	1		470	33.50		503.50	570
0160	2'-8" x 7'-0"			17	.941		495	31.50		526.50	600
0180	3'-0" x 7'-0"			16	1		480	33.50		513.50	585
0200	4'-0" x 7'-0"			15	1.067		620	36		656	740
0220	For "A" label, 3 hour, 18 ga., use same price as "B" label										
0240	For vision lite, add					Ea.	149			149	164
0520	Flush, "B" label 90 min., egress core, 20 ga., 2'-0" x 6'-8"		2 Carp	18	.889		625	30		655	735
0540	2'-8" x 6'-8"			17	.941		635	31.50		666.50	750
0560	3'-0" x 6'-8"			16	1		635	33.50		668.50	750
0580	3'-0" x 7'-0"			16	1		655	33.50		688.50	775
0640	Flush, "A" label 3 hour, egress core, 18 ga., 3'-0" x 6'-8"			15	1.067		695	36		731	820
0660	2'-8" x 7'-0"			16	1		715	33.50		748.50	840
0680	3'-0" x 7'-0"			15	1.067		705	36		741	835
0700	4'-0" x 7'-0"			14	1.143		840	38.50		878.50	990

08 13 13.20 Residential Steel Doors		Crew	Daily Output	Labor-Hours	Unit	Material	Labor	Equipment	Total	Total Incl O&P
0010	**RESIDENTIAL STEEL DOORS**									
2510	Bi-passing closet, incl. hardware, no frame or trim incl.									
2511	Mirrored, metal frame, 6'-8" x 4'-0" wide	2 Carp	10	1.600	Opng.	184	53.50		237.50	293
2512	5'-0" wide		10	1.600		214	53.50		267.50	325
2513	6'-0" wide		10	1.600		244	53.50		297.50	360
2514	7'-0" wide		9	1.778		264	59.50		323.50	390
2515	8'-0" wide		9	1.778		425	59.50		484.50	570
2611	Mirrored, metal, 8'-0" x 4'-0" wide		10	1.600		310	53.50		363.50	430
2612	5'-0" wide		10	1.600		320	53.50		373.50	445
2613	6'-0" wide		10	1.600		375	53.50		428.50	505
2614	7'-0" wide		9	1.778		395	59.50		454.50	535
2615	8'-0" wide		9	1.778		435	59.50		494.50	575

08 13 13.25 Doors Hollow Metal		Crew	Daily Output	Labor-Hours	Unit	Material	Labor	Equipment	Total	Total Incl O&P
0010	**DOORS HOLLOW METAL**									
0500	Exterior, commercial, flush, 20 ga., 1-3/4" x 7'-0" x 2'-6" wide	2 Carp	15	1.067	Ea.	375	36		411	475
0530	2'-8" wide		15	1.067		415	36		451	520
0560	3'-0" wide		14	1.143		415	38.50		453.50	525

08 13 Metal Doors

08 13 13 – Hollow Metal Doors

08 13 13.25 Doors Hollow Metal		Crew	Daily Output	Labor-Hours	Unit	2011 Bare Costs Material	Labor	Equipment	Total	Total Incl O&P
1000	18 ga., 1-3/4" x 7'-0" x 2'-6" wide	2 Carp	15	1.067	Ea.	475	36		511	585
1030	2'-8" wide		15	1.067		480	36		516	590
1060	3'-0" wide		14	1.143		465	38.50		503.50	580
1500	16 ga., 1-3/4 x 7'-0" x 2'-6" wide		15	1.067		540	36		576	655
1530	2'-8" wide		15	1.067		545	36		581	660
1560	3'-0" wide		14	1.143		530	38.50		568.50	650
1590	3'-6" wide		14	1.143		625	38.50		663.50	750
2900	Fire door, "A" label, 1-3/4" x 7'-0" x 2'-6" wide		15	1.067		630	36		666	750
2930	2'-8" wide		15	1.067		630	36		666	755
2960	3'-0" wide		14	1.143		620	38.50		658.50	745
2990	3'-6" wide		14	1.143		705	38.50		743.50	840
3100	"B" label, 2'-6" wide		15	1.067		565	36		601	680
3130	2'-8" wide		15	1.067		565	36		601	685
3160	3'-0" wide		14	1.143		560	38.50		598.50	680

08 13 16 – Aluminum Doors

08 13 16.10 Commercial Aluminum Doors		Crew	Daily Output	Labor-Hours	Unit	Material	Labor	Equipment	Total	Total Incl O&P
0010	**COMMERCIAL ALUMINUM DOORS**, no glazing									
0020	Incl. hinges, push/pull, deadlock, cyl., threshold									
1000	Narrow stile, no glazing, standard hardware, 3'-0" x 7'-0", single	2 Carp	3	5.333	Ea.	535	179		714	885
1050	Bronze finish		3	5.333		465	179		644	815
1100	Black finish		3	5.333		555	179		734	910
1200	Pair of 3'-0" x 7'-0"		1.70	9.412	Pr.	1,200	315		1,515	1,850
1500	3'-6" x 7'-0", single		3	5.333	Ea.	700	179		879	1,075
1550	Bronze finish		3	5.333		645	179		824	1,000
1600	Black finish		3	5.333		705	179		884	1,075
2100	Medium stile, 3'-0" x 7'-0", single		3	5.333		740	179		919	1,125
2200	Pair of 3'-0" x 7'-0"		1.70	9.412	Pr.	1,475	315		1,790	2,150
2300	3'-6" x 7'-0", single		3	5.333	Ea.	955	179		1,134	1,350
5000	Flush panel doors, pair of 2'-6" x 7'-0"	2 Sswk	2	8	Pr.	1,375	291		1,666	2,050
5050	3'-0" x 7'-0", single		2.50	6.400	Ea.	780	233		1,013	1,300
5100	Pair of 3'-0" x 7'-0"		2	8	Pr.	1,450	291		1,741	2,150
5150	3'-6" x 7'-0", single		2.50	6.400	Ea.	930	233		1,163	1,475

08 14 Wood Doors

08 14 13 – Carved Wood Doors

08 14 13.10 Types of Wood Doors, Carved		Crew	Daily Output	Labor-Hours	Unit	Material	Labor	Equipment	Total	Total Incl O&P
0010	**TYPES OF WOOD DOORS, CARVED**									
3000	Solid wood, 1-3/4" thick stile and rail									
3020	Mahogany, 3'-0" x 7'-0", minimum	2 Carp	14	1.143	Ea.	945	38.50		983.50	1,125
3030	Maximum		10	1.600		1,500	53.50		1,553.50	1,750
3040	3'-6" x 8'-0", minimum		10	1.600		1,250	53.50		1,303.50	1,475
3050	Maximum		8	2		1,850	67		1,917	2,150
3100	Pine, 3'-0" x 7'-0", minimum		14	1.143		490	38.50		528.50	600
3120	3'-6" x 8'-0", minimum		10	1.600		885	53.50		938.50	1,075
3130	Maximum		8	2		1,800	67		1,867	2,075
3200	Red oak, 3'-0" x 7'-0", minimum		14	1.143		1,625	38.50		1,663.50	1,875
3210	Maximum		10	1.600		2,025	53.50		2,078.50	2,325
3220	3'-6" x 8'-0", minimum		10	1.600		2,200	53.50		2,253.50	2,525
3230	Maximum		8	2		3,200	67		3,267	3,625
4000	Hand carved door, mahogany									

08 14 Wood Doors

08 14 13 – Carved Wood Doors

08 14 13.10 Types of Wood Doors, Carved		Crew	Daily Output	Labor-Hours	Unit	2011 Bare Costs Material	Labor	Equipment	Total	Total Incl O&P
4020	3'-0" x 7'-0", minimum	2 Carp	14	1.143	Ea.	1,600	38.50		1,638.50	1,825
4040	3'-6" x 8'-0", minimum		10	1.600		2,500	53.50		2,553.50	2,850
4050	Maximum		8	2		3,500	67		3,567	3,950
4200	Rose wood, 3'-0" x 7'-0", minimum		14	1.143		5,000	38.50		5,038.50	5,575
4210	Maximum		11	1.455		13,600	49		13,649	15,100
4220	3'-6" x 8'-0", minimum		10	1.600		5,700	53.50		5,753.50	6,375
4225	maximum		10	1.600		8,700	53.50		8,753.50	9,675
4280	For 6'-8" high door, deduct from 7'-0" door					34			34	37.50
4400	For custom finish, add					400			400	440
4600	Side light, mahogany, 7'-0" x 1'-6" wide, minimum	2 Carp	18	.889		900	30		930	1,050
4610	Maximum		14	1.143		2,625	38.50		2,663.50	2,975
4620	8'-0" x 1'-6" wide, minimum		14	1.143		1,700	38.50		1,738.50	1,950
4630	Maximum		10	1.600		1,950	53.50		2,003.50	2,250
4640	Side light, oak, 7'-0" x 1'-6" wide, minimum		18	.889		1,100	30		1,130	1,250
4650	Maximum		14	1.143		1,900	38.50		1,938.50	2,175
4660	8'-0" x 1'-6" wide, minimum		14	1.143		1,000	38.50		1,038.50	1,175
4670	Maximum		10	1.600		1,900	53.50		1,953.50	2,200

08 14 16 – Flush Wood Doors

08 14 16.09 Smooth Wood Doors		Crew	Daily Output	Labor-Hours	Unit	Material	Labor	Equipment	Total	Total Incl O&P
0010	**SMOOTH WOOD DOORS**									
0085	3'-0" x 6'-8"	1 Carp	16	.500	Ea.	52.50	16.80		69.30	86
0108	3'-0" x 7'-0"	2 Carp	16	1	"	95.50	33.50		129	161
0112	Pair of 3'-0" x 7'-0"		9	1.778	Pr.	182	59.50		241.50	300
0206	2'-6" x 7'-0"		16	1	Ea.	92.50	33.50		126	158
0210	3'-0" x 7'-0"		16	1	"	103	33.50		136.50	169
0214	Pair of 3'-0" x 7'-0"		9	1.778	Pr.	197	59.50		256.50	315
4045	3'-0" x 8'-0"	1 Carp	8	1	Ea.	420	33.50		453.50	515
5000	Wood doors, for vision lite, add					90			90	99
5010	Wood doors, for narrow lite, add					97.50			97.50	107
5015	Wood doors, for bottom (or top) louver, add					264			264	291

08 14 16.10 Wood Doors Decorator		Crew	Daily Output	Labor-Hours	Unit	Material	Labor	Equipment	Total	Total Incl O&P
0010	**WOOD DOORS DECORATOR**									
1800	Exterior, flush, solid wood core, birch 1-3/4" x 7'-0" x 2'-6" wide	2 Carp	15	1.067	Ea.	274	36		310	360
1820	2'-8" wide		15	1.067		282	36		318	370
1840	3'-0" wide		14	1.143		293	38.50		331.50	385
1900	Oak faced, 1-3/4" x 7'-0" x 2'-6" wide		15	1.067		282	36		318	370
1920	2'-8" wide		15	1.067		289	36		325	380
1940	3'-0" wide		14	1.143		300	38.50		338.50	395
2100	Walnut faced, 1-3/4" x 7'-0" x 2'-6" wide		15	1.067		345	36		381	435
2120	2'-8" wide		15	1.067		355	36		391	450
2140	3'-0" wide		14	1.143		365	38.50		403.50	470

08 14 23 – Clad Wood Doors

08 14 23.13 Metal-Faced Wood Doors		Crew	Daily Output	Labor-Hours	Unit	Material	Labor	Equipment	Total	Total Incl O&P
0010	**METAL-FACED WOOD DOORS**									
0020	Interior, flush type, 3' x 7'	2 Carp	4.30	3.721	Opng.	340	125		465	585

08 14 23.20 Tin Clad Wood Doors		Crew	Daily Output	Labor-Hours	Unit	Material	Labor	Equipment	Total	Total Incl O&P
0010	**TIN CLAD WOOD DOORS**									
0020	3 ply, 6' x 7', double sliding, doors only	2 Carp	1	16	Opng.	1,900	535		2,435	3,000
1000	For electric operator, add	1 Elec	2	4	"	3,250	161		3,411	3,825

08 14 Wood Doors

08 14 33 – Stile and Rail Wood Doors

08 14 33.10 Wood Doors Paneled		Crew	Daily Output	Labor-Hours	Unit	2011 Bare Costs Material	Labor	Equipment	Total	Total Incl O&P
0010	**WOOD DOORS PANELED**									
0020	Interior, six panel, hollow core, 1-3/8" thick									
0040	Molded hardboard, 2'-0" x 6'-8"	2 Carp	17	.941	Ea.	60	31.50		91.50	119
0060	2'-6" x 6'-8"		17	.941		62	31.50		93.50	121
0070	2'-8" x 6'-8"		17	.941		65	31.50		96.50	125
0080	3'-0" x 6'-8"		17	.941		68	31.50		99.50	128
0140	Embossed print, molded hardboard, 2'-0" x 6'-8"		17	.941		62	31.50		93.50	121
0160	2'-6" x 6'-8"		17	.941		62	31.50		93.50	121
0180	3'-0" x 6'-8"		17	.941		68	31.50		99.50	128
0540	Six panel, solid, 1-3/8" thick, pine, 2'-0" x 6'-8"		15	1.067		150	36		186	225
0560	2'-6" x 6'-8"		14	1.143		165	38.50		203.50	246
0580	3'-0" x 6'-8"		13	1.231		140	41.50		181.50	223
1020	Two panel, bored rail, solid, 1-3/8" thick, pine, 1'-6" x 6'-8"		16	1		265	33.50		298.50	350
1040	2'-0" x 6'-8"		15	1.067		350	36		386	445
1060	2'-6" x 6'-8"		14	1.143		395	38.50		433.50	500
1340	Two panel, solid, 1-3/8" thick, fir, 2'-0" x 6'-8"		15	1.067		155	36		191	231
1360	2'-6" x 6'-8"		14	1.143		205	38.50		243.50	290
1380	3'-0" x 6'-8"		13	1.231		410	41.50		451.50	520
1740	Five panel, solid, 1-3/8" thick, fir, 2'-0" x 6'-8"		15	1.067		275	36		311	365
1760	2'-6" x 6'-8"		14	1.143		415	38.50		453.50	520
1780	3'-0" x 6'-8"		13	1.231		415	41.50		456.50	525

08 14 33.20 Wood Doors Residential		Crew	Daily Output	Labor-Hours	Unit	Material	Labor	Equipment	Total	Total Incl O&P
0010	**WOOD DOORS RESIDENTIAL**									
1420	6'-8" x 3'-0" wide, fir	2 Carp	16	1	Ea.	400	33.50		433.50	490
1800	Lauan, solid core, 1-3/4" x 7'-0" x 2'-4" wide		16	1		106	33.50		139.50	172
1810	2'-6" wide		15	1.067		109	36		145	179
1820	2'-8" wide		9	1.778		113	59.50		172.50	224
1830	3'-0" wide		16	1		124	33.50		157.50	192
1840	3'-4" wide		16	1		207	33.50		240.50	283
1850	Pair of 3'-0" wide		15	1.067	Pr.	247	36		283	330
2920	Hardboard, primed 7'-0" x 4'-0", wide		12	1.333	Ea.	175	44.50		219.50	268
2930	6'-0" wide		10	1.600		171	53.50		224.50	278
3100	7'-0" x 3'-0" wide		12	1.333		231	44.50		275.50	330
3120	6'-0" wide		10	1.600		405	53.50		458.50	535
3140	For oak or ash, add					125%				
9000	Passage doors, flush, no frame, birch, solid core, 1-3/8" x 7'-0" x 2'-4"	2 Carp	16	1	Ea.	109	33.50		142.50	176
9020	2'-8" wide		16	1		116	33.50		149.50	184
9040	3'-0" wide		16	1		125	33.50		158.50	194
9060	3'-4" wide		15	1.067		144	36		180	219
9080	Pair of 3'-0" wide		9	1.778	Pr.	248	59.50		307.50	375
9100	Lauan, solid core, 1-3/8" x 7'-0" x 2'-4" wide		16	1	Ea.	101	33.50		134.50	167
9120	2'-8" wide		16	1		107	33.50		140.50	174
9140	3'-0" wide		16	1		114	33.50		147.50	181
9160	3'-4" wide		15	1.067		122	36		158	194
9180	Pair of 3'-0" wide		9	1.778	Pr.	207	59.50		266.50	330
9200	Hardboard, solid core, 1-3/8" x 7'-0" x 2'-4" wide		16	1	Ea.	124	33.50		157.50	193
9220	2'-8" wide		16	1		128	33.50		161.50	197
9240	3'-0" wide		16	1		133	33.50		166.50	202
9260	3'-4" wide		15	1.067		148	36		184	223

08 14 Wood Doors

08 14 40 – Interior Cafe Doors

08 14 40.10 Cafe Style Doors		Crew	Daily Output	Labor-Hours	Unit	Material	2011 Bare Costs Labor	Equipment	Total	Total Incl O&P
0010	**CAFE STYLE DOORS**									
6520	Interior cafe doors, 2'-6" opening, stock, panel pine	2 Carp	16	1	Ea.	191	33.50		224.50	266
6540	3'-0" opening	"	16	1	"	199	33.50		232.50	275
6550	Louvered pine									
6560	2'-6" opening	2 Carp	16	1	Ea.	165	33.50		198.50	238
8000	3'-0" opening		16	1		176	33.50		209.50	250
8010	2'-6" opening, hardwood		16	1		242	33.50		275.50	320
8020	3'-0" opening		16	1		268	33.50		301.50	350

08 16 Composite Doors

08 16 13 – Fiberglass Doors

08 16 13.10 Entrance Doors, Fiberous Glass			Crew	Daily Output	Labor-Hours	Unit	Material	Labor	Equipment	Total	Total Incl O&P
0010	**ENTRANCE DOORS, FIBEROUS GLASS**										
0020	Exterior, fiberglass, door, 2'-8" wide x 6'-8" high	G	2 Carp	15	1.067	Ea.	252	36		288	335
0040	3'-0" wide x 6'-8" high	G		15	1.067		252	36		288	335
0060	3'-0" wide x 7'-0" high	G		15	1.067		455	36		491	560
0080	3'-0" wide x 6'-8" high, with two lites	G		15	1.067		294	36		330	385
0100	3'-0" wide x 8'-0" high, with two lites	G		15	1.067		490	36		526	595
0110	Half glass, 3'-0" wide x 6'-8" high	G		15	1.067		294	36		330	385
0120	3'-0" wide x 6'-8" high, low e	G		15	1.067		325	36		361	415
0130	3'-0" wide x 8'-0" high	G		15	1.067		580	36		616	695
0140	3'-0" wide x 8'-0" high, low e	G		15	1.067		640	36		676	760
0150	Side lights, 1'-0" wide x 6'-8" high,	G					247			247	271
0160	1'-0" wide x 6'-8" high, low e	G					256			256	281
0180	1'-0" wide x 6'-8" high, full glass	G					275			275	305
0190	1'-0" wide x 6'-8" high, low e	G					284			284	315

08 16 14 – French Doors

08 16 14.10 Exterior Doors With Glass Lites		Crew	Daily Output	Labor-Hours	Unit	Material	Labor	Equipment	Total	Total Incl O&P
0010	**EXTERIOR DOORS WITH GLASS LITES**									
0020	French, Fir, 1-3/4", 3'-0"wide x 6'-8" high	2 Carp	12	1.333	Ea.	600	44.50		644.50	735
0025	double		12	1.333		1,200	44.50		1,244.50	1,400
0030	Maple, 1-3/4", 3'-0"wide x 6'-8" high		12	1.333		675	44.50		719.50	820
0035	double		12	1.333		1,350	44.50		1,394.50	1,550
0040	Cherry, 1-3/4", 3'-0"wide x 6'-8" high		12	1.333		790	44.50		834.50	940
0045	double		12	1.333		1,575	44.50		1,619.50	1,800
0100	Mahogany, 1-3/4", 3'-0"wide x 8'-0" high		10	1.600		700	53.50		753.50	860
0105	double		10	1.600		1,400	53.50		1,453.50	1,650
0110	Fir, 1-3/4", 3'-0"wide x 8'-0" high		10	1.600		1,100	53.50		1,153.50	1,300
0115	double		10	1.600		2,200	53.50		2,253.50	2,525
0120	Oak, 1-3/4", 3'-0"wide x 8'-0" high		10	1.600		1,825	53.50		1,878.50	2,100
0125	double		10	1.600		3,650	53.50		3,703.50	4,100

08 17 Integrated Door Opening Assemblies

08 17 23 – Integrated Wood Door Opening Assemblies

08 17 23.10 Pre-Hung Doors		Crew	Daily Output	Labor-Hours	Unit	2011 Bare Costs Material	Labor	Equipment	Total	Total Incl O&P
0010	**PRE-HUNG DOORS**									
0300	Exterior, wood, comb. storm & screen, 6′-9″ x 2′-6″ wide	2 Carp	15	1.067	Ea.	289	36		325	380
0320	2′-8″ wide		15	1.067		289	36		325	380
0340	3′-0″ wide		15	1.067		296	36		332	385
0360	For 7′-0″ high door, add					24			24	26.50
1600	Entrance door, flush, birch, solid core									
1620	4-5/8″ solid jamb, 1-3/4″ x 6′-8″ x 2′-8″ wide	2 Carp	16	1	Ea.	285	33.50		318.50	370
1640	3′-0″ wide		16	1		325	33.50		358.50	415
1642	5-5/8″ jamb		16	1		325	33.50		358.50	410
1680	For 7′-0″ high door, add					21			21	23
2000	Entrance door, colonial, 6 panel pine									
2020	4-5/8″ solid jamb, 1-3/4″ x 6′-8″ x 2′-8″ wide	2 Carp	16	1	Ea.	635	33.50		668.50	755
2040	3′-0″ wide	″	16	1		665	33.50		698.50	790
2060	For 7′-0″ high door, add					52.50			52.50	58
2200	For 5-5/8″ solid jamb, add					40.50			40.50	44.50
2230	French style, exterior, 1 lite, 1-3/4″ x 3′-0″ x 6′-8″	1 Carp	14	.571		590	19.15		609.15	675
2235	9 lites	″	14	.571		655	19.15		674.15	750
2245	15 lites	2 Carp	14	1.143		625	38.50		663.50	755
2250	Two, 2′- 0″, 15 lites, 4′ opening		7	2.286	Pr.	1,175	76.50		1,251.50	1,425
2260	Two 2′-6″ 15 lites, 5′ opening		7	2.286		1,300	76.50		1,376.50	1,550
2280	Two 3′-0″ 15 lites, 6′ opening		7	2.286		1,325	76.50		1,401.50	1,575
2430	3′-0″ x 7′-0″, 15 lites		14	1.143	Ea.	920	38.50		958.50	1,075
2432	Two 3′-0″ x 7′-0″		7	2.286	Pr.	1,900	76.50		1,976.50	2,225
2435	3′-0″ x 8′-0″		14	1.143	Ea.	980	38.50		1,018.50	1,150
2437	Two, 3′-0″ x 8′-0″		7	2.286	Pr.	2,025	76.50		2,101.50	2,350
4000	Interior, passage door, 4-5/8″ solid jamb									
4400	Lauan, flush, solid core, 1-3/8″ x 6′-8″ x 2′-6″ wide	2 Carp	17	.941	Ea.	184	31.50		215.50	255
4420	2′-8″ wide		17	.941		184	31.50		215.50	255
4440	3′-0″ wide		16	1		199	33.50		232.50	275
4600	Hollow core, 1-3/8″ x 6′-8″ x 2′-6″ wide		17	.941		124	31.50		155.50	189
4620	2′-8″ wide		17	.941		124	31.50		155.50	189
4640	3′-0″ wide		16	1		139	33.50		172.50	209
4700	For 7′-0″ high door, add					30			30	33
5000	Birch, flush, solid core, 1-3/8″ x 6′-8″ x 2′-6″ wide	2 Carp	17	.941		238	31.50		269.50	315
5020	2′-8″ wide		17	.941		199	31.50		230.50	272
5040	3′-0″ wide		16	1		270	33.50		303.50	355
5200	Hollow core, 1-3/8″ x 6′-8″ x 2′-6″ wide		17	.941		182	31.50		213.50	253
5220	2′-8″ wide		17	.941		158	31.50		189.50	226
5240	3′-0″ wide		16	1		195	33.50		228.50	270
5280	For 7′-0″ high door, add					25			25	27.50
5500	Hardboard paneled, 1-3/8″ x 6′-8″ x 2′-6″ wide	2 Carp	17	.941		143	31.50		174.50	211
5520	2′-8″ wide		17	.941		152	31.50		183.50	220
5540	3′-0″ wide		16	1		149	33.50		182.50	220
6000	Pine paneled, 1-3/8″ x 6′-8″ x 2′-6″ wide		17	.941		247	31.50		278.50	325
6020	2′-8″ wide		17	.941		267	31.50		298.50	345
6500	For 5-5/8″ solid jamb, add					20			20	22
6520	For split jamb, deduct					21.50			21.50	23.50
7600	Oak, 6 panel, 1-3/4″ x 6′-8″ x 3′-0″	1 Carp	17	.471		860	15.80		875.80	975
8200	Birch, flush, solid core, 1-3/4″ x 6′-8″ x 2′-4″ wide		17	.471		219	15.80		234.80	268
8220	2′-6″ wide		17	.471		215	15.80		230.80	264
8240	2′-8″ wide		17	.471		200	15.80		215.80	246
8260	3′-0″ wide		16	.500		208	16.80		224.80	257
8280	3′-6″ wide		15	.533		350	17.90		367.90	415

08 17 Integrated Door Opening Assemblies

08 17 23 – Integrated Wood Door Opening Assemblies

08 17 23.10 Pre-Hung Doors		Crew	Daily Output	Labor-Hours	Unit	Material	2011 Bare Costs Labor	Equipment	Total	Total Incl O&P
8500	Pocket door frame with lauan, flush, hollow core , 1-3/8" x 3'-0" x 6'-8"	1 Carp	17	.471	Ea.	250	15.80		265.80	300

08 31 Access Doors and Panels

08 31 13 – Access Doors and Frames

08 31 13.10 Types of Framed Access Doors		Crew	Daily Output	Labor-Hours	Unit	Material	Labor	Equipment	Total	Total Incl O&P
0010	**TYPES OF FRAMED ACCESS DOORS**									
7000	Ceiling hatches, 2'-6" x 2'-6", swing up, single leaf, st fr & cover	G-3	11	2.909	Ea.	570	94		664	780
7010	Aluminum cover		11	2.909		550	94		644	760
7020	2'-6" x 3'-0", swing up, single leaf, steel frame & cover		11	2.909		575	94		669	790
7040	Aluminum cover		11	2.909		585	94		679	795
7060	2'-6" x 2'-6", swing down model, steel frame and cover		11	2.909		725	94		819	955
7080	2'-6" x 3'-0"		11	2.909		765	94		859	1,000
7100	Aluminum cover		11	2.909		750	94		844	980

08 31 13.20 Bulkhead/Cellar Doors		Crew	Daily Output	Labor-Hours	Unit	Material	Labor	Equipment	Total	Total Incl O&P
0010	**BULKHEAD/CELLAR DOORS**									
0020	Steel, not incl. sides, 44" x 62"	1 Carp	5.50	1.455	Ea.	490	49		539	620
0100	52" x 73"		5.10	1.569		670	52.50		722.50	825
0500	With sides and foundation plates, 57" x 45" x 24"		4.70	1.702		700	57		757	865
0600	42" x 49" x 51"		4.30	1.860		790	62.50		852.50	975

08 31 13.30 Commercial Floor Doors		Crew	Daily Output	Labor-Hours	Unit	Material	Labor	Equipment	Total	Total Incl O&P
0010	**COMMERCIAL FLOOR DOORS**									
0020	Aluminum tile, steel frame, one leaf, 2' x 2' opng.	2 Sswk	3.50	4.571	Opng.	775	166		941	1,175
0050	3'-6" x 3'-6" opening		3.50	4.571		1,475	166		1,641	1,950
0500	Double leaf, 4' x 4' opening		3	5.333		1,575	194		1,769	2,100
0550	5' x 5' opening		3	5.333		2,800	194		2,994	3,450

08 31 13.35 Industrial Floor Doors		Crew	Daily Output	Labor-Hours	Unit	Material	Labor	Equipment	Total	Total Incl O&P
0010	**INDUSTRIAL FLOOR DOORS**									
0020	Steel 300 psf L.L., single leaf, 2' x 2', 175#	2 Sswk	6	2.667	Opng.	735	97		832	995
0050	3' x 3' opening, 300#		5.50	2.909		1,000	106		1,106	1,300
0300	Double leaf, 4' x 4' opening, 455#		5	3.200		1,625	116		1,741	2,025
0350	5' x 5' opening, 645#		4.50	3.556		2,450	129		2,579	2,950

08 32 Sliding Glass Doors

08 32 13 – Sliding Aluminum-Framed Glass Doors

08 32 13.10 Sliding Aluminum Doors		Crew	Daily Output	Labor-Hours	Unit	Material	Labor	Equipment	Total	Total Incl O&P
0010	**SLIDING ALUMINUM DOORS**									
0350	Aluminum, 5/8" tempered insulated glass, 6' wide									
0400	Premium	2 Carp	4	4	Ea.	1,550	134		1,684	1,950
0450	Economy		4	4		885	134		1,019	1,200
0500	8' wide, premium		3	5.333		1,625	179		1,804	2,075
0550	Economy		3	5.333		1,400	179		1,579	1,825
0600	12' wide, premium		2.50	6.400		2,950	215		3,165	3,600
0650	Economy		2.50	6.400		1,375	215		1,590	1,850
0700	Non-insulated, 6' wide		4	4		1,100	134		1,234	1,425
4000	Aluminum, baked on enamel, temp glass, 6'-8" x 10'-0" wide		4	4		1,025	134		1,159	1,350
4020	Insulating glass, 6'-8" x 6'-0" wide		4	4		895	134		1,029	1,200
4040	8'-0" wide		3	5.333		1,050	179		1,229	1,450
4060	10'-0" wide		2	8		1,300	268		1,568	1,875

08 32 Sliding Glass Doors

08 32 13 – Sliding Aluminum-Framed Glass Doors

08 32 13.10 Sliding Aluminum Doors		Crew	Daily Output	Labor-Hours	Unit	2011 Bare Costs Material	Labor	Equipment	Total	Total Incl O&P
4080	Anodized, temp glass, 6'-8" x 6'-0" wide	2 Carp	4	4	Ea.	450	134		584	720
4100	8'-0" wide		3	5.333		570	179		749	930
4120	10'-0" wide		2	8		645	268		913	1,150
4200	Vinyl clad, anodized, temp glass, 6'-8" x 6'-0" wide		4	4		420	134		554	685
4240	8'-0" wide		3	5.333		580	179		759	940
4260	10'-0" wide		2	8		630	268		898	1,150
4280	Insulating glass, 6'-8" x 6'-0' wide		4	4		1,100	134		1,234	1,425
4300	8'-0" wide		3	5.333		920	179		1,099	1,300
4320	10'-0" wide		4	4		1,625	134		1,759	2,000

08 32 19 – Sliding Wood-Framed Glass Doors

08 32 19.10 Sliding Wood Doors		Crew	Daily Output	Labor-Hours	Unit	Material	Labor	Equipment	Total	Total Incl O&P
0010	SLIDING WOOD DOORS									
0020	Wood, tempered insul. glass, 6' wide, premium	2 Carp	4	4	Ea.	1,400	134		1,534	1,750
0100	Economy		4	4		1,125	134		1,259	1,450
0150	8' wide, wood, premium		3	5.333		1,775	179		1,954	2,250
0200	Economy		3	5.333		1,475	179		1,654	1,925
0235	10' wide, wood,		2.50	6.400		2,525	215		2,740	3,125
0240	Economy		2.50	6.400		2,150	215		2,365	2,700
0250	12' wide, wood,		2.50	6.400		2,925	215		3,140	3,575
0300	Economy		2.50	6.400		2,400	215		2,615	3,000

08 32 19.15 Sliding Glass Vinyl-Clad Wood Doors		Crew	Daily Output	Labor-Hours	Unit	Material	Labor	Equipment	Total	Total Incl O&P
0010	SLIDING GLASS VINYL-CLAD WOOD DOORS									
0020	Glass, sliding vinyl clad, insul. glass, 6'-0" x 6'-8" G	2 Carp	4	4	Ea.	1,500	134		1,634	1,875
0025	6'-0" x 6'-10" high G		4	4	Opng.	1,625	134		1,759	2,000
0030	6'-0" x 8'-0" high G		4	4	Ea.	1,925	134		2,059	2,350
0050	5'-0" x 6'-8" high G		4	4	"	1,475	134		1,609	1,850
0100	8'-0" x 6'-10" high G		4	4	Opng.	2,000	134		2,134	2,425
0150	8'-0" x 8'-0" high G		4	4		2,175	134		2,309	2,625
0500	4 leaf, 9'-0" x 6'-10" high G		3	5.333		3,175	179		3,354	3,800
0550	9'-0" x 8'-0" high G		3	5.333	Ea.	3,675	179		3,854	4,350
0600	12'-0" x 6'-10" high G		3	5.333	Opng.	3,800	179		3,979	4,475
0650	12'-0" x 8'-0" high		3	5.333	"	3,800	179		3,979	4,475

08 33 Coiling Doors and Grilles

08 33 13 – Coiling Counter Doors

08 33 13.10 Counter Doors, Coiling Type		Crew	Daily Output	Labor-Hours	Unit	Material	Labor	Equipment	Total	Total Incl O&P
0010	COUNTER DOORS, COILING TYPE									
0020	Manual, incl. frm and hdwe, galv. stl., 4' roll-up, 6' long	2 Carp	2	8	Opng.	1,150	268		1,418	1,725
0300	Galvanized steel, UL label		1.80	8.889		1,200	298		1,498	1,825
0600	Stainless steel, 4' high roll-up, 6' long		2	8		2,100	268		2,368	2,750
0700	10' long		1.80	8.889		2,450	298		2,748	3,200
2000	Aluminum, 4' high, 4' long		2.20	7.273		1,325	244		1,569	1,875
2020	6' long		2	8		1,475	268		1,743	2,075
2040	8' long		1.90	8.421		1,675	283		1,958	2,325
2060	10' long		1.80	8.889		1,775	298		2,073	2,450
2080	14' long		1.40	11.429		2,500	385		2,885	3,400
2100	6' high, 4' long		2	8		1,500	268		1,768	2,100
2120	6' long		1.60	10		1,550	335		1,885	2,275
2140	10' long		1.40	11.429		2,050	385		2,435	2,900

08 33 Coiling Doors and Grilles

08 33 16 – Coiling Counter Grilles

08 33 16.10 Coiling Grilles		Crew	Daily Output	Labor-Hours	Unit	Material	2011 Bare Costs Labor	Equipment	Total	Total Incl O&P
0010	**COILING GRILLES**									
2020	Aluminum, manual operated, mill finish	2 Sswk	82	.195	S.F.	27.50	7.10		34.60	43.50
2040	Bronze anodized		82	.195	"	43.50	7.10		50.60	61.50
2060	Steel, manual operated, 10′ x 10′ high		1	16	Opng.	2,450	580		3,030	3,800
2080	15′ x 8′ high		.80	20	"	2,850	730		3,580	4,525
3000	For safety edge bottom bar, electric, add				L.F.	49			49	54
8000	For motor operation, add	2 Sswk	5	3.200	Opng.	1,250	116		1,366	1,600

08 33 23 – Overhead Coiling Doors

08 33 23.10 Rolling Service Doors		Crew	Daily Output	Labor-Hours	Unit	Material	Labor	Equipment	Total	Total Incl O&P
0010	**ROLLING SERVICE DOORS** Steel, manual, 20 ga., incl. hardware									
0050	8′ x 8′ high	2 Sswk	1.60	10	Ea.	1,075	365		1,440	1,875
0100	10′ x 10′ high		1.40	11.429		1,775	415		2,190	2,750
0130	12′ x 12′ high, standard		1.20	13.333		1,775	485		2,260	2,875
0160	10′ x 20′ high, standard		.50	32		2,350	1,175		3,525	4,800
2000	Class A fire doors, manual, 20 ga., 8′ x 8′ high		1.40	11.429		1,425	415		1,840	2,375
2100	10′ x 10′ high		1.10	14.545		1,950	530		2,480	3,150
2200	20′ x 10′ high		.80	20		3,975	730		4,705	5,775
2300	12′ x 12′ high		1	16		3,050	580		3,630	4,475
2304	Overhead door, roll up, fire rated 12′ x 14′		.90	17.778		3,500	645		4,145	5,075
2400	20′ x 12′ high		.80	20		4,325	730		5,055	6,150
2500	14′ x 14′ high		.60	26.667		3,350	970		4,320	5,550
2600	20′ x 16′ high		.50	32		5,400	1,175		6,575	8,175
2700	10′ x 20′ high		.40	40		4,175	1,450		5,625	7,375
3000	For 18 ga. doors, add				S.F.	1.40			1.40	1.54
3300	For enamel finish, add				"	1.65			1.65	1.82
3600	For safety edge bottom bar, pneumatic, add				L.F.	20.50			20.50	22.50
3700	Electric, add					38.50			38.50	42
4000	For weatherstripping, extruded rubber, jambs, add					12.90			12.90	14.20
4100	Hood, add					7.90			7.90	8.70
4200	Sill, add					4.55			4.55	5
4500	Motor operators, to 14′ x 14′ opening	2 Sswk	5	3.200	Ea.	1,025	116		1,141	1,375
4700	For fire door, additional fusible link, add				"	25			25	27.50

08 34 Special Function Doors

08 34 53 – Security Doors and Frames

08 34 53.20 Steel Door		Crew	Daily Output	Labor-Hours	Unit	Material	Labor	Equipment	Total	Total Incl O&P
0010	**STEEL DOOR** with ballistic core and welded frame both 14 gauge									
0050	Flush, UL 752 Level 3, 1-3/4″, 3′-0″ x 6′-8″	2 Carp	1.50	10.667	Opng.	2,150	360		2,510	2,950
0055	1-3/4″, 3′-6″ x 6′-8″		1.50	10.667		2,250	360		2,610	3,075
0060	1-3/4″, 4′-0″ x 6′-8″		1.20	13.333		2,425	445		2,870	3,425
0100	UL 752 Level 8, 1-3/4″, 3′-0″ x 6′-8″		1.50	10.667		9,475	360		9,835	11,000
0105	1-3/4″, 3′-6″ x 6′-8″		1.50	10.667		9,575	360		9,935	11,100
0110	1-3/4″, 4′-0″ x 6′-8″		1.20	13.333		10,900	445		11,345	12,800
0120	UL 752 Level 3, 1-3/4″, 3′-0″ x 7′-0″		1.50	10.667		2,175	360		2,535	2,975
0125	1-3/4″, 3′-6″ x 7′-0″		1.50	10.667		2,275	360		2,635	3,100
0130	1-3/4″, 4′-0″ x 7′-0″		1.20	13.333		2,475	445		2,920	3,450
0150	UL 752 Level 8, 1-3/4″, 3′-0″ x 7′-0″		1.50	10.667		9,500	360		9,860	11,100
0155	1-3/4″, 3′-6″ x 7′-0″		1.50	10.667		9,600	360		9,960	11,200
0160	1-3/4″, 4′-0″ x 7′-0″		1.20	13.333		10,900	445		11,345	12,800
1000	Safe Room sliding door and hardware, 1-3/4″, 3′-0″ x 7′-0″ UL 752 Level 3		.50	32		22,700	1,075		23,775	26,700

08 34 Special Function Doors

08 34 53 – Security Doors and Frames

08 34 53.20 Steel Door		Crew	Daily Output	Labor-Hours	Unit	2011 Bare Costs Material	Labor	Equipment	Total	Total Incl O&P
1050	Safe Room swinging door and hardware, 1-3/4", 3'-0" x 7'-0" UL 752 Level 3	2 Carp	.50	32	Opng.	26,800	1,075		27,875	31,300

08 34 53.30 Wood Ballistic Doors

		Crew	Daily Output	Labor-Hours	Unit	Material	Labor	Equipment	Total	Total Incl O&P
0010	**WOOD BALLISTIC DOORS** with frames and hardware									
0050	Wood, 1-3/4", 3'-0" x 7'-0" UL 752 Level 3	2 Carp	1.50	10.667	Opng.	2,050	360		2,410	2,875

08 34 56 – SECURITY GATES

08 34 56.10 Gates

		Crew	Daily Output	Labor-Hours	Unit	Material	Labor	Equipment	Total	Total Incl O&P
0010	**GATES**									
0015	Gates include mounting hardware									
0500	Wood, security gate, driveway, dual, 10' wide	H-4	.80	25	Opng.	3,675	790		4,465	5,350
0505	12' wide		.80	25		3,975	790		4,765	5,675
0510	15' wide		.80	25		4,575	790		5,365	6,325
0600	Steel, security gate, driveway, single, 10' wide		.80	25		2,150	790		2,940	3,650
0605	12' wide		.80	25		2,225	790		3,015	3,725
0620	Steel, security gate, driveway, steel, dual, 12' wide		.80	25		2,050	790		2,840	3,575
0625	14' wide		.80	25		2,200	790		2,990	3,725
0630	16' wide		.80	25		2,525	790		3,315	4,075
0700	Aluminum, security gate, driveway, dual, 10' wide		.80	25		2,075	790		2,865	3,600
0705	12' wide		.80	25		2,475	790		3,265	4,025
0710	16' wide		.80	25		3,300	790		4,090	4,950
1000	Security gate, driveway, opener 12 VDC				Ea.	650			650	715
1010	Wireless					1,600			1,600	1,750
1020	Security gate, driveway, opener 24 VDC					1,000			1,000	1,100
1030	Wireless					1,100			1,100	1,200
1040	Security gate, driveway, opener 12 VDC, solar panel 10 watt	1 Elec	2	4		160	161		321	435
1050	20 watt	″	2	4		300	161		461	590

08 34 73 – Sound Control Door Assemblies

08 34 73.10 Acoustical Doors

		Crew	Daily Output	Labor-Hours	Unit	Material	Labor	Equipment	Total	Total Incl O&P
0010	**ACOUSTICAL DOORS**									
0020	Including framed seals, 3' x 7', wood, 27 STC rating	2 Carp	1.50	10.667	Ea.	900	360		1,260	1,600
0100	Steel, 40 STC rating		1.50	10.667		2,900	360		3,260	3,775
0200	45 STC rating		1.50	10.667		3,400	360		3,760	4,350
0300	48 STC rating		1.50	10.667		3,925	360		4,285	4,900
0400	52 STC rating		1.50	10.667		4,425	360		4,785	5,475

08 36 Panel Doors

08 36 13 – Sectional Doors

08 36 13.10 Overhead Commercial Doors

		Crew	Daily Output	Labor-Hours	Unit	Material	Labor	Equipment	Total	Total Incl O&P
0010	**OVERHEAD COMMERCIAL DOORS**									
1000	Stock, sectional, heavy duty, wood, 1-3/4" thick, 8' x 8' high	2 Carp	2	8	Ea.	810	268		1,078	1,350
1100	10' x 10' high		1.80	8.889		1,175	298		1,473	1,800
1200	12' x 12' high		1.50	10.667		1,700	360		2,060	2,450
1300	Chain hoist, 14' x 14' high		1.30	12.308		2,600	415		3,015	3,575
1400	12' x 16' high		1	16		2,575	535		3,110	3,725
1500	20' x 8' high		1.30	12.270		2,325	410		2,735	3,275
1600	20' x 16' high		.65	24.615		4,700	825		5,525	6,550
1800	Center mullion openings, 8' high		4	4		1,200	134		1,334	1,550
1900	20' high		2	8		2,000	268		2,268	2,650
2100	For medium duty custom door, deduct					5%	5%			
2150	For medium duty stock doors, deduct					10%	5%			
2300	Fiberglass and aluminum, heavy duty, sectional, 12' x 12' high	2 Carp	1.50	10.667	Ea.	2,625	360		2,985	3,475

08 36 Panel Doors

08 36 13 – Sectional Doors

08 36 13.10 Overhead Commercial Doors		Crew	Daily Output	Labor-Hours	Unit	Material	2011 Bare Costs Labor	Equipment	Total	Total Incl O&P
2450	Chain hoist, 20′ x 20′ high	2 Carp	.50	32	Ea.	6,550	1,075		7,625	9,000
2600	Steel, 24 ga. sectional, manual, 8′ x 8′ high		2	8		805	268		1,073	1,325
2650	10′ x 10′ high		1.80	8.889		1,025	298		1,323	1,625
2700	12′ x 12′ high		1.50	10.667		1,250	360		1,610	1,975
2800	Chain hoist, 20′ x 14′ high		.70	22.857		3,450	765		4,215	5,075
2850	For 1-1/4″ rigid insulation and 26 ga. galv.									
2860	back panel, add				S.F.	4.50			4.50	4.95
2900	For electric trolley operator, 1/3 H.P., to 12′ x 12′, add	1 Carp	2	4	Ea.	905	134		1,039	1,225
2950	Over 12′ x 12′, 1/2 H.P., add	"	1	8	"	1,050	268		1,318	1,600

08 38 Traffic Doors

08 38 19 – Rigid Traffic Doors

08 38 19.20 Double Acting Swing Doors		Crew	Daily Output	Labor-Hours	Unit	Material	Labor	Equipment	Total	Total Incl O&P
0010	**DOUBLE ACTING SWING DOORS**									
1000	.063″ aluminum, 7′-0″ high, 4′-0″ wide	2 Carp	4.20	3.810	Pr.	1,900	128		2,028	2,325
1025	6′-0″ wide		4	4		2,100	134		2,234	2,525
1050	6′-8″ wide		4	4		2,300	134		2,434	2,750
2000	Solid core wood, 3/4″ thick, metal frame, stainless steel									
2010	base plate, 7′ high opening, 4′ wide	2 Carp	4	4	Pr.	2,200	134		2,334	2,650
2050	7′ wide	"	3.80	4.211	"	2,500	141		2,641	2,975
08 38 19.30 Shock Absorbing Doors										
0010	**SHOCK ABSORBING DOORS**									
0020	Rigid, no frame, 1-1/2″ thick, 5′ x 7′	2 Sswk	1.90	8.421	Opng.	1,400	305		1,705	2,125
0100	8′ x 8′		1.80	8.889		2,000	325		2,325	2,825
0500	Flexible, no frame, insulated, .16″ thick, economy, 5′ x 7′		2	8		1,750	291		2,041	2,475
0600	Deluxe		1.90	8.421		2,625	305		2,930	3,450
1000	8′ x 8′ opening, economy		2	8		2,750	291		3,041	3,575
1100	Deluxe		1.90	8.421		3,500	305		3,805	4,425

08 41 Entrances and Storefronts

08 41 13 – Aluminum-Framed Entrances and Storefronts

08 41 13.10 Aluminum Swing Doors		Crew	Daily Output	Labor-Hours	Unit	Material	Labor	Equipment	Total	Total Incl O&P
0010	**ALUMINUM SWING DOORS**									
0015	Aluminum entrance 6′ x 7′	2 Sswk	.70	22.857	Opng.	6,800	830		7,630	9,050
08 41 13.20 Tube Framing										
0010	**TUBE FRAMING**, For window walls and store fronts, aluminum stock									
0050	Plain tube frame, mill finish, 1-3/4″ x 1-3/4″	2 Glaz	103	.155	L.F.	9.05	5.15		14.20	18.40
0150	1-3/4″ x 4″		98	.163		12.50	5.40		17.90	22.50
0200	1-3/4″ x 4-1/2″		95	.168		14.35	5.60		19.95	25
0250	2″ x 6″		89	.180		21.50	5.95		27.45	33.50
0350	4″ x 4″		87	.184		24.50	6.10		30.60	36.50
0400	4-1/2″ x 4-1/2″		85	.188		25.50	6.25		31.75	38.50
0450	Glass bead		240	.067		2.73	2.21		4.94	6.65
1000	Flush tube frame, mill finish, 1/4″ glass, 1-3/4″ x 4″, open header		80	.200		12.05	6.65		18.70	24
1050	Open sill		82	.195		9.75	6.50		16.25	21.50
1100	Closed back header		83	.193		17.20	6.40		23.60	29.50
1150	Closed back sill		85	.188		16.40	6.25		22.65	28.50
1160	Tube fmg, spandrel cover both sides, alum 1″ wide	1 Sswk	85	.094	S.F.	91.50	3.43		94.93	107
1170	Tube fmg, spandrel cover both sides, alum 2″ wide	"	85	.094	"	35	3.43		38.43	45

08 41 Entrances and Storefronts

08 41 13 – Aluminum-Framed Entrances and Storefronts

08 41 13.20 Tube Framing		Crew	Daily Output	Labor-Hours	Unit	Material	2011 Bare Costs Labor	Equipment	Total	Total Incl O&P
1200	Vertical mullion, one piece	2 Glaz	75	.213	L.F.	18	7.10		25.10	31.50
1250	Two piece		73	.219		19.15	7.30		26.45	33
1300	90° or 180° vertical corner post		75	.213		29	7.10		36.10	43.50
1400	1-3/4" x 4-1/2", open header		80	.200		14.60	6.65		21.25	27
1450	Open sill		82	.195		12.20	6.50		18.70	24
1500	Closed back header		83	.193		17.45	6.40		23.85	29.50
1550	Closed back sill		85	.188		17.15	6.25		23.40	29
1600	Vertical mullion, one piece		75	.213		19.20	7.10		26.30	32.50
1650	Two piece		73	.219		20	7.30		27.30	34.50
1700	90° or 180° vertical corner post		75	.213		20.50	7.10		27.60	34.50
2000	Flush tube frame, mil fin.,ins. glass w/thml brk, 2" x 4-1/2", open header		75	.213		15	7.10		22.10	28
2050	Open sill		77	.208		12.60	6.90		19.50	25
2100	Closed back header		78	.205		14.15	6.80		20.95	27
2150	Closed back sill		80	.200		15.15	6.65		21.80	27.50
2200	Vertical mullion, one piece		70	.229		15.70	7.60		23.30	30
2250	Two piece		68	.235		16.95	7.80		24.75	31.50
2300	90° or 180° vertical corner post		70	.229		16.20	7.60		23.80	30.50
5000	Flush tube frame, mill fin., thermal brk., 2-1/4" x 4-1/2", open header		74	.216		15.60	7.20		22.80	29
5050	Open sill		75	.213		13.90	7.10		21	27
5100	Vertical mullion, one piece		69	.232		17.05	7.70		24.75	31.50
5150	Two piece		67	.239		19.75	7.95		27.70	34.50
5200	90° or 180° vertical corner post		69	.232		17.40	7.70		25.10	32
6980	Door stop (snap in)		380	.042		3	1.40		4.40	5.60
7000	For joints, 90°, clip type, add				Ea.	23.50			23.50	26
7050	Screw spline joint, add					22			22	24
7100	For joint other than 90°, add					45.50			45.50	50
8000	For bronze anodized aluminum, add					15%				
8050	For stainless steel materials, add					350%				
8100	For monumental grade, add					52%				
8150	For steel stiffener, add	2 Glaz	200	.080	L.F.	10.50	2.66		13.16	15.90
8200	For 2 to 5 stories, add per story				Story		6%			

08 41 19 – Stainless-Steel-Framed Entrances and Storefronts

08 41 19.10 Stainless-Steel and Glass Entrance Unit		Crew	Daily Output	Labor-Hours	Unit	Material	Labor	Equipment	Total	Total Incl O&P
0010	**STAINLESS-STEEL AND GLASS ENTRANCE UNIT**, narrow stiles									
0020	3' x 7' opening, including hardware, minimum	2 Sswk	1.60	10	Opng.	6,400	365		6,765	7,750
0050	Average		1.40	11.429		6,800	415		7,215	8,275
0100	Maximum		1.20	13.333		7,350	485		7,835	9,000
1000	For solid bronze entrance units, statuary finish, add					62%				
1100	Without statuary finish, add					45%				
2000	Balanced doors, 3' x 7', economy	2 Sswk	.90	17.778	Ea.	8,600	645		9,245	10,700
2100	Premium	"	.70	22.857	"	14,800	830		15,630	17,900

08 42 Entrances

08 42 26 – All-Glass Entrances

08 42 26.10 Swinging Glass Doors		Crew	Daily Output	Labor-Hours	Unit	Material	2011 Bare Costs Labor	Equipment	Total	Total Incl O&P
0010	**SWINGING GLASS DOORS**									
0020	Including hardware, 1/2" thick, tempered, 3' x 7' opening	2 Glaz	2	8	Opng.	2,100	266		2,366	2,750
0100	6' x 7' opening	"	1.40	11.429	"	4,025	380		4,405	5,050

08 42 29 – Automatic Entrances

08 42 29.23 Sliding Automatic Entrances		Crew	Daily Output	Labor-Hours	Unit	Material	2011 Bare Costs Labor	Equipment	Total	Total Incl O&P
0010	**SLIDING AUTOMATIC ENTRANCES** 12' x 7'-6" opng., 5' x 7' door, 2 way traffic									
0020	Mat or electronic activated, panic pushout, incl. operator & hardware,									
0030	not including glass or glazing	2 Glaz	.70	22.857	Opng.	8,150	760		8,910	10,200

08 42 36 – Balanced Door Entrances

08 42 36.10 Balanced Entrance Doors		Crew	Daily Output	Labor-Hours	Unit	Material	2011 Bare Costs Labor	Equipment	Total	Total Incl O&P
0010	**BALANCED ENTRANCE DOORS**									
0020	Hardware & frame, alum. & glass, 3' x 7', econ.	2 Sswk	.90	17.778	Ea.	6,050	645		6,695	7,875
0150	Premium	"	.70	22.857	"	7,400	830		8,230	9,725

08 43 Storefronts

08 43 13 – Aluminum-Framed Storefronts

08 43 13.10 Aluminum-Framed Entrance Doors		Crew	Daily Output	Labor-Hours	Unit	Material	2011 Bare Costs Labor	Equipment	Total	Total Incl O&P
0010	**ALUMINUM-FRAMED ENTRANCE DOORS**									
0015	Standard hardware and glass stops but no glazing									
0020	Entrance door, 3' x 7' opening, clear anodized finish	2 Sswk	7	2.286	Opng.	355	83		438	555
0040	Bronze finish		7	2.286		375	83		458	575
0060	Black finish		7	2.286		420	83		503	620
0200	3'-6" x 7'-0", mill finish		7	2.286		435	83		518	635
0220	Bronze finish		7	2.286		515	83		598	725
0240	Black finish		7	2.286		550	83		633	765
0500	6' x 7' opening, clear finish		6	2.667		720	97		817	975
0520	Bronze finish		6	2.667		755	97		852	1,025
0540	Black finish		6	2.667		845	97		942	1,100
0600	Door Frame 3'-0" x 7'-0", mill finish		6	2.667		310	97		407	525
0620	Bronze finish		6	2.667		360	97		457	585
0640	Black finish		6	2.667		395	97		492	620
0700	3'-6" x 7'-0", mill finish		6	2.667		272	97		369	485
0720	Bronze finish		6	2.667		272	97		369	485
0740	Black finish		6	2.667		272	97		369	485
0800	6'-0" x 7'-0", mill finish		6	2.667		272	97		369	485
0820	Bronze finish		6	2.667		278	97		375	490
0840	Black finish		6	2.667		298	97		395	515
8000	For 8' high doors add				Ea.	200			200	220

08 43 13.20 Storefront Systems		Crew	Daily Output	Labor-Hours	Unit	Material	2011 Bare Costs Labor	Equipment	Total	Total Incl O&P
0010	**STOREFRONT SYSTEMS,** aluminum frame clear 3/8" plate glass									
0020	incl. 3' x 7' door with hardware (400 sq. ft. max. wall)									
0500	Wall height to 12' high, commercial grade	2 Glaz	150	.107	S.F.	20	3.54		23.54	28
0600	Institutional grade		130	.123		25	4.09		29.09	34
0700	Monumental grade		115	.139		38	4.62		42.62	49
1000	6' x 7' door with hardware, commercial grade		135	.119		20	3.93		23.93	28.50
1100	Institutional grade		115	.139		28	4.62		32.62	38
1200	Monumental grade		100	.160		51	5.30		56.30	65
1500	For bronze anodized finish, add					15%				
1600	For black anodized finish, add					35%				
1700	For stainless steel framing, add to monumental					76%				

08 43 Storefronts

08 43 29 – Sliding Storefronts

08 43 29.10 Sliding Panels		Crew	Daily Output	Labor-Hours	Unit	Material	2011 Bare Costs Labor	Equipment	Total	Total Incl O&P
0010	**SLIDING PANELS**									
0020	Mall fronts, aluminum & glass, 15' x 9' high	2 Glaz	1.30	12.308	Opng.	3,250	410		3,660	4,250
0100	24' x 9' high		.70	22.857		4,625	760		5,385	6,325
0200	48' x 9' high, with fixed panels		.90	17.778		8,600	590		9,190	10,400
0500	For bronze finish, add					17%				

08 51 Metal Windows

08 51 13 – Aluminum Windows

08 51 13.10 Aluminum Sash		Crew	Daily Output	Labor-Hours	Unit	Material	Labor	Equipment	Total	Total Incl O&P
0010	**ALUMINUM SASH**									
0020	Stock, grade C, glaze & trim not incl., casement	2 Sswk	200	.080	S.F.	37.50	2.91		40.41	46.50
0050	Double hung		200	.080		38	2.91		40.91	47
0100	Fixed casement		200	.080		16.55	2.91		19.46	24
0150	Picture window		200	.080		17.55	2.91		20.46	25
0200	Projected window		200	.080		34	2.91		36.91	42.50
0250	Single hung		200	.080		15.80	2.91		18.71	23
0300	Sliding		200	.080		20.50	2.91		23.41	28
1000	Mullions for above, tubular		240	.067	L.F.	5.90	2.43		8.33	11.10
3000	Double glazing for above, add	2 Glaz	200	.080	S.F.	11.60	2.66		14.26	17.10
3100	Triple glazing for above, add	"	85	.188	"	13.75	6.25		20	25.50

08 51 13.20 Aluminum Windows		Crew	Daily Output	Labor-Hours	Unit	Material	Labor	Equipment	Total	Total Incl O&P
0010	**ALUMINUM WINDOWS**, incl. frame and glazing, commercial grade									
1000	Stock units, casement, 3'-1" x 3'-2" opening	2 Sswk	10	1.600	Ea.	355	58		413	500
1040	Insulating glass	"	10	1.600		485	58		543	645
1050	Add for storms					115			115	126
1600	Projected, with screen, 3'-1" x 3'-2" opening	2 Sswk	10	1.600		335	58		393	480
1650	Insulating glass	"	10	1.600		345	58		403	485
1700	Add for storms					112			112	123
2000	4'-5" x 5'-3" opening	2 Sswk	8	2		380	73		453	560
2050	Insulating glass	"	8	2		445	73		518	630
2100	Add for storms					120			120	132
2500	Enamel finish windows, 3'-1" x 3'-2"	2 Sswk	10	1.600		340	58		398	485
2550	Insulating glass		10	1.600		293	58		351	430
2600	4'-5" x 5'-3"		8	2		385	73		458	565
2700	Insulating glass		8	2		385	73		458	565
3000	Single hung, 2' x 3' opening, enameled, standard glazed		10	1.600		198	58		256	330
3100	Insulating glass		10	1.600		240	58		298	375
3300	2'-8" x 6'-8" opening, standard glazed		8	2		350	73		423	525
3400	Insulating glass		8	2		450	73		523	635
3700	3'-4" x 5'-0" opening, standard glazed		9	1.778		286	64.50		350.50	440
3800	Insulating glass		9	1.778		315	64.50		379.50	475
4000	Sliding aluminum, 3' x 2' opening, standard glazed		10	1.600		209	58		267	340
4100	Insulating glass		10	1.600		224	58		282	355
4300	5' x 3' opening, standard glazed		9	1.778		320	64.50		384.50	475
4400	Insulating glass		9	1.778		370	64.50		434.50	535
4600	8' x 4' opening, standard glazed		6	2.667		335	97		432	555
4700	Insulating glass		6	2.667		545	97		642	780
5000	9' x 5' opening, standard glazed		4	4		510	146		656	840
5100	Insulating glass		4	4		820	146		966	1,175
5500	Sliding, with thermal barrier and screen, 6' x 4', 2 track		8	2		695	73		768	905

08 51 Metal Windows

08 51 13 – Aluminum Windows

08 51 13.20 Aluminum Windows		Crew	Daily Output	Labor-Hours	Unit	Material	2011 Bare Costs Labor	Equipment	Total	Total Incl O&P
5700	4 track	2 Sswk	8	2	Ea.	880	73		953	1,100
6000	For above units with bronze finish, add					12%				
6200	For installation in concrete openings, add				↓	6%				

08 51 23 – Steel Windows

08 51 23.10 Steel Sash

		Crew	Daily Output	Labor-Hours	Unit	Material	Labor	Equipment	Total	Total Incl O&P
0010	**STEEL SASH** Custom units, glazing and trim not included									
0100	Casement, 100% vented	2 Sswk	200	.080	S.F.	62	2.91		64.91	73.50
0200	50% vented		200	.080		51	2.91		53.91	61.50
0300	Fixed		200	.080		27.50	2.91		30.41	35.50
1000	Projected, commercial, 40% vented		200	.080		48.50	2.91		51.41	59
1100	Intermediate, 50% vented	↓	200	.080	↓	54.50	2.91		57.41	65.50
1200	Basement/utility window, 1'-11" x 2'-8"	1 Carp	14	.571	Ea.	139	19.15		158.15	185
1500	Industrial, horizontally pivoted	2 Sswk	200	.080	S.F.	50.50	2.91		53.41	61
1600	Fixed		200	.080		29.50	2.91		32.41	37.50
2000	Industrial security sash, 50% vented		200	.080		54.50	2.91		57.41	65.50
2100	Fixed		200	.080		44.50	2.91		47.41	54.50
2500	Picture window		200	.080		28.50	2.91		31.41	37
3000	Double hung		200	.080	↓	56	2.91		58.91	67
5000	Mullions for above, open interior face		240	.067	L.F.	9.80	2.43		12.23	15.40
5100	With interior cover	↓	240	.067	"	16.35	2.43		18.78	22.50
6000	Double glazing for above, add	2 Glaz	200	.080	S.F.	12.40	2.66		15.06	18
6100	Triple glazing for above, add	"	85	.188	"	11.85	6.25		18.10	23.50

08 51 23.20 Steel Windows

		Crew	Daily Output	Labor-Hours	Unit	Material	Labor	Equipment	Total	Total Incl O&P
0010	**STEEL WINDOWS** Stock, including frame, trim and insul. glass									
1000	Custom units, double hung, 2'-8" x 4'-6" opening	2 Sswk	12	1.333	Ea.	680	48.50		728.50	845
1100	2'-4" x 3'-9" opening		12	1.333		565	48.50		613.50	715
1500	Commercial projected, 3'-9" x 5'-5" opening		10	1.600		1,200	58		1,258	1,400
1600	6'-9" x 4'-1" opening		7	2.286		1,575	83		1,658	1,875
2000	Intermediate projected, 2'-9" x 4'-1" opening		12	1.333		670	48.50		718.50	830
2100	4'-1" x 5'-5" opening	↓	10	1.600	↓	1,350	58		1,408	1,575

08 51 23.40 Basement Utility Windows

		Crew	Daily Output	Labor-Hours	Unit	Material	Labor	Equipment	Total	Total Incl O&P
0010	**BASEMENT UTILITY WINDOWS**									
0015	1'-3" x 2'-8"	1 Carp	16	.500	Ea.	123	16.80		139.80	163
1100	1'-7" x 2'-8"	"	16	.500	"	137	16.80		153.80	178

08 51 66 – Metal Window Screens

08 51 66.10 Screens

		Crew	Daily Output	Labor-Hours	Unit	Material	Labor	Equipment	Total	Total Incl O&P
0010	**SCREENS**									
0020	For metal sash, aluminum or bronze mesh, flat screen	2 Sswk	1200	.013	S.F.	4.10	.49		4.59	5.45
0500	Wicket screen, inside window	"	1000	.016	"	6.30	.58		6.88	8.05
0600	Residential, aluminum mesh and frame, 2' x 3'	2 Carp	32	.500	Ea.	14	16.80		30.80	43.50
0610	Rescreen		50	.320		12	10.75		22.75	31
0620	3' x 5'		32	.500		48	16.80		64.80	81
0630	Rescreen		45	.356		30	11.95		41.95	53
0640	4' x 8'		25	.640		77.50	21.50		99	122
0650	Rescreen		40	.400		46	13.40		59.40	73
0660	Patio door		25	.640	↓	140	21.50		161.50	190
0680	Rescreening	↓	1600	.010	S.F.	2.10	.34		2.44	2.87
0800	Security screen, aluminum frame with stainless steel cloth	2 Sswk	1200	.013		22.50	.49		22.99	26
0900	Steel grate, painted, on steel frame		1600	.010		12.50	.36		12.86	14.45
1000	For solar louvers, add	↓	160	.100	↓	23.50	3.64		27.14	32.50

08 52 Wood Windows

08 52 10 – Plain Wood Windows

08 52 10.10 Wood Windows		Crew	Daily Output	Labor-Hours	Unit	2011 Bare Costs Material	Labor	Equipment	Total	Total Incl O&P
0010	**WOOD WINDOWS**, including frame,screens and grilles									
0020	Residential, stock units									
0050	Awning type, double insulated glass, 2′-10″ x 1′-9″ opening	2 Carp	12	1.333	Opng.	224	44.50		268.50	320
0100	2′-10″ x 6′-0″ opening	1 Carp	8	1		530	33.50		563.50	640
0200	4′-0″ x 3′-6″ single pane		10	.800		360	27		387	440
0300	6′ x 5′ single pane		8	1	Ea.	490	33.50		523.50	590
1000	Casement, 2′-0″ x 3′-4″ high	2 Carp	20	.800		258	27		285	330
1020	2′-0″ x 4′-0″		18	.889		276	30		306	355
1040	2′-0″ x 5′-0″		17	.941		305	31.50		336.50	390
1060	2′-0″ x 6′-0″		16	1		300	33.50		333.50	385
1080	4′-0″ x 3′-4″		15	1.067		575	36		611	695
1100	4′-0″ x 4′-0″		15	1.067		645	36		681	770
1120	4′-0″ x 5′-0″		14	1.143		730	38.50		768.50	870
1140	4′-0″ x 6′-0″		12	1.333		825	44.50		869.50	980
1600	Casement units, 8′ x 5′, with screens, double insulated glass		2.50	6.400	Opng.	1,375	215		1,590	1,875
1700	Low E glass		2.50	6.400		1,575	215		1,790	2,075
2300	Casements, including screens, 2′-0″ x 3′-4″, dbl. insulated glass		11	1.455		262	49		311	370
2400	Low E glass		11	1.455		262	49		311	370
2600	2 lite, 4′-0″ x 4′-0″, double insulated glass		9	1.778		485	59.50		544.50	630
2700	Low E glass		9	1.778		495	59.50		554.50	640
2900	3 lite, 5′-2″ x 5′-0″, double insulated glass		7	2.286		730	76.50		806.50	935
3000	Low E glass		7	2.286		775	76.50		851.50	985
3200	4 lite, 7′-0″ x 5′-0″, double insulated glass		6	2.667		1,050	89.50		1,139.50	1,300
3300	Low E glass		6	2.667		1,100	89.50		1,189.50	1,375
3500	5 lite, 8′-6″ x 5′-0″, double insulated glass		5	3.200		1,375	107		1,482	1,700
3600	Low E glass		5	3.200		1,475	107		1,582	1,775
3800	For removable wood grilles, diamond pattern, add				Leaf	34			34	37.50
3900	Rectangular pattern, add				″	34			34	37.50
4000	Bow, fixed lites, 8′ x 5′, double insulated glass	2 Carp	3	5.333	Opng.	1,375	179		1,554	1,825
4100	Low E glass	″	3	5.333	″	1,875	179		2,054	2,375
4150	6′-0″ x 5′-0″	1 Carp	8	1	Ea.	1,250	33.50		1,283.50	1,425
4300	Fixed lites, 9′-9″ x 5′-0″, double insulated glass	2 Carp	2	8	Opng.	950	268		1,218	1,500
4400	Low E glass	″	2	8	″	1,050	268		1,318	1,600
5000	Bow, casement, 8′-1″ x 4′-8″ high	3 Carp	8	3	Ea.	1,550	101		1,651	1,900
5020	9′-6″ x 4′-8″		8	3		1,650	101		1,751	2,000
5040	8′-1″ x 5′-1″		8	3		1,875	101		1,976	2,250
5060	9′-6″ x 5′-1″		6	4		1,900	134		2,034	2,300
5080	8′-1″ x 6′-0″		6	4		1,875	134		2,009	2,300
5100	9′-6″ x 6′-0″		6	4		1,975	134		2,109	2,400
5800	Skylights, hatches, vents, and sky roofs, see Section 08 62 13.00									

08 52 10.20 Awning Window		Crew	Daily Output	Labor-Hours	Unit	Material	Labor	Equipment	Total	Total Incl O&P
0010	**AWNING WINDOW**, Including frame, screens and grilles									
0100	Average quality, 34″ x 22″, double insulated glass	1 Carp	10	.800	Ea.	256	27		283	325
0200	Low E glass		10	.800		265	27		292	335
0300	40″ x 28″, double insulated glass		9	.889		310	30		340	390
0400	Low E Glass		9	.889		335	30		365	420
0500	48″ x 36″, double insulated glass		8	1		460	33.50		493.50	560
0600	Low E glass		8	1		485	33.50		518.50	585
1000	Vinyl clad, 34″ x 22″		10	.800		260	27		287	330
1100	40″ x 22″		10	.800		284	27		311	355
1200	36″ x 28″		9	.889		300	30		330	380
1300	36″ x 36″		9	.889		335	30		365	420

08 52 Wood Windows

08 52 10 – Plain Wood Windows

08 52 10.20 Awning Window			Crew	Daily Output	Labor-Hours	Unit	Material	2011 Bare Costs Labor	Equipment	Total	Total Incl O&P
1400	48" x 28"		1 Carp	8	1	Ea.	360	33.50		393.50	455
1500	60" x 36"			8	1		510	33.50		543.50	620
2200	Metal clad, 36" x 25"			9	.889		269	30		299	345
2300	40" x 30"			9	.889		335	30		365	420
2400	48" x 28"			8	1		345	33.50		378.50	435
2500	60" x 36"			8	1		365	33.50		398.50	460
4000	Impact windows, minimum, add						60%				
4010	Impact windows, maximum, add						160%				

08 52 10.40 Casement Window			Crew	Daily Output	Labor-Hours	Unit	Material	Labor	Equipment	Total	Total Incl O&P
0010	**CASEMENT WINDOW**, including frame, screen and grilles										
0100	Avg. quality, bldrs. model, 2'-0" x 3'-0" H, dbl. insulated glass	G	1 Carp	10	.800	Ea.	276	27		303	350
0150	Low E glass	G		10	.800		225	27		252	293
0200	2'-0" x 4'-6" high, double insulated glass	G		9	.889		345	30		375	430
0250	Low E glass	G		9	.889		263	30		293	340
0260	Casement 4'-2" x 4'-2" double insulated glass	G		11	.727		855	24.50		879.50	980
0270	4'-0" x 4'-0" Low E glass	G		11	.727		465	24.50		489.50	555
0290	6'-4" x 5'-7" Low E glass	G		9	.889		1,100	30		1,130	1,250
0300	2'-4" x 6'-0" high, double insulated glass	G		8	1		400	33.50		433.50	495
0350	Low E glass	G		8	1		465	33.50		498.50	570
0600	3'-0" x 5'-0"			8	1		650	33.50		683.50	770
0700	4'-0" x 3'-0"			8	1		715	33.50		748.50	845
0710	4'-0" x 4'-0"	G		8	1		610	33.50		643.50	730
0720	4'-8" x 4'-0"			8	1		675	33.50		708.50	800
0730	4'-8" x 5'-0"			6	1.333		770	44.50		814.50	925
0740	4'-8" x 6'-0"			6	1.333		870	44.50		914.50	1,025
0750	6'-0" x 4'-0"			6	1.333		790	44.50		834.50	940
0800	6'-0" x 5'-0"			6	1.333		885	44.50		929.50	1,050
0900	5'-6" x 5'-6"	G	2 Carp	15	1.067		1,400	36		1,436	1,575
2000	Bay, casement units, 8' x 5', w/screens, dbl. insul. glass			2.50	6.400	Opng.	1,575	215		1,790	2,075
2100	Low E glass			2.50	6.400	"	1,650	215		1,865	2,150
8100	Metal clad, deluxe, dbl. insul. glass, 2'-0" x 3'-0" high	G	1 Carp	10	.800	Ea.	272	27		299	345
8120	2'-0" x 4'-0" high	G		9	.889		305	30		335	385
8140	2'-0" x 5'-0" high	G		8	1		320	33.50		353.50	405
8160	2'-0" x 6'-0" high	G		8	1		355	33.50		388.50	445
8190	For installation, add per leaf							15%			
8200	For multiple leaf units, deduct for stationary sash										
8220	2' high					Ea.	22.50			22.50	24.50
8240	4'-6" high						25.50			25.50	28
8260	6' high						33.50			33.50	37
8300	Impact windows, minimum, add						60%				
8310	Impact windows, maximum, add						160%				

08 52 10.50 Double Hung			Crew	Daily Output	Labor-Hours	Unit	Material	Labor	Equipment	Total	Total Incl O&P
0010	**DOUBLE HUNG**, Including frame, screens and grilles										
0100	Avg. quality, bldrs. model, 2'-0" x 3'-0" high, low e insul. glass	G	1 Carp	10	.800	Ea.	208	27		235	274
0200	3'-0" x 4'-0" high, double insulated glass	G		9	.889		264	30		294	340
0300	4'-0" x 4'-6" high, low e insulated glass	G		8	1		315	33.50		348.50	400
1000	Vinyl clad, premium, double insulated glass, 2'-6" x 3'-0"	G		10	.800		300	27		327	375
1005	2'-6" x 4'-0"	G		10	.800		355	27		382	435
1100	3'-0" x 3'-6"	G		10	.800		330	27		357	405
1200	3'-0" x 4'-0"	G		9	.889		375	30		405	465
1300	3'-0" x 4'-6"	G		9	.889		395	30		425	485
1400	3'-0" x 5'-0"	G		8	1		435	33.50		468.50	535

08 52 Wood Windows

08 52 10 – Plain Wood Windows

08 52 10.50 Double Hung			Crew	Daily Output	Labor-Hours	Unit	Material	2011 Bare Costs Labor	Equipment	Total	Total Incl O&P
1490	3'-4" x 5'-0"	G	1 Carp	8	1	Ea.	445	33.50		478.50	545
1500	3'-6" x 6'-0"	G		8	1		485	33.50		518.50	585
1520	4'-0" x 5'-0"	G		7	1.143		550	38.50		588.50	670
1530	4'-0" x 6'-0"	G		7	1.143		685	38.50		723.50	820
2000	Metal clad, deluxe, dbl. insul. glass, 2'-6" x 3'-0" high	G		10	.800		265	27		292	335
2100	3'-0" x 3'-6" high	G		10	.800		305	27		332	380
2200	3'-0" x 4'-0" high	G		9	.889		320	30		350	400
2300	3'-0" x 4'-6" high	G		9	.889		335	30		365	420
2400	3'-0" x 5'-0" high	G		8	1		365	33.50		398.50	455
2500	3'-6" x 6'-0" high	G		8	1		440	33.50		473.50	535
8000	Impact windows, minimum, add						60%				
8010	Impact windows, maximum, add						160%				

08 52 10.55 Picture Window		Crew	Daily Output	Labor-Hours	Unit	Material	Labor	Equipment	Total	Total Incl O&P
0010	**PICTURE WINDOW**, Including frame and grilles									
0100	Average quality, bldrs. model, 3'-6" x 4'-0" high, dbl. insulated glass	2 Carp	12	1.333	Ea.	395	44.50		439.50	510
0150	Low E glass		12	1.333		410	44.50		454.50	525
0200	4'-0" x 4'-6" high, double insulated glass		11	1.455		490	49		539	620
0250	Low E glass		11	1.455		520	49		569	650
0300	5'-0" x 4'-0" high, double insulated glass		11	1.455		570	49		619	705
0350	Low E glass		11	1.455		590	49		639	730
0400	6'-0" x 4'-6" high, double insulated glass		10	1.600		610	53.50		663.50	765
0450	Low E glass		10	1.600		620	53.50		673.50	775

08 52 10.65 Wood Sash		Crew	Daily Output	Labor-Hours	Unit	Material	Labor	Equipment	Total	Total Incl O&P
0010	**WOOD SASH**, Including glazing but not trim									
0050	Custom, 5'-0" x 4'-0", 1" dbl. glazed, 3/16" thick lites	2 Carp	3.20	5	Ea.	225	168		393	530
0100	1/4" thick lites		5	3.200		240	107		347	445
0200	1" thick, triple glazed		5	3.200		405	107		512	625
0300	7'-0" x 4'-6" high, 1" double glazed, 3/16" thick lites		4.30	3.721		415	125		540	665
0400	1/4" thick lites		4.30	3.721		465	125		590	720
0500	1" thick, triple glazed		4.30	3.721		530	125		655	795
0600	8'-6" x 5'-0" high, 1" double glazed, 3/16" thick lites		3.50	4.571		555	153		708	865
0700	1/4" thick lites		3.50	4.571		605	153		758	920
0800	1" thick, triple glazed		3.50	4.571		610	153		763	930
0900	Window frames only, based on perimeter length				L.F.	3.94			3.94	4.33

08 52 10.70 Sliding Windows			Crew	Daily Output	Labor-Hours	Unit	Material	Labor	Equipment	Total	Total Incl O&P
0010	**SLIDING WINDOWS**										
0100	Average quality, bldrs. model, 3'-0" x 3'-0" high, double insulated	G	1 Carp	10	.800	Ea.	272	27		299	345
0120	Low E glass	G		10	.800		293	27		320	365
0200	4'-0" x 3'-6" high, double insulated	G		9	.889		325	30		355	410
0220	Low E glass	G		9	.889		350	30		380	435
0300	6'-0" x 5'-0" high, double insulated	G		8	1		460	33.50		493.50	565
0320	Low E glass	G		8	1		500	33.50		533.50	605
6000	Sliding, insulating glass, including screens,										
6100	3'-0" x 3'-0"		2 Carp	6.50	2.462	Ea.	320	82.50		402.50	490
6200	4'-0" x 3'-6"			6.30	2.540		325	85		410	500
6300	5'-0" x 4'-0"			6	2.667		400	89.50		489.50	590

08 52 13 – Metal-Clad Wood Windows

08 52 13.10 Awning Windows, Metal-Clad		Crew	Daily Output	Labor-Hours	Unit	Material	Labor	Equipment	Total	Total Incl O&P
0010	**AWNING WINDOWS, METAL-CLAD**									
2000	Metal clad, awning deluxe, double insulated glass, 34" x 22"	1 Carp	10	.800	Ea.	243	27		270	310
2100	40" x 22"	"	10	.800	"	286	27		313	360

08 52 Wood Windows

08 52 13 – Metal-Clad Wood Windows

08 52 13.35 Picture and Sliding Windows Metal-Clad		Crew	Daily Output	Labor-Hours	Unit	2011 Bare Costs Material	Labor	Equipment	Total	Total Incl O&P
0010	**PICTURE AND SLIDING WINDOWS METAL-CLAD**									
2000	Metal clad, dlx picture, dbl. insul. glass, 4'-0" x 4'-0" high	2 Carp	12	1.333	Ea.	365	44.50		409.50	480
2100	4'-0" x 6'-0" high		11	1.455		535	49		584	670
2200	5'-0" x 6'-0" high		10	1.600		595	53.50		648.50	745
2300	6'-0" x 6'-0" high		10	1.600		680	53.50		733.50	840
2400	Metal clad, dlx sliding, double insulated glass, 3'-0" x 3'-0" high G	1 Carp	10	.800		330	27		357	405
2420	4'-0" x 3'-6" high G		9	.889		400	30		430	490
2440	5'-0" x 4'-0" high G		9	.889		480	30		510	575
2460	6'-0" x 5'-0" high G		8	1		705	33.50		738.50	830

08 52 16 – Plastic-Clad Wood Windows

08 52 16.10 Bow Window		Crew	Daily Output	Labor-Hours	Unit	Material	Labor	Equipment	Total	Total Incl O&P
0010	**BOW WINDOW** Including frames, screens, and grilles									
0020	End panels operable									
1000	Bow type, casement, wood, bldrs mdl, 8' x 5' dbl. insltd glass, 4 panel	2 Carp	10	1.600	Ea.	1,500	53.50		1,553.50	1,750
1050	Low E glass		10	1.600		1,325	53.50		1,378.50	1,550
1100	10'-0" x 5'-0", double insulated glass, 6 panels		6	2.667		1,350	89.50		1,439.50	1,650
1200	Low E glass, 6 panels		6	2.667		1,450	89.50		1,539.50	1,750
1300	Vinyl clad, bldrs model, double insulated glass, 6'-0" x 4'-0", 3 panel		10	1.600		1,025	53.50		1,078.50	1,225
1340	9'-0" x 4'-0", 4 panel		8	2		1,350	67		1,417	1,600
1380	10'-0" x 6'-0", 5 panels		7	2.286		2,250	76.50		2,326.50	2,600
1420	12'-0" x 6'-0", 6 panels		6	2.667		2,925	89.50		3,014.50	3,375
1600	Metal clad, casement, bldrs mdl, 6'-0" x 4'-0", dbl. insltd gls, 3 panels		10	1.600		1,200	53.50		1,253.50	1,425
1640	9'-0" x 4'-0", 4 panels		8	2		1,550	67		1,617	1,800
1680	10'-0" x 5'-0", 5 panels		7	2.286		2,100	76.50		2,176.50	2,450
1720	12'-0" x 6'-0", 6 panels		6	2.667		2,700	89.50		2,789.50	3,100
2000	Bay window, 8' x 5', dbl. insul glass		10	1.600		1,825	53.50		1,878.50	2,125
2050	Low E glass		10	1.600		2,225	53.50		2,278.50	2,550
2100	12'-0" x 6'-0", double insulated glass, 6 panels		6	2.667		2,300	89.50		2,389.50	2,675
2200	Low E glass		6	2.667		2,350	89.50		2,439.50	2,750
2280	6'-0" x 4'-0"		11	1.455		1,225	49		1,274	1,425
2300	Vinyl clad, premium, double insulated glass, 8'-0" x 5'-0"		10	1.600		1,750	53.50		1,803.50	2,025
2340	10'-0" x 5'-0"		8	2		2,275	67		2,342	2,600
2380	10'-0" x 6'-0"		7	2.286		2,600	76.50		2,676.50	2,975
2420	12'-0" x 6'-0"		6	2.667		3,200	89.50		3,289.50	3,675
2600	Metal clad, deluxe, dbl. insul. glass, 8'-0" x 5'-0" high, 4 panels		10	1.600		1,625	53.50		1,678.50	1,875
2640	10'-0" x 5'-0" high, 5 panels		8	2		1,750	67		1,817	2,025
2680	10'-0" x 6'-0" high, 5 panels		7	2.286		2,050	76.50		2,126.50	2,400
2720	12'-0" x 6'-0" high, 6 panels		6	2.667		2,850	89.50		2,939.50	3,300
3000	Double hung, bldrs. model, bay, 8' x 4' high, dbl. insulated glass		10	1.600		1,300	53.50		1,353.50	1,525
3050	Low E glass		10	1.600		1,375	53.50		1,428.50	1,625
3100	9'-0" x 5'-0" high, double insulated glass		6	2.667		1,400	89.50		1,489.50	1,675
3200	Low E glass		6	2.667		1,475	89.50		1,564.50	1,775
3300	Vinyl clad, premium, double insulated glass, 7'-0" x 4'-6"		10	1.600		1,325	53.50		1,378.50	1,575
3340	8'-0" x 4'-6"		8	2		1,375	67		1,442	1,600
3380	8'-0" x 5'-0"		7	2.286		1,425	76.50		1,501.50	1,700
3420	9'-0" x 5'-0"		6	2.667		1,475	89.50		1,564.50	1,775
3600	Metal clad, deluxe, dbl. insul. glass, 7'-0" x 4'-0" high		10	1.600		1,225	53.50		1,278.50	1,450
3640	8'-0" x 4'-0" high		8	2		1,275	67		1,342	1,500
3680	8'-0" x 5'-0" high		7	2.286		1,325	76.50		1,401.50	1,575
3720	9'-0" x 5'-0" high		6	2.667		1,400	89.50		1,489.50	1,700

08 52 Wood Windows

08 52 16 – Plastic-Clad Wood Windows

08 52 16.30 Palladian Windows		Crew	Daily Output	Labor-Hours	Unit	2011 Bare Costs Material	Labor	Equipment	Total	Total Incl O&P
0010	**PALLADIAN WINDOWS**									
0020	Vinyl clad, double insulated glass, including frame and grilles									
0040	3′-2″ x 2′-6″ high	2 Carp	11	1.455	Ea.	1,250	49		1,299	1,450
0060	3′-2″ x 4′-10″		11	1.455		1,750	49		1,799	1,975
0080	3′-2″ x 6′-4″		10	1.600		1,700	53.50		1,753.50	1,975
0100	4′-0″ x 4′-0″		10	1.600		1,525	53.50		1,578.50	1,775
0120	4′-0″ x 5′-4″	3 Carp	10	2.400		1,875	80.50		1,955.50	2,175
0140	4′-0″ x 6′-0″		9	2.667		1,950	89.50		2,039.50	2,300
0160	4′-0″ x 7′-4″		9	2.667		2,125	89.50		2,214.50	2,475
0180	5′-5″ x 4′-10″		9	2.667		2,275	89.50		2,364.50	2,650
0200	5′-5″ x 6′-10″		9	2.667		2,575	89.50		2,664.50	3,000
0220	5′-5″ x 7′-9″		9	2.667		2,800	89.50		2,889.50	3,225
0240	6′-0″ x 7′-11″		8	3		3,475	101		3,576	4,000
0260	8′-0″ x 6′-0″		8	3		3,075	101		3,176	3,550

08 52 16.40 Transom Windows

Line	Description	Crew	Daily Output	Labor-Hours	Unit	Material	Labor	Equipment	Total	Total Incl O&P
0010	**TRANSOM WINDOWS**									
1000	Vinyl clad, premium, dbl. insul. glass, 4′-0″ x 4′-0″	2 Carp	12	1.333	Ea.	480	44.50		524.50	605
1100	4′-0″ x 6′-0″		11	1.455		910	49		959	1,075
1200	5′-0″ x 6′-0″		10	1.600		1,000	53.50		1,053.50	1,225
1300	6′-0″ x 6′-0″		10	1.600		1,025	53.50		1,078.50	1,225

08 52 50 – Window Accessories

08 52 50.10 Window Grille or Muntin

Line	Description	Crew	Daily Output	Labor-Hours	Unit	Material	Labor	Equipment	Total	Total Incl O&P
0010	**WINDOW GRILLE OR MUNTIN**, snap in type									
0020	Standard pattern interior grilles									
2000	Wood, awning window, glass size 28″ x 16″ high	1 Carp	30	.267	Ea.	25.50	8.95		34.45	43
2060	44″ x 24″ high		32	.250		37	8.40		45.40	54.50
2100	Casement, glass size, 20″ x 36″ high		30	.267		29.50	8.95		38.45	47.50
2180	20″ x 56″ high		32	.250		41	8.40		49.40	59
2200	Double hung, glass size, 16″ x 24″ high		24	.333	Set	49	11.20		60.20	72
2280	32″ x 32″ high		34	.235	″	128	7.90		135.90	154
2500	Picture, glass size, 48″ x 48″ high		30	.267	Ea.	114	8.95		122.95	140
2580	60″ x 68″ high		28	.286	″	178	9.60		187.60	211
2600	Sliding, glass size, 14″ x 36″ high		24	.333	Set	34	11.20		45.20	55.50
2680	36″ x 36″ high		22	.364	″	41.50	12.20		53.70	66

08 52 69 – Wood Storm Windows

08 52 69.10 Storm Windows

Line	Description	Crew	Daily Output	Labor-Hours	Unit	Material	Labor	Equipment	Total	Total Incl O&P
0010	**STORM WINDOWS**, aluminum residential									
0300	Basement, mill finish, incl. fiberglass screen									
0320	1′-10″ x 1′-0″ high G	2 Carp	30	.533	Ea.	35	17.90		52.90	68.50
0340	2′-9″ x 1′-6″ high G		30	.533		38	17.90		55.90	72
0360	3′-4″ x 2′-0″ high G		30	.533		44	17.90		61.90	78.50
1600	Double-hung, combination, storm & screen									
1700	Custom, clear anodic coating, 2′-0″ x 3′-5″ high	2 Carp	30	.533	Ea.	90	17.90		107.90	129
1720	2′-6″ x 5′-0″ high		28	.571		110	19.15		129.15	153
1740	4′-0″ x 6′-0″ high		25	.640		226	21.50		247.50	285
1800	White painted, 2′-0″ x 3′-5″ high		30	.533		100	17.90		117.90	140
1820	2′-6″ x 5′-0″ high		28	.571		160	19.15		179.15	208
1840	4′-0″ x 6′-0″ high		25	.640		273	21.50		294.50	335
2000	Average quality, clear anodic coating, 2′-0″ x 3′-5″ high G		30	.533		85	17.90		102.90	124
2020	2′-6″ x 5′-0″ high G		28	.571		107	19.15		126.15	149
2040	4′-0″ x 6′-0″ high G		25	.640		126	21.50		147.50	175

08 52 Wood Windows

08 52 69 – Wood Storm Windows

08 52 69.10 Storm Windows			Crew	Daily Output	Labor-Hours	Unit	Material	2011 Bare Costs Labor	Equipment	Total	Total Incl O&P
2400	White painted, 2'-0" x 3'-5" high	G	2 Carp	30	.533	Ea.	85	17.90		102.90	124
2420	2'-6" x 5'-0" high	G		28	.571		90	19.15		109.15	131
2440	4'-0" x 6'-0" high	G		25	.640		98	21.50		119.50	144
2600	Mill finish, 2'-0" x 3'-5" high	G		30	.533		75	17.90		92.90	113
2620	2'-6" x 5'-0" high	G		28	.571		85	19.15		104.15	126
2640	4'-0" x 6'-8" high	G		25	.640		99	21.50		120.50	145
4000	Picture window, storm, 1 lite, white or bronze finish										
4020	4'-6" x 4'-6" high		2 Carp	25	.640	Ea.	120	21.50		141.50	168
4040	5'-8" x 4'-6" high			20	.800		144	27		171	203
4400	Mill finish, 4'-6" x 4'-6" high			25	.640		126	21.50		147.50	175
4420	5'-8" x 4'-6" high			20	.800		144	27		171	203
4600	3 lite, white or bronze finish										
4620	4'-6" x 4'-6" high		2 Carp	25	.640	Ea.	152	21.50		173.50	203
4640	5'-8" x 4'-6" high			20	.800		166	27		193	228
4800	Mill finish, 4'-6" x 4'-6" high			25	.640		144	21.50		165.50	194
4820	5'-8" x 4'-6" high			20	.800		152	27		179	212
5000	Sliding glass door, storm 6' x 6'-8", standard		1 Glaz	2	4		800	133		933	1,100
6000	Sliding window, storm, 2 lite, white or bronze finish										
6020	3'-4" x 2'-7" high		2 Carp	28	.571	Ea.	110	19.15		129.15	153
6040	4'-4" x 3'-3" high			25	.640		150	21.50		171.50	201
6060	5'-4" x 6'-0" high			20	.800		240	27		267	310
9000	Magnetic interior storm window										
9100	3/16" plate glass		1 Glaz	107	.075	S.F.	5.30	2.48		7.78	9.95

08 53 Plastic Windows

08 53 13 – Vinyl Windows

08 53 13.10 Solid Vinyl Windows			Crew	Daily Output	Labor-Hours	Unit	Material	Labor	Equipment	Total	Total Incl O&P
0010	**SOLID VINYL WINDOWS**										
0020	Double hung, including frame and screen, 2'-0" x 2'-6"		2 Carp	15	1.067	Ea.	197	36		233	277
0040	2'-0" x 3'-6"			14	1.143		209	38.50		247.50	294
0060	2'-6" x 4'-6"			13	1.231		247	41.50		288.50	340
0080	3'-0" x 4'-0"			10	1.600		214	53.50		267.50	325
0100	3'-0" x 4'-6"			9	1.778		275	59.50		334.50	400
0120	3'-6" x 4'-6"			8	2		289	67		356	430
0140	3'-6" x 6'-0"			7	2.286		335	76.50		411.50	495

08 53 13.20 Vinyl Single Hung Windows			Crew	Daily Output	Labor-Hours	Unit	Material	Labor	Equipment	Total	Total Incl O&P
0010	**VINYL SINGLE HUNG WINDOWS**, insulated glass										
0100	Grids, low E, J fin, ext. jambs, 21" x 53"	G	2 Carp	18	.889	Ea.	198	30		228	268
0110	21" x 57"	G		17	.941		195	31.50		226.50	268
0120	21" x 65"	G		16	1		200	33.50		233.50	276
0130	25" x 41"	G		20	.800		188	27		215	252
0140	25" x 49"	G		18	.889		198	30		228	268
0150	25" x 57"	G		17	.941		201	31.50		232.50	274
0160	25" x 65"	G		16	1		207	33.50		240.50	284
0170	29" x 41"	G		18	.889		192	30		222	261
0180	29" x 53"	G		18	.889		202	30		232	272
0190	29" x 57"	G		17	.941		206	31.50		237.50	280
0200	29" x 65"	G		16	1		212	33.50		245.50	289
0210	33" x 41"	G		20	.800		198	27		225	263
0220	33" x 53"	G		18	.889		209	30		239	280
0230	33" x 57"	G		17	.941		212	31.50		243.50	286

08 53 Plastic Windows

08 53 13 – Vinyl Windows

08 53 13.20 Vinyl Single Hung Windows		Crew	Daily Output	Labor-Hours	Unit	Material	2011 Bare Costs Labor	Equipment	Total	Total Incl O&P
0240	33" x 65" G	2 Carp	16	1	Ea.	218	33.50		251.50	296
0250	37" x 41" G		20	.800		206	27		233	272
0260	37" x 53" G		18	.889		218	30		248	290
0270	37" x 57" G		17	.941		221	31.50		252.50	296
0280	37" x 65" G	↓	16	1	↓	232	33.50		265.50	310

08 53 13.30 Vinyl Double Hung Windows		Crew	Daily Output	Labor-Hours	Unit	Material	Labor	Equipment	Total	Total Incl O&P
0010	**VINYL DOUBLE HUNG WINDOWS**, insulated glass									
0100	Grids, low E, J fin, ext. jambs, 21" x 53" G	2 Carp	18	.889	Ea.	215	30		245	287
0102	21" x 37" G		18	.889		197	30		227	267
0104	21" x 41" G		18	.889		201	30		231	271
0106	21" x 49" G		18	.889		221	30		251	293
0110	21" x 57" G		17	.941		218	31.50		249.50	293
0120	21" x 65" G		16	1		228	33.50		261.50	305
0128	25" x 37" G		20	.800		206	27		233	272
0130	25" x 41" G		20	.800		209	27		236	275
0140	25" x 49" G		18	.889		213	30		243	284
0145	25" x 53" G		18	.889		221	30		251	293
0150	25" x 57" G		17	.941		221	31.50		252.50	296
0160	25" x 65" G		16	1		231	33.50		264.50	310
0162	25" x 69" G		16	1		238	33.50		271.50	320
0164	25" x 77" G		16	1		248	33.50		281.50	330
0168	29" x 37" G		18	.889		211	30		241	282
0170	29" x 41" G		18	.889		214	30		244	285
0172	29" x 49" G		18	.889		221	30		251	293
0180	29" x 53" G		18	.889		225	30		255	298
0190	29" x 57" G		17	.941		229	31.50		260.50	305
0200	29" x 65" G		16	1		236	33.50		269.50	315
0202	29" x 69" G		16	1		243	33.50		276.50	325
0205	29" x 77" G		16	1		254	33.50		287.50	335
0208	33" x 37" G		20	.800		215	27		242	282
0210	33" x 41" G		20	.800		219	27		246	286
0215	33" x 49" G		20	.800		227	27		254	295
0220	33" x 53" G		18	.889		231	30		261	305
0230	33" x 57" G		17	.941		235	31.50		266.50	310
0240	33" x 65" G		16	1		239	33.50		272.50	320
0242	33" x 69" G		16	1		245	33.50		278.50	325
0246	33" x 77" G		16	1		251	33.50		284.50	330
0250	37" x 41" G		20	.800		223	27		250	290
0255	37" x 49" G		20	.800		231	27		258	299
0260	37" x 53" G		18	.889		239	30		269	315
0270	37" x 57" G		17	.941		243	31.50		274.50	320
0280	37" x 65" G		16	1		255	33.50		288.50	335
0282	37" x 69" G		16	1		262	33.50		295.50	345
0286	37" x 77" G	↓	16	1		270	33.50		303.50	355
0300	Solid vinyl, average quality, double insulated glass, 2'-0" x 3'-0" G	1 Carp	10	.800		287	27		314	360
0310	3'-0" x 4'-0" G		9	.889		198	30		228	267
0320	4'-0" x 4'-6" G		8	1		298	33.50		331.50	385
0330	Premium, double insulated glass, 2'-6" x 3'-0" G		10	.800		230	27		257	298
0340	3'-0" x 3'-6" G		9	.889		254	30		284	330
0350	3'-0" x 4'-0" G		9	.889		274	30		304	350
0360	3'-0" x 4'-6" G		9	.889		266	30		296	340
0370	3'-0" x 5'-0" G	↓	8	1	↓	286	33.50		319.50	370

08 53 Plastic Windows

08 53 13 – Vinyl Windows

08 53 13.30 Vinyl Double Hung Windows			Crew	Daily Output	Labor-Hours	Unit	2011 Bare Costs Material	Labor	Equipment	Total	Total Incl O&P
0380	3'-6" x 6'-0"	G	1 Carp	8	1	Ea.	320	33.50		353.50	405
08 53 13.40 Vinyl Casement Windows											
0010	**VINYL CASEMENT WINDOWS**, insulated glass										
0015	Grids, low E, J fin, extension jambs, screens										
0100	One lite, 21" x 41"	G	2 Carp	20	.800	Ea.	271	27		298	345
0110	21" x 47"	G		20	.800		294	27		321	370
0120	21" x 53"	G		20	.800		315	27		342	395
0128	24" x 35"	G		19	.842		263	28.50		291.50	335
0130	24" x 41"	G		19	.842		283	28.50		311.50	360
0140	24" x 47"	G		19	.842		305	28.50		333.50	385
0150	24" x 53"	G		19	.842		325	28.50		353.50	410
0158	28" x 35"	G		19	.842		278	28.50		306.50	355
0160	28" x 41"	G		19	.842		298	28.50		326.50	380
0170	28" x 47"	G		19	.842		320	28.50		348.50	400
0180	28" x 53"	G		19	.842		350	28.50		378.50	435
0184	28" x 59"	G		19	.842		350	28.50		378.50	435
0188	Two lites, 33" x 35"	G		18	.889		430	30		460	520
0190	33" x 41"	G		18	.889		455	30		485	555
0200	33" x 47"	G		18	.889		490	30		520	585
0210	33" x 53"	G		18	.889		520	30		550	620
0212	33" x 59"	G		18	.889		550	30		580	655
0215	33" x 72"	G		18	.889		570	30		600	675
0220	41" x 41"	G		18	.889		495	30		525	595
0230	41" x 47"	G		18	.889		530	30		560	630
0240	41" x 53"	G		17	.941		560	31.50		591.50	670
0242	41" x 59"	G		17	.941		590	31.50		621.50	700
0246	41" x 72"	G		17	.941		615	31.50		646.50	730
0250	47" x 41"	G		17	.941		500	31.50		531.50	605
0260	47" x 47"	G		17	.941		530	31.50		561.50	640
0270	47" x 53"	G		17	.941		560	31.50		591.50	675
0272	47" x 59"	G		17	.941		615	31.50		646.50	730
0280	56" x 41"	G		15	1.067		535	36		571	650
0290	56" x 47"	G		15	1.067		560	36		596	680
0300	56" x 53"	G		15	1.067		610	36		646	730
0302	56" x 59"	G		15	1.067		635	36		671	760
0310	56" x 72"	G		15	1.067		690	36		726	820
0340	Solid vinyl, premium, double insulated glass, 2'-0" x 3'-0" high	G	1 Carp	10	.800		259	27		286	330
0360	2'-0" x 4'-0" high	G		9	.889		289	30		319	370
0380	2'-0" x 5'-0" high	G		8	1		295	33.50		328.50	380
08 53 13.50 Vinyl Picture Windows											
0010	**VINYL PICTURE WINDOWS**, insulated glass										
0100	Grids, low E, J fin, ext. jambs, 33" x 47"		2 Carp	12	1.333	Ea.	238	44.50		282.50	335
0110	35" x 71"			12	1.333		310	44.50		354.50	415
0120	41" x 47"			12	1.333		252	44.50		296.50	350
0130	41" x 71"			12	1.333		320	44.50		364.50	430
0140	47" x 47"			12	1.333		277	44.50		321.50	380
0150	47" x 71"			11	1.455		350	49		399	465
0160	53" x 47"			11	1.455		300	49		349	410
0170	53" x 71"			11	1.455		315	49		364	425
0180	59" x 47"			11	1.455		340	49		389	455
0190	59" x 71"			11	1.455		365	49		414	480
0200	71" x 47"			10	1.600		375	53.50		428.50	505

08 53 Plastic Windows

08 53 13 – Vinyl Windows

08 53 13.50 Vinyl Picture Windows			Crew	Daily Output	Labor-Hours	Unit	2011 Bare Costs Material	Labor	Equipment	Total	Total Incl O&P
0210	71" x 71"		2 Carp	10	1.600	Ea.	395	53.50		448.50	525

08 54 Composite Windows

08 54 13 – Fiberglass Windows

08 54 13.10 Fiberglass Single Hung Windows			Crew	Daily Output	Labor-Hours	Unit	Material	Labor	Equipment	Total	Total Incl O&P
0010	**FIBERGLASS SINGLE HUNG WINDOWS**										
0100	Grids, low E, 18" x 24"	G	2 Carp	18	.889	Ea.	335	30		365	420
0110	18" x 40"	G		17	.941		300	31.50		331.50	385
0130	24" x 40"	G		20	.800		330	27		357	410
0230	36" x 36"	G		17	.941		370	31.50		401.50	460
0250	36" x 48"	G		20	.800		405	27		432	490
0260	36" x 60"	G		18	.889		445	30		475	540
0280	36" x 72"	G		16	1		470	33.50		503.50	570
0290	48" x 40"	G		16	1		470	33.50		503.50	570

08 61 Roof Windows

08 61 13 – Metal Roof Windows

08 61 13.10 Roof Windows		Crew	Daily Output	Labor-Hours	Unit	Material	Labor	Equipment	Total	Total Incl O&P
0010	**ROOF WINDOWS**, fixed high perf tmpd glazing, metallic framed									
0020	46" x 21-1/2", Flashed for shingled roof	1 Carp	8	1	Ea.	245	33.50		278.50	325
0100	46" x 28"		8	1		272	33.50		305.50	355
0125	57" x 44"		6	1.333		335	44.50		379.50	445
0130	72" x 28"		7	1.143		335	38.50		373.50	435
0150	Fixed, laminated tempered glazing, 46" x 21-1/2"		8	1		415	33.50		448.50	510
0175	46" x 28"		8	1		460	33.50		493.50	560
0200	57" x 44"		6	1.333		430	44.50		474.50	550
0500	Vented flashing set for shingled roof, 46" x 21-1/2"		7	1.143		415	38.50		453.50	520
0525	46" x 28"		6	1.333		460	44.50		504.50	580
0550	57" x 44"		5	1.600		575	53.50		628.50	725
0560	72" x 28"		5	1.600		575	53.50		628.50	725
0575	Flashing set for low pitched roof, 46" x 21-1/2"		7	1.143		475	38.50		513.50	590
0600	46" x 28"		7	1.143		525	38.50		563.50	640
0625	57" x 44"		5	1.600		650	53.50		703.50	805
0650	Flashing set for curb 46" x 21-1/2"		7	1.143		550	38.50		588.50	670
0675	46" x 28"		7	1.143		600	38.50		638.50	725
0700	57" x 44"		5	1.600		740	53.50		793.50	900

08 61 16 – Wood Roof Windows

08 61 16.16 Roof Windows, Wood Framed		Crew	Daily Output	Labor-Hours	Unit	Material	Labor	Equipment	Total	Total Incl O&P
0010	**ROOF WINDOWS, WOOD FRAMED**									
5600	Roof window incl. frame, flashing, double insulated glass & screens,									
5610	complete unit, 22" x 38"	2 Carp	3	5.333	Ea.	540	179		719	890
5650	2'-5" x 3'-8"		3.20	5		660	168		828	1,000
5700	3'-5" x 4'-9"		3.40	4.706		765	158		923	1,100

08 62 Unit Skylights

08 62 13 – Domed Unit Skylights

08 62 13.10 Domed Skylights			Crew	Daily Output	Labor-Hours	Unit	Material	Labor	Equipment	Total	Total Incl O&P
							2011 Bare Costs				
0010	**DOMED SKYLIGHTS**										
0020	Skylight, fixed dome type, 22" x 22"	G	G-3	12	2.667	Ea.	200	86		286	365
0030	22" x 46"	G		10	3.200		229	103		332	425
0040	30" x 30"	G		12	2.667		230	86		316	395
0050	30" x 46"	G		10	3.200		290	103		393	490
0110	Fixed, double glazed, 22" x 27"	G		12	2.667		213	86		299	375
0120	22" x 46"	G		10	3.200		269	103		372	470
0130	44" x 46"	G		10	3.200		400	103		503	610
0210	Operable, double glazed, 22" x 27"	G		12	2.667		310	86		396	485
0220	22" x 46"	G		10	3.200		365	103		468	570
0230	44" x 46"	G		10	3.200		795	103		898	1,050

08 62 13.20 Skylights			Crew	Daily Output	Labor-Hours	Unit	Material	Labor	Equipment	Total	Total Incl O&P
0010	**SKYLIGHTS**, Plastic domes, flush or curb mounted										
2120	Ventilating insulated plexiglass dome with										
2130	curb mounting, 36" x 36"	G	G-3	12	2.667	Ea.	460	86		546	650
2150	52" x 52"	G		12	2.667		640	86		726	850
2160	28" x 52"	G		10	3.200		470	103		573	685
2170	36" x 52"	G		10	3.200		520	103		623	740
2180	For electric opening system, add	G					300			300	330
2210	Operating skylight, with thermopane glass, 24" x 48"	G	G-3	10	3.200		560	103		663	790
2220	32" x 48"	G	"	9	3.556		590	115		705	835
2310	Non venting insulated plexiglass dome skylight with										
2320	Flush mount 22" x 46"	G	G-3	15.23	2.101	Ea.	320	68		388	465
2330	30" x 30"	G		16	2		292	64.50		356.50	425
2340	46" x 46"	G		13.91	2.301		540	74.50		614.50	720
2350	Curb mount 22" x 46"	G		15.23	2.101		350	68		418	500
2360	30" x 30"	G		16	2		390	64.50		454.50	535
2370	46" x 46"	G		13.91	2.301		600	74.50		674.50	785
2381	Non-insulated flush mount 22" x 46"			15.23	2.101		215	68		283	350
2382	30" x 30"			16	2		194	64.50		258.50	320
2383	46" x 46"			13.91	2.301		365	74.50		439.50	525
2384	Curb mount 22" x 46"			15.23	2.101		181	68		249	310
2385	30" x 30"			16	2		180	64.50		244.50	305
2400	Sandwich panels, fiberglass, for walls, 1-9/16" thick, to 250 S.F.	G		200	.160	S.F.	16.20	5.15		21.35	26.50
2700	As above, but for roofs, 2-3/4" thick, to 250 S.F.	G		295	.108		23.50	3.51		27.01	31.50
2800	250 S.F. and up	G		330	.097		19	3.14		22.14	26
4000	Skylight, solar tube kit, incl dome, flashing, diffuser, 1 pipe, 9" dia.	G	1 Carp	2	4	Ea.	230	134		364	480
4010	13" dia.	G		2	4		310	134		444	565
4020	21" dia.	G		2	4		490	134		624	765
4030	Accessories for, 1' long x 9" dia pipe	G		24	.333		30	11.20		41.20	51.50
4040	2' long x 9" dia pipe	G		24	.333		44	11.20		55.20	67
4050	4' long x 9" dia pipe	G		20	.400		72	13.40		85.40	102
4060	1' long x 13" dia pipe	G		24	.333		44	11.20		55.20	67
4070	2' long x 13" dia pipe	G		24	.333		57	11.20		68.20	81
4080	4' long x 13" dia pipe	G		20	.400		99	13.40		112.40	132
4090	2' long x 21" dia pipe	G		16	.500		100	16.80		116.80	138
4100	4' long x 21" dia pipe	G		12	.667		162	22.50		184.50	216
4110	45 degree elbow, 9"	G		16	.500		90	16.80		106.80	127
4120	13"	G		16	.500		83	16.80		99.80	120
4130	Interior decorative ring, 9"	G		20	.400		20	13.40		33.40	44.50
4140	13"	G		20	.400		22	13.40		35.40	46.50

08 71 Door Hardware

08 71 13 – Automatic Door Operators

08 71 13.10 Automatic Openers Commercial		Crew	Daily Output	Labor-Hours	Unit	Material	2011 Bare Costs Labor	Equipment	Total	Total Incl O&P
0010	**AUTOMATIC OPENERS COMMERCIAL**									
0020	Pneumatic, incl opener, motion sens, control box, tubing, compressor									
0050	For single swing door, per opening	2 Skwk	.80	20	Ea.	4,300	680		4,980	5,875
0100	Pair, per opening		.50	32	Opng.	7,000	1,100		8,100	9,525
1000	For single sliding door, per opening		.60	26.667		4,700	910		5,610	6,700
1300	Bi-parting pair		.50	32		7,100	1,100		8,200	9,625
1420	Electronic door opener incl motion sens, 12 V control box, motor									
1450	For single swing door, per opening	2 Skwk	.80	20	Opng.	3,500	680		4,180	5,000
1500	Pair, per opening		.50	32		5,700	1,100		6,800	8,100
1600	For single sliding door, per opening		.60	26.667		3,800	910		4,710	5,700
1700	Bi-parting pair		.50	32		5,800	1,100		6,900	8,200
1750	Handicap actuator buttons, 2, including 12 V DC wiring, add	1 Carp	1.50	5.333	Pr.	430	179		609	775

08 71 20 – Hardware

08 71 20.15 Hardware		Crew	Daily Output	Labor-Hours	Unit	Material	Labor	Equipment	Total	Total Incl O&P
0009	**HARDWARE**									
0010	Average percentage for hardware, total job cost									
0025	Minimum				Job					1%
0050	Maximum									4%
0500	Total hardware for building, average distribution					85%	15%			
1000	Door hardware, apartment, interior	1 Carp	4	2	Door	510	67		577	670
1300	Average, door hardware, motel/hotel interior, with access card		4	2		625	67		692	800
1500	Hospital bedroom, minimum		4	2		595	67		662	765
2000	Maximum		3	2.667		705	89.50		794.50	925
2100	Pocket door		6	1.333	Ea.	134	44.50		178.50	222
2250	School, single exterior, incl. lever, incl. panic device		3	2.667	Door	1,100	89.50		1,189.50	1,375
2500	Single interior, regular use, lever included		3	2.667		595	89.50		684.50	800
2550	Average, door hdwe, school, classroom, ANSI F84, lever handle		3	2.667		750	89.50		839.50	975
2600	Average, door hdwe, school, classroom, ANSI F88, incl. lever		3	2.667		790	89.50		879.50	1,025
2850	Stairway, single interior		3	2.667		525	89.50		614.50	725
3100	Double exterior, with panic device		2	4	Pr.	2,075	134		2,209	2,500
3600	Toilet, public, single interior				Door	210			210	231
6020	Add for fire alarm door holder, electro-magnetic	1 Elec	4	2	Ea.	99	80.50		179.50	240

08 71 20.30 Door Closers		Crew	Daily Output	Labor-Hours	Unit	Material	Labor	Equipment	Total	Total Incl O&P
0010	**DOOR CLOSERS**									
0020	Adjustable backcheck, 3 way mount, all sizes, regular arm	1 Carp	6	1.333	Ea.	182	44.50		226.50	275
0040	Hold open arm		6	1.333		230	44.50		274.50	330
0100	Fusible link		6.50	1.231		155	41.50		196.50	240
0200	Non sized, regular arm		6	1.333		165	44.50		209.50	257
0240	Hold open arm		6	1.333		200	44.50		244.50	295
0400	4 way mount, non sized, regular arm		6	1.333		220	44.50		264.50	315
0440	Hold open arm		6	1.333		230	44.50		274.50	330
8000	Surface mounted, stand. duty, parallel arm, primed, traditional		6	1.333		187	44.50		231.50	281
8030	Light duty		6	1.333		123	44.50		167.50	210
8050	Heavy duty		6	1.333		213	44.50		257.50	310
8100	Standard duty, parallel arm, modern		6	1.333		213	44.50		257.50	310
8150	Heavy duty		6	1.333		228	44.50		272.50	325

08 71 20.31 Door Closers		Crew	Daily Output	Labor-Hours	Unit	Material	Labor	Equipment	Total	Total Incl O&P
0010	**DOOR CLOSERS**									
0015	Door closer, rack and pinion	1 Carp	6.50	1.231	Ea.	168	41.50		209.50	253
1520	Door, single acting, standard arm		1	8		199	268		467	670
1522	Hold open arm		1	8		216	268		484	690
1526	Frame, single acting, standard arm		1	8		335	268		603	820

08 71 Door Hardware

08 71 20 – Hardware

08 71 20.31 Door Closers		Crew	Daily Output	Labor-Hours	Unit	Material	2011 Bare Costs Labor	Equipment	Total	Total Incl O&P
1530	Hold open arm	1 Carp	1	8	Ea.	355	268		623	840
1534	Double acting, standard arm		1	8		450	268		718	945
1536	Hold open arm		1	8		460	268		728	955
1554	Floor, center hung, single acting, bottom arm		1	8		320	268		588	800
1558	Double acting		1	8		330	268		598	815
1560	Offset hung, single acting, bottom arm	↓	1	8	↓	370	268		638	860
6500	Electro magnetic closer/holder									
6510	Single point, no detector	1 Carp	1	8	Ea.	350	268		618	835
6515	Including detector		1	8		670	268		938	1,200
6520	Multi-point, no detector		1	8		770	268		1,038	1,300
6524	Including detector	↓	1	8	↓	780	268		1,048	1,300
6550	Electric automatic operators									
6555	Operator	1 Carp	1	8	Ea.	1,525	268		1,793	2,125
6570	Wall plate actuator		1	8		195	268		463	665
8010	Light duty, regular arm		1	8		111	268		379	570
8032	Parallel arm		1	8		115	268		383	575
8034	Hold open arm		1	8		121	268		389	585
8036	Fusible link arm		1	8		176	268		444	645
8040	Medium duty, regular arm		1	8		140	268		408	605
8042	Extra duty parallel arm		1	8		150	268		418	615
8044	Hold open arm		1	8		157	268		425	620
8046	Positive stop arm		1	8		170	268		438	635
8052	Heavy duty, regular arm		1	8		172	268		440	640
8054	Top jamb mount		1	8		172	268		440	640
8056	Extra duty parallel arm		1	8		172	268		440	640
8058	Hold open arm		1	8		187	268		455	655
8060	Positive stop arm		1	8		189	268		457	660
8062	Fusible link arm		1	8		217	268		485	690
8080	Universal heavy duty, regular arm		1	8		186	268		454	655
8084	Parallel arm		1	8		186	268		454	655
8088	Extra duty, parallel arm		1	8		191	268		459	660
8090	Hold open arm		1	8		272	268		540	750
8094	Positive stop arm		1	8		277	268		545	755
8098	For delayed action add	↓	1	8	↓	30	268		298	485

08 71 20.35 Panic Devices		Crew	Daily Output	Labor-Hours	Unit	Material	Labor	Equipment	Total	Total Incl O&P
0010	**PANIC DEVICES**									
0015	For rim locks, single door exit only	1 Carp	6	1.333	Ea.	530	44.50		574.50	660
1000	Mortise, bar, exit only		4	2	"	550	67		617	715
4000	Double doors, exit only		2	4	Pr.	955	134		1,089	1,275
4500	Exit & entrance	↓	2	4	"	1,700	134		1,834	2,100

08 71 20.40 Lockset		Crew	Daily Output	Labor-Hours	Unit	Material	Labor	Equipment	Total	Total Incl O&P
0010	**LOCKSET**, Standard duty									
0020	Non-keyed, passage, w/sect.trim	1 Carp	12	.667	Ea.	54.50	22.50		77	97
0100	Privacy		12	.667		62	22.50		84.50	106
0400	Keyed, single cylinder function		10	.800		92.50	27		119.50	147
0420	Hotel (see also Section 08 71 20.15)		8	1		135	33.50		168.50	205
0500	Lever handled, keyed, single cylinder function		10	.800		159	27		186	220
0600	Bedroom, bathroom and inner office doors		10	.800		171	27		198	233
0900	Apartment, office and corridor doors		10	.800		230	27		257	298
1400	Keyed, single cylinder function		10	.800		220	27		247	287
1700	Residential, interior door, minimum		16	.500		25	16.80		41.80	55.50
1720	Maximum	↓	8	1	↓	70	33.50		103.50	133

08 71 Door Hardware

08 71 20 – Hardware

08 71 20.40 Lockset		Crew	Daily Output	Labor-Hours	Unit	2011 Bare Costs Material	Labor	Equipment	Total	Total Incl O&P
1800	Exterior, minimum	1 Carp	14	.571	Ea.	45	19.15		64.15	81.50
1810	Average		8	1		72	33.50		105.50	136
1820	Maximum		8	1		170	33.50		203.50	243
08 71 20.45 Peepholes										
0010	**PEEPHOLES**									
2010	Peephole	1 Carp	32	.250	Ea.	15.80	8.40		24.20	31.50
08 71 20.50 Door Stops										
0010	**DOOR STOPS**									
0020	Holder & bumper, floor or wall	1 Carp	32	.250	Ea.	35.50	8.40		43.90	53
1300	Wall bumper, 4" diameter, with rubber pad, aluminum		32	.250		10.95	8.40		19.35	26
1600	Door bumper, floor type, aluminum		32	.250		6.70	8.40		15.10	21.50
1900	Plunger type, door mounted		32	.250		25.50	8.40		33.90	42
08 71 20.55 Push-Pull Plates										
0010	**PUSH-PULL PLATES**									
0100	Push plate, .050 thick, 4" x 16", aluminum	1 Carp	12	.667	Ea.	10.20	22.50		32.70	49
0500	Bronze		12	.667		21.50	22.50		44	61
1500	Pull handle and push bar, aluminum		11	.727		131	24.50		155.50	185
2000	Bronze		10	.800		169	27		196	230
4000	Door pull, designer style, cast aluminum, minimum		12	.667		71.50	22.50		94	116
5000	Maximum		8	1		370	33.50		403.50	460
08 71 20.60 Entrance Locks										
0010	**ENTRANCE LOCKS**									
0015	Cylinder, grip handle deadlocking latch	1 Carp	9	.889	Ea.	132	30		162	195
0020	Deadbolt		8	1		160	33.50		193.50	232
0100	Push and pull plate, dead bolt		8	1		150	33.50		183.50	221
0900	For handicapped lever, add					170			170	187
08 71 20.65 Thresholds										
0010	**THRESHOLDS**									
0011	Threshold 3' long saddles aluminum	1 Carp	48	.167	L.F.	4.93	5.60		10.53	14.75
0100	Aluminum, 8" wide, 1/2" thick		12	.667	Ea.	39	22.50		61.50	80.50
0500	Bronze		60	.133	L.F.	41.50	4.47		45.97	53
0600	Bronze, panic threshold, 5" wide, 1/2" thick		12	.667	Ea.	67.50	22.50		90	112
0700	Rubber, 1/2" thick, 5-1/2" wide		20	.400		40	13.40		53.40	66.50
0800	2-3/4" wide		20	.400		29	13.40		42.40	54.50
08 71 20.75 Door Hardware Accessories										
0010	**DOOR HARDWARE ACCESSORIES**									
0050	Door closing coordinator, 36" (for paired openings up to 56")	1 Carp	8	1	Ea.	94	33.50		127.50	159
0060	48" (for paired openings up to 84")		8	1		101	33.50		134.50	167
0070	56" (for paired openings up to 96")		8	1		111	33.50		144.50	178
1000	Knockers, brass, standard		16	.500		44	16.80		60.80	76.50
1100	Deluxe		10	.800		118	27		145	174
4100	Deluxe		18	.444		44	14.90		58.90	73.50
4500	Rubber door silencers		540	.015		.33	.50		.83	1.19
08 71 20.80 Hasps										
0010	**HASPS**, steel assembly									
0015	3"	1 Carp	26	.308	Ea.	4.71	10.30		15.01	22.50
0020	4-1/2"		13	.615		6.50	20.50		27	41.50
0040	6"		12.50	.640		9.25	21.50		30.75	46

08 71 Door Hardware

08 71 20 – Hardware

08 71 20.90 Hinges		Crew	Daily Output	Labor-Hours	Unit	Material	2011 Bare Costs Labor	Equipment	Total	Total Incl O&P
0010	**HINGES**									
0012	Full mortise, avg. freq., steel base, USP, 4-1/2" x 4-1/2"				Pr.	27			27	29.50
0100	5" x 5", USP					43			43	47.50
0200	6" x 6", USP					91			91	100
0400	Brass base, 4-1/2" x 4-1/2", US10					53.50			53.50	59
0500	5" x 5", US10					77			77	85
0600	6" x 6", US10					130			130	143
0800	Stainless steel base, 4-1/2" x 4-1/2", US32					79.50			79.50	87.50
0900	For non removable pin, add (security item)				Ea.	4.72			4.72	5.20
0910	For floating pin, driven tips, add					3.25			3.25	3.58
0930	For hospital type tip on pin, add					14.10			14.10	15.50
0940	For steeple type tip on pin, add					13			13	14.30
0950	Full mortise, high frequency, steel base, 3-1/2" x 3-1/2", US26D				Pr.	27			27	29.50
1000	4-1/2" x 4-1/2", USP					62			62	68.50
1100	5" x 5", USP					58.50			58.50	64
1200	6" x 6", USP					142			142	156
1400	Brass base, 3-1/2" x 3-1/2", US4					48			48	52.50
1430	4-1/2" x 4-1/2", US10					84			84	92
1500	5" x 5", US10					123			123	135
1600	6" x 6", US10					178			178	196
1800	Stainless steel base, 4-1/2" x 4-1/2", US32					127			127	140
1810	5" x 4-1/2", US32					182			182	200
1930	For hospital type tip on pin, add				Ea.	12.60			12.60	13.85
1950	Full mortise, low frequency, steel base, 3-1/2" x 3-1/2", US26D				Pr.	11.55			11.55	12.75
2000	4-1/2" x 4-1/2", USP					12.80			12.80	14.05
2100	5" x 5", USP					31.50			31.50	34.50
2200	6" x 6", USP					63.50			63.50	69.50
2300	4-1/2" x 4-1/2", US3					18.40			18.40	20
2310	5" x 5", US3					44			44	48.50
2400	Brass bass, 4-1/2" x 4-1/2", US10					44.50			44.50	49
2500	5" x 5", US10					67			67	74
2800	Stainless steel base, 4-1/2" x 4-1/2", US32					75.50			75.50	83

08 71 20.92 Mortised Hinges		Crew	Daily Output	Labor-Hours	Unit	Material	Labor	Equipment	Total	Total Incl O&P
0010	**MORTISED HINGES**									
0200	Average frequency, steel plated, ball bearing, 3-1/2" x 3-1/2"				Pr.	25.50			25.50	28.50
0300	Bronze, ball bearing					32			32	35
0900	High frequency, steel plated, ball bearing					79.50			79.50	87.50
1100	Bronze, ball bearing					82			82	90.50
1300	Average frequency, steel plated, ball bearing, 4-1/2" x 4-1/2"					31.50			31.50	35
1500	Bronze, ball bearing, to 36" wide					34			34	37
1700	Low frequency, steel, plated, plain bearing					16.10			16.10	17.70
1900	Bronze, plain bearing					22.50			22.50	25

08 71 20.95 Kick Plates		Crew	Daily Output	Labor-Hours	Unit	Material	Labor	Equipment	Total	Total Incl O&P
0010	**KICK PLATES**									
0020	Stainless steel, .050, 16 ga, 8" x 28", US32	1 Carp	15	.533	Ea.	35	17.90		52.90	68.50
0030	8" x 30"		15	.533		36	17.90		53.90	69.50
0040	8" x 34"		15	.533		40	17.90		57.90	74
0050	10" x 28"		15	.533		67	17.90		84.90	104
0060	10" x 30"		15	.533		72	17.90		89.90	109
0070	10" x 34"		15	.533		81	17.90		98.90	119
0080	Mop/Kick, 4" x 28"		15	.533		30	17.90		47.90	63
0090	4" x 30"		15	.533		32	17.90		49.90	65

08 71 Door Hardware

08 71 20 – Hardware

08 71 20.95 Kick Plates		Crew	Daily Output	Labor-Hours	Unit	Material	2011 Bare Costs Labor	Equipment	Total	Total Incl O&P
0100	4" x 34"	1 Carp	15	.533	Ea.	36	17.90		53.90	69.50
0110	6" x 28"		15	.533		38	17.90		55.90	72
0120	6" x 30"		15	.533		41	17.90		58.90	75
0130	6" x 34"		15	.533		47	17.90		64.90	81.50
0500	Bronze, .050", 8" x 28"		15	.533		47	17.90		64.90	81.50
0510	8" x 30"		15	.533		63	17.90		80.90	99.50
0520	8" x 34"		15	.533		71	17.90		88.90	108
0530	10" x 28"		15	.533		73	17.90		90.90	111
0540	10" x 30"		15	.533		78	17.90		95.90	116
0550	10" x 34"		15	.533		89	17.90		106.90	128
0560	Mop/Kick, 4" x 28"		15	.533		31	17.90		48.90	64
0570	4" x 30"		15	.533		34	17.90		51.90	67.50
0580	4" x 34"		15	.533		35	17.90		52.90	68.50
0590	6" x 28"		15	.533		44	17.90		61.90	78.50
0600	6" x 30"		15	.533		50	17.90		67.90	85
0610	6" x 34"		15	.533		54	17.90		71.90	89.50
1000	Acrylic, .125", 8" x 26"		15	.533		28	17.90		45.90	61
1010	8" x 36"		15	.533		38	17.90		55.90	72
1020	8" x 42"		15	.533		45	17.90		62.90	79.50
1030	10" x 26"		15	.533		35	17.90		52.90	68.50
1040	10" x 36"		15	.533		48	17.90		65.90	83
1050	10" x 42"		15	.533		68.50	17.90		86.40	106
1060	Mop/Kick, 4" x 26"		15	.533		17	17.90		34.90	48.50
1070	4" x 36"		15	.533		24	17.90		41.90	56.50
1080	4" x 42"		15	.533		27	17.90		44.90	59.50
1090	6" x 26"		15	.533		23	17.90		40.90	55.50
1100	6" x 36"		15	.533		34	17.90		51.90	67.50
1110	6" x 42"		15	.533		39	17.90		56.90	73
1220	Brass, .050", 8" x 26"		15	.533		56	17.90		73.90	91.50
1230	8" x 36"		15	.533		75	17.90		92.90	113
1240	8" x 42"		15	.533		86	17.90		103.90	125
1250	10" x 26"		15	.533		71	17.90		88.90	108
1260	10" x 36"		15	.533		91	17.90		108.90	130
1270	10" x 42"		15	.533		105	17.90		122.90	146
1320	Mop/Kick, 4" x 26"		15	.533		28	17.90		45.90	61
1330	4" x 36"		15	.533		39	17.90		56.90	73
1340	4" x 42"		15	.533		44	17.90		61.90	78.50
1350	6" x 26"		15	.533		38	17.90		55.90	72
1360	6" x 36"		15	.533		48	17.90		65.90	83
1370	6" x 42"		15	.533		55	17.90		72.90	90.50
1800	Aluminum, .050", 8" x 26"		15	.533		22	17.90		39.90	54
1810	8" x 36"		15	.533		30.50	17.90		48.40	63.50
1820	8" x 42"		15	.533		35.50	17.90		53.40	69
1830	10" x 26"		15	.533		27.50	17.90		45.40	60.50
1840	10" x 36"		15	.533		38	17.90		55.90	72
1850	10" x 42"		15	.533		44.50	17.90		62.40	79
1860	Mop/Kick, 4" x 26"		15	.533		11	17.90		28.90	42
1870	4" x 36"		15	.533		15.50	17.90		33.40	47
1880	4" x 42"		15	.533		17	17.90		34.90	48.50
1890	6" x 26"		15	.533		16.50	17.90		34.40	48
1900	6" x 36"		15	.533		23	17.90		40.90	55.50
1910	6" x 42"	↓	15	.533	↓	27	17.90		44.90	59.50

08 71 Door Hardware

08 71 21 – Astragals

08 71 21.10 Exterior Mouldings, Astragals		Crew	Daily Output	Labor-Hours	Unit	Material	2011 Bare Costs Labor	Equipment	Total	Total Incl O&P
0010	**EXTERIOR MOULDINGS, ASTRAGALS**									
4170	Astragal for double doors, aluminum	1 Carp	4	2	Opng.	27.50	67		94.50	143
4174	Bronze	"	4	2	"	38	67		105	154

08 71 25 – Weatherstripping

08 71 25.10 Mechanical Seals, Weatherstripping		Crew	Daily Output	Labor-Hours	Unit	Material	2011 Bare Costs Labor	Equipment	Total	Total Incl O&P
0010	**MECHANICAL SEALS, WEATHERSTRIPPING**									
1000	Doors, wood frame, interlocking, for 3′ x 7′ door, zinc	1 Carp	3	2.667	Opng.	16	89.50		105.50	168
1100	Bronze		3	2.667		25	89.50		114.50	178
1300	6′ x 7′ opening, zinc		2	4		18	134		152	245
1400	Bronze		2	4		33	134		167	262
1700	Wood frame, spring type, bronze									
1800	3′ x 7′ door	1 Carp	7.60	1.053	Opng.	19.35	35.50		54.85	80.50
1900	6′ x 7′ door		7	1.143		28	38.50		66.50	94.50
1920	Felt, 3′ x 7′ door		14	.571		2.53	19.15		21.68	35
1930	6′ x 7′ door		13	.615		2.93	20.50		23.43	37.50
1950	Rubber, 3′ x 7′ door		7.60	1.053		5.45	35.50		40.95	65
1951	Rubber, 3′ x 7′ door		119	.067	L.F.	.32	2.26		2.58	4.13
1960	6′ x 7′ door		7	1.143	Opng.	6	38.50		44.50	70.50
2200	Metal frame, spring type, bronze									
2300	3′ x 7′ door	1 Carp	3	2.667	Opng.	37	89.50		126.50	191
2400	6′ x 7′ door	"	2.50	3.200	"	47	107		154	232
2500	For stainless steel, spring type, add					133%				
2700	Metal frame, extruded sections, 3′ x 7′ door, aluminum	1 Carp	3	2.667	Opng.	46	89.50		135.50	201
2800	Bronze		3	2.667		120	89.50		209.50	282
3100	6′ x 7′ door, aluminum		1.50	5.333		58	179		237	365
3200	Bronze		1.50	5.333		137	179		316	450
3500	Threshold weatherstripping									
3650	Door sweep, flush mounted, aluminum	1 Carp	25	.320	Ea.	13.55	10.75		24.30	33
3700	Vinyl		25	.320		16	10.75		26.75	35.50
5000	Garage door bottom weatherstrip, 12′ aluminum, clear		14	.571		21.50	19.15		40.65	56
5010	Bronze		14	.571		84	19.15		103.15	125
5050	Bottom protection, Rubber		14	.571		24.50	19.15		43.65	58.50
5100	Threshold		14	.571		91.50	19.15		110.65	133

08 74 Access Control Hardware

08 74 13 – Card Key Access Control Hardware

08 74 13.60 Entrance Card System		Crew	Daily Output	Labor-Hours	Unit	Material	2011 Bare Costs Labor	Equipment	Total	Total Incl O&P
0010	**ENTRANCE CARD SYSTEMS**									
0100	Entrance card, barium ferrite				Ea.	2.55			2.55	2.81
0120	Credential					2.65			2.65	2.92
0140	Proximity					4.22			4.22	4.64
0160	Weigand					6.55			6.55	7.20
0500	Entrance card reader, barium ferrite	R-19	4	5		188	202		390	530
0520	Credential		4	5		246	202		448	595
0540	Proximity		4	5		310	202		512	665
0560	Weigand		4	5		320	202		522	675
0600	Local processor for card system		4	5		1,675	202		1,877	2,175
0650	Scanner, eye retina		4	5		6,000	202		6,202	6,925
0700	Gate opener, cantilever	R-18	3	8.667		2,725	270		2,995	3,450
0750	Switch, tamper	R-19	6	3.333		15	135		150	235

08 74 Access Control Hardware

08 74 13 – Card Key Access Control Hardware

08 74 13.60 Entrance Card System		Crew	Daily Output	Labor-Hours	Unit	Material	Labor	Equipment	Total	Total Incl O&P
						2011 Bare Costs				
0900	Accessories, electric door strike/bolt	R-19	5	4	Ea.	70	161		231	340
0920	Electromagnetic lock		5	4		141	161		302	415
0940	Keypad for card reader	↓	4	5	↓	615	202		817	1,000

08 75 Window Hardware

08 75 10 – Window Handles and Latches

08 75 10.10 Handles and Latches

		Crew	Daily Output	Labor-Hours	Unit	Material	Labor	Equipment	Total	Total Incl O&P
0010	**HANDLES AND LATCHES**									
1000	Handles, surface mounted, aluminum	1 Carp	24	.333	Ea.	4.10	11.20		15.30	23
1020	Brass		24	.333		7.45	11.20		18.65	27
1040	Chrome		24	.333		6.25	11.20		17.45	25.50
1500	Recessed, aluminum		12	.667		2.27	22.50		24.77	40
1520	Brass		12	.667		2.76	22.50		25.26	40.50
1540	Chrome		12	.667		2.31	22.50		24.81	40
2000	Latches, aluminum		20	.400		3.01	13.40		16.41	26
2020	Brass		20	.400		4.02	13.40		17.42	27
2040	Chrome	↓	20	.400	↓	2.52	13.40		15.92	25.50

08 75 30 – Weatherstripping

08 75 30.10 Mechanical Weather Seals

		Crew	Daily Output	Labor-Hours	Unit	Material	Labor	Equipment	Total	Total Incl O&P
0010	**MECHANICAL WEATHER SEALS**, Window, double hung, 3' X 5'									
0020	Zinc	1 Carp	7.20	1.111	Opng.	12.80	37.50		50.30	76.50
0100	Bronze		7.20	1.111		28	37.50		65.50	93.50
0200	Vinyl V strip		7	1.143		5.75	38.50		44.25	70.50
0500	As above but heavy duty, zinc		4.60	1.739		18	58.50		76.50	117
0600	Bronze	↓	4.60	1.739	↓	32	58.50		90.50	133

08 79 Hardware Accessories

08 79 20 – Door Accessories

08 79 20.10 Door Hardware Accessories

		Crew	Daily Output	Labor-Hours	Unit	Material	Labor	Equipment	Total	Total Incl O&P
0010	**DOOR HARDWARE ACCESSORIES**									
0140	Door bolt, surface, 4"	1 Carp	32	.250	Ea.	9	8.40		17.40	24
0160	Door latch	"	12	.667	"	8.45	22.50		30.95	47
0200	Sliding closet door									
0220	Track and hanger, single	1 Carp	10	.800	Ea.	54	27		81	105
0240	Double		8	1		75	33.50		108.50	139
0260	Door guide, single		48	.167		25	5.60		30.60	37
0280	Double		48	.167		34	5.60		39.60	47
0600	Deadbolt and lock cover plate, brass or stainless steel		30	.267		28	8.95		36.95	46
0620	Hole cover plate, brass or chrome		35	.229		7.20	7.65		14.85	21
2240	Mortise lockset, passage, lever handle		9	.889		198	30		228	268
4000	Security chain, standard	↓	18	.444	↓	7.90	14.90		22.80	33.50

08 81 Glass Glazing

08 81 10 – Float Glass

08 81 10.10 Various Types and Thickness of Float Glass		Crew	Daily Output	Labor-Hours	Unit	Material	2011 Bare Costs Labor	Equipment	Total	Total Incl O&P
0010	VARIOUS TYPES AND THICKNESS OF FLOAT GLASS R088110-10									
0020	3/16" Plain	2 Glaz	130	.123	S.F.	4.63	4.09		8.72	11.80
0200	Tempered, clear		130	.123		6.45	4.09		10.54	13.80
0300	Tinted		130	.123		7.35	4.09		11.44	14.75
0600	1/4" thick, clear, plain		120	.133		5.40	4.43		9.83	13.20
0700	Tinted		120	.133		8	4.43		12.43	16.05
0800	Tempered, clear		120	.133		7.85	4.43		12.28	15.90
0900	Tinted		120	.133		10.05	4.43		14.48	18.30
1600	3/8" thick, clear, plain		75	.213		9.45	7.10		16.55	22
1700	Tinted		75	.213		14.80	7.10		21.90	28
1800	Tempered, clear		75	.213		15.60	7.10		22.70	29
1900	Tinted		75	.213		17.40	7.10		24.50	31
2200	1/2" thick, clear, plain		55	.291		16	9.65		25.65	33.50
2300	Tinted		55	.291		26	9.65		35.65	44.50
2400	Tempered, clear		55	.291		23	9.65		32.65	41.50
2500	Tinted		55	.291		24	9.65		33.65	42.50
2800	5/8" thick, clear, plain		45	.356		26	11.80		37.80	48
2900	Tempered, clear		45	.356		29.50	11.80		41.30	52
8900	For low emissivity coating for 3/16" & 1/4" only, add to above					16%				

08 81 17 – Fire Glass

08 81 17.10 Fire Resistant Glass		Crew	Daily Output	Labor-Hours	Unit	Material	Labor	Equipment	Total	Total Incl O&P
0010	FIRE RESISTANT GLASS									
0020	Fire Glass Minimum	2 Glaz	40	.400	S.F.	26	13.30		39.30	50.50
0030	Mid Range		40	.400		66	13.30		79.30	94.50
0050	High End		40	.400		300	13.30		313.30	350

08 81 20 – Vision Panels

08 81 20.10 Full Vision		Crew	Daily Output	Labor-Hours	Unit	Material	Labor	Equipment	Total	Total Incl O&P
0010	FULL VISION, window system with 3/4" glass mullions									
0020	Up to 10' high	H-2	130	.185	S.F.	59.50	5.70		65.20	75
0100	10' to 20' high, minimum		110	.218		63	6.75		69.75	80
0150	Average		100	.240		68	7.45		75.45	87.50
0200	Maximum		80	.300		76.50	9.30		85.80	99.50

08 81 25 – Glazing Variables

08 81 25.10 Applications of Glazing		Crew	Daily Output	Labor-Hours	Unit	Material	Labor	Equipment	Total	Total Incl O&P
0010	APPLICATIONS OF GLAZING									
0500	For high rise glazing, exterior, add per S.F. per story				S.F.					.40
0600	For glass replacement, add				"		100%			
0700	For gasket settings, add				L.F.	5.55			5.55	6.10
0900	For sloped glazing, add				S.F.		25%			
2000	Fabrication, polished edges, 1/4" thick				Inch	.50			.50	.55
2100	1/2" thick					1.23			1.23	1.35
2500	Mitered edges, 1/4" thick					1.23			1.23	1.35
2600	1/2" thick					2			2	2.20

08 81 30 – Insulating Glass

08 81 30.10 Reduce Heat Transfer Glass		Crew	Daily Output	Labor-Hours	Unit	Material	Labor	Equipment	Total	Total Incl O&P
0010	REDUCE HEAT TRANSFER GLASS									
0015	2 lites 1/8" float, 1/2" thk under 15 S.F.									
0100	Tinted G	2 Glaz	95	.168	S.F.	13.10	5.60		18.70	23.50
0280	Double glazed, 5/8" thk unit, 3/16" float, 15-30 S.F., clear		90	.178		12.95	5.90		18.85	24
0400	1" thk, dbl. glazed, 1/4" float, 30-70 S.F., clear G		75	.213		15.65	7.10		22.75	29

08 81 Glass Glazing

08 81 30 – Insulating Glass

08 81 30.10 Reduce Heat Transfer Glass			Crew	Daily Output	Labor-Hours	Unit	Material	2011 Bare Costs Labor	Equipment	Total	Total Incl O&P
0500	Tinted	G	2 Glaz	75	.213	S.F.	22	7.10		29.10	36
2000	Both lites, light & heat reflective	G		85	.188		29.50	6.25		35.75	43
2500	Heat reflective, film inside, 1" thick unit, clear	G		85	.188		26	6.25		32.25	39
2600	Tinted	G		85	.188		27	6.25		33.25	40
3000	Film on weatherside, clear, 1/2" thick unit	G		95	.168		18.60	5.60		24.20	29.50
3100	5/8" thick unit	G		90	.178		18.60	5.90		24.50	30
3200	1" thick unit	G		85	.188		25.50	6.25		31.75	38.50
5000	Spectrally selective film, on ext, blocks solar gain/allows 70% of light	G	↓	95	.168	↓	9.80	5.60		15.40	20

08 81 55 – Window Glass

08 81 55.10 Sheet Glass		Crew	Daily Output	Labor-Hours	Unit	Material	Labor	Equipment	Total	Total Incl O&P
0010	**SHEET GLASS** (window), clear float, stops, putty bed									
0015	1/8" thick, clear float	2 Glaz	480	.033	S.F.	3.40	1.11		4.51	5.55
0500	3/16" thick, clear		480	.033		5.40	1.11		6.51	7.75
0600	Tinted		480	.033		7.15	1.11		8.26	9.65
0700	Tempered	↓	480	.033	↓	8.65	1.11		9.76	11.30

08 83 Mirrors

08 83 13 – Mirrored Glass Glazing

08 83 13.10 Mirrors		Crew	Daily Output	Labor-Hours	Unit	Material	Labor	Equipment	Total	Total Incl O&P
0010	**MIRRORS**, No frames, wall type, 1/4" plate glass, polished edge									
0100	Up to 5 S.F.	2 Glaz	125	.128	S.F.	9.35	4.25		13.60	17.30
0200	Over 5 S.F.		160	.100		9.05	3.32		12.37	15.40
0500	Door type, 1/4" plate glass, up to 12 S.F.		160	.100		8.20	3.32		11.52	14.45
1000	Float glass, up to 10 S.F., 1/8" thick		160	.100		5.45	3.32		8.77	11.40
1100	3/16" thick		150	.107		6.65	3.54		10.19	13.10
1500	12" x 12" wall tiles, square edge, clear		195	.082		1.90	2.72		4.62	6.55
1600	Veined		195	.082		4.86	2.72		7.58	9.80
2010	Bathroom, unframed, laminated	↓	160	.100	↓	12.60	3.32		15.92	19.30

08 87 Glazing Surface Films

08 87 23 – Safety and Security Films

08 87 23.16 Security Films		Crew	Daily Output	Labor-Hours	Unit	Material	Labor	Equipment	Total	Total Incl O&P
0010	**SECURITY FILMS**, clear, 32000 psi tensile strength, adhered to glass									
0100	.002" thick, daylight installation	H-2	950	.025	S.F.	2.20	.78		2.98	3.71
0150	.004" thick, daylight installation		800	.030		2.30	.93		3.23	4.06
0200	.006" thick, daylight installation		700	.034		2.45	1.06		3.51	4.45
0210	Install for anchorage		600	.040		2.72	1.24		3.96	5.05
0400	.007" thick, daylight installation		600	.040		2.65	1.24		3.89	4.96
0410	Install for anchorage		500	.048		2.94	1.49		4.43	5.70
0500	.008" thick, daylight installation		500	.048		2.85	1.49		4.34	5.60
0510	Install for anchorage		500	.048		3.17	1.49		4.66	5.95
0600	.015" thick, daylight installation		400	.060		4.75	1.86		6.61	8.30
0610	Install for anchorage	↓	400	.060	↓	3.17	1.86		5.03	6.55
0900	Security film anchorage, mechanical attachment and cover plate	H-3	370	.043	L.F.	9.05	1.27		10.32	12.05
0950	Security film anchorage, wet glaze structural caulking	1 Glaz	225	.036	"	.83	1.18		2.01	2.85
1000	Adhered security film removal	1 Clab	275	.029	S.F.		.77		.77	1.29

08 88 Special Function Glazing

08 88 56 – Ballistics-Resistant Glazing

08 88 56.10 Laminated Glass		Crew	Daily Output	Labor-Hours	Unit	Material	2011 Bare Costs Labor	Equipment	Total	Total Incl O&P
0010	**LAMINATED GLASS**									
2700	Level 2 (.357 magnum)	2 Glaz	12	1.333	S.F.	65	44.50		109.50	144
2750	Level 3A (.44 magnum)		12	1.333		70	44.50		114.50	150
2800	Level 4 (AK-47)		12	1.333		91.50	44.50		136	174
2850	Level 5 (M-16)		12	1.333		102	44.50		146.50	185
2900	Level 3 (7.62 Armor Piercing)	↓	12	1.333	↓	122	44.50		166.50	207

08 91 Louvers

08 91 19 – Fixed Louvers

08 91 19.10 Aluminum Louvers		Crew	Daily Output	Labor-Hours	Unit	Material	Labor	Equipment	Total	Total Incl O&P
0010	**ALUMINUM LOUVERS**									
0020	Aluminum with screen, residential, 8" x 8"	1 Carp	38	.211	Ea.	10.55	7.05		17.60	23.50
0100	12" x 12"		38	.211		15	7.05		22.05	28.50
0200	12" x 18"		35	.229		18.85	7.65		26.50	33.50
0250	14" x 24"		30	.267		26	8.95		34.95	43.50
0300	18" x 24"		27	.296		29	9.95		38.95	48.50
0500	24" x 30"		24	.333		32	11.20		43.20	53.50
0700	Triangle, adjustable, small		20	.400		32	13.40		45.40	57.50
0800	Large		15	.533		41	17.90		58.90	75
2100	Midget, aluminum, 3/4" deep, 1" diameter		85	.094		.69	3.16		3.85	6.05
2150	3" diameter		60	.133		1.87	4.47		6.34	9.55
2200	4" diameter		50	.160		3.64	5.35		8.99	13
2250	6" diameter	↓	30	.267	↓	3.74	8.95		12.69	19.05

08 91 26 – Door Louvers

08 91 26.10 Steel Louvers, 18 Gauge, Fixed Blade		Crew	Daily Output	Labor-Hours	Unit	Material	Labor	Equipment	Total	Total Incl O&P
0010	**STEEL LOUVERS, 18 GAUGE, FIXED BLADE**									
0050	12" x 12", with enamel or powder coat	1 Carp	20	.400	Ea.	67.50	13.40		80.90	97
0055	18" x 12"		20	.400		76	13.40		89.40	106
0060	18" x 18"		20	.400		83	13.40		96.40	114
0065	24" x 12"		20	.400		91.50	13.40		104.90	124
0070	24" x 18"		20	.400		126	13.40		139.40	162
0075	24" x 24"		20	.400		130	13.40		143.40	166
0100	12" x 12", galvanized		20	.400		58	13.40		71.40	86.50
0105	18" x 12"		20	.400		65.50	13.40		78.90	94.50
0115	24" x 12"		20	.400		82	13.40		95.40	113
0125	24" x 24"	↓	20	.400	↓	119	13.40		132.40	153

08 95 Vents

08 95 13 – Soffit Vents

08 95 13.10 Wall Louvers		Crew	Daily Output	Labor-Hours	Unit	Material	Labor	Equipment	Total	Total Incl O&P
0010	**WALL LOUVERS**									
2400	Under eaves vent, aluminum, mill finish, 16" x 4"	1 Carp	48	.167	Ea.	1.37	5.60		6.97	10.85
2500	16" x 8"	"	48	.167	"	1.93	5.60		7.53	11.45

08 95 16 – Wall Vents

08 95 16.10 Louvers		Crew	Daily Output	Labor-Hours	Unit	Material	Labor	Equipment	Total	Total Incl O&P
0010	**LOUVERS**									
0020	Redwood, 2'-0" diameter, full circle	1 Carp	16	.500	Ea.	169	16.80		185.80	214
0100	Half circle	↓	16	.500	↓	189	16.80		205.80	236

08 95 Vents

08 95 16 – Wall Vents

08 95 16.10 Louvers		Crew	Daily Output	Labor-Hours	Unit	Material	2011 Bare Costs Labor	Equipment	Total	Total Incl O&P
0200	Octagonal	1 Carp	16	.500	Ea.	129	16.80		145.80	170
0300	Triangular, 5/12 pitch, 5'-0" at base		16	.500		189	16.80		205.80	236
7000	Vinyl gable vent, 8" x 8"		38	.211		11.55	7.05		18.60	24.50
7020	12" x 12"		38	.211		22.50	7.05		29.55	37
7080	12" x 18"		35	.229		29.50	7.65		37.15	45.50
7200	18" x 24"		30	.267		36.50	8.95		45.45	55

Estimating Tips

General

- Room Finish Schedule: A complete set of plans should contain a room finish schedule. If one is not available, it would be well worth the time and effort to obtain one.

09 20 00 Plaster and Gypsum Board

- Lath is estimated by the square yard plus a 5% allowance for waste. Furring, channels, and accessories are measured by the linear foot. An extra foot should be allowed for each accessory miter or stop.
- Plaster is also estimated by the square yard. Deductions for openings vary by preference, from zero deduction to 50% of all openings over 2 feet in width. The estimator should allow one extra square foot for each linear foot of horizontal interior or exterior angle located below the ceiling level. Also, double the areas of small radius work.
- Drywall accessories, studs, track, and acoustical caulking are all measured by the linear foot. Drywall taping is figured by the square foot. Gypsum wallboard is estimated by the square foot. No material deductions should be made for door or window openings under 32 S.F.

09 60 00 Flooring

- Tile and terrazzo areas are taken off on a square foot basis. Trim and base materials are measured by the linear foot. Accent tiles are listed per each. Two basic methods of installation are used. Mud set is approximately 30% more expensive than thin set. In terrazzo work, be sure to include the linear footage of embedded decorative strips, grounds, machine rubbing, and power cleanup.
- Wood flooring is available in strip, parquet, or block configuration. The latter two types are set in adhesives with quantities estimated by the square foot. The laying pattern will influence labor costs and material waste. In addition to the material and labor for laying wood floors, the estimator must make allowances for sanding and finishing these areas unless the flooring is prefinished.
- Sheet flooring is measured by the square yard. Roll widths vary, so consideration should be given to use the most economical width, as waste must be figured into the total quantity. Consider also the installation methods available, direct glue down or stretched.

09 70 00 Wall Finishes

- Wall coverings are estimated by the square foot. The area to be covered is measured, length by height of wall above baseboards, to calculate the square footage of each wall. This figure is divided by the number of square feet in the single roll which is being used. Deduct, in full, the areas of openings such as doors and windows. Where a pattern match is required allow 25%-30% waste.

09 80 00 Acoustic Treatment

- Acoustical systems fall into several categories. The takeoff of these materials should be by the square foot of area with a 5% allowance for waste. Do not forget about scaffolding, if applicable, when estimating these systems.

09 90 00 Painting and Coating

- A major portion of the work in painting involves surface preparation. Be sure to include cleaning, sanding, filling, and masking costs in the estimate.
- Protection of adjacent surfaces is not included in painting costs. When considering the method of paint application, an important factor is the amount of protection and masking required. These must be estimated separately and may be the determining factor in choosing the method of application.

Reference Numbers

Reference numbers are shown in shaded boxes at the beginning of some major classifications. These numbers refer to related items in the Reference Section. The reference information may be an estimating procedure, an alternate pricing method, or technical information.

Note: Not all subdivisions listed here necessarily appear in this publication.

09 05 Common Work Results for Finishes

09 05 05 – Selective Finishes Demolition

09 05 05.10 Selective Demolition, Ceilings

	09 05 05.10 Selective Demolition, Ceilings	Crew	Daily Output	Labor-Hours	Unit	Material	2011 Bare Costs Labor	Equipment	Total	Total Incl O&P
0010	SELECTIVE DEMOLITION, CEILINGS R024119-10									
0200	Ceiling, drywall, furred and nailed or screwed	2 Clab	800	.020	S.F.		.53		.53	.89
1000	Plaster, lime and horse hair, on wood lath, incl. lath		700	.023			.60		.60	1.01
1200	Suspended ceiling, mineral fiber, 2′ x 2′ or 2′ x 4′		1500	.011			.28		.28	.47
1250	On suspension system, incl. system		1200	.013			.35		.35	.59
1500	Tile, wood fiber, 12″ x 12″, glued		900	.018			.47		.47	.79
1540	Stapled		1500	.011			.28		.28	.47
2000	Wood, tongue and groove, 1″ x 4″		1000	.016			.42		.42	.71
2040	1″ x 8″		1100	.015			.38		.38	.64
2400	Plywood or wood fiberboard, 4′ x 8′ sheets		1200	.013			.35		.35	.59

09 05 05.20 Selective Demolition, Flooring

	09 05 05.20 Selective Demolition, Flooring	Crew	Daily Output	Labor-Hours	Unit	Material	Labor	Equipment	Total	Total Incl O&P
0010	SELECTIVE DEMOLITION, FLOORING									
0200	Brick with mortar	2 Clab	475	.034	S.F.		.89		.89	1.49
0400	Carpet, bonded, including surface scraping		2000	.008			.21		.21	.35
0480	Tackless		9000	.002			.05		.05	.08
0550	Carpet tile, releasable adhesive		5000	.003			.08		.08	.14
0560	Permanent adhesive		1850	.009			.23		.23	.38
0800	Resilient, sheet goods		1400	.011			.30		.30	.51
0850	Vinyl or rubber cove base	1 Clab	1000	.008	L.F.		.21		.21	.35
0860	Vinyl or rubber cove base, molded corner	"	1000	.008	Ea.		.21		.21	.35
0870	For glued and caulked installation, add to labor						50%			
0900	Vinyl composition tile, 12″ x 12″	2 Clab	1000	.016	S.F.		.42		.42	.71
2000	Tile, ceramic, thin set		675	.024			.63		.63	1.05
2020	Mud set		625	.026			.68		.68	1.13
3000	Wood, block, on end	1 Carp	400	.020			.67		.67	1.12
3200	Parquet		450	.018			.60		.60	1
3400	Strip flooring, interior, 2-1/4″ x 25/32″ thick		325	.025			.83		.83	1.38
3500	Exterior, porch flooring, 1″ x 4″		220	.036			1.22		1.22	2.04
3800	Subfloor, tongue and groove, 1″ x 6″		325	.025			.83		.83	1.38
3820	1″ x 8″		430	.019			.62		.62	1.04
3840	1″ x 10″		520	.015			.52		.52	.86
4000	Plywood, nailed		600	.013			.45		.45	.75
4100	Glued and nailed		400	.020			.67		.67	1.12
4200	Hardboard, 1/4″ thick		760	.011			.35		.35	.59

09 05 05.30 Selective Demolition, Walls and Partitions

	09 05 05.30 Selective Demolition, Walls and Partitions	Crew	Daily Output	Labor-Hours	Unit	Material	Labor	Equipment	Total	Total Incl O&P
0010	SELECTIVE DEMOLITION, WALLS AND PARTITIONS R024119-10									
0020	Walls, concrete, reinforced	B-39	120	.400	C.F.		10.70	1.62	12.32	19.75
0025	Plain	"	160	.300	"		8.05	1.22	9.27	14.80
1000	Drywall, nailed or screwed	1 Clab	1000	.008	S.F.		.21		.21	.35
1010	2 layers		400	.020			.53		.53	.89
1500	Fiberboard, nailed		900	.009			.24		.24	.39
1568	Plenum barrier, sheet lead		300	.027			.71		.71	1.18
2200	Metal or wood studs, finish 2 sides, fiberboard	B-1	520	.046			1.25		1.25	2.10
2250	Lath and plaster		260	.092			2.50		2.50	4.19
2300	Plasterboard (drywall)		520	.046			1.25		1.25	2.10
2350	Plywood		450	.053			1.45		1.45	2.42
2800	Paneling, 4′ x 8′ sheets	1 Clab	475	.017			.45		.45	.75
3000	Plaster, lime and horsehair, on wood lath		400	.020			.53		.53	.89
3020	On metal lath		335	.024			.63		.63	1.06
3450	Plaster, interior gypsum, acoustic, or cement		60	.133	S.Y.		3.53		3.53	5.90
3500	Stucco, on masonry		145	.055			1.46		1.46	2.44
3510	Commercial 3-coat		80	.100			2.65		2.65	4.43

09 05 Common Work Results for Finishes

09 05 05 – Selective Finishes Demolition

09 05 05.30 Selective Demolition, Walls and Partitions		Crew	Daily Output	Labor-Hours	Unit	2011 Bare Costs Material	Labor	Equipment	Total	Total Incl O&P
3520	Interior stucco	1 Clab	25	.320	S.Y.		8.45		8.45	14.20
3760	Tile, ceramic, on walls, thin set		300	.027	S.F.		.71		.71	1.18
3765	Mud set		250	.032	"		.85		.85	1.42

09 22 Supports for Plaster and Gypsum Board

09 22 03 – Fastening Methods for Finishes

09 22 03.20 Drilling Plaster/Drywall		Crew	Daily Output	Labor-Hours	Unit	2011 Bare Costs Material	Labor	Equipment	Total	Total Incl O&P
0010	**DRILLING PLASTER/DRYWALL**									
1100	Drilling & layout for drywall/plaster walls, up to 1" deep, no anchor									
1200	Holes, 1/4" diameter	1 Carp	150	.053	Ea.	.01	1.79		1.80	3
1300	3/8" diameter		140	.057		.01	1.92		1.93	3.22
1400	1/2" diameter		130	.062		.01	2.06		2.07	3.47
1500	3/4" diameter		120	.067		.01	2.24		2.25	3.75
1600	1" diameter		110	.073		.02	2.44		2.46	4.10
1700	1-1/4" diameter		100	.080		.03	2.68		2.71	4.52
1800	1-1/2" diameter		90	.089		.04	2.98		3.02	5.05
1900	For ceiling installations, add						40%			

09 22 13 – Metal Furring

09 22 13.13 Metal Channel Furring		Crew	Daily Output	Labor-Hours	Unit	2011 Bare Costs Material	Labor	Equipment	Total	Total Incl O&P
0010	**METAL CHANNEL FURRING**									
0030	Beams and columns, 7/8" channels, galvanized, 12" O.C.	1 Lath	155	.052	S.F.	.31	1.55		1.86	2.82
0050	16" O.C.		170	.047		.25	1.41		1.66	2.54
0070	24" O.C.		185	.043		.17	1.30		1.47	2.27
0100	Ceilings, on steel, 7/8" channels, galvanized, 12" O.C.		210	.038		.28	1.14		1.42	2.14
0300	16" O.C.		290	.028		.25	.83		1.08	1.61
0400	24" O.C.		420	.019		.17	.57		.74	1.11
0600	1-5/8" channels, galvanized, 12" O.C.		190	.042		.38	1.27		1.65	2.44
0700	16" O.C.		260	.031		.34	.92		1.26	1.85
0900	24" O.C.		390	.021		.23	.62		.85	1.24
0930	7/8" channels with sound isolation clips, 12" O.C.		120	.067		1.35	2		3.35	4.68
0940	16" O.C.		100	.080		1.88	2.40		4.28	5.90
0950	24" O.C.		165	.048		1.23	1.46		2.69	3.69
0960	1-5/8" channels, galvanized, 12" O.C.		110	.073		1.44	2.19		3.63	5.10
0970	16" O.C.		100	.080		1.97	2.40		4.37	6
0980	24" O.C.		155	.052		1.29	1.55		2.84	3.90
1000	Walls, 7/8" channels, galvanized, 12" O.C.		235	.034		.28	1.02		1.30	1.95
1200	16" O.C.		265	.030		.25	.91		1.16	1.73
1300	24" O.C.		350	.023		.17	.69		.86	1.29
1500	1-5/8" channels, galvanized, 12" O.C.		210	.038		.38	1.14		1.52	2.25
1600	16" O.C.		240	.033		.34	1		1.34	1.97
1800	24" O.C.		305	.026		.23	.79		1.02	1.51
1920	7/8" channels with sound isolation clips, 12" O.C.		125	.064		1.35	1.92		3.27	4.56
1940	16" O.C.		100	.080		1.88	2.40		4.28	5.90
1950	24" O.C.		150	.053		1.23	1.60		2.83	3.92
1960	1-5/8" channels, galvanized, 12" O.C.		115	.070		1.44	2.09		3.53	4.93
1970	16" O.C.		95	.084		1.97	2.53		4.50	6.20
1980	24" O.C.		140	.057		1.29	1.72		3.01	4.17

09 22 Supports for Plaster and Gypsum Board

09 22 16 – Non-Structural Metal Framing

09 22 16.13 Non-Structural Metal Stud Framing		Crew	Daily Output	Labor-Hours	Unit	2011 Bare Costs Material	Labor	Equipment	Total	Total Incl O&P
0010	**NON-STRUCTURAL METAL STUD FRAMING**									
1600	Non-load bearing, galv, 8′ high, 25 ga. 1-5/8″ wide, 16″ O.C.	1 Carp	619	.013	S.F.	.24	.43		.67	1
1610	24″ O.C.		950	.008		.18	.28		.46	.67
1620	2-1/2″ wide, 16″ O.C.		613	.013		.28	.44		.72	1.04
1630	24″ O.C.		938	.009		.21	.29		.50	.71
1640	3-5/8″ wide, 16″ O.C.		600	.013		.32	.45		.77	1.11
1650	24″ O.C.		925	.009		.24	.29		.53	.76
1660	4″ wide, 16″ O.C.		594	.013		.37	.45		.82	1.16
1670	24″ O.C.		925	.009		.27	.29		.56	.79
1680	6″ wide, 16″ O.C.		588	.014		.46	.46		.92	1.27
1690	24″ O.C.		906	.009		.35	.30		.65	.88
1700	20 ga. studs, 1-5/8″ wide, 16″ O.C.		494	.016		.37	.54		.91	1.31
1710	24″ O.C.		763	.010		.27	.35		.62	.89
1720	2-1/2″ wide, 16″ O.C.		488	.016		.42	.55		.97	1.38
1730	24″ O.C.		750	.011		.32	.36		.68	.95
1740	3-5/8″ wide, 16″ O.C.		481	.017		.45	.56		1.01	1.43
1750	24″ O.C.		738	.011		.34	.36		.70	.98
1760	4″ wide, 16″ O.C.		475	.017		.54	.57		1.11	1.54
1770	24″ O.C.		738	.011		.40	.36		.76	1.05
1780	6″ wide, 16″ O.C.		469	.017		.63	.57		1.20	1.65
1790	24″ O.C.		725	.011		.47	.37		.84	1.14
2000	Non-load bearing, galv, 10′ high, 25 ga. 1-5/8″ wide, 16″ O.C.		495	.016		.23	.54		.77	1.16
2100	24″ O.C.		760	.011		.17	.35		.52	.78
2200	2-1/2″ wide, 16″ O.C.		490	.016		.27	.55		.82	1.21
2250	24″ O.C.		750	.011		.20	.36		.56	.81
2300	3-5/8″ wide, 16″ O.C.		480	.017		.31	.56		.87	1.28
2350	24″ O.C.		740	.011		.23	.36		.59	.86
2400	4″ wide, 16″ O.C.		475	.017		.35	.57		.92	1.33
2450	24″ O.C.		740	.011		.25	.36		.61	.89
2500	6″ wide, 16″ O.C.		470	.017		.44	.57		1.01	1.44
2550	24″ O.C.		725	.011		.32	.37		.69	.97
2600	20 ga. studs, 1-5/8″ wide, 16″ O.C.		395	.020		.35	.68		1.03	1.52
2650	24″ O.C.		610	.013		.25	.44		.69	1.02
2700	2-1/2″ wide, 16″ O.C.		390	.021		.40	.69		1.09	1.59
2750	24″ O.C.		600	.013		.29	.45		.74	1.07
2800	3-5/8″ wide, 16″ OC		385	.021		.43	.70		1.13	1.64
2850	24″ O.C.		590	.014		.32	.45		.77	1.11
2900	4″ wide, 16″ O.C.		380	.021		.51	.71		1.22	1.74
2950	24″ O.C.		590	.014		.37	.45		.82	1.17
3000	6″ wide, 16″ O.C.		375	.021		.60	.72		1.32	1.86
3050	24″ O.C.		580	.014		.44	.46		.90	1.25
3060	Non-load bearing, galv, 12′ high, 25 ga. 1-5/8″ wide, 16″ O.C.		413	.019		.22	.65		.87	1.33
3070	24″ O.C.		633	.013		.16	.42		.58	.89
3080	2-1/2″ wide, 16″ O.C.		408	.020		.25	.66		.91	1.38
3090	24″ O.C.		625	.013		.18	.43		.61	.92
3100	3-5/8″ wide, 16″ O.C.		400	.020		.29	.67		.96	1.44
3110	24″ O.C.		617	.013		.21	.44		.65	.97
3120	4″ wide, 16″ O.C.		396	.020		.33	.68		1.01	1.50
3130	24″ O.C.		617	.013		.24	.44		.68	1
3140	6″ wide, 16″ O.C.		392	.020		.42	.68		1.10	1.61
3150	24″ O.C.		604	.013		.30	.44		.74	1.07
3160	20 ga. studs, 1-5/8″ wide, 16″ O.C.	↓	329	.024	↓	.33	.82		1.15	1.73

09 22 Supports for Plaster and Gypsum Board

09 22 16 – Non-Structural Metal Framing

09 22 16.13 Non-Structural Metal Stud Framing		Crew	Daily Output	Labor-Hours	Unit	Material	2011 Bare Costs Labor	Equipment	Total	Total Incl O&P
3170	24" O.C.	1 Carp	508	.016	S.F.	.24	.53		.77	1.15
3180	2-1/2" wide, 16" O.C.		325	.025		.38	.83		1.21	1.80
3190	24" O.C.		500	.016		.28	.54		.82	1.20
3200	3-5/8" wide, 16" O.C.		321	.025		.41	.84		1.25	1.85
3210	24" O.C.		492	.016		.30	.55		.85	1.24
3220	4" wide, 16" O.C.		317	.025		.49	.85		1.34	1.96
3230	24" O.C.		492	.016		.35	.55		.90	1.30
3240	6" wide, 16" O.C.		313	.026		.57	.86		1.43	2.07
3250	24" O.C.		483	.017		.42	.56		.98	1.39
5000	Load bearing studs, see Section 05 41 13.30									

09 22 26 – Suspension Systems

09 22 26.13 Ceiling Suspension Systems		Crew	Daily Output	Labor-Hours	Unit	Material	Labor	Equipment	Total	Total Incl O&P
0010	**CEILING SUSPENSION SYSTEMS** For gypsum board or plaster									
8000	Suspended ceilings, including carriers									
8200	1-1/2" carriers, 24" O.C. with:									
8300	7/8" channels, 16" O.C.	1 Lath	275	.029	S.F.	.43	.87		1.30	1.87
8320	24" O.C.		310	.026		.34	.78		1.12	1.62
8400	1-5/8" channels, 16" O.C.		205	.039		.51	1.17		1.68	2.44
8420	24" O.C.		250	.032		.40	.96		1.36	1.98
8600	2" carriers, 24" O.C. with:									
8700	7/8" channels, 16" O.C.	1 Lath	250	.032	S.F.	.50	.96		1.46	2.10
8720	24" O.C.		285	.028		.42	.84		1.26	1.81
8800	1-5/8" channels, 16" O.C.		190	.042		.59	1.27		1.86	2.67
8820	24" O.C.		225	.036		.48	1.07		1.55	2.23

09 22 36 – Lath

09 22 36.13 Gypsum Lath		Crew	Daily Output	Labor-Hours	Unit	Material	Labor	Equipment	Total	Total Incl O&P
0010	**GYPSUM LATH** R092000-50									
0020	Plain or perforated, nailed, 3/8" thick	1 Lath	85	.094	S.Y.	4.95	2.83		7.78	9.95
0100	1/2" thick		80	.100		4.86	3.01		7.87	10.15
0300	Clipped to steel studs, 3/8" thick		75	.107		4.95	3.21		8.16	10.60
0400	1/2" thick		70	.114		4.86	3.43		8.29	10.85
1500	For ceiling installations, add		216	.037			1.11		1.11	1.78
1600	For columns and beams, add		170	.047			1.41		1.41	2.26

09 22 36.23 Metal Lath		Crew	Daily Output	Labor-Hours	Unit	Material	Labor	Equipment	Total	Total Incl O&P
0010	**METAL LATH** R092000-50									
3600	2.5 lb. diamond painted, on wood framing, on walls	1 Lath	85	.094	S.Y.	3.62	2.83		6.45	8.50
3700	On ceilings		75	.107		3.62	3.21		6.83	9.15
4200	3.4 lb. diamond painted, wired to steel framing		75	.107		4	3.21		7.21	9.55
4300	On ceilings		60	.133		4	4.01		8.01	10.80
5100	Rib lath, painted, wired to steel, on walls, 2.5 lb.		75	.107		3.14	3.21		6.35	8.60
5200	3.4 lb.		70	.114		4.84	3.43		8.27	10.80
5700	Suspended ceiling system, incl. 3.4 lb. diamond lath, painted		15	.533		3.35	16.05		19.40	29
5800	Galvanized		15	.533		6.50	16.05		22.55	32.50

09 22 36.43 Security Mesh		Crew	Daily Output	Labor-Hours	Unit	Material	Labor	Equipment	Total	Total Incl O&P
0010	**SECURITY MESH**, expanded metal, flat, screwed to framing									
0100	On walls, 3/4", 1.76 lb./S.F.	2 Carp	1500	.011	S.F.	1.49	.36		1.85	2.24
0110	1-1/2", 1.14 lb./S.F.		1600	.010		1.21	.34		1.55	1.89
0200	On ceilings, 3/4", 1.76 lb./S.F.		1350	.012		1.49	.40		1.89	2.31
0210	1-1/2", 1.14 lb./S.F.		1450	.011		1.21	.37		1.58	1.95

09 22 Supports for Plaster and Gypsum Board

09 22 36 – Lath

09 22 36.83 Accessories, Plaster		Crew	Daily Output	Labor-Hours	Unit	2011 Bare Costs Material	Labor	Equipment	Total	Total Incl O&P
0010	**ACCESSORIES, PLASTER**									
0020	Casing bead, expanded flange, galvanized	1 Lath	2.70	2.963	C.L.F.	43.50	89		132.50	190
0200	Foundation weep screed, galvanized	″	2.70	2.963		58	89		147	206
0900	Channels, cold rolled, 16 ga., 3/4″ deep, galvanized					28.50			28.50	31
1200	1-1/2″ deep, 16 ga., galvanized					38			38	41.50
1620	Corner bead, expanded bullnose, 3/4″ radius, #10, galvanized	1 Lath	2.60	3.077		30.50	92.50		123	182
1650	#1, galvanized		2.55	3.137		48	94.50		142.50	204
1670	Expanded wing, 2-3/4″ wide, #1, galvanized		2.65	3.019		33.50	90.50		124	182
1700	Inside corner (corner rite), 3″ x 3″, painted		2.60	3.077		28.50	92.50		121	180
1750	Strip-ex, 4″ wide, painted		2.55	3.137		29	94.50		123.50	183
1800	Expansion joint, 3/4″ grounds, limited expansion, galv., 1 piece		2.70	2.963		98	89		187	250
2100	Extreme expansion, galvanized, 2 piece		2.60	3.077		157	92.50		249.50	320

09 23 Gypsum Plastering

09 23 13 – Acoustical Gypsum Plastering

09 23 13.10 Perlite or Vermiculite Plaster

		Crew	Daily Output	Labor-Hours	Unit	Material	Labor	Equipment	Total	Total Incl O&P
0010	**PERLITE OR VERMICULITE PLASTER** R092000-50									
0020	In 100 lb. bags, under 200 bags				Bag	14.40			14.40	15.80
0300	2 coats, no lath included, on walls	J-1	92	.435	S.Y.	3.78	12.75	1.34	17.87	26.50
0400	On ceilings		79	.506		3.78	14.85	1.56	20.19	30
0900	3 coats, no lath included, on walls		74	.541		6.30	15.85	1.67	23.82	35
1000	On ceilings		63	.635		6.30	18.65	1.96	26.91	39.50
1700	For irregular or curved surfaces, add to above						30%			
1800	For columns and beams, add to above						50%			
1900	For soffits, add to ceiling prices						40%			

09 23 20 – Gypsum Plaster

09 23 20.10 Gypsum Plaster On Walls and Ceilings

		Crew	Daily Output	Labor-Hours	Unit	Material	Labor	Equipment	Total	Total Incl O&P
0010	**GYPSUM PLASTER ON WALLS AND CEILINGS** R092000-50									
0020	80# bag, less than 1 ton				Bag	17.30			17.30	19
0300	2 coats, no lath included, on walls	J-1	105	.381	S.Y.	3.87	11.20	1.18	16.25	24
0400	On ceilings		92	.435		3.87	12.75	1.34	17.96	26.50
0900	3 coats, no lath included, on walls		87	.460		5.55	13.50	1.42	20.47	29.50
1000	On ceilings		78	.513		5.55	15.05	1.58	22.18	32.50
1600	For irregular or curved surfaces, add						30%			
1800	For columns & beams, add						50%			

09 24 Cement Plastering

09 24 23 – Cement Stucco

09 24 23.40 Stucco

		Crew	Daily Output	Labor-Hours	Unit	Material	Labor	Equipment	Total	Total Incl O&P
0010	**STUCCO** R092000-50									
0015	3 coats 1″ thick, float finish, with mesh, on wood frame	J-2	63	.762	S.Y.	5.30	22.50	1.97	29.77	44.50
0100	On masonry construction, no mesh incl.	J-1	67	.597		2.28	17.55	1.84	21.67	33
0150	2 coats, 3/4″ thick, float finish, no lath incl.	″	110	.364		2.16	10.70	1.12	13.98	21
0300	For trowel finish, add	1 Plas	170	.047			1.44		1.44	2.35
0600	For coloring and special finish, add, minimum	J-1	685	.058		.39	1.71	.18	2.28	3.42
0700	Maximum		200	.200		1.36	5.85	.62	7.83	11.75
1000	Exterior stucco, with bonding agent, 3 coats, on walls, no mesh incl.		200	.200		3.54	5.85	.62	10.01	14.10
1200	Ceilings		180	.222		3.54	6.50	.69	10.73	15.25

09 24 Cement Plastering

09 24 23 – Cement Stucco

09 24 23.40 Stucco		Crew	Daily Output	Labor-Hours	Unit	Material	2011 Bare Costs Labor	Equipment	Total	Total Incl O&P
1300	Beams	J-1	80	.500	S.Y.	3.54	14.70	1.55	19.79	29.50
1500	Columns		100	.400		3.54	11.75	1.24	16.53	24.50
1600	Mesh, painted, nailed to wood, 1.8 lb.	1 Lath	60	.133		5	4.01		9.01	11.90
1800	3.6 lb.		55	.145		3.03	4.37		7.40	10.35
1900	Wired to steel, painted, 1.8 lb.		53	.151		5	4.54		9.54	12.75
2100	3.6 lb.		50	.160		3.03	4.81		7.84	11.05

09 26 Veneer Plastering

09 26 13 – Gypsum Veneer Plastering

09 26 13.20 Blueboard

Code	Description	Crew	Daily Output	Labor-Hours	Unit	Material	Labor	Equipment	Total	Total Incl O&P
0010	**BLUEBOARD** For use with thin coat									
0100	plaster application (see Section 09 26 13.80)									
1000	3/8" thick, on walls or ceilings, standard, no finish included	2 Carp	1900	.008	S.F.	.26	.28		.54	.76
1100	With thin coat plaster finish		875	.018		.36	.61		.97	1.42
1400	On beams, columns, or soffits, standard, no finish included		675	.024		.30	.80		1.10	1.66
1450	With thin coat plaster finish		475	.034		.40	1.13		1.53	2.32
3000	1/2" thick, on walls or ceilings, standard, no finish included		1900	.008		.29	.28		.57	.79
3100	With thin coat plaster finish		875	.018		.39	.61		1	1.45
3300	Fire resistant, no finish included		1900	.008		.29	.28		.57	.79
3400	With thin coat plaster finish		875	.018		.39	.61		1	1.45
3450	On beams, columns, or soffits, standard, no finish included		675	.024		.33	.80		1.13	1.70
3500	With thin coat plaster finish		475	.034		.43	1.13		1.56	2.36
3700	Fire resistant, no finish included		675	.024		.33	.80		1.13	1.70
3800	With thin coat plaster finish		475	.034		.43	1.13		1.56	2.36
5000	5/8" thick, on walls or ceilings, fire resistant, no finish included		1900	.008		.30	.28		.58	.80
5100	With thin coat plaster finish		875	.018		.40	.61		1.01	1.47
5500	On beams, columns, or soffits, no finish included		675	.024		.35	.80		1.15	1.71
5600	With thin coat plaster finish		475	.034		.44	1.13		1.57	2.38
6000	For high ceilings, over 8' high, add		3060	.005			.18		.18	.29
6500	For over 3 stories high, add per story		6100	.003			.09		.09	.15

09 26 13.80 Thin Coat Plaster

Code	Description	Crew	Daily Output	Labor-Hours	Unit	Material	Labor	Equipment	Total	Total Incl O&P
0010	**THIN COAT PLASTER** R092000-50									
0012	1 coat veneer, not incl. lath	J-1	3600	.011	S.F.	.10	.33	.03	.46	.68
1000	In 50 lb. bags				Bag	12.95			12.95	14.25

09 29 Gypsum Board

09 29 10 – Gypsum Board Panels

09 29 10.30 Gypsum Board

Code	Description	Crew	Daily Output	Labor-Hours	Unit	Material	Labor	Equipment	Total	Total Incl O&P
0010	**GYPSUM BOARD** on walls & ceilings R092910-10									
0100	Nailed or screwed to studs unless otherwise noted									
0150	3/8" thick, on walls, standard, no finish included	2 Carp	2000	.008	S.F.	.24	.27		.51	.71
0200	On ceilings, standard, no finish included		1800	.009		.24	.30		.54	.76
0250	On beams, columns, or soffits, no finish included		675	.024		.24	.80		1.04	1.59
0300	1/2" thick, on walls, standard, no finish included		2000	.008		.23	.27		.50	.70
0350	Taped and finished (level 4 finish)		965	.017		.27	.56		.83	1.23
0390	With compound skim coat (level 5 finish)		775	.021		.32	.69		1.01	1.52
0400	Fire resistant, no finish included		2000	.008		.28	.27		.55	.76
0450	Taped and finished (level 4 finish)		965	.017		.32	.56		.88	1.29
0490	With compound skim coat (level 5 finish)		775	.021		.37	.69		1.06	1.57

09 29 Gypsum Board

09 29 10 – Gypsum Board Panels

09 29 10.30 Gypsum Board		Crew	Daily Output	Labor-Hours	Unit	2011 Bare Costs Material	Labor	Equipment	Total	Total Incl O&P
0500	Water resistant, no finish included	2 Carp	2000	.008	S.F.	.36	.27		.63	.85
0550	Taped and finished (level 4 finish)		965	.017		.40	.56		.96	1.38
0590	With compound skim coat (level 5 finish)		775	.021		.45	.69		1.14	1.66
0600	Prefinished, vinyl, clipped to studs		900	.018		.46	.60		1.06	1.51
0700	Mold resistant, no finish included		2000	.008		.37	.27		.64	.86
0710	Taped and finished (level 4 finish)		965	.017		.41	.56		.97	1.39
0720	With compound skim coat (level 5 finish)		775	.021		.46	.69		1.15	1.67
1000	On ceilings, standard, no finish included		1800	.009		.23	.30		.53	.75
1050	Taped and finished (level 4 finish)		765	.021		.27	.70		.97	1.47
1090	With compound skim coat (level 5 finish)		610	.026		.32	.88		1.20	1.83
1100	Fire resistant, no finish included		1800	.009		.28	.30		.58	.81
1150	Taped and finished (level 4 finish)		765	.021		.32	.70		1.02	1.53
1195	With compound skim coat (level 5 finish)		610	.026		.37	.88		1.25	1.88
1200	Water resistant, no finish included		1800	.009		.36	.30		.66	.90
1250	Taped and finished (level 4 finish)		765	.021		.40	.70		1.10	1.62
1290	With compound skim coat (level 5 finish)		610	.026		.45	.88		1.33	1.97
1310	Mold resistant, no finish included		1800	.009		.37	.30		.67	.91
1320	Taped and finished (level 4 finish)		765	.021		.41	.70		1.11	1.63
1330	With compound skim coat (level 5 finish)		610	.026		.46	.88		1.34	1.98
1500	On beams, columns, or soffits, standard, no finish included		675	.024		.26	.80		1.06	1.62
1550	Taped and finished (level 4 finish)		475	.034		.27	1.13		1.40	2.19
1590	With compound skim coat (level 5 finish)		540	.030		.32	.99		1.31	2.02
1600	Fire resistant, no finish included		675	.024		.28	.80		1.08	1.64
1650	Taped and finished (level 4 finish)		475	.034		.32	1.13		1.45	2.25
1690	With compound skim coat (level 5 finish)		540	.030		.37	.99		1.36	2.07
1700	Water resistant, no finish included		675	.024		.41	.80		1.21	1.79
1750	Taped and finished (level 4 finish)		475	.034		.40	1.13		1.53	2.34
1790	With compound skim coat (level 5 finish)		540	.030		.45	.99		1.44	2.16
1800	Mold resistant, no finish included		675	.024		.43	.80		1.23	1.80
1810	Taped and finished (level 4 finish)		475	.034		.41	1.13		1.54	2.35
1820	With compound skim coat (level 5 finish)		540	.030		.46	.99		1.45	2.17
2000	5/8" thick, on walls, standard, no finish included		2000	.008		.28	.27		.55	.76
2050	Taped and finished (level 4 finish)		965	.017		.32	.56		.88	1.29
2090	With compound skim coat (level 5 finish)		775	.021		.37	.69		1.06	1.57
2100	Fire resistant, no finish included		2000	.008		.28	.27		.55	.76
2150	Taped and finished (level 4 finish)		965	.017		.32	.56		.88	1.29
2195	With compound skim coat (level 5 finish)		775	.021		.37	.69		1.06	1.57
2200	Water resistant, no finish included		2000	.008		.40	.27		.67	.89
2250	Taped and finished (level 4 finish)		965	.017		.44	.56		1	1.42
2290	With compound skim coat (level 5 finish)		775	.021		.49	.69		1.18	1.70
2300	Prefinished, vinyl, clipped to studs		900	.018		.72	.60		1.32	1.79
2510	Mold resistant, no finish included		2000	.008		.42	.27		.69	.91
2520	Taped and finished (level 4 finish)		965	.017		.46	.56		1.02	1.44
2530	With compound skim coat (level 5 finish)		775	.021		.51	.69		1.20	1.72
3000	On ceilings, standard, no finish included		1800	.009		.28	.30		.58	.81
3050	Taped and finished (level 4 finish)		765	.021		.32	.70		1.02	1.53
3090	With compound skim coat (level 5 finish)		615	.026		.37	.87		1.24	1.87
3100	Fire resistant, no finish included		1800	.009		.28	.30		.58	.81
3150	Taped and finished (level 4 finish)		765	.021		.32	.70		1.02	1.53
3190	With compound skim coat (level 5 finish)		615	.026		.37	.87		1.24	1.87
3200	Water resistant, no finish included		1800	.009		.40	.30		.70	.94
3250	Taped and finished (level 4 finish)		765	.021		.44	.70		1.14	1.66
3290	With compound skim coat (level 5 finish)		615	.026		.49	.87		1.36	2

09 29 Gypsum Board

09 29 10 – Gypsum Board Panels

09 29 10.30 Gypsum Board		Crew	Daily Output	Labor-Hours	Unit	Material	2011 Bare Costs Labor	Equipment	Total	Total Incl O&P
3300	Mold resistant, no finish included	2 Carp	1800	.009	S.F.	.42	.30		.72	.96
3310	Taped and finished (level 4 finish)		765	.021		.46	.70		1.16	1.68
3320	With compound skim coat (level 5 finish)		615	.026		.51	.87		1.38	2.02
3500	On beams, columns, or soffits, no finish included		675	.024		.32	.80		1.12	1.68
3550	Taped and finished (level 4 finish)		475	.034		.37	1.13		1.50	2.30
3590	With compound skim coat (level 5 finish)		380	.042		.43	1.41		1.84	2.83
3600	Fire resistant, no finish included		675	.024		.32	.80		1.12	1.68
3650	Taped and finished (level 4 finish)		475	.034		.37	1.13		1.50	2.30
3690	With compound skim coat (level 5 finish)		380	.042		.37	1.41		1.78	2.77
3700	Water resistant, no finish included		675	.024		.46	.80		1.26	1.84
3750	Taped and finished (level 4 finish)		475	.034		.51	1.13		1.64	2.45
3790	With compound skim coat (level 5 finish)		380	.042		.49	1.41		1.90	2.90
3800	Mold resistant, no finish included		675	.024		.48	.80		1.28	1.86
3810	Taped and finished (level 4 finish)		475	.034		.53	1.13		1.66	2.48
3820	With compound skim coat (level 5 finish)		380	.042		.51	1.41		1.92	2.92
4000	Fireproofing, beams or columns, 2 layers, 1/2" thick, incl finish		330	.048		.60	1.63		2.23	3.39
4010	Mold resistant		330	.048		.78	1.63		2.41	3.58
4050	5/8" thick		300	.053		.65	1.79		2.44	3.70
4060	Mold resistant		300	.053		.93	1.79		2.72	4.01
4100	3 layers, 1/2" thick		225	.071		.88	2.39		3.27	4.96
4110	Mold resistant		225	.071		1.15	2.39		3.54	5.25
4150	5/8" thick		210	.076		.97	2.56		3.53	5.35
4160	Mold resistant		210	.076		1.39	2.56		3.95	5.80
5200	For work over 8' high, add		3060	.005			.18		.18	.29
5270	For textured spray, add	2 Lath	1600	.010		.05	.30		.35	.54
5300	For distribution cost over 3 stories high, add per story	2 Carp	6100	.003			.09		.09	.15
5350	For finishing inner corners, add		950	.017	L.F.	.10	.57		.67	1.06
5355	For finishing outer corners, add		1250	.013		.21	.43		.64	.96
5500	For acoustical sealant, add per bead	1 Carp	500	.016		.04	.54		.58	.95
5550	Sealant, 1 quart tube				Ea.	7.05			7.05	7.80
5600	Sound deadening board, 1/4" gypsum	2 Carp	1800	.009	S.F.	.34	.30		.64	.87
5650	1/2" wood fiber	"	1800	.009	"	.34	.30		.64	.87

09 29 10.50 High Abuse Gypsum Board		Crew	Daily Output	Labor-Hours	Unit	Material	Labor	Equipment	Total	Total Incl O&P
0010	**HIGH ABUSE GYPSUM BOARD**, fiber reinforced, nailed or									
0100	screwed to studs unless otherwise noted									
0110	1/2" thick, on walls, no finish included	2 Carp	1800	.009	S.F.	.86	.30		1.16	1.45
0120	Taped and finished (level 4 finish)		870	.018		.90	.62		1.52	2.03
0130	With compound skim coat (level 5 finish)		700	.023		.95	.77		1.72	2.33
0150	On ceilings, no finish included		1620	.010		.86	.33		1.19	1.50
0160	Taped and finished (level 4 finish)		690	.023		.90	.78		1.68	2.30
0170	With compound skim coat (level 5 finish)		550	.029		.95	.98		1.93	2.68
0210	5/8" thick, on walls, no finish included		1800	.009		.90	.30		1.20	1.49
0220	Taped and finished (level 4 finish)		870	.018		.94	.62		1.56	2.07
0230	With compound skim coat (level 5 finish)		700	.023		.99	.77		1.76	2.37
0250	On ceilings, no finish included		1620	.010		.90	.33		1.23	1.54
0260	Taped and finished (level 4 finish)		690	.023		.94	.78		1.72	2.34
0270	With compound skim coat (level 5 finish)		550	.029		.99	.98		1.97	2.72
0310	5/8" thick, on walls, very high impact, no finish included		1800	.009		.98	.30		1.28	1.58
0320	Taped and finished (level 4 finish)		870	.018		1.02	.62		1.64	2.16
0330	With compound skim coat (level 5 finish)		700	.023		1.07	.77		1.84	2.46
0350	On ceilings, no finish included		1620	.010		.98	.33		1.31	1.63
0360	Taped and finished (level 4 finish)		690	.023		1.02	.78		1.80	2.43

09 29 Gypsum Board

09 29 10 – Gypsum Board Panels

09 29 10.50 High Abuse Gypsum Board		Crew	Daily Output	Labor-Hours	Unit	Material	2011 Bare Costs Labor	Equipment	Total	Total Incl O&P
0370	With compound skim coat (level 5 finish)	2 Carp	550	.029	S.F.	1.07	.98		2.05	2.81
0400	High abuse, gypsum core, paper face									
0410	1/2″ thick, on walls, no finish included	2 Carp	1800	.009	S.F.	.79	.30		1.09	1.37
0420	Taped and finished (level 4 finish)		870	.018		.83	.62		1.45	1.95
0430	With compound skim coat (level 5 finish)		700	.023		.88	.77		1.65	2.25
0450	On ceilings, no finish included		1620	.010		.79	.33		1.12	1.42
0460	Taped and finished (level 4 finish)		690	.023		.83	.78		1.61	2.22
0470	With compound skim coat (level 5 finish)		550	.029		.88	.98		1.86	2.60
0510	5/8″ thick, on walls, no finish included		1800	.009		.79	.30		1.09	1.37
0520	Taped and finished (level 4 finish)		870	.018		.83	.62		1.45	1.95
0530	With compound skim coat (level 5 finish)		700	.023		.88	.77		1.65	2.25
0550	On ceilings, no finish included		1620	.010		.79	.33		1.12	1.42
0560	Taped and finished (level 4 finish)		690	.023		.83	.78		1.61	2.22
0570	With compound skim coat (level 5 finish)		550	.029		.88	.98		1.86	2.60
1000	For high ceilings, over 8′ high, add		2750	.006			.20		.20	.33
1010	For distribution cost over 3 stories high, add per story		5500	.003			.10		.10	.16

09 29 15 – Gypsum Board Accessories

09 29 15.10 Accessories, Gypsum Board		Crew	Daily Output	Labor-Hours	Unit	Material	Labor	Equipment	Total	Total Incl O&P
0010	ACCESSORIES, GYPSUM BOARD									
0020	Casing bead, galvanized steel	1 Carp	2.90	2.759	C.L.F.	27	92.50		119.50	185
0100	Vinyl		3	2.667		23	89.50		112.50	176
0300	Corner bead, galvanized steel, 1″ x 1″		4	2		14.35	67		81.35	128
0400	1-1/4″ x 1-1/4″		3.50	2.286		15.30	76.50		91.80	145
0600	Vinyl		4	2		16.75	67		83.75	130
0900	Furring channel, galv. steel, 7/8″ deep, standard		2.60	3.077		24.50	103		127.50	200
1000	Resilient		2.55	3.137		27	105		132	206
1100	J trim, galvanized steel, 1/2″ wide		3	2.667		29	89.50		118.50	182
1120	5/8″ wide		2.95	2.712		28	91		119	183
1160	Screws #6 x 1″ A				M	8.45			8.45	9.30
1170	#6 x 1-5/8″ A				″	12.40			12.40	13.65
1500	Z stud, galvanized steel, 1-1/2″ wide	1 Carp	2.60	3.077	C.L.F.	38.50	103		141.50	216

09 30 Tiling

09 30 13 – Ceramic Tiling

09 30 13.10 Ceramic Tile		Crew	Daily Output	Labor-Hours	Unit	Material	Labor	Equipment	Total	Total Incl O&P
0010	CERAMIC TILE									
0020	Backsplash, thinset, average grade tiles	1 Tilf	50	.160	S.F.	2.32	5.05		7.37	10.60
0022	Custom grade tiles		50	.160		4.65	5.05		9.70	13.15
0024	Luxury grade tiles		50	.160		9.30	5.05		14.35	18.30
0026	Economy grade tiles		50	.160		2.12	5.05		7.17	10.40
0050	Base, using 1′ x 4″ high pc. with 1″ x 1″ tiles, mud set	D-7	82	.195	L.F.	4.81	5.50		10.31	14.10
0100	Thin set	″	128	.125		4.46	3.53		7.99	10.55
0300	For 6″ high base, 1″ x 1″ tile face, add					.72			.72	.79
0400	For 2″ x 2″ tile face, add to above					.38			.38	.42
0600	Cove base, 4-1/4″ x 4-1/4″ high, mud set	D-7	91	.176		3.75	4.96		8.71	12.05
0700	Thin set		128	.125		3.81	3.53		7.34	9.85
0900	6″ x 4-1/4″ high, mud set		100	.160		3.20	4.51		7.71	10.70
1000	Thin set		137	.117		3.07	3.29		6.36	8.65
1200	Sanitary cove base, 6″ x 4-1/4″ high, mud set		93	.172		3.85	4.85		8.70	12
1300	Thin set		124	.129		3.75	3.64		7.39	9.95

09 30 Tiling

09 30 13 – Ceramic Tiling

09 30 13.10 Ceramic Tile		Crew	Daily Output	Labor-Hours	Unit	2011 Bare Costs Material	Labor	Equipment	Total	Total Incl O&P
1500	6" x 6" high, mud set	D-7	84	.190	L.F.	4.34	5.35		9.69	13.40
1600	Thin set		117	.137		4.21	3.86		8.07	10.80
1800	Bathroom accessories, average (soap dish, tooth brush holder)		82	.195	Ea.	11	5.50		16.50	21
1900	Bathtub, 5', rec. 4-1/4" x 4-1/4" tile wainscot, adhesive set 6' high		2.90	5.517		156	156		312	420
2100	7' high wainscot		2.50	6.400		179	180		359	485
2200	8' high wainscot		2.20	7.273		190	205		395	540
2400	Bullnose trim, 4-1/4" x 4-1/4", mud set		82	.195	L.F.	3.52	5.50		9.02	12.65
2500	Thin set		128	.125		3.37	3.53		6.90	9.35
2700	6" x 4-1/4" bullnose trim, mud set		84	.190		2.70	5.35		8.05	11.55
2800	Thin set		124	.129		2.61	3.64		6.25	8.65
3000	Floors, natural clay, random or uniform, thin set, color group 1		183	.087	S.F.	3.83	2.47		6.30	8.15
3100	Color group 2		183	.087		5.15	2.47		7.62	9.60
3255	Floors, glazed, thin set, 6" x 6", color group 1		200	.080		4.43	2.26		6.69	8.50
3260	8" x 8" tile		250	.064		4.43	1.80		6.23	7.75
3270	12" x 12" tile		325	.049		5.25	1.39		6.64	7.95
3280	16" x 16" tile		550	.029		6.05	.82		6.87	7.95
3281	18" x 18" tile		600	.027		6.05	.75		6.80	7.85
3283	24" x 24" tile		650	.025		6.05	.69		6.74	7.75
3285	Border, 6" x 12" tile		275	.058		11.55	1.64		13.19	15.35
3290	3" x 12" tile		200	.080		34	2.26		36.26	40.50
3300	Porcelain type, 1 color, color group 2, 1" x 1"		183	.087		4.39	2.47		6.86	8.75
3310	2" x 2" or 2" x 1", thin set		190	.084		4.26	2.37		6.63	8.50
3350	For random blend, 2 colors, add					.88			.88	.97
3360	4 colors, add					1.24			1.24	1.36
4300	Specialty tile, 4-1/4" x 4-1/4" x 1/2", decorator finish	D-7	183	.087		9.90	2.47		12.37	14.85
4500	Add for epoxy grout, 1/16" joint, 1" x 1" tile		800	.020		.62	.56		1.18	1.58
4600	2" x 2" tile		820	.020		.59	.55		1.14	1.53
4800	Pregrouted sheets, walls, 4-1/4" x 4-1/4", 6" x 4-1/4"									
4810	and 8-1/2" x 4-1/4", 4 S.F. sheets, silicone grout	D-7	240	.067	S.F.	4.73	1.88		6.61	8.20
5100	Floors, unglazed, 2 S.F. sheets,									
5110	Urethane adhesive	D-7	180	.089	S.F.	4.71	2.51		7.22	9.20
5400	Walls, interior, thin set, 4-1/4" x 4-1/4" tile		190	.084		2.14	2.37		4.51	6.15
5500	6" x 4-1/4" tile		190	.084		2.73	2.37		5.10	6.80
5700	8-1/2" x 4-1/4" tile		190	.084		3.80	2.37		6.17	7.95
5800	6" x 6" tile		200	.080		2.38	2.26		4.64	6.20
5810	8" x 8" tile		225	.071		4.08	2.01		6.09	7.70
5820	12" x 12" tile		300	.053		3.53	1.50		5.03	6.30
5830	16" x 16" tile		500	.032		4.13	.90		5.03	6
6000	Decorated wall tile, 4-1/4" x 4-1/4", minimum		270	.059		2.64	1.67		4.31	5.60
6100	Maximum		180	.089		44	2.51		46.51	52
6600	Crystalline glazed, 4-1/4" x 4-1/4", mud set, plain		100	.160		3.78	4.51		8.29	11.35
6700	4-1/4" x 4-1/4", scored tile		100	.160		4.95	4.51		9.46	12.65
6900	6" x 6" plain		93	.172		5.15	4.85		10	13.40
7000	For epoxy grout, 1/16" joints, 4-1/4" tile, add		800	.020		.39	.56		.95	1.33
7200	For tile set in dry mortar, add		1735	.009			.26		.26	.42
7300	For tile set in Portland cement mortar, add		290	.055		.16	1.56		1.72	2.67
9300	Ceramic tiles, recycled glass, standard colors, 2" x 2" thru 6" x 6" G		190	.084		17	2.37		19.37	22.50
9310	6" x 6" G		200	.080		17	2.26		19.26	22.50
9320	8" x 8" G		225	.071		18.45	2.01		20.46	23.50
9330	12" x 12" G		300	.053		18.45	1.50		19.95	23
9340	Earthtones, 2" x 2" to 4" x 8" G		190	.084		21.50	2.37		23.87	27.50
9350	6" x 6" G		200	.080		21.50	2.26		23.76	27
9360	8" x 8" G		225	.071		23	2.01		25.01	28

09 30 Tiling

09 30 13 – Ceramic Tiling

09 30 13.10 Ceramic Tile		Crew	Daily Output	Labor-Hours	Unit	2011 Bare Costs Material	Labor	Equipment	Total	Total Incl O&P
9370	12" x 12" G	D-7	300	.053	S.F.	23	1.50		24.50	27.50
9380	Deep colors, 2" x 2" to 4" x 8" G		190	.084		31	2.37		33.37	38
9390	6" x 6" G		200	.080		31	2.26		33.26	37.50
9400	8" x 8" G		225	.071		32	2.01		34.01	38.50
9410	12" x 12" G	↓	300	.053	↓	32	1.50		33.50	38

09 30 16 – Quarry Tiling

09 30 16.10 Quarry Tile		Crew	Daily Output	Labor-Hours	Unit	Material	Labor	Equipment	Total	Total Incl O&P
0010	**QUARRY TILE**									
0100	Base, cove or sanitary, mud set, to 5" high, 1/2" thick	D-7	110	.145	L.F.	5.30	4.10		9.40	12.35
0300	Bullnose trim, red, mud set, 6" x 6" x 1/2" thick		120	.133		4.34	3.76		8.10	10.80
0400	4" x 4" x 1/2" thick		110	.145		4.82	4.10		8.92	11.85
0600	4" x 8" x 1/2" thick, using 8" as edge		130	.123	↓	4.26	3.47		7.73	10.25
0700	Floors, mud set, 1,000 S.F. lots, red, 4" x 4" x 1/2" thick		120	.133	S.F.	6.10	3.76		9.86	12.70
0900	6" x 6" x 1/2" thick		140	.114		5.50	3.22		8.72	11.20
1000	4" x 8" x 1/2" thick	↓	130	.123		6.10	3.47		9.57	12.25
1300	For waxed coating, add					.71			.71	.78
1500	For non-standard colors, add					.42			.42	.46
1600	For abrasive surface, add					.49			.49	.54
1800	Brown tile, imported, 6" x 6" x 3/4"	D-7	120	.133		7.30	3.76		11.06	14
1900	8" x 8" x 1"		110	.145		7.95	4.10		12.05	15.30
2100	For thin set mortar application, deduct		700	.023			.64		.64	1.03
2700	Stair tread, 6" x 6" x 3/4", plain		50	.320		5.15	9		14.15	20
2800	Abrasive		47	.340		5.60	9.60		15.20	21.50
3000	Wainscot, 6" x 6" x 1/2", thin set, red		105	.152		4.31	4.30		8.61	11.60
3100	Non-standard colors		105	.152	↓	4.78	4.30		9.08	12.10
3300	Window sill, 6" wide, 3/4" thick		90	.178	L.F.	5.20	5		10.20	13.75
3400	Corners	↓	80	.200	Ea.	6	5.65		11.65	15.60

09 30 29 – Metal Tiling

09 30 29.10 Metal Tile		Crew	Daily Output	Labor-Hours	Unit	Material	Labor	Equipment	Total	Total Incl O&P
0010	**METAL TILE** 4' x 4' sheet, 24 ga., tile pattern, nailed									
0200	Stainless steel	2 Carp	512	.031	S.F.	25.50	1.05		26.55	30
0400	Aluminized steel	"	512	.031	"	13.80	1.05		14.85	16.95

09 34 Waterproofing-Membrane Tiling

09 34 13 – Waterproofing-Membrane Ceramic Tiling

09 34 13.10 Ceramic Tile Waterproofing Membrane		Crew	Daily Output	Labor-Hours	Unit	Material	Labor	Equipment	Total	Total Incl O&P
0010	**CERAMIC TILE WATERPROOFING MEMBRANE**									
0020	On floors, including thinset									
0030	Fleece laminated polyethylene grid, 1/8" thick	D-7	250	.064	S.F.	2.24	1.80		4.04	5.35
0040	5/16" thick	"	250	.064	"	2.60	1.80		4.40	5.75
0050	On walls, including thinset									
0060	Fleece laminated polyethylene sheet, 8 mil thick	D-7	480	.033	S.F.	2.24	.94		3.18	3.96
0070	Accessories, including thinset									
0080	Joint and corner sheet, 4 mils thick, 5" wide	1 Tilf	240	.033	L.F.	1.45	1.05		2.50	3.27
0090	7-1/4" wide		180	.044		1.84	1.40		3.24	4.26
0100	10" wide		120	.067	↓	2.24	2.10		4.34	5.80
0110	Pre-formed corners, inside		32	.250	Ea.	3.93	7.90		11.83	16.90
0120	Outside		32	.250		7.40	7.90		15.30	21
0130	2" flanged floor drain with 6" stainless steel grate		16	.500	↓	355	15.80		370.80	415
0140	EPS, sloped shower floor	↓	480	.017	S.F.	4.82	.53		5.35	6.15

09 34 Waterproofing-Membrane Tiling

09 34 13 – Waterproofing-Membrane Ceramic Tiling

09 34 13.10 Ceramic Tile Waterproofing Membrane		Crew	Daily Output	Labor-Hours	Unit	2011 Bare Costs Material	Labor	Equipment	Total	Total Incl O&P
0150	Curb	1 Tilf	32	.250	L.F.	13.55	7.90		21.45	27.50

09 51 Acoustical Ceilings

09 51 23 – Acoustical Tile Ceilings

09 51 23.10 Suspended Acoustic Ceiling Tiles

		Crew	Daily Output	Labor-Hours	Unit	Material	Labor	Equipment	Total	Total Incl O&P
0010	**SUSPENDED ACOUSTIC CEILING TILES**, not including									
0100	suspension system									
0300	Fiberglass boards, film faced, 2' x 2' or 2' x 4', 5/8" thick	1 Carp	625	.013	S.F.	.72	.43		1.15	1.51
0400	3/4" thick		600	.013		1.66	.45		2.11	2.58
0500	3" thick, thermal, R11		450	.018		1.53	.60		2.13	2.68
0600	Glass cloth faced fiberglass, 3/4" thick		500	.016		1.49	.54		2.03	2.54
0700	1" thick		485	.016		2.18	.55		2.73	3.33
0820	1-1/2" thick, nubby face		475	.017		2.66	.57		3.23	3.88
1110	Mineral fiber tile, lay-in, 2' x 2' or 2' x 4', 5/8" thick, fine texture		625	.013		.59	.43		1.02	1.37
1115	Rough textured		625	.013		1.44	.43		1.87	2.30
1125	3/4" thick, fine textured		600	.013		1.50	.45		1.95	2.40
1130	Rough textured		600	.013		1.79	.45		2.24	2.72
1135	Fissured		600	.013		2.45	.45		2.90	3.45
1150	Tegular, 5/8" thick, fine textured		470	.017		1.45	.57		2.02	2.56
1155	Rough textured		470	.017		1.90	.57		2.47	3.05
1165	3/4" thick, fine textured		450	.018		2.08	.60		2.68	3.29
1170	Rough textured		450	.018		2.35	.60		2.95	3.59
1175	Fissured		450	.018		3.63	.60		4.23	4.99
1180	For aluminum face, add					6.05			6.05	6.65
1185	For plastic film face, add					.94			.94	1.03
1190	For fire rating, add					.45			.45	.50
1900	Eggcrate, acrylic, 1/2" x 1/2" x 1/2" cubes	1 Carp	500	.016		1.82	.54		2.36	2.90
2100	Polystyrene eggcrate, 3/8" x 3/8" x 1/2" cubes		510	.016		1.53	.53		2.06	2.56
2200	1/2" x 1/2" x 1/2" cubes		500	.016		1.75	.54		2.29	2.83
2400	Luminous panels, prismatic, acrylic		400	.020		2.22	.67		2.89	3.56
2500	Polystyrene		400	.020		1.14	.67		1.81	2.37
2700	Flat white acrylic		400	.020		3.86	.67		4.53	5.35
2800	Polystyrene		400	.020		2.65	.67		3.32	4.04
3000	Drop pan, white, acrylic		400	.020		5.65	.67		6.32	7.35
3100	Polystyrene		400	.020		4.73	.67		5.40	6.30
3600	Perforated aluminum sheets, .024" thick, corrugated, painted		490	.016		2.25	.55		2.80	3.40
3700	Plain		500	.016		3.76	.54		4.30	5.05
5020	66 – 78% recycled content, 3/4" thick G		600	.013		1.82	.45		2.27	2.75
5040	Mylar, 42% recycled content, 3/4" thick G		600	.013		4.28	.45		4.73	5.45

09 51 23.30 Suspended Ceilings, Complete

		Crew	Daily Output	Labor-Hours	Unit	Material	Labor	Equipment	Total	Total Incl O&P
0010	**SUSPENDED CEILINGS, COMPLETE** Including standard									
0100	suspension system but not incl. 1-1/2" carrier channels									
0600	Fiberglass ceiling board, 2' x 4' x 5/8", plain faced	1 Carp	500	.016	S.F.	1.37	.54		1.91	2.41
0700	Offices, 2' x 4' x 3/4"		380	.021		2.31	.71		3.02	3.72
1800	Tile, Z bar suspension, 5/8" mineral fiber tile		150	.053		2.16	1.79		3.95	5.35
1900	3/4" mineral fiber tile		150	.053		2.31	1.79		4.10	5.55

09 51 53 – Direct-Applied Acoustical Ceilings

09 51 53.10 Ceiling Tile

		Crew	Daily Output	Labor-Hours	Unit	Material	Labor	Equipment	Total	Total Incl O&P
0010	**CEILING TILE**, Stapled or cemented									
0100	12" x 12" or 12" x 24", not including furring									

09 51 Acoustical Ceilings

09 51 53 – Direct-Applied Acoustical Ceilings

09 51 53.10 Ceiling Tile		Crew	Daily Output	Labor-Hours	Unit	2011 Bare Costs Material	Labor	Equipment	Total	Total Incl O&P
0600	Mineral fiber, vinyl coated, 5/8" thick	1 Carp	300	.027	S.F.	1.79	.89		2.68	3.47
0700	3/4" thick		300	.027		1.91	.89		2.80	3.60
0900	Fire rated, 3/4" thick, plain faced		300	.027		1.40	.89		2.29	3.04
1000	Plastic coated face		300	.027		1.18	.89		2.07	2.80
1200	Aluminum faced, 5/8" thick, plain		300	.027		1.51	.89		2.40	3.16
3300	For flameproofing, add					.10			.10	.11
3400	For sculptured 3 dimensional, add					.29			.29	.32
3900	For ceiling primer, add					.13			.13	.14
4000	For ceiling cement, add					.37			.37	.41

09 53 Acoustical Ceiling Suspension Assemblies

09 53 23 – Metal Acoustical Ceiling Suspension Assemblies

09 53 23.30 Ceiling Suspension Systems		Crew	Daily Output	Labor-Hours	Unit	Material	Labor	Equipment	Total	Total Incl O&P
0010	**CEILING SUSPENSION SYSTEMS** for boards and tile									
0050	Class A suspension system, 15/16" T bar, 2' x 4' grid	1 Carp	800	.010	S.F.	.65	.34		.99	1.28
0300	2' x 2' grid		650	.012		.81	.41		1.22	1.58
0310	25% recycled steel, 2' x 4' grid G		800	.010		.74	.34		1.08	1.37
0320	2' x 2' grid G		650	.012		.93	.41		1.34	1.71
0350	For 9/16" grid, add					.16			.16	.18
0360	For fire rated grid, add					.09			.09	.10
0370	For colored grid, add					.21			.21	.23
0400	Concealed Z bar suspension system, 12" module	1 Carp	520	.015		.66	.52		1.18	1.59
0600	1-1/2" carrier channels, 4' O.C., add		470	.017		.10	.57		.67	1.06
0650	1-1/2" x 3-1/2" channels		470	.017		.25	.57		.82	1.24
0700	Carrier channels for ceilings with									
0900	recessed lighting fixtures, add	1 Carp	460	.017	S.F.	.17	.58		.75	1.17
5000	Wire hangers, #12 wire	"	300	.027	Ea.	1.22	.89		2.11	2.84

09 54 Specialty Ceilings

09 54 33 – Decorative Panel Ceilings

09 54 33.20 Metal Panel Ceilings		Crew	Daily Output	Labor-Hours	Unit	Material	Labor	Equipment	Total	Total Incl O&P
0010	**METAL PANEL CEILINGS**									
0020	Lay-in or screwed to furring, not including grid									
0100	Tin ceilings, 2' x 2' or 2' x 4', bare steel finish	2 Carp	300	.053	S.F.	2.37	1.79		4.16	5.60
0120	Painted white finish		300	.053	"	3.66	1.79		5.45	7
0140	Copper, chrome or brass finish		300	.053	L.F.	6.40	1.79		8.19	10.05
0200	Cornice molding, 2-1/2" to 3-1/2" wide, 4' long, bare steel finish		200	.080	S.F.	2.25	2.68		4.93	6.95
0220	Painted white finish		200	.080		2.94	2.68		5.62	7.70
0240	Copper, chrome or brass finish		200	.080		3.78	2.68		6.46	8.65
0320	5" to 6-1/2" wide, 4' long, bare steel finish		150	.107		3.19	3.58		6.77	9.50
0340	Painted white finish		150	.107		4.20	3.58		7.78	10.60
0360	Copper, chrome or brass finish		150	.107		6.40	3.58		9.98	13.05
0420	Flat molding, 3-1/2" to 5" wide, 4' long, bare steel finish		250	.064		3.43	2.15		5.58	7.35
0440	Painted white finish		250	.064		4.09	2.15		6.24	8.10
0460	Copper, chrome or brass finish		250	.064		7.35	2.15		9.50	11.70

09 62 Specialty Flooring

09 62 19 – Laminate Flooring

09 62 19.10 Floating Floor			Crew	Daily Output	Labor-Hours	Unit	2011 Bare Costs Material	Labor	Equipment	Total	Total Incl O&P
0010	FLOATING FLOOR										
8300	Floating floor, laminate, wood pattern strip, complete		1 Clab	133	.060	S.F.	4.57	1.59		6.16	7.70
8310	Components, T & G wood composite strips						3.66			3.66	4.03
8320	Film						.15			.15	.16
8330	Foam						.24			.24	.27
8340	Adhesive						.50			.50	.55
8350	Installation kit						.17			.17	.19
8360	Trim, 2″ wide x 3′ long					L.F.	3.85			3.85	4.24
8370	Reducer moulding					″	5.20			5.20	5.70

09 62 23 – Bamboo Flooring

09 62 23.10 Flooring, Bamboo			Crew	Daily Output	Labor-Hours	Unit	Material	Labor	Equipment	Total	Total Incl O&P
0010	FLOORING, BAMBOO										
8600	Flooring, wood, bamboo strips, unfinished, 5/8″ x 4″ x 3′	G	1 Carp	255	.031	S.F.	4.44	1.05		5.49	6.65
8610	5/8″ x 4″ x 4′	G		275	.029		4.61	.98		5.59	6.70
8620	5/8″ x 4″ x 6′	G		295	.027		5.05	.91		5.96	7.05
8630	Finished, 5/8″ x 4″ x 3′	G		255	.031		4.89	1.05		5.94	7.15
8640	5/8″ x 4″ x 4′	G		275	.029		5.10	.98		6.08	7.30
8650	5/8″ x 4″ x 6′	G		295	.027		4.74	.91		5.65	6.70
8660	Stair treads, unfinished, 1-1/16″ x 11-1/2″ x 4′	G		18	.444	Ea.	42.50	14.90		57.40	72
8670	Finished, 1-1/16″ x 11-1/2″ x 4′	G		18	.444		71.50	14.90		86.40	104
8680	Stair risers, unfinished, 5/8″ x 7-1/2″ x 4′	G		18	.444		15.75	14.90		30.65	42.50
8690	Finished, 5/8″ x 7-1/2″ x 4′	G		18	.444		30	14.90		44.90	58
8700	Stair nosing, unfinished, 6′ long	G		16	.500		35	16.80		51.80	66.50
8710	Finished, 6′ long	G		16	.500		39.50	16.80		56.30	71.50

09 63 Masonry Flooring

09 63 13 – Brick Flooring

09 63 13.10 Miscellaneous Brick Flooring			Crew	Daily Output	Labor-Hours	Unit	Material	Labor	Equipment	Total	Total Incl O&P
0010	MISCELLANEOUS BRICK FLOORING										
0020	Acid-proof shales, red, 8″ x 3-3/4″ x 1-1/4″ thick		D-7	.43	37.209	M	860	1,050		1,910	2,625
0050	2-1/4″ thick		D-1	.40	40		1,000	1,225		2,225	3,100
0200	Acid-proof clay brick, 8″ x 3-3/4″ x 2-1/4″ thick	G		.40	40		895	1,225		2,120	2,975
0250	9″ x 4-1/2″ x 3″	G		95	.168	S.F.	3.91	5.10		9.01	12.70
0260	Cast ceramic, pressed, 4″ x 8″ x 1/2″, unglazed		D-7	100	.160		6.25	4.51		10.76	14.10
0270	Glazed			100	.160		8.35	4.51		12.86	16.35
0280	Hand molded flooring, 4″ x 8″ x 3/4″, unglazed			95	.168		8.25	4.75		13	16.70
0290	Glazed			95	.168		10.35	4.75		15.10	19
0300	8″ hexagonal, 3/4″ thick, unglazed			85	.188		9.05	5.30		14.35	18.45
0310	Glazed			85	.188		16.35	5.30		21.65	26.50
0450	Acid-proof joints, 1/4″ wide		D-1	65	.246		1.44	7.50		8.94	13.90
0500	Pavers, 8″ x 4″, 1″ to 1-1/4″ thick, red		D-7	95	.168		3.64	4.75		8.39	11.60
0510	Ironspot		″	95	.168		5.15	4.75		9.90	13.25
0540	1-3/8″ to 1-3/4″ thick, red		D-1	95	.168		3.51	5.10		8.61	12.25
0560	Ironspot			95	.168		5.10	5.10		10.20	14
0580	2-1/4″ thick, red			90	.178		3.57	5.40		8.97	12.80
0590	Ironspot			90	.178		5.55	5.40		10.95	14.95
0700	Paver, adobe brick, 6″ x 12″, 1/2″ joint	G		42	.381		.84	11.60		12.44	19.90
0710	Mexican red, 12″ x 12″	G	1 Tilf	48	.167		1.38	5.25		6.63	9.90
0720	Saltillo, 12″ x 12″	G	″	48	.167		1.38	5.25		6.63	9.90
0800	For sidewalks and patios with pavers, see Section 32 14 16.10										

09 63 Masonry Flooring

09 63 13 – Brick Flooring

	09 63 13.10 Miscellaneous Brick Flooring	Crew	Daily Output	Labor-Hours	Unit	Material	Labor	Equipment	Total	Total Incl O&P
						2011 Bare Costs				
0870	For epoxy joints, add	D-1	600	.027	S.F.	2.72	.81		3.53	4.32
0880	For Furan underlayment, add	"	600	.027		2.25	.81		3.06	3.81
0890	For waxed surface, steam cleaned, add	A-1H	1000	.008		.20	.21	.06	.47	.64

09 63 40 – Stone Flooring

09 63 40.10 Marble

		Crew	Daily Output	Labor-Hours	Unit	Material	Labor	Equipment	Total	Total Incl O&P
0010	**MARBLE**									
0020	Thin gauge tile, 12" x 6", 3/8", white Carara	D-7	60	.267	S.F.	11.25	7.50		18.75	24.50
0100	Travertine		60	.267		12.30	7.50		19.80	25.50
0200	12" x 12" x 3/8", thin set, floors		60	.267		6.05	7.50		13.55	18.65
0300	On walls		52	.308		9.50	8.70		18.20	24.50
1000	Marble threshold, 4" wide x 36" long x 5/8" thick, white		60	.267	Ea.	6.05	7.50		13.55	18.65

09 63 40.20 Slate Tile

		Crew	Daily Output	Labor-Hours	Unit	Material	Labor	Equipment	Total	Total Incl O&P
0010	**SLATE TILE**									
0020	Vermont, 6" x 6" x 1/4" thick, thin set	D-7	180	.089	S.F.	6.80	2.51		9.31	11.45

09 64 Wood Flooring

09 64 16 – Wood Block Flooring

09 64 16.10 End Grain Block Flooring

		Crew	Daily Output	Labor-Hours	Unit	Material	Labor	Equipment	Total	Total Incl O&P
0010	**END GRAIN BLOCK FLOORING**									
0020	End grain flooring, coated, 2" thick	1 Carp	295	.027	S.F.	3.29	.91		4.20	5.15
0400	Natural finish, 1" thick, fir		125	.064		3.40	2.15		5.55	7.35
0600	1-1/2" thick, pine		125	.064		3.34	2.15		5.49	7.25
0700	2" thick, pine		125	.064		4.64	2.15		6.79	8.70

09 64 23 – Wood Parquet Flooring

09 64 23.10 Wood Parquet

		Crew	Daily Output	Labor-Hours	Unit	Material	Labor	Equipment	Total	Total Incl O&P
0010	**WOOD PARQUET** flooring									
5200	Parquetry, standard, 5/16" thick, not incl. finish, oak, minimum	1 Carp	160	.050	S.F.	4.55	1.68		6.23	7.80
5300	Maximum		100	.080		5.55	2.68		8.23	10.60
5500	Teak, minimum		160	.050		4.86	1.68		6.54	8.15
5600	Maximum		100	.080		8.50	2.68		11.18	13.85
5650	13/16" thick, select grade oak, minimum		160	.050		9.35	1.68		11.03	13.10
5700	Maximum		100	.080		14.20	2.68		16.88	20
5800	Custom parquetry, including finish, minimum		100	.080		15.65	2.68		18.33	21.50
5900	Maximum		50	.160		21	5.35		26.35	32
6700	Parquetry, prefinished white oak, 5/16" thick, minimum		160	.050		3.71	1.68		5.39	6.90
6800	Maximum		100	.080		5.50	2.68		8.18	10.55
7000	Walnut or teak, parquetry, minimum		160	.050		5.35	1.68		7.03	8.70
7100	Maximum		100	.080		9.35	2.68		12.03	14.80
7200	Acrylic wood parquet blocks, 12" x 12" x 5/16",									
7210	Irradiated, set in epoxy	1 Carp	160	.050	S.F.	7.80	1.68		9.48	11.40

09 64 29 – Wood Strip and Plank Flooring

09 64 29.10 Wood

		Crew	Daily Output	Labor-Hours	Unit	Material	Labor	Equipment	Total	Total Incl O&P
0010	**WOOD**									
0020	Fir, vertical grain, 1" x 4", not incl. finish, grade B & better	1 Carp	255	.031	S.F.	2.74	1.05		3.79	4.77
0100	C grade & better		255	.031		2.58	1.05		3.63	4.60
0300	Flat grain, 1" x 4", not incl. finish, B & better		255	.031		3.13	1.05		4.18	5.20
0400	C & better		255	.031		3.01	1.05		4.06	5.05
4000	Maple, strip, 25/32" x 2-1/4", not incl. finish, select		170	.047		4.37	1.58		5.95	7.45
4100	#2 & better		170	.047		3.49	1.58		5.07	6.50

09 64 Wood Flooring

09 64 29 – Wood Strip and Plank Flooring

09 64 29.10 Wood		Crew	Daily Output	Labor-Hours	Unit	Material	2011 Bare Costs Labor	Equipment	Total	Total Incl O&P
4300	33/32" x 3-1/4", not incl. finish, #1 grade	1 Carp	170	.047	S.F.	4.19	1.58		5.77	7.25
4400	#2 & better		170	.047		3.73	1.58		5.31	6.75
4600	Oak, white or red, 25/32" x 2-1/4", not incl. finish									
4700	#1 common	1 Carp	170	.047	S.F.	2.49	1.58		4.07	5.40
4900	Select quartered, 2-1/4" wide		170	.047		3.09	1.58		4.67	6.05
5000	Clear		170	.047		3.95	1.58		5.53	7
6100	Prefinished, white oak, prime grade, 2-1/4" wide		170	.047		4.49	1.58		6.07	7.60
6200	3-1/4" wide		185	.043		4.99	1.45		6.44	7.95
6400	Ranch plank		145	.055		6.75	1.85		8.60	10.50
6500	Hardwood blocks, 9" x 9", 25/32" thick		160	.050		5.75	1.68		7.43	9.15
7400	Yellow pine, 3/4" x 3-1/8", T & G, C & better, not incl. finish		200	.040		1.49	1.34		2.83	3.89
7500	Refinish wood floor, sand, 2 cts poly, wax, soft wood, min.	1 Clab	400	.020		.78	.53		1.31	1.75
7600	Hard wood, max		130	.062		1.16	1.63		2.79	4.01
7800	Sanding and finishing, 2 coats polyurethane		295	.027		.78	.72		1.50	2.06
7900	Subfloor and underlayment, see Section 06 16									
8015	Transition molding, 2 1/4" wide, 5' long	1 Carp	19.20	.417	Ea.	15	14		29	40

09 65 Resilient Flooring

09 65 10 – Resilient Tile Underlayment

09 65 10.10 Latex Underlayment		Crew	Daily Output	Labor-Hours	Unit	Material	Labor	Equipment	Total	Total Incl O&P
0010	**LATEX UNDERLAYMENT**									
3600	Latex underlayment, 1/8" thk., cementitious for resilient flooring	1 Tilf	160	.050	S.F.	.82	1.58		2.40	3.42
4000	Liquid, fortified				Gal.	40.50			40.50	44.50

09 65 13 – Resilient Base and Accessories

09 65 13.13 Resilient Base		Crew	Daily Output	Labor-Hours	Unit	Material	Labor	Equipment	Total	Total Incl O&P
0010	**RESILIENT BASE**									
0800	Base, cove, rubber or vinyl									
1100	Standard colors, 0.080" thick, 2-1/2" high	1 Tilf	315	.025	L.F.	1	.80		1.80	2.38
1150	4" high		315	.025		1.08	.80		1.88	2.47
1200	6" high		315	.025		1.50	.80		2.30	2.93
1450	1/8" thick, 2-1/2" high		315	.025		1	.80		1.80	2.38
1500	4" high		315	.025		.90	.80		1.70	2.27
1550	6" high		315	.025		1.45	.80		2.25	2.88
1600	Corners, 2-1/2" high		315	.025	Ea.	1.60	.80		2.40	3.04
1630	4" high		315	.025		3	.80		3.80	4.58
1660	6" high		315	.025		3	.80		3.80	4.58

09 65 16 – Resilient Sheet Flooring

09 65 16.10 Rubber and Vinyl Sheet Flooring		Crew	Daily Output	Labor-Hours	Unit	Material	Labor	Equipment	Total	Total Incl O&P
0010	**RUBBER AND VINYL SHEET FLOORING**									
5500	Linoleum, sheet goods G	1 Tilf	360	.022	S.F.	4.28	.70		4.98	5.85
5900	Rubber, sheet goods, 36" wide, 1/8" thick		120	.067		6.85	2.10		8.95	10.90
5950	3/16" thick		100	.080		9.25	2.52		11.77	14.25
6000	1/4" thick		90	.089		10.95	2.80		13.75	16.55
8000	Vinyl sheet goods, backed, .065" thick, minimum		250	.032		3.75	1.01		4.76	5.75
8050	Maximum		200	.040		3.69	1.26		4.95	6.10
8100	.080" thick, minimum		230	.035		3.80	1.10		4.90	5.95
8150	Maximum		200	.040		5.40	1.26		6.66	7.95
8200	.125" thick, minimum		230	.035		4.35	1.10		5.45	6.55
8250	Maximum		200	.040		6.50	1.26		7.76	9.15
8700	Adhesive cement, 1 gallon per 200 to 300 S.F.				Gal.	25			25	27.50

09 65 Resilient Flooring

09 65 16 – Resilient Sheet Flooring

09 65 16.10 Rubber and Vinyl Sheet Flooring		Crew	Daily Output	Labor-Hours	Unit	2011 Bare Costs Material	Labor	Equipment	Total	Total Incl O&P
8800	Asphalt primer, 1 gallon per 300 S.F.				Gal.	12.75			12.75	14.05
8900	Emulsion, 1 gallon per 140 S.F.				↓	16.25			16.25	17.90

09 65 19 – Resilient Tile Flooring

09 65 19.10 Miscellaneous Resilient Tile Flooring		Crew	Daily Output	Labor-Hours	Unit	Material	Labor	Equipment	Total	Total Incl O&P
0010	**MISCELLANEOUS RESILIENT TILE FLOORING**									
2200	Cork tile, standard finish, 1/8" thick G	1 Tilf	315	.025	S.F.	6.25	.80		7.05	8.20
2250	3/16" thick G		315	.025		6.30	.80		7.10	8.20
2300	5/16" thick G		315	.025		7.75	.80		8.55	9.85
2350	1/2" thick G		315	.025		9.20	.80		10	11.40
2500	Urethane finish, 1/8" thick G		315	.025		7.45	.80		8.25	9.45
2550	3/16" thick G		315	.025		7.80	.80		8.60	9.90
2600	5/16" thick G		315	.025		10	.80		10.80	12.30
2650	1/2" thick G		315	.025		13.50	.80		14.30	16.15
6050	Rubber tile, marbleized colors, 12" x 12", 1/8" thick		400	.020		6.65	.63		7.28	8.35
6100	3/16" thick		400	.020		10	.63		10.63	12
6300	Special tile, plain colors, 1/8" thick		400	.020		10	.63		10.63	12
6350	3/16" thick		400	.020		10	.63		10.63	12
7000	Vinyl composition tile, 12" x 12", 1/16" thick		500	.016		.93	.50		1.43	1.83
7050	Embossed		500	.016		2.17	.50		2.67	3.20
7100	Marbleized		500	.016		2.17	.50		2.67	3.20
7150	Solid		500	.016		2.79	.50		3.29	3.88
7200	3/32" thick, embossed		500	.016		1.35	.50		1.85	2.30
7250	Marbleized		500	.016		2.49	.50		2.99	3.55
7300	Solid		500	.016		2.31	.50		2.81	3.35
7350	1/8" thick, marbleized		500	.016		1.64	.50		2.14	2.61
7400	Solid		500	.016		1.73	.50		2.23	2.71
7450	Conductive		500	.016		5.95	.50		6.45	7.35
7500	Vinyl tile, 12" x 12", .050" thick, minimum		500	.016		3	.50		3.50	4.11
7550	Maximum		500	.016		6	.50		6.50	7.40
7600	1/8" thick, minimum		500	.016		4.50	.50		5	5.75
7650	Solid colors		500	.016		6.20	.50		6.70	7.60
7700	Marbleized or Travertine pattern		500	.016		5.65	.50		6.15	7
7750	Florentine pattern		500	.016		6	.50		6.50	7.40
7800	Maximum	↓	500	.016	↓	13.50	.50		14	15.65

09 65 33 – Conductive Resilient Flooring

09 65 33.10 Conductive Rubber and Vinyl Flooring		Crew	Daily Output	Labor-Hours	Unit	Material	Labor	Equipment	Total	Total Incl O&P
0010	**CONDUCTIVE RUBBER AND VINYL FLOORING**									
1700	Conductive flooring, rubber tile, 1/8" thick	1 Tilf	315	.025	S.F.	6	.80		6.80	7.90
1800	Homogeneous vinyl tile, 1/8" thick	"	315	.025	"	8	.80		8.80	10.10

09 66 Terrazzo Flooring

09 66 13 – Portland Cement Terrazzo Flooring

09 66 13.10 Portland Cement Terrazzo		Crew	Daily Output	Labor-Hours	Unit	Material	2011 Bare Costs Labor	Equipment	Total	Total Incl O&P
0010	**PORTLAND CEMENT TERRAZZO**, cast-in-place									
0020	Cove base, 6" high, 16 ga. zinc	1 Mstz	20	.400	L.F.	2.98	12.50		15.48	23.50
0101	Curb, 6" high and 6" wide	J-3	12	1.333		5.55	38.50	20	64.05	89
0300	Divider strip for floors, 14 ga., 1-1/4" deep, zinc	1 Mstz	375	.021		1.39	.67		2.06	2.60
0400	Brass		375	.021		2.43	.67		3.10	3.74
0600	Heavy top strip 1/4" thick, 1-1/4" deep, zinc		300	.027		1.90	.83		2.73	3.42
1200	For thin set floors, 16 ga., 1/2" x 1/2", zinc		350	.023		.82	.72		1.54	2.04
1500	Floor, bonded to concrete, 1-3/4" thick, gray cement	J-3	75	.213	S.F.	2.92	6.10	3.22	12.24	16.55
1600	White cement, mud set		75	.213		3.32	6.10	3.22	12.64	17
1800	Not bonded, 3" total thickness, gray cement		70	.229		3.65	6.55	3.45	13.65	18.30
1900	White cement, mud set		70	.229		3.98	6.55	3.45	13.98	18.70
4300	Stone chips, onyx gemstone, per 50 lb. bag				Bag	16.30			16.30	17.95
09 66 13.30 Terrazzo, Precast										
0010	**TERRAZZO, PRECAST**									
0020	Base, 6" high, straight	1 Mstz	70	.114	L.F.	10.55	3.58		14.13	17.35
0100	Cove		60	.133		12.15	4.17		16.32	20
0300	8" high, straight		60	.133		10.85	4.17		15.02	18.55
0400	Cove		50	.160		15.95	5		20.95	25.50
0600	For white cement, add					.43			.43	.47
0700	For 16 ga. zinc toe strip, add					1.63			1.63	1.79
0900	Curbs, 4" x 4" high	1 Mstz	40	.200		30.50	6.25		36.75	43.50
1000	8" x 8" high	"	30	.267		35.50	8.35		43.85	52.50
1200	Floor tiles, non-slip, 1" thick, 12" x 12"	D-1	60	.267	S.F.	18.60	8.10		26.70	34
1300	1-1/4" thick, 12" x 12"		60	.267		19.60	8.10		27.70	35
1500	16" x 16"		50	.320		21.50	9.75		31.25	39.50
1600	1-1/2" thick, 16" x 16"		45	.356		19.45	10.80		30.25	39.50
4800	Wainscot, 12" x 12" x 1" tiles	1 Mstz	12	.667		6.50	21		27.50	40.50
4900	16" x 16" x 1-1/2" tiles	"	8	1		13.70	31.50		45.20	65

09 66 16 – Terrazzo Floor Tile

09 66 16.10 Tile or Terrazzo Base		Crew	Daily Output	Labor-Hours	Unit	Material	Labor	Equipment	Total	Total Incl O&P
0010	**TILE OR TERRAZZO BASE**									
0020	Scratch coat only	1 Mstz	150	.053	S.F.	.41	1.67		2.08	3.12
0500	Scratch and brown coat only	"	75	.107	"	.78	3.34		4.12	6.20

09 67 Fluid-Applied Flooring

09 67 20 – Epoxy-Marble Chip Flooring

09 67 20.13 Elastomeric Liquid Flooring		Crew	Daily Output	Labor-Hours	Unit	Material	Labor	Equipment	Total	Total Incl O&P
0010	**ELASTOMERIC LIQUID FLOORING**									
0020	Cementitious acrylic, 1/4" thick	C-6	520	.092	S.F.	1.47	2.56	.09	4.12	5.95
0200	Methyl methachrylate, 1/4" thick	C-8A	3000	.016	"	5.40	.46		5.86	6.70
09 67 20.16 Epoxy Terrazzo										
0010	**EPOXY TERRAZZO**									
1800	Epoxy terrazzo, 1/4" thick, chemical resistant, minimum	J-3	200	.080	S.F.	5.15	2.30	1.21	8.66	10.70
1900	Maximum	"	150	.107	"	7.85	3.06	1.61	12.52	15.30
09 67 20.19 Polyacrylate Terrazzo										
0010	**POLYACRYLATE TERRAZZO**									
3150	Polyacrylate, 1/4" thick, minimum	C-6	735	.065	S.F.	3.07	1.81	.06	4.94	6.45
3170	Maximum		480	.100		3.91	2.77	.10	6.78	9
3200	3/8" thick, minimum		620	.077		4	2.14	.08	6.22	8.05
3220	Maximum		480	.100		5.65	2.77	.10	8.52	10.95

09 67 Fluid-Applied Flooring

09 67 20 – Epoxy-Marble Chip Flooring

09 67 20.19 Polyacrylate Terrazzo		Crew	Daily Output	Labor-Hours	Unit	2011 Bare Costs Material	Labor	Equipment	Total	Total Incl O&P
3300	Conductive, 1/4" thick, minimum	C-6	450	.107	S.F.	6.40	2.95	.10	9.45	12.05
3330	Maximum		305	.157		8.80	4.36	.15	13.31	17.10
3350	3/8" thick, minimum		365	.132		8.40	3.64	.13	12.17	15.45
3370	Maximum		255	.188		11.25	5.20	.18	16.63	21.50
3450	Granite, conductive, 1/4" thick, minimum		695	.069		8.15	1.91	.07	10.13	12.20
3470	Maximum		420	.114		10.50	3.16	.11	13.77	16.90
3500	3/8" thick, minimum		695	.069		11.90	1.91	.07	13.88	16.35
3520	Maximum		380	.126		14.40	3.50	.12	18.02	22
09 67 20.26 Quartz Flooring										
0010	**QUARTZ FLOORING**									
0600	Epoxy, with colored quartz chips, broadcast, minimum	C-6	675	.071	S.F.	2.57	1.97	.07	4.61	6.20
0700	Maximum		490	.098		3.12	2.71	.10	5.93	8.05
0900	Trowelled, minimum		560	.086		3.31	2.37	.08	5.76	7.65
1000	Maximum		480	.100		4.84	2.77	.10	7.71	10
1200	Heavy duty epoxy topping, 1/4" thick,									
1300	500 to 1,000 S.F.	C-6	420	.114	S.F.	4.89	3.16	.11	8.16	10.75
1500	1,000 to 2,000 S.F.		450	.107		4.06	2.95	.10	7.11	9.50
1600	Over 10,000 S.F.		480	.100		3.83	2.77	.10	6.70	8.90

09 68 Carpeting

09 68 05 – Carpet Accessories

09 68 05.11 Flooring Transition Strip		Crew	Daily Output	Labor-Hours	Unit	Material	Labor	Equipment	Total	Total Incl O&P
0010	**FLOORING TRANSITION STRIP**									
0107	Clamp down brass divider, 12' strip, vinyl to carpet	1 Tilf	31.25	.256	Ea.	10.45	8.10		18.55	24.50
0117	Vinyl to hard surface	"	31.25	.256	"	10.45	8.10		18.55	24.50

09 68 10 – Carpet Pad

09 68 10.10 Commercial Grade Carpet Pad		Crew	Daily Output	Labor-Hours	Unit	Material	Labor	Equipment	Total	Total Incl O&P
0010	**COMMERCIAL GRADE CARPET PAD**									
9000	Sponge rubber pad, minimum	1 Tilf	150	.053	S.Y.	4.06	1.68		5.74	7.15
9100	Maximum		150	.053		10	1.68		11.68	13.70
9200	Felt pad, minimum		150	.053		4.14	1.68		5.82	7.25
9300	Maximum		150	.053		7.10	1.68		8.78	10.50
9400	Bonded urethane pad, minimum		150	.053		4.57	1.68		6.25	7.75
9500	Maximum		150	.053		7.80	1.68		9.48	11.30
9600	Prime urethane pad, minimum		150	.053		3	1.68		4.68	6
9700	Maximum		150	.053		5	1.68		6.68	8.20

09 68 13 – Tile Carpeting

09 68 13.10 Carpet Tile		Crew	Daily Output	Labor-Hours	Unit	Material	Labor	Equipment	Total	Total Incl O&P
0010	**CARPET TILE**									
0100	Tufted nylon, 18" x 18", hard back, 20 oz.	1 Tilf	80	.100	S.Y.	22	3.16		25.16	29.50
0110	26 oz.		80	.100		38	3.16		41.16	46.50
0200	Cushion back, 20 oz.		80	.100		28	3.16		31.16	35.50
0210	26 oz.		80	.100		43.50	3.16		46.66	53

09 68 16 – Sheet Carpeting

09 68 16.10 Sheet Carpet		Crew	Daily Output	Labor-Hours	Unit	Material	Labor	Equipment	Total	Total Incl O&P
0010	**SHEET CARPET**									
0700	Nylon, level loop, 26 oz., light to medium traffic	1 Tilf	75	.107	S.Y.	28	3.37		31.37	36.50
0720	28 oz., light to medium traffic		75	.107		29	3.37		32.37	37.50
0900	32 oz., medium traffic		75	.107		35	3.37		38.37	43.50

09 68 Carpeting

09 68 16 – Sheet Carpeting

09 68 16.10 Sheet Carpet		Crew	Daily Output	Labor-Hours	Unit	Material	2011 Bare Costs Labor	Equipment	Total	Total Incl O&P
1100	40 oz., medium to heavy traffic	1 Tilf	75	.107	S.Y.	51	3.37		54.37	61.50
2920	Nylon plush, 30 oz., medium traffic		57	.140		27	4.43		31.43	36.50
3000	36 oz., medium traffic		75	.107		35.50	3.37		38.87	44.50
3100	42 oz., medium to heavy traffic		70	.114		40	3.61		43.61	50
3200	46 oz., medium to heavy traffic		70	.114		46.50	3.61		50.11	57.50
3300	54 oz., heavy traffic		70	.114		51	3.61		54.61	62
4110	Wool, level loop, 40 oz., medium traffic		75	.107		119	3.37		122.37	136
4500	50 oz., medium to heavy traffic		75	.107		120	3.37		123.37	137
4700	Patterned, 32 oz., medium to heavy traffic		70	.114		119	3.61		122.61	137
4900	48 oz., heavy traffic		70	.114		122	3.61		125.61	140
5000	For less than full roll (approx. 1500 S.F.), add					25%				
5100	For small rooms, less than 12′ wide, add						25%			
5200	For large open areas (no cuts), deduct						25%			
5600	For bound carpet baseboard, add	1 Tilf	300	.027	L.F.	2.75	.84		3.59	4.37
5610	For stairs, not incl. price of carpet, add	"	30	.267	Riser		8.40		8.40	13.45
5620	For borders and patterns, add to labor						18%			
8950	For tackless, stretched installation, add padding from 09 68 10.10 to above									
9850	For brand-named specific fiber, add				S.Y.	25%				

09 68 20 – Athletic Carpet

09 68 20.10 Indoor Athletic Carpet		Crew	Daily Output	Labor-Hours	Unit	Material	Labor	Equipment	Total	Total Incl O&P
0010	**INDOOR ATHLETIC CARPET**									
3700	Polyethylene, in rolls, no base incl., landscape surfaces	1 Tilf	275	.029	S.F.	3.30	.92		4.22	5.10
3800	Nylon action surface, 1/8″ thick		275	.029		3.46	.92		4.38	5.30
3900	1/4″ thick		275	.029		4.99	.92		5.91	6.95
4000	3/8″ thick		275	.029		6.25	.92		7.17	8.35

09 72 Wall Coverings

09 72 23 – Wallpapering

09 72 23.10 Wallpaper		Crew	Daily Output	Labor-Hours	Unit	Material	Labor	Equipment	Total	Total Incl O&P
0010	**WALLPAPER** including sizing; add 10-30 percent waste @ takeoff									
0050	Aluminum foil	1 Pape	275	.029	S.F.	1.02	.87		1.89	2.53
0100	Copper sheets, .025″ thick, vinyl backing		240	.033		5.45	1		6.45	7.60
0300	Phenolic backing		240	.033		7.05	1		8.05	9.40
0600	Cork tiles, light or dark, 12″ x 12″ x 3/16″		240	.033		4.47	1		5.47	6.55
0700	5/16″ thick		235	.034		3.80	1.02		4.82	5.85
0900	1/4″ basketweave		240	.033		5.85	1		6.85	8.05
1000	1/2″ natural, non-directional pattern		240	.033		7.85	1		8.85	10.25
1100	3/4″ natural, non-directional pattern		240	.033		10.80	1		11.80	13.45
1200	Granular surface, 12″ x 36″, 1/2″ thick		385	.021		1.26	.62		1.88	2.40
1300	1″ thick		370	.022		1.63	.65		2.28	2.84
1500	Polyurethane coated, 12″ x 12″ x 3/16″ thick		240	.033		3.95	1		4.95	5.95
1600	5/16″ thick		235	.034		5.65	1.02		6.67	7.85
1800	Cork wallpaper, paperbacked, natural		480	.017		1.99	.50		2.49	3
1900	Colors		480	.017		2.78	.50		3.28	3.87
2100	Flexible wood veneer, 1/32″ thick, plain woods		100	.080		2.36	2.39		4.75	6.50
2200	Exotic woods		95	.084		3.58	2.52		6.10	8.05
2400	Gypsum-based, fabric-backed, fire resistant									
2500	for masonry walls, minimum, 21 oz./S.Y.	1 Pape	800	.010	S.F.	.82	.30		1.12	1.39
2600	Average		720	.011		1.23	.33		1.56	1.89
2700	Maximum (small quantities)		640	.013		1.37	.37		1.74	2.12

09 72 Wall Coverings

09 72 23 – Wallpapering

09 72 23.10 Wallpaper		Crew	Daily Output	Labor-Hours	Unit	Material	Labor	Equipment	Total	Total Incl O&P
						2011 Bare Costs				
2750	Acrylic, modified, semi-rigid PVC, .028″ thick	2 Carp	330	.048	S.F.	1.13	1.63		2.76	3.96
2800	.040″ thick	″	320	.050		1.49	1.68		3.17	4.45
3000	Vinyl wall covering, fabric-backed, lightweight, type 1 (12-15 oz./S.Y.)	1 Pape	640	.013		.69	.37		1.06	1.37
3300	Medium weight, type 2 (20-24 oz./S.Y.)		480	.017		.82	.50		1.32	1.71
3400	Heavy weight, type 3 (28 oz./S.Y.)		435	.018		1.64	.55		2.19	2.69
3600	Adhesive, 5 gal. lots (18 S.Y./gal.)				Gal.	10.40			10.40	11.45
3700	Wallpaper, average workmanship, solid pattern, low cost paper	1 Pape	640	.013	S.F.	.37	.37		.74	1.02
3900	basic patterns (matching required), avg. cost paper		535	.015		.78	.45		1.23	1.59
4000	Paper at $85 per double roll, quality workmanship		435	.018		2.47	.55		3.02	3.61
4100	Linen wall covering, paper backed									
4150	Flame treatment, minimum				S.F.	1.04			1.04	1.14
4180	Maximum					1.55			1.55	1.71
4200	Grass cloths with lining paper, minimum (G)	1 Pape	400	.020		.78	.60		1.38	1.83
4300	Maximum (G)	″	350	.023		2.49	.68		3.17	3.85

09 77 Special Wall Surfacing

09 77 33 – Fiberglass Reinforced Panels

09 77 33.10 Fiberglass Reinforced Plastic Panels		Crew	Daily Output	Labor-Hours	Unit	Material	Labor	Equipment	Total	Total Incl O&P
0010	**FIBERGLASS REINFORCED PLASTIC PANELS**, .090″ thick									
0020	On walls, adhesive mounted, embossed surface	2 Carp	640	.025	S.F.	1.23	.84		2.07	2.75
0030	Smooth surface		640	.025		1.51	.84		2.35	3.06
0040	Fire rated, embossed surface		640	.025		2.19	.84		3.03	3.81
0050	Nylon rivet mounted, on drywall, embossed surface		480	.033		1.15	1.12		2.27	3.14
0060	Smooth surface		480	.033		1.43	1.12		2.55	3.44
0070	Fire rated, embossed surface		480	.033		2.11	1.12		3.23	4.19
0080	On masonry, embossed surface		320	.050		1.15	1.68		2.83	4.08
0090	Smooth surface		320	.050		1.43	1.68		3.11	4.38
0100	Fire rated, embossed surface		320	.050		2.11	1.68		3.79	5.15
0110	Nylon rivet and adhesive mounted, on drywall, embossed surface		240	.067		1.30	2.24		3.54	5.15
0120	Smooth surface		240	.067		1.58	2.24		3.82	5.50
0130	Fire rated, embossed surface		240	.067		2.26	2.24		4.50	6.25
0140	On masonry, embossed surface		190	.084		1.30	2.83		4.13	6.15
0150	Smooth surface		190	.084		1.58	2.83		4.41	6.45
0160	Fire rated, embossed surface		190	.084		2.26	2.83		5.09	7.20
0170	For moldings add	1 Carp	250	.032	L.F.	.28	1.07		1.35	2.11
0180	On ceilings, for lay in grid system, embossed surface		400	.020	S.F.	1.23	.67		1.90	2.47
0190	Smooth surface		400	.020		1.51	.67		2.18	2.78
0200	Fire rated, embossed surface		400	.020		2.19	.67		2.86	3.53

09 77 43 – Panel Systems

09 77 43.20 Slatwall Panels and Accessories		Crew	Daily Output	Labor-Hours	Unit	Material	Labor	Equipment	Total	Total Incl O&P
0010	**SLATWALL PANELS AND ACCESSORIES**									
0100	Slatwall panel, 4′ x 8′ x 3/4″ T, MDF, paint grade	1 Carp	500	.016	S.F.	1.32	.54		1.86	2.35
0110	Melamine finish		500	.016		1.87	.54		2.41	2.96
0120	High pressure plastic laminate finish		500	.016		3.15	.54		3.69	4.37
0130	Aluminum channel inserts, add					2.94			2.94	3.23
0200	Accessories, corner forms, 8′ L				L.F.	4.28			4.28	4.71
0210	T-connector, 8′ L					5.15			5.15	5.70
0220	J-mold, 8′ L					1.33			1.33	1.46
0230	Edge cap, 8′ L					2.05			2.05	2.26
0240	Finish end cap, 8′ L					3.18			3.18	3.50

09 77 Special Wall Surfacing

09 77 43 – Panel Systems

09 77 43.20 Slatwall Panels and Accessories		Crew	Daily Output	Labor-Hours	Unit	Material	2011 Bare Costs Labor	Equipment	Total	Total Incl O&P
0300	Display hook, metal, 4" L				Ea.	.44			.44	.48
0310	6" L					.50			.50	.55
0320	8" L					.54			.54	.59
0330	10" L					.58			.58	.64
0340	12" L					.63			.63	.69
0350	Acrylic, 4" L					.91			.91	1
0360	6" L					1.05			1.05	1.16
0370	8" L					1.10			1.10	1.21
0380	10" L					1.12			1.12	1.23
0400	Waterfall hanger, metal, 12" - 16"					4.52			4.52	4.97
0410	Acrylic					9.85			9.85	10.85
0500	Shelf bracket, metal, 8"					4.43			4.43	4.87
0510	10"					4.60			4.60	5.05
0520	12"					4.49			4.49	4.94
0530	14"					4.79			4.79	5.25
0540	16"					6.05			6.05	6.65
0550	Acrylic, 8"					3.04			3.04	3.34
0560	10"					3.37			3.37	3.71
0570	12"					3.72			3.72	4.09
0580	14"					3.99			3.99	4.39
0600	Shelf, acrylic, 12" x 16" x 1/4"					19.90			19.90	22
0610	12" x 24" x 1/4"				↓	30			30	33

09 81 Acoustic Insulation

09 81 16 – Acoustic Blanket Insulation

09 81 16.10 Sound Attenuation Blanket		Crew	Daily Output	Labor-Hours	Unit	Material	Labor	Equipment	Total	Total Incl O&P
0010	**SOUND ATTENUATION BLANKET**									
0020	Blanket, 1" thick	1 Carp	925	.009	S.F.	.25	.29		.54	.77
0500	1-1/2" thick		920	.009		.25	.29		.54	.77
1000	2" thick		915	.009		.34	.29		.63	.86
1500	3" thick	↓	910	.009	↓	.50	.29		.79	1.04

09 84 Acoustic Room Components

09 84 13 – Fixed Sound-Absorptive Panels

09 84 13.10 Fixed Panels		Crew	Daily Output	Labor-Hours	Unit	Material	Labor	Equipment	Total	Total Incl O&P
0010	**FIXED PANELS** Perforated steel facing, painted with									
0100	Fiberglass or mineral filler, no backs, 2-1/4" thick, modular									
0200	space units, ceiling or wall hung, white or colored	1 Carp	100	.080	S.F.	7.35	2.68		10.03	12.60
0300	Fiberboard sound deadening panels, 1/2" thick	"	600	.013	"	.32	.45		.77	1.10
0500	Fiberglass panels, 4' x 8' x 1" thick, with									
0600	glass cloth face for walls, cemented	1 Carp	155	.052	S.F.	6.50	1.73		8.23	10.05
0700	1-1/2" thick, dacron covered, inner aluminum frame,									
0710	wall mounted	1 Carp	300	.027	S.F.	8.75	.89		9.64	11.10

09 91 Painting

09 91 03 – Paint Restoration

09 91 03.20 Sanding		Crew	Daily Output	Labor-Hours	Unit	Material	2011 Bare Costs Labor	Equipment	Total	Total Incl O&P
0010	**SANDING** and puttying interior trim, compared to									
0100	Painting 1 coat, on quality work				L.F.		100%			
0300	Medium work						50%			
0400	Industrial grade				↓		25%			
0500	Surface protection, placement and removal									
0510	Basic drop cloths	1 Pord	6400	.001	S.F.		.04		.04	.06
0520	Masking with paper		800	.010		.05	.30		.35	.54
0530	Volume cover up (using plastic sheathing, or building paper)	↓	16000	.001	↓		.01		.01	.02

09 91 03.30 Exterior Surface Preparation		Crew	Daily Output	Labor-Hours	Unit	Material	Labor	Equipment	Total	Total Incl O&P
0010	**EXTERIOR SURFACE PREPARATION**									
0015	Doors, per side, not incl. frames or trim									
0020	Scrape & sand									
0030	Wood, flush	1 Pord	616	.013	S.F.		.39		.39	.63
0040	Wood, detail		496	.016			.48		.48	.78
0050	Wood, louvered		280	.029			.85		.85	1.38
0060	Wood, overhead	↓	616	.013	↓		.39		.39	.63
0070	Wire brush									
0080	Metal, flush	1 Pord	640	.013	S.F.		.37		.37	.60
0090	Metal, detail		520	.015			.46		.46	.74
0100	Metal, louvered		360	.022			.66		.66	1.07
0110	Metal or fibr., overhead		640	.013			.37		.37	.60
0120	Metal, roll up		560	.014			.42		.42	.69
0130	Metal, bulkhead	↓	640	.013	↓		.37		.37	.60
0140	Power wash, based on 2500 lb. operating pressure									
0150	Metal, flush	A-1H	2240	.004	S.F.		.09	.03	.12	.19
0160	Metal, detail		2120	.004			.10	.03	.13	.20
0170	Metal, louvered		2000	.004			.11	.03	.14	.21
0180	Metal or fibr., overhead		2400	.003			.09	.03	.12	.18
0190	Metal, roll up		2400	.003			.09	.03	.12	.18
0200	Metal, bulkhead	↓	2200	.004	↓		.10	.03	.13	.19
0400	Windows, per side, not incl. trim									
0410	Scrape & sand									
0420	Wood, 1-2 lite	1 Pord	320	.025	S.F.		.74		.74	1.20
0430	Wood, 3-6 lite		280	.029			.85		.85	1.38
0440	Wood, 7-10 lite		240	.033			.99		.99	1.60
0450	Wood, 12 lite		200	.040			1.19		1.19	1.93
0460	Wood, Bay/Bow	↓	320	.025	↓		.74		.74	1.20
0470	Wire brush									
0480	Metal, 1-2 lite	1 Pord	480	.017	S.F.		.49		.49	.80
0490	Metal, 3-6 lite		400	.020			.59		.59	.96
0500	Metal, Bay/Bow	↓	480	.017	↓		.49		.49	.80
0510	Power wash, based on 2500 lb. operating pressure									
0520	1-2 lite	A-1H	4400	.002	S.F.		.05	.01	.06	.10
0530	3-6 lite		4320	.002			.05	.01	.06	.10
0540	7-10 lite		4240	.002			.05	.01	.06	.10
0550	12 lite		4160	.002			.05	.01	.06	.11
0560	Bay/Bow	↓	4400	.002	↓		.05	.01	.06	.10
0600	Siding, scrape and sand, light=10-30%, med.=30-70%									
0610	Heavy=70-100% of surface to sand									
0650	Texture 1-11, light	1 Pord	480	.017	S.F.		.49		.49	.80
0660	Med.		440	.018			.54		.54	.88
0670	Heavy	↓	360	.022	↓		.66		.66	1.07

09 91 Painting

09 91 03 – Paint Restoration

09 91 03.30 Exterior Surface Preparation		Crew	Daily Output	Labor-Hours	Unit	Material	2011 Bare Costs Labor	Equipment	Total	Total Incl O&P
0680	Wood shingles, shakes, light	1 Pord	440	.018	S.F.		.54		.54	.88
0690	Med.		360	.022			.66		.66	1.07
0700	Heavy		280	.029			.85		.85	1.38
0710	Clapboard, light		520	.015			.46		.46	.74
0720	Med.		480	.017			.49		.49	.80
0730	Heavy		400	.020			.59		.59	.96
0740	Wire brush									
0750	Aluminum, light	1 Pord	600	.013	S.F.		.40		.40	.64
0760	Med.		520	.015			.46		.46	.74
0770	Heavy		440	.018			.54		.54	.88
0780	Pressure wash, based on 2500 lb. operating pressure									
0790	Stucco	A-1H	3080	.003	S.F.		.07	.02	.09	.14
0800	Aluminum or vinyl		3200	.003			.07	.02	.09	.13
0810	Siding, masonry, brick & block		2400	.003			.09	.03	.12	.18
1300	Miscellaneous, wire brush									
1310	Metal, pedestrian gate	1 Pord	100	.080	S.F.		2.37		2.37	3.85
8000	For chemical washing, see Section 04 01 30									
8010	For steam cleaning, see Section 04 01 30.20									
8020	For sand blasting, see Section 05 01 10.51 and 03 35 29.60									

09 91 03.40 Interior Surface Preparation		Crew	Daily Output	Labor-Hours	Unit	Material	Labor	Equipment	Total	Total Incl O&P
0010	**INTERIOR SURFACE PREPARATION**									
0020	Doors, per side, not incl. frames or trim									
0030	Scrape & sand									
0040	Wood, flush	1 Pord	616	.013	S.F.		.39		.39	.63
0050	Wood, detail		496	.016			.48		.48	.78
0060	Wood, louvered		280	.029			.85		.85	1.38
0070	Wire brush									
0080	Metal, flush	1 Pord	640	.013	S.F.		.37		.37	.60
0090	Metal, detail		520	.015			.46		.46	.74
0100	Metal, louvered		360	.022			.66		.66	1.07
0110	Hand wash									
0120	Wood, flush	1 Pord	2160	.004	S.F.		.11		.11	.18
0130	Wood, detailed		2000	.004			.12		.12	.19
0140	Wood, louvered		1360	.006			.17		.17	.28
0150	Metal, flush		2160	.004			.11		.11	.18
0160	Metal, detail		2000	.004			.12		.12	.19
0170	Metal, louvered		1360	.006			.17		.17	.28
0400	Windows, per side, not incl. trim									
0410	Scrape & sand									
0420	Wood, 1-2 lite	1 Pord	360	.022	S.F.		.66		.66	1.07
0430	Wood, 3-6 lite		320	.025			.74		.74	1.20
0440	Wood, 7-10 lite		280	.029			.85		.85	1.38
0450	Wood, 12 lite		240	.033			.99		.99	1.60
0460	Wood, Bay/Bow		360	.022			.66		.66	1.07
0470	Wire brush									
0480	Metal, 1-2 lite	1 Pord	520	.015	S.F.		.46		.46	.74
0490	Metal, 3-6 lite		440	.018			.54		.54	.88
0500	Metal, Bay/Bow		520	.015			.46		.46	.74
0600	Walls, sanding, light=10-30%, medium - 30-70%,									
0610	heavy=70-100% of surface to sand									
0650	Walls, sand									
0660	Gypsum board or plaster, light	1 Pord	3077	.003	S.F.		.08		.08	.13

09 91 Painting

09 91 03 – Paint Restoration

09 91 03.40 Interior Surface Preparation		Crew	Daily Output	Labor-Hours	Unit	Material	2011 Bare Costs Labor	Equipment	Total	Total Incl O&P
0670	Gypsum board or plaster, medium	1 Pord	2160	.004	S.F.		.11		.11	.18
0680	Gypsum board or plaster, heavy		923	.009			.26		.26	.42
0690	Wood, T&G, light		2400	.003			.10		.10	.16
0700	Wood, T&G, med.		1600	.005			.15		.15	.24
0710	Wood, T&G, heavy		800	.010			.30		.30	.48
0720	Walls, wash									
0730	Gypsum board or plaster	1 Pord	3200	.003	S.F.		.07		.07	.12
0740	Wood, T&G		3200	.003			.07		.07	.12
0750	Masonry, brick & block, smooth		2800	.003			.08		.08	.14
0760	Masonry, brick & block, coarse		2000	.004			.12		.12	.19
8000	For chemical washing, see Section 04 01 30									
8020	For sand blasting, see Section 03 35 29.60 and 05 01 10.51									

09 91 03.41 Scrape After Fire Damage		Crew	Daily Output	Labor-Hours	Unit	Material	Labor	Equipment	Total	Total Incl O&P
0010	**SCRAPE AFTER FIRE DAMAGE**									
0050	Boards, 1" x 4"	1 Pord	336	.024	L.F.		.71		.71	1.15
0060	1" x 6"		260	.031			.91		.91	1.48
0070	1" x 8"		207	.039			1.15		1.15	1.86
0080	1" x 10"		174	.046			1.36		1.36	2.21
0500	Framing, 2" x 4"		265	.030			.90		.90	1.45
0510	2" x 6"		221	.036			1.07		1.07	1.74
0520	2" x 8"		190	.042			1.25		1.25	2.03
0530	2" x 10"		165	.048			1.44		1.44	2.33
0540	2" x 12"		144	.056			1.65		1.65	2.68
1000	Heavy framing, 3" x 4"		226	.035			1.05		1.05	1.70
1010	4" x 4"		210	.038			1.13		1.13	1.83
1020	4" x 6"		191	.042			1.24		1.24	2.02
1030	4" x 8"		165	.048			1.44		1.44	2.33
1040	4" x 10"		144	.056			1.65		1.65	2.68
1060	4" x 12"		131	.061			1.81		1.81	2.94
2900	For sealing, minimum		825	.010	S.F.	.14	.29		.43	.62
2920	Maximum		460	.017	"	.29	.52		.81	1.16

09 91 13 – Exterior Painting

09 91 13.30 Fences		Crew	Daily Output	Labor-Hours	Unit	Material	Labor	Equipment	Total	Total Incl O&P
0010	**FENCES** R099100-20									
0100	Chain link or wire metal, one side, water base									
0110	Roll & brush, first coat	1 Pord	960	.008	S.F.	.06	.25		.31	.47
0120	Second coat		1280	.006		.06	.19		.25	.36
0130	Spray, first coat		2275	.004		.06	.10		.16	.24
0140	Second coat		2600	.003		.06	.09		.15	.22
0150	Picket, water base									
0160	Roll & brush, first coat	1 Pord	865	.009	S.F.	.06	.27		.33	.52
0170	Second coat		1050	.008		.06	.23		.29	.44
0180	Spray, first coat		2275	.004		.06	.10		.16	.24
0190	Second coat		2600	.003		.06	.09		.15	.22
0200	Stockade, water base									
0210	Roll & brush, first coat	1 Pord	1040	.008	S.F.	.06	.23		.29	.44
0220	Second coat		1200	.007		.06	.20		.26	.39
0230	Spray, first coat		2275	.004		.06	.10		.16	.24
0240	Second coat		2600	.003		.06	.09		.15	.22

09 91 Painting

09 91 13 – Exterior Painting

09 91 13.42 Miscellaneous, Exterior		Crew	Daily Output	Labor-Hours	Unit	Material	2011 Bare Costs Labor	Equipment	Total	Total Incl O&P
0010	**MISCELLANEOUS, EXTERIOR**									
0015	For painting metals, see Section 09 97 13.23									
0100	Railing, ext., decorative wood, incl. cap & baluster									
0110	Newels & spindles @ 12" O.C.									
0120	Brushwork, stain, sand, seal & varnish									
0130	First coat	1 Pord	90	.089	L.F.	.69	2.64		3.33	5.05
0140	Second coat	"	120	.067	"	.69	1.98		2.67	3.97
0150	Rough sawn wood, 42" high, 2" x 2" verticals, 6" O.C.									
0160	Brushwork, stain, each coat	1 Pord	90	.089	L.F.	.22	2.64		2.86	4.53
0170	Wrought iron, 1" rail, 1/2" sq. verticals									
0180	Brushwork, zinc chromate, 60" high, bars 6" O.C.									
0190	Primer	1 Pord	130	.062	L.F.	.59	1.82		2.41	3.61
0200	Finish coat		130	.062		.18	1.82		2	3.16
0210	Additional coat		190	.042		.21	1.25		1.46	2.26
0220	Shutters or blinds, single panel, 2' x 4', paint all sides									
0230	Brushwork, primer	1 Pord	20	.400	Ea.	.58	11.85		12.43	19.90
0240	Finish coat, exterior latex		20	.400		.49	11.85		12.34	19.80
0250	Primer & 1 coat, exterior latex		13	.615		.94	18.25		19.19	30.50
0260	Spray, primer		35	.229		.85	6.80		7.65	11.95
0270	Finish coat, exterior latex		35	.229		1.04	6.80		7.84	12.15
0280	Primer & 1 coat, exterior latex		20	.400		.91	11.85		12.76	20.50
0290	For louvered shutters, add				S.F.	10%				
0300	Stair stringers, exterior, metal									
0310	Roll & brush, zinc chromate, to 14", each coat	1 Pord	320	.025	L.F.	.06	.74		.80	1.27
0320	Rough sawn wood, 4" x 12"									
0330	Roll & brush, exterior latex, each coat	1 Pord	215	.037	L.F.	.07	1.10		1.17	1.87
0340	Trellis/lattice, 2" x 2" @ 3" O.C. with 2" x 8" supports									
0350	Spray, latex, per side, each coat	1 Pord	475	.017	S.F.	.07	.50		.57	.89
0450	Decking, ext., sealer, alkyd, brushwork, sealer coat		1140	.007		.09	.21		.30	.44
0460	1st coat		1140	.007		.09	.21		.30	.43
0470	2nd coat		1300	.006		.06	.18		.24	.37
0500	Paint, alkyd, brushwork, primer coat		1140	.007		.08	.21		.29	.42
0510	1st coat		1140	.007		.16	.21		.37	.51
0520	2nd coat		1300	.006		.12	.18		.30	.43
0600	Sand paint, alkyd, brushwork, 1 coat		150	.053		.12	1.58		1.70	2.70

09 91 13.60 Siding Exterior		Crew	Daily Output	Labor-Hours	Unit	Material	Labor	Equipment	Total	Total Incl O&P
0010	**SIDING EXTERIOR**, Alkyd (oil base)									
0450	Steel siding, oil base, paint 1 coat, brushwork	2 Pord	2015	.008	S.F.	.13	.24		.37	.52
0500	Spray		4550	.004		.19	.10		.29	.38
0800	Paint 2 coats, brushwork		1300	.012		.25	.37		.62	.87
1000	Spray		2750	.006		.17	.17		.34	.47
1200	Stucco, rough, oil base, paint 2 coats, brushwork		1300	.012		.25	.37		.62	.87
1400	Roller		1625	.010		.27	.29		.56	.76
1600	Spray		2925	.005		.28	.16		.44	.57
1800	Texture 1-11 or clapboard, oil base, primer coat, brushwork		1300	.012		.10	.37		.47	.70
2000	Spray		4550	.004		.10	.10		.20	.28
2100	Paint 1 coat, brushwork		1300	.012		.18	.37		.55	.79
2200	Spray		4550	.004		.18	.10		.28	.37
2400	Paint 2 coats, brushwork		810	.020		.37	.59		.96	1.35
2600	Spray		2600	.006		.41	.18		.59	.75
3000	Stain 1 coat, brushwork		1520	.011		.07	.31		.38	.59
3200	Spray		5320	.003		.08	.09		.17	.23

09 91 Painting

09 91 13 – Exterior Painting

09 91 13.60 Siding Exterior		Crew	Daily Output	Labor-Hours	Unit	Material	2011 Bare Costs Labor	Equipment	Total	Total Incl O&P
3400	Stain 2 coats, brushwork	2 Pord	950	.017	S.F.	.15	.50		.65	.97
4000	Spray		3050	.005		.17	.16		.33	.43
4200	Wood shingles, oil base primer coat, brushwork		1300	.012		.09	.37		.46	.69
4400	Spray		3900	.004		.09	.12		.21	.29
4600	Paint 1 coat, brushwork		1300	.012		.15	.37		.52	.76
4800	Spray		3900	.004		.19	.12		.31	.41
5000	Paint 2 coats, brushwork		810	.020		.31	.59		.90	1.29
5200	Spray		2275	.007		.29	.21		.50	.66
5800	Stain 1 coat, brushwork		1500	.011		.07	.32		.39	.59
6000	Spray		3900	.004		.07	.12		.19	.28
6500	Stain 2 coats, brushwork		950	.017		.15	.50		.65	.97
7000	Spray		2660	.006		.21	.18		.39	.52
8000	For latex paint, deduct					10%				
8100	For work over 12′ H, from pipe scaffolding, add						15%			
8200	For work over 12′ H, from extension ladder, add						25%			
8300	For work over 12′ H, from swing staging, add						35%			

09 91 13.62 Siding, Misc.		Crew	Daily Output	Labor-Hours	Unit	Material	Labor	Equipment	Total	Total Incl O&P
0010	**SIDING, MISC.**, latex paint									
0100	Aluminum siding									
0110	Brushwork, primer	2 Pord	2275	.007	S.F.	.05	.21		.26	.40
0120	Finish coat, exterior latex		2275	.007		.04	.21		.25	.39
0130	Primer & 1 coat exterior latex		1300	.012		.10	.37		.47	.70
0140	Primer & 2 coats exterior latex		975	.016		.14	.49		.63	.95
0150	Mineral fiber shingles									
0160	Brushwork, primer	2 Pord	1495	.011	S.F.	.10	.32		.42	.63
0170	Finish coat, industrial enamel		1495	.011		.15	.32		.47	.68
0180	Primer & 1 coat enamel		810	.020		.25	.59		.84	1.22
0190	Primer & 2 coats enamel		540	.030		.40	.88		1.28	1.86
0200	Roll, primer		1625	.010		.11	.29		.40	.59
0210	Finish coat, industrial enamel		1625	.010		.16	.29		.45	.65
0220	Primer & 1 coat enamel		975	.016		.27	.49		.76	1.09
0230	Primer & 2 coats enamel		650	.025		.43	.73		1.16	1.66
0240	Spray, primer		3900	.004		.09	.12		.21	.29
0250	Finish coat, industrial enamel		3900	.004		.13	.12		.25	.35
0260	Primer & 1 coat enamel		2275	.007		.22	.21		.43	.58
0270	Primer & 2 coats enamel		1625	.010		.35	.29		.64	.86
0280	Waterproof sealer, first coat		4485	.004		.07	.11		.18	.24
0290	Second coat		5235	.003		.06	.09		.15	.22
0300	Rough wood incl. shingles, shakes or rough sawn siding									
0310	Brushwork, primer	2 Pord	1280	.013	S.F.	.12	.37		.49	.73
0320	Finish coat, exterior latex		1280	.013		.07	.37		.44	.68
0330	Primer & 1 coat exterior latex		960	.017		.19	.49		.68	1.01
0340	Primer & 2 coats exterior latex		700	.023		.26	.68		.94	1.39
0350	Roll, primer		2925	.005		.16	.16		.32	.43
0360	Finish coat, exterior latex		2925	.005		.09	.16		.25	.36
0370	Primer & 1 coat exterior latex		1790	.009		.24	.27		.51	.70
0380	Primer & 2 coats exterior latex		1300	.012		.33	.37		.70	.95
0390	Spray, primer		3900	.004		.13	.12		.25	.35
0400	Finish coat, exterior latex		3900	.004		.07	.12		.19	.27
0410	Primer & 1 coat exterior latex		2600	.006		.20	.18		.38	.52
0420	Primer & 2 coats exterior latex		2080	.008		.27	.23		.50	.66
0430	Waterproof sealer, first coat		4485	.004		.12	.11		.23	.30

09 91 Painting

09 91 13 – Exterior Painting

09 91 13.62 Siding, Misc.		Crew	Daily Output	Labor-Hours	Unit	Material	2011 Bare Costs Labor	Equipment	Total	Total Incl O&P
0440	Second coat	2 Pord	4485	.004	S.F.	.07	.11		.18	.24
0450	Smooth wood incl. butt, T&G, beveled, drop or B&B siding									
0460	Brushwork, primer	2 Pord	2325	.007	S.F.	.08	.20		.28	.42
0470	Finish coat, exterior latex		1280	.013		.07	.37		.44	.68
0480	Primer & 1 coat exterior latex		800	.020		.16	.59		.75	1.13
0490	Primer & 2 coats exterior latex		630	.025		.23	.75		.98	1.47
0500	Roll, primer		2275	.007		.09	.21		.30	.44
0510	Finish coat, exterior latex		2275	.007		.08	.21		.29	.43
0520	Primer & 1 coat exterior latex		1300	.012		.17	.37		.54	.78
0530	Primer & 2 coats exterior latex		975	.016		.25	.49		.74	1.07
0540	Spray, primer		4550	.004		.07	.10		.17	.25
0550	Finish coat, exterior latex		4550	.004		.07	.10		.17	.24
0560	Primer & 1 coat exterior latex		2600	.006		.14	.18		.32	.45
0570	Primer & 2 coats exterior latex		1950	.008		.21	.24		.45	.63
0580	Waterproof sealer, first coat		5230	.003		.07	.09		.16	.22
0590	Second coat		5980	.003		.07	.08		.15	.20
0600	For oil base paint, add					10%				
09 91 13.70 Doors and Windows, Exterior										
0010	**DOORS AND WINDOWS, EXTERIOR** R099100-10									
0100	Door frames & trim, only									
0110	Brushwork, primer	1 Pord	512	.016	L.F.	.05	.46		.51	.81
0120	Finish coat, exterior latex		512	.016		.06	.46		.52	.82
0130	Primer & 1 coat, exterior latex		300	.027		.11	.79		.90	1.41
0135	2 coats, exterior latex, both sides		15	.533	Ea.	5.45	15.80		21.25	31.50
0140	Primer & 2 coats, exterior latex		265	.030	L.F.	.18	.90		1.08	1.64
0150	Doors, flush, both sides, incl. frame & trim									
0160	Roll & brush, primer	1 Pord	10	.800	Ea.	4.03	23.50		27.53	43
0170	Finish coat, exterior latex		10	.800		4.65	23.50		28.15	43.50
0180	Primer & 1 coat, exterior latex		7	1.143		8.70	34		42.70	64.50
0190	Primer & 2 coats, exterior latex		5	1.600		13.30	47.50		60.80	91.50
0200	Brushwork, stain, sealer & 2 coats polyurethane		4	2		24	59.50		83.50	123
0210	Doors, French, both sides, 10-15 lite, incl. frame & trim									
0220	Brushwork, primer	1 Pord	6	1.333	Ea.	2.02	39.50		41.52	66
0230	Finish coat, exterior latex		6	1.333		2.32	39.50		41.82	66.50
0240	Primer & 1 coat, exterior latex		3	2.667		4.34	79		83.34	133
0250	Primer & 2 coats, exterior latex		2	4		6.55	119		125.55	200
0260	Brushwork, stain, sealer & 2 coats polyurethane		2.50	3.200		8.70	95		103.70	164
0270	Doors, louvered, both sides, incl. frame & trim									
0280	Brushwork, primer	1 Pord	7	1.143	Ea.	4.03	34		38.03	59.50
0290	Finish coat, exterior latex		7	1.143		4.65	34		38.65	60
0300	Primer & 1 coat, exterior latex		4	2		8.70	59.50		68.20	106
0310	Primer & 2 coats, exterior latex		3	2.667		13.05	79		92.05	142
0320	Brushwork, stain, sealer & 2 coats polyurethane		4.50	1.778		24	52.50		76.50	112
0330	Doors, panel, both sides, incl. frame & trim									
0340	Roll & brush, primer	1 Pord	6	1.333	Ea.	4.03	39.50		43.53	68.50
0350	Finish coat, exterior latex		6	1.333		4.65	39.50		44.15	69
0360	Primer & 1 coat, exterior latex		3	2.667		8.70	79		87.70	138
0370	Primer & 2 coats, exterior latex		2.50	3.200		13.05	95		108.05	168
0380	Brushwork, stain, sealer & 2 coats polyurethane		3	2.667		24	79		103	154
0400	Windows, per ext. side, based on 15 S.F.									
0410	1 to 6 lite									
0420	Brushwork, primer	1 Pord	13	.615	Ea.	.80	18.25		19.05	30.50

09 91 Painting

09 91 13 – Exterior Painting

09 91 13.70 Doors and Windows, Exterior		Crew	Daily Output	Labor-Hours	Unit	2011 Bare Costs Material	Labor	Equipment	Total	Total Incl O&P
0430	Finish coat, exterior latex	1 Pord	13	.615	Ea.	.92	18.25		19.17	30.50
0440	Primer & 1 coat, exterior latex		8	1		1.71	29.50		31.21	50
0450	Primer & 2 coats, exterior latex		6	1.333		2.58	39.50		42.08	67
0460	Stain, sealer & 1 coat varnish		7	1.143		3.44	34		37.44	59
0470	7 to 10 lite									
0480	Brushwork, primer	1 Pord	11	.727	Ea.	.80	21.50		22.30	36
0490	Finish coat, exterior latex		11	.727		.92	21.50		22.42	36
0500	Primer & 1 coat, exterior latex		7	1.143		1.71	34		35.71	57
0510	Primer & 2 coats, exterior latex		5	1.600		2.58	47.50		50.08	80
0520	Stain, sealer & 1 coat varnish		6	1.333		3.44	39.50		42.94	68
0530	12 lite									
0540	Brushwork, primer	1 Pord	10	.800	Ea.	.80	23.50		24.30	39.50
0550	Finish coat, exterior latex		10	.800		.92	23.50		24.42	39.50
0560	Primer & 1 coat, exterior latex		6	1.333		1.71	39.50		41.21	66
0570	Primer & 2 coats, exterior latex		5	1.600		2.58	47.50		50.08	80
0580	Stain, sealer & 1 coat varnish		6	1.333		3.37	39.50		42.87	67.50
0590	For oil base paint, add					10%				

09 91 13.80 Trim, Exterior		Crew	Daily Output	Labor-Hours	Unit	Material	Labor	Equipment	Total	Total Incl O&P
0010	**TRIM, EXTERIOR**									
0100	Door frames & trim (see Doors, interior or exterior)									
0110	Fascia, latex paint, one coat coverage									
0120	1" x 4", brushwork	1 Pord	640	.013	L.F.	.02	.37		.39	.62
0130	Roll		1280	.006		.02	.19		.21	.32
0140	Spray		2080	.004		.02	.11		.13	.21
0150	1" x 6" to 1" x 10", brushwork		640	.013		.06	.37		.43	.67
0160	Roll		1230	.007		.07	.19		.26	.39
0170	Spray		2100	.004		.05	.11		.16	.24
0180	1" x 12", brushwork		640	.013		.06	.37		.43	.67
0190	Roll		1050	.008		.07	.23		.30	.45
0200	Spray		2200	.004		.05	.11		.16	.24
0210	Gutters & downspouts, metal, zinc chromate paint									
0220	Brushwork, gutters, 5", first coat	1 Pord	640	.013	L.F.	.06	.37		.43	.67
0230	Second coat		960	.008		.06	.25		.31	.47
0240	Third coat		1280	.006		.05	.19		.24	.35
0250	Downspouts, 4", first coat		640	.013		.06	.37		.43	.67
0260	Second coat		960	.008		.06	.25		.31	.47
0270	Third coat		1280	.006		.05	.19		.24	.35
0280	Gutters & downspouts, wood									
0290	Brushwork, gutters, 5", primer	1 Pord	640	.013	L.F.	.05	.37		.42	.66
0300	Finish coat, exterior latex		640	.013		.05	.37		.42	.66
0310	Primer & 1 coat exterior latex		400	.020		.11	.59		.70	1.09
0320	Primer & 2 coats exterior latex		325	.025		.18	.73		.91	1.38
0330	Downspouts, 4", primer		640	.013		.05	.37		.42	.66
0340	Finish coat, exterior latex		640	.013		.05	.37		.42	.66
0350	Primer & 1 coat exterior latex		400	.020		.11	.59		.70	1.09
0360	Primer & 2 coats exterior latex		325	.025		.09	.73		.82	1.29
0370	Molding, exterior, up to 14" wide									
0380	Brushwork, primer	1 Pord	640	.013	L.F.	.06	.37		.43	.67
0390	Finish coat, exterior latex		640	.013		.07	.37		.44	.67
0400	Primer & 1 coat exterior latex		400	.020		.14	.59		.73	1.11
0410	Primer & 2 coats exterior latex		315	.025		.14	.75		.89	1.37
0420	Stain & fill		1050	.008		.09	.23		.32	.47

09 91 Painting

09 91 13 – Exterior Painting

09 91 13.80 Trim, Exterior		Crew	Daily Output	Labor-Hours	Unit	Material	2011 Bare Costs Labor	Equipment	Total	Total Incl O&P
0430	Shellac	1 Pord	1850	.004	L.F.	.09	.13		.22	.31
0440	Varnish	↓	1275	.006	↓	.10	.19		.29	.41

09 91 13.90 Walls, Masonry (CMU), Exterior		Crew	Daily Output	Labor-Hours	Unit	Material	Labor	Equipment	Total	Total Incl O&P
0350	**WALLS, MASONRY (CMU), EXTERIOR**									
0360	Concrete masonry units (CMU), smooth surface									
0370	Brushwork, latex, first coat	1 Pord	640	.013	S.F.	.06	.37		.43	.66
0380	Second coat		960	.008		.05	.25		.30	.45
0390	Waterproof sealer, first coat		736	.011		.25	.32		.57	.79
0400	Second coat		1104	.007		.25	.22		.47	.62
0410	Roll, latex, paint, first coat		1465	.005		.07	.16		.23	.34
0420	Second coat		1790	.004		.05	.13		.18	.28
0430	Waterproof sealer, first coat		1680	.005		.25	.14		.39	.50
0440	Second coat		2060	.004		.25	.12		.37	.46
0450	Spray, latex, paint, first coat		1950	.004		.05	.12		.17	.26
0460	Second coat		2600	.003		.04	.09		.13	.20
0470	Waterproof sealer, first coat		2245	.004		.25	.11		.36	.44
0480	Second coat	↓	2990	.003	↓	.25	.08		.33	.40
0490	Concrete masonry unit (CMU), porous									
0500	Brushwork, latex, first coat	1 Pord	640	.013	S.F.	.12	.37		.49	.73
0510	Second coat		960	.008		.06	.25		.31	.46
0520	Waterproof sealer, first coat		736	.011		.25	.32		.57	.79
0530	Second coat		1104	.007		.25	.22		.47	.62
0540	Roll latex, first coat		1465	.005		.09	.16		.25	.36
0550	Second coat		1790	.004		.06	.13		.19	.28
0560	Waterproof sealer, first coat		1680	.005		.25	.14		.39	.50
0570	Second coat		2060	.004		.25	.12		.37	.46
0580	Spray latex, first coat		1950	.004		.06	.12		.18	.27
0590	Second coat		2600	.003		.04	.09		.13	.20
0600	Waterproof sealer, first coat		2245	.004		.25	.11		.36	.44
0610	Second coat	↓	2990	.003	↓	.25	.08		.33	.40

09 91 23 – Interior Painting

09 91 23.20 Cabinets and Casework		Crew	Daily Output	Labor-Hours	Unit	Material	Labor	Equipment	Total	Total Incl O&P
0010	**CABINETS AND CASEWORK**									
1000	Primer coat, oil base, brushwork	1 Pord	650	.012	S.F.	.06	.37		.43	.66
2000	Paint, oil base, brushwork, 1 coat		650	.012		.08	.37		.45	.68
2500	2 coats		400	.020		.16	.59		.75	1.14
3000	Stain, brushwork, wipe off		650	.012		.07	.37		.44	.67
4000	Shellac, 1 coat, brushwork		650	.012		.07	.37		.44	.67
4500	Varnish, 3 coats, brushwork, sand after 1st coat	↓	325	.025		.24	.73		.97	1.45
5000	For latex paint, deduct				↓	10%				

09 91 23.33 Doors and Windows, Interior Alkyd (Oil Base)		Crew	Daily Output	Labor-Hours	Unit	Material	Labor	Equipment	Total	Total Incl O&P
0010	**DOORS AND WINDOWS, INTERIOR ALKYD (OIL BASE)**									
0500	Flush door & frame, 3' x 7', oil, primer, brushwork	1 Pord	10	.800	Ea.	3.04	23.50		26.54	42
1000	Paint, 1 coat		10	.800		3.62	23.50		27.12	42.50
1200	2 coats		6	1.333		4.35	39.50		43.85	69
1400	Stain, brushwork, wipe off		18	.444		1.57	13.20		14.77	23
1600	Shellac, 1 coat, brushwork		25	.320		1.54	9.50		11.04	17.10
1800	Varnish, 3 coats, brushwork, sand after 1st coat		9	.889		5	26.50		31.50	48.50
2000	Panel door & frame, 3' x 7', oil, primer, brushwork		6	1.333		2.31	39.50		41.81	66.50
2200	Paint, 1 coat		6	1.333		3.62	39.50		43.12	68
2400	2 coats		3	2.667		9.55	79		88.55	139
2600	Stain, brushwork, panel door, 3' x 7', not incl. frame	↓	16	.500	↓	1.57	14.85		16.42	25.50

09 91 Painting

09 91 23 – Interior Painting

09 91 23.33 Doors and Windows, Interior Alkyd (Oil Base)		Crew	Daily Output	Labor-Hours	Unit	2011 Bare Costs Material	Labor	Equipment	Total	Total Incl O&P
2800	Shellac, 1 coat, brushwork	1 Pord	22	.364	Ea.	1.54	10.80		12.34	19.20
3000	Varnish, 3 coats, brushwork, sand after 1st coat		7.50	1.067		5	31.50		36.50	57
3020	French door, incl. 3′ x 7′, 6 lites, frame & trim									
3022	Paint, 1 coat, over existing paint	1 Pord	5	1.600	Ea.	7.25	47.50		54.75	85
3024	2 coats, over existing paint		5	1.600		14.05	47.50		61.55	92.50
3026	Primer & 1 coat		3.50	2.286		11.85	68		79.85	123
3028	Primer & 2 coats		3	2.667		19.10	79		98.10	149
3032	Varnish or polyurethane, 1 coat		5	1.600		6.40	47.50		53.90	84
3034	2 coats, sanding between		3	2.667		12.80	79		91.80	142
4400	Windows, including frame and trim, per side									
4600	Colonial type, 6/6 lites, 2′ x 3′, oil, primer, brushwork	1 Pord	14	.571	Ea.	.36	16.95		17.31	28
5800	Paint, 1 coat		14	.571		.57	16.95		17.52	28
6000	2 coats		9	.889		1.11	26.50		27.61	44
6200	3′ x 5′ opening, 6/6 lites, primer coat, brushwork		12	.667		.91	19.75		20.66	33
6400	Paint, 1 coat		12	.667		1.43	19.75		21.18	33.50
6600	2 coats		7	1.143		2.78	34		36.78	58
6800	4′ x 8′ opening, 6/6 lites, primer coat, brushwork		8	1		1.95	29.50		31.45	50
7000	Paint, 1 coat		8	1		3.05	29.50		32.55	51.50
7200	2 coats		5	1.600		5.90	47.50		53.40	83.50
8000	Single lite type, 2′ x 3′, oil base, primer coat, brushwork		33	.242		.36	7.20		7.56	12.05
8200	Paint, 1 coat		33	.242		.57	7.20		7.77	12.30
8400	2 coats		20	.400		1.11	11.85		12.96	20.50
8600	3′ x 5′ opening, primer coat, brushwork		20	.400		.91	11.85		12.76	20.50
8800	Paint, 1 coat		20	.400		1.43	11.85		13.28	21
8900	2 coats		13	.615		2.78	18.25		21.03	32.50
9200	4′ x 8′ opening, primer coat, brushwork		14	.571		1.95	16.95		18.90	29.50
9400	Paint, 1 coat		14	.571		3.05	16.95		20	31
9600	2 coats		8	1		5.90	29.50		35.40	54.50

09 91 23.35 Doors and Windows, Interior Latex		Crew	Daily Output	Labor-Hours	Unit	Material	Labor	Equipment	Total	Total Incl O&P
0010	**DOORS & WINDOWS, INTERIOR LATEX**									
0100	Doors, flush, both sides, incl. frame & trim									
0110	Roll & brush, primer	1 Pord	10	.800	Ea.	3.73	23.50		27.23	42.50
0120	Finish coat, latex		10	.800		4.13	23.50		27.63	43
0130	Primer & 1 coat latex		7	1.143		7.85	34		41.85	63.50
0140	Primer & 2 coats latex		5	1.600		11.75	47.50		59.25	90
0160	Spray, both sides, primer		20	.400		3.93	11.85		15.78	23.50
0170	Finish coat, latex		20	.400		4.33	11.85		16.18	24
0180	Primer & 1 coat latex		11	.727		8.30	21.50		29.80	44
0190	Primer & 2 coats latex		8	1		12.45	29.50		41.95	61.50
0200	Doors, French, both sides, 10-15 lite, incl. frame & trim									
0210	Roll & brush, primer	1 Pord	6	1.333	Ea.	1.87	39.50		41.37	66
0220	Finish coat, latex		6	1.333		2.06	39.50		41.56	66.50
0230	Primer & 1 coat latex		3	2.667		3.93	79		82.93	132
0240	Primer & 2 coats latex		2	4		5.90	119		124.90	199
0260	Doors, louvered, both sides, incl. frame & trim									
0270	Roll & brush, primer	1 Pord	7	1.143	Ea.	3.73	34		37.73	59
0280	Finish coat, latex		7	1.143		4.13	34		38.13	59.50
0290	Primer & 1 coat, latex		4	2		7.65	59.50		67.15	105
0300	Primer & 2 coats, latex		3	2.667		12	79		91	141
0320	Spray, both sides, primer		20	.400		3.93	11.85		15.78	23.50
0330	Finish coat, latex		20	.400		4.33	11.85		16.18	24
0340	Primer & 1 coat, latex		11	.727		8.30	21.50		29.80	44

09 91 Painting

09 91 23 – Interior Painting

09 91 23.35 Doors and Windows, Interior Latex		Crew	Daily Output	Labor-Hours	Unit	2011 Bare Costs Material	Labor	Equipment	Total	Total Incl O&P
0350	Primer & 2 coats, latex	1 Pord	8	1	Ea.	12.70	29.50		42.20	62
0360	Doors, panel, both sides, incl. frame & trim									
0370	Roll & brush, primer	1 Pord	6	1.333	Ea.	3.93	39.50		43.43	68.50
0380	Finish coat, latex		6	1.333		4.13	39.50		43.63	68.50
0390	Primer & 1 coat, latex		3	2.667		7.85	79		86.85	137
0400	Primer & 2 coats, latex		2.50	3.200		12	95		107	167
0420	Spray, both sides, primer		10	.800		3.93	23.50		27.43	43
0430	Finish coat, latex		10	.800		4.33	23.50		27.83	43.50
0440	Primer & 1 coat, latex		5	1.600		8.30	47.50		55.80	86
0450	Primer & 2 coats, latex		4	2		12.70	59.50		72.20	110
0460	Windows, per interior side, based on 15 S.F.									
0470	1 to 6 lite									
0480	Brushwork, primer	1 Pord	13	.615	Ea.	.74	18.25		18.99	30.50
0490	Finish coat, enamel		13	.615		.81	18.25		19.06	30.50
0500	Primer & 1 coat enamel		8	1		1.55	29.50		31.05	49.50
0510	Primer & 2 coats enamel		6	1.333		2.37	39.50		41.87	66.50
0530	7 to 10 lite									
0540	Brushwork, primer	1 Pord	11	.727	Ea.	.74	21.50		22.24	36
0550	Finish coat, enamel		11	.727		.81	21.50		22.31	36
0560	Primer & 1 coat enamel		7	1.143		1.55	34		35.55	56.50
0570	Primer & 2 coats enamel		5	1.600		2.37	47.50		49.87	79.50
0590	12 lite									
0600	Brushwork, primer	1 Pord	10	.800	Ea.	.74	23.50		24.24	39.50
0610	Finish coat, enamel		10	.800		.81	23.50		24.31	39.50
0620	Primer & 1 coat enamel		6	1.333		1.55	39.50		41.05	65.50
0630	Primer & 2 coats enamel		5	1.600		2.37	47.50		49.87	79.50
0650	For oil base paint, add					10%				

09 91 23.39 Doors and Windows, Interior Latex, Zero Voc		Crew	Daily Output	Labor-Hours	Unit	Material	Labor	Equipment	Total	Total Incl O&P
0010	DOORS & WINDOWS, INTERIOR LATEX, ZERO VOC									
0100	Doors flush, both sides, incl. frame & trim									
0110	Roll & brush, primer G	1 Pord	10	.800	Ea.	5	23.50		28.50	44
0120	Finish coat, latex G		10	.800		5.25	23.50		28.75	44.50
0130	Primer & 1 coat latex G		7	1.143		10.25	34		44.25	66.50
0140	Primer & 2 coats latex G		5	1.600		15.20	47.50		62.70	94
0160	Spray, both sides, primer G		20	.400		5.25	11.85		17.10	25
0170	Finish coat, latex G		20	.400		5.50	11.85		17.35	25.50
0180	Primer & 1 coat latex G		11	.727		10.85	21.50		32.35	47
0190	Primer & 2 coats latex G		8	1		16.10	29.50		45.60	65.50
0200	Doors, French, both sides, 10-15 lite, incl. frame & trim									
0210	Roll & brush, primer G	1 Pord	6	1.333	Ea.	2.50	39.50		42	67
0220	Finish coat, latex G		6	1.333		2.63	39.50		42.13	67
0230	Primer & 1 coat latex G		3	2.667		5.15	79		84.15	134
0240	Primer & 2 coats latex G		2	4		7.60	119		126.60	201
0360	Doors, panel, both sides, incl. frame & trim									
0370	Roll & brush, primer G	1 Pord	6	1.333	Ea.	5.25	39.50		44.75	70
0380	Finish coat, latex G		6	1.333		5.25	39.50		44.75	70
0390	Primer & 1 coat, latex G		3	2.667		10.25	79		89.25	139
0400	Primer & 2 coats, latex G		2.50	3.200		15.50	95		110.50	171
0420	Spray, both sides, primer G		10	.800		5.25	23.50		28.75	44.50
0430	Finish coat, latex G		10	.800		5.50	23.50		29	44.50
0440	Primer & 1 coat, latex G		5	1.600		10.85	47.50		58.35	89
0450	Primer & 2 coats, latex G		4	2		16.45	59.50		75.95	115

09 91 Painting

09 91 23 – Interior Painting

09 91 23.39 Doors and Windows, Interior Latex, Zero Voc			Crew	Daily Output	Labor-Hours	Unit	2011 Bare Costs Material	Labor	Equipment	Total	Total Incl O&P
0460	Windows, per interior side, based on 15 S.F.										
0470	1 to 6 lite										
0480	Brushwork, primer	G	1 Pord	13	.615	Ea.	.99	18.25		19.24	30.50
0490	Finish coat, enamel	G		13	.615		1.04	18.25		19.29	30.50
0500	Primer & 1 coat enamel	G		8	1		2.02	29.50		31.52	50
0510	Primer & 2 coats enamel	G		6	1.333		3.06	39.50		42.56	67.50

09 91 23.40 Floors, Interior		Crew	Daily Output	Labor-Hours	Unit	Material	Labor	Equipment	Total	Total Incl O&P
0010	**FLOORS, INTERIOR**									
0100	Concrete paint, latex									
0110	Brushwork									
0120	1st coat	1 Pord	975	.008	S.F.	.15	.24		.39	.57
0130	2nd coat		1150	.007		.10	.21		.31	.45
0140	3rd coat		1300	.006		.08	.18		.26	.39
0150	Roll									
0160	1st coat	1 Pord	2600	.003	S.F.	.20	.09		.29	.37
0170	2nd coat		3250	.002		.12	.07		.19	.25
0180	3rd coat		3900	.002		.09	.06		.15	.20
0190	Spray									
0200	1st coat	1 Pord	2600	.003	S.F.	.17	.09		.26	.34
0210	2nd coat		3250	.002		.09	.07		.16	.22
0220	3rd coat		3900	.002		.08	.06		.14	.18
0300	Acid stain and sealer									
0310	Stain, one coat	1 Pord	650	.012	S.F.	.11	.37		.48	.71
0320	Two coats		570	.014		.22	.42		.64	.92
0330	Acrylic sealer, one coat		2600	.003		.18	.09		.27	.35
0340	Two coats		1400	.006		.36	.17		.53	.66

09 91 23.52 Miscellaneous, Interior		Crew	Daily Output	Labor-Hours	Unit	Material	Labor	Equipment	Total	Total Incl O&P
0010	**MISCELLANEOUS, INTERIOR**									
2400	Floors, conc./wood, oil base, primer/sealer coat, brushwork	2 Pord	1950	.008	S.F.	.06	.24		.30	.47
2450	Roller		5200	.003		.06	.09		.15	.22
2600	Spray		6000	.003		.06	.08		.14	.20
2650	Paint 1 coat, brushwork		1950	.008		.10	.24		.34	.51
2800	Roller		5200	.003		.10	.09		.19	.26
2850	Spray		6000	.003		.11	.08		.19	.25
3000	Stain, wood floor, brushwork, 1 coat		4550	.004		.07	.10		.17	.25
3200	Roller		5200	.003		.08	.09		.17	.24
3250	Spray		6000	.003		.08	.08		.16	.22
3400	Varnish, wood floor, brushwork		4550	.004		.08	.10		.18	.26
3450	Roller		5200	.003		.08	.09		.17	.24
3600	Spray		6000	.003		.09	.08		.17	.23
3800	Grilles, per side, oil base, primer coat, brushwork	1 Pord	520	.015		.12	.46		.58	.87
3850	Spray		1140	.007		.13	.21		.34	.48
3880	Paint 1 coat, brushwork		520	.015		.19	.46		.65	.95
3900	Spray		1140	.007		.21	.21		.42	.57
3920	Paint 2 coats, brushwork		325	.025		.37	.73		1.10	1.60
3940	Spray		650	.012		.42	.37		.79	1.06
4500	Louvers, one side, primer, brushwork		524	.015		.06	.45		.51	.81
4520	Paint one coat, brushwork		520	.015		.09	.46		.55	.84
4530	Spray		1140	.007		.10	.21		.31	.45
4540	Paint two coats, brushwork		325	.025		.17	.73		.90	1.38
4550	Spray		650	.012		.19	.37		.56	.80
4560	Paint three coats, brushwork		270	.030		.25	.88		1.13	1.71

09 91 Painting

09 91 23 – Interior Painting

09 91 23.52 Miscellaneous, Interior		Crew	Daily Output	Labor-Hours	Unit	Material	2011 Bare Costs Labor	Equipment	Total	Total Incl O&P
4570	Spray	1 Pord	500	.016	S.F.	.28	.47		.75	1.08
5000	Pipe, 1" - 4" diameter, primer or sealer coat, oil base, brushwork	2 Pord	1250	.013	L.F.	.06	.38		.44	.69
5100	Spray		2165	.007		.06	.22		.28	.43
5200	Paint 1 coat, brushwork		1250	.013		.10	.38		.48	.73
5300	Spray		2165	.007		.09	.22		.31	.46
5350	Paint 2 coats, brushwork		775	.021		.18	.61		.79	1.19
5400	Spray		1240	.013		.20	.38		.58	.84
5450	5" - 8" diameter, primer or sealer coat, brushwork		620	.026		.13	.77		.90	1.38
5500	Spray		1085	.015		.21	.44		.65	.94
5550	Paint 1 coat, brushwork		620	.026		.27	.77		1.04	1.54
5600	Spray		1085	.015		.30	.44		.74	1.04
5650	Paint 2 coats, brushwork		385	.042		.36	1.23		1.59	2.39
5700	Spray		620	.026		.40	.77		1.17	1.68
5750	9" - 12" diameter, primer or sealer coat, brushwork		415	.039		.19	1.14		1.33	2.07
5800	Spray		725	.022		.33	.65		.98	1.42
5850	Paint 1 coat, brushwork		415	.039		.28	1.14		1.42	2.16
6000	Spray		725	.022		.31	.65		.96	1.40
6200	Paint 2 coats, brushwork		260	.062		.54	1.82		2.36	3.55
6250	Spray		415	.039		.60	1.14		1.74	2.52
6300	13" - 16" diameter, primer or sealer coat, brushwork		310	.052		.26	1.53		1.79	2.78
6350	Spray		540	.030		.29	.88		1.17	1.75
6400	Paint 1 coat, brushwork		310	.052		.37	1.53		1.90	2.90
6450	Spray		540	.030		.41	.88		1.29	1.88
6500	Paint 2 coats, brushwork		195	.082		.72	2.43		3.15	4.74
6550	Spray		310	.052		.80	1.53		2.33	3.37
6600	Radiators, per side, primer, brushwork	1 Pord	520	.015	S.F.	.06	.46		.52	.81
6620	Paint, one coat		520	.015		.06	.46		.52	.80
6640	Two coats		340	.024		.17	.70		.87	1.32
6660	Three coats		283	.028		.25	.84		1.09	1.64
7000	Trim, wood, incl. puttying, under 6" wide									
7200	Primer coat, oil base, brushwork	1 Pord	650	.012	L.F.	.03	.37		.40	.62
7250	Paint, 1 coat, brushwork		650	.012		.05	.37		.42	.64
7400	2 coats		400	.020		.09	.59		.68	1.06
7450	3 coats		325	.025		.14	.73		.87	1.34
7500	Over 6" wide, primer coat, brushwork		650	.012		.06	.37		.43	.66
7550	Paint, 1 coat, brushwork		650	.012		.10	.37		.47	.69
7600	2 coats		400	.020		.19	.59		.78	1.16
7650	3 coats		325	.025		.27	.73		1	1.49
8000	Cornice, simple design, primer coat, oil base, brushwork		650	.012	S.F.	.06	.37		.43	.66
8250	Paint, 1 coat		650	.012		.10	.37		.47	.69
8300	2 coats		400	.020		.19	.59		.78	1.16
8350	Ornate design, primer coat		350	.023		.06	.68		.74	1.17
8400	Paint, 1 coat		350	.023		.10	.68		.78	1.20
8450	2 coats		400	.020		.19	.59		.78	1.16
8600	Balustrades, primer coat, oil base, brushwork		520	.015		.06	.46		.52	.81
8650	Paint, 1 coat		520	.015		.10	.46		.56	.84
8700	2 coats		325	.025		.19	.73		.92	1.39
8900	Trusses and wood frames, primer coat, oil base, brushwork		800	.010		.06	.30		.36	.55
8950	Spray		1200	.007		.06	.20		.26	.39
9000	Paint 1 coat, brushwork		750	.011		.10	.32		.42	.61
9200	Spray		1200	.007		.11	.20		.31	.44
9220	Paint 2 coats, brushwork		500	.016		.19	.47		.66	.97
9240	Spray		600	.013		.21	.40		.61	.87

09 91 Painting

09 91 23 – Interior Painting

09 91 23.52 Miscellaneous, Interior

Code	Description	Crew	Daily Output	Labor-Hours	Unit	2011 Bare Costs Material	Labor	Equipment	Total	Total Incl O&P
9260	Stain, brushwork, wipe off	1 Pord	600	.013	S.F.	.07	.40		.47	.72
9280	Varnish, 3 coats, brushwork		275	.029		.24	.86		1.10	1.66
9350	For latex paint, deduct					10%				

09 91 23.62 Electrostatic Painting

Code	Description	Crew	Daily Output	Labor-Hours	Unit	2011 Bare Costs Material	Labor	Equipment	Total	Total Incl O&P
0010	**ELECTROSTATIC PAINTING**									
0100	In Shop									
0200	Flat Surfaces (lockers, casework, elev doors. etc)									
0300	One coat	1 Pord	200	.040	S.F.	.46	1.19		1.65	2.44
0400	Two coats	"	120	.067	"	.67	1.98		2.65	3.95
0500	Irregular Surfaces (furniture, door frames, etc)									
0600	One coat	1 Pord	150	.053	S.F.	.46	1.58		2.04	3.08
0700	Two coats	"	100	.080	"	.67	2.37		3.04	4.59
0800	On Site									
0900	Flat Surfaces (lockers, casework, elev doors. etc)									
1000	One coat	1 Pord	150	.053	S.F.	.46	1.58		2.04	3.08
1100	Two coats	"	100	.080	"	.67	2.37		3.04	4.59
1200	Irregular Surfaces (furniture, door frames, etc)									
1300	One coat	1 Pord	115	.070	S.F.	.46	2.06		2.52	3.86
1400	Two coats	"	70	.114	"	.67	3.39		4.06	6.25

09 91 23.72 Walls and Ceilings, Interior

Code	Description	Crew	Daily Output	Labor-Hours	Unit	2011 Bare Costs Material	Labor	Equipment	Total	Total Incl O&P
0010	**WALLS AND CEILINGS, INTERIOR**									
0100	Concrete, drywall or plaster, latex, primer or sealer coat									
0200	Smooth finish, brushwork	1 Pord	1150	.007	S.F.	.06	.21		.27	.40
0240	Roller		1350	.006		.06	.18		.24	.35
0280	Spray		2750	.003		.05	.09		.14	.19
0300	Sand finish, brushwork		975	.008		.06	.24		.30	.46
0340	Roller		1150	.007		.06	.21		.27	.40
0380	Spray		2275	.004		.05	.10		.15	.22
0400	Paint 1 coat, smooth finish, brushwork		1200	.007		.10	.20		.30	.43
0440	Roller		1300	.006		.10	.18		.28	.41
0480	Spray		2275	.004		.09	.10		.19	.27
0500	Sand finish, brushwork		1050	.008		.10	.23		.33	.48
0540	Roller		1600	.005		.10	.15		.25	.35
0580	Spray		2100	.004		.09	.11		.20	.28
0800	Paint 2 coats, smooth finish, brushwork		680	.012		.11	.35		.46	.69
0840	Roller		800	.010		.11	.30		.41	.60
0880	Spray		1625	.005		.10	.15		.25	.35
0900	Sand finish, brushwork		605	.013		.11	.39		.50	.76
0940	Roller		1020	.008		.11	.23		.34	.50
0980	Spray		1700	.005		.10	.14		.24	.34
1200	Paint 3 coats, smooth finish, brushwork		510	.016		.16	.47		.63	.94
1240	Roller		650	.012		.16	.37		.53	.77
1280	Spray		1625	.005		.15	.15		.30	.41
1300	Sand finish, brushwork		454	.018		.29	.52		.81	1.17
1340	Roller		680	.012		.31	.35		.66	.91
1380	Spray		1133	.007		.26	.21		.47	.63
1600	Glaze coating, 2 coats, spray, clear		1200	.007		.49	.20		.69	.86
1640	Multicolor		1200	.007		.95	.20		1.15	1.37
1660	Painting walls, complete, including surface prep, primer &									
1670	2 coats finish, on drywall or plaster, with roller	1 Pord	325	.025	S.F.	.17	.73		.90	1.38
1700	For oil base paint, add					10%				
1800	For ceiling installations, add						25%			

09 91 Painting

09 91 23 – Interior Painting

09 91 23.72 Walls and Ceilings, Interior		Crew	Daily Output	Labor-Hours	Unit	Material	2011 Bare Costs Labor	Equipment	Total	Total Incl O&P
2000	Masonry or concrete block, primer/sealer, latex paint									
2100	Primer, smooth finish, brushwork	1 Pord	1000	.008	S.F.	.10	.24		.34	.50
2110	Roller		1150	.007		.10	.21		.31	.45
2180	Spray		2400	.003		.09	.10		.19	.26
2200	Sand finish, brushwork		850	.009		.10	.28		.38	.56
2210	Roller		975	.008		.10	.24		.34	.51
2280	Spray		2050	.004		.09	.12		.21	.29
2400	Finish coat, smooth finish, brush		1100	.007		.08	.22		.30	.43
2410	Roller		1300	.006		.08	.18		.26	.38
2480	Spray		2400	.003		.07	.10		.17	.23
2500	Sand finish, brushwork		950	.008		.08	.25		.33	.49
2510	Roller		1090	.007		.08	.22		.30	.43
2580	Spray		2040	.004		.07	.12		.19	.26
2800	Primer plus one finish coat, smooth brush		525	.015		.27	.45		.72	1.03
2810	Roller		615	.013		.17	.39		.56	.82
2880	Spray		1200	.007		.15	.20		.35	.49
2900	Sand finish, brushwork		450	.018		.17	.53		.70	1.05
2910	Roller		515	.016		.17	.46		.63	.94
2980	Spray		1025	.008		.15	.23		.38	.55
3200	Primer plus 2 finish coats, smooth, brush		355	.023		.25	.67		.92	1.37
3210	Roller		415	.019		.25	.57		.82	1.21
3280	Spray		800	.010		.22	.30		.52	.72
3300	Sand finish, brushwork		305	.026		.25	.78		1.03	1.54
3301	Sand finish, brushwork, ° 600 S.F. per loss site		440	.018		.30	.54		.84	1.21
3310	Roller		350	.023		.25	.68		.93	1.38
3380	Spray		675	.012		.22	.35		.57	.81
3600	Glaze coating, 3 coats, spray, clear		900	.009		.70	.26		.96	1.20
3620	Multicolor		900	.009		1.10	.26		1.36	1.64
4000	Block filler, 1 coat, brushwork		425	.019		.13	.56		.69	1.06
4100	Silicone, water repellent, 2 coats, spray		2000	.004		.26	.12		.38	.47
4120	For oil base paint, add					10%				
8200	For work 8' - 15' H, add						10%			
8300	For work over 15' H, add						20%			
8400	For light textured surfaces, add						10%			
8410	Heavy textured, add						25%			

09 91 23.74 Walls and Ceilings, Interior, Zero VOC Latex		Crew	Daily Output	Labor-Hours	Unit	Material	Labor	Equipment	Total	Total Incl O&P
0010	**WALLS AND CEILINGS, INTERIOR, ZERO VOC LATEX**									
0100	Concrete, dry wall or plaster, latex, primer or sealer coat									
0200	Smooth finish, brushwork G	1 Pord	1150	.007	S.F.	.07	.21		.28	.41
0240	Roller G		1350	.006		.06	.18		.24	.36
0280	Spray G		2750	.003		.05	.09		.14	.19
0300	Sand finish, brushwork G		975	.008		.06	.24		.30	.47
0340	Roller G		1150	.007		.07	.21		.28	.42
0380	Spray G		2275	.004		.05	.10		.15	.23
0400	Paint 1 coat, smooth finish, brushwork G		1200	.007		.06	.20		.26	.39
0440	Roller G		1300	.006		.06	.18		.24	.37
0480	Spray G		2275	.004		.06	.10		.16	.23
0500	Sand finish, brushwork G		1050	.008		.06	.23		.29	.44
0540	Roller G		1600	.005		.06	.15		.21	.31
0580	Spray G		2100	.004		.06	.11		.17	.24
0800	Paint 2 coats, smooth finish, brushwork G		680	.012		.13	.35		.48	.71
0840	Roller G		800	.010		.13	.30		.43	.62

09 91 Painting

09 91 23 – Interior Painting

09 91 23.74 Walls and Ceilings, Interior, Zero VOC Latex		Crew	Daily Output	Labor-Hours	Unit	Material	2011 Bare Costs Labor	Equipment	Total	Total Incl O&P
0880	Spray G	1 Pord	1625	.005	S.F.	.11	.15		.26	.36
0900	Sand finish, brushwork G		605	.013		.12	.39		.51	.78
0940	Roller G		1020	.008		.13	.23		.36	.52
0980	Spray G		1700	.005		.11	.14		.25	.35
1200	Paint 3 coats, smooth finish, brushwork G		510	.016		.18	.47		.65	.96
1240	Roller G		650	.012		.19	.37		.56	.80
1280	Spray G	↓	1625	.005		.17	.15		.32	.42
1800	For ceiling installations, add G						25%			
8200	For work 8' - 15' H, add						10%			
8300	For work over 15' H, add				↓		20%			

09 91 23.75 Dry Fall Painting		Crew	Daily Output	Labor-Hours	Unit	Material	Labor	Equipment	Total	Total Incl O&P
0010	**DRY FALL PAINTING** R099100-20									
0100	Sprayed on walls, gypsum board or plaster									
0220	One coat	1 Pord	2600	.003	S.F.	.05	.09		.14	.20
0250	Two coats		1560	.005		.10	.15		.25	.36
0280	Concrete or textured plaster, one coat		1560	.005		.05	.15		.20	.30
0310	Two coats		1300	.006		.10	.18		.28	.41
0340	Concrete block, one coat		1560	.005		.05	.15		.20	.30
0370	Two coats		1300	.006		.10	.18		.28	.41
0400	Wood, one coat		877	.009		.05	.27		.32	.49
0430	Two coats	↓	650	.012	↓	.10	.37		.47	.70
0440	On ceilings, gypsum board or plaster									
0470	One coat	1 Pord	1560	.005	S.F.	.05	.15		.20	.30
0500	Two coats		1300	.006		.10	.18		.28	.41
0530	Concrete or textured plaster, one coat		1560	.005		.05	.15		.20	.30
0560	Two coats		1300	.006		.10	.18		.28	.41
0570	Structural steel, bar joists or metal deck, one coat		1560	.005		.05	.15		.20	.30
0580	Two coats	↓	1040	.008	↓	.10	.23		.33	.48

09 93 Staining and Transparent Finishing

09 93 23 – Interior Staining and Finishing

09 93 23.10 Varnish		Crew	Daily Output	Labor-Hours	Unit	Material	Labor	Equipment	Total	Total Incl O&P
0010	**VARNISH**									
0012	1 coat + sealer, on wood trim, brush, no sanding included	1 Pord	400	.020	S.F.	.07	.59		.66	1.04
0100	Hardwood floors, 2 coats, no sanding included, roller	"	1890	.004	"	.15	.13		.28	.37

09 96 High-Performance Coatings

09 96 23 – Graffiti-Resistant Coatings

09 96 23.10 Graffiti Resistant Treatments		Crew	Daily Output	Labor-Hours	Unit	Material	Labor	Equipment	Total	Total Incl O&P
0010	**GRAFFITI RESISTANT TREATMENTS**, sprayed on walls									
0100	Non-sacrificial, permanent non-stick coating, clear, on metals	1 Pord	2000	.004	S.F.	2.06	.12		2.18	2.46
0200	Concrete		2000	.004		2.35	.12		2.47	2.77
0300	Concrete block		2000	.004		3.03	.12		3.15	3.53
0400	Brick		2000	.004		3.44	.12		3.56	3.97
0500	Stone		2000	.004		3.44	.12		3.56	3.97
0600	Unpainted wood		2000	.004		3.97	.12		4.09	4.56
2000	Semi-permanent cross linking polymer primer, on metals		2000	.004		.40	.12		.52	.63
2100	Concrete		2000	.004		.48	.12		.60	.72
2200	Concrete block	↓	2000	.004	↓	.60	.12		.72	.85

09 96 High-Performance Coatings

09 96 23 – Graffiti-Resistant Coatings

09 96 23.10 Graffiti Resistant Treatments		Crew	Daily Output	Labor-Hours	Unit	Material	2011 Bare Costs Labor	Equipment	Total	Total Incl O&P
2300	Brick	1 Pord	2000	.004	S.F.	.48	.12		.60	.72
2400	Stone		2000	.004		.48	.12		.60	.72
2500	Unpainted wood		2000	.004		.67	.12		.79	.92
3000	Top coat, on metals		2000	.004		.43	.12		.55	.66
3100	Concrete		2000	.004		.49	.12		.61	.72
3200	Concrete block		2000	.004		.68	.12		.80	.94
3300	Brick		2000	.004		.57	.12		.69	.81
3400	Stone		2000	.004		.57	.12		.69	.81
3500	Unpainted wood		2000	.004		.68	.12		.80	.94
5000	Sacrificial, water based, on metal		2000	.004		.24	.12		.36	.46
5100	Concrete		2000	.004		.24	.12		.36	.46
5200	Concrete block		2000	.004		.24	.12		.36	.46
5300	Brick		2000	.004		.24	.12		.36	.46
5400	Stone		2000	.004		.24	.12		.36	.46
5500	Unpainted wood	↓	2000	.004	↓	.24	.12		.36	.46
8000	Cleaner for use after treatment									
8100	Towels or wipes, per package of 30				Ea.	.63			.63	.70
8200	Aerosol spray, 24 oz. can				"	18			18	19.80

09 96 56 – Epoxy Coatings

09 96 56.20 Wall Coatings		Crew	Daily Output	Labor-Hours	Unit	Material	2011 Bare Costs Labor	Equipment	Total	Total Incl O&P
0010	**WALL COATINGS**									
0100	Acrylic glazed coatings, minimum	1 Pord	525	.015	S.F.	.29	.45		.74	1.05
0200	Maximum		305	.026		.62	.78		1.40	1.94
0300	Epoxy coatings, minimum		525	.015		.38	.45		.83	1.15
0400	Maximum		170	.047		1.15	1.40		2.55	3.54
0600	Exposed aggregate, troweled on, 1/16" to 1/4", minimum		235	.034		.58	1.01		1.59	2.28
0700	Maximum (epoxy or polyacrylate)		130	.062		1.24	1.82		3.06	4.32
0900	1/2" to 5/8" aggregate, minimum		130	.062		1.14	1.82		2.96	4.21
1000	Maximum		80	.100		1.96	2.97		4.93	7
1200	1" aggregate size, minimum		90	.089		1.99	2.64		4.63	6.45
1300	Maximum		55	.145		3.04	4.31		7.35	10.35
1500	Exposed aggregate, sprayed on, 1/8" aggregate, minimum		295	.027		.53	.80		1.33	1.89
1600	Maximum		145	.055		.99	1.64		2.63	3.75
1800	High build epoxy, 50 mil, minimum		390	.021		.64	.61		1.25	1.69
1900	Maximum		95	.084		1.09	2.50		3.59	5.25
2100	Laminated epoxy with fiberglass, minimum		295	.027		.69	.80		1.49	2.07
2200	Maximum		145	.055		1.23	1.64		2.87	4.01
2400	Sprayed perlite or vermiculite, 1/16" thick, minimum		2935	.003		.25	.08		.33	.41
2500	Maximum		640	.013		.70	.37		1.07	1.37
2700	Vinyl plastic wall coating, minimum		735	.011		.31	.32		.63	.86
2800	Maximum		240	.033		.77	.99		1.76	2.45
3000	Urethane on smooth surface, 2 coats, minimum		1135	.007		.25	.21		.46	.62
3100	Maximum		665	.012		.55	.36		.91	1.19
3300	3 coat, minimum		840	.010		.33	.28		.61	.82
3400	Maximum		470	.017		.74	.50		1.24	1.63
3600	Ceramic-like glazed coating, cementitious, minimum		440	.018		.45	.54		.99	1.38
3700	Maximum		345	.023		.76	.69		1.45	1.96
3900	Resin base, minimum		640	.013		.31	.37		.68	.94
4000	Maximum	↓	330	.024	↓	.50	.72		1.22	1.72

09 97 Special Coatings

09 97 13 – Steel Coatings

09 97 13.23 Exterior Steel Coatings		Crew	Daily Output	Labor-Hours	Unit	2011 Bare Costs Material	Labor	Equipment	Total	Total Incl O&P
0010	**EXTERIOR STEEL COATINGS**									
6101	Cold galvanizing, brush in field	1 Pord	1100	.007	S.F.	.07	.22		.29	.42
6510	Paints & protective coatings, sprayed in field									
6520	Alkyds, primer	2 Psst	3600	.004	S.F.	.06	.14		.20	.33
6540	Gloss topcoats		3200	.005		.07	.15		.22	.37
6560	Silicone alkyd		3200	.005		.12	.15		.27	.42
6610	Epoxy, primer		3000	.005		.24	.16		.40	.57
6630	Intermediate or topcoat		2800	.006		.21	.17		.38	.58
6650	Enamel coat		2800	.006		.27	.17		.44	.63
6700	Epoxy ester, primer		2800	.006		.52	.17		.69	.91
6720	Topcoats		2800	.006		.18	.17		.35	.54
6810	Latex primer		3600	.004		.05	.14		.19	.32
6830	Topcoats		3200	.005		.06	.15		.21	.35
6910	Universal primers, one part, phenolic, modified alkyd		2000	.008		.11	.24		.35	.59
6940	Two part, epoxy spray		2000	.008		.28	.24		.52	.78
7000	Zinc rich primers, self cure, spray, inorganic		1800	.009		.59	.27		.86	1.17
7010	Epoxy, spray, organic	↓	1800	.009	↓	.21	.27		.48	.76
7020	Above one story, spray painting simple structures, add						25%			
7030	Intricate structures, add						50%			

Estimating Tips

31 05 00 Common Work Results for Earthwork

- Estimating the actual cost of performing earthwork requires careful consideration of the variables involved. This includes items such as type of soil, whether water will be encountered, dewatering, whether banks need bracing, disposal of excavated earth, and length of haul to fill or spoil sites, etc. If the project has large quantities of cut or fill, consider raising or lowering the site to reduce costs, while paying close attention to the effect on site drainage and utilities.
- If the project has large quantities of fill, creating a borrow pit on the site can significantly lower the costs.
- It is very important to consider what time of year the project is scheduled for completion. Bad weather can create large cost overruns from dewatering, site repair, and lost productivity from cold weather.
- New lines have been added for marine piling.

Reference Numbers

Reference numbers are shown in shaded boxes at the beginning of some major classifications. These numbers refer to related items in the Reference Section. The reference information may be an estimating procedure, an alternate pricing method, or technical information.

Note: Not all subdivisions listed here necessarily appear in this publication.

31 05 Common Work Results for Earthwork

31 05 13 – Soils for Earthwork

31 05 13.10 Borrow		Crew	Daily Output	Labor-Hours	Unit	2011 Bare Costs Material	Labor	Equipment	Total	Total Incl O&P
0010	**BORROW**									
0020	Spread, 200 H.P. dozer, no compaction, 2 mi. RT haul									
0200	Common borrow	B-15	600	.047	C.Y.	15	1.37	3.44	19.81	22.50

31 05 16 – Aggregates for Earthwork

31 05 16.10 Borrow		Crew	Daily Output	Labor-Hours	Unit	Material	Labor	Equipment	Total	Total Incl O&P
0010	**BORROW**									
0020	Spread, with 200 H.P. dozer, no compaction, 2 mi. RT haul									
0100	Bank run gravel	B-15	600	.047	C.Y.	22	1.37	3.44	26.81	30
0300	Crushed stone (1.40 tons per CY), 1-1/2"		600	.047		38	1.37	3.44	42.81	48
0320	3/4"		600	.047		38	1.37	3.44	42.81	48
0340	1/2"		600	.047		47	1.37	3.44	51.81	58
0360	3/8"		600	.047		32	1.37	3.44	36.81	41.50
0400	Sand, washed, concrete		600	.047		41	1.37	3.44	45.81	51
0500	Dead or bank sand	↓	600	.047	↓	15	1.37	3.44	19.81	22.50

31 11 Clearing and Grubbing

31 11 10 – Clearing and Grubbing Land

31 11 10.10 Clear and Grub Site		Crew	Daily Output	Labor-Hours	Unit	Material	Labor	Equipment	Total	Total Incl O&P
0010	**CLEAR AND GRUB SITE**									
0020	Cut & chip light trees to 6" diam.	B-7	1	48	Acre		1,350	1,300	2,650	3,700
0150	Grub stumps and remove	B-30	2	12			355	980	1,335	1,675
0200	Cut & chip medium, trees to 12" diam.	B-7	.70	68.571			1,925	1,875	3,800	5,275
0250	Grub stumps and remove	B-30	1	24			715	1,950	2,665	3,325
0300	Cut & chip heavy, trees to 24" diam.	B-7	.30	160			4,500	4,350	8,850	12,300
0350	Grub stumps and remove	B-30	.50	48			1,425	3,925	5,350	6,650
0400	If burning is allowed, reduce cut & chip				↓					40%

31 13 Selective Tree and Shrub Removal and Trimming

31 13 13 – Selective Tree and Shrub Removal

31 13 13.10 Selective Clearing		Crew	Daily Output	Labor-Hours	Unit	Material	Labor	Equipment	Total	Total Incl O&P
0010	**SELECTIVE CLEARING**									
0020	Clearing brush with brush saw	A-1C	.25	32	Acre		845	103	948	1,550
0100	By hand	1 Clab	.12	66.667			1,775		1,775	2,950
0300	With dozer, ball and chain, light clearing	B-11A	2	8			245	525	770	985
0400	Medium clearing	"	1.50	10.667	↓		325	700	1,025	1,300

31 22 Grading

31 22 16 – Fine Grading

31 22 16.10 Finish Grading		Crew	Daily Output	Labor-Hours	Unit	Material	Labor	Equipment	Total	Total Incl O&P
0010	**FINISH GRADING**									
0012	Finish grading area to be paved with grader, small area	B-11L	400	.040	S.Y.		1.23	1.36	2.59	3.52
1050	For small irregular areas	B-32C	2000	.024			.74	.91	1.65	2.23
1100	Fine grade for slab on grade, machine	B-11L	1040	.015	↓		.47	.52	.99	1.35
1160	Hand grading	1 Clab	500	.016	S.F.		.42		.42	.71
3300	Finishing grading slopes, gentle	B-11L	8900	.002	S.Y.		.06	.06	.12	.16
3500	Finish grading lagoon bottoms	"	4	4	M.S.F.		123	136	259	350

31 23 Excavation and Fill

31 23 16 – Excavation

31 23 16.13 Excavating, Trench

	31 23 16.13 Excavating, Trench	Crew	Daily Output	Labor-Hours	Unit	Material	2011 Bare Costs Labor	Equipment	Total	Total Incl O&P
0010	**EXCAVATING, TRENCH**									
0011	Or continuous footing									
0050	1' to 4' deep, 3/8 C.Y. excavator	B-11C	150	.107	B.C.Y.		3.27	2.14	5.41	7.75
0060	1/2 C.Y. excavator	B-11M	200	.080			2.45	1.75	4.20	5.95
0090	4' to 6' deep, 1/2 C.Y. excavator	"	200	.080			2.45	1.75	4.20	5.95
0100	5/8 C.Y. excavator	B-12Q	250	.064			1.99	2.12	4.11	5.60
0300	1/2 C.Y. excavator, truck mounted	B-12J	200	.080			2.49	4.29	6.78	8.85
1352	4' to 6' deep, 1/2 C.Y. excavator w/trench box	B-13H	188	.085			2.65	5.15	7.80	10.05
1354	5/8 C.Y. excavator	"	235	.068			2.12	4.13	6.25	8.05
1400	By hand with pick and shovel 2' to 6' deep, light soil	1 Clab	8	1			26.50		26.50	44.50
1500	Heavy soil	"	4	2			53		53	88.50
5050	1' to 4' deep, 3/8 C.Y. tractor loader/backhoe	B-11C	162	.099			3.03	1.98	5.01	7.20
5060	1/2 C.Y. excavator	B-11M	216	.074			2.27	1.62	3.89	5.55
5080	4' to 6' deep, 1/2 C.Y. excavator	"	216	.074			2.27	1.62	3.89	5.55
5090	5/8 C.Y. excavator	B-12Q	276	.058			1.80	1.92	3.72	5.10
5130	1/2 C.Y. excavator, truck mounted	B-12J	216	.074			2.31	3.97	6.28	8.20
5352	4' to 6' deep, 1/2 C.Y. excavator w/trench box	B-13H	205	.078			2.43	4.74	7.17	9.20
5354	5/8 C.Y. excavator	"	257	.062			1.94	3.78	5.72	7.35
6050	1' to 4' deep, 3/8 C.Y. excavator	B-11C	165	.097			2.98	1.94	4.92	7.05
6060	1/2 C.Y. excavator	B-11M	220	.073			2.23	1.59	3.82	5.45
6080	4' to 6' deep, 1/2 C.Y. excavator	"	220	.073			2.23	1.59	3.82	5.45
6090	5/8 C.Y. excavator	B-12Q	275	.058			1.81	1.93	3.74	5.10
6130	1/2 C.Y. excavator, truck mounted	B-12J	220	.073			2.26	3.90	6.16	8.05
6352	4' to 6' deep, 1/2 C.Y. excavator w/trench box	B-13H	209	.077			2.38	4.65	7.03	9.05
6354	5/8 C.Y. excavator	"	261	.061			1.91	3.72	5.63	7.25
7050	1' to 4' deep, 3/8 C.Y. excavator	B-11C	132	.121			3.72	2.43	6.15	8.80
7060	1/2 C.Y. excavator	B-11M	176	.091			2.79	1.99	4.78	6.80
7080	4' to 6' deep, 1/2 C.Y. excavator	"	176	.091			2.79	1.99	4.78	6.80
7090	5/8 C.Y. excavator	B-12Q	220	.073			2.26	2.41	4.67	6.40
7130	1/2 C.Y. excavator, truck mounted	B-12J	176	.091			2.83	4.87	7.70	10

31 23 16.14 Excavating, Utility Trench

	31 23 16.14 Excavating, Utility Trench	Crew	Daily Output	Labor-Hours	Unit	Material	Labor	Equipment	Total	Total Incl O&P
0010	**EXCAVATING, UTILITY TRENCH**									
0011	Common earth									
0050	Trenching with chain trencher, 12 H.P., operator walking									
0100	4" wide trench, 12" deep	B-53	800	.010	L.F.		.26	.07	.33	.52
1000	Backfill by hand including compaction, add									
1050	4" wide trench, 12" deep	A-1G	800	.010	L.F.		.26	.06	.32	.51

31 23 16.16 Structural Excavation for Minor Structures

	31 23 16.16 Structural Excavation for Minor Structures	Crew	Daily Output	Labor-Hours	Unit	Material	Labor	Equipment	Total	Total Incl O&P
0010	**STRUCTURAL EXCAVATION FOR MINOR STRUCTURES**									
0015	Hand, pits to 6' deep, sandy soil	1 Clab	8	1	B.C.Y.		26.50		26.50	44.50
0100	Heavy soil or clay		4	2			53		53	88.50
1100	Hand loading trucks from stock pile, sandy soil		12	.667			17.65		17.65	29.50
1300	Heavy soil or clay		8	1			26.50		26.50	44.50
1500	For wet or muck hand excavation, add to above				%				50%	50%

31 23 16.42 Excavating, Bulk Bank Measure

	31 23 16.42 Excavating, Bulk Bank Measure	Crew	Daily Output	Labor-Hours	Unit	Material	Labor	Equipment	Total	Total Incl O&P
0010	**EXCAVATING, BULK BANK MEASURE** R312316-40									
0011	Common earth piled									
0020	For loading onto trucks, add								15%	15%
0200	Excavator, hydraulic, crawler mtd., 1 C.Y. cap. = 100 C.Y./hr.	B-12A	800	.020	B.C.Y.		.62	.91	1.53	2.03
0310	Wheel mounted, 1/2 C.Y. cap. = 40 C.Y./hr.	B-12E	320	.050			1.56	1.14	2.70	3.82
1200	Front end loader, track mtd., 1-1/2 C.Y. cap. = 70 C.Y./hr.	B-10N	560	.014			.50	.83	1.33	1.74
1500	Wheel mounted, 3/4 C.Y. cap. = 45 C.Y./hr.	B-10R	360	.022			.78	.66	1.44	1.99

31 23 Excavation and Fill

31 23 16 – Excavation

31 23 16.42 Excavating, Bulk Bank Measure		Crew	Daily Output	Labor-Hours	Unit	Material	2011 Bare Costs Labor	Equipment	Total	Total Incl O&P
5000	Excavating, bulk bank measure, sandy clay & loam piled									
5020	For loading onto trucks, add								15%	15%
5100	Excavator, hydraulic, crawler mtd., 1 C.Y. cap. = 120 C.Y./hr.	B-12A	960	.017	B.C.Y.		.52	.76	1.28	1.70
5610	Wheel mounted, 1/2 C.Y. cap. = 44 C.Y./hr.	B-12E	352	.045	"		1.41	1.04	2.45	3.48
8000	For hauling excavated material, see Section 31 23 23.20									

31 23 19 – Dewatering

31 23 19.20 Dewatering Systems

Code	Description	Crew	Daily Output	Labor-Hours	Unit	Material	Labor	Equipment	Total	Total Incl O&P
0010	**DEWATERING SYSTEMS**									
0020	Excavate drainage trench, 2′ wide, 2′ deep	B-11C	90	.178	C.Y.		5.45	3.56	9.01	12.90
0100	2′ wide, 3′ deep, with backhoe loader	"	135	.119	↓		3.64	2.37	6.01	8.60
0200	Excavate sump pits by hand, light soil	1 Clab	7.10	1.127	↓		30		30	50
0500	Pumping 8 hr., attended 2 hrs. per day, including 20 L.F.									
0550	of suction hose & 100 L.F. discharge hose									
0600	2″ diaphragm pump used for 8 hours	B-10H	4	2	Day		70	17.60	87.60	133
0650	4″ diaphragm pump used for 8 hours	B-10I	4	2	"		70	29.50	99.50	147
1600	Sump hole construction, incl. excavation and gravel, pit	B-6	1250	.019	C.F.	.99	.55	.26	1.80	2.29
1700	With 12″ gravel collar, 12″ pipe, corrugated, 16 ga.	"	70	.343	L.F.	23	9.90	4.58	37.48	47

31 23 23 – Fill

31 23 23.13 Backfill

Code	Description	Crew	Daily Output	Labor-Hours	Unit	Material	Labor	Equipment	Total	Total Incl O&P
0010	**BACKFILL**									
0015	By hand, no compaction, light soil	1 Clab	14	.571	L.C.Y.		15.10		15.10	25.50
0100	Heavy soil	↓	11	.727	"		19.25		19.25	32
0300	Compaction in 6″ layers, hand tamp, add to above	↓	20.60	.388	E.C.Y.		10.25		10.25	17.20
0500	Air tamp, add	B-9D	190	.211	↓		5.65	1.20	6.85	10.75
0600	Vibrating plate, add	A-1D	60	.133	↓		3.53	.51	4.04	6.45
0800	Compaction in 12″ layers, hand tamp, add to above	1 Clab	34	.235	↓		6.20		6.20	10.40
1300	Dozer backfilling, bulk, up to 300′ haul, no compaction	B-10B	1200	.007	L.C.Y.		.23	.88	1.11	1.34
1400	Air tamped, add	B-11B	80	.200	E.C.Y.		6	3.11	9.11	13.30

31 23 23.16 Fill By Borrow and Utility Bedding

Code	Description	Crew	Daily Output	Labor-Hours	Unit	Material	Labor	Equipment	Total	Total Incl O&P
0010	**FILL BY BORROW AND UTILITY BEDDING**									
0049	Utility bedding, for pipe & conduit, not incl. compaction									
0050	Crushed or screened bank run gravel	B-6	150	.160	L.C.Y.	25	4.61	2.14	31.75	37.50
0100	Crushed stone 3/4″ to 1/2″	↓	150	.160	↓	38	4.61	2.14	44.75	52
0200	Sand, dead or bank	↓	150	.160	↓	15	4.61	2.14	21.75	26.50
0500	Compacting bedding in trench	A-1D	90	.089	E.C.Y.		2.35	.34	2.69	4.32
0600	If material source exceeds 2 miles, add for extra mileage.									

31 23 23.17 General Fill

Code	Description	Crew	Daily Output	Labor-Hours	Unit	Material	Labor	Equipment	Total	Total Incl O&P
0010	**GENERAL FILL**									
0011	Spread dumped material, no compaction									
0020	By dozer, no compaction	B-10B	1000	.008	L.C.Y.		.28	1.05	1.33	1.62
0100	By hand	1 Clab	12	.667	"		17.65		17.65	29.50
0500	Gravel fill, compacted, under floor slabs, 4″ deep	B-37	10000	.005	S.F.	.40	.13	.01	.54	.68
0600	6″ deep	↓	8600	.006	↓	.60	.16	.02	.78	.94
0700	9″ deep	↓	7200	.007	↓	1	.19	.02	1.21	1.43
0800	12″ deep	↓	6000	.008	↓	1.40	.22	.02	1.64	1.94
1000	Alternate pricing method, 4″ deep	↓	120	.400	E.C.Y.	30	11.20	1.21	42.41	53
1100	6″ deep	↓	160	.300	↓	30	8.40	.91	39.31	48
1200	9″ deep	↓	200	.240	↓	30	6.70	.73	37.43	45
1300	12″ deep	↓	220	.218	↓	30	6.10	.66	36.76	44

Chapter 7

Assemblies Estimating

Assemblies (or systems) estimating provides a fast and reasonably accurate way to develop construction costs. During the design phases of a project many decisions must be made in the process of developing the preliminary design into actual building components. Actual building layout is developed, the structural system defined and its components sized, and the materials and building systems are selected. Every decision has financial impact. Using assemblies cost data allows the estimator to evaluate the impact of each decision on the total project budget.

The assemblies approach to preliminary cost estimating involves grouping several trades and/or work items and tasks into a building element. An assembly is the grouping of unit price line items, with appropriate quantities, to form a building component such as an exterior wall, partition, roof, or footing system. For example, a foundation wall usually requires formwork, reinforcing, concrete, concrete placement, and finish. A unit price estimate of the foundation wall would require the estimator to evaluate and compile five separate line items, each with its own quantity and unit, to arrive at a final cost for the wall. In an assemblies estimate, the separate line items are combined in a single package. The assembly cost of a foundation wall, then, is represented as one line item that includes the cost of materials and installation (labor plus equipment) applied to one unit/quantity.

Assemblies data is organized differently than the unit price estimates discussed earlier in this book. Because these estimates are typically performed during the schematic and design development phases of the project, the materials-based MasterFormat organizational format doesn't work well. The UNIFORMAT II structure, based on major building elements such as walls, flooring, or HVAC systems, is more appropriate for this work. In UNIFORMAT II, data is organized by where it occurs in the building, in the general order of construction. Hence, Element A is Substructure and contains information on foundations. Element B is Superstructure, and covers the framing, exterior closure, and roofing. Element C includes interior finish, and so on through Element G, Building Site work.

Assemblies estimates do not require a completed design or detailed drawings. Instead, they are based on the general size of the structure and other known parameters of the project. The degree of accuracy of an assemblies estimate is generally +/- 10%.

Chapter 7 Exercises

Using the plans for the medical office found at **https://rsmeansonline.com/academic**, calculate the cost of the following items using RSMeans assemblies data wherever possible.

1. Excavation for the foundation, 8′ deep. Sandy soil.
2. Foundation: strip and spread footings, piers, foundation wall and slab on grade.

3. Floor Framing: 11-7/8″ TJI @ 16″ on center. No RSMeans assembly so use 2 × 12 @ 12″ on center instead. R30 fiberglass batt insulation.
4. Exterior wall: 2 x 6 @ 16″ on center w/R21 batt and beveled wood siding.
5. Roof framing, insulation and shingles.
6. Metal roof at dormers.
7. Exterior doors and windows.
8. Partitions, 2 x 4 wood with skim coat plaster.
9. Ceilings.
10. Create an assemblies estimate summary for this project. Add general requirements at 10%, CG overhead at 5%, and GC profit at 10%. Add an architect fee of 9%. Add contingency at 15%. Adjust the location to Plymouth, Massachusetts 02360.

Assemblies Section

Table of Contents

How to Use the Assemblies Cost Tables

The following is a detailed explanation of a sample Assemblies Cost Table. Most Assembly Tables are separated into three parts: 1) an illustration of the system to be estimated; 2) the components and related costs of a typical system; and 3) the costs for similar systems with dimensional and/or size variations. For costs of the components that comprise these systems, or assemblies, refer to the Unit Price Section. Next to each bold number below is the described item with the appropriate component of the sample entry following in parentheses. In most cases, if the work is to be subcontracted, the general contractor will need to add an additional markup (RSMeans suggests using 10%) to the "Total" figures.

1 System/Line Numbers (C1010 124 2400)

Each Assemblies Cost Line has been assigned a unique identification number based on the UNIFORMAT II classification system.

UNIFORMAT II Major Group

C1010 124 2400

UNIFORMAT II Level 3

Means Major Classification

Means Individual Line Number

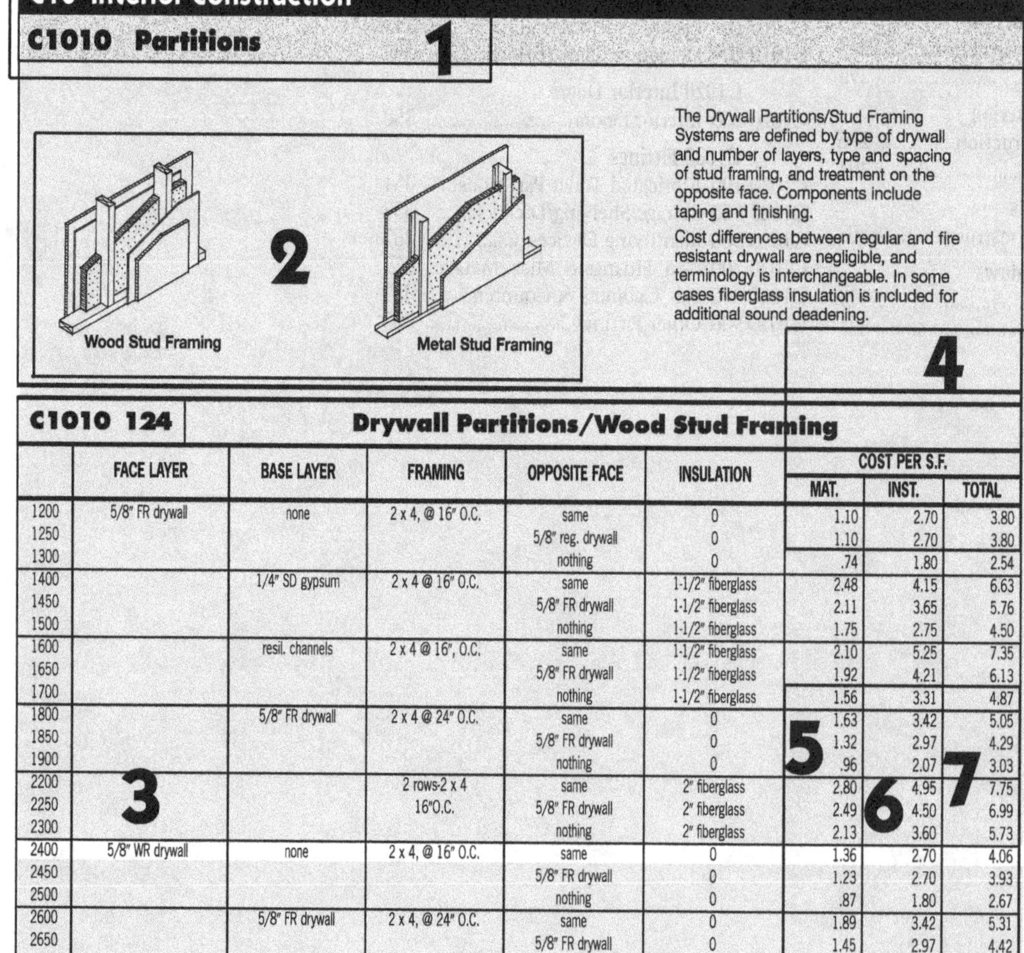

C10 Interior Construction

C1010 Partitions 1

The Drywall Partitions/Stud Framing Systems are defined by type of drywall and number of layers, type and spacing of stud framing, and treatment on the opposite face. Components include taping and finishing.

Cost differences between regular and fire resistant drywall are negligible, and terminology is interchangeable. In some cases fiberglass insulation is included for additional sound deadening.

4

C1010 124 Drywall Partitions/Wood Stud Framing

	FACE LAYER	BASE LAYER	FRAMING	OPPOSITE FACE	INSULATION	COST PER S.F. MAT.	INST.	TOTAL
1200	5/8" FR drywall	none	2 x 4, @ 16" O.C.	same	0	1.10	2.70	3.80
1250				5/8" reg. drywall	0	1.10	2.70	3.80
1300				nothing	0	.74	1.80	2.54
1400		1/4" SD gypsum	2 x 4 @ 16" O.C.	same	1-1/2" fiberglass	2.48	4.15	6.63
1450				5/8" FR drywall	1-1/2" fiberglass	2.11	3.65	5.76
1500				nothing	1-1/2" fiberglass	1.75	2.75	4.50
1600		resil. channels	2 x 4 @ 16", O.C.	same	1-1/2" fiberglass	2.10	5.25	7.35
1650				5/8" FR drywall	1-1/2" fiberglass	1.92	4.21	6.13
1700				nothing	1-1/2" fiberglass	1.56	3.31	4.87
1800		5/8" FR drywall	2 x 4 @ 24" O.C.	same	0	1.63	3.42	5.05
1850				5/8" FR drywall	0	1.32	2.97	4.29
1900				nothing	0	.96	2.07	3.03
2200	3		2 rows-2 x 4 16"O.C.	same	2" fiberglass	2.80	4.95	7.75
2250				5/8" FR drywall	2" fiberglass	2.49	4.50	6.99
2300				nothing	2" fiberglass	2.13	3.60	5.73
2400	5/8" WR drywall	none	2 x 4, @ 16" O.C.	same	0	1.36	2.70	4.06
2450				5/8" FR drywall	0	1.23	2.70	3.93
2500				nothing	0	.87	1.80	2.67
2600		5/8" FR drywall	2 x 4, @ 24" O.C.	same	0	1.89	3.42	5.31
2650				5/8" FR drywall	0	1.45	2.97	4.42

5 6 7

2 *Illustration*

At the top of most assembly pages are an illustration, a brief description, and the design criteria used to develop the cost.

3 *System Description*

The components of a typical system are listed in the description to show what has been included in the development of the total system price. The rest of the table contains prices for other similar systems with dimensional and/or size variations.

4 *Unit of Measure for Each System (S.F.)*

Costs shown in the three right-hand columns have been adjusted by the component quantity and unit of measure for the entire system. In this example, "S.F. Cost" is the unit of measure for this system, or assembly.

5 *Materials (1.36)*

This column contains the Materials Cost of each component. These cost figures are bare costs plus 10% for profit.

6 *Installation (2.70)*

Installation includes labor and equipment plus the installing contractor's overhead and profit. Equipment costs are the bare rental costs plus 10% for profit. The labor overhead and profit is defined on the inside back cover of this book.

7 *Total (4.06)*

The figure in this column is the sum of the material and installation costs.

Material Cost	+	Installation Cost	=	Total
$1.36	+	$2.70	=	$4.06

A10 Foundations

A1010 Standard Foundations

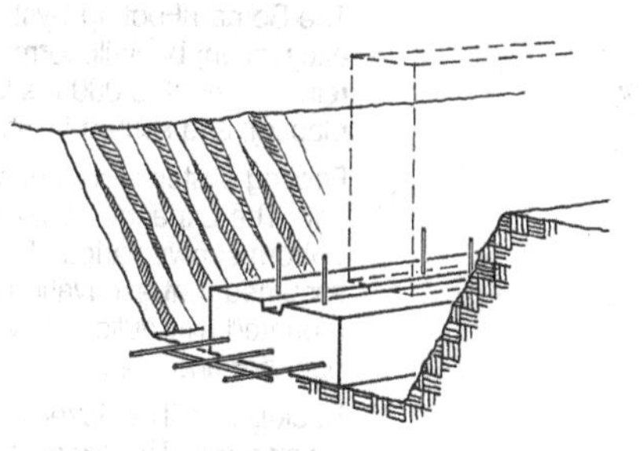

The Strip Footing System includes: excavation; hand trim; all forms needed for footing placement; forms for 2″ x 6″ keyway (four uses); dowels; and 3,000 p.s.i. concrete.

The footing size required varies for different soils. Soil bearing capacities are listed for 3 KSF and 6 KSF. Depths of the system range from 8″ and deeper. Widths range from 16″ and wider. Smaller strip footings may not require reinforcement.

Please see the reference section for further design and cost information.

A1010 110	Strip Footings	COST PER L.F.		
		MAT.	INST.	TOTAL
2100	Strip footing, load 2.6 KLF, soil capacity 3 KSF, 16″ wide x 8″ deep plain	7.55	10.75	18.30
2300	Load 3.9 KLF, soil capacity, 3 KSF, 24″ wide x 8″ deep, plain	9.40	11.80	21.20
2500	Load 5.1KLF, soil capacity 3 KSF, 24″ wide x 12″ deep, reinf.	14.95	17.95	32.90
2700	Load 11.1KLF, soil capacity 6 KSF, 24″ wide x 12″ deep, reinf.	14.95	17.95	32.90
2900	Load 6.8 KLF, soil capacity 3 KSF, 32″ wide x 12″ deep, reinf.	18.25	19.60	37.85
3100	Load 14.8 KLF, soil capacity 6 KSF, 32″ wide x 12″ deep, reinf.	18.25	19.60	37.85
3300	Load 9.3 KLF, soil capacity 3 KSF, 40″ wide x 12″ deep, reinf.	21.50	21	42.50
3500	Load 18.4 KLF, soil capacity 6 KSF, 40″ wide x 12″ deep, reinf.	21.50	21.50	43
4500	Load 10KLF, soil capacity 3 KSF, 48″ wide x 16″ deep, reinf.	30.50	26.50	57
4700	Load 22KLF, soil capacity 6 KSF, 48″ wide, 16″ deep, reinf.	31	27	58
5700	Load 15KLF, soil capacity 3 KSF, 72″ wide x 20″ deep, reinf.	53	38.50	91.50
5900	Load 33KLF, soil capacity 6 KSF, 72″ wide x 20″ deep, reinf.	56	41	97

A10 Foundations

A1010 Standard Foundations

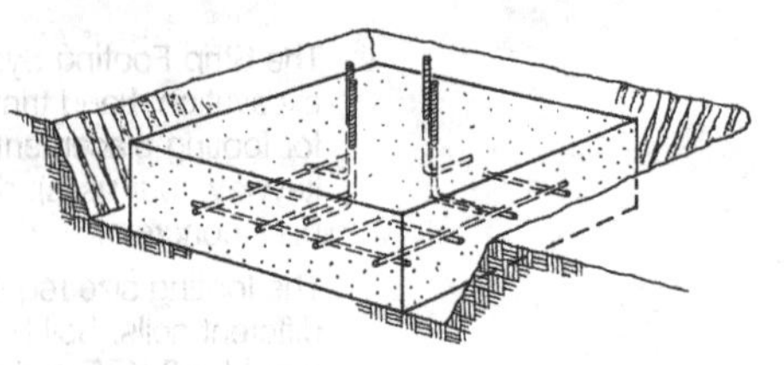

The Spread Footing System includes: excavation; backfill; forms (four uses); all reinforcement; 3,000 p.s.i. concrete (chute placed); and screed finish.

Footing systems are priced per individual unit. The Expanded System Listing at the bottom shows various footing sizes. It is assumed that excavation is done by a truck mounted hydraulic excavator with an operator and oiler.

Backfill is with a dozer, and compaction by air tamp. The excavation and backfill equipment is assumed to operate at 30 C.Y. per hour.

Please see the reference section for further design and cost information.

A1010 210	Spread Footings	COST EACH		
		MAT.	INST.	TOTAL
7090	Spread footings, 3000 psi concrete, chute delivered			
7100	Load 25K, soil capacity 3 KSF, 3'-0" sq. x 12" deep	54	92	146
7150	Load 50K, soil capacity 3 KSF, 4'-6" sq. x 12" deep	117	159	276
7200	Load 50K, soil capacity 6 KSF, 3'-0" sq. x 12" deep	54	92	146
7250	Load 75K, soil capacity 3 KSF, 5'-6" sq. x 13" deep	186	225	411
7300	Load 75K, soil capacity 6 KSF, 4'-0" sq. x 12" deep	94.50	136	230.50
7350	Load 100K, soil capacity 3 KSF, 6'-0" sq. x 14" deep	235	270	505
7410	Load 100K, soil capacity 6 KSF, 4'-6" sq. x 15" deep	144	187	331
7450	Load 125K, soil capacity 3 KSF, 7'-0" sq. x 17" deep	375	390	765
7500	Load 125K, soil capacity 6 KSF, 5'-0" sq. x 16" deep	186	224	410
7550	Load 150K, soil capacity 3 KSF 7'-6" sq. x 18" deep	455	455	910
7610	Load 150K, soil capacity 6 KSF, 5'-6" sq. x 18" deep	248	284	532
7650	Load 200K, soil capacity 3 KSF, 8'-6" sq. x 20" deep	645	600	1,245
7700	Load 200K, soil capacity 6 KSF, 6'-0" sq. x 20" deep	325	350	675
7750	Load 300K, soil capacity 3 KSF, 10'-6" sq. x 25" deep	1,200	985	2,185
7810	Load 300K, soil capacity 6 KSF, 7'-6" sq. x 25" deep	620	595	1,215
7850	Load 400K, soil capacity 3 KSF, 12'-6" sq. x 28" deep	1,900	1,450	3,350
7900	Load 400K, soil capacity 6 KSF, 8'-6" sq. x 27" deep	860	770	1,630
8010	Load 500K, soil capacity 6 KSF, 9'-6" sq. x 30" deep	1,200	1,000	2,200
8100	Load 600K, soil capacity 6 KSF, 10'-6" sq. x 33" deep	1,600	1,300	2,900
8200	Load 700K, soil capacity 6 KSF, 11'-6" sq. x 36" deep	2,050	1,575	3,625
8300	Load 800K, soil capacity 6 KSF, 12'-0" sq. x 37" deep	2,300	1,725	4,025
8400	Load 900K, soil capacity 6 KSF, 13'-0" sq. x 39" deep	2,850	2,075	4,925
8500	Load 1000K, soil capacity 6 KSF, 13'-6" sq. x 41" deep	3,225	2,300	5,525

A1010 320	Foundation Dampproofing	COST PER L.F.		
		MAT.	INST.	TOTAL
1000	Foundation dampproofing, bituminous, 1 coat, 4' high	.88	3.51	4.39
1400	8' high	1.76	7	8.76
1800	12' high	2.64	10.85	13.49
2000	2 coats, 4' high	1.72	4.31	6.03
2400	8' high	3.44	8.60	12.04
2800	12' high	5.15	13.25	18.40
3000	Asphalt with fibers, 1/16" thick, 4' high	2.04	4.31	6.35
3400	8' high	4.08	8.60	12.68
3800	12' high	6.10	13.25	19.35
4000	1/8" thick, 4' high	3.64	5.15	8.79

A10 Foundations

A1010 Standard Foundations

A1010 320	Foundation Dampproofing	COST PER L.F.		
		MAT.	INST.	TOTAL
4400	8′ high	7.30	10.30	17.60
4800	12′ high	10.90	15.80	26.70
5000	Asphalt coated board and mastic, 1/4″ thick, 4′ high	3.56	4.67	8.23
5400	8′ high	7.10	9.35	16.45
5800	12′ high	10.70	14.35	25.05
6000	1/2″ thick, 4′ high	5.35	6.45	11.80
6400	8′ high	10.70	12.95	23.65
6800	12′ high	16	19.75	35.75
7000	Cementitious coating, on walls, 1/8″ thick coating, 4′ high	8.80	6.85	15.65
7400	8′ high	17.60	13.70	31.30
7800	12′ high	26.50	20.50	47
8000	Cementitious/metallic slurry, 4 coat, 2′ high	28.50	340	368.50
8400	4′ high	57	680	737
8800	6′ high	85.50	1,025	1,110.50

A10 Foundations

A1030 Slab on Grade

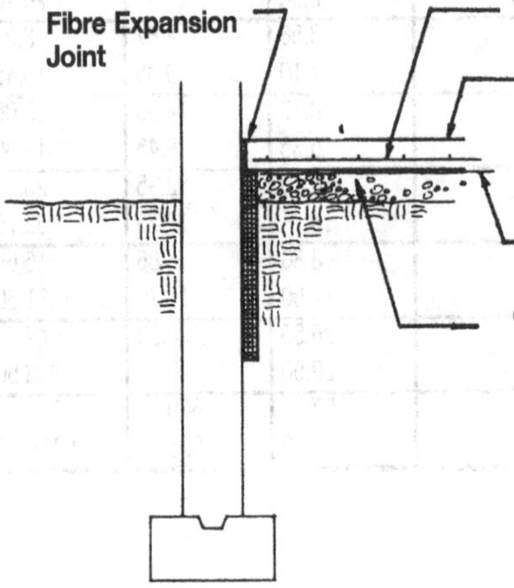

Reinforced Slab on Grade

A Slab on Grade system includes fine grading; 6″ of compacted gravel; vapor barrier; 3500 p.s.i. concrete; bituminous fiber expansion joint; all necessary edge forms 4 uses; steel trowel finish; and sprayed on membrane curing compound. Wire mesh reinforcing used in all reinforced slabs.

Non-industrial slabs are for foot traffic only with negligible abrasion. Light industrial slabs are for pneumatic wheels and light abrasion. Industrial slabs are for solid rubber wheels and moderate abrasion. Heavy industrial slabs are for steel wheels and severe abrasion.

A1030 120	Plain & Reinforced	COST PER S.F.		
		MAT.	INST.	TOTAL
2220	Slab on grade, 4″ thick, non industrial, non reinforced	1.99	2.10	4.09
2240	Reinforced	2.13	2.39	4.52
2260	Light industrial, non reinforced	2.59	2.58	5.17
2280	Reinforced	2.73	2.87	5.60
2300	Industrial, non reinforced	3.22	5.40	8.62
2320	Reinforced	3.36	5.70	9.06
3340	5″ thick, non industrial, non reinforced	2.32	2.16	4.48
3360	Reinforced	2.46	2.45	4.91
3380	Light industrial, non reinforced	2.93	2.64	5.57
3400	Reinforced	3.07	2.93	6
3420	Heavy industrial, non reinforced	4.19	6.50	10.69
3440	Reinforced	4.31	6.80	11.11
4460	6″ thick, non industrial, non reinforced	2.76	2.11	4.87
4480	Reinforced	3.01	2.50	5.51
4500	Light industrial, non reinforced	3.38	2.59	5.97
4520	Reinforced	3.78	3.11	6.89
4540	Heavy industrial, non reinforced	4.65	6.55	11.20
4560	Reinforced	4.90	6.95	11.85
5580	7″ thick, non industrial, non reinforced	3.10	2.18	5.28
5600	Reinforced	3.38	2.60	5.98
5620	Light industrial, non reinforced	3.73	2.66	6.39
5640	Reinforced	4.01	3.08	7.09
5660	Heavy industrial, non reinforced	5	6.50	11.50
5680	Reinforced	5.25	6.85	12.10
6700	8″ thick, non industrial, non reinforced	3.43	2.22	5.65
6720	Reinforced	3.67	2.57	6.24
6740	Light industrial, non reinforced	4.07	2.70	6.77
6760	Reinforced	4.31	3.05	7.36
6780	Heavy industrial, non reinforced	5.35	6.55	11.90
6800	Reinforced	5.70	6.95	12.65

A20 Basement Construction

A2010 Basement Excavation

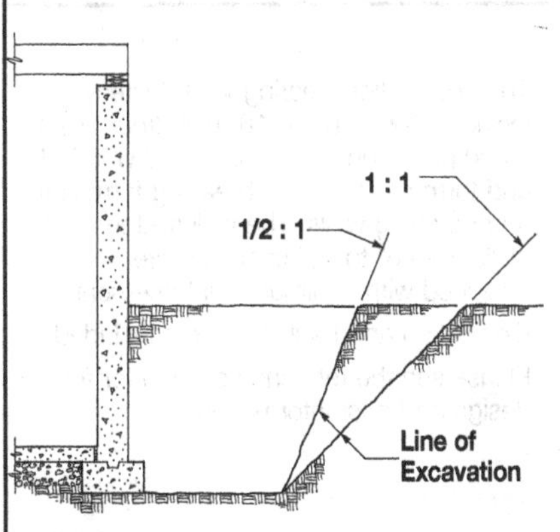

Pricing Assumptions: Two-thirds of excavation is by 2-1/2 C.Y. wheel mounted front end loader and one-third by 1-1/2 C.Y. hydraulic excavator.

Two-mile round trip haul by 12 C.Y. tandem trucks is included for excavation wasted and storage of suitable fill from excavated soil. For excavation in clay, all is wasted and the cost of suitable backfill with two-mile haul is included.

Sand and gravel assumes 15% swell and compaction; common earth assumes 25% swell and 15% compaction; clay assumes 40% swell and 15% compaction (non-clay).

In general, the following items are accounted for in the costs in the table below.

1. Excavation for building or other structure to depth and extent indicated.
2. Backfill compacted in place.
3. Haul of excavated waste.
4. Replacement of unsuitable material with bank run gravel.

Note: Additional excavation and fill beyond this line of general excavation for the building (as required for isolated spread footings, strip footings, etc.) are included in the cost of the appropriate component systems.

A2010 110	**Building Excavation & Backfill**	COST PER S.F.		
		MAT.	INST.	TOTAL
2220	Excav & fill, 1000 S.F. 4′ sand, gravel, or common earth, on site storage		.78	.78
2240	Off site storage		1.18	1.18
2260	Clay excavation, bank run gravel borrow for backfill	2.40	1.83	4.23
2280	8′ deep, sand, gravel, or common earth, on site storage		4.72	4.72
2300	Off site storage		10.45	10.45
2320	Clay excavation, bank run gravel borrow for backfill	9.75	8.50	18.25
2340	16′ deep, sand, gravel, or common earth, on site storage		12.90	12.90
2350	Off site storage		26	26
2360	Clay excavation, bank run gravel borrow for backfill	27	22	49
3380	4000 S.F.,4′ deep, sand, gravel, or common earth, on site storage		.41	.41
3400	Off site storage		.82	.82
3420	Clay excavation, bank run gravel borrow for backfill	1.22	.93	2.15
3440	8′ deep, sand, gravel, or common earth, on site storage		3.32	3.32
3460	Off site storage		5.85	5.85
3480	Clay excavation, bank run gravel borrow for backfill	4.44	5.10	9.54
3500	16′ deep, sand, gravel, or common earth, on site storage		8	8
3520	Off site storage		15.95	15.95
3540	Clay, excavation, bank run gravel borrow for backfill	11.90	12.15	24.05
4560	10,000 S.F., 4′ deep, sand gravel, or common earth, on site storage		.24	.24
4580	Off site storage		.49	.49
4600	Clay excavation, bank run gravel borrow for backfill	.74	.56	1.30
4620	8′ deep, sand, gravel, or common earth, on site storage		2.90	2.90
4640	Off site storage		4.41	4.41
4660	Clay excavation, bank run gravel borrow for backfill	2.69	3.97	6.66
4680	16′ deep, sand, gravel, or common earth, on site storage		6.55	6.55
4700	Off site storage		11.15	11.15
4720	Clay excavation, bank run gravel borrow for backfill	7.15	9.10	16.25
5740	30,000 S.F., 4′ deep, sand, gravel, or common earth, on site storage		.15	.15
5760	Off site storage		.30	.30
5780	Clay excavation, bank run gravel borrow for backfill	.43	.32	.75
5860	16′ deep, sand, gravel, or common earth, on site storage		5.60	5.60
5880	Off site storage		8.10	8.10
5900	Clay excavation, bank run gravel borrow for backfill	3.96	7	10.96
6910	100,000 S.F., 4′ deep, sand, gravel, or common earth, on site storage		.08	.08
6920	Off site storage		.17	.17
6930	Clay excavation, bank run gravel borrow for backfill	.22	.17	.39
6940	8′ deep, sand, gravel, or common earth, on site storage		2.43	2.43
6950	Off site storage		2.89	2.89
6960	Clay excavation, bank run gravel borrow for backfill	.82	2.77	3.59
6970	16′ deep, sand, gravel, or common earth, on site storage		5.10	5.10
6980	Off site storage		6.40	6.40
6990	Clay excavation, bank run gravel borrow for backfill	2.14	5.85	7.99

A20 Basement Construction

A2020 Basement Walls

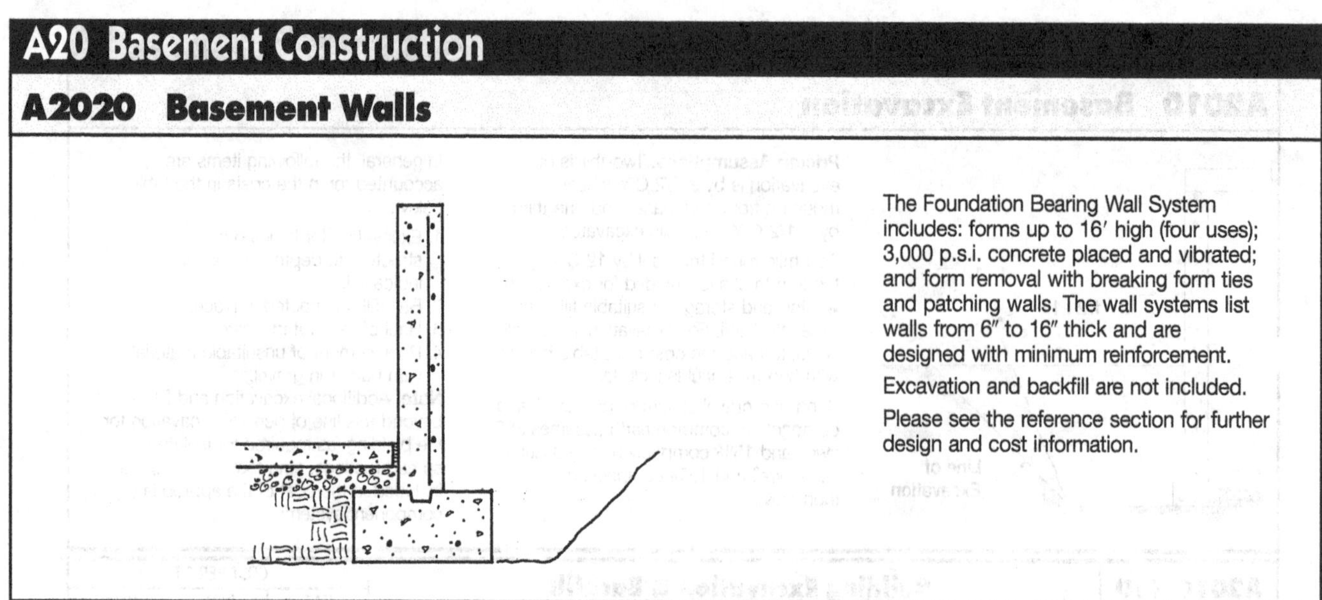

The Foundation Bearing Wall System includes: forms up to 16′ high (four uses); 3,000 p.s.i. concrete placed and vibrated; and form removal with breaking form ties and patching walls. The wall systems list walls from 6″ to 16″ thick and are designed with minimum reinforcement.

Excavation and backfill are not included.

Please see the reference section for further design and cost information.

A2020 110 Walls, Cast in Place

	WALL HEIGHT (FT.)	PLACING METHOD	CONCRETE (C.Y. per L.F.)	REINFORCING (LBS. per L.F.)	WALL THICKNESS (IN.)	COST PER L.F. MAT.	INST.	TOTAL
1500	4′	direct chute	.074	3.3	6	14.65	38	52.65
1520			.099	4.8	8	18.10	39	57.10
1540			.123	6.0	10	21.50	40	61.50
1561			.148	7.2	12	24.50	41	65.50
1580			.173	8.1	14	27.50	41.50	69
1600			.197	9.44	16	31	42.50	73.50
3000	6′	direct chute	.111	4.95	6	22	57	79
3020			.149	7.20	8	27	59	86
3040			.184	9.00	10	32	60	92
3061			.222	10.8	12	37	61.50	98.50
5000	8′	direct chute	.148	6.6	6	29.50	76	105.50
5020			.199	9.6	8	36.50	78.50	115
5040			.250	12	10	43	80	123
5061			.296	14.39	12	49	81.50	130.50
6020	10′	direct chute	.248	12	8	45.50	98	143.50
6040			.307	14.99	10	53	99.50	152.50
6061			.370	17.99	12	61.50	102	163.50
7220	12′	pumped	.298	14.39	8	54.50	122	176.50
7240			.369	17.99	10	64	125	189
7262			.444	21.59	12	73.50	129	202.50
9220	16′	pumped	.397	19.19	8	72.50	163	235.50
9240			.492	23.99	10	85	166	251
9260			.593	28.79	12	98.50	172	270.50

B10 Superstructure

B1010 Floor Construction

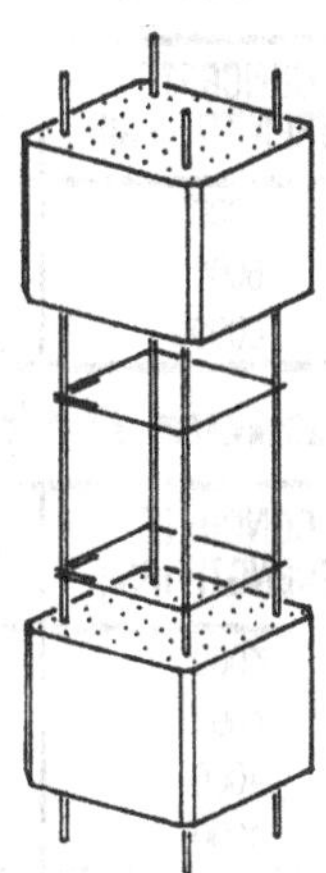

General: It is desirable for purposes of consistency and simplicity to maintain constant column sizes throughout building height. To do this, concrete strength may be varied (higher strength concrete at lower stories and lower strength concrete at upper stories), as well as varying the amount of reinforcing.

The table provides probably minimum column sizes with related costs and weight per lineal foot of story height.

B1010 203 C.I.P. Column, Square Tied

	LOAD (KIPS)	STORY HEIGHT (FT.)	COLUMN SIZE (IN.)	COLUMN WEIGHT (P.L.F.)	CONCRETE STRENGTH (PSI)	COST PER V.L.F.		
						MAT.	INST.	TOTAL
0640	100	10	10	96	4000	10	36.50	46.50
0680		12	10	97	4000	10.15	36.50	46.65
0720		14	12	142	4000	13.30	44	57.30
0840	200	10	12	140	4000	14.55	46	60.55
0860		12	12	142	4000	14.80	46	60.80
0900		14	14	196	4000	17.25	51.50	68.75
0920	300	10	14	192	4000	18.25	52.50	70.75
0960		12	14	194	4000	18.60	53	71.60
0980		14	16	253	4000	21.50	58.50	80
1020	400	10	16	248	4000	23	60.50	83.50
1060		12	16	251	4000	23.50	61	84.50
1080		14	16	253	4000	24	61.50	85.50
1200	500	10	18	315	4000	27	67.50	94.50
1250		12	20	394	4000	32.50	77	109.50
1300		14	20	397	4000	33	78	111
1350	600	10	20	388	4000	34	78.50	112.50
1400		12	20	394	4000	34.50	79.50	114
1600		14	20	397	4000	35	80.50	115.50
3400	900	10	24	560	4000	46	98	144
3800		12	24	567	4000	47	99.50	146.50
4000		14	24	571	4000	48	101	149
7300	300	10	14	192	6000	19.25	52	71.25
7500		12	14	194	6000	19.60	52.50	72.10
7600		14	14	196	6000	19.90	53	72.90
8000	500	10	16	248	6000	24.50	60.50	85
8050		12	16	251	6000	25	61	86
8100		14	16	253	6000	25.50	61.50	87
8200	600	10	18	315	6000	28.50	68	96.50
8300		12	18	319	6000	29	68.50	97.50
8400		14	18	321	6000	29.50	69.50	99
8800	800	10	20	388	6000	35	77	112
8900		12	20	394	6000	36	78	114
9000		14	20	397	6000	36.50	78.50	115

B10 Superstructure

B1010 Floor Construction

B1010 203 C.I.P. Column, Square Tied

	LOAD (KIPS)	STORY HEIGHT (FT.)	COLUMN SIZE (IN.)	COLUMN WEIGHT (P.L.F.)	CONCRETE STRENGTH (PSI)	COST PER V.L.F.		
						MAT.	INST.	TOTAL
9100	900	10	20	388	6000	38	80.50	118.50
9300		12	20	394	6000	38.50	81.50	120
9600		14	20	397	6000	39.50	82.50	122

B1010 204 C.I.P. Column, Square Tied-Minimum Reinforcing

	LOAD (KIPS)	STORY HEIGHT (FT.)	COLUMN SIZE (IN.)	COLUMN WEIGHT (P.L.F.)	CONCRETE STRENGTH (PSI)	COST PER V.L.F.		
						MAT.	INST.	TOTAL
9913	150	10-14	12	135	4000	12.05	42.50	54.55
9918	300	10-14	16	240	4000	19.30	55.50	74.80
9924	500	10-14	20	375	4000	29	72.50	101.50
9930	700	10-14	24	540	4000	40	90.50	130.50
9936	1000	10-14	28	740	4000	52.50	110	162.50
9942	1400	10-14	32	965	4000	64.50	123	187.50
9948	1800	10-14	36	1220	4000	80	144	224

B10 Superstructure

B1010 Floor Construction

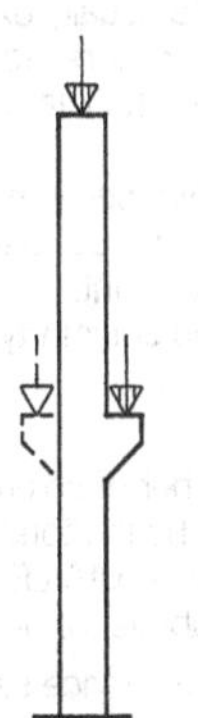

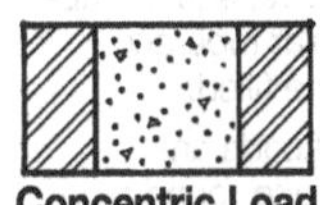

Concentric Load

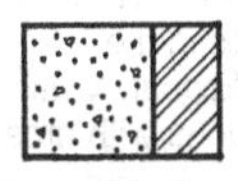

Eccentric Load

General: Data presented here is for plant produced members transported 50 miles to 100 miles to the site and erected.

Design and pricing assumptions:
Normal wt. concrete, f'c = 5 KSI

Main reinforcement, fy = 60 KSI
Ties, fy = 40 KSI

Minimum design eccentricity, 0.1t.

Concrete encased structural steel haunches are assumed where practical; otherwise galvanized rebar haunches are assumed.

Base plates are integral with columns.

Foundation anchor bolts, nuts and washers are included in price.

B1010 206 Tied, Concentric Loaded Precast Concrete Columns

	LOAD (KIPS)	STORY HEIGHT (FT.)	COLUMN SIZE (IN.)	COLUMN WEIGHT (P.L.F.)	LOAD LEVELS	COST PER V.L.F.		
						MAT.	INST.	TOTAL
0560	100	10	12x12	164	2	122	9.35	131.35
0570		12	12x12	162	2	118	7.75	125.75
0580		14	12x12	161	2	117	7.60	124.60
0590	150	10	12x12	166	3	117	7.55	124.55
0600		12	12x12	169	3	117	6.75	123.75
0610		14	12x12	162	3	117	6.65	123.65
0620	200	10	12x12	168	4	117	8	125
0630		12	12x12	170	4	117	7.25	124.25
0640		14	12x12	220	4	117	7.15	124.15

B1010 207 Tied, Eccentric Loaded Precast Concrete Columns

	LOAD (KIPS)	STORY HEIGHT (FT.)	COLUMN SIZE (IN.)	COLUMN WEIGHT (P.L.F.)	LOAD LEVELS	COST PER V.L.F.		
						MAT.	INST.	TOTAL
1130	100	10	12x12	161	2	115	9.35	124.35
1140		12	12x12	159	2	115	7.75	122.75
1150		14	12x12	159	2	115	7.60	122.60
1390	600	10	18x18	385	4	195	8	203
1400		12	18x18	380	4	194	7.25	201.25
1410		14	18x18	375	4	193	7.15	200.15
1480	800	10	20x20	490	4	195	8	203
1490		12	20x20	480	4	194	7.25	201.25
1500		14	20x20	475	4	195	7.15	202.15

B10 Superstructure

B1010 Floor Construction

(A) Wide Flange

(B) Pipe

(C) Pipe, Concrete Filled

(D) Square Tube

(E) Square Tube Concrete Filled

(F) Rectangular Tube

(G) Rectangular Tube, Concrete Filled

General: The following pages provide data for seven types of steel columns: wide flange, round pipe, round pipe concrete filled, square tube, square tube concrete filled, rectangular tube and rectangular tube concrete filled.

Design Assumptions: Loads are concentric; wide flange and round pipe bearing capacity is for 36 KSI steel. Square and rectangular tubing bearing capacity is for 46 KSI steel.

The effective length factor K=1.1 is used for determining column values in the tables. K=1.1 is within a frequently used range for pinned connections with cross bracing.

How To Use Tables:

a. Steel columns usually extend through two or more stories to minimize splices. Determine floors with splices.

b. Enter Table No. below with load to column at the splice. Use the unsupported height.

c. Determine the column type desired by price or design.

Cost:

a. Multiply number of columns at the desired level by the total height of the column by the cost/VLF.

b. Repeat the above for all tiers.

Please see the reference section for further design and cost information.

B1010 208 Steel Columns

	LOAD (KIPS)	UNSUPPORTED HEIGHT (FT.)	WEIGHT (P.L.F.)	SIZE (IN.)	TYPE	COST PER V.L.F.		
						MAT.	INST.	TOTAL
1000	25	10	13	4	A	19.50	8.40	27.90
1020			7.58	3	B	11.35	8.40	19.75
1040			15	3-1/2	C	13.90	8.40	22.30
1120			20	4x3	G	16.05	8.40	24.45
1200		16	16	5	A	22	6.30	28.30
1220			10.79	4	B	15	6.30	21.30
1240			36	5-1/2	C	21	6.30	27.30
1320			64	8x6	G	32.50	6.30	38.80
1600	50	10	16	5	A	24	8.40	32.40
1620			14.62	5	B	22	8.40	30.40
1640			24	4-1/2	C	16.60	8.40	25
1720			28	6x3	G	21.50	8.40	29.90
1800		16	24	8	A	33.50	6.30	39.80
1840			36	5-1/2	C	21	6.30	27.30
1920			64	8x6	G	32.50	6.30	38.80
2000		20	28	8	A	37	6.30	43.30
2040			49	6-5/8	C	26	6.30	32.30
2120			64	8x6	G	31	6.30	37.30
2200	75	10	20	6	A	30	8.40	38.40
2240			36	4-1/2	C	41.50	8.40	49.90
2320			35	6x4	G	24	8.40	32.40
2400		16	31	8	A	43	6.30	49.30
2440			49	6-5/8	C	27.50	6.30	33.80
2520			64	8x6	G	32.50	6.30	38.80
2600		20	31	8	A	41	6.30	47.30
2640			81	8-5/8	C	39	6.30	45.30
2720			64	8x6	G	31	6.30	37.30
2800	100	10	24	8	A	36	8.40	44.40
2840			35	4-1/2	C	41.50	8.40	49.90
2920			46	8x4	G	29.50	8.40	37.90
3000		16	31	8	A	43	6.30	49.30
3040			56	6-5/8	C	40.50	6.30	46.80
3120			64	8x6	G	32.50	6.30	38.80

B10 Superstructure

B1010 Floor Construction

B1010 208 Steel Columns

	LOAD (KIPS)	UNSUPPORTED HEIGHT (FT.)	WEIGHT (P.L.F.)	SIZE (IN.)	TYPE	COST PER V.L.F.		
						MAT.	INST.	TOTAL
3200	100	20	40	8	A	52.50	6.30	58.80
3240			81	8-5/8	C	39	6.30	45.30
3320			70	8x6	G	44	6.30	50.30
3400	125	10	31	8	A	46.50	8.40	54.90
3440			81	8	C	44.50	8.40	52.90
3520			64	8x6	G	35	8.40	43.40
3600		16	40	8	A	55.50	6.30	61.80
3640			81	8	C	41	6.30	47.30
3720			64	8x6	G	32.50	6.30	38.80
3800		20	48	8	A	63	6.30	69.30
3840			81	8	C	39	6.30	45.30
3920			60	8x6	G	44	6.30	50.30
4000	150	10	35	8	A	52.50	8.40	60.90
4040			81	8-5/8	C	44.50	8.40	52.90
4120			64	8x6	G	35	8.40	43.40
4200		16	45	10	A	62.50	6.30	68.80
4240			81	8-5/8	C	41	6.30	47.30
4320			70	8x6	G	46.50	6.30	52.80
4400		20	49	10	A	64.50	6.30	70.80
4440			123	10-3/4	C	55.50	6.30	61.80
4520			86	10x6	G	43.50	6.30	49.80
4600	200	10	45	10	A	67.50	8.40	75.90
4640			81	8-5/8	C	44.50	8.40	52.90
4720			70	8x6	G	50	8.40	58.40
4800		16	49	10	A	68	6.30	74.30
4840			123	10-3/4	C	58.50	6.30	64.80
4920			85	10x6	G	54	6.30	60.30
5200	300	10	61	14	A	91.50	8.40	99.90
5240			169	12-3/4	C	78	8.40	86.40
5320			86	10x6	G	74.50	8.40	82.90
5400		16	72	12	A	100	6.30	106.30
5440			169	12-3/4	C	72.50	6.30	78.80
5600		20	79	12	A	104	6.30	110.30
5640			169	12-3/4	C	68.50	6.30	74.80
5800	400	10	79	12	A	119	8.40	127.40
5840			178	12-3/4	C	102	8.40	110.40
6000		16	87	12	A	121	6.30	127.30
6040			178	12-3/4	C	94.50	6.30	100.80
6400	500	10	99	14	A	149	8.40	157.40
6600		16	109	14	A	151	6.30	157.30
6800		20	120	12	A	158	6.30	164.30
7000	600	10	120	12	A	180	8.40	188.40
7200		16	132	14	A	183	6.30	189.30
7400		20	132	14	A	174	6.30	180.30
7600	700	10	136	12	A	204	8.40	212.40
7800		16	145	14	A	201	6.30	207.30
8000		20	145	14	A	191	6.30	197.30
8200	800	10	145	14	A	218	8.40	226.40
8300		16	159	14	A	221	6.30	227.30
8400		20	176	14	A	231	6.30	237.30
8800	900	10	159	14	A	239	8.40	247.40
8900		16	176	14	A	244	6.30	250.30
9000		20	193	14	A	254	6.30	260.30

B10 Superstructure

B1010 Floor Construction

B1010 208 Steel Columns

	LOAD (KIPS)	UNSUPPORTED HEIGHT (FT.)	WEIGHT (P.L.F.)	SIZE (IN.)	TYPE	COST PER V.L.F.		
						MAT.	INST.	TOTAL
9100	1000	10	176	14	A	264	8.40	272.40
9200		16	193	14	A	268	6.30	274.30
9300		20	211	14	A	277	6.30	283.30

B10 Superstructure

B1010 Floor Construction

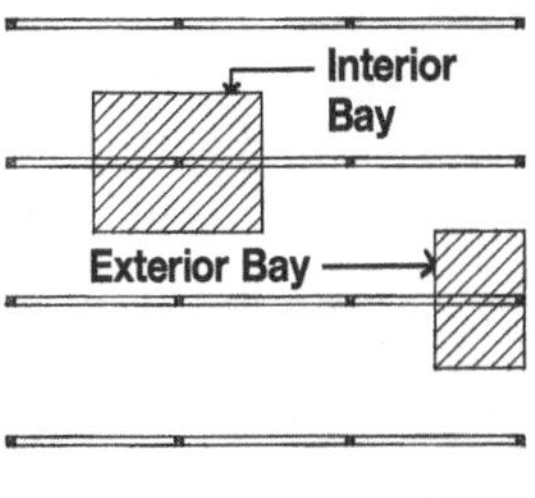

Description: Table below lists costs of columns per S.F. of bay for wood columns of various sizes and unsupported heights and the maximum allowable total load per S.F. by bay size.

Design Assumptions: Columns are concentrically loaded and are not subject to bending.

Fiber stress (f) is 1200 psi maximum.

Modulus of elasticity is 1,760,000. Use table to factor load capacity figures for modulus of elasticity other than 1,760,000.

The cost of columns per S.F. of exterior bay is proportional to the area supported. For exterior bays, multiply the costs below by two. For corner bays, multiply the cost by four.

Modulus of Elasticity	Factor
1,210,000 psi	0.69
1,320,000 psi	0.75
1,430,000 psi	0.81
1,540,000 psi	0.87
1,650,000 psi	0.94
1,760,000 psi	1.00

B1010 210 Wood Columns

	NOMINAL COLUMN SIZE (IN.)	BAY SIZE (FT.)	UNSUPPORTED HEIGHT (FT.)	MATERIAL (BF per M.S.F.)	TOTAL LOAD (P.S.F.)	COST PER S.F.		
						MAT.	INST.	TOTAL
1000	4 x 4	10 x 8	8	133	100	.19	.20	.39
1050			10	167	60	.23	.25	.48
1200		10 x 10	8	106	80	.15	.16	.31
1250			10	133	50	.19	.20	.39
1400		10 x 15	8	71	50	.10	.11	.21
1450			10	88	30	.12	.13	.25
1600		15 x 15	8	47	30	.07	.07	.14
1650			10	59	15	.08	.09	.17
2000	6 x 6	10 x 15	8	160	230	.23	.22	.45
2050			10	200	210	.29	.28	.57
2200		15 x 15	8	107	150	.15	.15	.30
2250			10	133	140	.19	.18	.37
2400		15 x 20	8	80	110	.11	.11	.22
2450			10	100	100	.14	.14	.28
2600		20 x 20	8	60	80	.09	.08	.17
2650			10	75	70	.11	.10	.21
2800		20 x 25	8	48	60	.07	.07	.14
2850			10	60	50	.09	.08	.17
3400	8 x 8	20 x 20	8	107	160	.16	.14	.30
3450			10	133	160	.20	.17	.37
3600		20 x 25	8	85	130	.12	.13	.25
3650			10	107	130	.15	.16	.31
3800		25 x 25	8	68	100	.11	.08	.19
3850			10	85	100	.14	.10	.24
4200	10 x 10	20 x 25	8	133	210	.21	.16	.37
4250			10	167	210	.27	.20	.47
4400		25 x 25	8	107	160	.17	.13	.30
4450			10	133	160	.21	.16	.37
4700	12 x 12	20 x 25	8	192	310	.32	.22	.54
4750			10	240	310	.40	.27	.67
4900		25 x 25	8	154	240	.26	.17	.43
4950			10	192	240	.32	.22	.54

B10 Superstructure

B1010 Floor Construction

"T" Shaped Precast Beams

"L" Shaped Precast Beams

B1010 214 "T" Shaped Precast Beams

	SPAN (FT.)	SUPERIMPOSED LOAD (K.L.F.)	SIZE W X D (IN.)	BEAM WEIGHT (P.L.F.)	TOTAL LOAD (K.L.F.)	COST PER L.F.		
						MAT.	INST.	TOTAL
2300	15	2.8	12x16	260	3.06	121	13.20	134.20
2500		8.37	12x28	515	8.89	175	14.20	189.20
8900	45	3.34	12x60	1165	4.51	345	8.50	353.50
9900		6.5	24x60	1915	8.42	465	11.85	476.85

B1010 215 "L" Shaped Precast Beams

	SPAN (FT.)	SUPERIMPOSED LOAD (K.L.F.)	SIZE W X D (IN.)	BEAM WEIGHT (P.L.F.)	TOTAL LOAD (K.L.F.)	COST PER L.F.		
						MAT.	INST.	TOTAL
2250	15	2.58	12x16	230	2.81	93	13.20	106.20
2400		5.92	12x24	370	6.29	126	13.20	139.20
4000	25	2.64	12x28	435	3.08	142	8.50	150.50
4450		6.44	18x36	790	7.23	203	10.15	213.15
5300	30	2.80	12x36	565	3.37	170	7.70	177.70
6400		8.66	24x44	1245	9.90	279	11.10	290.10

B10 Superstructure

B1010 Floor Construction

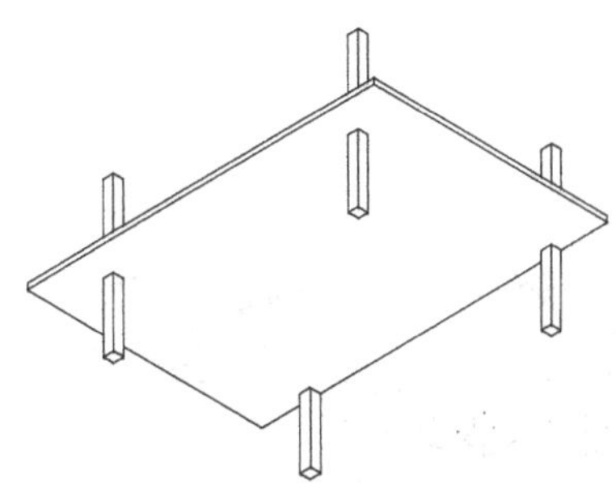

General: Flat Plates: Solid uniform depth concrete two way slab without drops or interior beams. Primary design limit is shear at columns.

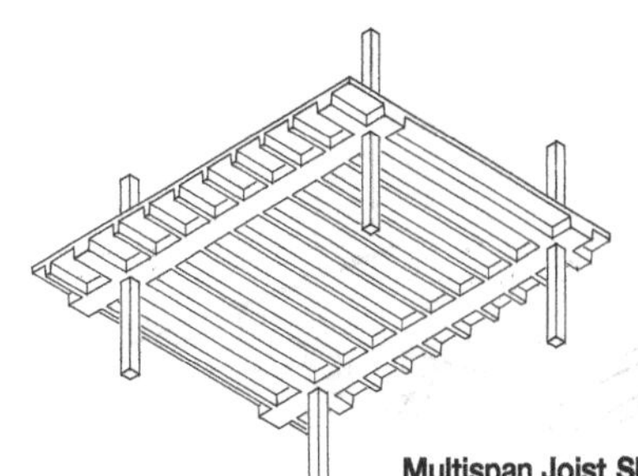

Multispan Joist Slab

General: Combination of thin concrete slab and monolithic ribs at uniform spacing to reduce dead weight and increase rigidity.

B1010 223 Cast in Place Flat Plate

	BAY SIZE (FT.)	SUPERIMPOSED LOAD (P.S.F.)	MINIMUM COL. SIZE (IN.)	SLAB THICKNESS (IN.)	TOTAL LOAD (P.S.F.)	COST PER S.F.		
						MAT.	INST.	TOTAL
3000	15 x 20	40	14	7	127	4.57	6.50	11.07
3400		75	16	7-1/2	169	4.89	6.70	11.59
3600		125	22	8-1/2	231	5.40	6.90	12.30
3800		175	24	8-1/2	281	5.45	6.90	12.35
4200	20 x 20	40	16	7	127	4.58	6.50	11.08
4400		75	20	7-1/2	175	4.93	6.70	11.63
4600		125	24	8-1/2	231	5.40	6.85	12.25
5000		175	24	8-1/2	281	5.45	6.90	12.35
5600	20 x 25	40	18	8-1/2	146	5.35	6.85	12.20
6000		75	20	9	188	5.55	6.95	12.50
6400		125	26	9-1/2	244	6.05	7.15	13.20
6600		175	30	10	300	6.30	7.30	13.60
7000	25 x 25	40	20	9	152	5.60	6.90	12.50
7400		75	24	9-1/2	194	5.90	7.10	13
7600		125	30	10	250	6.30	7.30	13.60

B1010 226 Cast in Place Multispan Joist Slab

	BAY SIZE (FT.)	SUPERIMPOSED LOAD (P.S.F.)	MINIMUM COL. SIZE (IN.)	RIB DEPTH (IN.)	TOTAL LOAD (P.S.F.)	COST PER S.F.		
						MAT.	INST.	TOTAL
2000	15 x 15	40	12	8	115	5.85	7.95	13.80
2100		75	12	8	150	5.90	7.95	13.85
2200		125	12	8	200	6	8.05	14.05
2300		200	14	8	275	6.20	8.35	14.55
2600	15 x 20	40	12	8	115	6	7.90	13.90
2800		75	12	8	150	6.10	8.40	14.50
3000		125	14	8	200	6.30	8.50	14.80
3300		200	16	8	275	6.60	8.65	15.25
3600	20 x 20	40	12	10	120	6.10	7.80	13.90
3900		75	14	10	155	6.35	8.35	14.70
4000		125	16	10	205	6.40	8.45	14.85
4100		200	18	10	280	6.70	8.85	15.55
6200	30 x 30	40	14	14	131	6.80	8.25	15.05
6400		75	18	14	166	7	8.50	15.50
6600		125	20	14	216	7.40	8.95	16.35
6700		200	24	16	297	7.95	9.30	17.25

B10 Superstructure

B1010 Floor Construction

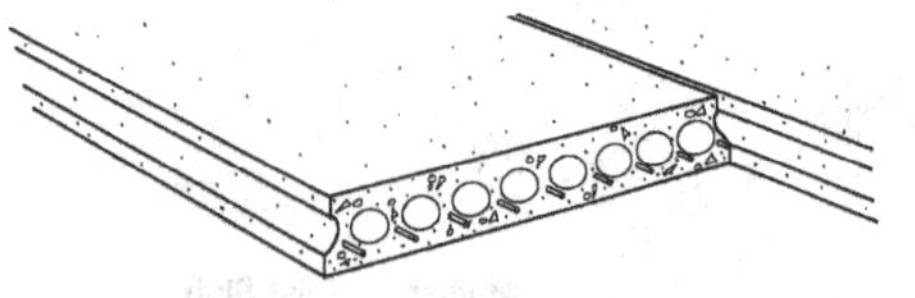

Precast Plank with No Topping

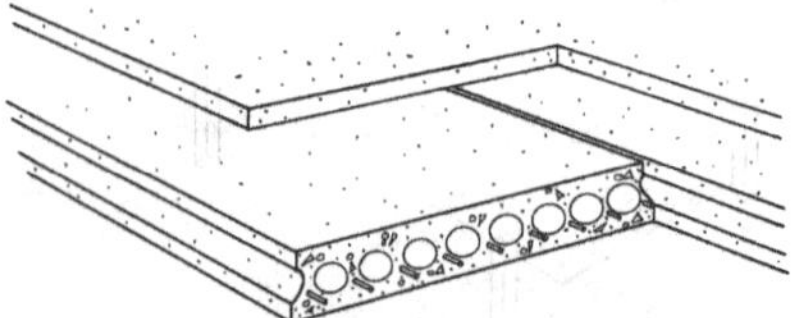

Precast Plank with 2″ Concrete Topping

B1010 229 Precast Plank with No Topping

	SPAN (FT.)	SUPERIMPOSED LOAD (P.S.F.)	TOTAL DEPTH (IN.)	DEAD LOAD (P.S.F.)	TOTAL LOAD (P.S.F.)	COST PER S.F.		
						MAT.	INST.	TOTAL
0720	10	40	4	50	90	6.10	2.22	8.32
0750		75	6	50	125	7.15	1.90	9.05
0770		100	6	50	150	7.15	1.90	9.05
0800	15	40	6	50	90	7.15	1.90	9.05
0820		75	6	50	125	7.15	1.90	9.05
0850		100	6	50	150	7.15	1.90	9.05
0950	25	40	6	50	90	7.15	1.90	9.05
0970		75	8	55	130	7.90	1.67	9.57
1000		100	8	55	155	7.90	1.67	9.57
1200	30	40	8	55	95	7.90	1.67	9.57
1300		75	8	55	130	7.90	1.67	9.57
1400		100	10	70	170	8.25	1.48	9.73
1500	40	40	10	70	110	8.25	1.48	9.73
1600		75	12	70	145	8.80	1.33	10.13
1700	45	40	12	70	110	8.80	1.33	10.13

B1010 230 Precast Plank with 2″ Concrete Topping

	SPAN (FT.)	SUPERIMPOSED LOAD (P.S.F.)	TOTAL DEPTH (IN.)	DEAD LOAD (P.S.F.)	TOTAL LOAD (P.S.F.)	COST PER S.F.		
						MAT.	INST.	TOTAL
2000	10	40	6	75	115	7	4.04	11.04
2100		75	8	75	150	8.05	3.72	11.77
2200		100	8	75	175	8.05	3.72	11.77
2500	15	40	8	75	115	8.05	3.72	11.77
2600		75	8	75	150	8.05	3.72	11.77
2700		100	8	75	175	8.05	3.72	11.77
3100	25	40	8	75	115	8.05	3.72	11.77
3200		75	8	75	150	8.05	3.72	11.77
3300		100	10	80	180	8.80	3.49	12.29
3400	30	40	10	80	120	8.80	3.49	12.29
3500		75	10	80	155	8.80	3.49	12.29
3600		100	10	80	180	8.80	3.49	12.29
4000	40	40	12	95	135	9.15	3.30	12.45
4500		75	14	95	170	9.70	3.15	12.85
5000	45	40	14	95	135	9.70	3.15	12.85

B10 Superstructure

B1010 Floor Construction

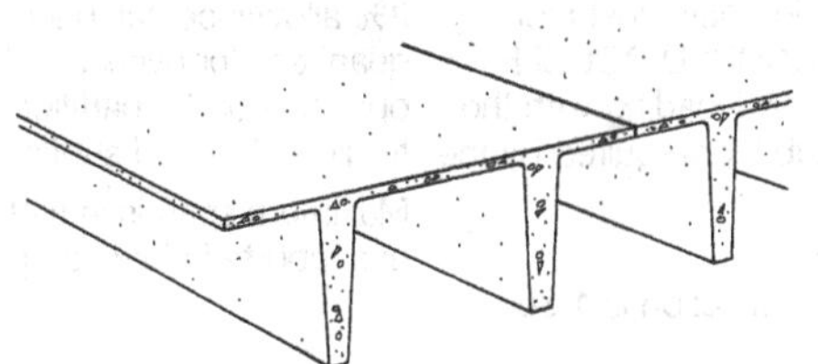

Most widely used for moderate span floors and roofs. At shorter spans, they tend to be competitive with hollow core slabs. They are also used as wall panels.

B1010 234 Precast Double "T" Beams with No Topping

	SPAN (FT.)	SUPERIMPOSED LOAD (P.S.F.)	DBL. "T" SIZE D (IN.) W (FT.)	CONCRETE "T" TYPE	TOTAL LOAD (P.S.F.)	COST PER S.F. MAT.	INST.	TOTAL
4300	50	30	20x8	Lt. Wt.	66	9.85	1.05	10.90
4400		40	20x8	Lt. Wt.	76	10.55	.95	11.50
4500		50	20x8	Lt. Wt.	86	10.65	1.16	11.81
4600		75	20x8	Lt. Wt.	111	10.70	1.34	12.04
5600	70	30	32x10	Lt. Wt.	78	11.55	.72	12.27
5750		40	32x10	Lt. Wt.	88	11.65	.89	12.54
5900		50	32x10	Lt. Wt.	98	11.70	1.04	12.74
6000		75	32x10	Lt. Wt.	123	11.80	1.27	13.07
6100		100	32x10	Lt. Wt.	148	12.05	1.75	13.80
6200	80	30	32x10	Lt. Wt.	78	11.70	1.04	12.74
6300		40	32x10	Lt. Wt.	88	11.80	1.27	13.07
6400		50	32x10	Lt. Wt.	98	11.95	1.51	13.46

B1010 235 Precast Double "T" Beams With 2" Topping

	SPAN (FT.)	SUPERIMPOSED LOAD (P.S.F.)	DBL. "T" SIZE D (IN.) W (FT.)	CONCRETE "T" TYPE	TOTAL LOAD (P.S.F.)	COST PER S.F. MAT.	INST.	TOTAL
7100	40	30	18x8	Reg. Wt.	120	8.85	2.31	11.16
7200		40	20x8	Reg. Wt.	130	8.90	2.18	11.08
7300		50	20x8	Reg. Wt.	140	9	2.39	11.39
7400		75	20x8	Reg. Wt.	165	9.05	2.49	11.54
7500		100	20x8	Reg. Wt.	190	9.15	2.79	11.94
7550	50	30	24x8	Reg. Wt.	120	9.55	2.31	11.86
7600		40	24x8	Reg. Wt.	130	9.60	2.39	11.99
7750		50	24x8	Reg. Wt.	140	9.60	2.41	12.01
7800		75	24x8	Reg. Wt.	165	9.75	2.70	12.45
7900		100	32x10	Reg. Wt.	189	11	2.24	13.24

B10 Superstructure

B1010 Floor Construction

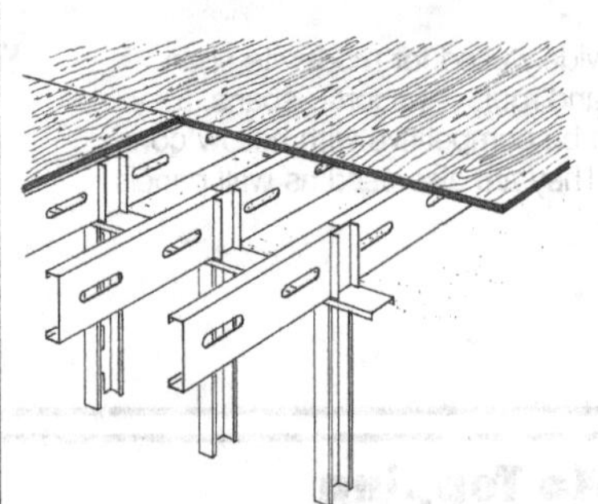

Description: Table below lists costs for light gauge CEE or PUNCHED DOUBLE joists to suit the span and loading with the minimum thickness subfloor required by the joist spacing.

Design Assumptions:
Maximum live load deflection is 1/360 of the clear span.

Maximum total load deflection is 1/240 of the clear span.

Bending strength is 20,000 psi.

8% allowance has been added to framing quantities for overlaps, double joists at openings under partitions, etc.; 5% added to glued & nailed subfloor for waste.

Maximum span is in feet and is the unsupported clear span.

B1010 244 Light Gauge Steel Floor Systems

B1010 244	SPAN (FT.)	SUPERIMPOSED LOAD (P.S.F.)	FRAMING DEPTH (IN.)	FRAMING SPAC. (IN.)	TOTAL LOAD (P.S.F.)	COST PER S.F.		
						MAT.	INST.	TOTAL
1500	15	40	8	16	54	2.40	1.44	3.84
1550			8	24	54	2.51	1.49	4
1600		65	10	16	80	3.10	1.72	4.82
1650			10	24	80	3.07	1.71	4.78
1700		75	10	16	90	3.10	1.72	4.82
1750			10	24	90	3.07	1.71	4.78
1800		100	10	16	116	4	2.13	6.13
1850			10	24	116	3.07	1.71	4.78
1900		125	10	16	141	4	2.13	6.13
1950			10	24	141	3.07	1.71	4.78
2500	20	40	8	16	55	2.74	1.61	4.35
2550			8	24	55	2.51	1.49	4
2600		65	8	16	80	3.16	1.80	4.96
2650			10	24	80	3.08	1.71	4.79
2700		75	10	16	90	2.93	1.65	4.58
2750			12	24	90	3.42	1.55	4.97
2800		100	10	16	115	2.93	1.65	4.58
2850			10	24	116	3.89	1.90	5.79
2900		125	12	16	142	4.52	1.88	6.40
2950			12	24	141	4.21	1.99	6.20
3500	25	40	10	16	55	3.10	1.72	4.82
3550			10	24	55	3.07	1.71	4.78
3600		65	10	16	81	4	2.13	6.13
3650			12	24	81	3.77	1.66	5.43
3700		75	12	16	92	4.51	1.87	6.38
3750			12	24	91	4.21	1.99	6.20
3800		100	12	16	117	5.05	2.05	7.10
3850		125	12	16	143	5.75	2.55	8.30
4500	30	40	12	16	57	4.51	1.87	6.38
4550			12	24	56	3.76	1.65	5.41
4600		65	12	16	82	5.05	2.04	7.09
4650		75	12	16	92	5.05	2.04	7.09

B10 Superstructure

B1010 Floor Construction

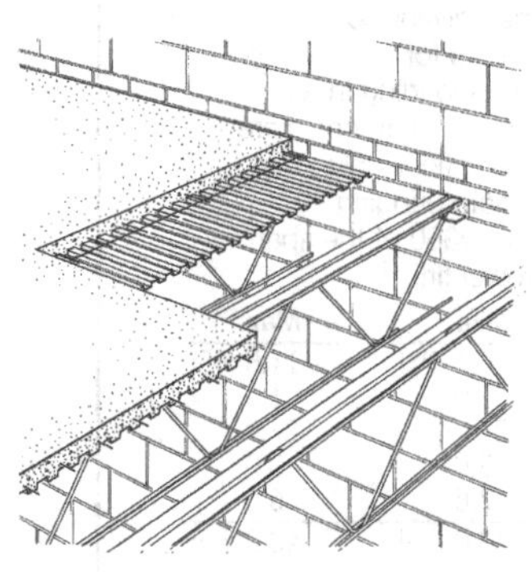

Description: Table below lists cost per S.F. for a floor system on bearing walls using open web steel joists, galvanized steel slab form and 2-1/2" concrete slab reinforced with welded wire fabric.

Design and Pricing Assumptions:
Concrete f'c = 3 KSI placed by pump.
WWF 6 x 6 – W1.4 x W1.4 (10 x 10)
Joists are spaced as shown.
Slab form is 28 gauge galvanized.
Joists costs include appropriate bridging. Deflection is limited to 1/360 of the span. Screeds and steel trowel finish.

Design Loads	Min.	Max.
Joists	3.0 PSF	7.6 PSF
Slab Form	1.0	1.0
2-1/2" Concrete	27.0	27.0
Ceiling	3.0	3.0
Misc.	9.0	9.4
	43.0 PSF	48.0 PSF

B1010 246 Deck & Joists on Bearing Walls

	SPAN (FT.)	SUPERIMPOSED LOAD (P.S.F.)	JOIST SPACING FT. - IN.	DEPTH (IN.)	TOTAL LOAD (P.S.F.)	COST PER S.F. MAT.	COST PER S.F. INST.	COST PER S.F. TOTAL
1050	20	40	2-0	14-1/2	83	4.59	3.03	7.62
1070		65	2-0	16-1/2	109	4.96	3.22	8.18
1100		75	2-0	16-1/2	119	4.96	3.22	8.18
1120		100	2-0	18-1/2	145	4.99	3.23	8.22
1150		125	1-9	18-1/2	170	6	3.78	9.78
1170	25	40	2-0	18-1/2	84	5.45	3.53	8.98
1200		65	2-0	20-1/2	109	5.65	3.61	9.26
1220		75	2-0	20-1/2	119	5.50	3.44	8.94
1250		100	2-0	22-1/2	145	5.60	3.47	9.07
1270		125	1-9	22-1/2	170	6.35	3.78	10.13
1300	30	40	2-0	22-1/2	84	5.60	3.15	8.75
1320		65	2-0	24-1/2	110	6.05	3.29	9.34
1350		75	2-0	26-1/2	121	6.25	3.36	9.61
1370		100	2-0	26-1/2	146	6.70	3.48	10.18
1400		125	2-0	24-1/2	172	7.20	4.45	11.65
1420	35	40	2-0	26-1/2	85	6.85	4.27	11.12
1450		65	2-0	28-1/2	111	7.05	4.38	11.43
1470		75	2-0	28-1/2	121	7.05	4.38	11.43
1500		100	1-11	28-1/2	147	7.65	4.66	12.31
1520		125	1-8	28-1/2	172	8.40	5	13.40
1550	5/8" gyp. fireproof.							
1560	On metal furring, add					.67	2.34	3.01

B10 Superstructure

B1010 Floor Construction

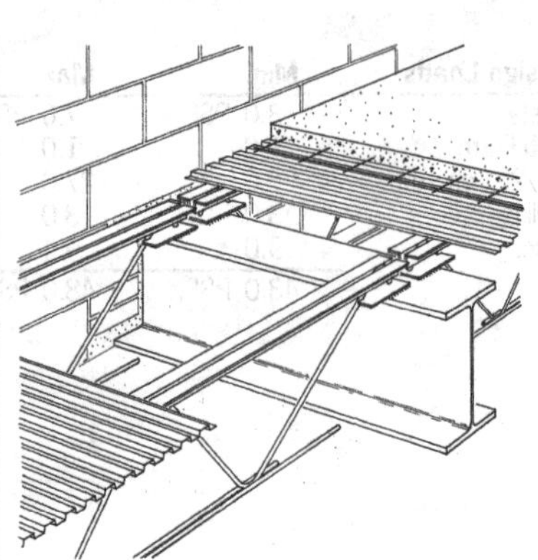

Table below lists costs for a floor system on exterior bearing walls and interior columns and beams using open web steel joists, galvanized steel slab form, 2-1/2" concrete slab reinforced with welded wire fabric.

Design and Pricing Assumptions:
Structural Steel is A36.
Concrete f'c = 3 KSI placed by pump.
WWF 6 x 6 – W1.4 x W1.4 (10 x 10)
Columns are 12' high.
Building is 4 bays long by 4 bays wide.
Joists are 2' O.C. ± and span the long direction of the bay.

Joists at columns have bottom chords extended and are connected to columns.

Slab form is 28 gauge galvanized.
Column costs in table are for columns to support 1 floor plus roof loading in a 2-story building; however, column costs are from ground floor to 2nd floor only. Joist costs include appropriate bridging.
Deflection is limited to 1/360 of the span.
Screeds and steel trowel finish.

Design Loads	Min.	Max.
S.S & Joists	4.4 PSF	11.5 PSF
Slab Form	1.0	1.0
2-1/2" Concrete	27.0	27.0
Ceiling	3.0	3.0
Misc.	7.6	5.5
	43.0 PSF	48.0 PSF

B1010 248 Steel Joists on Beam & Wall

	BAY SIZE (FT.)	SUPERIMPOSED LOAD (P.S.F.)	DEPTH (IN.)	TOTAL LOAD (P.S.F.)	COLUMN ADD	COST PER S.F. MAT.	INST.	TOTAL
1720	25x25	40	23	84		6.95	3.76	10.71
1730					columns	.35	.10	.45
1750	25x25	65	29	110		7.30	3.90	11.20
1760					columns	.35	.10	.45
1770	25x25	75	26	120		7.60	3.84	11.44
1780					columns	.41	.11	.52
1800	25x25	100	29	145		8.55	4.20	12.75
1810					columns	.41	.11	.52
1820	25x25	125	29	170		9	4.34	13.34
1830					columns	.46	.12	.58
2020	30x30	40	29	84		7.45	3.49	10.94
2030					columns	.32	.08	.40
2050	30x30	65	29	110		8.45	3.77	12.22
2060					columns	.32	.08	.40
2070	30x30	75	32	120		8.65	3.84	12.49
2080					columns	.36	.10	.46
2100	30x30	100	35	145		9.60	4.10	13.70
2110					columns	.43	.11	.54
2120	30x30	125	35	172		10.45	5.20	15.65
2130					columns	.51	.14	.65
2170	30x35	40	29	85		8.40	3.77	12.17
2180					columns	.31	.08	.39
2200	30x35	65	29	111		9.35	4.84	14.19
2210					columns	.35	.10	.45
2220	30x35	75	32	121		9.35	4.84	14.19
2230					columns	.36	.10	.46
2250	30x35	100	35	148		10.15	4.28	14.43
2260					columns	.44	.11	.55
2270	30x35	125	38	173		11.40	4.64	16.04
2280					columns	.45	.12	.57
2320	35x35	40	32	85		8.65	3.84	12.49
2330					columns	.31	.08	.39
2350	35x35	65	35	111		9.90	5	14.90
2360					columns	.38	.10	.48
2370	35x35	75	35	121		10.15	5.10	15.25
2380					columns	.38	.10	.48

B10 Superstructure

B1010 Floor Construction

B1010 248 Steel Joists on Beam & Wall

	BAY SIZE (FT.)	SUPERIMPOSED LOAD (P.S.F.)	DEPTH (IN.)	TOTAL LOAD (P.S.F.)	COLUMN ADD	COST PER S.F.		
						MAT.	INST.	TOTAL
2400	35x35	100	38	148		10.55	5.25	15.80
2410					columns	.46	.12	.58
2460	5/8 gyp. fireproof.							
2475	On metal furring, add					.67	2.34	3.01

B10 Superstructure

B1010 Floor Construction

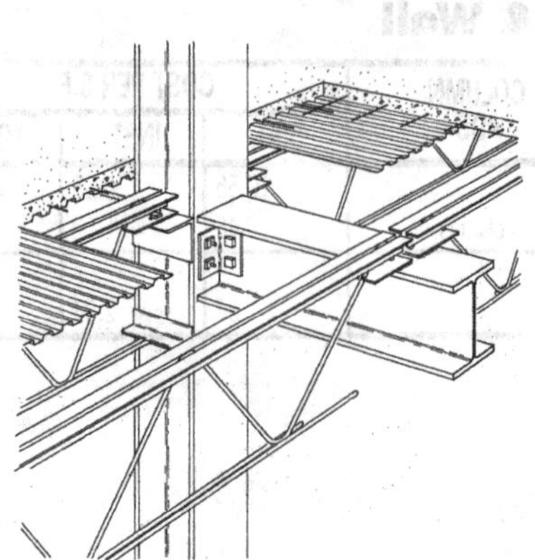

Table below lists costs for a floor system on steel columns and beams using open web steel joists, galvanized steel slab form, and 2-1/2″ concrete slab reinforced with welded wire fabric.

Design and Pricing Assumptions:

Structural Steel is A36.
Concrete f′c = 3 KSI placed by pump.
WWF 6 x 6 – W1.4 x W1.4 (10 x 10)
Columns are 12′ high.
Building is 4 bays long by 4 bays wide.
Joists are 2′ O.C. ± and span the long direction of the bay.

Joists at columns have bottom chords extended and are connected to columns.

Slab form is 28 gauge galvanized.
Column costs in table are for columns to support 1 floor plus roof loading in a 2-story building; however, column costs are from ground floor to 2nd floor only. Joist costs include appropriate bridging.
Deflection is limited to 1/360 of the span.
Screeds and steel trowel finish.

Design Loads	Min.	Max.
S.S. & Joists	6.3 PSF	15.3 PSF
Slab Form	1.0	1.0
2-1/2″ Concrete	27.0	27.0
Ceiling	3.0	3.0
Misc.	5.7	1.7
	43.0 PSF	48.0 PSF

B1010 250 Steel Joists, Beams & Slab on Columns

	BAY SIZE (FT.)	SUPERIMPOSED LOAD (P.S.F.)	DEPTH (IN.)	TOTAL LOAD (P.S.F.)	COLUMN ADD	COST PER S.F.		
						MAT.	INST.	TOTAL
2350	15x20	40	17	83		6.75	3.49	10.24
2400					column	1.17	.32	1.49
2450	15x20	65	19	108		7.55	3.71	11.26
2500					column	1.17	.32	1.49
2550	15x20	75	19	119		7.80	3.83	11.63
2600					column	1.28	.34	1.62
2650	15x20	100	19	144		8.35	3.98	12.33
2700					column	1.28	.34	1.62
2750	15x20	125	19	170		9.30	4.44	13.74
2800					column	1.71	.45	2.16
2850	20x20	40	19	83		7.45	3.64	11.09
2900					column	.96	.26	1.22
2950	20x20	65	23	109		8.20	3.89	12.09
3000					column	1.28	.34	1.62
3100	20x20	75	26	119		8.70	4.02	12.72
3200					column	1.28	.34	1.62
3400	20x20	100	23	144		9.05	4.13	13.18
3450					column	1.28	.34	1.62
3500	20x20	125	23	170		10	4.46	14.46
3600					column	1.54	.41	1.95
3700	20x25	40	44	83		8.40	4.14	12.54
3800					column	1.02	.27	1.29
3900	20x25	65	26	110		9.20	4.39	13.59
4000					column	1.02	.27	1.29
4100	20x25	75	26	120		9.05	4.20	13.25
4200					column	1.23	.33	1.56
4300	20x25	100	26	145		9.65	4.38	14.03
4400					column	1.23	.33	1.56
4500	20x25	125	29	170		10.70	4.77	15.47
4600					column	1.43	.38	1.81
4700	25x25	40	23	84		9.10	4.31	13.41
4800					column	.98	.26	1.24
4900	25x25	65	29	110		9.65	4.49	14.14
5000					column	.98	.26	1.24
5100	25x25	75	26	120		10.15	4.49	14.64
5200					column	1.15	.30	1.45

B10 Superstructure

B1010 Floor Construction

B1010 250	Steel Joists, Beams & Slab on Columns							
	BAY SIZE (FT.)	SUPERIMPOSED LOAD (P.S.F.)	DEPTH (IN.)	TOTAL LOAD (P.S.F.)	COLUMN ADD	COST PER S.F.		
						MAT.	INST.	TOTAL
5300	25x25	100	29	145		11.25	4.89	16.14
5400					column	1.15	.30	1.45
5500	25x25	125	32	170		11.90	5.10	17
5600					column	1.27	.34	1.61
5700	25x30	40	29	84		9.45	4.54	13.99
5800					column	.95	.25	1.20
5900	25x30	65	29	110		9.75	4.71	14.46
6000					column	.95	.25	1.20
6050	25x30	75	29	120		10.60	4.33	14.93
6100					column	1.06	.28	1.34
6150	25x30	100	29	145		11.45	4.57	16.02
6200					column	1.06	.28	1.34
6250	25x30	125	32	170		12.30	5.55	17.85
6300					column	1.22	.33	1.55
6350	30x30	40	29	84		9.90	4.11	14.01
6400					column	.88	.23	1.11
6500	30x30	65	29	110		11.25	4.49	15.74
6600					column	.88	.23	1.11
6700	30x30	75	32	120		11.50	4.56	16.06
6800					column	1.01	.27	1.28
6900	30x30	100	35	145		12.75	4.91	17.66
7000					column	1.18	.32	1.50
7100	30x30	125	35	172		13.90	6.05	19.95
7200					column	1.31	.35	1.66
7300	30x35	40	29	85		11.15	4.48	15.63
7400					column	.75	.20	.95
7500	30x35	65	29	111		12.30	5.60	17.90
7600					column	.97	.26	1.23
7700	30x35	75	32	121		12.35	5.60	17.95
7800					column	.99	.26	1.25
7900	30x35	100	35	148		13.40	5.10	18.50
8000					column	1.22	.33	1.55
8100	30x35	125	38	173		14.80	5.50	20.30
8200					column	1.24	.33	1.57
8300	35x35	40	32	85		11.45	4.55	16
8400					column	.87	.23	1.10
8500	35x35	65	35	111		13	5.80	18.80
8600					column	1.04	.28	1.32
9300	35x35	75	38	121		13.35	5.90	19.25
9400					column	1.04	.28	1.32
9500	35x35	100	38	148		14.40	6.25	20.65
9600					column	1.29	.34	1.63
9750	35x35	125	41	173		15.65	5.70	21.35
9800					column	1.31	.35	1.66
9810	5/8 gyp. fireproof.							
9815	On metal furring, add					.67	2.34	3.01

B10 Superstructure

B1010 Floor Construction

Description: Table B1010 258 lists S.F. costs for steel deck and concrete slabs for various spans.

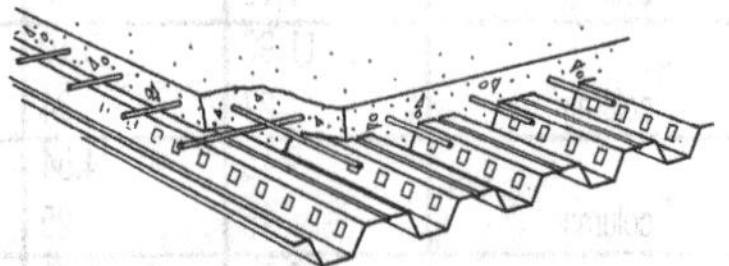

Metal Deck/Concrete Fill

Description: Table B1010 261 lists the S.F. costs for wood joists and a minimum thickness plywood subfloor.

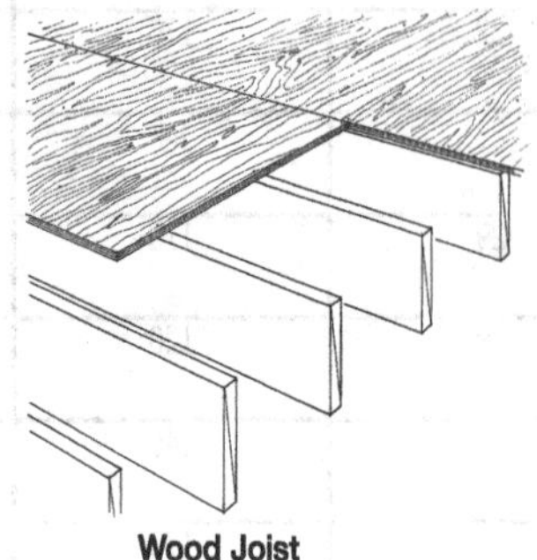

Wood Joist

Description: Table B1010 264 lists the S.F. costs, total load, and member sizes, for various bay sizes and loading conditions.

Wood Beam and Joist

B1010 258 Metal Deck/Concrete Fill

	SUPERIMPOSED LOAD (P.S.F.)	DECK SPAN (FT.)	DECK GAGE DEPTH	SLAB THICKNESS (IN.)	TOTAL LOAD (P.S.F.)	COST PER S.F.		
						MAT.	INST.	TOTAL
0900	125	6	22 1-1/2	4	164	2.37	1.95	4.32
0920		7	20 1-1/2	4	164	2.60	2.08	4.68
0950		8	20 1-1/2	4	165	2.60	2.08	4.68
0970		9	18 1-1/2	4	165	2.99	2.08	5.07
1000		10	18 2	4	165	2.92	2.19	5.11
1020		11	18 3	5	169	3.51	2.37	5.88

B1010 261 Wood Joist

		COST PER S.F.		
		MAT.	INST.	TOTAL
2902	Wood joist, 2"x8", 12" O.C.	1.55	1.53	3.08
2950	16" O.C.	1.33	1.31	2.64
3000	24" O.C.	1.39	1.22	2.61
3300	2"x10", 12" O.C.	2.02	1.74	3.76
3350	16" O.C.	1.68	1.46	3.14
3400	24" O.C.	1.62	1.31	2.93
3700	2"x12", 12" O.C.	2.32	1.76	4.08
3750	16" O.C.	1.91	1.48	3.39
3800	24" O.C.	1.78	1.33	3.11
4100	2"x14", 12" O.C.	2.72	1.91	4.63
4150	16" O.C.	2.21	1.60	3.81
4200	24" O.C.	1.98	1.41	3.39
7101	Note: Subfloor cost is included in these prices.			

B1010 264 Wood Beam & Joist

	BAY SIZE (FT.)	SUPERIMPOSED LOAD (P.S.F.)	GIRDER BEAM (IN.)	JOISTS (IN.)	TOTAL LOAD (P.S.F.)	COST PER S.F.		
						MAT.	INST.	TOTAL
2000	15x15	40	8 x 12 4 x 12	2 x 6 @ 16	53	9.10	3.54	12.64
2050		75	8 x 16 4 x 16	2 x 8 @ 16	90	11.90	3.87	15.77
2100		125	12 x 16 6 x 16	2 x 8 @ 12	144	18.45	4.97	23.42
2150		200	14 x 22 12 x 16	2 x 10 @ 12	227	37	7.60	44.60
3000	20x20	40	10 x 14 10 x 12	2 x 8 @ 16	63	11.05	3.57	14.62
3050		75	12 x 16 8 x 16	2 x 10 @ 16	102	15.90	4	19.90
3100		125	14 x 22 12 x 16	2 x 10 @ 12	163	31	6.35	37.35

B10 Superstructure

B1010 Floor Construction

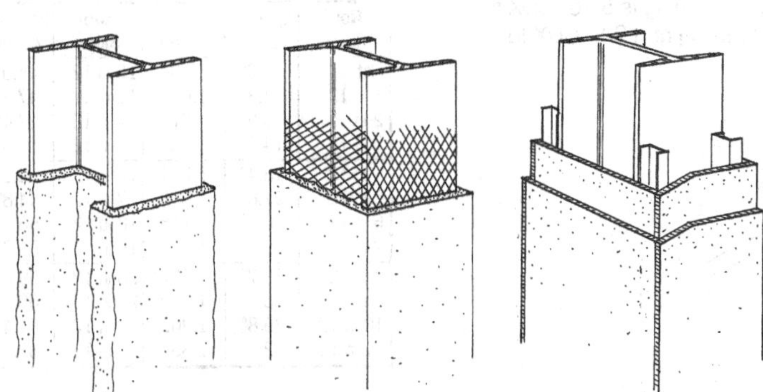

Listed below are costs per V.L.F. for fireproofing by material, column size, thickness and fire rating. Weights listed are for the fireproofing material only.

B1010 720 Steel Column Fireproofing

	ENCASEMENT SYSTEM	COLUMN SIZE (IN.)	THICKNESS (IN.)	FIRE RATING (HRS.)	WEIGHT (P.L.F.)	COST PER V.L.F.		
						MAT.	INST.	TOTAL
3000	Concrete	8	1	1	110	5.90	27	32.90
3300		14	1	1	258	9.95	39	48.95
3400			2	3	325	11.90	44.50	56.40
3450	Gypsum board	8	1/2	2	8	2.92	17.40	20.32
3550	1 layer	14	1/2	2	18	3.16	18.60	21.76
3600	Gypsum board	8	1	3	14	4.04	22	26.04
3650	1/2" fire rated	10	1	3	17	4.32	23.50	27.82
3700	2 layers	14	1	3	22	4.48	24.50	28.98
3750	Gypsum board	8	1-1/2	3	23	5.35	28	33.35
3800	1/2" fire rated	10	1-1/2	3	27	6	31	37
3850	3 layers	14	1-1/2	3	35	6.65	34	40.65
3900	Sprayed fiber	8	1-1/2	2	6.3	3.93	5.60	9.53
3950	Direct application		2	3	8.3	5.40	7.75	13.15
4050		10	1-1/2	2	7.9	4.75	6.80	11.55
4200		14	1-1/2	2	10.8	5.90	8.45	14.35

B10 Superstructure

B1020 Roof Construction

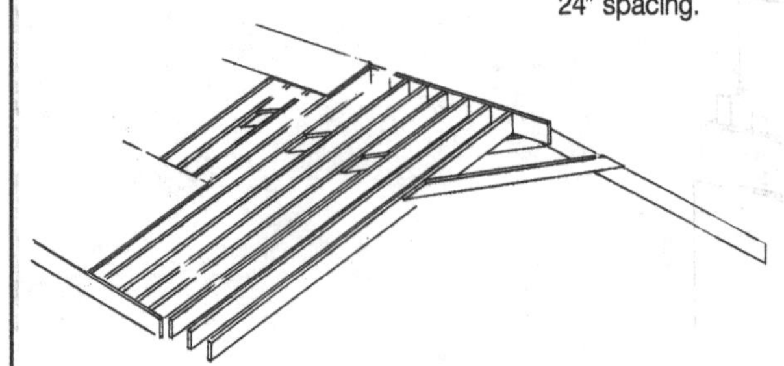

The table below lists prices per S.F. for roof rafters and sheathing by nominal size and spacing. Sheathing is 5/16" CDX for 12" and 16" spacing and 3/8" CDX for 24" spacing.

Factors for Converting Inclined to Horizontal

Roof Slope	Approx. Angle	Factor	Roof Slope	Approx. Angle	Factor
Flat	0°	1.000	12 in 12	45.0°	1.414
1 in 12	4.8°	1.003	13 in 12	47.3°	1.474
2 in 12	9.5°	1.014	14 in 12	49.4°	1.537
3 in 12	14.0°	1.031	15 in 12	51.3°	1.601
4 in 12	18.4°	1.054	16 in 12	53.1°	1.667
5 in 12	22.6°	1.083	17 in 12	54.8°	1.734
6 in 12	26.6°	1.118	18 in 12	56.3°	1.803
7 in 12	30.3°	1.158	19 in 12	57.7°	1.873
8 in 12	33.7°	1.202	20 in 12	59.0°	1.943
9 in 12	36.9°	1.250	21 in 12	60.3°	2.015
10 in 12	39.8°	1.302	22 in 12	61.4°	2.088
11 in 12	42.5°	1.357	23 in 12	62.4°	2.162

B1020 102	Wood/Flat or Pitched	COST PER S.F. MAT.	INST.	TOTAL
2500	Flat rafter, 2"x4", 12" O.C.	1.03	1.37	2.40
2550	16" O.C.	.93	1.18	2.11
2600	24" O.C.	.75	1.02	1.77
2900	2"x6", 12" O.C.	1.27	1.38	2.65
2950	16" O.C.	1.11	1.19	2.30
3000	24" O.C.	.87	1.02	1.89
3300	2"x8", 12" O.C.	1.51	1.49	3
3350	16" O.C.	1.29	1.27	2.56
3400	24" O.C.	.99	1.08	2.07
3700	2"x10", 12" O.C.	1.98	1.70	3.68
3750	16" O.C.	1.64	1.42	3.06
3800	24" O.C.	1.22	1.17	2.39
4100	2"x12", 12" O.C.	2.28	1.72	4
4150	16" O.C.	1.87	1.44	3.31
4200	24" O.C.	1.38	1.19	2.57
4500	2"x14", 12" O.C.	2.68	2.51	5.19
4550	16" O.C.	2.17	2.04	4.21
4600	24" O.C.	1.58	1.59	3.17
4900	3"x6", 12" O.C.	2.53	1.46	3.99
4950	16" O.C.	2.06	1.25	3.31
5000	24" O.C.	1.50	1.06	2.56
5300	3"x8", 12" O.C.	3.22	1.64	4.86
5350	16" O.C.	2.57	1.37	3.94
5400	24" O.C.	1.84	1.14	2.98
5700	3"x10", 12" O.C.	3.86	1.86	5.72
5750	16" O.C.	3.06	1.54	4.60
5800	24" O.C.	2.17	1.25	3.42
6100	3"x12", 12" O.C.	4.59	2.24	6.83
6150	16" O.C.	3.61	1.83	5.44
6200	24" O.C.	2.53	1.45	3.98
7001	Wood truss, 4 in 12 slope, 24" O.C., 24' to 29' span	3.44	2.04	5.48
7100	30' to 43' span	4.19	2.04	6.23
7200	44' to 60' span	4.38	2.04	6.42

B10 Superstructure

B1020 Roof Construction

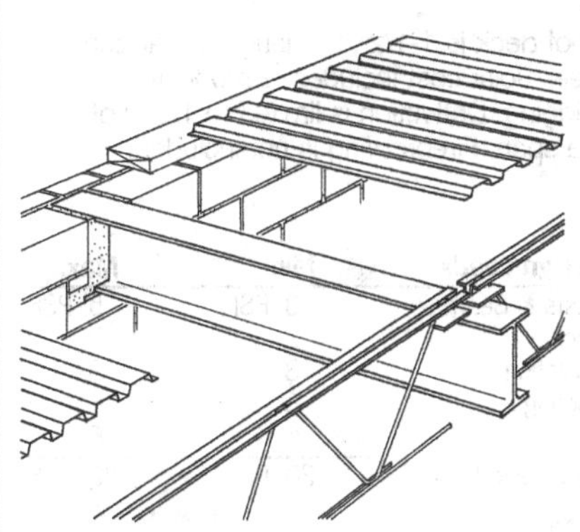

Table below lists the cost per S.F. for a roof system with steel columns, beams, and deck using open web steel joists and 1-1/2" galvanized metal deck. Perimeter of system is supported on bearing walls.

Design and Pricing Assumptions:
Columns are 18' high.
Joists are 5'-0" O.C. and span the long direction of the bay.

Joists at columns have bottom chords extended and are connected to columns. Column costs are not included but are listed separately per S.F. of floor.

Roof deck is 1-1/2", 22 gauge galvanized steel. Joist cost includes appropriate bridging. Deflection is limited to 1/240 of the span. Fireproofing is not included.

Costs/S.F. are based on a building 4 bays long and 4 bays wide.

Design Loads	Min.	Max.
Joists & Beams	3 PSF	5 PSF
Deck	2	2
Insulation	3	3
Roofing	6	6
Misc.	6	6
Total Dead Load	20 PSF	22 PSF

B1020 108 Steel Joists, Beams & Deck on Columns & Walls

	BAY SIZE (FT.)	SUPERIMPOSED LOAD (P.S.F.)	DEPTH (IN.)	TOTAL LOAD (P.S.F.)	COLUMN ADD	COST PER S.F.		
						MAT.	INST.	TOTAL
3000	25x25	20	18	40		3.29	1.10	4.39
3100					columns	.33	.07	.40
3200		30	22	50		3.73	1.29	5.02
3300					columns	.44	.09	.53
3400		40	20	60		3.95	1.29	5.24
3500					columns	.44	.09	.53
3600	25x30	20	22	40		3.47	1.07	4.54
3700					columns	.37	.08	.45
3800		30	20	50		3.78	1.37	5.15
3900					columns	.37	.08	.45
4000		40	25	60		4.12	1.22	5.34
4100					columns	.44	.09	.53
4200	30x30	20	25	42		3.85	1.14	4.99
4300					columns	.31	.07	.38
4400		30	22	52		4.22	1.25	5.47
4500					columns	.37	.08	.45
4600		40	28	62		4.40	1.29	5.69
4700					columns	.37	.08	.45
4800	30x35	20	22	42		3.99	1.19	5.18
4900					columns	.31	.07	.38
5000		30	28	52		4.23	1.25	5.48
5100					columns	.31	.07	.38
5200		40	25	62		4.59	1.35	5.94
5300					columns	.37	.08	.45
5400	35x35	20	28	42		4	1.20	5.20
5500					columns	.27	.06	.33
5600		30	25	52		4.91	1.42	6.33
5700					columns	.31	.07	.38
5800		40	28	62		5	1.44	6.44
5900					columns	.35	.07	.42

B10 Superstructure

B1020 Roof Construction

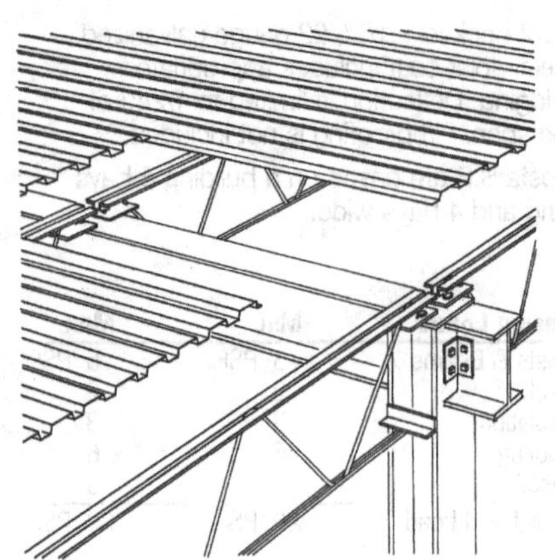

Description: Table below lists the cost per S.F. for a roof system with steel columns, beams, and deck, using open web steel joists and 1-1/2" galvanized metal deck.

Design and Pricing Assumptions:
Columns are 18′ high.
Building is 4 bays long by 4 bays wide.
Joists are 5′-0″ O.C. and span the long direction of the bay.
Joists at columns have bottom chords extended and are connected to columns.
Column costs are not included but are listed separately per S.F. of floor.

Roof deck is 1-1/2", 22 gauge galvanized steel. Joist cost includes appropriate bridging. Deflection is limited to 1/240 of the span. Fireproofing is not included.

Design Loads	Min.	Max.
Joists & Beams	3 PSF	5 PSF
Deck	2	2
Insulation	3	3
Roofing	6	6
Misc.	6	6
Total Dead Load	20 PSF	22 PSF

B1020 112 Steel Joists, Beams, & Deck on Columns

	BAY SIZE (FT.)	SUPERIMPOSED LOAD (P.S.F.)	DEPTH (IN.)	TOTAL LOAD (P.S.F.)	COLUMN ADD	COST PER S.F.		
						MAT.	INST.	TOTAL
1100	15x20	20	16	40		3.24	1.09	4.33
1200					columns	1.92	.40	2.32
1500		40	18	60		3.65	1.22	4.87
1600					columns	1.92	.40	2.32
2300	20x25	20	18	40		3.72	1.22	4.94
2400					columns	1.15	.24	1.39
2700		40	20	60		4.20	1.37	5.57
2800					columns	1.54	.33	1.87
2900	25x25	20	18	40		4.44	1.38	5.82
3000					columns	.92	.19	1.11
3100		30	22	50		4.80	1.55	6.35
3200					columns	1.23	.26	1.49
3300		40	20	60		5.15	1.56	6.71
3400					columns	1.23	.26	1.49
3500	25x30	20	22	40		4.33	1.25	5.58
3600					columns	1.02	.22	1.24
3900		40	25	60		5.25	1.49	6.74
4000					columns	1.23	.26	1.49
4100	30x30	20	25	42		4.94	1.39	6.33
4200					columns	.85	.18	1.03
4300		30	22	52		5.45	1.53	6.98
4400					columns	1.02	.22	1.24
4500		40	28	62		5.75	1.59	7.34
4600					columns	1.02	.22	1.24
5300	35x35	20	28	42		5.40	1.51	6.91
5400					columns	.75	.16	.91
5500		30	25	52		6.10	1.67	7.77
5600					columns	.88	.18	1.06
5700		40	28	62		6.60	1.79	8.39
5800					columns	.97	.20	1.17

B10 Superstructure

B1020 Roof Construction

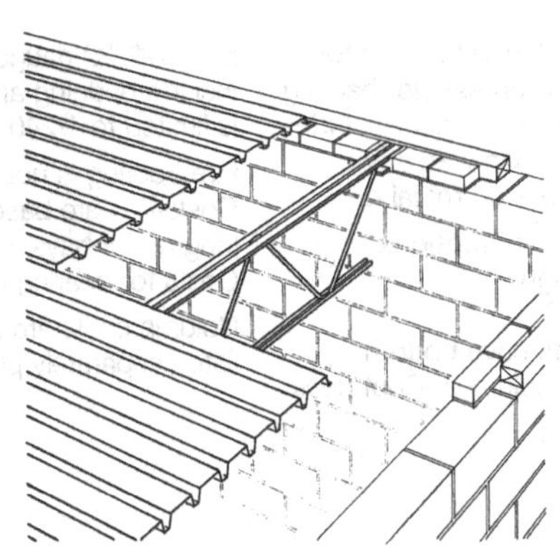

Description: Table below lists cost per S.F. for a roof system using open web steel joists and 1-1/2" galvanized metal deck. The system is assumed supported on bearing walls or other suitable support. Costs for the supports are not included.

Design and Pricing Assumptions:
Joists are 5'-0" O.C.
Roof deck is 1-1/2", 22 gauge galvanized.

B1020 116 Steel Joists & Deck on Bearing Walls

	BAY SIZE (FT.)	SUPERIMPOSED LOAD (P.S.F.)	DEPTH (IN.)	TOTAL LOAD (P.S.F.)		COST PER S.F.		
						MAT.	INST.	TOTAL
1100	20	20	13-1/2	40		1.96	.80	2.76
1200		30	15-1/2	50		2	.82	2.82
1300		40	15-1/2	60		2.15	.88	3.03
1400	25	20	17-1/2	40		2.16	.88	3.04
1500		30	17-1/2	50		2.34	.96	3.30
1600		40	19-1/2	60		2.37	.98	3.35
1700	30	20	19-1/2	40		2.36	.97	3.33
1800		30	21-1/2	50		2.42	.99	3.41
1900		40	23-1/2	60		2.61	1.07	3.68
2000	35	20	23-1/2	40		2.59	.91	3.50
2100		30	25-1/2	50		2.67	.93	3.60
2200		40	25-1/2	60		2.84	.98	3.82
2300	40	20	25-1/2	41		2.88	1	3.88
2400		30	25-1/2	51		3.07	1.05	4.12
2500		40	25-1/2	61		3.18	1.09	4.27
2600	45	20	27-1/2	41		3.22	1.46	4.68
2700		30	31-1/2	51		3.41	1.55	4.96
2800		40	31-1/2	61		3.60	1.64	5.24
2900	50	20	29-1/2	42		3.60	1.64	5.24
3000		30	31-1/2	52		3.94	1.81	5.75
3100		40	31-1/2	62		4.18	1.92	6.10
3200	60	20	37-1/2	42		4.67	1.69	6.36
3300		30	37-1/2	52		5.20	1.87	7.07
3400		40	37-1/2	62		5.20	1.87	7.07
3500	70	20	41-1/2	42		5.20	1.86	7.06
3600		30	41-1/2	52		5.55	1.99	7.54
3700		40	41-1/2	64		6.75	2.41	9.16
3800	80	20	45-1/2	44		7.55	2.62	10.17
3900		30	45-1/2	54		7.55	2.62	10.17
4000		40	45-1/2	64		8.35	2.90	11.25
4400	100	20	57-1/2	44		6.45	2.22	8.67
4500		30	57-1/2	54		7.75	2.64	10.39
4600		40	57-1/2	65		8.55	2.91	11.46
4700	125	20	69-1/2	44		9.10	2.96	12.06
4800		30	69-1/2	56		10.60	3.44	14.04
4900		40	69-1/2	67		12.15	3.92	16.07

B10 Superstructure

B1020 Roof Construction

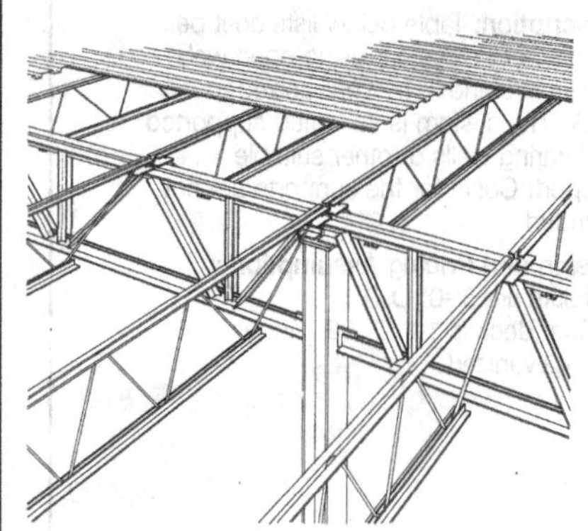

Description: Table below lists costs for a roof system supported on exterior bearing walls and interior columns. Costs include bracing, joist girders, open web steel joists and 1-1/2" galvanized metal deck.

Design and Pricing Assumptions:
Columns are 18' high.
Joists are 5'-0" O.C.
Joist girders and joists have bottom chords connected to columns. Roof deck is 1-1/2", 22 gauge galvanized steel. Costs include bridging and bracing. Deflection is limited to 1/240 of the span.

Fireproofing is not included.
Costs/S.F. are based on a building 4 bays long and 4 bays wide.
Costs for bearing walls are not included.

Column costs are not included but are listed separately per S.F. of floor.

B1020 120 Steel Joists, Joist Girders & Deck on Columns & Walls

	BAY SIZE (FT.) GIRD X JOISTS	SUPERIMPOSED LOAD (P.S.F.)	DEPTH (IN.)	TOTAL LOAD (P.S.F.)	COLUMN ADD	COST PER S.F.		
						MAT.	INST.	TOTAL
2350	30x35	20	32-1/2	40		2.89	1.15	4.04
2400					columns	.32	.07	.39
2550		40	36-1/2	60		3.25	1.27	4.52
2600					columns	.37	.08	.45
3000	35x35	20	36-1/2	40		3.17	1.41	4.58
3050					columns	.27	.06	.33
3200		40	36-1/2	60		3.58	1.54	5.12
3250					columns	.35	.07	.42
3300	35x40	20	36-1/2	40		3.21	1.29	4.50
3350					columns	.28	.06	.34
3500		40	36-1/2	60		3.62	1.44	5.06
3550					columns	.30	.07	.37
3900	40x40	20	40-1/2	41		3.55	1.64	5.19
3950					columns	.27	.06	.33
4100		40	40-1/2	61		3.98	1.81	5.79
4150					columns	.27	.06	.33
5100	45x50	20	52-1/2	41		3.99	1.89	5.88
5150					columns	.19	.04	.23
5300		40	52-1/2	61		4.85	2.28	7.13
5350					columns	.28	.06	.34
5400	50x45	20	56-1/2	41		3.87	1.83	5.70
5450					columns	.19	.04	.23
5600		40	56-1/2	61		4.97	2.32	7.29
5650					columns	.26	.06	.32
5700	50x50	20	56-1/2	42		4.34	2.05	6.39
5750					columns	.19	.04	.23
5900		40	59	64		5.95	2.31	8.26
5950					columns	.27	.06	.33
6300	60x50	20	62-1/2	43		4.38	2.08	6.46
6350					columns	.31	.07	.38
6500		40	71	65		6.10	2.38	8.48
6550					columns	.41	.08	.49

B10 Superstructure

B1020 Roof Construction

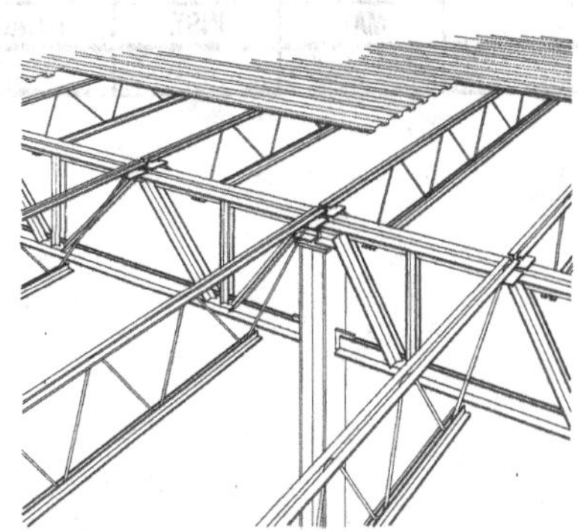

Description: Table below lists the cost per S.F. for a roof system supported on columns. Costs include joist girders, open web steel joists and 1-1/2" galvanized metal deck.

Design and Pricing Assumptions:
Columns are 18′ high.
Joists are 5′-0″ O.C.
Joist girders and joists have bottom chords connected to columns. Roof deck is 1-1/2", 22 gauge galvanized steel. Costs include bridging and bracing. Deflection is limited to 1/240 of the span. Fireproofing is not included.

Costs/S.F. are based on a building 4 bays long and 4 bays wide.

Costs for columns are not included, but are listed separately per S.F. of area.

B1020 124 Steel Joists & Joist Girders on Columns

	BAY SIZE (FT.) GIRD X JOISTS	SUPERIMPOSED LOAD (P.S.F.)	DEPTH (IN.)	TOTAL LOAD (P.S.F.)	COLUMN ADD	COST PER S.F.		
						MAT.	INST.	TOTAL
2000	30x30	20	17-1/2	40		2.95	1.32	4.27
2050					columns	1.02	.22	1.24
2100		30	17-1/2	50		3.21	1.42	4.63
2150					columns	1.02	.22	1.24
2200		40	21-1/2	60		3.29	1.45	4.74
2250					columns	1.02	.22	1.24
2300	30x35	20	32-1/2	40		3.11	1.27	4.38
2350					columns	.88	.19	1.07
2500		40	36-1/2	60		3.51	1.41	4.92
2550					columns	1.02	.22	1.24
3200	35x40	20	36-1/2	40		3.89	1.95	5.84
3250					columns	.77	.17	.94
3400		40	36-1/2	60		4.38	2.16	6.54
3450					columns	.85	.18	1.03
3800	40x40	20	40-1/2	41		3.94	1.95	5.89
3850					columns	.74	.16	.90
4000		40	40-1/2	61		4.42	2.16	6.58
4050					columns	.74	.16	.90
4100	40x45	20	40-1/2	41		4.24	2.20	6.44
4150					columns	.66	.14	.80
4200		30	40-1/2	51		4.61	2.36	6.97
4250					columns	.66	.14	.80
4300		40	40-1/2	61		5.10	2.59	7.69
4350					columns	.74	.16	.90
5000	45x50	20	52-1/2	41		4.46	2.26	6.72
5050					columns	.53	.11	.64
5200		40	52-1/2	61		5.45	2.70	8.15
5250					columns	.68	.15	.83
5300	50x45	20	56-1/2	41		4.35	2.19	6.54
5350					columns	.53	.11	.64
5500		40	56-1/2	61		5.60	2.75	8.35
5550					columns	.68	.15	.83
5600	50x50	20	56-1/2	42		4.85	2.42	7.27
5650					columns	.54	.11	.65
5800		40	59	64		5.35	2.21	7.56
5850					columns	.75	.16	.91

B10 Superstructure

B1020 Roof Construction

B1020 310	Canopies	COST PER S.F. MAT.	COST PER S.F. INST.	COST PER S.F. TOTAL
0100	Canopies, wall hung, prefinished aluminum, 8′ x 10′	2,350	1,325	3,675

B20 Exterior Enclosure

B2010 Exterior Walls

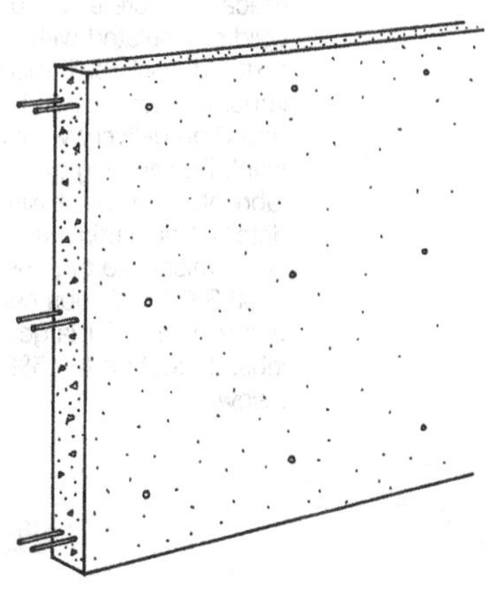

The table below describes a concrete wall system for exterior closure. There are several types of wall finishes priced from plain finish to a finish with 3/4″ rustication strip.

Design Assumptions:
Conc. f'c = 3000 to 5000 psi
Reinf. fy = 60,000 psi

B2010 101	Cast In Place Concrete	COST PER S.F.		
		MAT.	INST.	TOTAL
2100	Conc wall reinforced, 8′ high, 6″ thick, plain finish, 3000 PSI	4.22	11.95	16.17
2700	Aged wood liner, 3000 PSI	5.25	13.70	18.95
3000	Sand blast light 1 side, 3000 PSI	5	13.75	18.75
3700	3/4″ bevel rustication strip, 3000 PSI	4.36	12.75	17.11
4000	8″ thick, plain finish, 3000 PSI	5.05	12.30	17.35
4100	4000 PSI	5.15	12.30	17.45
4300	Rub concrete 1 side, 3000 PSI	5.05	14.65	19.70
4550	8″ thick, aged wood liner, 3000 PSI	6.10	14.05	20.15
4750	Sand blast light 1 side, 3000 PSI	5.85	14.10	19.95
5300	3/4″ bevel rustication strip, 3000 PSI	5.20	13.10	18.30
5600	10″ thick, plain finish, 3000 PSI	5.85	12.60	18.45
5900	Rub concrete 1 side, 3000 PSI	5.85	14.95	20.80
6200	Aged wood liner, 3000 PSI	6.85	14.35	21.20
6500	Sand blast light 1 side, 3000 PSI	6.65	14.40	21.05
7100	3/4″ bevel rustication strip, 3000 PSI	5.95	13.40	19.35
7400	12″ thick, plain finish, 3000 PSI	6.75	13	19.75
7700	Rub concrete 1 side, 3000 PSI	6.75	15.30	22.05
8000	Aged wood liner, 3000 PSI	7.80	14.75	22.55
8300	Sand blast light 1 side, 3000 PSI	7.55	14.75	22.30
8900	3/4″ bevel rustication strip, 3000 PSI	6.90	13.75	20.65

B20 Exterior Enclosure

B2010 Exterior Walls

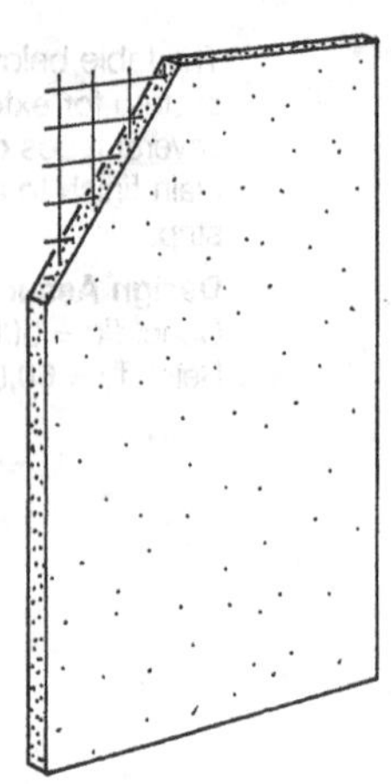

Precast concrete wall panels are either solid or insulated with plain, colored or textured finishes. Transportation is an important cost factor. Prices below are based on delivery within fifty miles of a plant. Engineering data is available from fabricators to assist with construction details. Usual minimum job size for economical use of panels is about 5000 S.F. Small jobs can double the prices below. For large, highly repetitive jobs, deduct up to 15% from the prices below.

B2010 102 Flat Precast Concrete

	THICKNESS (IN.)	PANEL SIZE (FT.)	FINISHES	RIGID INSULATION (IN)	TYPE	COST PER S.F.		
						MAT.	INST.	TOTAL
3000	4	5x18	smooth gray	none	low rise	12.95	5.10	18.05
3050		6x18				10.80	4.25	15.05
3100		8x20				21.50	2.06	23.56
3150		12x20				20	1.95	21.95
3200	6	5x18	smooth gray	2	low rise	13.80	5.55	19.35
3250		6x18				11.70	4.71	16.41
3300		8x20				22.50	2.58	25.08
3350		12x20				20.50	2.37	22.87
3400	8	5x18	smooth gray	2	low rise	29	3.22	32.22
3450		6x18				27.50	3.07	30.57
3500		8x20				25.50	2.84	28.34
3550		12x20				23	2.61	25.61
3600	4	4x8	white face	none	low rise	49.50	2.95	52.45
3650		8x8				37	2.21	39.21
3700		10x10				32.50	1.94	34.44
3750		20x10				29.50	1.75	31.25
3800	5	4x8	white face	none	low rise	50.50	3	53.50
3850		8x8				38	2.27	40.27
3900		10x10				33.50	2.01	35.51
3950		20x20				31	1.85	32.85
4000	6	4x8	white face	none	low rise	52	3.12	55.12
4050		8x8				39.50	2.37	41.87
4100		10x10				35	2.08	37.08
4150		20x10				32	1.91	33.91
4200	6	4x8	white face	2	low rise	53	3.62	56.62
4250		8x8				40.50	2.87	43.37
4300		10x10				36	2.58	38.58
4350		20x10				32	1.91	33.91
4400	7	4x8	white face	none	low rise	53.50	3.19	56.69
4450		8x8				41	2.46	43.46
4500		10x10				37	2.20	39.20
4550		20x10				33.50	2.01	35.51

B20 Exterior Enclosure

B2010 Exterior Walls

B2010 102 Flat Precast Concrete

	THICKNESS (IN.)	PANEL SIZE (FT.)	FINISHES	RIGID INSULATION (IN)	TYPE	COST PER S.F.		
						MAT.	INST.	TOTAL
4600	7	4x8	white face	2	low rise	54.50	3.69	58.19
4650		8x8				42	2.96	44.96
4700		10x10				38	2.70	40.70
4750		20x10				34.50	2.51	37.01
4800	8	4x8	white face	none	low rise	54.50	3.27	57.77
4850		8x8				42	2.51	44.51
4900		10x10				38	2.26	40.26
4950		20x10				35	2.08	37.08
5000		4x8	white face	2	low rise	55.50	3.77	59.27
5050		8x8				43	3.01	46.01
5100		10x10				39	2.76	41.76
5150		20x10				36	2.58	38.58

B2010 103 Fluted Window or Mullion Precast Concrete

	THICKNESS (IN.)	PANEL SIZE (FT.)	FINISHES	RIGID INSULATION (IN)	TYPE	COST PER S.F.		
						MAT.	INST.	TOTAL
5200	4	4x8	smooth gray	none	high rise	29.50	11.65	41.15
5250		8x8				21	8.35	29.35
5300		10x10				40.50	3.93	44.43
5350		20x10				35.50	3.45	38.95
5400	5	4x8	smooth gray	none	high rise	30	11.80	41.80
5450		8x8				22	8.60	30.60
5500		10x10				42.50	4.09	46.59
5550		20x10				37.50	3.63	41.13
5600	6	4x8	smooth gray	none	high rise	31	12.15	43.15
5650		8x8				22.50	8.90	31.40
5700		10x10				43.50	4.21	47.71
5750		20x10				39	3.74	42.74
5800		4x8	smooth gray	2	high rise	32	12.65	44.65
5850		8x8				23.50	9.40	32.90
5900		10x10				44.50	4.71	49.21
5950		20x10				40	4.24	44.24
6000	7	4x8	smooth gray	none	high rise	31.50	12.40	43.90
6050		8x8				23.50	9.15	32.65
6100		10x10				45.50	4.40	49.90
6150		20x10				40	3.86	43.86
6200		4x8	smooth gray	2	high rise	32.50	12.90	45.40
6250		8x8				24.50	9.65	34.15
6300		10x10				46.50	4.90	51.40
6350		20x10				41	4.36	45.36
6400	8	4x8	smooth gray	none	high rise	32	12.60	44.60
6450		8x8				24	9.40	33.40
6500		10x10				47	4.54	51.54
6550		20x10				42	4.05	46.05
6600		4x8	smooth gray	2	high rise	33	13.10	46.10
6650		8x8				25	9.90	34.90
6700		10x10				48	5.05	53.05
6750		20x10				43	4.55	47.55

B20 Exterior Enclosure

B2010 Exterior Walls

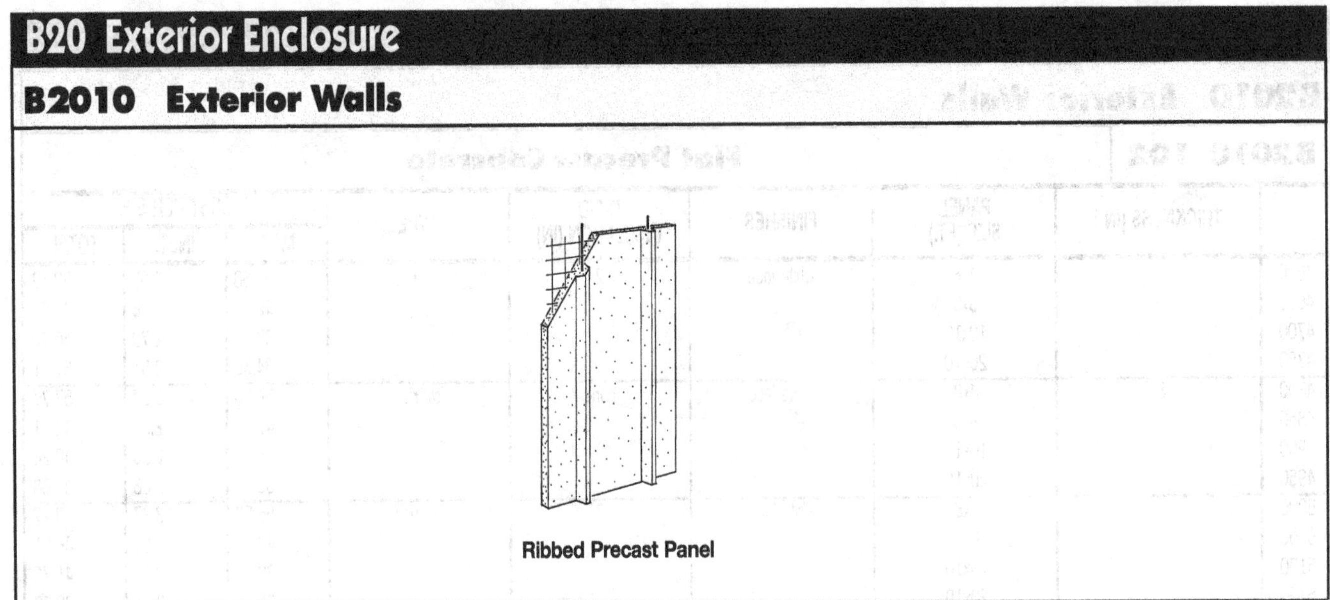
Ribbed Precast Panel

B2010 104 Ribbed Precast Concrete

	THICKNESS (IN.)	PANEL SIZE (FT.)	FINISHES	RIGID INSULATION(IN.)	TYPE	COST PER S.F.		
						MAT.	INST.	TOTAL
6800	4	4x8	aggregate	none	high rise	34.50	11.65	46.15
6850		8x8				25.50	8.50	34
6900		10x10				43	3.07	46.07
6950		20x10				38	2.72	40.72
7000	5	4x8	aggregate	none	high rise	35	11.80	46.80
7050		8x8				26	8.75	34.75
7100		10x10				45	3.18	48.18
7150		20x10				40	2.85	42.85
7200	6	4x8	aggregate	none	high rise	36	12.05	48.05
7250		8x8				26.50	9	35.50
7300		10x10				46.50	3.29	49.79
7350		20x10				41.50	2.95	44.45
7400		4x8	aggregate	2	high rise	37	12.55	49.55
7450		8x8				27.50	9.50	37
7500		10x10				47.50	3.79	51.29
7550		20x10				42.50	3.45	45.95
7600	7	4x8	aggregate	none	high rise	36.50	12.35	48.85
7650		8x8				27.50	9.15	36.65
7700		10x10				47.50	3.39	50.89
7750		20x10				42.50	3.03	45.53
7800		4x8	aggregate	2	high rise	37.50	12.85	50.35
7850		8x8				28.50	9.65	38.15
7900		10x10				48.50	3.89	52.39
7950		20x10				43.50	3.53	47.03
8000	8	4x8	aggregate	none	high rise	37.50	12.55	50.05
8050		8x8				28	9.50	37.50
8100		10x10				49.50	3.51	53.01
8150		20x10				44.50	3.17	47.67
8200		4x8	aggregate	2	high rise	38.50	13.05	51.55
8250		8x8				29	10	39
8300		10x10				50.50	4.01	54.51
8350		20x10				45.50	3.67	49.17

B20 Exterior Enclosure

B2010 Exterior Walls

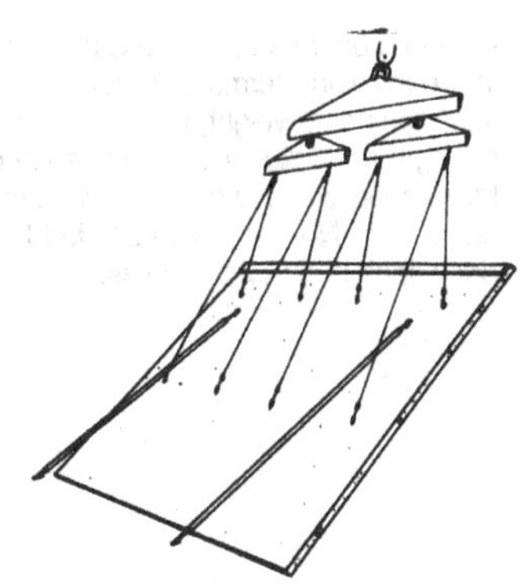

The advantage of tilt up construction is in the low cost of forms and placing of concrete and reinforcing. Tilt up has been used for several types of buildings, including warehouses, stores, offices, and schools. The panels are cast in forms on the ground, or floor slab. Most jobs use 5-1/2" thick solid reinforced concrete panels.

Design Assumptions:
Conc. f'c = 3000 psi
Reinf. fy = 60,000

B2010 106	Tilt-Up Concrete Panel	COST PER S.F.		
		MAT.	INST.	TOTAL
3200	Tilt up conc panels, broom finish, 5-1/2" thick, 3000 PSI	3.75	4.70	8.45
3250	5000 PSI	3.88	4.65	8.53
3300	6" thick, 3000 PSI	4.13	4.85	8.98
3350	5000 PSI	4.28	4.77	9.05
3400	7-1/2" thick, 3000 PSI	5.25	5.05	10.30
3450	5000 PSI	5.45	4.94	10.39
3500	8" thick, 3000 PSI	5.65	5.15	10.80
3550	5000 PSI	5.90	5.05	10.95
3700	Steel trowel finish, 5-1/2" thick, 3000 PSI	3.75	4.81	8.56
3750	5000 PSI	3.88	4.72	8.60
3800	6" thick, 3000 PSI	4.13	4.92	9.05
3850	5000 PSI	4.28	4.84	9.12
3900	7-1/2" thick, 3000 PSI	5.25	5.10	10.35
3950	5000 PSI	5.45	5	10.45
4000	8" thick, 3000 PSI	5.65	5.25	10.90
4050	5000 PSI	5.90	5.15	11.05
4200	Exp. aggregate finish, 5-1/2" thick, 3000 PSI	4.07	4.84	8.91
4250	5000 PSI	4.21	4.76	8.97
4300	6" thick, 3000 PSI	4.46	4.96	9.42
4350	5000 PSI	4.60	4.87	9.47
4400	7-1/2" thick, 3000 PSI	5.60	5.15	10.75
4450	5000 PSI	5.80	5.05	10.85
4500	8" thick, 3000 PSI	5.95	5.25	11.20
4550	5000 PSI	6.20	5.20	11.40
4600	Exposed aggregate & vert. rustication 5-1/2" thick, 3000 PSI	6.05	6	12.05
4650	5000 PSI	6.20	5.90	12.10
4700	6" thick, 3000 PSI	6.45	6.10	12.55
4750	5000 PSI	6.60	6.05	12.65
4800	7-1/2" thick, 3000 PSI	7.55	6.30	13.85
4850	5000 PSI	7.75	6.20	13.95
4900	8" thick, 3000 PSI	7.95	6.45	14.40
4950	5000 PSI	8.20	6.35	14.55
5000	Vertical rib & light sandblast, 5-1/2" thick, 3000 PSI	6.15	7.95	14.10
5100	6" thick, 3000 PSI	6.55	8.10	14.65
5200	7-1/2" thick, 3000 PSI	7.65	8.25	15.90
5300	8" thick, 3000 PSI	8.05	8.40	16.45
6000	Broom finish w/2" polystyrene insulation, 6" thick, 3000 PSI	3.61	5.85	9.46
6100	Broom finish 2" fiberplank insulation, 6" thick, 3000 PSI	4.06	5.80	9.86
6200	Exposed aggregate w/2" polystyrene insulation, 6" thick, 3000 PSI	3.84	5.85	9.69
6300	Exposed aggregate 2" fiberplank insulation, 6" thick, 3000 PSI	4.29	5.80	10.09

B20 Exterior Enclosure

B2010 Exterior Walls

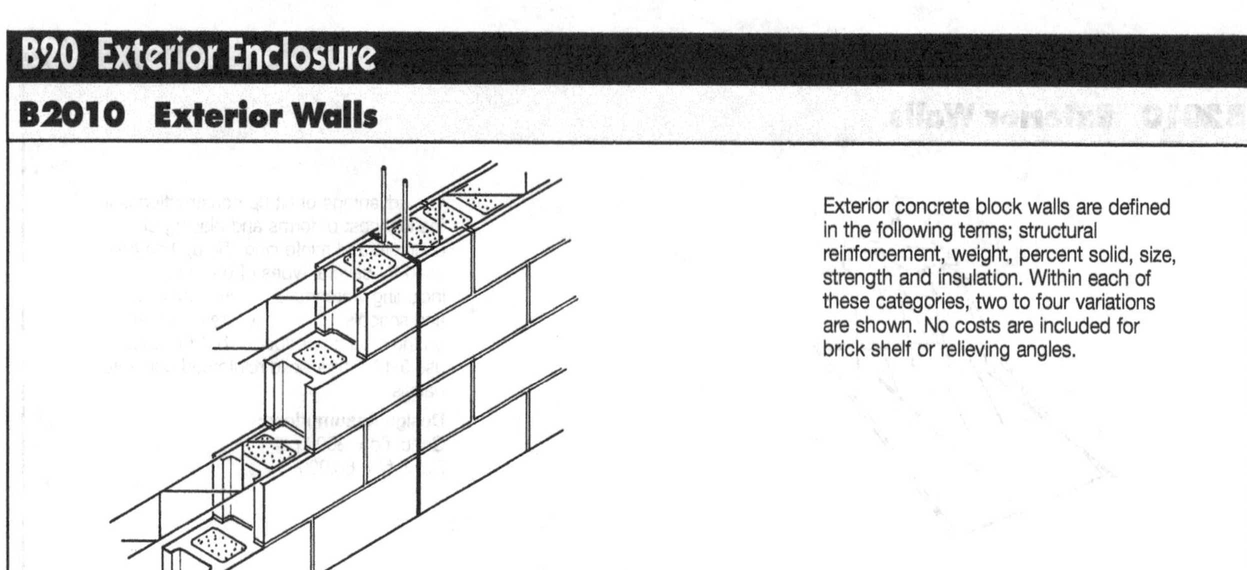

Exterior concrete block walls are defined in the following terms; structural reinforcement, weight, percent solid, size, strength and insulation. Within each of these categories, two to four variations are shown. No costs are included for brick shelf or relieving angles.

B2010 109 Concrete Block Wall - Regular Weight

	TYPE	SIZE (IN.)	STRENGTH (P.S.I.)	CORE FILL		COST PER S.F.		
						MAT.	INST.	TOTAL
1200	Hollow	4x8x16	2,000	none		1.80	4.92	6.72
1250			4,500	none		2.23	4.92	7.15
1400		8x8x16	2,000	perlite		3.36	6	9.36
1410				styrofoam		4.80	5.65	10.45
1440				none		2.81	5.65	8.46
1450			4,500	perlite		3.89	6	9.89
1460				styrofoam		5.35	5.65	11
1490				none		3.34	5.65	8.99
2000	75% solid	4x8x16	2,000	none		2.17	4.97	7.14
2100		6x8x16	2,000	perlite		3.19	5.45	8.64
2500	Solid	4x8x16	2,000	none		2.17	5.10	7.27
2700		8x8x16	2,000	none		3.75	5.85	9.60

B2010 110 Concrete Block Wall - Lightweight

	TYPE	SIZE (IN.)	WEIGHT (P.C.F.)	CORE FILL		COST PER S.F.		
						MAT.	INST.	TOTAL
3100	Hollow	8x4x16	105	perlite		3.09	5.65	8.74
3110				styrofoam		4.53	5.30	9.83
3200		4x8x16	105	none		1.97	4.81	6.78
3250			85	none		2.42	4.71	7.13
3300		6x8x16	105	perlite		3.17	5.40	8.57
3310				styrofoam		4.78	5.15	9.93
3340				none		2.79	5.15	7.94
3400		8x8x16	105	perlite		3.89	5.85	9.74
3410				styrofoam		5.35	5.50	10.85
3440				none		3.34	5.50	8.84
3450			85	perlite		4.33	5.70	10.03
4000	75% solid	4x8x16	105	none		2.85	4.86	7.71
4050			85	none		2.98	4.76	7.74
4100		6x8x16	105	perlite		4.27	5.35	9.62
4500	Solid	4x8x16	105	none		2.75	5.05	7.80
4700		8x8x16	105	none		3.89	5.80	9.69

B20 Exterior Enclosure

B2010 Exterior Walls

B2010 111 — Reinforced Concrete Block Wall - Regular Weight

	TYPE	SIZE (IN.)	STRENGTH (P.S.I.)	VERT. REINF & GROUT SPACING		COST PER S.F. MAT.	INST.	TOTAL
5200	Hollow	4x8x16	2,000	#4 @ 48"		1.94	5.45	7.39
5300		6x8x16	2,000	#4 @ 48"		2.72	5.90	8.62
5330				#5 @ 32"		2.93	6.15	9.08
5340				#5 @ 16"		3.36	7	10.36
5350			4,500	#4 @ 28"		2.99	5.90	8.89
5390				#5 @ 16"		3.63	7	10.63
5400		8x8x16	2,000	#4 @ 48"		3.11	6.30	9.41
5430				#5 @ 32"		3.33	6.70	10.03
5440				#5 @ 16"		3.84	7.70	11.54
5450			4,500	#4 @ 48"		3.63	6.40	10.03
5490				#5 @ 16"		4.37	7.70	12.07
5500		12x8x16	2,000	#4 @ 48"		4.38	8.05	12.43
5540				#5 @ 16"		5.40	9.45	14.85
6100	75% solid	6x8x16	2,000	#4 @ 48"		3.13	5.80	8.93
6140				#5 @ 16"		3.55	6.60	10.15
6150			4,500	#4 @ 48"		3.49	5.80	9.29
6190				#5 @ 16"		3.91	6.60	10.51
6200		8x8x16	2,000	#4 @ 48"		3.52	6.25	9.77
6230				#5 @ 32"		3.68	6.45	10.13
6240				#5 @ 16"		3.97	7.20	11.17
6250			4,500	#4 @ 48"		4.23	6.25	10.48
6280				#5 @ 32"		4.39	6.45	10.84
6290				#5 @ 16"		4.68	7.20	11.88
6500	Solid-double	2-4x8x16	2,000	#4 @ 48" E.W.		5.20	11.65	16.85
6530	Wythe			#5 @ 16" E.W.		5.90	12.40	18.30
6550			4,500	#4 @ 48" E.W.		7	11.55	18.55
6580				#5 @ 16" E.W.		7.70	12.25	19.95

B2010 112 — Reinforced Concrete Block Wall - Lightweight

	TYPE	SIZE (IN.)	WEIGHT (P.C.F.)	VERT REINF. & GROUT SPACING		COST PER S.F. MAT.	INST.	TOTAL
7100	Hollow	8x4x16	105	#4 @ 48"		2.83	6.05	8.88
7140				#5 @ 16"		3.57	7.35	10.92
7150			85	#4 @ 48"		4.04	5.90	9.94
7190				#5 @ 16"		4.78	7.20	11.98
7400		8x8x16	105	#4 @ 48"		3.63	6.25	9.88
7440				#5 @ 16"		4.37	7.55	11.92
7450			85	#4 @ 48"		4.07	6.10	10.17
7490				#5 @ 16"		4.81	7.40	12.21
7800		8x8x24	105	#4 @ 48"		2.79	6.70	9.49
7840				#5 @ 16"		3.53	8	11.53
7850			85	#4 @ 48"		6.30	5.75	12.05
7890				#5 @ 16"		7.05	7.05	14.10
8100	75% solid	6x8x16	105	#4 @ 48"		4.21	5.70	9.91
8130				#5 @ 32"		4.36	5.85	10.21
8150			85	#4 @ 48"		4.28	5.55	9.83
8180				#5 @ 32"		4.43	5.75	10.18
8200		8x8x16	105	#4 @ 48"		5	6.10	11.10
8230				#5 @ 32"		5.20	6.30	11.50
8250			85	#4 @ 48"		4.80	5.95	10.75
8280				#5 @ 32"		4.96	6.15	11.11

B20 Exterior Enclosure

B2010 Exterior Walls

B2010 112 Reinforced Concrete Block Wall - Lightweight

	TYPE	SIZE (IN.)	WEIGHT (P.C.F.)	VERT REINF. & GROUT SPACING		COST PER S.F.		
						MAT.	INST.	TOTAL
8500	Solid-double	2-4x8x16	105	#4 @ 48"		6.35	11.55	17.90
8530	Wythe		105	#5 @ 16"		7.05	12.25	19.30
8600		2-6x8x16	105	#4 @ 48"		7	12.30	19.30
8630				#5 @ 16"		7.65	13	20.65
8650			85	#4 @ 48"		10.50	11.85	22.35

B20 Exterior Enclosure

B2010 Exterior Walls

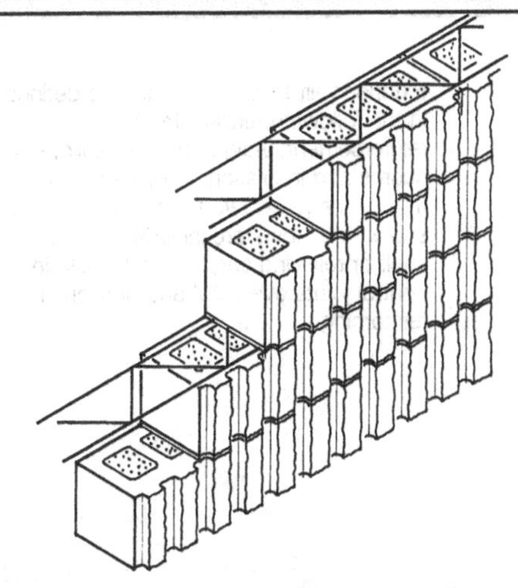

Exterior split ribbed block walls are defined in the following terms; structural reinforcement, weight, percent solid, size, number of ribs and insulation. Within each of these categories two to four variations are shown. No costs are included for brick shelf or relieving angles. Costs include control joints every 20′ and horizontal reinforcing.

B2010 113 Split Ribbed Block Wall - Regular Weight

	TYPE	SIZE (IN.)	RIBS	CORE FILL		COST PER S.F.		
						MAT.	INST.	TOTAL
1220	Hollow	4x8x16	4	none		3.51	6.10	9.61
1250			8	none		3.81	6.10	9.91
1280			16	none		4.11	6.20	10.31
1430		8x8x16	8	perlite		5.55	7.25	12.80
1440				styrofoam		7	6.90	13.90
1450				none		4.99	6.90	11.89
1530		12x8x16	8	perlite		6.85	9.60	16.45
1540				styrofoam		8.40	8.95	17.35
1550				none		5.95	8.95	14.90
2120	75% solid	4x8x16	4	none		4.40	6.20	10.60
2150			8	none		4.79	6.20	10.99
2180			16	none		5.15	6.30	11.45
2520	Solid	4x8x16	4	none		5	6.30	11.30
2550			8	none		5.90	6.30	12.20
2580			16	none		5.90	6.35	12.25

B2010 115 Reinforced Split Ribbed Block Wall - Regular Weight

	TYPE	SIZE (IN.)	RIBS	VERT. REINF. & GROUT SPACING		COST PER S.F.		
						MAT.	INST.	TOTAL
5200	Hollow	4x8x16	4	#4 @ 48″		3.65	6.60	10.25
5230			8	#4 @ 48″		3.95	6.60	10.55
5260			16	#4 @ 48″		4.25	6.70	10.95
5430		8x8x16	8	#4 @ 48″		5.30	7.65	12.95
5440				#5 @ 32″		5.50	7.95	13.45
5450				#5 @ 16″		6	8.95	14.95
5530		12x8x16	8	#4 @ 48″		6.35	9.70	16.05
5540				#5 @ 32″		6.65	10	16.65
5550				#5 @ 16″		7.35	11.10	18.45
6230	75% solid	6x8x16	8	#4 @ 48″		5.65	7	12.65
6240				#5 @ 32″		5.80	7.15	12.95
6250				#5 @ 16″		6.10	7.80	13.90
6330		8x8x16	8	#4 @ 48″		6.35	7.55	13.90
6340				#5 @ 32″		6.50	7.75	14.25
6350				#5 @ 16″		6.80	8.50	15.30

B20 Exterior Enclosure

B2010 Exterior Walls

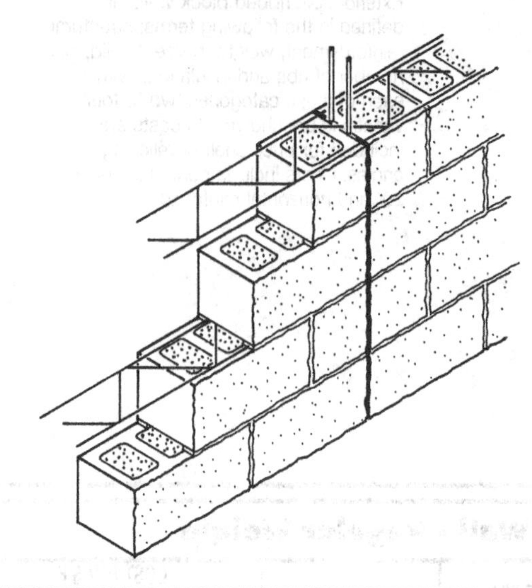

Exterior split face block walls are defined in the following terms; structural reinforcement, weight, percent solid, size, scores and insulation. Within each of these categories two to four variations are shown. No costs are included for brick shelf or relieving angles. Costs include control joints every 20′ and horizontal reinforcing.

B2010 117 Split Face Block Wall - Regular Weight

	TYPE	SIZE (IN.)	SCORES	CORE FILL		COST PER S.F.		
						MAT.	INST.	TOTAL
1200	Hollow	8x4x16	0	perlite		5.40	7.95	13.35
1210				styrofoam		6.80	7.60	14.40
1240				none		4.83	7.60	12.43
1250			1	perlite		5.70	7.95	13.65
1260				styrofoam		7.15	7.60	14.75
1290				none		5.15	7.60	12.75
1300		12x4x16	0	perlite		6.95	9.05	16
1310				styrofoam		8.50	8.40	16.90
1340				none		6.05	8.40	14.45
1350			1	perlite		7.35	9.05	16.40
1360				styrofoam		8.90	8.40	17.30
1390				none		6.45	8.40	14.85
1400		4x8x16	0	none		3.17	6	9.17
1450			1	none		3.43	6.10	9.53
1500		6x8x16	0	perlite		4.05	6.90	10.95
1510				styrofoam		5.65	6.65	12.30
1540				none		3.67	6.65	10.32
1550			1	perlite		4.35	7	11.35
1560				styrofoam		5.95	6.75	12.70
1590				none		3.97	6.75	10.72
1600		8x8x16	0	perlite		4.71	7.45	12.16
1610				styrofoam		6.15	7.10	13.25
1640				none		4.16	7.10	11.26
1650			1	perlite		5.05	7.60	12.65
1660				styrofoam		6.50	7.25	13.75
1690				none		4.49	7.25	11.74
1700		12x8x16	0	perlite		6	9.75	15.75
1705			0	perlite		7.35	11.65	19
1710				styrofoam		7.55	9.10	16.65
1740				none		5.10	9.10	14.20

B20 Exterior Enclosure

B2010 Exterior Walls

B2010 117 Split Face Block Wall - Regular Weight

	TYPE	SIZE (IN.)	SCORES	CORE FILL		COST PER S.F.		
						MAT.	INST.	TOTAL
1750	Hollow	8x4x16	1	perlite		6.40	9.95	16.35
1760				styrofoam		7.95	9.30	17.25
1790				none		5.50	9.30	14.80
1800	75% solid	8x4x16	0	perlite		6.10	7.90	14
1840				none		5.80	7.75	13.55
1852			1	perlite		6.55	7.90	14.45
1890				none		6.25	7.75	14
2000		4x8x16	0	none		3.95	6.10	10.05
2050			1	none		4.29	6.20	10.49
2400	Solid	8x4x16	0	none		6.45	7.90	14.35
2450			1	none		6.95	7.90	14.85
2900		12x8x16	0	none		7.10	9.45	16.55
2950			1	none		7.65	9.65	17.30

B2010 119 Reinforced Split Face Block Wall - Regular Weight

	TYPE	SIZE (IN.)	SCORES	VERT. REINF. & GROUT SPACING		COST PER S.F.		
						MAT.	INST.	TOTAL
5200	Hollow	8x4x16	0	#4 @ 48"		5.10	8.35	13.45
5210				#5 @ 32"		5.35	8.65	14
5240				#5 @ 16"		5.85	9.65	15.50
5250			1	#4 @ 48"		5.45	8.35	13.80
5260				#5 @ 32"		5.70	8.65	14.35
5290				#5 @ 16"		6.20	9.65	15.85
5700		12x8x16	0	#4 @ 48"		5.50	9.85	15.35
5710				#5 @ 32"		5.80	10.15	15.95
5740				#5 @ 16"		6.50	11.25	17.75
5750			1	#4 @ 48"		5.90	10.05	15.95
5760				#5 @ 32"		6.20	10.35	16.55
5790				#5 @ 16"		6.90	11.45	18.35
6000	75% solid	8x4x16	0	#4 @ 48"		5.95	8.30	14.25
6010				#5 @ 32"		6.10	8.50	14.60
6040				#5 @ 16"		6.40	9.25	15.65
6050			1	#4 @ 48"		6.40	8.30	14.70
6060				#5 @ 32"		6.55	8.50	15.05
6090				#5 @ 16"		6.85	9.25	16.10
6100		12x4x16	0	#4 @ 48"		7.45	9.15	16.60
6110				#5 @ 32"		7.70	9.40	17.10
6140				#5 @ 16"		8.05	10.05	18.10
6150			1	#4 @ 48"		7.95	9.15	17.10
6160				#5 @ 32"		8.20	9.40	17.60
6190				#5 @ 16"		8.60	10.25	18.85
6700	Solid-double Wythe	2-4x8x16	0	#4 @ 48" E.W.		9.85	13.85	23.70
6710				#5 @ 32" E.W.		10.10	14	24.10
6740				#5 @ 16" E.W.		10.50	14.55	25.05
6750			1	#4 @ 48" E.W.		10.60	14.05	24.65
6760				#5 @ 32" E.W.		10.85	14.20	25.05
6790				#5 @ 16" E.W.		11.30	14.75	26.05
6800		2-6x8x16	0	#4 @ 48" E.W.		11.15	15.30	26.45
6810				#5 @ 32" E.W.		11.40	15.40	26.80
6840				#5 @ 16" E.W.		11.80	16	27.80
6850			1	#4 @ 48" E.W.		12	15.50	27.50
6860				#5 @ 32" E.W.		12.25	15.60	27.85
6890				#5 @ 16" E.W.		12.65	16.20	28.85

B20 Exterior Enclosure

B2010 Exterior Walls

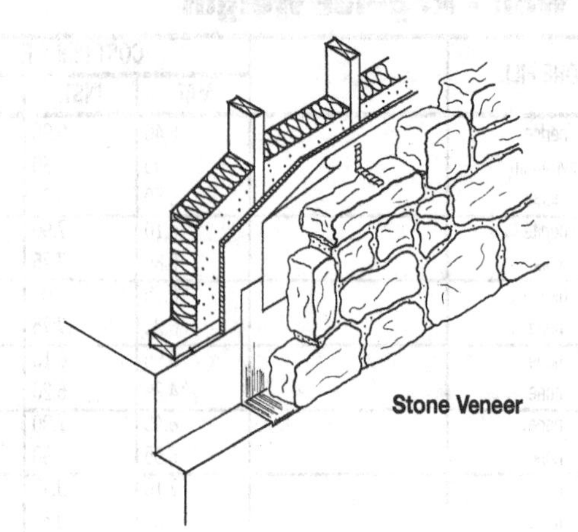

The table below lists costs per S.F. for stone veneer walls on various backup using different stone.

B2010 128	Stone Veneer	COST PER S.F.		
		MAT.	INST.	TOTAL
2000	Ashlar veneer, 4", 2"x4" stud backup, 16" O.C., 8' high, low priced stone	12.45	17.35	29.80
2100	Metal stud backup, 8' high, 16" O.C.	13.10	17.60	30.70
2150	24" O.C.	12.80	17.20	30
2200	Conc. block backup, 4" thick	13.75	20.50	34.25
2300	6" thick	14.45	21	35.45
2350	8" thick	14.75	21.50	36.25
2400	10" thick	15.45	22.50	37.95
2500	12" thick	15.90	24	39.90
3100	High priced stone, wood stud backup, 10' high, 16" O.C.	18	19.80	37.80
3200	Metal stud backup, 10' high, 16" O.C.	18.60	20	38.60
3250	24" O.C.	18.30	19.65	37.95
3300	Conc. block backup, 10' high, 4" thick	19.25	23	42.25
3350	6" thick	19.95	23	42.95
3400	8" thick	20.50	24	44.50
3450	10" thick	21	25	46
3500	12" thick	21.50	26.50	48
4000	Indiana limestone 2" thick,sawn finish,wood stud backup,10' high,16" O.C.	40.50	10.85	51.35
4100	Metal stud backup, 10' high, 16" O.C.	41	11.10	52.10
4150	24" O.C.	41	10.70	51.70
4200	Conc. block backup, 4" thick	42	14.15	56.15
4250	6" thick	42.50	14.35	56.85
4300	8" thick	43	14.85	57.85
4350	10" thick	43.50	15.90	59.40
4400	12" thick	44	17.45	61.45
4450	2" thick, smooth finish, wood stud backup, 8' high, 16" O.C.	40.50	10.85	51.35
4550	Metal stud backup, 8' high, 16" O.C.	41	10.95	51.95
4600	24" O.C.	41	10.55	51.55
4650	Conc. block backup, 4" thick	42	14	56
4700	6" thick	42.50	14.20	56.70
4750	8" thick	43	14.70	57.70
4800	10" thick	43.50	15.90	59.40
4850	12" thick	44	17.45	61.45
5350	4" thick, smooth finish, wood stud backup, 8' high, 16" O.C.	47	10.85	57.85
5450	Metal stud backup, 8' high, 16" O.C.	47.50	11.10	58.60
5500	24" O.C.	47.50	10.70	58.20
5550	Conc. block backup, 4" thick	48.50	14.15	62.65
5600	6" thick	49	14.35	63.35
5650	8" thick	49.50	14.85	64.35

B20 Exterior Enclosure

B2010 Exterior Walls

B2010 128	Stone Veneer	COST PER S.F. MAT.	COST PER S.F. INST.	COST PER S.F. TOTAL
5700	10" thick	50	15.90	65.90
5750	12" thick	50.50	17.45	67.95
6000	Granite, gray or pink, 2" thick, wood stud backup, 8' high, 16" O.C.	24	21.50	45.50
6100	Metal studs, 8' high, 16" O.C.	24.50	22	46.50
6150	24" O.C.	24.50	21.50	46
6200	Conc. block backup, 4" thick	25.50	25	50.50
6250	6" thick	26	25	51
6300	8" thick	26.50	25.50	52
6350	10" thick	27	26.50	53.50
6400	12" thick	27.50	28	55.50
6900	4" thick, wood stud backup, 8' high, 16" O.C.	38.50	24.50	63
7000	Metal studs, 8' high, 16" O.C.	39	24.50	63.50
7050	24" O.C.	39	24	63
7100	Conc. block backup, 4" thick	40	27.50	67.50
7150	6" thick	40.50	28	68.50
7200	8" thick	41	28.50	69.50
7250	10" thick	41.50	29.50	71
7300	12" thick	42	31	73

B20 Exterior Enclosure

B2010 Exterior Walls

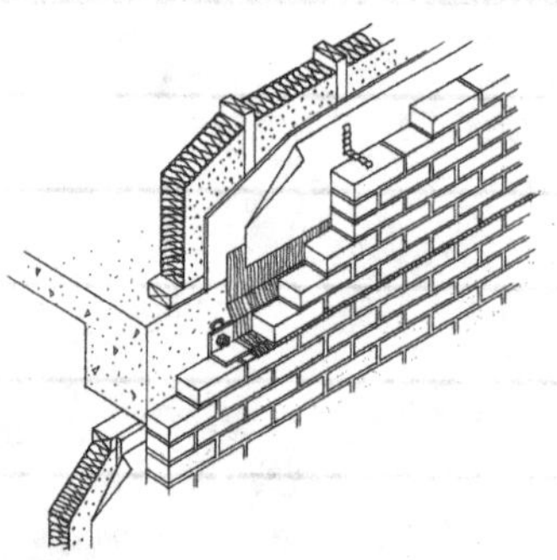

Exterior brick veneer/stud backup walls are defined in the following terms: type of brick and studs, stud spacing and bond. All systems include a back-up wall, a control joint every 20′, a brick shelf every 12′ of height, ties to the backup and the necessary dampproofing, flashing and insulation.

B2010 129 Brick Veneer/Wood Stud Backup

	FACE BRICK	STUD BACKUP	STUD SPACING (IN.)	BOND		COST PER S.F.		
						MAT.	INST.	TOTAL
1100	Standard	2x4-wood	16	running		7.35	13.55	20.90
1120				common		8.45	15.30	23.75
1140				Flemish		9.15	17.90	27.05
1160				English		10.10	18.85	28.95
1400		2x6-wood	16	running		7.60	13.65	21.25
1420				common		8.65	15.40	24.05
1440				Flemish		9.35	18	27.35
1460				English		10.30	18.95	29.25
1700	Glazed	2x4-wood	16	running		21	14	35
1720				common		25	15.95	40.95
1740				Flemish		27.50	18.85	46.35
1760				English		30.50	19.95	50.45
2300	Engineer	2x4-wood	16	running		6.90	12.15	19.05
2320				common		7.85	13.55	21.40
2340				Flemish		8.50	15.95	24.45
2360				English		9.35	16.65	26
2900	Roman	2x4-wood	16	running		8.90	12.45	21.35
2920				common		10.30	14	24.30
2940				Flemish		11.20	16.30	27.50
2960				English		12.45	17.45	29.90
4100	Norwegian	2x4-wood	16	running		6.80	9.75	16.55
4120				common		7.70	10.75	18.45
4140				Flemish		8.35	12.45	20.80
4160				English		9.15	12.95	22.10

B2010 130 Brick Veneer/Metal Stud Backup

	FACE BRICK	STUD BACKUP	STUD SPACING (IN.)	BOND		COST PER S.F.		
						MAT.	INST.	TOTAL
5100	Standard	25ga.x6"NLB	24	running		6.95	13.25	20.20
5120				common		8	15	23
5140				Flemish		8.40	17.05	25.45
5160				English		9.65	18.55	28.20
5200		20ga.x3-5/8"NLB	16	running		7.05	13.80	20.85
5220				common		8.15	15.55	23.70
5240				Flemish		8.85	18.15	27
5260				English		9.80	19.10	28.90

B20 Exterior Enclosure

B2010 Exterior Walls

B2010 130 Brick Veneer/Metal Stud Backup

	FACE BRICK	STUD BACKUP	STUD SPACING (IN.)	BOND		COST PER S.F.		
						MAT.	INST.	TOTAL
5400	Standard	16ga.x3-5/8"LB	16	running		7.65	14	21.65
5420				common		8.75	15.75	24.50
5440				Flemish		9.45	18.35	27.80
5460				English		10.40	19.30	29.70
5700	Glazed	25ga.x6"NLB	24	running		20.50	13.70	34.20
5720				common		24.50	15.65	40.15
5740				Flemish		27	18.55	45.55
5760				English		30	19.65	49.65
5800		20ga.x3-5/8"NLB	24	running		20.50	13.85	34.35
5820				common		24.50	15.80	40.30
5840				Flemish		27	18.70	45.70
5860				English		30	19.80	49.80
6000		16ga.x3-5/8"LB	16	running		21	14.45	35.45
6020				common		25	16.40	41.40
6040				Flemish		27.50	19.30	46.80
6060				English		30.50	20.50	51
6300	Engineer	25ga.x6"NLB	24	running		6.50	11.85	18.35
6320				common		7.40	13.25	20.65
6340				Flemish		8.05	15.65	23.70
6360				English		8.90	16.35	25.25
6400		20ga.x3-5/8"NLB	16	running		6.60	12.40	19
6420				common		7.55	13.80	21.35
6440				Flemish		8.20	16.20	24.40
6460				English		9.05	16.90	25.95
6900	Roman	25ga.x6"NLB	24	running		8.45	12.15	20.60
6920				common		9.85	13.65	23.50
6940				Flemish		10.75	16	26.75
6960				English		12	17.15	29.15
7000		20ga.x3-5/8"NLB	16	running		8.60	12.70	21.30
7020				common		10	14.25	24.25
7040				Flemish		10.90	16.55	27.45
7060				English		12.15	17.70	29.85
7500	Norman	25ga.x6"NLB	24	running		7.50	10.35	17.85
7520				common		8.65	11.55	20.20
7540				Flemish		18.15	13.50	31.65
7560				English		10.45	14.20	24.65
7600		20ga.x3-5/8"NLB	24	running		7.50	10.50	18
7620				common		8.65	11.70	20.35
7640				Flemish		18.15	13.65	31.80
7660				English		10.45	14.35	24.80
8100	Norwegian	25ga.x6"NLB	24	running		6.40	9.45	15.85
8120				common		7.30	10.45	17.75
8140				Flemish		7.90	12.15	20.05
8160				English		8.70	12.65	21.35
8400		16ga.x3-5/8"LB	16	running		7.10	10.20	17.30
8420				common		8	11.20	19.20
8440				Flemish		8.65	12.90	21.55
8460				English		9.45	13.40	22.85

B20 Exterior Enclosure

B2010 Exterior Walls

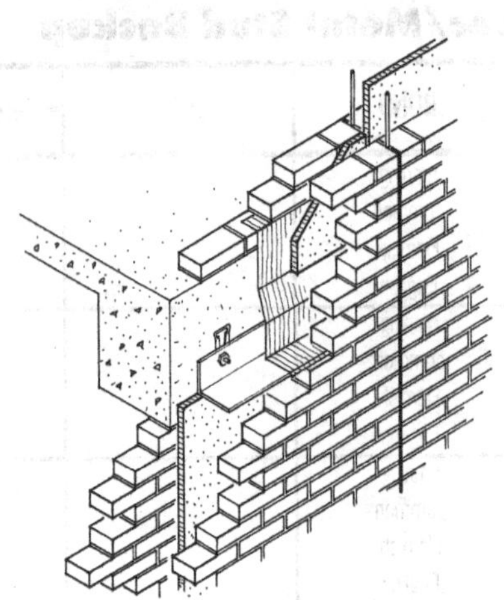

Exterior brick face cavity walls are defined in the following terms: cavity treatment, type of face brick, backup masonry, total thickness and insulation. Seven types of face brick are shown with various types of backup. All systems include a brick shelf, ties to the backups and necessary dampproofing, flashing, and control joints every 20′.

B2010 134 Brick Face Cavity Wall

	FACE BRICK	BACKUP MASONRY	TOTAL THICKNESS (IN.)	CAVITY INSULATION		COST PER S.F.		
						MAT.	INST.	TOTAL
1000	Standard	4″ common brick	10	polystyrene		9.75	21	30.75
1020				none		9.45	20.50	29.95
1040		6″ SCR brick	12	polystyrene		12	18.70	30.70
1060				none		11.70	18.15	29.85
1080		4″ conc. block	10	polystyrene		7.85	16.90	24.75
1100				none		7.50	16.35	23.85
1120		6″ conc. block	12	polystyrene		8.50	17.25	25.75
1140				none		8.20	16.70	24.90
1160		4″ L.W. block	10	polystyrene		8	16.80	24.80
1180				none		7.70	16.25	23.95
1200		6″ L.W. block	12	polystyrene		8.80	17.15	25.95
1220				none		8.50	16.55	25.05
1240		4″ glazed block	10	polystyrene		15.55	18.05	33.60
1260				none		15.20	17.50	32.70
1280		6″ glazed block	12	polystyrene		15.55	18.05	33.60
1300				none		15.25	17.50	32.75
1320		4″ clay tile	10	polystyrene		8.35	4.08	12.43
1340				none		8.35	4.08	12.43
1360		4″ glazed tile	10	polystyrene		18.25	21.50	39.75
1380				none		17.90	21	38.90
1500	Glazed	4″ common brick	10	polystyrene		23	21.50	44.50
1520				none		23	21	44
1580		4″ conc. block	10	polystyrene		21.50	17.35	38.85
1600				none		21	16.80	37.80
1660		4″ L.W. block	10	polystyrene		21.50	17.25	38.75
1680				none		21	16.70	37.70
1740		4″ glazed block	10	polystyrene		29	18.50	47.50
1760				none		28.50	17.95	46.45
1820		4″ clay tile	10	polystyrene		8.35	4.08	12.43
1840				none		8.35	4.08	12.43
1860		4″ glazed tile	10	polystyrene		29	17.15	46.15
1880				none		29	16.60	45.60
2000	Engineer	4″ common brick	10	polystyrene		9.30	19.60	28.90
2020				none		9	19.05	28.05

B20 Exterior Enclosure

B2010 Exterior Walls

B2010 134 Brick Face Cavity Wall

	FACE BRICK	BACKUP MASONRY	TOTAL THICKNESS (IN.)	CAVITY INSULATION		COST PER S.F.		
						MAT.	INST.	TOTAL
2080	Engineer	4" conc. block	10	polystyrene		7.40	15.50	22.90
2100				none		7.05	14.95	22
2162		4" L.W. block	10	polystyrene		7.55	15.40	22.95
2180				none		7.25	14.85	22.10
2240		4" glazed block	10	polystyrene		15.10	16.65	31.75
2260				none		14.75	16.10	30.85
2320		4" clay tile	10	polystyrene		8.35	4.08	12.43
2340				none		8.35	4.08	12.43
2360		4" glazed tile	10	polystyrene		17.80	20	37.80
2380				none		17.45	19.45	36.90
2500	Roman	4" common brick	10	polystyrene		11.30	19.90	31.20
2520				none		10.95	19.35	30.30
2580		4" conc. block	10	polystyrene		9.35	15.80	25.15
2600				none		9.05	15.25	24.30
2660		4" L.W. block	10	polystyrene		9.50	15.70	25.20
2680				none		9.20	15.15	24.35
2740		4" glazed block	10	polystyrene		17.05	16.95	34
2760				none		16.75	16.40	33.15
2820		4" clay tile	10	polystyrene		8.35	4.08	12.43
2840				none		8.35	4.08	12.43
2860		4" glazed tile	10	polystyrene		19.75	20.50	40.25
2880				none		19.45	19.75	39.20
3000	Norman	4" common brick	10	polystyrene		10.35	18.10	28.45
3020				none		10	17.55	27.55
3080		4" conc. block	10	polystyrene		8.40	14	22.40
3100				none		8.10	13.45	21.55
3160		4" L.W. block	10	polystyrene		8.55	13.90	22.45
3180				none		8.25	13.35	21.60
3240		4" glazed block	10	polystyrene		16.10	15.15	31.25
3260				none		15.80	14.60	30.40
3320		4" clay tile	10	polystyrene		8.35	4.08	12.43
3340				none		8.35	4.08	12.43
3360		4" glazed tile	10	polystyrene		18.80	18.50	37.30
3380				none		18.50	17.95	36.45
3500	Norwegian	4" common brick	10	polystyrene		9.20	17.20	26.40
3520				none		8.85	16.65	25.50
3580		4" conc. block	10	polystyrene		7.25	13.10	20.35
3600				none		6.95	12.55	19.50
3660		4" L.W. block	10	polystyrene		7.45	13	20.45
3680				none		7.10	12.45	19.55
3740		4" glazed block	10	polystyrene		14.95	14.25	29.20
3760				none		14.65	13.70	28.35
3820		4" clay tile	10	polystyrene		8.35	4.08	12.43
3840				none		8.35	4.08	12.43
3860		4" glazed tile	10	polystyrene		17.65	17.60	35.25
3880				none		17.35	17.05	34.40
4000	Utility	4" common brick	10	polystyrene		8.85	16.30	25.15
4020				none		8.50	15.75	24.25
4080		4" conc. block	10	polystyrene		6.90	12.20	19.10
4100				none		6.60	11.60	18.20
4160		4" L.W. block	10	polystyrene		7.05	12.05	19.10
4180				none		6.75	11.50	18.25

B20 Exterior Enclosure

B2010 Exterior Walls

B2010 134 Brick Face Cavity Wall

	FACE BRICK	BACKUP MASONRY	TOTAL THICKNESS (IN.)	CAVITY INSULATION		COST PER S.F.		
						MAT.	INST.	TOTAL
4240	Utility	4" glazed block	10	polystyrene		14.60	13.35	27.95
4260				none		14.30	12.80	27.10
4320		4" clay tile	10	polystyrene		8.35	4.08	12.43
4340				none		8.35	4.08	12.43
4360		4" glazed tile	10	polystyrene		17.30	16.70	34
4380				none		17	16.15	33.15

B2010 135 Brick Face Cavity Wall - Insulated Backup

	FACE BRICK	BACKUP MASONRY	TOTAL THICKNESS (IN.)	BACKUP CORE FILL		COST PER S.F.		
						MAT.	INST.	TOTAL
5100	Standard	6" conc. block	10	perlite		8.55	16.95	25.50
5120				styrofoam		10.15	16.70	26.85
5180		6" L.W. block	10	perlite		8.85	16.85	25.70
5200				styrofoam		10.45	16.55	27
5260		6" glazed block	10	perlite		16.30	18	34.30
5280				styrofoam		17.90	17.75	35.65
5340		6" clay tile	10	none		15.40	16.15	31.55
5360		8" clay tile	12	none		18.10	16.70	34.80
5600	Glazed	6" conc. block	10	perlite		22	17.40	39.40
5620				styrofoam		23.50	17.15	40.65
5680		6" L.W. block	10	perlite		22.50	17.30	39.80
5700				styrofoam		24	17	41
5760		6" glazed block	10	perlite		30	18.45	48.45
5780				styrofoam		31.50	18.20	49.70
5840		6" clay tile	10	none		29	16.60	45.60
5860		8" clay tile	8	none		31.50	17.15	48.65
6100	Engineer	6" conc. block	10	perlite		8.10	15.55	23.65
6120				styrofoam		9.70	15.30	25
6180		6" L.W. block	10	perlite		8.40	15.45	23.85
6200				styrofoam		10	15.15	25.15
6260		6" glazed block	10	perlite		15.85	16.60	32.45
6280				styrofoam		17.45	16.35	33.80
6340		6" clay tile	10	none		14.95	14.75	29.70
6360		8" clay tile	12	none		17.65	15.30	32.95
6600	Roman	6" conc. block	10	perlite		10.05	15.85	25.90
6620				styrofoam		11.70	15.60	27.30
6680		6" L.W. block	10	perlite		10.40	15.75	26.15
6700				styrofoam		12	15.45	27.45
6760		6" glazed block	10	perlite		17.80	16.90	34.70
6780				styrofoam		19.45	16.65	36.10
6840		6" clay tile	10	none		16.95	15.05	32
6860		8" clay tile	12	none		19.60	15.60	35.20
7100	Norman	6" conc. block	10	perlite		9.10	14.05	23.15
7120				styrofoam		10.75	13.80	24.55
7180		6" L.W. block	10	perlite		9.45	13.95	23.40
7200				styrofoam		11.05	13.65	24.70
7260		6" glazed block	10	perlite		16.85	15.10	31.95
7280				styrofoam		18.50	14.85	33.35
7340		6" clay tile	10	none		16	13.25	29.25
7360		8" clay tile	12	none		18.65	13.80	32.45

B20 Exterior Enclosure

B2010 Exterior Walls

B2010 135 Brick Face Cavity Wall - Insulated Backup

	FACE BRICK	BACKUP MASONRY	TOTAL THICKNESS (IN.)	BACKUP CORE FILL		COST PER S.F.		
						MAT.	INST.	TOTAL
7600	Norwegian	6″ conc. block	10	perlite		8	13.15	21.15
7620				styrofoam		9.60	12.90	22.50
7680		6″ L.W. block	10	perlite		8.30	13.05	21.35
7700				styrofoam		9.90	12.75	22.65
7760		6″ glazed block	10	perlite		15.75	14.20	29.95
7780				styrofoam		17.35	13.95	31.30
7840		6″ clay tile	10	none		14.85	12.35	27.20
7860		8″ clay tile	12	none		17.50	12.90	30.40
8100	Utility	6″ conc. block	10	perlite		7.60	12.25	19.85
8120				styrofoam		9.25	12	21.25
8180		6″ L.W. block	10	perlite		7.95	12.10	20.05
8200				styrofoam		9.55	11.85	21.40
8260		6″ glazed block	10	perlite		15.35	13.30	28.65
8280				styrofoam		17	13.05	30.05
8340		6″ clay tile	10	none		14.50	11.40	25.90
8360		8″ clay tile	12	none		17.15	12	29.15

B20 Exterior Enclosure

B2010 Exterior Walls

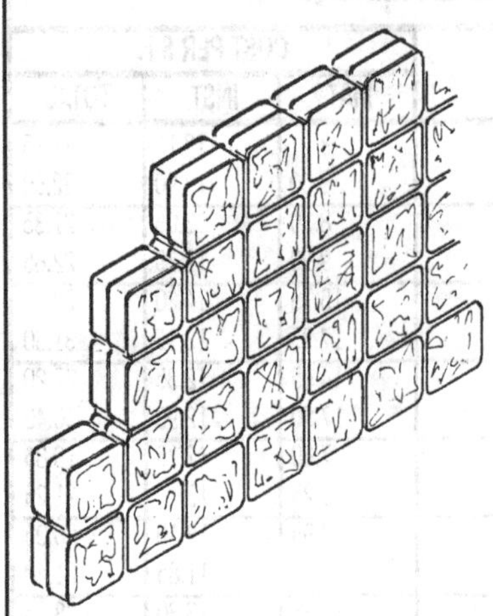

The table below lists costs per S.F. for glass block walls. Included in the costs are the following special accessories required for glass block walls.

Glass block accessories required for proper installation.

Wall ties: Galvanized double steel mesh full length of joint.

Fiberglass expansion joint at sides and top.

Silicone caulking: One gallon does 95 L.F.

Oakum: One lb. does 30 L.F.

Asphalt emulsion: One gallon does 600 L.F.

If block are not set in wall chase, use 2'-0" long wall anchors at 2'-0" O.C.

B2010 140	Glass Block	COST PER S.F.		
		MAT.	INST.	TOTAL
2300	Glass block 4" thick, 6"x6" plain, under 1,000 S.F.	28	19.75	47.75
2400	1,000 to 5,000 S.F.	27.50	17.15	44.65
2500	Over 5,000 S.F.	27	16.10	43.10
2600	Solar reflective, under 1,000 S.F.	39.50	27	66.50
2700	1,000 to 5,000 S.F.	38.50	23	61.50
2800	Over 5,000 S.F.	38	21.50	59.50
3500	8"x8" plain, under 1,000 S.F.	16.60	14.80	31.40
3600	1,000 to 5,000 S.F.	16.25	12.75	29
3700	Over 5,000 S.F.	15.75	11.55	27.30
3800	Solar reflective, under 1,000 S.F.	23	19.90	42.90
3900	1,000 to 5,000 S.F.	22.50	17	39.50
4000	Over 5,000 S.F.	22	15.35	37.35
5000	12"x12" plain, under 1,000 S.F.	23.50	13.70	37.20
5100	1,000 to 5,000 S.F.	23.50	11.55	35.05
5200	Over 5,000 S.F.	22.50	10.55	33.05
5300	Solar reflective, under 1,000 S.F.	33	18.35	51.35
5400	1,000 to 5,000 S.F.	33	15.35	48.35
5600	Over 5,000 S.F.	31.50	13.95	45.45
5800	3" thinline, 6"x6" plain, under 1,000 S.F.	19.35	19.75	39.10
5900	Over 5,000 S.F.	19.35	19.75	39.10
6000	Solar reflective, under 1,000 S.F.	27	27	54
6100	Over 5,000 S.F.	25.50	21.50	47
6200	8"x8" plain, under 1,000 S.F.	10.85	14.80	25.65
6300	Over 5,000 S.F.	10.20	11.55	21.75
6400	Solar reflective, under 1,000 S.F.	15.15	19.90	35.05
6500	Over 5,000 S.F.	14.25	15.35	29.60

B20 Exterior Enclosure

B2010 Exterior Walls

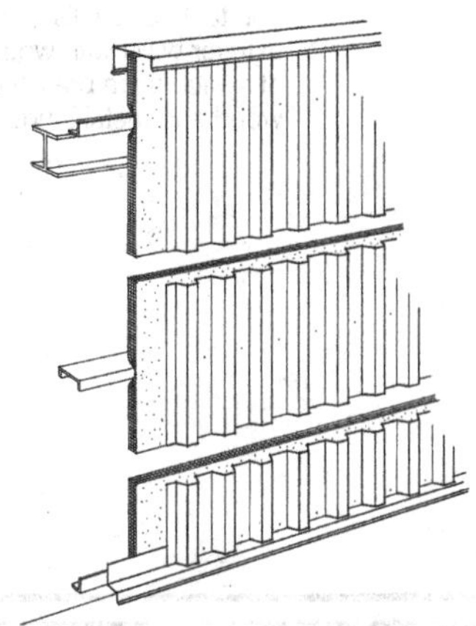

The table below lists costs for metal siding of various descriptions, not including the steel frame, or the structural steel, of a building. Costs are per S.F. including all accessories and insulation.

For steel frame support see System B2010 154.

B2010 146	Metal Siding Panel	COST PER S.F.		
		MAT.	INST.	TOTAL
1400	Metal siding aluminum panel, corrugated, .024" thick, natural	2.88	3.16	6.04
1450	Painted	3.05	3.16	6.21
1500	.032" thick, natural	3.22	3.16	6.38
1550	Painted	3.79	3.16	6.95
1600	Ribbed 4" pitch, .032" thick, natural	3.15	3.16	6.31
1650	Painted	3.83	3.16	6.99
1700	.040" thick, natural	3.62	3.16	6.78
1750	Painted	4.15	3.16	7.31
1800	.050" thick, natural	4.08	3.16	7.24
1850	Painted	4.62	3.16	7.78
1900	8" pitch panel, .032" thick, natural	3	3.03	6.03
1950	Painted	3.66	3.05	6.71
2000	.040" thick, natural	3.48	3.05	6.53
2050	Painted	3.98	3.08	7.06
2100	.050" thick, natural	3.93	3.06	6.99
2150	Painted	4.50	3.09	7.59
3000	Steel, corrugated or ribbed, 29 Ga. .0135" thick, galvanized	2.09	2.89	4.98
3050	Colored	2.97	2.93	5.90
3100	26 Ga. .0179" thick, galvanized	2.16	2.90	5.06
3150	Colored	3.06	2.94	6
3200	24 Ga. .0239" thick, galvanized	2.77	2.92	5.69
3250	Colored	3.45	2.96	6.41
3300	22 Ga. .0299" thick, galvanized	2.79	2.93	5.72
3350	Colored	4.19	2.97	7.16
3400	20 Ga. .0359" thick, galvanized	2.79	2.93	5.72
3450	Colored	4.46	3.14	7.60
4100	Sandwich panels, factory fab., 1" polystyrene, steel core, 26 Ga., galv.	5.80	4.52	10.32
4200	Colored, 1 side	7.05	4.52	11.57
4300	2 sides	9.10	4.52	13.62
4400	2" polystyrene, steel core, 26 Ga., galvanized	6.85	4.52	11.37
4500	Colored, 1 side	8.10	4.52	12.62
4600	2 sides	10.15	4.52	14.67
4700	22 Ga., baked enamel exterior	13.45	4.77	18.22
4800	Polyvinyl chloride exterior	14.15	4.77	18.92
5100	Textured aluminum, 4' x 8' x 5/16" plywood backing, single face	4.04	2.69	6.73
5200	Double face	5.50	2.69	8.19

B20 Exterior Enclosure

B2010 Exterior Walls

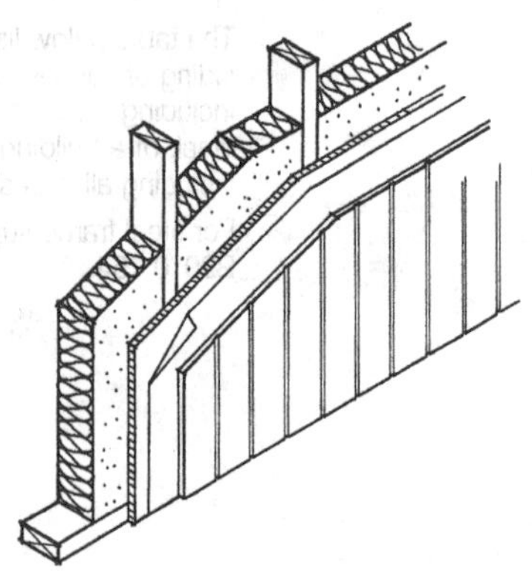

The table below lists costs per S.F. for exterior walls with wood siding. A variety of systems are presented using both wood and metal studs at 16" and 24" O.C.

B2010 148	Panel, Shingle & Lap Siding	COST PER S.F.		
		MAT.	INST.	TOTAL
1400	Wood siding w/2"x4"studs, 16"O.C., insul. wall, 5/8"text 1-11 fir plywood	3.41	4.62	8.03
1450	5/8" text 1-11 cedar plywood	4.57	4.46	9.03
1500	1" x 4" vert T.&G. redwood	4.69	5.35	10.04
1600	1" x 8" vert T.&G. redwood	6.40	4.65	11.05
1650	1" x 5" rabbetted cedar bev. siding	6.75	4.81	11.56
1700	1" x 6" cedar drop siding	6.85	4.84	11.69
1750	1" x 12" rough sawn cedar	3.26	4.66	7.92
1800	1" x 12" sawn cedar, 1" x 4" battens	6.30	4.36	10.66
1850	1" x 10" redwood shiplap siding	6.45	4.49	10.94
1900	18" no. 1 red cedar shingles, 5-1/2" exposed	4.39	6.10	10.49
1950	6" exposed	4.21	5.90	10.11
2000	6-1/2" exposed	4.04	5.70	9.74
2100	7" exposed	3.86	5.50	9.36
2150	7-1/2" exposed	3.68	5.30	8.98
3000	8" wide aluminum siding	4.09	4.12	8.21
3150	8" plain vinyl siding	2.74	4.19	6.93
3250	8" insulated vinyl siding	3.02	4.64	7.66
3300				
3400	2" x 6" studs, 16" O.C., insul. wall, w/ 5/8" text 1-11 fir plywood	3.91	4.74	8.65
3500	5/8" text 1-11 cedar plywood	5.05	4.74	9.79
3600	1" x 4" vert T.&G. redwood	5.20	5.45	10.65
3700	1" x 8" vert T.&G. redwood	6.90	4.77	11.67
3800	1" x 5" rabbetted cedar bev siding	7.25	4.93	12.18
3900	1" x 6" cedar drop siding	7.35	4.96	12.31
4000	1" x 12" rough sawn cedar	3.76	4.78	8.54
4200	1" x 12" sawn cedar, 1" x 4" battens	6.80	4.48	11.28
4500	1" x 10" redwood shiplap siding	6.95	4.61	11.56
4550	18" no. 1 red cedar shingles, 5-1/2" exposed	4.89	6.20	11.09
4600	6" exposed	4.71	6	10.71
4650	6-1/2" exposed	4.54	5.80	10.34
4700	7" exposed	4.36	5.60	9.96
4750	7-1/2" exposed	4.18	5.40	9.58
4800	8" wide aluminum siding	4.59	4.24	8.83
4850	8" plain vinyl siding	3.24	4.31	7.55
4900	8" insulated vinyl siding	3.52	4.76	8.28
4910				
5000	2" x 6" studs, 24" O.C., insul. wall, 5/8" text 1-11, fir plywood	3.77	4.47	8.24
5050	5/8" text 1-11 cedar plywood	4.93	4.47	9.40
5100	1" x 4" vert T.&G. redwood	5.05	5.20	10.25
5150	1" x 8" vert T.&G. redwood	6.75	4.50	11.25

B20 Exterior Enclosure

B2010 Exterior Walls

B2010 148	Panel, Shingle & Lap Siding	COST PER S.F.		
		MAT.	INST.	TOTAL
5200	1" x 5" rabbetted cedar bev siding	7.10	4.66	11.76
5250	1" x 6" cedar drop siding	7.20	4.69	11.89
5300	1" x 12" rough sawn cedar	3.62	4.51	8.13
5400	1" x 12" sawn cedar, 1" x 4" battens	6.70	4.21	10.91
5450	1" x 10" redwood shiplap siding	6.80	4.34	11.14
5500	18" no. 1 red cedar shingles, 5-1/2" exposed	4.75	5.95	10.70
5550	6" exposed	4.57	5.75	10.32
5650	7" exposed	4.22	5.35	9.57
5700	7-1/2" exposed	4.04	5.15	9.19
5750	8" wide aluminum siding	4.45	3.97	8.42
5800	8" plain vinyl siding	3.10	4.04	7.14
5850	8" insulated vinyl siding	3.38	4.49	7.87
5900	3-5/8" metal studs, 16 Ga., 16" O.C. insul.wall, 5/8"text 1-11 fir plywood	4.07	4.85	8.92
5950	5/8" text 1-11 cedar plywood	5.25	4.85	10.10
6000	1" x 4" vert T.&G. redwood	5.35	5.55	10.90
6050	1" x 8" vert T.&G. redwood	7.05	4.88	11.93
6100	1" x 5" rabbetted cedar bev siding	7.40	5.05	12.45
6150	1" x 6" cedar drop siding	7.50	5.05	12.55
6200	1" x 12" rough sawn cedar	3.92	4.89	8.81
6250	1" x 12" sawn cedar, 1" x 4" battens	7	4.59	11.59
6300	1" x 10" redwood shiplap siding	7.10	4.72	11.82
6350	18" no. 1 red cedar shingles, 5-1/2" exposed	5.05	6.30	11.35
6500	6" exposed	4.87	6.10	10.97
6550	6-1/2" exposed	4.70	5.90	10.60
6600	7" exposed	4.52	5.70	10.22
6650	7-1/2" exposed	4.52	5.55	10.07
6700	8" wide aluminum siding	4.75	4.35	9.10
6750	8" plain vinyl siding	3.40	4.26	7.66
6800	8" insulated vinyl siding	3.68	4.71	8.39
7000	3-5/8" metal studs, 16 Ga. 24" O.C. insul wall, 5/8" text 1-11 fir plywood	3.78	4.31	8.09
7050	5/8" text 1-11 cedar plywood	4.94	4.31	9.25
7100	1" x 4" vert T.&G. redwood	5.05	5.20	10.25
7150	1" x 8" vert T.&G. redwood	6.80	4.50	11.30
7200	1" x 5" rabbetted cedar bev siding	7.10	4.66	11.76
7250	1" x 6" cedar drop siding	7.20	4.69	11.89
7300	1" x 12" rough sawn cedar	3.63	4.51	8.14
7350	1" x 12" sawn cedar 1" x 4" battens	6.70	4.21	10.91
7400	1" x 10" redwood shiplap siding	6.80	4.34	11.14
7450	18" no. 1 red cedar shingles, 5-1/2" exposed	4.94	6.15	11.09
7500	6" exposed	4.76	5.95	10.71
7550	6-1/2" exposed	4.58	5.75	10.33
7600	7" exposed	4.41	5.55	9.96
7650	7-1/2" exposed	4.05	5.15	9.20
7700	8" wide aluminum siding	4.46	3.97	8.43
7750	8" plain vinyl siding	3.11	4.04	7.15
7800	8" insul. vinyl siding	3.39	4.49	7.88

B20 Exterior Enclosure

B2010 Exterior Walls

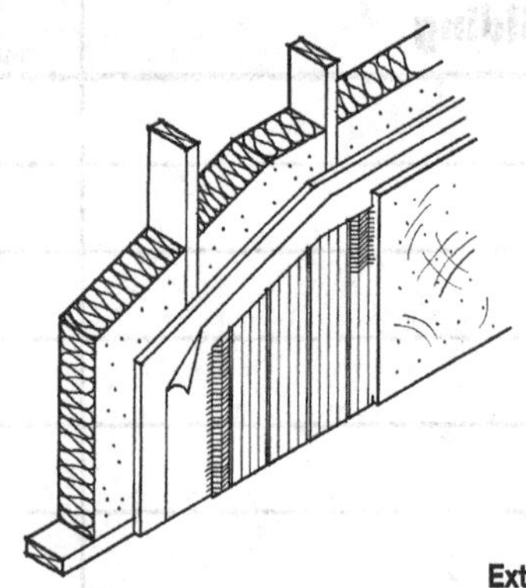

Exterior Stucco Wall

The table below lists costs for some typical stucco walls including all the components as demonstrated in the component block below. Prices are presented for backup walls using wood studs, metal studs and CMU.

B2010 151	Stucco Wall	COST PER S.F.		
		MAT.	INST.	TOTAL
2100	Cement stucco, 7/8" th., plywood sheathing, stud wall, 2" x 4", 16" O.C.	2.67	7.50	10.17
2200	24" O.C.	2.58	7.30	9.88
2300	2" x 6", 16" O.C.	3.17	7.65	10.82
2400	24" O.C.	3.03	7.35	10.38
2500	No sheathing, metal lath on stud wall, 2" x 4", 16" O.C.	1.84	6.55	8.39
2600	24" O.C.	1.75	6.30	8.05
2700	2" x 6", 16" O.C.	2.34	6.65	8.99
2800	24" O.C.	2.20	6.40	8.60
2900	1/2" gypsum sheathing, 3-5/8" metal studs, 16" O.C.	2.73	6.80	9.53
2950	24" O.C.	2.44	6.40	8.84
3000	Cement stucco, 5/8" th., 2 coats on std. CMU block, 8"x 16", 8" thick	3.21	9.15	12.36
3100	10" thick	4.20	9.40	13.60
3200	12" thick	4.34	10.80	15.14
3300	Std. light Wt. block 8" x 16", 8" thick	3.70	9	12.70
3400	10" thick	4.37	9.25	13.62
3500	12" thick	5.30	10.60	15.90
3600	3 coat stucco, self furring metal lath 3.4 Lb/SY, on 8" x 16", 8" thick	3.25	9.70	12.95
3700	10" thick	4.24	9.95	14.19
3800	12" thick	4.38	11.35	15.73
3900	Lt. Wt. block, 8" thick	3.74	9.55	13.29
4000	10" thick	4.41	9.80	14.21
4100	12" thick	5.35	11.15	16.50

B2010 152	E.I.F.S.	COST PER S.F.		
		MAT.	INST.	TOTAL
5100	E.I.F.S., plywood sheathing, stud wall, 2" x 4", 16" O.C., 1" EPS	4.07	7.55	11.62
5110	2" EPS	4.38	7.55	11.93
5120	3" EPS	4.70	7.55	12.25
5130	4" EPS	5	7.55	12.55
5140	2" x 6", 16" O.C., 1" EPS	4.57	7.70	12.27
5150	2" EPS	4.88	7.70	12.58
5160	3" EPS	5.20	7.70	12.90
5170	4" EPS	5.50	7.70	13.20
5180	Cement board sheathing, 3-5/8" metal studs, 16" O.C., 1" EPS	4.70	9.35	14.05
5190	2" EPS	5	9.35	14.35
5200	3" EPS	5.35	9.35	14.70
5210	4" EPS	6.40	9.35	15.75
5220	6" metal studs, 16" O.C., 1" EPS	5.30	9.40	14.70
5230	2" EPS	5.60	9.40	15
5240	3" EPS	5.95	9.40	15.35
5250	4" EPS	7	9.40	16.40

B20 Exterior Enclosure

B2010 Exterior Walls

B2010 152	E.I.F.S.	COST PER S.F.		
		MAT.	INST.	TOTAL
5260	CMU block, 8" x 8" x 16", 1" EPS	5.10	10.75	15.85
5270	2" EPS	5.40	10.75	16.15
5280	3" EPS	5.70	10.75	16.45
5290	4" EPS	6.05	10.75	16.80
5300	8" x 10" x 16", 1" EPS	6.05	11	17.05
5310	2" EPS	6.40	11	17.40
5320	3" EPS	6.70	11	17.70
5330	4" EPS	7	11	18
5340	8" x 12" x 16", 1" EPS	6.20	12.40	18.60
5350	2" EPS	6.50	12.40	18.90
5360	3" EPS	6.85	12.40	19.25
5370	4" EPS	7.15	12.40	19.55

B20 Exterior Enclosure

B2010 Exterior Walls

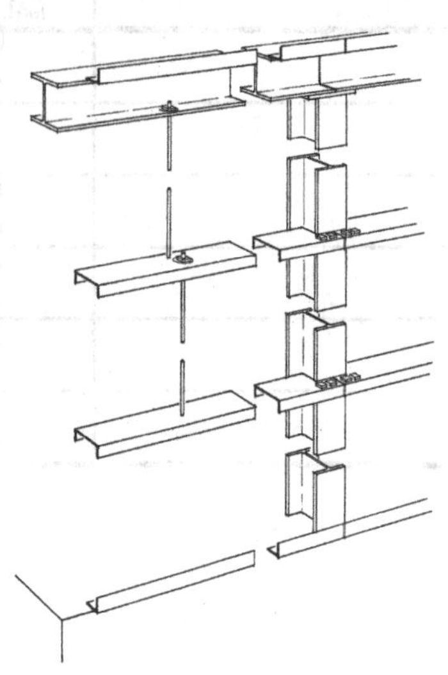

Description: The table below lists costs, $/S.F., for channel girts with sag rods and connector angles top and bottom for various column spacings, building heights and wind loads. Additive costs are shown for wind columns.

How to Use this Table: Add the cost of girts, sag rods, and angles to the framing costs of steel buildings clad in metal or composition siding. If the column spacing is in excess of the column spacing shown, use intermediate wind columns. Additive costs are shown under "Wind Columns".

B2010 154 Metal Siding Support

	BLDG. HEIGHT (FT.)	WIND LOAD (P.S.F.)	COL. SPACING (FT.)		INTERMEDIATE COLUMNS	COST PER S.F.		
						MAT.	INST.	TOTAL
3000	18	20	20			2.06	3.11	5.17
3100					wind cols.	1.06	.22	1.28
3200		20	25			2.25	3.15	5.40
3300					wind cols.	.85	.18	1.03
3400		20	30			2.48	3.19	5.67
3500					wind cols.	.71	.15	.86
3600		20	35			2.73	3.25	5.98
3700					wind cols.	.61	.12	.73
3800		30	20			2.28	3.15	5.43
3900					wind cols.	1.06	.22	1.28
4000		30	25			2.48	3.19	5.67
4100					wind cols.	.85	.18	1.03
4200		30	30			2.74	3.25	5.99
4300					wind cols.	.96	.20	1.16
4600	30	20	20			2	2.19	4.19
4700					wind cols.	1.50	.31	1.81
4800		20	25			2.24	2.24	4.48
4900					wind cols.	1.20	.25	1.45
5000		20	30			2.53	2.30	4.83
5100					wind cols.	1.18	.25	1.43
5200		20	35			2.82	2.36	5.18
5300					wind cols.	1.17	.25	1.42
5400		30	20			2.27	2.25	4.52
5500					wind cols.	1.77	.37	2.14
5600		30	25			2.52	2.30	4.82
5700					wind cols.	1.69	.36	2.05
5800		30	30			2.84	2.36	5.20
5900					wind cols.	1.59	.33	1.92

B20 Exterior Enclosure

B2020 Exterior Windows

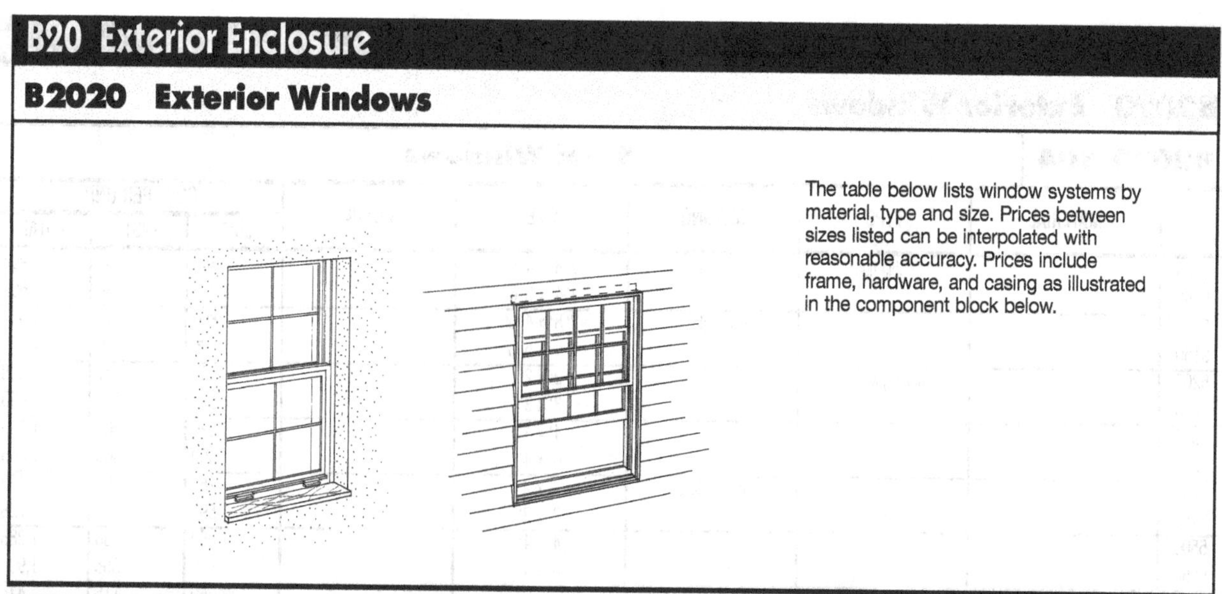

The table below lists window systems by material, type and size. Prices between sizes listed can be interpolated with reasonable accuracy. Prices include frame, hardware, and casing as illustrated in the component block below.

B2020 102	Wood Windows							
	MATERIAL	TYPE	GLAZING	SIZE	DETAIL	COST PER UNIT		
						MAT.	INST.	TOTAL
3000	Wood	double hung	std. glass	2'-8" x 4'-6"		239	196	435
3050				3'-0" x 5'-6"		315	225	540
3100			insul. glass	2'-8" x 4'-6"		256	196	452
3150				3'-0" x 5'-6"		340	225	565
3200		sliding	std. glass	3'-4" x 2'-7"		315	163	478
3250				4'-4" x 3'-3"		330	179	509
3300				5'-4" x 6'-0"		420	216	636
3350			insul. glass	3'-4" x 2'-7"		385	191	576
3400				4'-4" x 3'-3"		400	208	608
3450				5'-4" x 6'-0"		505	246	751
3500		awning	std. glass	2'-10" x 1'-9"		224	98.50	322.50
3600				4'-4" x 2'-8"		355	95.50	450.50
3700			insul. glass	2'-10" x 1'-9"		274	113	387
3800				4'-4" x 2'-8"		435	105	540
3900		casement	std. glass	1'-10" x 3'-2"	1 lite	320	129	449
3950				4'-2" x 4'-2"	2 lite	580	172	752
4000				5'-11" x 5'-2"	3 lite	870	221	1,091
4050				7'-11" x 6'-3"	4 lite	1,225	266	1,491
4100				9'-11" x 6'-3"	5 lite	1,600	310	1,910
4150			insul. glass	1'-10" x 3'-2"	1 lite	320	129	449
4200				4'-2" x 4'-2"	2 lite	590	172	762
4250				5'-11" x 5'-2"	3 lite	920	221	1,141
4300				7'-11" x 6'-3"	4 lite	1,300	266	1,566
4350				9'-11" x 6'-3"	5 lite	1,675	310	1,985
4400		picture	std. glass	4'-6" x 4'-6"		460	222	682
4450				5'-8" x 4'-6"		515	247	762
4500			insul. glass	4'-6" x 4'-6"		560	258	818
4550				5'-8" x 4'-6"		630	287	917
4600		fixed bay	std. glass	8' x 5'		1,600	405	2,005
4650				9'-9" x 5'-4"		1,125	570	1,695
4700			insul. glass	8' x 5'		2,150	405	2,555
4750				9'-9" x 5'-4"		1,225	570	1,795
4800		casement bay	std. glass	8' x 5'		1,600	465	2,065
4850			insul. glass	8' x 5'		1,800	465	2,265
4900		vert. bay	std. glass	8' x 5'		1,800	465	2,265
4950			insul. glass	8' x 5'		1,875	465	2,340

B20 Exterior Enclosure

B2020 Exterior Windows

B2020 104 Steel Windows

	MATERIAL	TYPE	GLAZING	SIZE	DETAIL	COST PER UNIT		
						MAT.	INST.	TOTAL
5000	Steel	double hung	1/4" tempered	2'-8" x 4'-6"		840	154	994
5050				3'-4" x 5'-6"		1,275	234	1,509
5100			insul. glass	2'-8" x 4'-6"		860	177	1,037
5150				3'-4" x 5'-6"		1,325	270	1,595
5202		horiz. pivoted	std. glass	2' x 2'		257	51	308
5250				3' x 3'		575	115	690
5300				4' x 4'		1,025	205	1,230
5350				6' x 4'		1,550	305	1,855
5400			insul. glass	2' x 2'		263	59	322
5450				3' x 3'		595	133	728
5500				4' x 4'		1,050	236	1,286
5550				6' x 4'		1,575	355	1,930
5600		picture window	std. glass	3' x 3'		360	115	475
5650				6' x 4'		965	305	1,270
5700			insul. glass	3' x 3'		375	133	508
5750				6' x 4'		1,000	355	1,355
5800		industrial security	std. glass	2'-9" x 4'-1"		770	144	914
5850				4'-1" x 5'-5"		1,525	283	1,808
5900			insul. glass	2'-9" x 4'-1"		790	165	955
5950				4'-1" x 5'-5"		1,550	325	1,875
6000		comm. projected	std. glass	3'-9" x 5'-5"		1,275	260	1,535
6050				6'-9" x 4'-1"		1,700	355	2,055
6100			insul. glass	3'-9" x 5'-5"		1,300	300	1,600
6150				6'-9" x 4'-1"		1,750	405	2,155
6200		casement	std. glass	4'-2" x 4'-2"	2 lite	1,125	222	1,347
6250			insul. glass	4'-2" x 4'-2"		1,150	256	1,406
6300			std. glass	5'-11" x 5'-2"	3 lite	2,350	390	2,740
6350			insul. glass	5'-11" x 5'-2"		2,400	450	2,850

B2020 106 Aluminum Windows

	MATERIAL	TYPE	GLAZING	SIZE	DETAIL	COST PER UNIT		
						MAT.	INST.	TOTAL
6400	Aluminum	awning	std. glass	3'-1" x 3'-2"		370	111	481
6450				4'-5" x 5'-3"		420	139	559
6500			insul. glass	3'-1" x 3'-2"		445	133	578
6550				4'-5" x 5'-3"		505	167	672
6600		sliding	std. glass	3' x 2'		230	111	341
6650				5' x 3'		350	123	473
6700				8' x 4'		370	185	555
6750				9' x 5'		560	278	838
6800			insul. glass	3' x 2'		246	111	357
6850				5' x 3'		410	123	533
6900				8' x 4'		595	185	780
6950				9' x 5'		900	278	1,178
7000		single hung	std. glass	2' x 3'		218	111	329
7050				2'-8" x 6'-8"		385	139	524
7100				3'-4" x 5'-0"		315	123	438
7150			insul. glass	2' x 3'		264	111	375
7200				2'-8" x 6'-8"		495	139	634
7250				3'-4" x 5'		350	123	473
7300		double hung	std. glass	2' x 3'		300	77	377
7350				2'-8" x 6'-8"		895	228	1,123

B20 Exterior Enclosure

B2020 Exterior Windows

B2020 106 Aluminum Windows

	MATERIAL	TYPE	GLAZING	SIZE	DETAIL	COST PER UNIT		
						MAT.	INST.	TOTAL
7400	Aluminum	Double Hung		3'-4" x 5'		835	213	1,048
7450			insul. glass	2' x 3'		310	88.50	398.50
7500				2'-8" x 6'-8"		925	263	1,188
7550				3'-4" x 5'-0"		865	246	1,111
7600		casement	std. glass	3'-1" x 3'-2"		262	125	387
7650				4'-5" x 5'-3"		635	300	935
7700			insul. glass	3'-1" x 3'-2"		278	144	422
7750				4'-5" x 5'-3"		675	350	1,025
7800		hinged swing	std. glass	3' x 4'		550	154	704
7850				4' x 5'		915	256	1,171
7900			insul. glass	3' x 4'		570	177	747
7950				4' x 5'		945	295	1,240
8200		picture unit	std. glass	2'-0" x 3'-0"		168	77	245
8250				2'-8" x 6'-8"		500	228	728
8300				3'-4" x 5'-0"		465	213	678
8350			insul. glass	2'-0" x 3'-0"		178	88.50	266.50
8400				2'-8" x 6'-8"		530	263	793
8450				3'-4" x 5'-0"		495	246	741
8500		awning type	std. glass	3'-0" x 3'-0"	2 lite	445	79.50	524.50
8550				3'-0" x 4'-0"	3 lite	520	111	631
8600				3'-0" x 5'-4"	4 lite	625	111	736
8650				4'-0" x 5'-4"	4 lite	685	123	808
8700			insul. glass	3'-0" x 3'-0"	2 lite	475	79.50	554.50
8750				3'-0" x 4'-0"	3 lite	595	111	706
8800				3'-0" x 5'-4"	4 lite	735	111	846
8850				4'-0" x 5'-4"	4 lite	825	123	948

B2020 210 Tubular Aluminum Framing

		COST/S.F. OPNG.		
		MAT.	INST.	TOTAL
1100	Alum flush tube frame, for 1/4" glass, 1-3/4"x4", 5'x6'opng, no inter horiz	13.20	9.35	22.55
1150	One intermediate horizontal	18	11.15	29.15
1200	Two intermediate horizontals	23	13	36
1250	5' x 20' opening, three intermediate horizontals	12.15	7.90	20.05
1400	1-3/4" x 4-1/2", 5' x 6' opening, no intermediate horizontals	15.20	9.35	24.55
1450	One intermediate horizontal	20	11.15	31.15
1500	Two intermediate horizontals	25	13	38
1550	5' x 20' opening, three intermediate horizontals	13.65	7.90	21.55
1700	For insulating glass, 2"x4-1/2", 5'x6' opening, no intermediate horizontals	15.55	9.85	25.40
1750	One intermediate horizontal	20	11.80	31.80
1800	Two intermediate horizontals	25	13.75	38.75
1850	5' x 20' opening, three intermediate horizontals	13.55	8.40	21.95
2000	Thermal break frame, 2-1/4"x4-1/2", 5'x6'opng, no intermediate horizontals	16.20	10	26.20
2050	One intermediate horizontal	21.50	12.30	33.80
2100	Two intermediate horizontals	27	14.60	41.60
2150	5' x 20' opening, three intermediate horizontals	14.65	8.80	23.45

B20 Exterior Enclosure

B2020 Exterior Windows

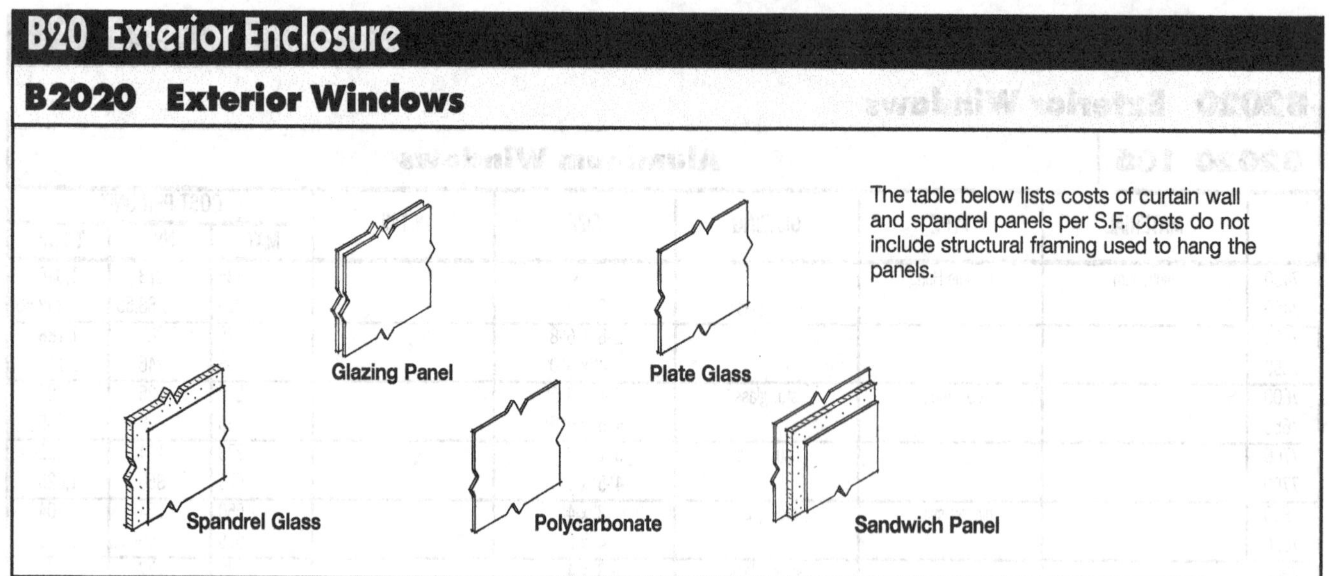

The table below lists costs of curtain wall and spandrel panels per S.F. Costs do not include structural framing used to hang the panels.

B2020 220	Curtain Wall Panels	COST PER S.F.		
		MAT.	INST.	TOTAL
1000	Glazing panel, insulating, 1/2" thick, 2 lites 1/8" float, clear	10.35	9.20	19.55
1100	Tinted	14.45	9.20	23.65
1200	5/8" thick units, 2 lites 3/16" float, clear	14.25	9.70	23.95
1300	Tinted	14.65	9.70	24.35
1400	1" thick units, 2 lites, 1/4" float, clear	17.20	11.65	28.85
1500	Tinted	24.50	11.65	36.15
1600	Heat reflective film inside	28.50	10.25	38.75
1700	Light and heat reflective glass, tinted	32.50	10.25	42.75
2000	Plate glass, 1/4" thick, clear	5.95	7.25	13.20
2050	Tempered	8.65	7.25	15.90
2100	Tinted	8.80	7.25	16.05
2200	3/8" thick, clear	10.40	11.65	22.05
2250	Tempered	17.15	11.65	28.80
2300	Tinted	16.30	11.65	27.95
2400	1/2" thick, clear	17.60	15.85	33.45
2450	Tempered	25.50	15.85	41.35
2500	Tinted	28.50	15.85	44.35
2600	3/4" thick, clear	36.50	25	61.50
2650	Tempered	43	25	68
3000	Spandrel glass, panels, 1/4" plate glass insul w/fiberglass, 1" thick	17.50	7.25	24.75
3100	2" thick	21	7.25	28.25
3200	Galvanized steel backing, add	6.05		6.05
3300	3/8" plate glass, 1" thick	29.50	7.25	36.75
3400	2" thick	33	7.25	40.25
4000	Polycarbonate, masked, clear or colored, 1/8" thick	10.65	5.15	15.80
4100	3/16" thick	12.20	5.30	17.50
4200	1/4" thick	13.75	5.65	19.40
4300	3/8" thick	24.50	5.80	30.30
5000	Facing panel, textured al, 4' x 8' x 5/16" plywood backing, sgl face	4.04	2.69	6.73
5100	Double face	5.50	2.69	8.19
5200	4' x 10' x 5/16" plywood backing, single face	4.32	2.69	7.01
5300	Double face	5.80	2.69	8.49
5400	4' x 12' x 5/16" plywood backing, single face	3.66	2.69	6.35
5500	Sandwich panel, 22 Ga. galv., both sides 2" insulation, enamel exterior	13.45	4.77	18.22
5600	Polyvinylidene fluoride exterior finish	14.15	4.77	18.92
5700	26 Ga., galv. both sides, 1" insulation, colored 1 side	7.05	4.52	11.57
5800	Colored 2 sides	9.10	4.52	13.62

B20 Exterior Enclosure

B2030 Exterior Doors

B2030 110 Glazed Doors, Steel or Aluminum

	MATERIAL	TYPE	DOORS	SPECIFICATION	OPENING	COST PER OPNG. MAT.	INST.	TOTAL
5600	St. Stl. & glass	revolving	stock unit	manual oper.	6′-0″ x 7′-0″	42,900	7,350	50,250
5650				auto Cntrls.	6′-10″ x 7′-0″	59,000	7,875	66,875
5700	Bronze	revolving	stock unit	manual oper.	6′-10″ x 7′-0″	50,500	14,800	65,300
5750				auto Cntrls.	6′-10″ x 7′-0″	66,500	15,300	81,800
5800	St. Stl. & glass	balanced	standard	economy	3′-0″ x 7′-0″	9,450	1,225	10,675
5850				premium	3′-0″ x 7′-0″	16,400	1,625	18,025
6300	Alum. & glass	w/o transom	narrow stile	w/panic Hrdwre.	3′-0″ x 7′-0″	1,675	825	2,500
6350				dbl. door, Hrdwre.	6′-0″ x 7′-0″	2,875	1,350	4,225
6400			wide stile	hdwre.	3′-0″ x 7′-0″	1,900	805	2,705
6450				dbl. door, Hdwre.	6′-0″ x 7′-0″	3,675	1,625	5,300
6500			full vision	hdwre.	3′-0″ x 7′-0″	2,475	1,300	3,775
6550				dbl. door, Hdwre.	6′-0″ x 7′-0″	3,450	1,825	5,275
6600			non-standard	hdwre.	3′-0″ x 7′-0″	2,075	805	2,880
6650				dbl. door, Hdwre.	6′-0″ x 7′-0″	4,125	1,625	5,750
6700			bronze fin.	hdwre.	3′-0″ x 7′-0″	1,525	805	2,330
6750				dbl. door, Hrdwre.	6′-0″ x 7′-0″	3,075	1,625	4,700
6800			black fin.	hdwre.	3′-0″ x 7′-0″	2,275	805	3,080
6850				dbl. door, Hdwre.	6′-0″ x 7′-0″	4,525	1,625	6,150
6900		w/transom	narrow stile	hdwre.	3′-0″ x 10′-0″	2,200	930	3,130
6950				dbl. door, Hdwre.	6′-0″ x 10′-0″	3,475	1,625	5,100
7000			wide stile	hdwre.	3′-0″ x 10′-0″	2,500	1,125	3,625
7050				dbl. door, Hdwre.	6′-0″ x 10′-0″	3,800	1,925	5,725
7100			full vision	hdwre.	3′-0″ x 10′-0″	2,775	1,250	4,025
7150				dbl. door, Hdwre.	6′-0″ x 10′-0″	4,150	2,125	6,275
7200			non-standard	hdwre.	3′-0″ x 10′-0″	2,150	870	3,020
7250				dbl. door, Hdwre.	6′-0″ x 10′-0″	4,300	1,750	6,050
7300			bronze fin.	hdwre.	3′-0″ x 10′-0″	1,625	870	2,495
7350				dbl. door, Hdwre.	6′-0″ x 10′-0″	3,225	1,750	4,975
7400			black fin.	hdwre.	3′-0″ x 10′-0″	2,350	870	3,220
7450				dbl. door, Hdwre.	6′-0″ x 10′-0″	4,700	1,750	6,450
7500		revolving	stock design	minimum	6′-10″ x 7′-0″	21,000	2,950	23,950
7550				average	6′-0″ x 7′-0″	25,400	3,700	29,100
7600				maximum	6′-10″ x 7′-0″	43,600	4,925	48,525
7650				min., automatic	6′-10″ x 7′-0″	36,900	3,475	40,375
7700				avg., automatic	6′-10″ x 7′-0″	41,300	4,225	45,525
7750				max., automatic	6′-10″ x 7′-0″	59,500	5,450	64,950
7800		balanced	standard	economy	3′-0″ x 7′-0″	6,650	1,225	7,875
7850				premium	3′-0″ x 7′-0″	8,150	1,575	9,725
7900		mall front	sliding panels	alum. fin.	16′-0″ x 9′-0″	3,575	670	4,245
7950					24′-0″ x 9′-0″	5,075	1,250	6,325
8000				bronze fin.	16′-0″ x 9′-0″	4,175	780	4,955
8050					24′-0″ x 9′-0″	5,900	1,450	7,350
8100			fixed panels	alum. fin.	48′-0″ x 9′-0″	9,475	970	10,445
8150				bronze fin.	48′-0″ x 9′-0″	11,000	1,125	12,125
8200		sliding entrance	5′ x 7′ door	electric oper.	12′-0″ x 7′-6″	8,975	1,250	10,225
8250		sliding patio	temp. glass	economy	6′-0″ x 7′-0″	1,775	225	2,000
8300			temp. glass	economy	12′-0″ x 7′-0″	4,175	299	4,474
8350				premium	6′-0″ x 7′-0″	2,675	340	3,015
8400					12′-0″ x 7′-0″	6,275	450	6,725

B20 Exterior Enclosure

B2030 Exterior Doors

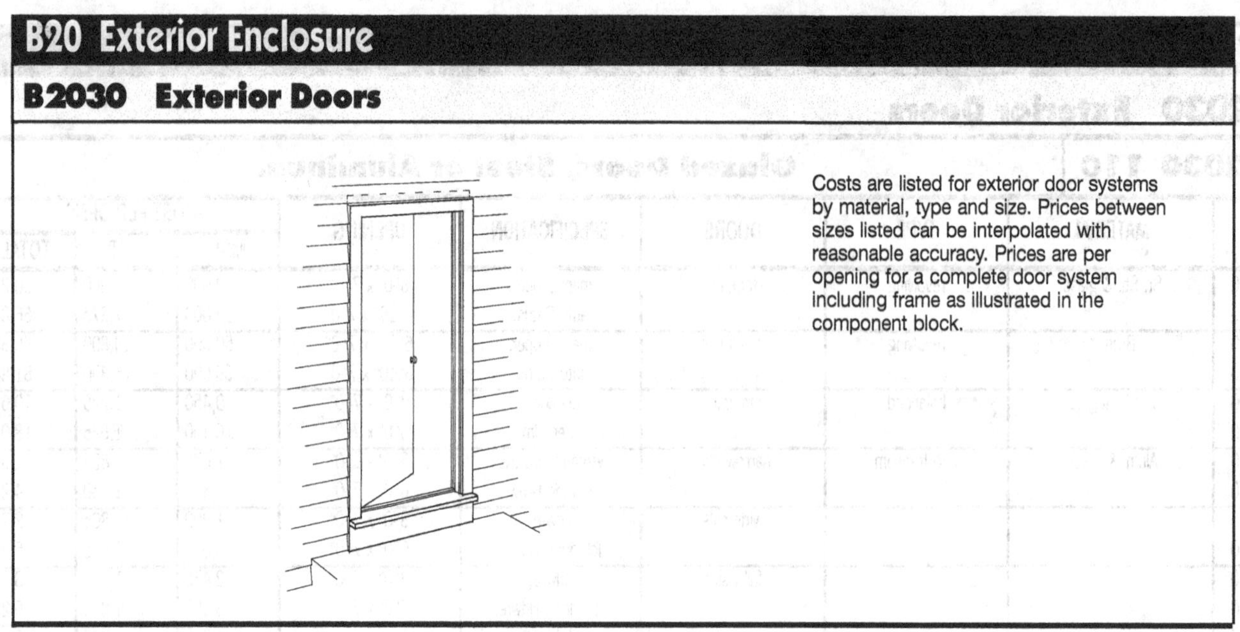

Costs are listed for exterior door systems by material, type and size. Prices between sizes listed can be interpolated with reasonable accuracy. Prices are per opening for a complete door system including frame as illustrated in the component block.

B2030 210 Wood Doors

	MATERIAL	TYPE	DOORS	SPECIFICATION	OPENING	COST PER OPNG.		
						MAT.	INST.	TOTAL
2350	Birch	solid core	single door	hinged	2'-6" x 6'-8"	1,250	243	1,493
2400					2'-6" x 7'-0"	1,225	245	1,470
2450					2'-8" x 7'-0"	1,225	245	1,470
2500					3'-0" x 7'-0"	1,250	249	1,499
2550			double door	hinged	2'-6" x 6'-8"	2,225	440	2,665
2600					2'-6" x 7'-0"	2,250	445	2,695
2650					2'-8" x 7'-0"	2,225	445	2,670
2700					3'-0" x 7'-0"	2,325	455	2,780
2750	Wood	combination	storm & screen	hinged	3'-0" x 6'-8"	325	60	385
2800					3'-0" x 7'-0"	360	66	426
2850		overhead	panels, H.D.	manual oper.	8'-0" x 8'-0"	890	450	1,340
2900					10'-0" x 10'-0"	1,300	500	1,800
2950					12'-0" x 12'-0"	1,850	600	2,450
3000					14'-0" x 14'-0"	2,875	690	3,565
3050					20'-0" x 16'-0"	5,175	1,375	6,550
3100				electric oper.	8'-0" x 8'-0"	1,875	675	2,550
3150					10'-0" x 10'-0"	2,300	725	3,025
3200					12'-0" x 12'-0"	2,850	825	3,675
3250					14'-0" x 14'-0"	3,875	915	4,790
3300					20'-0" x 16'-0"	6,325	1,825	8,150

B2030 220 Steel Doors

	MATERIAL	TYPE	DOORS	SPECIFICATION	OPENING	COST PER OPNG.		
						MAT.	INST.	TOTAL
3350	Steel 18 Ga.	hollow metal	1 door w/frame	no label	2'-6" x 7'-0"	1,500	253	1,753
3400					2'-8" x 7'-0"	1,500	253	1,753
3450					3'-0" x 7'-0"	1,500	253	1,753
3500					3'-6" x 7'-0"	1,650	266	1,916
3550					4'-0" x 8'-0"	1,925	267	2,192
3600			2 doors w/frame	no label	5'-0" x 7'-0"	2,950	470	3,420
3650					5'-4" x 7'-0"	2,950	470	3,420
3700					6'-0" x 7'-0"	2,950	470	3,420
3750					7'-0" x 7'-0"	3,225	495	3,720
3800					8'-0" x 8'-0"	3,750	495	4,245

B20 Exterior Enclosure

B2030 Exterior Doors

B2030 220 Steel Doors

	MATERIAL	TYPE	DOORS	SPECIFICATION	OPENING	COST PER OPNG. MAT.	COST PER OPNG. INST.	COST PER OPNG. TOTAL
3850	Steel 18 Ga.	Hollow Metal	1 door w/frame	"A" label	2'-6" x 7'-0"	1,875	305	2,180
3900					2'-8" x 7'-0"	1,875	310	2,185
3950					3'-0" x 7'-0"	1,875	310	2,185
4000					3'-6" x 7'-0"	2,025	315	2,340
4050					4'-0" x 8'-0"	2,250	330	2,580
4100			2 doors w/frame	"A" label	5'-0" x 7'-0"	3,625	565	4,190
4150					5'-4" x 7'-0"	3,675	575	4,250
4200					6'-0" x 7'-0"	3,675	575	4,250
4250					7'-0" x 7'-0"	3,975	585	4,560
4300					8'-0" x 8'-0"	3,975	585	4,560
4350	Steel 24 Ga.	overhead	sectional	manual oper.	8'-0" x 8'-0"	885	450	1,335
4400					10'-0" x 10'-0"	1,125	500	1,625
4450					12'-0" x 12'-0"	1,375	600	1,975
4500					20'-0" x 14'-0"	3,800	1,275	5,075
4550				electric oper.	8'-0" x 8'-0"	1,875	675	2,550
4600					10'-0" x 10'-0"	2,125	725	2,850
4650					12'-0" x 12'-0"	2,375	825	3,200
4700					20'-0" x 14'-0"	4,950	1,725	6,675
4750	Steel	overhead	rolling	manual oper.	8'-0" x 8'-0"	1,175	695	1,870
4800					10'-0" x 10'-0"	1,950	795	2,745
4850					12'-0" x 12'-0"	2,000	925	2,925
4900					14'-0" x 14'-0"	3,225	1,400	4,625
4950					20'-0" x 12'-0"	2,275	1,225	3,500
5000					20'-0" x 16'-0"	3,650	1,850	5,500
5050				electric oper.	8'-0" x 8'-0"	2,325	915	3,240
5100					10'-0" x 10'-0"	3,100	1,025	4,125
5150					12'-0" x 12'-0"	3,150	1,150	4,300
5200					14'-0" x 14'-0"	4,375	1,625	6,000
5250					20'-0" x 12'-0"	3,475	1,450	4,925
5300					20'-0" x 16'-0"	4,850	2,075	6,925
5350				fire rated	10'-0" x 10'-0"	2,150	1,000	3,150
5400			rolling grille	manual oper.	10'-0" x 10'-0"	2,700	1,100	3,800
5450					15'-0" x 8'-0"	3,125	1,400	4,525
5500		vertical lift	1 door w/frame	motor operator	16'-0" x 16'-0"	23,700	4,675	28,375
5550					32'-0" x 24'-0"	51,000	3,100	54,100

B2030 230 Aluminum Doors

	MATERIAL	TYPE	DOORS	SPECIFICATION	OPENING	COST PER OPNG. MAT.	COST PER OPNG. INST.	COST PER OPNG. TOTAL
6000	Aluminum	combination	storm & screen	hinged	3'-0" x 6'-8"	325	64	389
6050					3'-0" x 7'-0"	360	70.50	430.50
6100		overhead	rolling grille	manual oper.	12'-0" x 12'-0"	4,325	1,950	6,275
6150				motor oper.	12'-0" x 12'-0"	5,700	2,175	7,875
6200	Alum. & Fbrgls.	overhead	heavy duty	manual oper.	12'-0" x 12'-0"	2,875	600	3,475
6250				electric oper.	12'-0" x 12'-0"	3,875	825	4,700

B30 Roofing

B3010 Roof Coverings

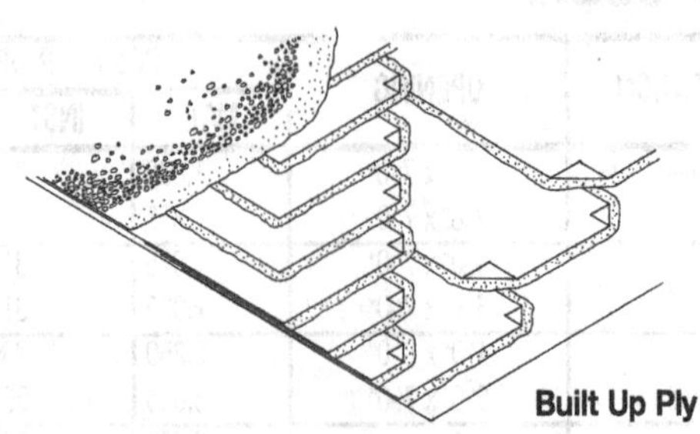

Multiple ply roofing is the most popular covering for minimum pitch roofs.

B3010 105	Built-Up	COST PER S.F.		
		MAT.	INST.	TOTAL
1200	Asphalt flood coat w/gravel; not incl. insul, flash., nailers			
1300				
1400	Asphalt base sheets & 3 plies #15 asphalt felt, mopped	.91	1.42	2.33
1500	On nailable deck	.96	1.49	2.45
1600	4 plies #15 asphalt felt, mopped	1.26	1.56	2.82
1700	On nailable deck	1.13	1.64	2.77
1800	Coated glass base sheet, 2 plies glass (type IV), mopped	.95	1.42	2.37
1900	For 3 plies	1.13	1.56	2.69
2000	On nailable deck	1.06	1.64	2.70
2300	4 plies glass fiber felt (type IV), mopped	1.38	1.56	2.94
2400	On nailable deck	1.25	1.64	2.89
2500	Organic base sheet & 3 plies #15 organic felt, mopped	.96	1.57	2.53
2600	On nailable deck	.95	1.64	2.59
2700	4 plies #15 organic felt, mopped	1.19	1.42	2.61
2750				
2800	Asphalt flood coat, smooth surface, not incl. insul, flash., nailers			
2900	Asphalt base sheet & 3 plies #15 asphalt felt, mopped	.97	1.30	2.27
3000	On nailable deck	.90	1.36	2.26
3100	Coated glass fiber base sheet & 2 plies glass fiber felt, mopped	.89	1.25	2.14
3200	On nailable deck	.84	1.30	2.14
3300	For 3 plies, mopped	1.07	1.36	2.43
3400	On nailable deck	1	1.42	2.42
3700	4 plies glass fiber felt (type IV), mopped	1.25	1.36	2.61
3800	On nailable deck	1.18	1.42	2.60
3900	Organic base sheet & 3 plies #15 organic felt, mopped	.96	1.30	2.26
4000	On nailable decks	.89	1.36	2.25
4100	4 plies #15 organic felt, mopped	1.12	1.42	2.54
4200	Coal tar pitch with gravel surfacing			
4300	4 plies #15 tarred felt, mopped	1.94	1.49	3.43
4400	3 plies glass fiber felt (type IV), mopped	1.60	1.64	3.24
4500	Coated glass fiber base sheets 2 plies glass fiber felt, mopped	1.66	1.64	3.30
4600	On nailable decks	1.45	1.73	3.18
4800	3 plies glass fiber felt (type IV), mopped	2.22	1.49	3.71
4900	On nailable decks	2.02	1.56	3.58

B30 Roofing

B3010 Roof Coverings

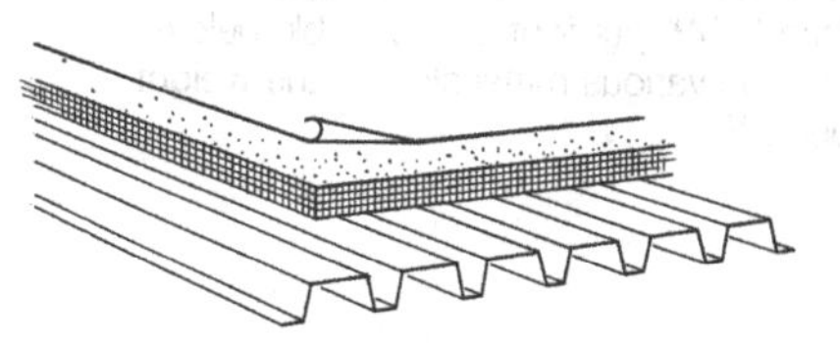

Fully Adhered

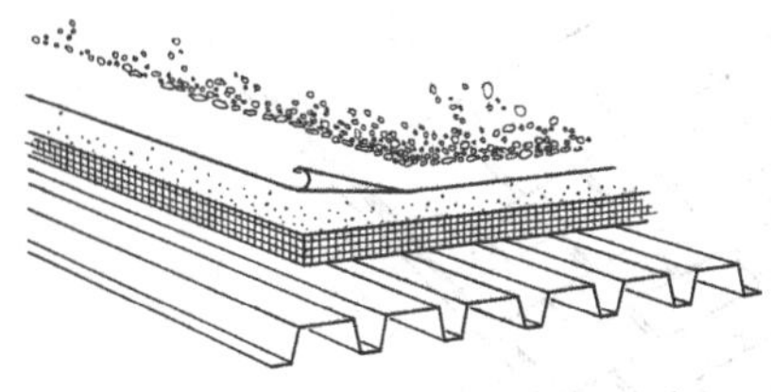

Ballasted

The systems listed below reflect only the cost for the single ply membrane.

B3010 120	Single Ply Membrane	COST PER S.F. MAT.	COST PER S.F. INST.	COST PER S.F. TOTAL
1000	CSPE (Chlorosulfonated polyethylene), 35 mils, fully adhered	2.04	.78	2.82
2000	EPDM (Ethylene propylene diene monomer), 45 mils, fully adhered	1.11	.78	1.89
4000	Modified bit., SBS modified, granule surface cap sheet, mopped, 150 mils	.64	1.56	2.20
4500	APP modified, granule surface cap sheet, torched, 180 mils	.65	1.02	1.67
6000	Reinforced PVC, 48 mils, loose laid and ballasted with stone	1.12	.41	1.53
6200	Fully adhered with adhesive	1.39	.78	2.17

B3010 130	Preformed Metal Roofing	COST PER S.F. MAT.	COST PER S.F. INST.	COST PER S.F. TOTAL
0200	Corrugated roofing, aluminum, mill finish, .0175" thick, .272 P.S.F.	1.06	1.43	2.49
0250	.0215" thick, .334 P.S.F.	1.40	1.43	2.83

B3010 135	Formed Metal	COST PER S.F. MAT.	COST PER S.F. INST.	COST PER S.F. TOTAL
1000	Batten seam, formed copper roofing, 3" min slope, 16 oz., 1.2 P.S.F.	10.35	4.67	15.02
1100	18 oz., 1.35 P.S.F.	11.60	5.10	16.70
2000	Zinc copper alloy, 3" min slope, .020" thick, .88 P.S.F.	12.35	4.27	16.62
3000	Flat seam, copper, 1/4" min. slope, 16 oz., 1.2 P.S.F.	7.60	4.27	11.87
3100	18 oz., 1.35 P.S.F.	8.50	4.47	12.97
5000	Standing seam, copper, 2-1/2" min. slope, 16 oz., 1.25 P.S.F.	8.20	3.97	12.17
5100	18 oz., 1.40 P.S.F.	9.15	4.27	13.42
6000	Zinc copper alloy, 2-1/2" min. slope, .020" thick, .87 P.S.F.	12.35	4.27	16.62
6100	.032" thick, 1.39 P.S.F.	21	4.67	25.67

B30 Roofing

B3010 Roof Coverings

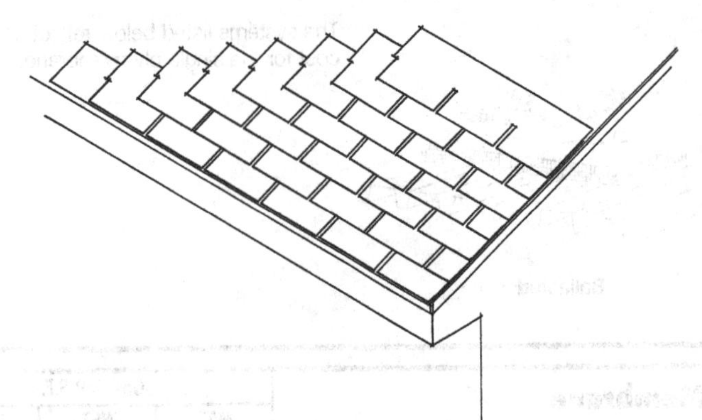

Shingles and tiles are practical in applications where the roof slope is more than 3-1/2" per foot of rise. Table below lists the various materials and the weight per S.F.

B3010 140	Shingle & Tile	COST PER S.F. MAT.	COST PER S.F. INST.	COST PER S.F. TOTAL
1095	Asphalt roofing			
1100	Strip shingles, 4" slope, inorganic class A 210-235 lb./sq.	.88	.86	1.74
1150	Organic, class C, 235-240 lb./sq.	.86	.94	1.80
1200	Premium laminated multi-layered, class A, 260-300 lb./sq.	1.79	1.28	3.07
1545	Metal roofing			
1550	Alum., shingles, colors, 3" min slope, .019" thick, 0.4 PSF	2.42	.97	3.39
1850	Steel, colors, 3" min slope, 26 gauge, 1.0 PSF	3.46	1.93	5.39
2795	Slate roofing			
2800	Slate roofing, 4" min. slope, shingles, 3/16" thick, 8.0 PSF	5.35	2.41	7.76
3495	Wood roofing			
3500	4" min slope, cedar shingles, 16" x 5", 5" exposure 1.6 PSF	2.48	1.94	4.42
4000	Shakes, 18", 8-1/2" exposure, 2.8 PSF	1.87	2.39	4.26
5095	Tile roofing			
5100	Aluminum, mission, 3" min slope, .019" thick, 0.65 PSF	9.05	1.87	10.92
6000	Clay, Americana, 3" minimum slope, 8 PSF	4.95	2.55	7.50
6002				

B30 Roofing

B3010 Roof Coverings

B3010 320	Roof Deck Rigid Insulation	COST PER S.F.		
		MAT.	INST.	TOTAL
0100	Fiberboard low density, 1/2" thick, R1.39			
0150	1" thick R2.78	.51	.51	1.02
0300	1 1/2" thick R4.17	.77	.51	1.28
0350	2" thick R5.56	1.02	.51	1.53
0370	Fiberboard high density, 1/2" thick R1.3	.26	.41	.67
0380	1" thick R2.5	.53	.51	1.04
0390	1 1/2" thick R3.8	.79	.51	1.30
0410	Fiberglass, 3/4" thick R2.78	.58	.41	.99
0450	15/16" thick R3.70	.78	.41	1.19
0550	1-5/16" thick R5.26	1.32	.41	1.73
0650	2 7/16" thick R10	1.62	.51	2.13
1510	Polyisocyanurate 2#/CF density, 1" thick	.56	.29	.85
1550	1 1/2" thick	.73	.33	1.06
1600	2" thick	.86	.37	1.23
1650	2 1/2" thick	1.11	.39	1.50
1700	3" thick	1.34	.41	1.75
1750	3 1/2" thick	2.15	.41	2.56
1800	Tapered for drainage	2.15	.29	2.44
1810	Expanded polystyrene, 1#/CF density, 3/4" thick R2.89	.24	.27	.51
1820	2" thick R7.69	.64	.33	.97
1830	Extruded polystyrene, 15 PSI compressive strength, 1" thick, R5	.48	.27	.75
1835	2" thick R10	.62	.33	.95
1840	3" thick R15	1.27	.41	1.68
2550	40 PSI compressive strength, 1" thick R5	.52	.27	.79
2600	2" thick R10	.97	.33	1.30
2650	3" thick R15	1.41	.41	1.82
2700	4" thick R20	1.87	.41	2.28
2750	Tapered for drainage	.74	.29	1.03
2810	60 PSI compressive strength, 1" thick R5	.72	.28	1
2850	2" thick R10	1.38	.34	1.72
2900	Tapered for drainage	.94	.29	1.23

B3010 410	Base Flashing							
	TYPE	DESCRIPTION	SPECIFICATION	REGLET	COUNTER FLASHING	COST PER L.F.		
						MAT.	INST.	TOTAL
1000	Aluminum	mill finish	.019" thick	alum. .025	alum. .032	11.50	12.45	23.95
1100			.050" thick	.025	.032	12.80	12.45	25.25
1200		fabric, 2 sides	.004" thick	.025	.032	11.90	10.90	22.80
1300			.016" thick	.025	.032	12.20	10.90	23.10
1400		mastic, 2 sides	.004" thick	.025	.032	11.90	10.90	22.80
1500			.016" thick	.025	.032	12.35	10.90	23.25
1700	Copper	sheets, plain	16 oz.	copper 10 oz.	copper 10 oz.	21.50	13.20	34.70
1800			20 oz.	10 oz.	10 oz.	23	13.35	36.35
1900			24 oz.	10 oz.	10 oz.	28	13.50	41.50
2000			32 oz.	10 oz.	10 oz.	30	13.70	43.70
2200	Sheets, Metal	galvanized	20 Ga.	galv. 24 Ga.	galv. 24 Ga.	11.45	28.50	39.95
2300			24 Ga.	24 Ga.	24 Ga.	10.75	23	33.75
2400			30 Ga.	24 Ga.	24 Ga.	10.05	17.20	27.25
2600	Rubber	butyl	1/32" thick	galv. 24 Ga.	galv. 24 Ga.	11	11.10	22.10
2700			1/16" thick	24 Ga.	24 Ga.	11.70	11.10	22.80
2800		neoprene	1/16" thick	24 Ga.	24 Ga.	11.20	11.10	22.30
2900			1/8" thick	24 Ga.	24 Ga.	13.50	11.10	24.60

B30 Roofing

B3010 Roof Coverings

B3010 420 Roof Edges

	EDGE TYPE	DESCRIPTION	SPECIFICATION	FACE HEIGHT		COST PER L.F.		
						MAT.	INST.	TOTAL
1000	Aluminum	mill finish	.050″ thick	4″		13.50	7.75	21.25
1100				6″		14	8	22
1300		duranodic	.050″ thick	4″		13.25	7.75	21
1400				6″		14.40	8	22.40
1600		painted	.050″ thick	4″		14.25	7.75	22
2000	Copper	plain	16 oz.	4″		9.70	7.75	17.45
2100				6″		11.20	8	19.20
2300			20 oz.	4″		10.65	8.90	19.55
2400				6″		11.90	8.60	20.50
2700	Sheet Metal	galvanized	20 Ga.	4″		10.95	9.25	20.20
2800				6″		13.10	9.25	22.35
3000			24 Ga.	4″		9.70	7.75	17.45
3100				6″		11.15	7.75	18.90

B3010 430 Flashing

	MATERIAL	BACKING	SIDES	SPECIFICATION	QUANTITY	COST PER S.F.		
						MAT.	INST.	TOTAL
0040	Aluminum	none		.019″		1.18	2.80	3.98
0050				.032″		1.38	2.80	4.18
0300		fabric	2	.004″		1.55	1.23	2.78
0400		mastic		.004″		1.55	1.23	2.78
0700	Copper	none		16 oz.	<500 lbs.	6.50	3.53	10.03
0800				24 oz.	<500 lbs.	10.20	3.87	14.07
2000	Copper lead	fabric	1	2 oz.		2.66	1.23	3.89
3500	PVC black	none		.010″		.23	1.43	1.66
3700				.030″		.43	1.43	1.86
4200	Neoprene			1/16″		2.11	1.43	3.54
4500	Stainless steel	none		.015″	<500 lbs.	6.40	3.53	9.93
4600	Copper clad				>2000 lbs.	6.10	2.62	8.72
5000	Plain			32 ga.		4.40	2.62	7.02
5009								

B30 Roofing

B3010 Roof Coverings

B3010 610 Gutters

	SECTION	MATERIAL	THICKNESS	SIZE	FINISH	COST PER L.F.		
						MAT.	INST.	TOTAL
0050	Box	aluminum	.027″	5″	enameled	2.76	4.21	6.97
0100					mill	2.84	4.21	7.05
0200			.032″	5″	enameled	3.62	4.21	7.83
0500		copper	16 Oz.	4″	lead coated	12.50	4.21	16.71
0600					mill	7.55	4.21	11.76
1000		steel galv.	28 Ga.	5″	enameled	1.66	4.21	5.87
1200			26 Ga.	5″	mill	1.75	4.21	5.96
1800		vinyl		4″	colors	1.19	4.08	5.27
1900				5″	colors	1.52	4.08	5.60
2300		hemlock or fir		4″x5″	treated	12.70	4.49	17.19
3000	Half round	copper	16 Oz.	4″	lead coated	10.40	4.21	14.61
3100					mill	9	4.21	13.21
3600		steel galv.	28 Ga.	5″	enameled	1.66	4.21	5.87
4102		stainless steel		5″	mill	6.95	4.21	11.16
5000		vinyl		4″	white	1.19	4.08	5.27
5002								

B3010 620 Downspouts

	MATERIALS	SECTION	SIZE	FINISH	THICKNESS	COST PER V.L.F.		
						MAT.	INST.	TOTAL
0100	Aluminum	rectangular	2″x3″	embossed mill	.020″	1.13	2.66	3.79
0150				enameled	.020″	1.39	2.66	4.05
0250			3″x4″	enameled	.024″	3.03	3.61	6.64
0300		round corrugated	3″	enameled	.020″	1.76	2.66	4.42
0350			4″	enameled	.025″	2.72	3.61	6.33
0500	Copper	rectangular corr.	2″x3″	mill	16 Oz.	8.30	2.66	10.96
0600		smooth		mill	16 Oz.	7.35	2.66	10.01
0700		rectangular corr.	3″x4″	mill	16 Oz.	10.95	3.48	14.43
1300	Steel	rectangular corr.	2″x3″	galvanized	28 Ga.	1.41	2.66	4.07
1350				epoxy coated	24 Ga.	2.20	2.66	4.86
1400		smooth		galvanized	28 Ga.	1.41	2.66	4.07
1450		rectangular corr.	3″x4″	galvanized	28 Ga.	1.80	3.48	5.28
1500				epoxy coated	24 Ga.	2.75	3.48	6.23
1550		smooth		galvanized	28 Ga.	2.56	3.48	6.04
1652		round corrugated	3″	galvanized	28 Ga.	1.89	2.66	4.55
1700			4″	galvanized	28 Ga.	2.59	3.48	6.07
1750			5″	galvanized	28 Ga.	3.83	3.88	7.71
2552	S.S. tubing sch.5	rectangular	3″x4″	mill		62	3.48	65.48
2702								

B3010 630 Gravel Stop

	MATERIALS	SECTION	SIZE	FINISH	THICKNESS	COST PER L.F.		
						MAT.	INST.	TOTAL
5100	Aluminum	extruded	4″	mill	.050″	6.75	3.48	10.23
5200			4″	duranodic	.050″	6.50	3.48	9.98
5300			8″	mill	.050″	9.05	4.04	13.09
5400			8″	duranodic	.050″	8.80	4.04	12.84
6000			12″-2 pc.	duranodic	.050″	11.05	5.05	16.10
6100	Stainless	formed	6″	mill	24 Ga.	13.85	3.74	17.59

B30 Roofing

B3020 Roof Openings

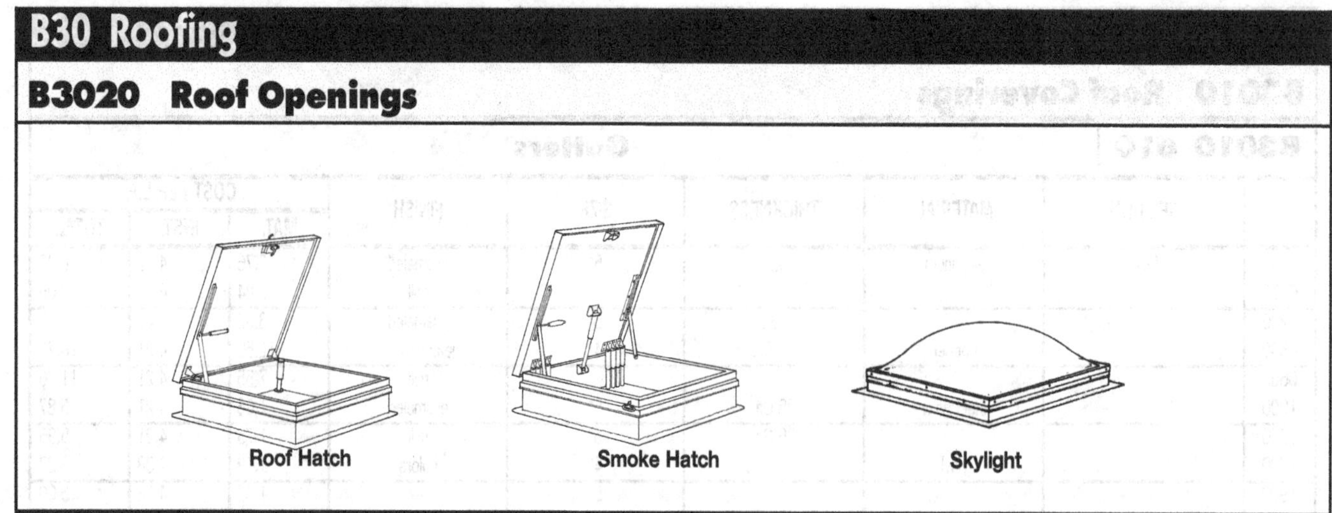

B3020 110	Skylights	COST PER S.F.		
		MAT.	INST.	TOTAL
5100	Skylights, plastic domes, insul curbs, nom. size to 10 S.F., single glaze	28	10.75	38.75
5200	Double glazing	32	13.20	45.20
5300	10 S.F. to 20 S.F., single glazing	27	4.35	31.35
5400	Double glazing	27.50	5.45	32.95
5500	20 S.F. to 30 S.F., single glazing	22.50	3.70	26.20
5600	Double glazing	25.50	4.35	29.85
5700	30 S.F. to 65 S.F., single glazing	19.15	2.82	21.97
5800	Double glazing	25.50	3.70	29.20
6000	Sandwich panels fiberglass, 1-9/16" thick, 2 S.F. to 10 S.F.	17.80	8.60	26.40
6100	10 S.F. to 18 S.F.	16.05	6.50	22.55
6200	2-3/4" thick, 25 S.F. to 40 S.F.	25.50	5.80	31.30
6300	40 S.F. to 70 S.F.	21	5.20	26.20

B3020 210	Hatches	COST PER OPNG.		
		MAT.	INST.	TOTAL
0200	Roof hatches with curb and 1" fiberglass insulation, 2'-6"x3'-0", aluminum	1,075	172	1,247
0300	Galvanized steel 165 lbs.	655	172	827
0400	Primed steel 164 lbs.	605	172	777
0500	2'-6"x4'-6" aluminum curb and cover, 150 lbs.	935	191	1,126
0600	Galvanized steel 220 lbs.	660	191	851
0650	Primed steel 218 lbs.	910	191	1,101
0800	2'x6"x8'-0" aluminum curb and cover, 260 lbs.	1,825	260	2,085
0900	Galvanized steel, 360 lbs.	1,750	260	2,010
0950	Primed steel 358 lbs.	1,325	260	1,585
1200	For plexiglass panels, add to the above	440		440
2100	Smoke hatches, unlabeled not incl. hand winch operator, 2'-6"x3', galv	795	208	1,003
2200	Plain steel, 160 lbs.	730	208	938
2400	2'-6"x8'-0",galvanized steel, 360 lbs.	1,925	284	2,209
2500	Plain steel, 350 lbs.	1,450	284	1,734
3000	4'-0"x8'-0", double leaf low profile, aluminum cover, 359 lb.	2,850	215	3,065
3100	Galvanized steel 475 lbs.	2,425	215	2,640
3200	High profile, aluminum cover, galvanized curb, 361 lbs.	2,750	215	2,965

C10 Interior Construction

C1010 Partitions

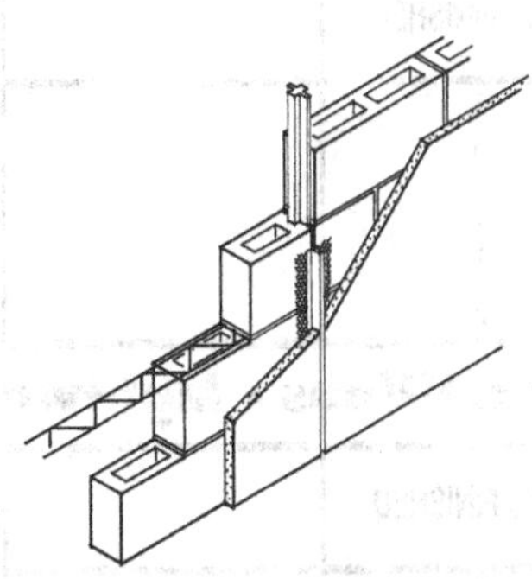

The Concrete Block Partition Systems are defined by weight and type of block, thickness, type of finish and number of sides finished. System components include joint reinforcing on alternate courses and vertical control joints.

C1010 102 Concrete Block Partitions - Regular Weight

	TYPE	THICKNESS (IN.)	TYPE FINISH	SIDES FINISHED		COST PER S.F.		
						MAT.	INST.	TOTAL
1000	Hollow	4	none	0		1.80	4.92	6.72
1010			gyp. plaster 2 coat	1		2.18	7.20	9.38
1020				2		2.56	9.45	12.01
1200			portland - 3 coat	1		2.07	7.50	9.57
1400			5/8" drywall	1		2.30	6.45	8.75
1500		6	none	0		2.48	5.30	7.78
1510			gyp. plaster 2 coat	1		2.86	7.55	10.41
1520				2		3.24	9.80	13.04
1700			portland - 3 coat	1		2.75	7.85	10.60
1900			5/8" drywall	1		2.98	6.85	9.83
1910				2		3.48	8.40	11.88
2000		8	none	0		2.81	5.65	8.46
2010			gyp. plaster 2 coat	1		3.19	7.90	11.09
2020			gyp. plaster 2 coat	2		3.57	10.20	13.77
2200			portland - 3 coat	1		3.08	8.25	11.33
2400			5/8" drywall	1		3.31	7.20	10.51
2410				2		3.81	8.75	12.56
2500		10	none	0		3.52	5.90	9.42
2510			gyp. plaster 2 coat	1		3.90	8.15	12.05
2520				2		4.28	10.45	14.73
2700			portland - 3 coat	1		3.79	8.50	12.29
2900			5/8" drywall	1		4.02	7.45	11.47
2910				2		4.52	9	13.52
3000	Solid	2	none	0		1.63	4.86	6.49
3010			gyp. plaster	1		2.01	7.15	9.16
3020				2		2.39	9.40	11.79
3200			portland - 3 coat	1		1.90	7.45	9.35
3400			5/8" drywall	1		2.13	6.40	8.53
3410				2		2.63	7.95	10.58
3500		4	none	0		2.17	5.10	7.27
3510			gyp. plaster	1		2.64	7.40	10.04
3520				2		2.93	9.65	12.58
3700			portland - 3 coat	1		2.44	7.70	10.14
3900			5/8" drywall	1		2.67	6.65	9.32
3910				2		3.17	8.20	11.37

C10 Interior Construction

C1010 Partitions

C1010 102 Concrete Block Partitions - Regular Weight

	TYPE	THICKNESS (IN.)	TYPE FINISH	SIDES FINISHED		COST PER S.F. MAT.	INST.	TOTAL
4010	Solid	6	gyp. plaster	1		2.94	7.75	10.69
4020				2		3.32	10	13.32
4200			portland - 3 coat	1		2.83	8.05	10.88
4400			5/8" drywall	1		3.06	7.05	10.11
4410				2		3.56	8.60	12.16

C1010 104 Concrete Block Partitions - Lightweight

	TYPE	THICKNESS (IN.)	TYPE FINISH	SIDES FINISHED		COST PER S.F. MAT.	INST.	TOTAL
5000	Hollow	4	none	0		1.97	4.81	6.78
5010			gyp. plaster	1		2.35	7.10	9.45
5020				2		2.73	9.35	12.08
5200			portland - 3 coat	1		2.24	7.40	9.64
5400			5/8" drywall	1		2.47	6.35	8.82
5410				2		2.97	7.90	10.87
5500		6	none	0		2.79	5.15	7.94
5520			gyp. plaster	2		3.55	9.70	13.25
5700			portland - 3 coat	1		3.06	7.75	10.81
5900			5/8" drywall	1		3.29	6.70	9.99
5910				2		3.79	8.25	12.04
6000		8	none	0		3.34	5.50	8.84
6010			gyp. plaster	1		3.72	7.75	11.47
6020				2		4.10	10.05	14.15
6200			portland - 3 coat	1		3.61	8.10	11.71
6400			5/8" drywall	1		3.84	7.05	10.89
6410				2		4.34	8.60	12.94
6500		10	none	0		4.01	5.75	9.76
6700			portland - 3 coat	1		4.28	8.35	12.63
6900			5/8" drywall	1		4.51	7.30	11.81
6910				2		5	8.85	13.85
7000	Solid	4	none	0		2.75	5.05	7.80
7010			gyp. plaster	1		3.13	7.30	10.43
7020				2		3.51	9.55	13.06
7200			portland - 3 coat	1		3.02	7.60	10.62
7400			5/8" drywall	1		3.25	6.60	9.85
7410				2		3.75	8.15	11.90
7500		6	none	0		3.08	5.40	8.48
7510			gyp. plaster	1		3.54	7.70	11.24
7520				2		3.84	9.90	13.74
7700			portland - 3 coat	1		3.35	7.95	11.30
7900			5/8" drywall	1		3.58	6.95	10.53
7910				2		4.08	8.50	12.58
8000		8	none	0		3.89	5.80	9.69
8010			gyp. plaster	1		4.27	8.05	12.32
8020				2		4.65	10.35	15
8200			portland - 3 coat	1		4.16	8.40	12.56
8400			5/8" drywall	1		4.39	7.35	11.74
8410				2		4.89	8.90	13.79

C10 Interior Construction

C1010 Partitions

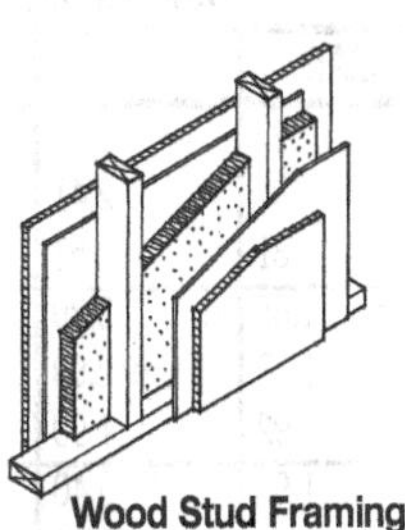
Wood Stud Framing

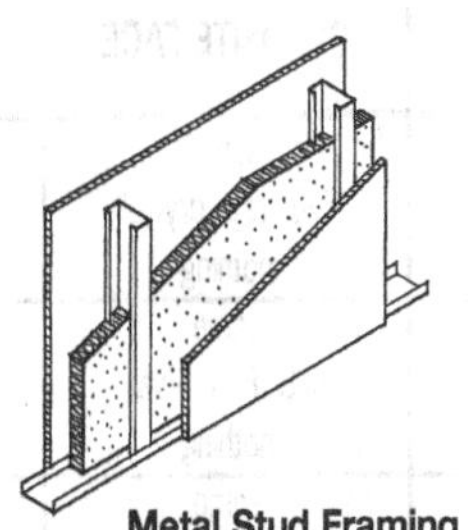
Metal Stud Framing

The Drywall Partitions/Stud Framing Systems are defined by type of drywall and number of layers, type and spacing of stud framing, and treatment on the opposite face. Components include taping and finishing.

Cost differences between regular and fire resistant drywall are negligible, and terminology is interchangeable. In some cases fiberglass insulation is included for additional sound deadening.

C1010 124 Drywall Partitions/Wood Stud Framing

	FACE LAYER	BASE LAYER	FRAMING	OPPOSITE FACE	INSULATION	COST PER S.F.		
						MAT.	INST.	TOTAL
1200	5/8" FR drywall	none	2 x 4, @ 16" O.C.	same	0	1.10	2.70	3.80
1250				5/8" reg. drywall	0	1.10	2.70	3.80
1300				nothing	0	.74	1.80	2.54
1400		1/4" SD gypsum	2 x 4 @ 16" O.C.	same	1-1/2" fiberglass	2.48	4.15	6.63
1450				5/8" FR drywall	1-1/2" fiberglass	2.11	3.65	5.76
1500				nothing	1-1/2" fiberglass	1.75	2.75	4.50
1600		resil. channels	2 x 4 @ 16", O.C.	same	1-1/2" fiberglass	2.10	5.25	7.35
1650				5/8" FR drywall	1-1/2" fiberglass	1.92	4.21	6.13
1700				nothing	1-1/2" fiberglass	1.56	3.31	4.87
1800		5/8" FR drywall	2 x 4 @ 24" O.C.	same	0	1.63	3.42	5.05
1850				5/8" FR drywall	0	1.32	2.97	4.29
1900				nothing	0	.96	2.07	3.03
2200			2 rows-2 x 4 16"O.C.	same	2" fiberglass	2.80	4.95	7.75
2250				5/8" FR drywall	2" fiberglass	2.49	4.50	6.99
2300				nothing	2" fiberglass	2.13	3.60	5.73
2400	5/8" WR drywall	none	2 x 4, @ 16" O.C.	same	0	1.36	2.70	4.06
2450				5/8" FR drywall	0	1.23	2.70	3.93
2500				nothing	0	.87	1.80	2.67
2600		5/8" FR drywall	2 x 4, @ 24" O.C.	same	0	1.89	3.42	5.31
2650				5/8" FR drywall	0	1.45	2.97	4.42
2700				nothing	0	1.09	2.07	3.16
2800	5/8 VF drywall	none	2 x 4, @ 16" O.C.	same	0	1.96	2.90	4.86
2850				5/8" FR drywall	0	1.53	2.80	4.33
2900				nothing	0	1.17	1.90	3.07
3000		5/8" FR drywall	2 x 4 , 24" O.C.	same	0	2.49	3.62	6.11
3050				5/8" FR drywall	0	1.75	3.07	4.82
3100				nothing	0	1.39	2.17	3.56
3200	1/2" reg drywall	3/8" reg drywall	2 x 4, @ 16" O.C.	same	0	1.50	3.60	5.10
3252				5/8"FR drywall	0	1.30	3.15	4.45
3300				nothing	0	.94	2.25	3.19

C1010 126 Drywall Partitions/Metal Stud Framing

	FACE LAYER	BASE LAYER	FRAMING	OPPOSITE FACE	INSULATION	COST PER S.F.		
						MAT.	INST.	TOTAL
5200	5/8" FR drywall	none	1-5/8" @ 24" O.C.	same	0	.91	2.39	3.30
5250				5/8" reg. drywall	0	.91	2.39	3.30
5300				nothing	0	.55	1.49	2.04

C10 Interior Construction

C1010 Partitions

C1010 126 — Drywall Partitions/Metal Stud Framing

	FACE LAYER	BASE LAYER	FRAMING	OPPOSITE FACE	INSULATION	COST PER S.F. MAT.	COST PER S.F. INST.	COST PER S.F. TOTAL
5400			3-5/8" @ 24" O.C.	same	0	.97	2.41	3.38
5450				5/8" reg. drywall	0	.97	2.41	3.38
5500				nothing	0	.61	1.51	2.12
5600		1/4" SD gypsum	1-5/8" @ 24" O.C.	same	0	1.65	3.39	5.04
5650				5/8" FR drywall	0	1.28	2.89	4.17
5700				nothing	0	.92	1.99	2.91
5800			2-1/2" @ 24" O.C.	same	0	1.67	3.40	5.07
5850				5/8" FR drywall	0	1.30	2.90	4.20
5900				nothing	0	.94	2	2.94
6000		5/8" FR drywall	2-1/2" @ 16" O.C.	same	0	1.78	3.85	5.63
6050				5/8" FR drywall	0	1.47	3.40	4.87
6100				nothing	0	1.11	2.50	3.61
6200			3-5/8" @ 24" O.C.	same	0	1.59	3.31	4.90
6250				5/8"FR drywall	3-1/2" fiberglass	1.87	3.19	5.06
6300				nothing	0	.92	1.96	2.88
6400	5/8" WR drywall	none	1-5/8" @ 24" O.C.	same	0	1.17	2.39	3.56
6450				5/8" FR drywall	0	1.04	2.39	3.43
6500				nothing	0	.68	1.49	2.17
6600			3-5/8" @ 24" O.C.	same	0	1.23	2.41	3.64
6650				5/8" FR drywall	0	1.10	2.41	3.51
6700				nothing	0	.74	1.51	2.25
6800		5/8" FR drywall	2-1/2" @ 16" O.C.	same	0	2.04	3.85	5.89
6850				5/8" FR drywall	0	1.60	3.40	5
6900				nothing	0	1.24	2.50	3.74
7000			3-5/8" @ 24" O.C.	same	0	1.85	3.31	5.16
7050				5/8"FR drywall	3-1/2" fiberglass	2	3.19	5.19
7100				nothing	0	1.05	1.96	3.01
7200	5/8" VF drywall	none	1-5/8" @ 24" O.C.	same	0	1.77	2.59	4.36
7250				5/8" FR drywall	0	1.34	2.49	3.83
7300				nothing	0	.98	1.59	2.57
7400			3-5/8" @ 24" O.C.	same	0	1.83	2.61	4.44
7450				5/8" FR drywall	0	1.40	2.51	3.91
7500				nothing	0	1.04	1.61	2.65
7600		5/8" FR drywall	2-1/2" @ 16" O.C.	same	0	2.64	4.05	6.69
7650				5/8" FR drywall	0	1.90	3.50	5.40
7700				nothing	0	1.54	2.60	4.14
7800			3-5/8" @ 24" O.C.	same	0	2.45	3.51	5.96
7850				5/8"FR drywall	3-1/2" fiberglass	2.30	3.29	5.59
7900				nothing	0	1.35	2.06	3.41

C10 Interior Construction

C1010 Partitions

C1010 128	Drywall Components	COST PER S.F.		
		MAT.	INST.	TOTAL
0060	Metal studs, 24" O.C. including track, load bearing, 20 gage, 2-1/2"	.56	.84	1.40
0080	3-5/8"	.67	.86	1.53
0100	4"	.60	.87	1.47
0120	6"	.89	.89	1.78
0140	Metal studs, 24" O.C. including track, load bearing, 18 gage, 2-1/2"	.56	.84	1.40
0160	3-5/8"	.67	.86	1.53
0180	4"	.60	.87	1.47
0200	6"	.89	.89	1.78
0220	16 gage, 2-1/2"	.66	.96	1.62
0240	3-5/8"	.78	.98	1.76
0260	4"	.82	1	1.82
0280	6"	1.03	1.02	2.05
0300	Non load bearing, 25 gage, 1-5/8"	.19	.59	.78
0340	3-5/8"	.25	.61	.86
0360	4"	.28	.61	.89
0380	6"	.35	.62	.97
0400	20 gage, 2-1/2"	.32	.75	1.07
0420	3-5/8"	.35	.76	1.11
0440	4"	.41	.76	1.17
0460	6"	.48	.77	1.25
0540	Wood studs including blocking, shoe and double top plate, 2"x4", 12" O.C.	.47	1.13	1.60
0560	16" O.C.	.38	.90	1.28
0580	24" O.C.	.29	.72	1.01
0600	2"x6", 12" O.C.	.76	1.29	2.05
0620	16" O.C.	.61	1	1.61
0640	24" O.C.	.46	.78	1.24
0642	Furring one side only, steel channels, 3/4", 12" O.C.	.31	1.64	1.95
0644	16" O.C.	.28	1.45	1.73
0646	24" O.C.	.19	1.10	1.29
0647	1-1/2" , 12" O.C.	.42	1.83	2.25
0648	16" O.C.	.37	1.60	1.97
0649	24" O.C.	.25	1.26	1.51
0650	Wood strips, 1" x 3", on wood, 12" O.C.	.38	.82	1.20
0651	16" O.C.	.29	.62	.91
0652	On masonry, 12" O.C.	.38	.91	1.29
0653	16" O.C.	.29	.68	.97
0654	On concrete, 12" O.C.	.38	1.73	2.11
0655	16" O.C.	.29	1.30	1.59
0665	Gypsum board, one face only, exterior sheathing, 1/2"	.42	.80	1.22
0680	Interior, fire resistant, 1/2"	.31	.45	.76
0700	5/8"	.31	.45	.76
0720	Sound deadening board 1/4"	.37	.50	.87
0740	Standard drywall 3/8"	.26	.45	.71
0760	1/2"	.25	.45	.70
0780	5/8"	.31	.45	.76
0800	Tongue & groove coreboard 1"	.62	1.87	2.49
0820	Water resistant, 1/2"	.40	.45	.85
0840	5/8"	.44	.45	.89
0860	Add for the following:, foil backing	.15		.15
0880	Fiberglass insulation, 3-1/2"	.59	.33	.92
0900	6"	.85	.33	1.18
0920	Rigid insulation 1"	.51	.45	.96
0940	Resilient furring @ 16" O.C.	.24	1.41	1.65
0960	Taping and finishing	.05	.45	.50
0980	Texture spray	.06	.48	.54
1000	Thin coat plaster	.11	.57	.68
1040	2"x4" staggered studs 2"x6" plates & blocking	.55	.98	1.53
1050				

C10 Interior Construction

C1010 Partitions

C1010 144	Plaster Partition Components	MAT.	INST.	TOTAL
		COST PER S.F.		
0060	Metal studs, 16" O.C., including track, non load bearing, 25 gage, 1-5/8"	.29	.92	1.21
0080	2-1/2"	.29	.92	1.21
0100	3-1/4"	.34	.94	1.28
0120	3-5/8"	.34	.94	1.28
0140	4"	.38	.95	1.33
0160	6"	.48	.96	1.44
0180	Load bearing, 20 gage, 2-1/2"	.76	1.17	1.93
0200	3-5/8"	.90	1.18	2.08
0220	4"	.95	1.22	2.17
0240	6"	1.20	1.23	2.43
0260	16 gage 2-1/2"	.90	1.32	2.22
0280	3-5/8"	1.07	1.36	2.43
0300	4"	1.13	1.38	2.51
0320	6"	1.41	1.41	2.82
0340	Wood studs, including blocking, shoe and double plate, 2"x4" , 12" O.C.	.47	1.13	1.60
0360	16" O.C.	.38	.90	1.28
0380	24" O.C.	.29	.72	1.01
0400	2"x6" , 12" O.C.	.76	1.29	2.05
0420	16" O.C.	.61	1	1.61
0440	24" O.C.	.46	.78	1.24
0460	Furring one face only, steel channels, 3/4" , 12" O.C.	.31	1.64	1.95
0480	16" O.C.	.28	1.45	1.73
0500	24" O.C.	.19	1.10	1.29
0520	1-1/2" , 12" O.C.	.42	1.83	2.25
0540	16" O.C.	.37	1.60	1.97
0560	24"O.C.	.25	1.26	1.51
0580	Wood strips 1"x3", on wood., 12" O.C.	.38	.82	1.20
0600	16"O.C.	.29	.62	.91
0620	On masonry, 12" O.C.	.38	.91	1.29
0640	16" O.C.	.29	.68	.97
0660	On concrete, 12" O.C.	.38	1.73	2.11
0680	16" O.C.	.29	1.30	1.59
0700	Gypsum lath. plain or perforated, nailed to studs, 3/8" thick	.60	.50	1.10
0720	1/2" thick	.59	.53	1.12
0740	Clipped to studs, 3/8" thick	.60	.57	1.17
0760	1/2" thick	.59	.61	1.20
0780	Metal lath, diamond painted, nailed to wood studs, 2.5 lb.	.44	.50	.94
0800	3.4 lb.	.49	.53	1.02
0820	Screwed to steel studs, 2.5 lb.	.44	.53	.97
0840	3.4 lb.	.54	.57	1.11
0860	Rib painted, wired to steel, 2.75 lb	.38	.57	.95
0880	3.4 lb	.59	.61	1.20
0900	4.0 lb	.66	.65	1.31
0910				
0920	Gypsum plaster, 2 coats	.47	2.16	2.63
0940	3 coats	.68	2.61	3.29
0960	Perlite or vermiculite plaster, 2 coats	.46	2.49	2.95
0980	3 coats	.77	3.09	3.86
1000	Stucco, 3 coats, 1" thick, on wood framing	.65	4.29	4.94
1020	On masonry	.28	3.39	3.67
1100	Metal base galvanized and painted 2-1/2" high	.72	1.60	2.32

C10 Interior Construction

C1010 Partitions

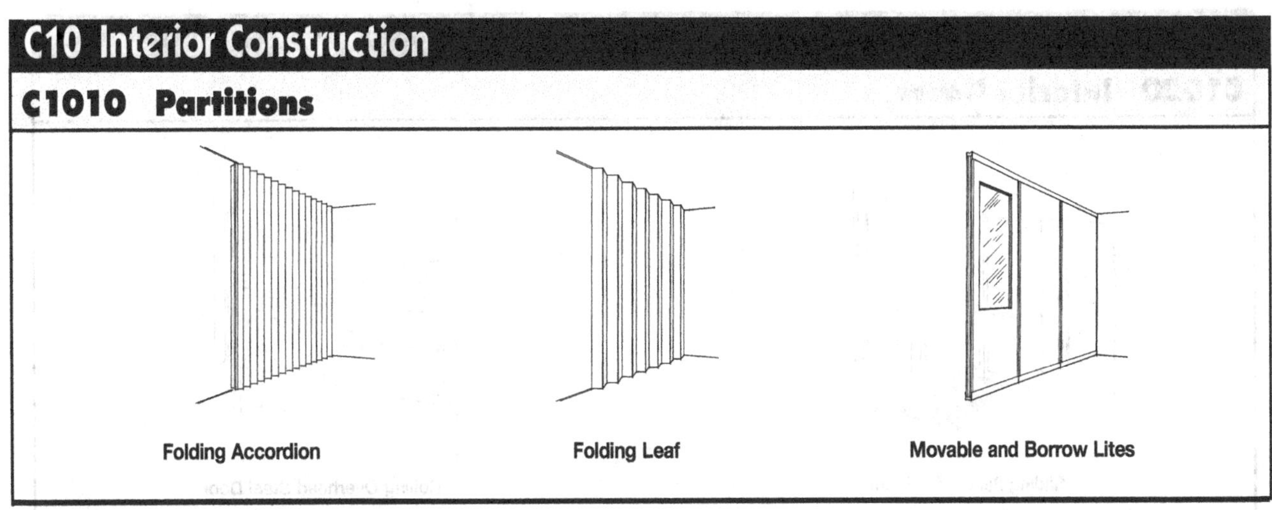

C1010 205	Partitions	COST PER S.F.		
		MAT.	INST.	TOTAL
0360	Folding accordion, vinyl covered, acoustical, 3 lb. S.F., 17 ft max. hgt	29	9	38
0380	5 lb. per S.F. 27 ft max height	40	9.45	49.45
0400	5.5 lb. per S.F., 17 ft. max height	47	10	57
0420	Commercial, 1.75 lb per S.F., 8 ft. max height	25	3.99	28.99
0440	2.0 Lb per S.F., 17 ft. max height	25.50	6	31.50
0460	Industrial, 4.0 lb. per S.F. 27 ft. max height	36.50	12	48.50
0480	Vinyl clad wood or steel, electric operation 6 psf	54	5.60	59.60
0500	Wood, non acoustic, birch or mahogany	29	2.99	31.99
0560	Folding leaf, aluminum framed acoustical 12 ft.high.,5.5 lb per S.F., min.	40.50	14.95	55.45
0580	Maximum	49	30	79
0600	6.5 lb. per S.F., minimum	43	14.95	57.95
0620	Maximum	53	30	83
0640	Steel acoustical, 7.5 per S.F., vinyl faced, minimum	60.50	14.95	75.45
0660	Maximum	73.50	30	103.50
0680	Wood acoustic type, vinyl faced to 18' high 6 psf, minimum	56.50	14.95	71.45
0700	Average	67.50	19.95	87.45
0720	Maximum	87.50	30	117.50
0740	Formica or hardwood faced, minimum	58.50	14.95	73.45
0760	Maximum	62.50	30	92.50
0780	Wood, low acoustical type to 12 ft. high 4.5 psf	42.50	17.95	60.45
0840	Demountable, trackless wall, cork finish, semi acous, 1-5/8" th, min	42.50	2.76	45.26
0860	Maximum	40	4.73	44.73
0880	Acoustic, 2" thick, minimum	34	2.95	36.95
0900	Maximum	59	3.99	62.99
0920	In-plant modular office system, w/prehung steel door			
0940	3" thick honeycomb core panels			
0960	12' x 12', 2 wall	12.25	.56	12.81
0970	4 wall	10.35	.75	11.10
0980	16' x 16', 2 wall	10.80	.39	11.19
0990	4 wall	7.30	.40	7.70
1000	Gypsum, demountable, 3" to 3-3/4" thick x 9' high, vinyl clad	7.05	2.08	9.13
1020	Fabric clad	17.55	2.28	19.83
1040	1.75 system, vinyl clad hardboard, paper honeycomb core panel			
1060	1-3/4" to 2-1/2" thick x 9' high	11.90	2.08	13.98
1080	Unitized gypsum panel system, 2" to 2-1/2" thick x 9' high			
1100	Vinyl clad gypsum	15.20	2.08	17.28
1120	Fabric clad gypsum	25	2.28	27.28
1140	Movable steel walls, modular system			
1160	Unitized panels, 48" wide x 9' high			
1180	Baked enamel, pre-finished	17.20	1.66	18.86
1200	Fabric clad	25	1.78	26.78
1203				

C10 Interior Construction

C1020 Interior Doors

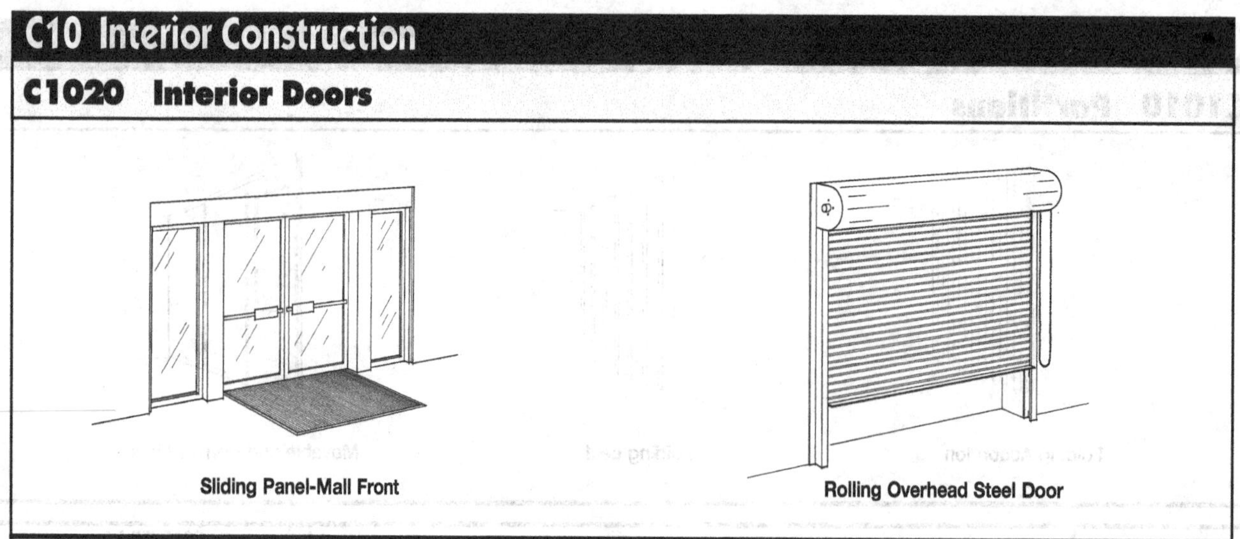

Sliding Panel-Mall Front

Rolling Overhead Steel Door

C1020 102	Special Doors	COST PER OPNG.		
		MAT.	INST.	TOTAL
2500	Single leaf, wood, 3'-0"x7'-0"x1 3/8", birch, solid core	395	168	563
2510	Hollow core	325	161	486
2530	Hollow core, lauan	315	161	476
2540	Louvered pine	460	161	621
2550	Paneled pine	455	161	616
2600	Hollow metal, comm. quality, flush, 3'-0"x7'-0"x1-3/8"	845	177	1,022
2650	3'-0"x10'-0" openings with panel	1,350	229	1,579
2700	Metal fire, comm. quality, 3'-0"x7'-0"x1-3/8"	1,075	184	1,259
2800	Kalamein fire, comm. quality, 3'-0"x7'-0"x1-3/4"	740	335	1,075
3200	Double leaf, wood, hollow core, 2 - 3'-0"x7'-0"x1-3/8"	570	310	880
3300	Hollow metal, comm. quality, B label, 2'-3'-0"x7'-0"x1-3/8"	2,100	385	2,485
3400	6'-0"x10'-0" opening, with panel	2,650	475	3,125
3500	Double swing door system, 12'-0"x7'-0", mill finish	8,625	2,475	11,100
3700	Black finish	8,725	2,550	11,275
3800	Sliding entrance door and system mill finish	10,200	2,200	12,400
3900	Bronze finish	11,200	2,375	13,575
4000	Black finish	11,700	2,475	14,175
4100	Sliding panel mall front, 16'x9' opening, mill finish	3,575	670	4,245
4200	Bronze finish	4,650	870	5,520
4300	Black finish	5,725	1,075	6,800
4400	24'x9' opening mill finish	5,075	1,250	6,325
4500	Bronze finish	6,600	1,625	8,225
4600	Black finish	8,125	2,000	10,125
4700	48'x9' opening mill finish	9,475	970	10,445
4800	Bronze finish	12,300	1,250	13,550
4900	Black finish	15,200	1,550	16,750
5000	Rolling overhead steel door, manual, 8' x 8' high	1,175	695	1,870
5100	10' x 10' high	1,950	795	2,745
5200	20' x 10' high	3,300	1,100	4,400
5300	12' x 12' high	2,000	925	2,925
5400	Motor operated, 8' x 8' high	2,325	915	3,240
5500	10' x 10' high	3,100	1,025	4,125
5600	20' x 10' high	4,450	1,325	5,775
5700	12' x 12' high	3,150	1,150	4,300
5800	Roll up grille, aluminum, manual, 10' x 10' high, mill finish	3,000	1,350	4,350
5900	Bronze anodized	4,800	1,350	6,150
6000	Motor operated, 10' x 10' high, mill finish	4,375	1,575	5,950
6100	Bronze anodized	6,175	1,575	7,750
6200	Steel, manual, 10' x 10' high	2,700	1,100	3,800
6300	15' x 8' high	3,125	1,400	4,525
6400	Motor operated, 10' x 10' high	4,075	1,325	5,400
6500	15' x 8' high	4,500	1,625	6,125

C10 Interior Construction

C1030 Fittings

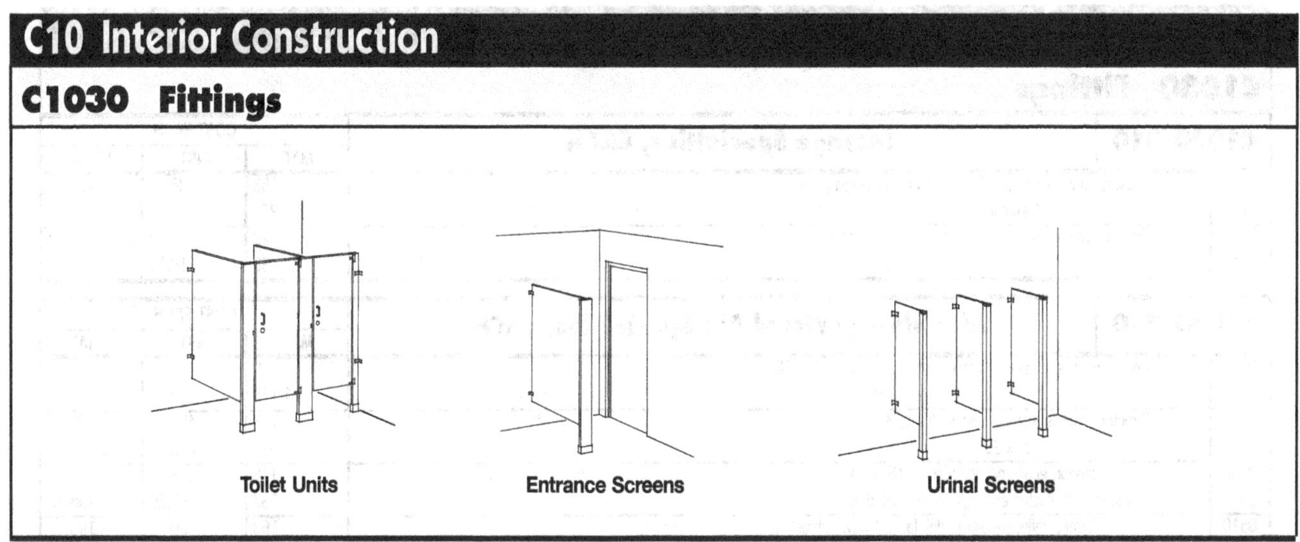

C1030 110	Toilet Partitions	MAT.	INST.	TOTAL
		COST PER UNIT		
0380	Toilet partitions, cubicles, ceiling hung, marble	1,600	415	2,015
0400	Painted metal	530	225	755
0420	Plastic laminate	630	225	855
0460	Stainless steel	830	225	1,055
0480	Handicap addition	450		450
0520	Floor and ceiling anchored, marble	2,000	335	2,335
0540	Painted metal	645	180	825
0560	Plastic laminate	825	180	1,005
0600	Stainless steel	1,150	180	1,330
0620	Handicap addition	315		315
0660	Floor mounted marble	1,175	278	1,453
0680	Painted metal	610	128	738
0700	Plastic laminate	585	128	713
0740	Stainless steel	1,375	128	1,503
0760	Handicap addition	315		315
0780	Juvenile deduction	45.50		45.50
0820	Floor mounted with handrail marble	1,125	278	1,403
0840	Painted metal	535	150	685
0860	Plastic laminate	880	150	1,030
0900	Stainless steel	1,300	150	1,450
0920	Handicap addition	365		365
0960	Wall hung, painted metal	685	128	813
1020	Stainless steel	1,825	128	1,953
1040	Handicap addition	365		365
1080	Entrance screens, floor mounted, 54″ high, marble	745	92.50	837.50
1100	Painted metal	262	60	322
1140	Stainless steel	985	60	1,045
1300	Urinal screens, floor mounted, 24″ wide, laminated plastic	198	112	310
1320	Marble	630	133	763
1340	Painted metal	268	112	380
1380	Stainless steel	630	112	742
1428	Wall mounted wedge type, painted metal	148	90	238
1460	Stainless steel	620	90	710
1500	Partitions, shower stall, single wall, painted steel, 2′-8″ x 2′-8″	1,025	202	1,227
1510	Fiberglass, 2′-8″ x 2′-8″	800	224	1,024
1520	Double wall, enameled steel, 2′-8″ x 2′-8″	1,050	202	1,252
1530	Stainless steel, 2′-8″ x 2′-8″	2,700	202	2,902
1560	Tub enclosure, sliding panels, tempered glass, aluminum frame	415	252	667
1570	Chrome/brass frame-deluxe	1,225	335	1,560

C10 Interior Construction

C1030 Fittings

C1030 310	Storage Specialties, EACH	COST EACH		
		MAT.	INST.	TOTAL
0200	Lockers, steel, single tier, 5′ to 6′ high, per opening, min.	153	36	189
0210	Maximum	271	42	313
0600	Shelving, metal industrial, braced, 3′ wide, 1′ deep	22	9.50	31.50
0610	2′ deep	30	10.10	40.10

C1030 510	Identifying/Visual Aid Specialties, EACH	COST EACH		
		MAT.	INST.	TOTAL
0100	Control boards, magnetic, porcelain finish, framed, 24″ x 18″	219	112	331
0110	96″ x 48″	1,175	180	1,355
0120	Directory boards, outdoor, black plastic, 36″ x 24″	790	450	1,240
0130	36″ x 36″	915	600	1,515
0140	Indoor, economy, open faced, 18″ x 24″	148	128	276
0500	Signs, interior electric exit sign, wall mounted, 6″	133	65.50	198.50
0510	Street, reflective alum., dbl. face, 4 way, w/bracket	163	30	193
0520	Letters, cast aluminum, 1/2″ deep, 4″ high	25	25	50
0530	1″ deep, 10″ high	57	25	82
0540	Plaques, cast aluminum, 20″ x 30″	1,250	225	1,475
0550	Cast bronze, 36″ x 48″	4,700	450	5,150

C1030 520	Identifying/Visual Aid Specialties, S.F.	COST PER S.F.		
		MAT.	INST.	TOTAL
0100	Bulletin board, cork sheets, no frame, 1/4″ thick	2.13	3.10	5.23
0120	Aluminum frame, 1/4″ thick, 3′ x 5′	8.30	3.75	12.05
0200	Chalkboards, wall hung, alum, frame & chalktrough	12.10	1.98	14.08
0210	Wood frame & chalktrough	11.95	2.14	14.09
0220	Sliding board, one board with back panel	51	1.92	52.92
0230	Two boards with back panel	86	1.92	87.92
0240	Liquid chalk type, alum. frame & chalktrough	11.45	1.98	13.43
0250	Wood frame & chalktrough	26	1.98	27.98

C1030 710	Bath and Toilet Accessories, EACH	COST EACH		
		MAT.	INST.	TOTAL
0100	Specialties, bathroom accessories, st. steel, curtain rod, 5′ long, 1″ diam	38	34.50	72.50
0120	Dispenser, towel, surface mounted	36.50	28	64.50
0140	Grab bar, 1-1/4″ diam., 12″ long	32	18.70	50.70
0160	Mirror, framed with shelf, 18″ x 24″	207	22.50	229.50
0170	72″ x 24″	291	75	366
0180	Toilet tissue dispenser, surface mounted, single roll	18.65	14.95	33.60
0200	Towel bar, 18″ long	43.50	19.55	63.05
0300	Medicine cabinets, sliding mirror doors, 20″ x 16″ x 4-3/4″, unlighted	121	64	185
0310	24″ x 19″ x 8-1/2″, lighted	175	90	265

C1030 730	Bath and Toilet Accessories, L.F.	COST PER L.F.		
		MAT.	INST.	TOTAL
0100	Partitions, hospital curtain, ceiling hung, polyester oxford cloth	19.80	4.39	24.19
0110	Designer oxford cloth	37	5.55	42.55

C1030 830	Fabricated Cabinets & Counters, L.F.	COST PER L.F.		
		MAT.	INST.	TOTAL
0110	Household, base, hardwood, one top drawer & one door below x 12″ wide	252	36	288
0120	Four drawer x 24″ wide	335	40.50	375.50
0130	Wall, hardwood, 30″ high with one door x 12″ wide	204	41	245
0140	Two doors x 48″ wide	475	49	524
0150	Counter top-laminated plastic, stock, economy	10.80	14.95	25.75
0160	Custom-square edge, 7/8″ thick	16.30	33.50	49.80
0170	School, counter, wood, 32″ high	236	45	281
0180	Metal, 84″ high	425	60	485

C10 Interior Construction

C1030 Fittings

C1030 910	Other Fittings, EACH	COST EACH		
		MAT.	INST.	TOTAL
0500	Mail boxes, horizontal, rear loaded, aluminum, 5" x 6" x 15" deep	38.50	13.20	51.70
0510	Front loaded, aluminum, 10" x 12" x 15" deep	105	22.50	127.50
0520	Vertical, front loaded, aluminum, 15" x 5" x 6" deep	33	13.20	46.20
0530	Bronze, duranodic finish	38.50	13.20	51.70
0540	Letter slot, post office	116	56	172
0550	Mail counter, window, post office, with grille	605	225	830
0700	Turnstiles, one way, 4' arm, 46" diam., manual	885	180	1,065
0710	Electric	1,450	750	2,200
0720	3 arm, 5'-5" diam. & 7' high, manual	4,375	900	5,275
0730	Electric	4,950	1,500	6,450

C20 Stairs

C2010 Stair Construction

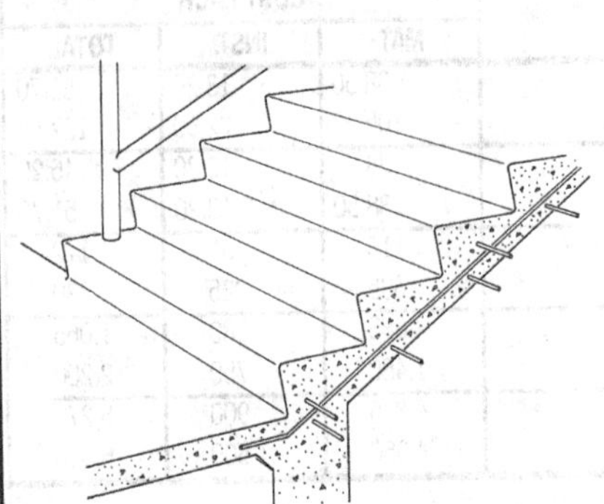

General Design: See reference section for code requirements. Maximum height between landings is 12′; usual stair angle is 20° to 50° with 30° to 35° best. Usual relation of riser to treads is:

Riser + tread = 17.5.

2x (Riser) + tread = 25.

Riser x tread = 70 or 75.

Maximum riser height is 7″ for commercial, 8-1/4″ for residential.

Usual riser height is 6-1/2″ to 7-1/4″.

Minimum tread width is 11″ for commercial and 9″ for residential.

For additional information please see reference section.

Cost Per Flight: Table below lists the cost per flight for 4′-0″ wide stairs. Side walls are not included. Railings are included.

C2010 110	Stairs	COST PER FLIGHT		
		MAT.	INST.	TOTAL
0470	Stairs, C.I.P. concrete, w/o landing, 12 risers, w/o nosing	970	1,875	2,845
0480	With nosing	1,650	2,150	3,800
0550	W/landing, 12 risers, w/o nosing	1,100	2,300	3,400
0560	With nosing	1,775	2,550	4,325
0570	16 risers, w/o nosing	1,350	2,900	4,250
0580	With nosing	2,250	3,250	5,500
0590	20 risers, w/o nosing	1,600	3,475	5,075
0600	With nosing	2,725	3,925	6,650
0610	24 risers, w/o nosing	1,850	4,075	5,925
0620	With nosing	3,200	4,600	7,800
0630	Steel, grate type w/nosing & rails, 12 risers, w/o landing	4,800	950	5,750
0640	With landing	6,275	1,300	7,575
0660	16 risers, with landing	7,875	1,600	9,475
0680	20 risers, with landing	9,475	1,925	11,400
0700	24 risers, with landing	11,100	2,250	13,350
0710	Concrete fill metal pan & picket rail, 12 risers, w/o landing	6,300	950	7,250
0720	With landing	8,275	1,425	9,700
0740	16 risers, with landing	10,400	1,750	12,150
0760	20 risers, with landing	12,500	2,050	14,550
0780	24 risers, with landing	14,600	2,375	16,975
0790	Cast iron tread & pipe rail, 12 risers, w/o landing	6,300	950	7,250
0800	With landing	8,275	1,425	9,700
1120	Wood, prefab box type, oak treads, wood rails 3′-6″ wide, 14 risers	2,100	375	2,475
1150	Prefab basement type, oak treads, wood rails 3′-0″ wide, 14 risers	1,050	92.50	1,142.50

C30 Interior Finishes

C3010 Wall Finishes

C3010 230	Paint & Covering	COST PER S.F.		
		MAT.	INST.	TOTAL
0060	Painting, interior on plaster and drywall, brushwork, primer & 1 coat	.11	.59	.70
0080	Primer & 2 coats	.17	.79	.96
0100	Primer & 3 coats	.23	.96	1.19
0120	Walls & ceilings, roller work, primer & 1 coat	.11	.40	.51
0140	Primer & 2 coats	.17	.51	.68
0160	Woodwork incl. puttying, brushwork, primer & 1 coat	.11	.86	.97
0180	Primer & 2 coats	.17	1.13	1.30
0200	Primer & 3 coats	.23	1.54	1.77
0260	Cabinets and casework, enamel, primer & 1 coat	.12	.96	1.08
0280	Primer & 2 coats	.18	1.19	1.37
0300	Masonry or concrete, latex, brushwork, primer & 1 coat	.24	.80	1.04
0320	Primer & 2 coats	.31	1.15	1.46
0340	Addition for block filler	.19	.99	1.18
0380	Fireproof paints, intumescent, 1/8" thick 3/4 hour	2.14	.79	2.93
0400	3/16" thick 1 hour	4.78	1.19	5.97
0420	7/16" thick 2 hour	6.15	2.75	8.90
0440	1-1/16" thick 3 hour	10.05	5.50	15.55
0500	Gratings, primer & 1 coat	.29	1.20	1.49
0600	Pipes over 12" diameter	.66	3.85	4.51
0700	Structural steel, brushwork, light framing 300-500 S.F./Ton	.09	1.47	1.56
0720	Heavy framing 50-100 S.F./Ton	.09	.74	.83
0740	Spraywork, light framing 300-500 S.F./Ton	.10	.33	.43
0760	Heavy framing 50-100 S.F./Ton	.10	.36	.46
0800	Varnish, interior wood trim, no sanding sealer & 1 coat	.08	.96	1.04
0820	Hardwood floor, no sanding 2 coats	.17	.20	.37
0840	Wall coatings, acrylic glazed coatings, minimum	.32	.73	1.05
0860	Maximum	.68	1.26	1.94
0880	Epoxy coatings, minimum	.42	.73	1.15
0900	Maximum	1.27	2.27	3.54
0940	Exposed epoxy aggregate, troweled on, 1/16" to 1/4" aggregate, minimum	.64	1.64	2.28
0960	Maximum	1.36	2.96	4.32
0980	1/2" to 5/8" aggregate, minimum	1.25	2.96	4.21
1000	Maximum	2.16	4.82	6.98
1020	1" aggregate, minimum	2.19	4.28	6.47
1040	Maximum	3.34	7	10.34
1060	Sprayed on, minimum	.58	1.31	1.89
1080	Maximum	1.09	2.66	3.75
1100	High build epoxy 50 mil, minimum	.70	.99	1.69
1120	Maximum	1.20	4.05	5.25
1140	Laminated epoxy with fiberglass minimum	.76	1.31	2.07
1160	Maximum	1.35	2.66	4.01
1180	Sprayed perlite or vermiculite 1/16" thick, minimum	.28	.13	.41
1200	Maximum	.77	.60	1.37
1260	Wall coatings, vinyl plastic, minimum	.34	.52	.86
1280	Maximum	.85	1.60	2.45
1300	Urethane on smooth surface, 2 coats, minimum	.28	.34	.62
1320	Maximum	.61	.58	1.19
1340	3 coats, minimum	.36	.46	.82
1360	Maximum	.81	.82	1.63
1380	Ceramic-like glazed coating, cementitious, minimum	.50	.88	1.38
1400	Maximum	.84	1.12	1.96
1420	Resin base, minimum	.34	.60	.94
1440	Maximum	.55	1.17	1.72
1460	Wall coverings, aluminum foil	1.12	1.41	2.53
1480	Copper sheets, .025" thick, phenolic backing	7.80	1.62	9.42
1500	Vinyl backing	6	1.62	7.62
1520	Cork tiles, 12"x12", light or dark, 3/16" thick	4.92	1.62	6.54
1540	5/16" thick	4.18	1.65	5.83

C30 Interior Finishes

C3010 Wall Finishes

C3010 230	Paint & Covering	COST PER S.F.		
		MAT.	INST.	TOTAL
1560	Basketweave, 1/4" thick	6.45	1.62	8.07
1580	Natural, non-directional, 1/2" thick	8.65	1.62	10.27
1600	12"x36", granular, 3/16" thick	1.39	1.01	2.40
1620	1" thick	1.79	1.05	2.84
1640	12"x12", polyurethane coated, 3/16" thick	4.35	1.62	5.97
1660	5/16" thick	6.20	1.65	7.85
1661	Paneling, prefinished plywood, birch	1.49	2.14	3.63
1662	Mahogany, African	2.76	2.25	5.01
1663	Philippine (lauan)	1.19	1.80	2.99
1664	Oak or cherry	3.54	2.25	5.79
1665	Rosewood	5.05	2.81	7.86
1666	Teak	3.54	2.25	5.79
1667	Chestnut	5.25	2.40	7.65
1668	Pecan	2.27	2.25	4.52
1669	Walnut	5.75	2.25	8
1670	Wood board, knotty pine, finished	2.40	3.72	6.12
1671	Rough sawn cedar	3.47	3.72	7.19
1672	Redwood	5.75	3.72	9.47
1673	Aromatic cedar	2.52	4	6.52
1680	Cork wallpaper, paper backed, natural	2.19	.81	3
1700	Color	3.06	.81	3.87
1720	Gypsum based, fabric backed, minimum	.90	.49	1.39
1740	Average	1.35	.54	1.89
1760	Maximum	1.51	.61	2.12
1780	Vinyl wall covering, fabric back, light weight	.76	.61	1.37
1800	Medium weight	.90	.81	1.71
1820	Heavy weight	1.80	.89	2.69
1840	Wall paper, double roll, solid pattern, avg. workmanship	.41	.61	1.02
1860	Basic pattern, avg. workmanship	.86	.73	1.59
1880	Basic pattern, quality workmanship	2.72	.89	3.61
1900	Grass cloths with lining paper, minimum	.86	.97	1.83
1920	Maximum	2.74	1.11	3.85
1940	Ceramic tile, thin set, 4-1/4" x 4-1/4"	2.36	3.79	6.15
1960	12" x 12"	3.89	2.40	6.29

C3010 235	Paint Trim	COST PER L.F.		
		MAT.	INST.	TOTAL
2040	Painting, wood trim, to 6" wide, enamel, primer & 1 coat	.12	.48	.60
2060	Primer & 2 coats	.18	.61	.79
2080	Misc. metal brushwork, ladders	.59	4.82	5.41
2100	Pipes, to 4" dia.	.07	1.01	1.08
2120	6" to 8" dia.	.14	2.03	2.17
2140	10" to 12" dia.	.52	3.03	3.55
2160	Railings, 2" pipe	.21	2.41	2.62
2180	Handrail, single	.16	.96	1.12
2185	Caulking & Sealants, Polyureathane, In place, 1 or 2 component, 1/2" X 1/4"	.36	1.53	1.89

C30 Interior Finishes

C3020 Floor Finishes

C3020 410	Tile & Covering	COST PER S.F. MAT.	INST.	TOTAL
0060	Carpet tile, nylon, fusion bonded, 18" x 18" or 24" x 24", 24 oz.	3.50	.56	4.06
0080	35 oz.	4.16	.56	4.72
0100	42 oz.	5.55	.56	6.11
0140	Carpet, tufted, nylon, roll goods, 12' wide, 26 oz.	5.70	.64	6.34
0160	36 oz.	8.55	.64	9.19
0180	Woven, wool, 36 oz.	14.55	.64	15.19
0200	42 oz.	14.85	.64	15.49
0220	Padding, add to above, minimum	.56	.30	.86
0240	Maximum	.95	.30	1.25
0260	Composition flooring, acrylic, 1/4" thick	1.62	4.34	5.96
0280	3/8" thick	2.10	5	7.10
0300	Epoxy, minimum	2.83	3.35	6.18
0320	Maximum	3.43	4.60	8.03
0340	Epoxy terrazzo, minimum	5.70	5	10.70
0360	Maximum	8.65	6.65	15.30
0380	Mastic, hot laid, 1-1/2" thick, minimum	4.15	3.27	7.42
0400	Maximum	5.30	4.34	9.64
0420	Neoprene 1/4" thick, minimum	4.08	4.14	8.22
0440	Maximum	5.60	5.25	10.85
0460	Polyacrylate with ground granite 1/4", minimum	3.38	3.07	6.45
0480	Maximum	6.25	4.71	10.96
0500	Polyester with colored quart 2 chips 1/16", minimum	3.01	2.12	5.13
0520	Maximum	4.65	3.35	8
0540	Polyurethane with vinyl chips, minimum	7.95	2.12	10.07
0560	Maximum	11.45	2.62	14.07
0600	Concrete topping, granolithic concrete, 1/2" thick	.34	3.61	3.95
0620	1" thick	.67	3.70	4.37
0640	2" thick	1.34	4.25	5.59
0660	Heavy duty 3/4" thick, minimum	.45	5.60	6.05
0680	Maximum	.79	6.65	7.44
0700	For colors, add to above, minimum	.38	1.29	1.67
0720	Maximum	.63	1.42	2.05
0740	Exposed aggregate finish, minimum	.23	.65	.88
0760	Maximum	.70	.87	1.57
0780	Abrasives, .25 P.S.F. add to above, minimum	.47	.48	.95
0800	Maximum	.71	.48	1.19
0820	Dust on coloring, add, minimum	.38	.31	.69
0840	Maximum	.63	.65	1.28
0860	Floor coloring using 0.6 psf powdered color, 1/2" integral, minimum	4.85	3.61	8.46
0880	Maximum	5.10	3.61	8.71
0900	Dustproofing, add, minimum	.15	.21	.36
0920	Maximum	.55	.31	.86
0930	Paint	.31	1.15	1.46
0940	Hardeners, metallic add, minimum	.56	.48	1.04
0960	Maximum	1.67	.71	2.38
0980	Non-metallic, minimum	.25	.48	.73
1000	Maximum	.75	.71	1.46
1020	Integral topping and finish, 1:1:2 mix, 3/16" thick	.11	2.13	2.24
1040	1/2" thick	.30	2.24	2.54
1060	3/4" thick	.45	2.51	2.96
1080	1" thick	.60	2.83	3.43
1100	Terrazzo, minimum	3.21	13.35	16.56
1120	Maximum	6	16.70	22.70
1320	On wood, add for felt underlay	.22		.22
1340	Cork tile, minimum	6.90	1.28	8.18
1360	Maximum	14.85	1.28	16.13

C30 Interior Finishes

C3020 Floor Finishes

C3020 410	Tile & Covering	COST PER S.F. MAT.	INST.	TOTAL
1380	Polyethylene, in rolls, minimum	3.63	1.47	5.10
1400	Maximum	6.90	1.47	8.37
1420	Polyurethane, thermoset, minimum	5.45	4.03	9.48
1440	Maximum	6.45	8.05	14.50
1460	Rubber, sheet goods, minimum	7.55	3.36	10.91
1480	Maximum	12.05	4.48	16.53
1500	Tile, minimum	7.35	1.01	8.36
1520	Maximum	13.75	1.47	15.22
1540	Synthetic turf, minimum	5.25	1.92	7.17
1560	Maximum	13.20	2.12	15.32
1580	Vinyl, composition tile, minimum	1.02	.81	1.83
1600	Maximum	1.90	.81	2.71
1620	Vinyl tile, minimum	3.30	.81	4.11
1640	Maximum	6.80	.81	7.61
1660	Sheet goods, minimum	4.13	1.61	5.74
1680	Maximum	7.15	2.02	9.17
1720	Tile, ceramic natural clay	5.65	3.94	9.59
1730	Marble, synthetic 12"x12"x5/8"	6.65	12	18.65
1740	Porcelain type, minimum	4.83	3.94	8.77
1760	Maximum	5.50	4.65	10.15
1800	Quarry tile, mud set, minimum	6.05	5.15	11.20
1820	Maximum	8.75	6.55	15.30
1840	Thin set, deduct		1.03	1.03
1850	Tile, natural stone, marble, in mortar bed, 12" x 12" x 3/8" thick	201	19.05	220.05
1860	Terrazzo precast, minimum	4.88	5.40	10.28
1880	Maximum	9.85	5.40	15.25
1900	Non-slip, minimum	20.50	13.30	33.80
1920	Maximum	21.50	17.75	39.25
2020	Wood block, end grain factory type, natural finish, 2" thick	5.10	3.59	8.69
2040	Fir, vertical grain, 1"x4", no finish, minimum	2.84	1.76	4.60
2060	Maximum	3.01	1.76	4.77
2080	Prefinished white oak, prime grade, 2-1/4" wide	4.94	2.64	7.58
2100	3-1/4" wide	5.50	2.43	7.93
2120	Maple strip, sanded and finished, minimum	4.70	3.84	8.54
2140	Maximum	5.45	3.84	9.29
2160	Oak strip, sanded and finished, minimum	3.60	3.84	7.44
2180	Maximum	5.20	3.84	9.04
2200	Parquetry, sanded and finished, minimum	5.85	4.01	9.86
2220	Maximum	6.95	5.70	12.65
2260	Add for sleepers on concrete, treated, 24" O.C., 1"x2"	1.88	2.30	4.18
2280	1"x3"	2.30	1.80	4.10
2300	2"x4"	.60	.91	1.51
2340	Underlayment, plywood, 3/8" thick	.85	.60	1.45
2350	1/2" thick	.88	.62	1.50
2360	5/8" thick	1.11	.64	1.75
2370	3/4" thick	1.35	.69	2.04
2380	Particle board, 3/8" thick	.36	.60	.96
2390	1/2" thick	.40	.62	1.02
2400	5/8" thick	.51	.64	1.15
2410	3/4" thick	.72	.69	1.41
2420	Hardboard, 4' x 4', .215" thick	.63	.60	1.23

C30 Interior Finishes

C3030 Ceiling Finishes

2 Coats of Plaster on Gypsum Lath on Wood Furring

Fiberglass Board on Exposed Suspended Grid System

Plaster and Metal Lath on Metal Furring

C3030 105 Plaster Ceilings

	TYPE	LATH	FURRING	SUPPORT		COST PER S.F. MAT.	INST.	TOTAL
2400	2 coat gypsum	3/8" gypsum	1"x3" wood, 16" O.C.	wood		1.47	4.55	6.02
2500	Painted			masonry		1.47	4.64	6.11
2600				concrete		1.47	5.20	6.67
2700	3 coat gypsum	3.4# metal	1"x3" wood, 16" O.C.	wood		1.57	4.88	6.45
2800	Painted			masonry		1.57	4.97	6.54
2900				concrete		1.57	5.55	7.12
3000	2 coat perlite	3/8" gypsum	1"x3" wood, 16" O.C.	wood		1.46	4.71	6.17
3100	Painted			masonry		1.46	4.80	6.26
3200				concrete		1.46	5.35	6.81
3300	3 coat perlite	3.4# metal	1"x3" wood, 16" O.C.	wood		1.35	4.82	6.17
3400	Painted			masonry		1.35	4.91	6.26
3500				concrete		1.35	5.50	6.85
3600	2 coat gypsum	3/8" gypsum	3/4" CRC, 12" O.C.	1-1/2" CRC, 48"O.C.		1.49	5.40	6.89
3700	Painted		3/4" CRC, 16" O.C.	1-1/2" CRC, 48"O.C.		1.46	4.92	6.38
3800			3/4" CRC, 24" O.C.	1-1/2" CRC, 48"O.C.		1.37	4.51	5.88
3900	2 coat perlite	3/8" gypsum	3/4" CRC, 12" O.C.	1-1/2" CRC, 48"O.C		1.48	5.80	7.28
4000	Painted		3/4" CRC, 16" O.C.	1-1/2" CRC, 48"O.C.		1.45	5.30	6.75
4100			3/4" CRC, 24" O.C.	1-1/2" CRC, 48"O.C.		1.36	4.87	6.23
4200	3 coat gypsum	3.4# metal	3/4" CRC, 12" O.C.	1-1/2" CRC, 36" O.C.		1.80	7.10	8.90
4300	Painted		3/4" CRC, 16" O.C.	1-1/2" CRC, 36" O.C.		1.77	6.60	8.37
4400			3/4" CRC, 24" O.C.	1-1/2" CRC, 36" O.C.		1.68	6.20	7.88
4500	3 coat perlite	3.4# metal	3/4" CRC, 12" O.C.	1-1/2" CRC,36" O.C.		1.89	7.85	9.74
4600	Painted		3/4" CRC, 16" O.C.	1-1/2" CRC, 36" O.C.		1.86	7.35	9.21
4700			3/4" CRC, 24" O.C.	1-1/2" CRC, 36" O.C.		1.77	6.90	8.67

C3030 110 Drywall Ceilings

	TYPE	FINISH	FURRING	SUPPORT		COST PER S.F. MAT.	INST.	TOTAL
4800	1/2" F.R. drywall	painted and textured	1"x3" wood, 16" O.C.	wood		.82	2.77	3.59
4900				masonry		.82	2.86	3.68
5000				concrete		.82	3.42	4.24
5100	5/8" F.R. drywall	painted and textured	1"x3" wood, 16" O.C.	wood		.82	2.77	3.59
5200				masonry		.82	2.86	3.68
5300				concrete		.82	3.42	4.24

C30 Interior Finishes

C3030 Ceiling Finishes

C3030 110 Drywall Ceilings

	TYPE	FINISH	FURRING	SUPPORT		COST PER S.F. MAT.	INST.	TOTAL
5400	1/2" F.R. drywall	painted and textured	7/8"resil. channels	24" O.C.		.68	2.69	3.37
5500			1"x2" wood	stud clips		.86	2.59	3.45
5602		painted	1-5/8" metal studs	24" O.C.		.72	2.40	3.12
5700	5/8" F.R. drywall	painted and textured	1-5/8"metal studs	24" O.C.		.72	2.40	3.12
5702								

C3030 140 Plaster Ceiling Components

		COST PER S.F. MAT.	INST.	TOTAL
0060	Plaster, gypsum incl. finish, 2 coats			
0080	3 coats	.68	2.91	3.59
0100	Perlite, incl. finish, 2 coats	.46	2.85	3.31
0120	3 coats	.77	3.63	4.40
0140	Thin coat on drywall	.11	.57	.68
0200	Lath, gypsum, 3/8" thick	.60	.70	1.30
0220	1/2" thick	.59	.73	1.32
0240	5/8" thick	.31	.85	1.16
0260	Metal, diamond, 2.5 lb.	.44	.57	1.01
0280	3.4 lb.	.49	.61	1.10
0300	Flat rib, 2.75 lb.	.38	.57	.95
0320	3.4 lb.	.59	.61	1.20
0440	Furring, steel channels, 3/4" galvanized , 12" O.C.	.31	1.83	2.14
0460	16" O.C.	.28	1.33	1.61
0480	24" O.C.	.19	.92	1.11
0500	1-1/2" galvanized , 12" O.C.	.42	2.02	2.44
0520	16" O.C.	.37	1.48	1.85
0540	24" O.C.	.25	.99	1.24
0560	Wood strips, 1"x3", on wood, 12" O.C.	.38	1.28	1.66
0580	16" O.C.	.29	.96	1.25
0600	24" O.C.	.19	.64	.83
0620	On masonry, 12" O.C.	.38	1.40	1.78
0640	16" O.C.	.29	1.05	1.34
0660	24" O.C.	.19	.70	.89
0680	On concrete, 12" O.C.	.38	2.14	2.52
0700	16" O.C.	.29	1.61	1.90
0720	24" O.C.	.19	1.07	1.26
0722				
0740				
0940	Paint on plaster or drywall, roller work, primer + 1 coat	.11	.40	.51
0960	Primer + 2 coats	.17	.51	.68

C3030 210 Acoustical Ceilings

	TYPE	TILE	GRID	SUPPORT		COST PER S.F. MAT.	INST.	TOTAL
5800	5/8" fiberglass board	24" x 48"	tee	suspended		1.73	1.35	3.08
5900		24" x 24"	tee	suspended		1.90	1.48	3.38
6000	3/4" fiberglass board	24" x 48"	tee	suspended		2.77	1.38	4.15
6100		24" x 24"	tee	suspended		2.94	1.51	4.45
6500	5/8" mineral fiber	12" x 12"	1"x3" wood, 12" O.C.	wood		2.35	2.78	5.13
6600				masonry		2.35	2.90	5.25
6700				concrete		2.35	3.64	5.99
6800	3/4" mineral fiber	12" x 12"	1"x3" wood, 12" O.C.	wood		2.48	2.78	5.26
6900				masonry		2.48	2.78	5.26
7000				concrete		2.48	2.78	5.26

C30 Interior Finishes

C3030 Ceiling Finishes

C3030 210 Acoustical Ceilings

	TYPE	TILE	GRID	SUPPORT		COST PER S.F.		
						MAT.	INST.	TOTAL
7100 7102	3/4"mineral fiber on 5/8" F.R. drywall	12" x 12"	25 ga. channels	runners		2.83	3.44	6.27
7200 7201 7202	5/8" plastic coated Mineral fiber	12" x 12"		adhesive backed		2.38	1.50	3.88
7300 7301 7302	3/4" plastic coated Mineral fiber	12" x 12"		adhesive backed		2.51	1.50	4.01
7400 7401 7402	3/4" mineral fiber	12" x 12"	conceal 2" bar & channels	suspended		2.59	3.39	5.98

C3030 240 Acoustical Ceiling Components

		COST PER S.F.		
		MAT.	INST.	TOTAL
2480	Ceiling boards, eggcrate, acrylic, 1/2" x 1/2" x 1/2" cubes	2	.90	2.90
2500	Polystyrene, 3/8" x 3/8" x 1/2" cubes	1.68	.88	2.56
2520	1/2" x 1/2" x 1/2" cubes	1.93	.90	2.83
2540	Fiberglass boards, plain, 5/8" thick	.79	.72	1.51
2560	3/4" thick	1.83	.75	2.58
2580	Grass cloth faced, 3/4" thick	1.64	.90	2.54
2600	1" thick	2.40	.93	3.33
2620	Luminous panels, prismatic, acrylic	2.44	1.12	3.56
2640	Polystyrene	1.25	1.12	2.37
2660	Flat or ribbed, acrylic	4.25	1.12	5.37
2680	Polystyrene	2.92	1.12	4.04
2700	Drop pan, white, acrylic	6.25	1.12	7.37
2720	Polystyrene	5.20	1.12	6.32
2740	Mineral fiber boards, 5/8" thick, standard	.86	.67	1.53
2760	Plastic faced	1.35	1.12	2.47
2780	2 hour rating	1.29	.67	1.96
2800	Perforated aluminum sheets, .024 thick, corrugated painted	2.48	.92	3.40
2820	Plain	4.14	.90	5.04
3080	Mineral fiber, plastic coated, 12" x 12" or 12" x 24", 5/8" thick	1.97	1.50	3.47
3100	3/4" thick	2.10	1.50	3.60
3120	Fire rated, 3/4" thick, plain faced	1.54	1.50	3.04
3140	Mylar faced	1.30	1.50	2.80
3160	Add for ceiling primer	.14		.14
3180	Add for ceiling cement	.41		.41
3240	Suspension system, furring, 1" x 3" wood 12" O.C.	.38	1.28	1.66
3260	T bar suspension system, 2' x 4' grid	.72	.56	1.28
3280	2' x 2' grid	.89	.69	1.58
3300	Concealed Z bar suspension system 12" module	.73	.86	1.59
3320	Add to above for 1-1/2" carrier channels 4' O.C.	.10	.96	1.06
3340	Add to above for carrier channels for recessed lighting	.19	.98	1.17

Chapter

8

Square Foot Cost Models

There is a thought process at the start of every construction project that defines the needs of the owner. Usually the owner and his design team work to develop a program that defines the space requirements, location, and physical attributes of the project. As with any major undertaking, this list of owner needs must be accompanied by a plan of how to pay for it. Thus, in the very early stages, a cost of construction must be forecast, often with very little definition beyond the building use and the square foot area required.

At this point, parametric cost models are useful, as they provide a reasonable cost per square foot, including contractor markups and design fees. This figure can be adjusted for design variations the anticipated cost escalation that will occur between development of the budget and the time of construction.

The RSMeans database includes models of 75 building use types, each with six variations of framing and exterior wall construction. Each model is built from approximately 75 assemblies from the RSMeans database. Each estimate is calculated using building parameters (area, perimeter, number of stories, etc.) and mathematical algorithms, to provide costs for a wide range of building configurations. The print versions of the models allow the user to adjust the cost for variations in perimeter and exterior wall height. The CD-ROM and On-Line versions of CostWorks allow the user to change a wider range of parameters to calculate costs for buildings with different numbers of stories, story heights, perimeters, partition and door densities, and partition heights.

The models are building construction costs only, for buildings with fairly simple standard finishes and systems. They do not include costs for any site work beyond the excavation for the building foundation. They do not include the cost for purchase of land, grading, or site improvements like roadways, parking areas, sidewalks, or site drainage and site lighting.

The cost models in this set of data are from *RSMeans Light Commercial Cost Data*, and are calculated using open shop labor rates. These cost models will provide a construction estimate with an anticipated accuracy of +/- 20%. Thus it is advised to add a contingency factor of at least 20% for an initial budget.

Chapter 8 Exercises

1. Develop a cost estimate for a 2-story apartment to be located in Anchorage, Alaska. The building will be 15,000 SF and constructed with steel joists and face brick with concrete block back-up.
2. Develop a cost estimate for the medical office shown in the plans found at **https://rsmeansonline.com/academic**

   Adjust the cost to Plymouth, Massachusetts (02360).
3. Develop a SF estimate for a 16,000 SF store in a strip mall in Billings, Montana. Assume the building is constructed with split-face concrete block bearing walls and open web steel joists with metal deck.

4. What is the approximate cost per SF of the electrical work in the store in example 3?
5. Develop a budget to construct a burger restaurant in Tyler, Texas. The building will be 5,800 SF with wood frame and wood siding. Include the cost of a 10′ × 10′ walk-in refrigerator, one range, a broiler, and a dishwasher. Put 20 LF of stainless steel counters in the dishwashing area.

   Note: Equipment items such as the kitchen equipment in this example are normally sold at the same price nationwide, and should not be adjusted with a location factor.
6. An automobile repair shop will be built in Denver, Colorado. The exterior wall will be concrete block and the roof will be framed with open web steel joists. Total area will be 10,000 SF. What is the total cost of this building? What will the general contractor earn for profit? What will the architectural fee be?
7. A local supermarket chain in Phoenix Arizona, would like to develop a prototype mini-market with a maximum construction cost of $750,000. How large a building can he build if the construction is stucco on CMU bearing walls? How large a building could he build with steel frame/precast concrete panels?
8. A local developer in Newark, New Jersey, is building a small shopping area, and would like to include a 6,900 SF free-standing branch bank near the entrance. The building will be face brick with steel frame, and will include a 5,000 SF basement area. The bank will have a 12′ story height and a perimeter of approximately 420 LF. What should the developer budget for this building?
9. How much will it cost to build a municipal complex in St. Paul, Minnesota, that includes an 18,000 SF town hall, an 11,000 SF police station, and a 7,000 S.F. fire station? The construction of the town hall will be stone with concrete block back-up and a steel frame. Both police and fire stations will be face brick with concrete block back-up and bearing wall construction.
10. A 26,500 SF library will be built in Los Angeles, California, to house a collection of rare books. It will be steel-framed with limestone veneer on concrete block. The developer is concerned that the HVAC and electrical systems in the model are not adequate for this special purpose, and requests the estimator to increase the budget allowance in these areas by 35%. Calculate a cost for the proposed building.

Commercial/Industrial/Institutional Section

Table of Contents

How to Use the Commercial/ Industrial/Institutional Section

The following is a detailed explanation of a sample entry in the Commercial/Industrial/Institutional Square Foot Cost Section. Each bold number below corresponds to the described item on the following page with the appropriate component or cost of the sample entry following in parentheses.

Prices listed are costs that include overhead and profit of the installing contractor and additional markups for General Conditions and Architects' Fees.

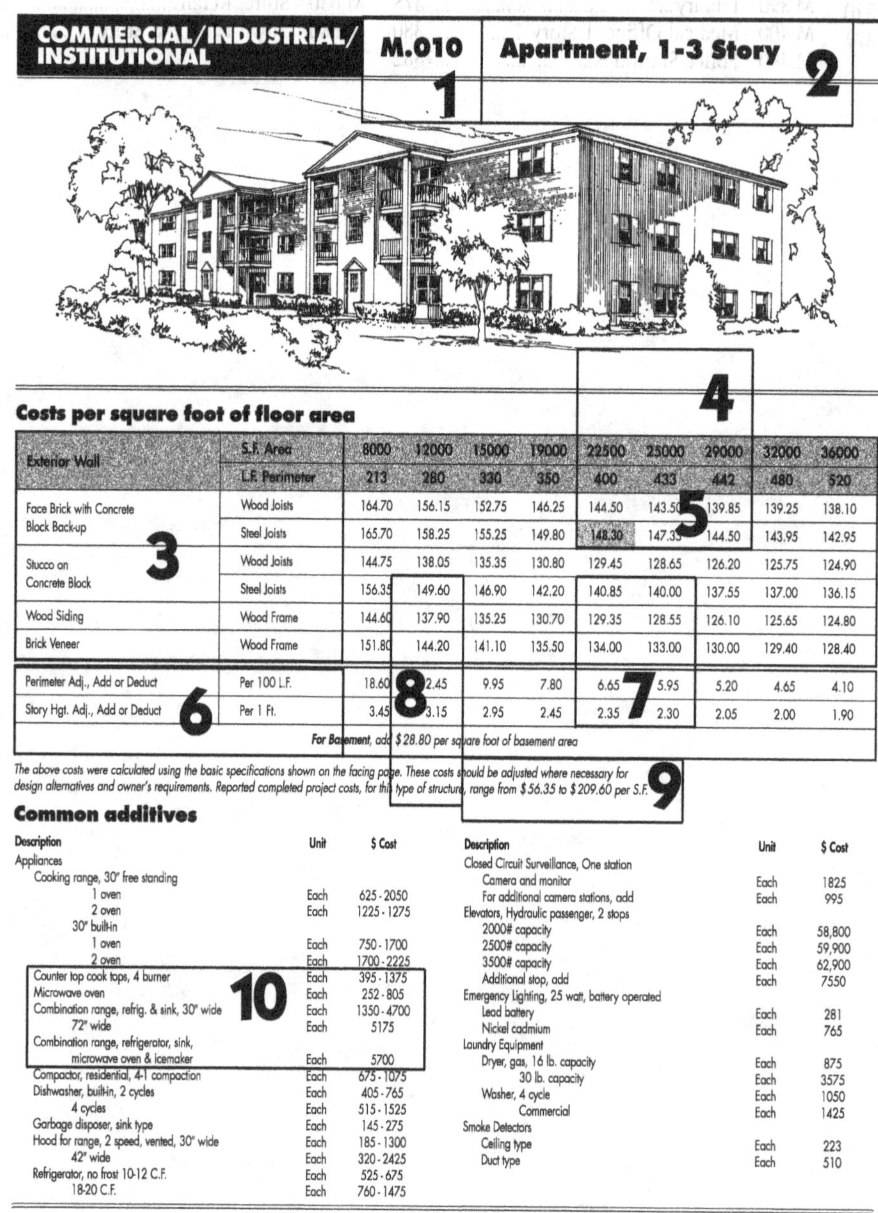

COMMERCIAL/INDUSTRIAL/ INSTITUTIONAL | **M.010** | **Apartment, 1-3 Story**

Costs per square foot of floor area

Exterior Wall	S.F. Area	8000	12000	15000	19000	22500	25000	29000	32000	36000
	L.F. Perimeter	213	280	330	350	400	433	442	480	520
Face Brick with Concrete Block Back-up	Wood Joists	164.70	156.15	152.75	146.25	144.50	143.50	139.85	139.25	138.10
	Steel Joists	165.70	158.25	155.25	149.80	148.30	147.35	144.50	143.95	142.95
Stucco on Concrete Block	Wood Joists	144.75	138.05	135.35	130.80	129.45	128.65	126.20	125.75	124.90
	Steel Joists	156.35	149.60	146.90	142.20	140.85	140.00	137.55	137.00	136.15
Wood Siding	Wood Frame	144.60	137.90	135.25	130.70	129.35	128.55	126.10	125.65	124.80
Brick Veneer	Wood Frame	151.80	144.20	141.10	135.50	134.00	133.00	130.00	129.40	128.40
Perimeter Adj., Add or Deduct	Per 100 L.F.	18.60	2.45	9.95	7.80	6.65	5.95	5.20	4.65	4.10
Story Hgt. Adj., Add or Deduct	Per 1 Ft.	3.45	3.15	2.95	2.45	2.35	2.30	2.05	2.00	1.90

For Basement, add $28.80 per square foot of basement area

The above costs were calculated using the basic specifications shown on the facing page. These costs should be adjusted where necessary for design alternatives and owner's requirements. Reported completed project costs, for this type of structure, range from $56.35 to $209.60 per S.F.

Common additives

Description	Unit	$ Cost
Appliances		
Cooking range, 30" free standing		
1 oven	Each	625 - 2050
2 oven	Each	1225 - 1275
30" built-in		
1 oven	Each	750 - 1700
2 oven	Each	1700 - 2225
Counter top cook tops, 4 burner	Each	395 - 1375
Microwave oven	Each	252 - 805
Combination range, refrig. & sink, 30" wide	Each	1350 - 4700
72" wide	Each	5175
Combination range, refrigerator, sink, microwave oven & icemaker	Each	5700
Compactor, residential, 4-1 compaction	Each	675 - 1075
Dishwasher, built-in, 2 cycles	Each	405 - 765
4 cycles	Each	515 - 1525
Garbage disposer, sink type	Each	145 - 275
Hood for range, 2 speed, vented, 30" wide	Each	185 - 1300
42" wide	Each	320 - 2425
Refrigerator, no frost 10-12 C.F.	Each	525 - 675
18-20 C.F.	Each	760 - 1475

Description	Unit	$ Cost
Closed Circuit Surveillance, One station		
Camera and monitor	Each	1825
For additional camera stations, add	Each	995
Elevators, Hydraulic passenger, 2 stops		
2000# capacity	Each	58,800
2500# capacity	Each	59,900
3500# capacity	Each	62,900
Additional stop, add	Each	7550
Emergency Lighting, 25 watt, battery operated		
Lead battery	Each	281
Nickel cadmium	Each	765
Laundry Equipment		
Dryer, gas, 16 lb. capacity	Each	875
30 lb. capacity	Each	3575
Washer, 4 cycle	Each	1050
Commercial	Each	1425
Smoke Detectors		
Ceiling type	Each	223
Duct type	Each	510

1 *Model Number (M.010)*

"M" distinguishes this section of the book and stands for model. The number designation is a sequential number.

2 *Type of Building (Apartment, 1-3 Story)*

There are 48 different types of commercial/industrial/institutional buildings highlighted in this section.

3 *Exterior Wall Construction and Building Framing Options (Face Brick with Concrete Block Back-up and Open Web Steel Bar Joists)*

Three or more commonly used exterior walls and, in most cases, two typical building framing systems are presented for each type of building. The model selected should be based on the actual characteristics of the building being estimated.

4 *Total Square Foot of Floor Area and Base Perimeter Used to Compute Base Costs (22,500 Square Feet and 400 Linear Feet)*

Square foot of floor area is the total gross area of all floors at grade, and above, and does not include a basement. The perimeter in linear feet used for the base cost is generally for a rectangular, economical building shape.

5 *Cost per Square Foot of Floor Area ($148.30)*

The highlighted cost is for a building of the selected exterior wall and framing system and floor area. Costs for buildings with floor areas other than those calculated may be interpolated between the costs shown.

6 *Building Perimeter and Story Height Adjustments*

Square-foot costs for a building with a perimeter or floor-to-floor story height significantly different from the model used to calculate the base cost may be adjusted (add or deduct) to reflect the actual building geometry.

7 *Cost per Square Foot of Floor Area for the Perimeter and/or Height Adjustment ($6.65 for Perimeter Difference and $2.35 for Story Height Difference)*

Add (or deduct) $6.65 to the base square-foot cost for each 100 feet of perimeter difference between the model and the actual building. Add (or deduct) $2.35 to the base square-foot cost for each 1 foot of story height difference between the model and the actual building.

8 *Optional Cost per Square Foot of Basement Floor Area ($28.80)*

The cost of an unfinished basement for the building being estimated is $28.80 times the gross floor area of the basement.

9 *Range of Cost per Square Foot of Floor Area for Similar Buildings ($56.35 to $209.60)*

Many different buildings of the same type have been built using similar materials and systems. RSMeans historical cost data of actual construction projects indicates a range of $56.35 to $209.60 for this type of building.

10 *Common Additives*

Common components and/or systems used in this type of building are listed. These costs should be added to the total building cost. Additional selections may be found in the Assemblies Section.

How to Use the Commercial/ Industrial/Institutional Section (Cont.)

The following is a detailed explanation of the specification and costs for a model building in the Commercial/Industrial/Institutional Square Foot Cost Section. Each bold number below corresponds to the described item on the following page with the appropriate component of the sample entry following in parentheses.

Prices listed are costs that include overhead and profit of the installing contractor.

Model costs calculated for a 3 story building with 10′ story height and 22,500 square feet of floor area **1**

2 **Apartment, 1-3 Story**

				Unit	Unit Cost	Cost Per S.F.	% Of Sub-Total
A. SUBSTRUCTURE							
1010	Standard Foundations	Poured concrete; strip and spread footings		S.F. Ground	5.34	1.78	
1020	Special Foundations	N/A		—	—	—	
1030	Slab on Grade	4" reinforced concrete with vapor barrier and granular base		S.F. Slab	4.52	1.51	4.2%
2010	Basement Excavation	Site preparation for slab and trench for foundation wall and footing		S.F. Ground	.15	.05	
2020	Basement Walls	4′ foundation wall		L.F. Wall	61	1.27	
B. SHELL							
	B10 Superstructure						
1010	Floor Construction	Open web steel joists, slab form, concrete, interior steel columns		S.F. Floor	16.97	11.31	12.8%
1020	Roof Construction	Open web steel joists with rib metal deck, interior steel columns		S.F. Roof	8.22	2.74	
	B20 Exterior Enclosure						
2010	Exterior Walls	Face brick with concrete block backup	*88% of wall*	S.F. Wall	20	9.55	
2020	Exterior Windows	Aluminum horizontal sliding	*12% of wall*	Each	473	2.01	10.8%
2030	Exterior Doors	Aluminum and glass		Each	1916	.34	
	B30 Roofing						
3010	Roof Coverings	Built-up tar and gravel with flashing; perlite/EPS composite insulation		S.F. Roof	5.52	1.84	1.7%
3020	Roof Openings	N/A		—	—	—	
C. INTERIORS	**3**	**4**	**5**	**6**			
1010	Partitions	Gypsum board and sound deadening board on metal studs	*10 S.F. of Floor/L.F. Partition*	S.F. Partition	5.57	4.95	
1020	Interior Doors	15% solid core wood, 85% hollow core wood	*80 S.F. Floor/Door*	Each	498	6.22	
1030	Fittings	Kitchen Cabinets		S.F. Floor	2.79	2.79	
2010	Stair Construction	Concrete filled metal pan		Flight	12,150	2.70	24.4%
3010	Wall Finishes	70% paint, 25% vinyl wall covering, 5% ceramic tile		S.F. Surface	1.21	2.15	
3020	Floor Finishes	60% carpet, 30% vinyl composition tile, 10% ceramic tile		S.F. Floor	4.92	4.92	
3030	Ceiling Finishes	Painted gypsum board on resilient channels		S.F. Ceiling	[illegible]	[illegible]	
D. SERVICES					**7**	**8**	**9**
	D10 Conveying						
1010	Elevators & Lifts	One hydraulic passenger elevator		Each	97,650	4.34	4.0%
1020	Escalators & Moving Walks	N/A		—	—	—	
	D20 Plumbing						
2010	Plumbing Fixtures	Kitchen, bathroom and service fixtures, supply and drainage	*1 Fixture/200 S.F. Floor*	Each	2558	12.79	
2020	Domestic Water Distribution	Gas fired water heater		S.F. Floor	4.04	4.04	15.7%
2040	Rain Water Drainage	Roof drains		S.F. Roof	1.20	.40	
	D30 HVAC						
3010	Energy Supply	Oil fired hot water, baseboard radiation		S.F. Floor	7.58	7.58	
3020	Heat Generating Systems	N/A		—	—	—	
3030	Cooling Generating Systems	Chilled water, air cooled condenser system		S.F. Floor	8.13	8.13	14.3%
3050	Terminal & Package Units	N/A		—	—	—	
3090	Other HVAC Sys. & Equipment	N/A		—	—	—	
	D40 Fire Protection						
4010	Sprinklers	Wet pipe sprinkler system		S.F. Floor	2.90	2.90	2.6%
4020	Standpipes	N/A		—	—	—	
	D50 Electrical						
5010	Electrical Service/Distribution	600 ampere service, panel board and feeders		S.F. Floor	2.43	2.43	
5020	Lighting & Branch Wiring	Incandescent fixtures, receptacles, switches, A.C. and misc. power		S.F. Floor	6.23	6.23	9.5%
5030	Communications & Security	Addressable alarm systems, internet wiring, and emergency lighting		S.F. Floor	1.56	1.56	
5090	Other Electrical Systems	Emergency generator, 11.5 kW		S.F. Floor	.20	.20	
E. EQUIPMENT & FURNISHINGS							
1010	Commercial Equipment	N/A		—	—	—	
1020	Institutional Equipment	N/A		—	**10**	—	0.0 %
1030	Vehicular Equipment	N/A		—		—	
1090	Other Equipment	N/A		—	—	—	
F. SPECIAL CONSTRUCTION							
1020	Integrated Construction	N/A		—	—	—	0.0 %
1040	Special Facilities	N/A		—	—	—	
G. BUILDING SITEWORK	**N/A**						
					Sub-Total	109.85	**100%**
	CONTRACTOR FEES (General Requirements: 10%, Overhead: 5%, Profit: 10%)	**11**	**12**		25%	27.47	
	ARCHITECT FEES				8%	10.98	
					Total Building Cost	148.30	

1 ***Building Description (Model costs are calculated for a 3-story apartment building with 10' story height and 22,500 square feet of floor area)***

The model highlighted is described in terms of building type, number of stories, typical story height, and square footage.

2 ***Type of Building (Apartment, 1-3 Story)***

3 ***Division C Interiors (C1020 Interior Doors)***

System costs are presented in divisions according to the 7-division UNIFORMAT II classifications. Each of the component systems is listed.

4 ***Specification Highlights (15% solid core wood; 85% hollow core wood)***

All systems in each subdivision are described with the material and proportions used.

5 ***Quality Criteria (80 S.F. Floor/Door)***

The criteria used in determining quantities for the calculations are shown.

6 ***Unit (Each)***

The unit of measure shown in this column is the unit of measure of the particular system shown that corresponds to the unit cost.

7 ***Unit Cost ($498)***

The cost per unit of measure of each system subdivision.

8 ***Cost per Square Foot ($6.22)***

The cost per square foot for each system is the unit cost of the system times the total number of units, divided by the total square feet of building area.

9 ***% of Sub-Total (24.4%)***

The percent of sub-total is the total cost per square foot of all systems in the division divided by the sub-total cost per square foot of the building.

10 ***Sub-Total ($109.85)***

The sub-total is the total of all the system costs per square foot.

11 ***Project Fees (Contractor Fees) (25%) (Architects' Fees) (8%)***

Contractor Fees to cover the general requirements, overhead, and profit of the General Contractor are added as a percentage of the sub-total. An Architect's Fee, also as a percentage of the sub-total, is also added. These values vary with the building type.

12 ***Total Building Cost ($148.30)***

The total building cost per square foot of building area is the sum of the square foot costs of all the systems, the General Contractor's general requirements, overhead and profit, and the Architect's fee. The total building cost is the amount which appears shaded in the Cost per Square Foot of Floor Area table shown previously.

COMMERCIAL/INDUSTRIAL/ INSTITUTIONAL	M.010	Apartment, 1-3 Story

Costs per square foot of floor area

Exterior Wall	S.F. Area	8000	12000	15000	19000	22500	25000	29000	32000	36000
	L.F. Perimeter	213	280	330	350	400	433	442	480	520
Face Brick with Concrete Block Back-up	Wood Joists	164.70	156.15	152.75	146.25	144.50	143.50	139.85	139.25	138.10
	Steel Joists	165.70	158.25	155.25	149.80	148.30	147.35	144.50	143.95	142.95
Stucco on Concrete Block	Wood Joists	144.75	138.05	135.35	130.80	129.45	128.65	126.20	125.75	124.90
	Steel Joists	156.35	149.60	146.90	142.20	140.85	140.00	137.55	137.00	136.15
Wood Siding	Wood Frame	144.60	137.90	135.25	130.70	129.35	128.55	126.10	125.65	124.80
Brick Veneer	Wood Frame	151.80	144.20	141.10	135.50	134.00	133.00	130.00	129.40	128.40
Perimeter Adj., Add or Deduct	Per 100 L.F.	18.60	12.45	9.95	7.80	6.65	5.95	5.20	4.65	4.10
Story Hgt. Adj., Add or Deduct	Per 1 Ft.	3.45	3.15	2.95	2.45	2.35	2.30	2.05	2.00	1.90
***For Basement**, add $28.80 per square foot of basement area*										

The above costs were calculated using the basic specifications shown on the facing page. These costs should be adjusted where necessary for design alternatives and owner's requirements. Reported completed project costs, for this type of structure, range from $56.35 to $209.60 per S.F.

Common additives

Description	Unit	$ Cost
Appliances		
Cooking range, 30" free standing		
1 oven	Each	625 - 2050
2 oven	Each	1225 - 1275
30" built-in		
1 oven	Each	750 - 1700
2 oven	Each	1700 - 2225
Counter top cook tops, 4 burner	Each	395 - 1375
Microwave oven	Each	252 - 805
Combination range, refrig. & sink, 30" wide	Each	1350 - 4700
72" wide	Each	5175
Combination range, refrigerator, sink, microwave oven & icemaker	Each	5700
Compactor, residential, 4-1 compaction	Each	675 - 1075
Dishwasher, built-in, 2 cycles	Each	405 - 765
4 cycles	Each	515 - 1525
Garbage disposer, sink type	Each	145 - 275
Hood for range, 2 speed, vented, 30" wide	Each	185 - 1300
42" wide	Each	320 - 2425
Refrigerator, no frost 10-12 C.F.	Each	525 - 675
18-20 C.F.	Each	760 - 1475

Description	Unit	$ Cost
Closed Circuit Surveillance, One station		
Camera and monitor	Each	1825
For additional camera stations, add	Each	995
Elevators, Hydraulic passenger, 2 stops		
2000# capacity	Each	58,800
2500# capacity	Each	59,900
3500# capacity	Each	62,900
Additional stop, add	Each	7550
Emergency Lighting, 25 watt, battery operated		
Lead battery	Each	281
Nickel cadmium	Each	765
Laundry Equipment		
Dryer, gas, 16 lb. capacity	Each	875
30 lb. capacity	Each	3575
Washer, 4 cycle	Each	1050
Commercial	Each	1425
Smoke Detectors		
Ceiling type	Each	223
Duct type	Each	510

Model costs calculated for a 3 story building with 10′ story height and 22,500 square feet of floor area

Apartment, 1-3 Story

			Unit	Unit Cost	Cost Per S.F.	% Of Sub-Total
A. SUBSTRUCTURE						
1010	Standard Foundations	Poured concrete; strip and spread footings	S.F. Ground	5.34	1.78	4.2%
1020	Special Foundations	N/A	–	–	–	
1030	Slab on Grade	4″ reinforced concrete with vapor barrier and granular base	S.F. Slab	4.52	1.51	
2010	Basement Excavation	Site preparation for slab and trench for foundation wall and footing	S.F. Ground	.15	.05	
2020	Basement Walls	4′ foundation wall	L.F. Wall	61	1.27	
B. SHELL						
	B10 Superstructure					
1010	Floor Construction	Open web steel joists, slab form, concrete, interior steel columns	S.F. Floor	16.97	11.31	12.8%
1020	Roof Construction	Open web steel joists with rib metal deck, interior steel columns	S.F. Roof	8.22	2.74	
	B20 Exterior Enclosure					
2010	Exterior Walls	Face brick with concrete block backup *88% of wall*	S.F. Wall	20	9.55	10.8%
2020	Exterior Windows	Aluminum horizontal sliding *12% of wall*	Each	473	2.01	
2030	Exterior Doors	Aluminum and glass	Each	1916	.34	
	B30 Roofing					
3010	Roof Coverings	Built-up tar and gravel with flashing; perlite/EPS composite insulation	S.F. Roof	5.52	1.84	1.7%
3020	Roof Openings	N/A	–	–	–	
C. INTERIORS						
1010	Partitions	Gypsum board and sound deadening board on metal studs *10 S.F. of Floor/L.F. Partition*	S.F. Partition	5.57	4.95	24.4%
1020	Interior Doors	15% solid core wood, 85% hollow core wood *80 S.F. Floor/Door*	Each	498	6.22	
1030	Fittings	Kitchen Cabinets	S.F. Floor	2.79	2.79	
2010	Stair Construction	Concrete filled metal pan	Flight	12,150	2.70	
3010	Wall Finishes	70% paint, 25% vinyl wall covering, 5% ceramic tile	S.F. Surface	1.21	2.15	
3020	Floor Finishes	60% carpet, 30% vinyl composition tile, 10% ceramic tile	S.F. Floor	4.92	4.92	
3030	Ceiling Finishes	Painted gypsum board on resilient channels	S.F. Ceiling	3.12	3.12	
D. SERVICES						
	D10 Conveying					
1010	Elevators & Lifts	One hydraulic passenger elevator	Each	97,650	4.34	4.0%
1020	Escalators & Moving Walks	N/A	–	–	–	
	D20 Plumbing					
2010	Plumbing Fixtures	Kitchen, bathroom and service fixtures, supply and drainage *1 Fixture/200 S.F. Floor*	Each	2558	12.79	15.7%
2020	Domestic Water Distribution	Gas fired water heater	S.F. Floor	4.04	4.04	
2040	Rain Water Drainage	Roof drains	S.F. Roof	1.20	.40	
	D30 HVAC					
3010	Energy Supply	Oil fired hot water, baseboard radiation	S.F. Floor	7.58	7.58	14.3%
3020	Heat Generating Systems	N/A	–	–	–	
3030	Cooling Generating Systems	Chilled water, air cooled condenser system	S.F. Floor	8.13	8.13	
3050	Terminal & Package Units	N/A	–	–	–	
3090	Other HVAC Sys. & Equipment	N/A	–	–	–	
	D40 Fire Protection					
4010	Sprinklers	Wet pipe sprinkler system	S.F. Floor	2.90	2.90	2.6%
4020	Standpipes	N/A	–	–	–	
	D50 Electrical					
5010	Electrical Service/Distribution	600 ampere service, panel board and feeders	S.F. Floor	2.43	2.43	9.5%
5020	Lighting & Branch Wiring	Incandescent fixtures, receptacles, switches, A.C. and misc. power	S.F. Floor	6.23	6.23	
5030	Communications & Security	Addressable alarm systems, internet wiring, and emergency lighting	S.F. Floor	1.56	1.56	
5090	Other Electrical Systems	Emergency generator, 11.5 kW	S.F. FLoor	.20	.20	
E. EQUIPMENT & FURNISHINGS						
1010	Commercial Equipment	N/A	–	–	–	0.0 %
1020	Institutional Equipment	N/A	–	–	–	
1030	Vehicular Equipment	N/A	–	–	–	
1090	Other Equipment	N/A	–	–	–	
F. SPECIAL CONSTRUCTION						
1020	Integrated Construction	N/A	–	–	–	0.0 %
1040	Special Facilities	N/A	–	–	–	
G. BUILDING SITEWORK		**N/A**				
				Sub-Total	109.85	**100%**
	CONTRACTOR FEES (General Requirements: 10%, Overhead: 5%, Profit: 10%)			*25%*	27.47	
	ARCHITECT FEES			*8%*	10.98	
				Total Building Cost	**148.30**	

COMMERCIAL/INDUSTRIAL/ INSTITUTIONAL | M.050 | Bank

Costs per square foot of floor area

Exterior Wall	S.F. Area	2000	2700	3400	4100	4800	5500	6200	6900	7600
	L.F. Perimeter	180	208	236	256	280	303	317	337	357
Face Brick with Concrete Block Back-up	Steel Frame	230.95	217.40	209.40	202.50	198.30	195.05	191.30	189.05	187.30
	R/Conc. Frame	244.30	230.70	222.75	**215.85**	211.70	208.40	204.70	202.45	200.60
Precast Concrete Panel	Steel Frame	244.00	228.55	219.40	211.45	206.65	202.90	198.55	196.00	193.85
	R/Conc. Frame	261.35	244.05	233.85	225.25	220.00	215.85	211.25	208.45	206.15
Limestone with Concrete Block Back-up	Steel Frame	273.50	253.80	242.25	232.00	225.90	221.10	215.50	212.15	209.50
	R/Conc. Frame	286.90	267.20	255.60	245.40	239.25	234.45	228.90	225.55	222.85
Perimeter Adj., Add or Deduct	Per 100 L.F.	41.75	30.90	24.55	20.35	17.45	15.20	13.50	12.15	11.00
Story Hgt. Adj., Add or Deduct	Per 1 Ft.	3.95	3.40	3.05	2.80	2.65	2.45	2.30	2.20	2.05

***For Basement**, add $27.30 per square foot of basement area*

The above costs were calculated using the basic specifications shown on the facing page. These costs should be adjusted where necessary for design alternatives and owner's requirements. Reported completed project costs, for this type of structure, range from $ 130.45 to $ 321.05 per S.F.

Common additives

Description	Unit	$ Cost
Bulletproof Teller Window, 44" x 60"	Each	4600
60" x 48"	Each	5900
Closed Circuit Surveillance, One station		
Camera and monitor	Each	1825
For additional camera stations, add	Each	995
Counters, Complete	Station	5550
Door & Frame, 3' x 6'-8", bullet resistant steel		
with vision panel	Each	6500 - 8475
Drive-up Window, Drawer & micr., not incl. glass	Each	7475 - 12,000
Emergency Lighting, 25 watt, battery operated		
Lead battery	Each	281
Nickel cadmium	Each	765
Night Depository	Each	9075 - 13,500
Package Receiver, painted	Each	1775
stainless steel	Each	2775
Partitions, Bullet resistant to 8' high	L.F.	298 - 485
Pneumatic Tube Systems, 2 station	Each	29,400
With TV viewer	Each	55,500

Description	Unit	$ Cost
Service Windows, Pass thru, steel		
24" x 36"	Each	3325
48" x 48"	Each	3725
72" x 40"	Each	5500
Smoke Detectors		
Ceiling type	Each	223
Duct type	Each	510
Twenty-four Hour Teller		
Automatic deposit cash & memo	Each	49,300
Vault Front, Door & frame		
1 hour test, 32"x 78"	Opening	3775
2 hour test, 32" door	Opening	4100
40" door	Opening	4775
4 hour test, 32" door	Opening	4500
40" door	Opening	5375
Time lock, two movement, add	Each	1925

Model costs calculated for a 1 story building with 14′ story height and 4,100 square feet of floor area

Bank

				Unit	Unit Cost	Cost Per S.F.	% Of Sub-Total
A. SUBSTRUCTURE							
1010	Standard Foundations	Poured concrete; strip and spread footings		S.F. Ground	3.53	3.53	8.5%
1020	Special Foundations	N/A		–	–	–	
1030	Slab on Grade	4″ reinforced concrete with vapor barrier and granular base		S.F. Slab	4.52	4.52	
2010	Basement Excavation	Site preparation for slab and trench for foundation wall and footing		S.F. Ground	.24	.24	
2020	Basement Walls	4′ Foundation wall		L.F. Wall	66	4.91	
B. SHELL							
	B10 Superstructure						
1010	Floor Construction	Cast-in-place columns		L.F. Column	99	4.83	11.5%
1020	Roof Construction	Cast-in-place concrete flat plate		S.F. Roof	13	13	
	B20 Exterior Enclosure						
2010	Exterior Walls	Face brick with concrete block backup	*80% of wall*	S.F. Wall	27	19.16	17.2%
2020	Exterior Windows	Horizontal aluminum sliding	*20% of wall*	Each	533	6.21	
2030	Exterior Doors	Double aluminum and glass and hollow metal		Each	2989	1.46	
	B30 Roofing						
3010	Roof Coverings	Built-up tar and gravel with flashing; perlite/EPS composite insulation		S.F. Roof	6.42	6.42	4.1%
3020	Roof Openings	N/A		–	–	–	
C. INTERIORS							
1010	Partitions	Gypsum board on metal studs	*20 S.F. of Floor/L.F. Partition*	S.F. Partition	8.54	4.27	13.9%
1020	Interior Doors	Single leaf hollow metal	*200 S.F. Floor/Door*	Each	1022	5.12	
1030	Fittings	N/A		–	–	–	
2010	Stair Construction	N/A		–	–	–	
3010	Wall Finishes	50% vinyl wall covering, 50% paint		S.F. Surface	1.21	1.21	
3020	Floor Finishes	50% carpet tile, 40% vinyl composition tile, 10% quarry tile		S.F. Floor	4.98	4.98	
3030	Ceiling Finishes	Mineral fiber tile on concealed zee bars		S.F. Ceiling	5.98	5.98	
D. SERVICES							
	D10 Conveying						
1010	Elevators & Lifts	N/A		–	–	–	0.0 %
1020	Escalators & Moving Walks	N/A		–	–	–	
	D20 Plumbing						
2010	Plumbing Fixtures	Toilet and service fixtures, supply and drainage	*1 Fixture/580 S.F. Floor*	Each	4501	7.76	6.5%
2020	Domestic Water Distribution	Gas fired water heater		S.F. Floor	1.12	1.12	
2040	Rain Water Drainage	Roof drains		S.F. Roof	1.27	1.27	
	D30 HVAC						
3010	Energy Supply	N/A		–	–	–	6.9 %
3020	Heat Generating Systems	Included in D3050		–	–	–	
3030	Cooling Generating Systems	N/A		–	–	–	
3050	Terminal & Package Units	Single zone rooftop unit, gas heating, electric cooling		S.F. Floor	10.67	10.67	
3090	Other HVAC Sys. & Equipment	N/A		–	–	–	
	D40 Fire Protection						
4010	Sprinklers	Wet pipe sprinkler system		S.F. Floor	3.97	3.97	4.0%
4020	Standpipes	Standpipe		S.F. Floor	2.31	2.31	
	D50 Electrical						
5010	Electrical Service/Distribution	200 ampere service, panel board and feeders		S.F. Floor	2.31	2.31	13.4%
5020	Lighting & Branch Wiring	Fluorescent fixtures, receptacles, switches, A.C. and misc. power		S.F. Floor	8.25	8.25	
5030	Communications & Security	Aarm systems, internet and phone wiring, emergency lighting, and security television		S.F. Floor	9.69	9.69	
5090	Other Electrical Systems	Emergency generator, 15 kW, Uninterruptible power supply		S.F. Floor	.60	.60	
E. EQUIPMENT & FURNISHINGS							
1010	Commercial Equipment	Automatic teller, drive up window, night depository		S.F. Floor	7.35	7.35	6.6%
1020	Institutional Equipment	Closed circuit TV monitoring system		S.F. Floor	2.87	2.87	
1030	Vehicular Equipment	N/A		–	–	–	
1090	Other Equipment	N/A		–	–	–	
F. SPECIAL CONSTRUCTION							
1020	Integrated Construction	N/A		–	–	–	7.4 %
1040	Special Facilities	Security vault door		S.F. Floor	11.56	11.56	
G. BUILDING SITEWORK	**N/A**						
					Sub-Total	155.57	**100%**
	CONTRACTOR FEES (General Requirements: 10%, Overhead: 5%, Profit: 10%)				*25%*	38.89	
	ARCHITECT FEES				*11%*	21.39	
					Total Building Cost	**215.85**	

COMMERCIAL/INDUSTRIAL/ INSTITUTIONAL | M.220 | Fire Station, 1 Story

Costs per square foot of floor area

Exterior Wall	S.F. Area	4000	4500	5000	5500	6000	6500	7000	7500	8000
	L.F. Perimeter	260	280	300	310	320	336	353	370	386
Face Brick Concrete Block Back-up	Steel Joists	159.10	155.95	153.45	150.20	147.45	145.70	144.40	143.20	142.10
	Bearing Walls	153.60	150.45	148.00	144.65	141.90	140.20	138.85	137.65	136.60
Decorative Concrete Block	Steel Joists	146.45	143.80	141.80	139.20	137.10	135.70	134.60	133.65	132.75
	Bearing Walls	140.90	138.35	136.35	133.75	131.60	130.15	129.10	128.10	127.25
Limestone with Concrete Block Back-up	Steel Joists	187.35	183.00	179.60	174.75	170.70	168.20	166.35	164.65	163.10
	Bearing Walls	181.90	177.50	174.10	169.20	165.15	162.70	160.85	159.15	157.65
Perimeter Adj., Add or Deduct	Per 100 L.F.	17.85	15.90	14.30	12.95	11.85	11.00	10.15	9.50	8.90
Story Hgt. Adj., Add or Deduct	Per 1 Ft.	2.35	2.25	2.20	2.00	1.90	1.90	1.80	1.80	1.75
***For Basement**, add $31.10 per square foot of basement area*										

The above costs were calculated using the basic specifications shown on the facing page. These costs should be adjusted where necessary for design alternatives and owner's requirements. Reported completed project costs, for this type of structure, range from $70.75 to $209.35 per S.F.

Common additives

Description	Unit	$ Cost
Appliances		
Cooking range, 30" free standing		
1 oven	Each	625 - 2050
2 oven	Each	1225 - 1275
30" built-in		
1 oven	Each	750 - 1700
2 oven	Each	1700 - 2225
Counter top cook tops, 4 burner	Each	395 - 1375
Microwave oven	Each	252 - 805
Combination range, refrig. & sink, 30" wide	Each	1350 - 4700
60" wide	Each	4000
72" wide	Each	5175
Combination range refrigerator, sink microwave oven & icemaker	Each	5700
Compactor, residential, 4-1 compaction	Each	675 - 1075
Dishwasher, built-in, 2 cycles	Each	405 - 765
4 cycles	Each	515 - 1525
Garbage disposer, sink type	Each	145 - 275
Hood for range, 2 speed, vented, 30" wide	Each	185 - 1300
42" wide	Each	320 - 2425

Description	Unit	$ Cost
Appliances, cont.		
Refrigerator, no frost 10-12 C.F.	Each	525 - 675
14-16 C.F.	Each	610 - 840
18-20 C.F.	Each	760 - 1475
Lockers, Steel, single tier, 60" or 72"	Opening	189 - 315
2 tier, 60" or 72" total	Opening	118 - 139
5 tier, box lockers	Opening	65 - 77.50
Locker bench, lam. maple top only	L.F.	21.50
Pedestals, steel pipe	Each	64
Sound System		
Amplifier, 250 watts	Each	1800
Speaker, ceiling or wall	Each	189
Trumpet	Each	360

Model costs calculated for a 1 story building with 14′ story height and 6,000 square feet of floor area

Fire Station, 1 Story

				Unit	Unit Cost	Cost Per S.F.	% Of Sub-Total
A. SUBSTRUCTURE							
1010	Standard Foundations	Poured concrete; strip and spread footings		S.F. Ground	2.81	2.81	11.6%
1020	Special Foundations	N/A		—	—	—	
1030	Slab on Grade	6″ reinforced concrete with vapor barrier and granular base		S.F. Slab	6	6	
2010	Basement Excavation	Site preparation for slab and trench for foundation wall and footing		S.F. Ground	.41	.41	
2020	Basement Walls	4′ foundation wall		L.F. Wall	66	3.50	
B. SHELL							
	B10 Superstructure						
1010	Floor Construction	N/A		—	—	—	7.5 %
1020	Roof Construction	Metal deck, open web steel joists, beams on columns		S.F. Roof	8.22	8.22	
	B20 Exterior Enclosure						
2010	Exterior Walls	Face brick with concrete block backup	*75% of wall*	S.F. Wall	27	15.34	19.4%
2020	Exterior Windows	Aluminum insulated glass	*10% of wall*	Each	780	1.82	
2030	Exterior Doors	Single aluminum and glass, overhead, hollow metal	*15% of wall*	S.F. Door	36	4.08	
	B30 Roofing						
3010	Roof Coverings	Built-up tar and gravel with flashing; perlite/EPS composite insulation		S.F. Roof	6.10	6.10	5.7%
3020	Roof Openings	Skylights, roof hatches		S.F. Roof	.14	.14	
C. INTERIORS							
1010	Partitions	Concrete block	*17 S.F. Floor/L.F. Partition*	S.F. Partition	7.94	4.67	13.0%
1020	Interior Doors	Single leaf hollow metal	*500 S.F. Floor/Door*	Each	1022	2.04	
1030	Fittings	Toilet partitions		S.F. Floor	.36	.36	
2010	Stair Construction	N/A		—	—	—	
3010	Wall Finishes	Paint		S.F. Surface	1.71	2.01	
3020	Floor Finishes	50% vinyl tile, 50% paint		S.F. Floor	2.10	2.10	
3030	Ceiling Finishes	Fiberglass board on exposed grid, suspended	*50% of area*	S.F. Ceiling	5.98	3	
D. SERVICES							
	D10 Conveying						
1010	Elevators & Lifts	N/A		—	—	—	0.0 %
1020	Escalators & Moving Walks	N/A		—	—	—	
	D20 Plumbing						
2010	Plumbing Fixtures	Kitchen, toilet and service fixtures, supply and drainage	*1 Fixture/375 S.F. Floor*	Each	3248	8.66	11.2%
2020	Domestic Water Distribution	Gast fired water heater		S.F. Floor	2.56	2.56	
2040	Rain Water Drainage	Roof drains		S.F. Roof	1.06	1.06	
	D30 HVAC						
3010	Energy Supply	N/A		—	—	—	18.5 %
3020	Heat Generating Systems	Included in D3050		—	—	—	
3030	Cooling Generating Systems	N/A		—	—	—	
3050	Terminal & Package Units	Rooftop multizone unit system		S.F. Floor	20	20.19	
3090	Other HVAC Sys. & Equipment	N/A		—	—	—	
	D40 Fire Protection						
4010	Sprinklers	Wet pipe sprinkler system		S.F. Floor	3.97	3.97	4.7%
4020	Standpipes	Standpipe, wet, Class III		S.F. Floor	1.13	1.13	
	D50 Electrical						
5010	Electrical Service/Distribution	200 ampere service, panel board and feeders		S.F. Floor	1.36	1.36	8.3%
5020	Lighting & Branch Wiring	High efficiency fluorescent fixtures, receptacles, switches, A.C. and misc. power		S.F. Floor	5.90	5.90	
5030	Communications & Security	Addressable alarm systems		S.F. Floor	1.79	1.79	
5090	Other Electrical Systems	N/A		—	—	—	
E. EQUIPMENT & FURNISHINGS							
1010	Commercial Equipment	N/A		—	—	—	0.0 %
1020	Institutional Equipment	N/A		—	—	—	
1030	Vehicular Equipment	N/A		—	—	—	
1090	Other Equipment	N/A		—	—	—	
F. SPECIAL CONSTRUCTION							
1020	Integrated Construction	N/A		—	—	—	0.0 %
1040	Special Facilities	N/A		—	—	—	
G. BUILDING SITEWORK		N/A					
					Sub-Total	109.22	**100%**
	CONTRACTOR FEES (General Requirements: 10%, Overhead: 5%, Profit: 10%)				*25%*	27.31	
	ARCHITECT FEES				*8%*	10.92	
					Total Building Cost	**147.45**	

COMMERCIAL/INDUSTRIAL/ INSTITUTIONAL	M.290	Garage, Repair

Costs per square foot of floor area

Exterior Wall	S.F. Area	2000	4000	6000	8000	10000	12000	14000	16000	18000
	L.F. Perimeter	180	260	340	420	500	580	586	600	610
Concrete Block	Wood Joists	134.70	115.25	108.75	105.45	103.55	102.30	98.90	96.65	94.75
	Steel Joists	135.15	116.30	109.95	106.80	**104.95**	103.70	100.40	98.15	96.30
Cast in Place Concrete	Wood Joists	142.40	121.05	113.95	110.30	108.25	106.85	102.90	100.20	98.00
	Steel Joists	143.90	122.55	115.45	111.85	109.80	108.40	104.35	101.70	99.45
Stucco	Wood Frame	139.55	117.25	109.80	106.00	103.90	102.40	98.20	95.30	92.90
Insulated Metal Panels	Steel Frame	132.55	115.40	109.60	106.70	105.05	103.90	101.00	99.00	97.40
Perimeter Adj., Add or Deduct	Per 100 L.F.	22.80	11.40	7.65	5.70	4.55	3.75	3.25	2.80	2.55
Story Hgt. Adj., Add or Deduct	Per 1 Ft.	1.75	1.25	1.10	1.05	1.00	0.90	0.85	0.70	0.65
***For Basement**, add $27.50 per square foot of basement area*										

The above costs were calculated using the basic specifications shown on the facing page. These costs should be adjusted where necessary for design alternatives and owner's requirements. Reported completed project costs, for this type of structure, range from $68.35 to $205.45 per S.F.

Common additives

Description	Unit	$ Cost
Air Compressors		
Electric 1-1/2 H.P., standard controls	Each	985
Dual controls	Each	1350
5 H.P. 115/230 Volt, standard controls	Each	3200
Dual controls	Each	3375
Product Dispenser		
with vapor recovery for 6 nozzles	Each	26,000
Hoists, Single post		
8000# cap., swivel arm	Each	7450
Two post, adjustable frames, 11,000# cap.	Each	10,200
24,000# cap.	Each	14,700
7500# Frame support	Each	9425
Four post, roll on ramp	Each	8650
Lockers, Steel, single tier, 60" or 72"	Opening	189 - 315
2 tier, 60" or 72" total	Opening	118 - 139
5 tier, box lockers	Opening	65 - 77.50
Locker bench, lam. maple top only	L.F.	21.50
Pedestals, steel pipe	Each	64
Lube Equipment		
3 reel type, with pumps, no piping	Each	10,300
Spray Painting Booth, 26' long, complete	Each	19,400

Model costs calculated for a 1 story building with 14′ story height and 10,000 square feet of floor area

Garage, Repair

			Unit	Unit Cost	Cost Per S.F.	% Of Sub-Total
A. SUBSTRUCTURE						
1010	Standard Foundations	Poured concrete; strip and spread footings	S.F. Ground	1.89	1.89	
1020	Special Foundations	N/A	—	—	—	
1030	Slab on Grade	6″ reinforced concrete with vapor barrier and granular base	S.F. Slab	6.89	6.89	15.8%
2010	Basement Excavation	Site preparation for slab and trench for foundation wall and footing	S.F. Ground	.24	.24	
2020	Basement Walls	4′ foundation wall	L.F. Wall	66	3.28	
B. SHELL						
	B10 Superstructure					
1010	Floor Construction	N/A	—	—	—	5.5 %
1020	Roof Construction	Metal deck on open web steel joists	S.F. Roof	4.27	4.27	
	B20 Exterior Enclosure					
2010	Exterior Walls	Concrete block *80% of wall*	S.F. Wall	10.84	6.07	
2020	Exterior Windows	Hopper type commercial steel *5% of wall*	Each	473	1.11	11.8%
2030	Exterior Doors	Steel overhead and hollow metal *15% of wall*	S.F. Door	18.76	1.97	
	B30 Roofing					
3010	Roof Coverings	Built-up tar and gravel; perlite/EPS composite insulation	S.F. Roof	5.77	5.77	7.4%
3020	Roof Openings	Skylight	S.F. Roof	.02	.02	
C. INTERIORS						
1010	Partitions	Concrete block *50 S.F. Floor/L.F. Partition*	S.F. Partition	6.45	1.29	
1020	Interior Doors	Single leaf hollow metal *3000 S.F. Floor/Door*	Each	1022	.34	
1030	Fittings	Toilet partitions	S.F. Floor	.10	.10	
2010	Stair Construction	N/A	—	—	—	7.3%
3010	Wall Finishes	Paint	S.F. Surface	6.35	2.54	
3020	Floor Finishes	90% metallic floor hardener, 10% vinyl composition tile	S.F. Floor	1.11	1.11	
3030	Ceiling Finishes	Gypsum board on wood joists in office and washrooms *10% of area*	S.F. Ceiling	3.08	.31	
D. SERVICES						
	D10 Conveying					
1010	Elevators & Lifts	N/A	—	—	—	0.0 %
1020	Escalators & Moving Walks	N/A	—	—	—	
	D20 Plumbing					
2010	Plumbing Fixtures	Toilet and service fixtures, supply and drainage *1 Fixture/500 S.F. Floor*	Each	1320	2.64	
2020	Domestic Water Distribution	Gas fired water heater	S.F. Floor	.46	.46	7.0%
2040	Rain Water Drainage	Roof drains	S.F. Roof	2.36	2.36	
	D30 HVAC					
3010	Energy Supply	N/A	—	—	—	
3020	Heat Generating Systems	N/A	—	—	—	
3030	Cooling Generating Systems	N/A	—	—	—	10.9 %
3050	Terminal & Package Units	Single zone AC unit	S.F. Floor	7.63	7.63	
3090	Other HVAC Sys. & Equipment	Garage exhaust system	S.F. Floor	.82	.82	
	D40 Fire Protection					
4010	Sprinklers	Sprinklers, ordinary hazard	S.F. Floor	3.85	3.85	5.9%
4020	Standpipes	Standpipe	S.F. Floor	.75	.75	
	D50 Electrical					
5010	Electrical Service/Distribution	200 ampere service, panel board and feeders	S.F. Floor	.49	.49	
5020	Lighting & Branch Wiring	T-8 fluorescent fixtures, receptacles, switches, A.C. and misc. power	S.F. Floor	7.13	7.13	14.0%
5030	Communications & Security	Addressable alarm systems, partial internet wiring and emergency lighting	S.F. Floor	3.17	3.17	
5090	Other Electrical Systems	Emergency generator, 15 kW	S.F. Floor	.08	.08	
E. EQUIPMENT & FURNISHINGS						
1010	Commercial Equipment	N/A	—	—	—	
1020	Institutional Equipment	N/A	—	—	—	14.4 %
1030	Vehicular Equipment	Hoists	S.F. Floor	11.17	11.17	
1090	Other Equipment	N/A	—	—	—	
F. SPECIAL CONSTRUCTION						
1020	Integrated Construction	N/A	—	—	—	0.0 %
1040	Special Facilities	N/A	—	—	—	
G. BUILDING SITEWORK		**N/A**				
				Sub-Total	77.75	**100%**
	CONTRACTOR FEES (General Requirements: 10%, Overhead: 5%, Profit: 10%)			25%	19.42	
	ARCHITECT FEES			8%	7.78	
				Total Building Cost	**104.95**	

COMMERCIAL/INDUSTRIAL/ INSTITUTIONAL | **M.390** | **Library**

Costs per square foot of floor area

Exterior Wall	S.F. Area	7000	10000	13000	16000	19000	22000	25000	28000	31000
	L.F. Perimeter	240	300	336	386	411	435	472	510	524
Face Brick with Concrete Block Back-up	R/Conc. Frame	165.70	157.35	150.25	147.10	143.00	139.90	138.40	137.20	135.20
	Steel Frame	166.30	157.95	150.85	147.70	143.60	140.55	139.00	137.85	135.80
Limestone with Concrete Block	R/Conc. Frame	210.15	194.80	181.75	175.90	168.45	162.90	160.05	157.90	154.20
	Steel Frame	206.10	192.85	180.90	175.75	168.70	163.50	160.95	159.00	155.45
Precast Concrete Panels	R/Conc. Frame	182.05	170.25	160.60	156.15	150.75	146.75	144.60	143.00	140.35
	Steel Frame	182.65	170.85	161.20	156.80	151.35	147.35	145.20	143.60	141.00
Perimeter Adj., Add or Deduct	Per 100 L.F.	20.45	14.35	11.00	8.95	7.55	6.60	5.75	5.10	4.60
Story Hgt. Adj., Add or Deduct	Per 1 Ft.	3.10	2.75	2.35	2.20	1.95	1.80	1.75	1.70	1.45
For Basement, add $38.40 per square foot of basement area										

The above costs were calculated using the basic specifications shown on the facing page. These costs should be adjusted where necessary for design alternatives and owner's requirements. Reported completed project costs, for this type of structure, range from $96.15 to $246.70 per S.F.

Common additives

Description	Unit	$ Cost
Carrels Hardwood	Each	665 - 1500
Closed Circuit Surveillance, One station		
Camera and monitor	Each	1825
For additional camera stations, add	Each	995
Elevators, Hydraulic passenger, 2 stops		
1500# capacity	Each	56,700
2500# capacity	Each	59,900
3500# capacity	Each	62,900
Emergency Lighting, 25 watt, battery operated		
Lead battery	Each	281
Nickel cadmium	Each	765
Flagpoles, Complete		
Aluminum, 20' high	Each	1500
40' high	Each	3450
70' high	Each	8550
Fiberglass, 23' high	Each	1725
39'-5" high	Each	3325
59' high	Each	6425

Description	Unit	$ Cost
Library Furnishings		
Bookshelf, 90" high, 10" shelf double face	L.F.	169
single face	L.F.	162
Charging desk, built-in with counter		
Plastic laminated top	L.F.	520
Reading table, laminated		
top 60" x 36"	Each	790

Model costs calculated for a 2 story building with 14' story height and 22,000 square feet of floor area

Library

				Unit	Unit Cost	Cost Per S.F.	% Of Sub-Total
A. SUBSTRUCTURE							
1010	Standard Foundations	Poured concrete; strip and spread footings		S.F. Ground	3.26	1.63	5.1%
1020	Special Foundations	N/A		–	–	–	
1030	Slab on Grade	4" reinforced concrete with vapor barrier and granular base		S.F. Slab	4.52	2.27	
2010	Basement Excavation	Site preparation for slab and trench for foundation wall and footing		S.F. Ground	.24	.12	
2020	Basement Walls	4' foundation wall		L.F. Wall	66	1.29	
B. SHELL							
	B10 Superstructure						
1010	Floor Construction	Concrete waffle slab		S.F. Floor	22	11.03	18.3%
1020	Roof Construction	Concrete waffle slab		S.F. Roof	15.96	7.98	
	B20 Exterior Enclosure						
2010	Exterior Walls	Face brick with concrete block backup	*90% of wall*	S.F. Wall	29	14.20	16.6%
2020	Exterior Windows	Window wall	*10% of wall*	Each	47	2.58	
2030	Exterior Doors	Double aluminum and glass, single leaf hollow metal		Each	5100	.47	
	B30 Roofing						
3010	Roof Coverings	Single ply membrane, EPDM, fully adhered; perlite/EPS composite insulation		S.F. Roof	4.72	2.36	2.3%
3020	Roof Openings	Roof hatches		S.F. Roof	.08	.04	
C. INTERIORS							
1010	Partitions	Gypsum board on metal studs	*30 S.F. Floor/L.F. Partition*	S.F. Partition	9.65	3.86	16.1%
1020	Interior Doors	Single leaf wood	*300 S.F. Floor/Door*	Each	563	1.88	
1030	Fittings	N/A		–	–	–	
2010	Stair Construction	Concrete filled metal pan		Flight	7800	.71	
3010	Wall Finishes	Paint		S.F. Surface	.69	.55	
3020	Floor Finishes	50% carpet, 50% vinyl tile		S.F. Floor	3.72	3.72	
3030	Ceiling Finishes	Mineral fiber on concealed zee bars		S.F. Ceiling	5.98	5.98	
D. SERVICES							
	D10 Conveying						
1010	Elevators & Lifts	One hydraulic passenger elevator		Each	76,560	3.48	3.4%
1020	Escalators & Moving Walks	N/A		–	–	–	
	D20 Plumbing						
2010	Plumbing Fixtures	Toilet and service fixtures, supply and drainage	*1 Fixture/1835 S.F. Floor*	Each	6202	3.38	4.8%
2020	Domestic Water Distribution	Gas fired water heater		S.F. Floor	1.11	1.11	
2040	Rain Water Drainage	Roof drains		S.F. Roof	.98	.49	
	D30 HVAC						
3010	Energy Supply	N/A		–	–	–	17.5 %
3020	Heat Generating Systems	Included in D3050		–	–	–	
3030	Cooling Generating Systems	N/A		–	–	–	
3050	Terminal & Package Units	Multizone unit, gas heating, electric cooling		S.F. Floor	18.10	18.10	
3090	Other HVAC Sys. & Equipment	N/A		–	–	–	
	D40 Fire Protection						
4010	Sprinklers	Wet pipe sprinkler system		S.F. Floor	2.64	2.64	3.3%
4020	Standpipes	Standpipe		S.F. Floor	.73	.73	
	D50 Electrical						
5010	Electrical Service/Distribution	400 ampere service, panel board and feeders		S.F. Floor	1.03	1.03	12.6%
5020	Lighting & Branch Wiring	High efficiency fluorescent fixtures, receptacles, switches, A.C. and misc. power		S.F. Floor	9.93	9.93	
5030	Communications & Security	Addressable alarm systems, internet wiring, and emergency lighting		S.F. Floor	1.86	1.86	
5090	Other Electrical Systems	Emergency generator, 7.5 kW, Uninterruptible power supply		S.F. Floor	.24	.24	
E. EQUIPMENT & FURNISHINGS							
1010	Commercial Equipment	N/A		–	–	–	0.0 %
1020	Institutional Equipment	N/A		–	–	–	
1030	Vehicular Equipment	N/A		–	–	–	
1090	Other Equipment	N/A		–	–	–	
F. SPECIAL CONSTRUCTION							
1020	Integrated Construction	N/A		–	–	–	0.0 %
1040	Special Facilities	N/A		–	–	–	
G. BUILDING SITEWORK		**N/A**					
					Sub-Total	103.66	**100%**
	CONTRACTOR FEES (General Requirements: 10%, Overhead: 5%, Profit: 10%)				*25%*	25.87	
	ARCHITECT FEES				*8%*	10.37	
					Total Building Cost	**139.90**	

COMMERCIAL/INDUSTRIAL/ INSTITUTIONAL | M.400 | Medical Office, 1 Story

Costs per square foot of floor area

Exterior Wall	S.F. Area	4000	5500	7000	8500	10000	11500	13000	14500	16000
	L.F. Perimeter	280	320	380	440	453	503	510	522	560
Face Brick with Concrete Block Back-up	Steel Joists	190.85	182.85	180.05	178.20	174.15	173.00	170.20	168.20	167.50
	Wood Truss	197.75	189.75	186.90	185.05	181.05	179.85	177.05	175.00	174.30
Stucco on Concrete Block	Steel Joists	180.05	173.95	171.70	170.25	167.20	166.25	164.15	162.65	162.10
	Wood Truss	186.95	180.80	178.55	177.10	174.05	173.15	171.00	169.50	169.00
Brick Veneer	Wood Truss	190.65	183.50	180.95	179.25	175.65	174.60	172.10	170.35	169.65
Wood Siding	Wood Frame	184.05	178.35	176.30	174.95	172.15	171.35	169.40	168.00	167.55
Perimeter Adj., Add or Deduct	Per 100 L.F.	14.95	10.85	8.50	7.00	5.95	5.20	4.60	4.10	3.70
Story Hgt. Adj., Add or Deduct	Per 1 Ft.	3.10	2.65	2.40	2.25	2.00	1.95	1.70	1.60	1.55
***For Basement**, add $25.00 per square foot of basement area*										

The above costs were calculated using the basic specifications shown on the facing page. These costs should be adjusted where necessary for design alternatives and owner's requirements. Reported completed project costs, for this type of structure, range from $81.00 to $208.40 per S.F.

Common additives

Description	Unit	$ Cost
Cabinets, Hospital, base		
Laminated plastic	L.F.	360
Stainless steel	L.F.	590
Counter top, laminated plastic	L.F.	69.50
Stainless steel	L.F.	174
For drop-in sink, add	Each	995
Nurses station, door type		
Laminated plastic	L.F.	405
Enameled steel	L.F.	390
Stainless steel	L.F.	690
Wall cabinets, laminated plastic	L.F.	263
Enameled steel	L.F.	310
Stainless steel	L.F.	555

Description	Unit	$ Cost
Directory Boards, Plastic, glass covered		
30" x 20"	Each	585
36" x 48"	Each	1275
Aluminum, 24" x 18"	Each	555
36" x 24"	Each	645
48" x 32"	Each	935
48" x 60"	Each	1900
Heat Therapy Unit		
Humidified, 26" x 78" x 28"	Each	3725
Smoke Detectors		
Ceiling type	Each	223
Duct type	Each	510
Tables, Examining, vinyl top		
with base cabinets	Each	1725 - 4700
Utensil Washer, Sanitizer	Each	12,300
X-Ray, Mobile	Each	14,700 - 83,000

Model costs calculated for a 1 story building with 10' story height and 7,000 square feet of floor area

Medical Office, 1 Story

				Unit	Unit Cost	Cost Per S.F.	% Of Sub-Total
A. SUBSTRUCTURE							
1010	Standard Foundations	Poured concrete; strip and spread footings		S.F. Ground	1.78	1.78	7.4%
1020	Special Foundations	N/A		—	—	—	
1030	Slab on Grade	4" reinforced concrete with vapor barrier and granular base		S.F. Slab	4.52	4.52	
2010	Basement Excavation	Site preparation for slab and trench for foundation wall and footing		S.F. Ground	.24	.24	
2020	Basement Walls	4' foundation wall		L.F. Wall	66	3.56	
B. SHELL							
	B10 Superstructure						
1010	Floor Construction	N/A		—	—	—	4.5 %
1020	Roof Construction	Plywood on wood trusses		S.F. Roof	6.23	6.23	
	B20 Exterior Enclosure						
2010	Exterior Walls	Face brick with concrete block backup	*70% of wall*	S.F. Wall	29	10.83	13.7%
2020	Exterior Windows	Wood double hung	*30% of wall*	Each	565	5.42	
2030	Exterior Doors	Aluminum and glass doors and entrance with transoms		Each	1499	2.57	
	B30 Roofing						
3010	Roof Coverings	Asphalt shingles with flashing (Pitched); rigid fiber glass insulation, gutters		S.F. Roof	3.82	3.82	2.8%
3020	Roof Openings	N/A		—	—	—	
C. INTERIORS							
1010	Partitions	Gypsum bd. & sound deadening bd. on wood studs w/insul.	*6 S.F. Floor/L.F. Partition*	S.F. Partition	7.68	10.24	25.2%
1020	Interior Doors	Single leaf wood	*60 S.F. Floor/Door*	Each	563	9.38	
1030	Fittings	N/A		—	—	—	
2010	Stair Construction	N/A		—	—	—	
3010	Wall Finishes	50% paint, 50% vinyl wall covering		S.F. Surface	1.20	3.19	
3020	Floor Finishes	50% carpet, 50% vinyl composition tile		S.F. Floor	6.59	6.59	
3030	Ceiling Finishes	Mineral fiber tile on concealed zee bars		S.F. Ceiling	5.13	5.13	
D. SERVICES							
	D10 Conveying						
1010	Elevators & Lifts	N/A		—	—	—	0.0 %
1020	Escalators & Moving Walks	N/A		—	—	—	
	D20 Plumbing						
2010	Plumbing Fixtures	Toilet, exam room and service fixtures, supply and drainage	*1 Fixture/195 S.F. Floor*	Each	3506	17.98	15.8%
2020	Domestic Water Distribution	Gas fired water heater		S.F. Floor	2.86	2.86	
2040	Rain Water Drainage	Roof drains		S.F. Roof	.84	.84	
	D30 HVAC						
3010	Energy Supply	N/A		—	—	—	9.5 %
3020	Heat Generating Systems	Included in D3050		—	—	—	
3030	Cooling Generating Systems	N/A		—	—	—	
3050	Terminal & Package Units	Multizone unit, gas heating, electric cooling		S.F. Floor	13.05	13.05	
3090	Other HVAC Sys. & Equipment	N/A		—	—	—	
	D40 Fire Protection						
4010	Sprinklers	Wet pipe sprinkler system		S.F. Floor	3.97	3.97	3.6%
4020	Standpipes	Standpipe		S.F. Floor	.97	.97	
	D50 Electrical						
5010	Electrical Service/Distribution	200 ampere service, panel board and feeders		S.F. Floor	1.37	1.37	11.5%
5020	Lighting & Branch Wiring	High efficiency fluorescent fixtures, receptacles, switches, A.C. and misc. power		S.F. Floor	6.58	6.58	
5030	Communications & Security	Alarm systems, internet and phone wiring, intercom system, and emergency lighting		S.F. Floor	6.94	6.94	
5090	Other Electrical Systems	Emergency generator, 7.5 kW		S.F. Floor	.85	.85	
E. EQUIPMENT & FURNISHINGS							
1010	Commercial Equipment	N/A		—	—	—	6.0 %
1020	Institutional Equipment	Exam room casework and countertops		S.F. Floor	8.26	8.26	
1030	Vehicular Equipment	N/A		—	—	—	
1090	Other Equipment	N/A		—	—	—	
F. SPECIAL CONSTRUCTION							
1020	Integrated Construction	N/A		—	—	—	0.0 %
1040	Special Facilities	N/A		—	—	—	
G. BUILDING SITEWORK	**N/A**						
					Sub-Total	137.17	**100%**
	CONTRACTOR FEES (General Requirements: 10%, Overhead: 5%, Profit: 10%)				25%	34.30	
	ARCHITECT FEES				9%	15.43	
					Total Building Cost	**186.90**	

COMMERCIAL/INDUSTRIAL/ INSTITUTIONAL	M.490	Police Station

Costs per square foot of floor area

Exterior Wall	S.F. Area	7000	9000	11000	13000	15000	17000	19000	21000	23000
	L.F. Perimeter	240	280	303	325	354	372	397	422	447
Limestone with Concrete Block Back-up	Bearing Walls	250.25	233.45	219.05	208.90	202.55	196.20	191.95	188.60	185.80
	R/Conc. Frame	260.20	244.05	230.40	220.75	214.75	208.75	204.75	201.65	198.90
Face Brick with Concrete Block Back-up	Bearing Walls	209.30	196.20	186.05	178.90	174.20	169.90	166.85	164.45	162.45
	R/Conc. Frame	227.05	213.95	203.80	196.65	191.95	187.60	184.60	182.20	180.15
Decorative Concrete Block	Bearing Walls	199.65	187.45	178.30	171.85	167.55	163.70	161.00	158.80	156.95
	R/Conc. Frame	218.30	206.05	196.95	190.50	186.20	182.30	179.60	177.40	175.55
Perimeter Adj., Add or Deduct	Per 100 L.F.	34.85	27.15	22.20	18.70	16.25	14.25	12.80	11.60	10.60
Story Hgt. Adj., Add or Deduct	Per 1 Ft.	6.35	5.75	5.10	4.55	4.35	4.00	3.85	3.75	3.60
***For Basement**, add $ 23.20 per square foot of basement area*										

The above costs were calculated using the basic specifications shown on the facing page. These costs should be adjusted where necessary for design alternatives and owner's requirements. Reported completed project costs, for this type of structure, range from $ 110.00 to $ 286.95 per S.F.

Common additives

Description	Unit	$ Cost
Cells Prefabricated, 5'-6' wide, 7'-8' high, 7'-8' deep	Each	12,200
Elevators, Hydraulic passenger, 2 stops		
1500# capacity	Each	56,700
2500# capacity	Each	59,900
3500# capacity	Each	62,900
Emergency Lighting, 25 watt, battery operated		
Lead battery	Each	281
Nickel cadmium	Each	765
Flagpoles, Complete		
Aluminum, 20' high	Each	1500
40' high	Each	3450
70' high	Each	8550
Fiberglass, 23' high	Each	1725
39'-5" high	Each	3325
59' high	Each	6425

Description	Unit	$ Cost
Lockers, Steel, Single tier, 60" to 72"	Opening	189 - 315
2 tier, 60" or 72" total	Opening	118 - 139
5 tier, box lockers	Opening	65 - 77.50
Locker bench, lam. maple top only	L.F.	21.50
Pedestals, steel pipe	Each	64
Safe, Office type, 1 hour rating		
30" x 18" x 18"	Each	2400
60" x 36" x 18", double door	Each	8975
Shooting Range, Incl. bullet traps, target provisions, and contols, not incl. structural shell	Each	43,000
Smoke Detectors		
Ceiling type	Each	223
Duct type	Each	510
Sound System		
Amplifier, 250 watts	Each	1800
Speaker, ceiling or wall	Each	189
Trumpet	Each	360

Model costs calculated for a 2 story building with 12′ story height and 11,000 square feet of floor area

Police Station

				Unit	Unit Cost	Cost Per S.F.	% Of Sub-Total
A. SUBSTRUCTURE							
1010	Standard Foundations	Poured concrete; strip and spread footings		S.F. Ground	2.42	1.21	
1020	Special Foundations	N/A		–	–	–	
1030	Slab on Grade	4″ reinforced concrete with vapor barrier and granular base		S.F. Slab	4.52	2.27	3.7%
2010	Basement Excavation	Site preparation for slab and trench for foundation wall and footing		S.F. Ground	.24	.12	
2020	Basement Walls	4′ foundation wall		L.F. Wall	66	2.40	
B. SHELL							
	B10 Superstructure						
1010	Floor Construction	Open web steel joists, slab form, concrete		S.F. Floor	9.62	4.81	4.1%
1020	Roof Construction	Metal deck on open web steel joists		S.F. Roof	3.70	1.85	
	B20 Exterior Enclosure						
2010	Exterior Walls	Limestone with concrete block backup	*80% of wall*	S.F. Wall	64	34.03	
2020	Exterior Windows	Metal horizontal sliding	*20% of wall*	Each	1048	9.24	28.0%
2030	Exterior Doors	Hollow metal		Each	2470	1.80	
	B30 Roofing						
3010	Roof Coverings	Built-up tar and gravel with flashing; perlite/EPS composite insulation		S.F. Roof	6.20	3.10	1.9%
3020	Roof Openings	N/A		–	–	–	
C. INTERIORS							
1010	Partitions	Concrete block	*20 S.F. Floor/L.F. Partition*	S.F. Partition	7.96	3.98	
1020	Interior Doors	Single leaf kalamein fire door	*200 S.F. Floor/Door*	Each	1022	5.12	
1030	Fittings	Toilet partitions		S.F. Floor	.73	.73	
2010	Stair Construction	Concrete filled metal pan		Flight	14,550	2.64	15.8%
3010	Wall Finishes	90% paint, 10% ceramic tile		S.F. Surface	3.08	3.08	
3020	Floor Finishes	70% vinyl composition tile, 20% carpet tile, 10% ceramic tile		S.F. Floor	3.80	3.80	
3030	Ceiling Finishes	Mineral fiber tile on concealed zee bars		S.F. Ceiling	5.98	5.98	
D. SERVICES							
	D10 Conveying						
1010	Elevators & Lifts	One hydraulic passenger elevator		Each	73,920	6.72	4.2%
1020	Escalators & Moving Walks	N/A		–	–	–	
	D20 Plumbing						
2010	Plumbing Fixtures	Toilet and service fixtures, supply and drainage	*1 Fixture/580 S.F. Floor*	Each	4025	6.94	
2020	Domestic Water Distribution	Oil fired water heater		S.F. Floor	2.81	2.81	7.1%
2040	Rain Water Drainage	Roof drains		S.F. Roof	3.42	1.71	
	D30 HVAC						
3010	Energy Supply	N/A		–	–	–	
3020	Heat Generating Systems	N/A		–	–	–	
3030	Cooling Generating Systems	N/A		–	–	–	11.4 %
3050	Terminal & Package Units	Multizone HVAC air cooled system		S.F. Floor	18.35	18.35	
3090	Other HVAC Sys. & Equipment	N/A		–	–	–	
	D40 Fire Protection						
4010	Sprinklers	Wet pipe sprinkler system		S.F. Floor	3.18	3.18	2.5%
4020	Standpipes	Standpipe		S.F. Floor	.81	.81	
	D50 Electrical						
5010	Electrical Service/Distribution	400 ampere service, panel board and feeders		S.F. Floor	1.75	1.75	
5020	Lighting & Branch Wiring	T-8 fluorescent fixtures, receptacles, switches, A.C. and misc. power		S.F. Floor	10.63	10.63	11.8%
5030	Communications & Security	Addressable alarm systems, internet wiring, intercom and emergency lighting		S.F. Floor	6.44	6.44	
5090	Other Electrical Systems	Emergency generator, 15 kW		S.F. Floor	.22	.22	
E. EQUIPMENT & FURNISHINGS							
1010	Commercial Equipment	N/A		–	–	–	
1020	Institutional Equipment	Lockers, detention rooms, cells, gasoline dispensers		S.F. Floor	12.69	12.69	9.4 %
1030	Vehicular Equipment	Gasoline dispenser system		S.F. Floor	2.36	2.36	
1090	Other Equipment	N/A		–	–	–	
F. SPECIAL CONSTRUCTION							
1020	Integrated Construction	N/A		–	–	–	0.0 %
1040	Special Facilities	N/A		–	–	–	
G. BUILDING SITEWORK	**N/A**						
					Sub-Total	160.77	**100%**
	CONTRACTOR FEES (General Requirements: 10%, Overhead: 5%, Profit: 10%)				*25%*	40.19	
	ARCHITECT FEES				*9%*	18.09	
					Total Building Cost	**219.05**	

COMMERCIAL/INDUSTRIAL/ INSTITUTIONAL	M.540	Restaurant, Fast Food

Costs per square foot of floor area

Exterior Wall	S.F. Area	2000	2800	3500	4000	5000	5800	6500	7200	8000
	L.F. Perimeter	180	212	240	260	300	314	336	344	368
Face Brick with Concrete Block Back-up	Bearing Walls	205.05	196.00	191.55	189.25	186.15	182.40	180.90	178.45	177.30
	Steel Frame	207.70	198.65	194.20	191.90	188.80	185.05	183.50	181.10	179.95
Concrete Block With Stucco	Bearing Walls	195.15	187.65	183.90	182.00	179.35	176.25	174.95	172.85	171.95
	Steel Frame	198.50	191.15	187.45	185.60	183.00	180.00	178.75	176.70	175.80
Wood Siding	Wood Frame	190.75	184.05	180.70	179.00	176.65	173.85	172.75	170.90	170.10
Brick Veneer	Steel Frame	203.10	194.95	190.90	188.90	186.00	182.70	181.35	179.10	178.10
Perimeter Adj., Add or Deduct	Per 100 L.F.	31.55	22.60	18.00	15.80	12.55	10.90	9.65	8.70	7.90
Story Hgt. Adj., Add or Deduct	Per 1 Ft.	4.15	3.55	3.15	3.05	2.75	2.55	2.35	2.20	2.15
Basement—Not Applicable										

The above costs were calculated using the basic specifications shown on the facing page. These costs should be adjusted where necessary for design alternatives and owner's requirements. Reported completed project costs, for this type of structure, range from $ 116.80 to $ 227.70 per S.F.

Common additives

Description	Unit	$ Cost
Bar, Front Bar	L.F.	365
Back bar	L.F.	290
Booth, Upholstered, custom straight	L.F.	209 - 385
"L" or "U" shaped	L.F.	216 - 365
Drive-up Window	Each	7475 - 12,000
Emergency Lighting, 25 watt, battery operated		
Lead battery	Each	281
Nickel cadmium	Each	765
Kitchen Equipment		
Broiler	Each	4025
Coffee urn, twin 6 gallon	Each	3250
Cooler, 6 ft. long	Each	5225
Dishwasher, 10-12 racks per hr.	Each	5025
Food warmer, counter, 1.2 KW	Each	530
Freezer, 44 C.F., reach-in	Each	3200
Ice cube maker, 50 lb. per day	Each	1775
Range with 1 oven	Each	3225

Description	Unit	$ Cost
Refrigerators, Prefabricated, walk-in		
7'-6" High, 6' x 6'	S.F.	170
10' x 10'	S.F.	135
12' x 14'	S.F.	119
12' x 20'	S.F.	105
Serving		
Counter top (Stainless steel)	L.F.	174
Base cabinets	L.F.	360 - 590
Sound System		
Amplifier, 250 watts	Each	1800
Speaker, ceiling or wall	Each	189
Trumpet	Each	360
Storage		
Shelving	S.F.	8.85
Washing		
Stainless steel counter	L.F.	174

Model costs calculated for a 1 story building with 10' story height and 4,000 square feet of floor area

Restaurant, Fast Food

			Unit	Unit Cost	Cost Per S.F.	% Of Sub-Total
A. SUBSTRUCTURE						
1010	Standard Foundations	Poured concrete; strip and spread footings	S.F. Ground	2.14	2.14	8.1%
1020	Special Foundations	N/A	—	—	—	
1030	Slab on Grade	4" reinforced concrete with vapor barrier and granular base	S.F. Slab	4.52	4.52	
2010	Basement Excavation	Site preparation for slab and trench for foundation wall and footing	S.F. Ground	.41	.41	
2020	Basement Walls	4' foundation wall	L.F. Wall	66	4.26	
B. SHELL						
	B10 Superstructure					
1010	Floor Construction	N/A	—	—	—	4.4 %
1020	Roof Construction	Metal deck on open web steel joists	S.F. Roof	6.14	6.14	
	B20 Exterior Enclosure					
2010	Exterior Walls	Face brick with concrete block backup *70% of wall*	S.F. Wall	27	12.47	20.1%
2020	Exterior Windows	Window wall *30% of wall*	Each	44	8.59	
2030	Exterior Doors	Aluminum and glass	Each	3527	7.06	
	B30 Roofing					
3010	Roof Coverings	Built-up tar and gravel with flashing; perlite/EPS composite insulation	S.F. Roof	5.86	5.86	4.5%
3020	Roof Openings	Roof hatches	S.F. Roof	.45	.45	
C. INTERIORS						
1010	Partitions	Gypsum board on metal studs *25 S.F. Floor/L.F. Partition*	S.F. Partition	5.06	1.82	16.6%
1020	Interior Doors	Hollow core wood *1000 S.F. Floor/Door*	Each	486	.49	
1030	Fittings	N/A	—	—	—	
2010	Stair Construction	N/A	—	—	—	
3010	Wall Finishes	Paint	S.F. Surface	2.35	1.69	
3020	Floor Finishes	Quarry tile	S.F. Floor	13.27	13.27	
3030	Ceiling Finishes	Mineral fiber tile on concealed zee bars	S.F. Ceiling	5.98	5.98	
D. SERVICES						
	D10 Conveying					
1010	Elevators & Lifts	N/A	—	—	—	0.0 %
1020	Escalators & Moving Walks	N/A	—	—	—	
	D20 Plumbing					
2010	Plumbing Fixtures	Kitchen, bathroom and service fixtures, supply and drainage *1 Fixture/400 S.F. Floor*	Each	3552	8.88	11.2%
2020	Domestic Water Distribution	Gas fired water heater	S.F. Floor	5.15	5.15	
2040	Rain Water Drainage	Roof drains	S.F. Roof	1.69	1.69	
	D30 HVAC					
3010	Energy Supply	N/A	—	—	—	11.9 %
3020	Heat Generating Systems	Included in D3050	—	—	—	
3030	Cooling Generating Systems	N/A	—	—	—	
3050	Terminal & Package Units	Multizone unit, gas heating, electric cooling; kitchen ventilation	S.F. Floor	16.72	16.72	
3090	Other HVAC Sys. & Equipment	N/A	—	—	—	
	D40 Fire Protection					
4010	Sprinklers	Sprinklers, light hazard	S.F. Floor	7.03	7.03	6.2%
4020	Standpipes	Standpipe	S.F. Floor	1.68	1.68	
	D50 Electrical					
5010	Electrical Service/Distribution	400 ampere service, panel board and feeders	S.F. Floor	4.17	4.17	12.6%
5020	Lighting & Branch Wiring	T-8 fluorescent fixtures, receptacles, switches, A.C. and misc. power	S.F. Floor	9.05	9.05	
5030	Communications & Security	Addressable alarm systems and emergency lighting	S.F. Floor	4	4	
5090	Other Electrical Systems	Emergency generator, 7.5 kW	S.F. Floor	.38	.38	
E. EQUIPMENT & FURNISHINGS						
1010	Commercial Equipment	N/A	—	—	—	4.5 %
1020	Institutional Equipment	N/A	—	—	—	
1030	Vehicular Equipment	N/A	—	—	—	
1090	Other Equipment	Walk-in refrigerator	S.F. Floor	6.31	6.31	
F. SPECIAL CONSTRUCTION						
1020	Integrated Construction	N/A	—	—	—	0.0 %
1040	Special Facilities	N/A	—	—	—	
G. BUILDING SITEWORK		**N/A**				
				Sub-Total	140.21	**100%**
	CONTRACTOR FEES (General Requirements: 10%, Overhead: 5%, Profit: 10%)			*25%*	35.02	
	ARCHITECT FEES			*8%*	14.02	
				Total Building Cost	**189.25**	

COMMERCIAL/INDUSTRIAL/ INSTITUTIONAL | M.600 | Store, Convenience

Costs per square foot of floor area

Exterior Wall	S.F. Area	1000	2000	3000	4000	6000	8000	10000	12000	15000
	L.F. Perimeter	126	179	219	253	310	358	400	438	490
Wood Siding	Wood Frame	133.15	113.10	105.00	100.50	95.40	92.50	90.55	89.15	87.65
Face Brick Veneer	Wood Frame	156.70	129.25	117.75	111.25	103.80	99.45	96.55	94.50	92.15
Stucco on Concrete Block	Steel Frame	150.80	126.15	115.85	110.15	103.55	99.70	97.15	95.35	93.30
	Bearing Walls	147.25	122.60	112.40	106.60	100.05	96.15	93.70	91.85	89.85
Metal Sandwich Panel	Steel Frame	164.30	135.85	123.90	117.15	109.40	104.85	101.85	99.70	97.25
Precast Concrete	Steel Frame	204.60	164.10	146.70	136.70	125.15	118.35	113.75	110.40	106.70
Perimeter Adj., Add or Deduct	Per 100 L.F.	32.10	16.05	10.70	8.05	5.40	4.00	3.25	2.65	2.05
Story Hgt. Adj., Add or Deduct	Per 1 Ft.	2.30	1.60	1.35	1.15	0.95	0.75	0.70	0.65	0.60
***For Basement**, add $21.00 per square foot of basement area*										

The above costs were calculated using the basic specifications shown on the facing page. These costs should be adjusted where necessary for design alternatives and owner's requirements. Reported completed project costs, for this type of structure, range from $65.85 to $228.20 per S.F.

Common additives

Description	Unit	$ Cost
Check Out Counter		
Single belt	Each	3275
Double belt	Each	4675
Emergency Lighting, 25 watt, battery operated		
Lead battery	Each	281
Nickel cadmium	Each	765
Refrigerators, Prefabricated, walk-in		
7'-6" high, 6' x 6'	S.F.	170
10' x 10'	S.F.	135
12' x 14'	S.F.	119
12' x 20'	S.F.	105
Refrigerated Food Cases		
Dairy, multi deck, 12' long	Each	12,100
Delicatessen case, single deck, 12' long	Each	8200
Multi deck, 18 S.F. shelf display	Each	7400
Freezer, self-contained chest type, 30 C.F.	Each	8400
Glass door upright, 78 C.F.	Each	11,300

Description	Unit	$ Cost
Refrigerated Food Cases, cont.		
Frozen food, chest type, 12' long	Each	8250
Glass door reach-in, 5 door	Each	15,600
Island case 12' long, single deck	Each	9300
Multi deck	Each	19,400
Meat cases, 12' long, single deck	Each	6825
Multi deck	Each	11,600
Produce, 12' long single deck	Each	8975
Multi deck	Each	9775
Safe, Office type, 1 hour rating		
30" x 18" x 18"	Each	2400
60" x 36" x 18", double door	Each	8975
Smoke Detectors		
Ceiling type	Each	223
Duct type	Each	510
Sound System		
Amplifier, 250 watts	Each	1800
Speaker, ceiling or wall	Each	189
Trumpet	Each	360

Model costs calculated for a 1 story building with 12′ story height and 4,000 square feet of floor area

Store, Convenience

				Unit	Unit Cost	Cost Per S.F.	% Of Sub-Total
A. SUBSTRUCTURE							
1010	Standard Foundations	Poured concrete; strip and spread footings		S.F. Ground	1.38	1.38	13.2%
1020	Special Foundations	N/A		–	–	–	
1030	Slab on Grade	4″ reinforced concrete with vapor barrier and granular base		S.F. Slab	4.52	4.52	
2010	Basement Excavation	Site preparation for slab and trench for foundation wall and footing		S.F. Ground	.41	.41	
2020	Basement Walls	4′ foundation wall		L.F. Wall	57	3.61	
B. SHELL							
	B10 Superstructure						
1010	Floor Construction	N/A		–	–	–	8.3 %
1020	Roof Construction	Wood truss with plywood sheathing		S.F. Roof	6.23	6.23	
	B20 Exterior Enclosure						
2010	Exterior Walls	Wood siding on wood studs, insulated	*80% of wall*	S.F. Wall	8.65	5.25	15.1%
2020	Exterior Windows	Storefront	*20% of wall*	Each	39	3.99	
2030	Exterior Doors	Double aluminum and glass, solid core wood		Each	2766	2.08	
	B30 Roofing						
3010	Roof Coverings	Asphalt shingles; rigid fiberglass insulation		S.F. Roof	4.58	4.58	6.1%
3020	Roof Openings	N/A		–	–	–	
C. INTERIORS							
1010	Partitions	Gypsum board on wood studs	*60 S.F. Floor/L.F. Partition*	S.F. Partition	10.86	1.81	14.4%
1020	Interior Doors	Single leaf wood, hollow metal	*1300 S.F. Floor/Door*	Each	1140	.87	
1030	Fittings	N/A		–	–	–	
2010	Stair Construction	N/A		–	–	–	
3010	Wall Finishes	Paint		S.F. Surface	.96	.32	
3020	Floor Finishes	Vinyl composition tile		S.F. Floor	2.71	2.71	
3030	Ceiling Finishes	Mineral fiber tile on wood furring		S.F. Ceiling	5.13	5.13	
D. SERVICES							
	D10 Conveying						
1010	Elevators & Lifts	N/A		–	–	–	0.0 %
1020	Escalators & Moving Walks	N/A		–	–	–	
	D20 Plumbing						
2010	Plumbing Fixtures	Toilet and service fixtures, supply and drainage	*1 Fixture/1000 S.F. Floor*	Each	3350	3.35	6.0%
2020	Domestic Water Distribution	Gas fired water heater		S.F. Floor	1.15	1.15	
2040	Rain Water Drainage	N/A		–	–	–	
	D30 HVAC						
3010	Energy Supply	N/A		–	–	–	9.6 %
3020	Heat Generating Systems	Included in D3050		–	–	–	
3030	Cooling Generating Systems	N/A		–	–	–	
3050	Terminal & Package Units	Single zone rooftop unit, gas heating, electric cooling		S.F. Floor	7.21	7.21	
3090	Other HVAC Sys. & Equipment	N/A		–	–	–	
	D40 Fire Protection						
4010	Sprinklers	Sprinkler, ordinary hazard		S.F. Floor	3.97	3.97	7.5%
4020	Standpipes	Standpipe		S.F. Floor	1.68	1.68	
	D50 Electrical						
5010	Electrical Service/Distribution	200 ampere service, panel board and feeders		S.F. Floor	2.56	2.56	19.8%
5020	Lighting & Branch Wiring	High efficiency fluorescent fixtures, receptacles, switches, A.C. and misc. power		S.F. Floor	9.42	9.42	
5030	Communications & Security	Addressable alarm systems and emergency lighting		S.F. Floor	2.55	2.55	
5090	Other Electrical Systems	Emergency generator, 7.5 kW		S.F. Floor	.35	.35	
E. EQUIPMENT & FURNISHINGS							
1010	Commercial Equipment	N/A		–	–	–	0.0 %
1020	Institutional Equipment	N/A		–	–	–	
1030	Vehicular Equipment	N/A		–	–	–	
1090	Other Equipment	N/A		–	–	–	
F. SPECIAL CONSTRUCTION							
1020	Integrated Construction	N/A		–	–	–	0.0 %
1040	Special Facilities	N/A		–	–	–	
G. BUILDING SITEWORK		**N/A**					
					Sub-Total	75.13	**100%**
	CONTRACTOR FEES (General Requirements: 10%, Overhead: 5%, Profit: 10%)				*25%*	18.80	
	ARCHITECT FEES				*7%*	6.57	
					Total Building Cost	**100.50**	

COMMERCIAL/INDUSTRIAL/ INSTITUTIONAL	M.630	Store, Retail

Costs per square foot of floor area

Exterior Wall	S.F. Area	4000	6000	8000	10000	12000	15000	18000	20000	22000
	L.F. Perimeter	260	340	360	410	440	490	540	565	594
Split Face Concrete Block	Steel Joists	127.40	116.25	106.50	102.35	98.60	95.15	92.80	91.40	90.30
Stucco on Concrete Block	Steel Joists	125.15	114.35	104.95	100.95	97.40	94.00	91.80	90.40	89.40
EIFS on Metal Studs	Steel Joists	127.90	116.70	106.95	102.70	98.95	95.45	93.05	91.65	90.60
Face Brick on Concrete Block	Steel Joists	142.35	129.35	116.90	111.80	107.10	102.70	99.70	97.85	96.55
Painted Reinforced Concrete	Steel Joists	132.65	120.85	110.15	105.65	101.60	97.75	95.20	93.65	92.50
Tilt-up Concrete Panels	Steel Joists	122.75	112.25	103.30	99.45	96.00	92.80	90.65	89.35	88.40
Perimeter Adj., Add or Deduct	Per 100 L.F.	13.90	9.30	6.95	5.60	4.65	3.75	3.05	2.75	2.55
Story Hgt. Adj., Add or Deduct	Per 1 Ft.	1.75	1.55	1.25	1.10	1.00	0.85	0.80	0.75	0.75
***For Basement**, add $29.10 per square foot of basement area*										

The above costs were calculated using the basic specifications shown on the facing page. These costs should be adjusted where necessary for design alternatives and owner's requirements. Reported completed project costs, for this type of structure, range from $57.25 to $199.15 per S.F.

Common additives

Description	Unit	$ Cost
Emergency Lighting, 25 watt, battery operated		
Lead battery	Each	281
Nickel cadmium	Each	765
Safe, Office type, 1 hour rating		
30" x 18" x 18"	Each	2400
60" x 36" x 18", double door	Each	8975
Smoke Detectors		
Ceiling type	Each	223
Duct type	Each	510
Sound System		
Amplifier, 250 watts	Each	1800
Speaker, ceiling or wall	Each	189
Trumpet	Each	360

Model costs calculated for a 1 story building with 14′ story height and 8,000 square feet of floor area

Store, Retail

				Unit	Unit Cost	Cost Per S.F.	% Of Sub-Total
A. SUBSTRUCTURE							
1010	Standard Foundations	Poured concrete; strip and spread footings		S.F. Ground	1.59	1.59	11.8%
1020	Special Foundations	N/A		–	–	–	
1030	Slab on Grade	4″ reinforced concrete with vapor barrier and granular base		S.F. Slab	4.52	4.52	
2010	Basement Excavation	Site preparation for slab and trench for foundation wall and footing		S.F. Ground	.24	.24	
2020	Basement Walls	4′ foundation wall		L.F. Wall	66	2.95	
B. SHELL							
	B10 Superstructure						
1010	Floor Construction	N/A		–	–	–	7.4 %
1020	Roof Construction	Metal deck, open web steel joists, beams, interior columns		S.F. Roof	5.87	5.87	
	B20 Exterior Enclosure						
2010	Exterior Walls	Decorative concrete block	*90% of wall*	S.F. Wall	14.94	8.47	14.7%
2020	Exterior Windows	Storefront windows	*10% of wall*	Each	42	2.65	
2030	Exterior Doors	Sliding entrance door and hollow metal service doors		Each	2042	.51	
	B30 Roofing						
3010	Roof Coverings	Built-up tar and gravel with flashing; perlite/EPS composite insulation		S.F. Roof	5.61	5.61	7.2%
3020	Roof Openings	Roof hatches		S.F. Roof	.10	.10	
C. INTERIORS							
1010	Partitions	Gypsum board on metal studs	*60 S.F. Floor/L.F. Partition*	S.F. Partition	5.10	.85	16.4%
1020	Interior Doors	Single leaf hollow metal	*600 S.F. Floor/Door*	Each	1022	1.71	
1030	Fittings	N/A		–	–	–	
2010	Stair Construction	N/A		–	–	–	
3010	Wall Finishes	Paint		S.F. Surface	5.16	1.72	
3020	Floor Finishes	Vinyl tile		S.F. Floor	2.71	2.71	
3030	Ceiling Finishes	Mineral fiber tile on concealed zee bars		S.F. Ceiling	5.98	5.98	
D. SERVICES							
	D10 Conveying						
1010	Elevators & Lifts	N/A		–	–	–	0.0 %
1020	Escalators & Moving Walks	N/A		–	–	–	
	D20 Plumbing						
2010	Plumbing Fixtures	Toilet and service fixtures, supply and drainage	*1 Fixture/890 S.F. Floor*	Each	2581	2.90	11.0%
2020	Domestic Water Distribution	Gas fired water heater		S.F. Floor	4.46	4.46	
2040	Rain Water Drainage	Roof drains		S.F. Roof	1.33	1.33	
	D30 HVAC						
3010	Energy Supply	N/A		–	–	–	8.5 %
3020	Heat Generating Systems	Included in D3050		–	–	–	
3030	Cooling Generating Systems	N/A		–	–	–	
3050	Terminal & Package Units	Single zone unit, gas heating, electric cooling		S.F. Floor	6.68	6.68	
3090	Other HVAC Sys. & Equipment	N/A		–	–	–	
	D40 Fire Protection						
4010	Sprinklers	Wet pipe sprinkler system		S.F. Floor	3.85	3.85	5.9%
4020	Standpipes	Standpipe, wet, Class III		S.F. Floor	.84	.84	
	D50 Electrical						
5010	Electrical Service/Distribution	400 ampere service, panel board and feeders		S.F. Floor	2.40	2.40	16.9%
5020	Lighting & Branch Wiring	High efficiency fluorescent fixtures, receptacles, switches, A.C. and misc. power		S.F. Floor	9.22	9.22	
5030	Communications & Security	Addressable alarm systems and emergency lighting		S.F. Floor	1.51	1.51	
5090	Other Electrical Systems	Emergency generator, 15 kW		S.F. Floor	.21	.21	
E. EQUIPMENT & FURNISHINGS							
1010	Commercial Equipment	N/A		–	–	–	0.0 %
1020	Institutional Equipment	N/A		–	–	–	
1030	Vehicular Equipment	N/A		–	–	–	
1090	Other Equipment	N/A		–	–	–	
F. SPECIAL CONSTRUCTION							
1020	Integrated Construction	N/A		–	–	–	0.0 %
1040	Special Facilities	N/A		–	–	–	
G. BUILDING SITEWORK		**N/A**					
					Sub-Total	78.88	**100%**
	CONTRACTOR FEES (General Requirements: 10%, Overhead: 5%, Profit: 10%)				*25%*	19.73	
	ARCHITECT FEES				*8%*	7.89	
					Total Building Cost	**106.50**	

COMMERCIAL/INDUSTRIAL/ INSTITUTIONAL	M.680	Town Hall, 2-3 Story

Costs per square foot of floor area

Exterior Wall	S.F. Area	8000	10000	12000	15000	18000	24000	28000	35000	40000
	L.F. Perimeter	206	233	260	300	320	360	393	451	493
Face Brick with Concrete Block Back-up	Steel Frame	184.50	174.85	168.35	161.80	155.60	147.85	144.90	141.40	139.65
	R/Conc. Frame	190.45	180.80	174.25	167.75	161.55	153.80	150.85	147.35	145.60
Stone with Concrete Block Back-up	Steel Frame	181.00	172.15	166.20	160.30	154.35	147.05	144.25	140.95	139.35
	R/Conc. Frame	198.30	187.90	180.85	173.85	167.00	158.35	155.10	151.20	149.35
Limestone with Concrete Block Back-up	Steel Frame	208.25	195.55	186.95	178.35	169.40	158.10	153.95	149.00	146.55
	R/Conc. Frame	214.20	201.45	192.90	184.30	175.35	164.05	159.90	154.95	152.45
Perimeter Adj., Add or Deduct	Per 100 L.F.	21.00	16.75	13.90	11.15	9.30	7.00	5.95	4.75	4.20
Story Hgt. Adj., Add or Deduct	Per 1 Ft.	3.20	2.90	2.65	2.55	2.25	1.85	1.75	1.55	1.55
***For Basement**, add $24.40 per square foot of basement area*										

The above costs were calculated using the basic specifications shown on the facing page. These costs should be adjusted where necessary for design alternatives and owner's requirements. Reported completed project costs, for this type of structure, range from $79.15 to $232.25 per S.F.

Common additives

Description	Unit	$ Cost
Directory Boards, Plastic, glass covered		
30" x 20"	Each	585
36" x 48"	Each	1275
Aluminum, 24" x 18"	Each	555
36" x 24"	Each	645
48" x 32"	Each	935
48" x 60"	Each	1900
Elevators, Hydraulic passenger, 2 stops		
1500# capacity	Each	56,700
2500# capacity	Each	59,900
3500# capacity	Each	62,900
Additional stop, add	Each	7550
Emergency Lighting, 25 watt, battery operated		
Lead battery	Each	281
Nickel cadmium	Each	765

Description	Unit	$ Cost
Flagpoles, Complete		
Aluminum, 20' high	Each	1500
40' high	Each	3450
70' high	Each	8550
Fiberglass, 23' high	Each	1725
39'-5" high	Each	3325
59' high	Each	6425
Safe, Office type, 1 hour rating		
30" x 18" x 18"	Each	2400
60" x 36" x 18", double door	Each	8975
Smoke Detectors		
Ceiling type	Each	223
Duct type	Each	510
Vault Front, Door & frame		
1 Hour test, 32" x 78"	Opening	3775
2 Hour test, 32" door	Opening	4100
40" door	Opening	4775
4 Hour test, 32" door	Opening	4500
40" door	Opening	5375
Time lock movement; two movement	Each	1925

Model costs calculated for a 3 story building with 12′ story height and 18,000 square feet of floor area

Town Hall, 2-3 Story

				Unit	Unit Cost	Cost Per S.F.	% Of Sub-Total
A. SUBSTRUCTURE							
1010	Standard Foundations	Poured concrete; strip and spread footings		S.F. Ground	3.45	1.15	3.5%
1020	Special Foundations	N/A		—	—	—	
1030	Slab on Grade	4″ reinforced concrete with vapor barrier and granular base		S.F. Slab	4.52	1.51	
2010	Basement Excavation	Site preparation for slab and trench for foundation wall and footing		S.F. Ground	.15	.05	
2020	Basement Walls	4′ foundation wall		L.F. Wall	61	1.30	
B. SHELL							
	B10 Superstructure						
1010	Floor Construction	Open web steel joists, slab form, concrete, wide flange steel columns		S.F. Floor	22	14.43	14.7%
1020	Roof Construction	Metal deck, open web steel joists, beams, interior columns		S.F. Roof	6.75	2.25	
	B20 Exterior Enclosure						
2010	Exterior Walls	Stone with concrete block backup	*70% of wall*	S.F. Wall	36	16.24	17.0%
2020	Exterior Windows	Metal outward projecting	*10% of wall*	Each	1025	2.37	
2030	Exterior Doors	Metal and glass with transoms		Each	2422	.67	
	B30 Roofing						
3010	Roof Coverings	Built-up tar and gravel with flashing; perlite/EPS composite insulation		S.F. Roof	5.52	1.84	1.6%
3020	Roof Openings	N/A		—	—	—	
C. INTERIORS							
1010	Partitions	Gypsum board on metal studs	*20 S.F. Floor/L.F. Partition*	S.F. Partition	8.36	4.18	23.4%
1020	Interior Doors	Wood solid core	*200 S.F. Floor/Door*	Each	563	2.82	
1030	Fittings	Toilet partitions		S.F. Floor	.29	.29	
2010	Stair Construction	Concrete filled metal pan		Flight	14,550	4.85	
3010	Wall Finishes	90% paint, 10% ceramic tile		S.F. Surface	1.23	1.23	
3020	Floor Finishes	70% carpet tile, 15% terrazzo, 15% vinyl composition tile		S.F. Floor	7.12	7.12	
3030	Ceiling Finishes	Mineral fiber tile on concealed zee bars		S.F. Ceiling	5.98	5.98	
D. SERVICES							
	D10 Conveying						
1010	Elevators & Lifts	Two hydraulic elevators		Each	108,720	12.08	10.7%
1020	Escalators & Moving Walks	N/A		—	—	—	
	D20 Plumbing						
2010	Plumbing Fixtures	Toilet and service fixtures, supply and drainage	*1 Fixture/1385 S.F. Floor*	Each	5388	3.89	4.2%
2020	Domestic Water Distribution	Gas fired water heater		S.F. Floor	.40	.40	
2040	Rain Water Drainage	Roof drains		S.F. Roof	1.41	.47	
	D30 HVAC						
3010	Energy Supply	N/A		—	—	—	6.4 %
3020	Heat Generating Systems	Included in D3050		—	—	—	
3030	Cooling Generating Systems	N/A		—	—	—	
3050	Terminal & Package Units	Multizone unit, gas heating, electric cooling		S.F. Floor	7.26	7.26	
3090	Other HVAC Sys. & Equipment	N/A		—	—	—	
	D40 Fire Protection						
4010	Sprinklers	Sprinklers, light hazard		S.F. Floor	2.77	2.77	4.7%
4020	Standpipes	Standpipe, wet, Class III		S.F. Floor	2.60	2.60	
	D50 Electrical						
5010	Electrical Service/Distribution	400 ampere service, panel board and feeders		S.F. Floor	1.50	1.50	13.7%
5020	Lighting & Branch Wiring	High efficiency fluorescent fixtures, receptacles, switches, A.C. and misc. power		S.F. Floor	10.87	10.87	
5030	Communications & Security	Addressable alarm systems, internet wiring, and emergency lighting		S.F. Floor	3.02	3.02	
5090	Other Electrical Systems	Emergency generator, 15 kW		S.F. Floor	.17	.17	
E. EQUIPMENT & FURNISHINGS							
1010	Commercial Equipment	N/A		—	—	—	0.0 %
1020	Institutional Equipment	N/A		—	—	—	
1030	Vehicular Equipment	N/A		—	—	—	
1090	Other Equipment	N/A		—	—	—	
F. SPECIAL CONSTRUCTION							
1020	Integrated Construction	N/A		—	—	—	0.0 %
1040	Special Facilities	N/A		—	—	—	
G. BUILDING SITEWORK		**N/A**					
					Sub-Total	113.31	**100%**
	CONTRACTOR FEES (General Requirements: 10%, Overhead: 5%, Profit: 10%)				25%	28.29	
	ARCHITECT FEES				9%	12.75	
					Total Building Cost	**154.35**	

Chapter 9

Project Costs, Order of Magnitude Estimates & RSMeans Indexes

The reference section in RSMeans books includes a great deal of information that is helpful in producing reliable approximate estimates.

Project Costs: The Square Foot Costs table in the Reference Section provides average costs for many types of structures. This data is derived from reports of the cost of actual buildings, by type, reported to RSMeans by its customers. This data is maintained in a database and updated annually to current cost using RSMeans historic cost index. This data can be used at the very early stages of preliminary design, and has a level of accuracy of +/- 25%. For each building type, the costs per SF of each building in the set are arranged from highest to lowest. The middle report is reported in the Median column. The report that is half way between the highest cost and the Median is shown in the 3/4 column. The report that is half way between the Median and the lowest report is shown in the 1/4 column. These figures then provide a reasonable range for the cost per SF of each building type. The same process is followed for the cost per cubic foot. Below the reports for SF and CF, there is data for various elements of the building, including masonry, finishes, plumbing, HVAC, electrical, etc. Data included for each building type varies, depending on the amount of statistically reliable data in the database. These estimates are most useful when the project is not yet defined, but the owner and designer need an indication of the cost of the project at completion, so that a decision can be made to proceed with the design process. This data is also useful as a reality check on the results of more detailed cost estimates.

The square foot data can be modified for economy of scale using the Square Foot Project Size Modifier table (located at the end of the Square Foot Costs Reference section). Follow the instructions at the top of the chart to compare a proposed project to the typical size for similar projects, and then develop a factor to adjust the project cost upward or downward, depending on the size comparison. Buildings larger than the typical size will have a lower cost per SF, and buildings smaller than the typical size will have a higher cost per SF.

Order of Magnitude Estimates: Order of Magnitude costs may be defined in relation to usable units that have been designed for a facility. For example, hospital administrators who plan to enlarge a hospital will need to know the projected cost per bed. If the number of beds in the proposed hospital (or the number of apartments in an apartment building, or parking spaces in a garage) is known, along with the number of units in the structure, an order of magnitude estimate can be generated. This data is available for many of the building types in the Square Foot Costs table, usually on line 9000.

For example, assume that a developer would like to develop an approximate cost for a proposed 20 story apartment building with 200 apartments. Using the Square Foot Cost table, the median cost for an apartment in a high rise (8- to 24-story) apartment building is $115,000. Thus, 200 apartments times $115,000 per apartment totals $23,000,000 for the building.

Square Foot and Order of Magnitude Comparison: Order of magnitude and square foot estimates may be used to develop preliminary costs only if certain limitations are taken into account. Always consider the following:

- Neither order of magnitude nor square foot estimates reflect actual floor and roof loading, structural systems, clear spans or column spacing. All of these items are important cost factors.

- At the time these estimates are assembled, the materials to be used for the exterior closure system as well as the extent and type of glass and glazing are not likely to be defined. Selection can have a major impact on cost.
- Neither of the preliminary estimates makes allowances for the type of heating and cooling systems, building automation systems, the extent of plumbing requirements, and the need for fire protection sprinklers. All have important cost considerations.
- Some building square foot cost reports include site work and others do not. If possible, determine whether or not this cost is part of the figures being used for the estimate.

Always carefully evaluate the parameters of an order of magnitude or square foot estimate to ensure its validity. At the conceptual stage, when there is little definition of the project, try to match the building type as closely as possible to the available reference data. When using RSMeans square foot and cubic foot costs, assume that they are for the building only and do not include the cost of site work or the purchase of land.

Means Cost Indexes: RSMeans publishes three cost indexes—the City Cost Index, Location Factors, and the Historical Cost Index. The information presented in the indexes is organized according to the Construction Specifications Institute MasterFormat 2010. To create a reliable index, RSMeans researched the building type most often constructed in the U.S. and Canada. Because it was concluded that no one type of building completely represented the building construction industry, nine different types of buildings (factory, office, store, school, etc.) were combined to create a composite model.

The exact material, labor and equipment quantities are based on detailed analysis of these nine building types; then each quantity is weighted in proportion to expected usage. These various material items, labor hours, and equipment rental rates are thus combined to form a composite building representing as closely as possible the actual usage of materials, labor and equipment used in the North American building construction industry.

For purposes of ensuring the timeliness of the data, the components of the index for the composite model have been streamlined. They currently consist of:

- Specific quantities of 66 commonly used construction materials.
- Specific labor-hours for 21 building construction trades.
- Specific days of equipment rental for 6 types of construction equipment. Fuel costs and routine maintenance costs are included in the equipment cost.

A sophisticated computer program handles the updating of all costs for each city on a quarterly basis. Material and equipment price quotations are gathered quarterly from 316 cities in the U.S. and Canada. These prices and the latest negotiated labor wage rates for 21 building trades are used to compile the quarterly update of the city cost indexes.

This ongoing research forms the basis of the three indexes. Each year, to create a U.S. National Average Cost, RSMeans averages the results for the 30 largest cities in the U.S. This average is set at a value of 100. All locations are then compared back to this average, and a numerical relationship is defined. Since the research covers all MasterFormat Divisions, the data can be reported back for both materials and installation costs in the City Cost Index (CCI) table by MasterFormat division.

The total weighted average for all divisions is shown at the bottom row on the CCI reports. The Location Factor Table is a summary table of these total weighted averages. Use Location Factors to adjust the costs of a complete building. Use the City Cost Index table for work done by a single trade, or where a more detailed evaluation is needed.

Means has been collecting data for its indexes since 1940, and has compiled that data into its Historical Index Table. This table allows the user to relate costs from location to location and year to year. The model was revised substantially in 1993, so the base year where the national average equals 100 is 1993.

Historical Cost Index: Often at the early stages of a project the estimator will be aware of a similar project that has been completed at some time in the past and/or in a different location. If the cost of that project is known, the Historical Cost Index table can be used to update that figure to current cost. For instance, if an office building cost $85.00 per square foot in Boston in 2006, and a similar building will be constructed in Philadelphia in 2011, use the Historical Cost Index to approximate a current cost:

(Index in Philadelphia in 2011 divided by Index in Boston in 2006) × Known Cost per SF in Boston in 2006 = Approx. Cost in Philadelphia in 2011

(211.6 divided by 180.4) × $85.00 = $99.70 per SF in Philadelphia in 2011.

Future Value: Often a preliminary estimate must be created to budget for a project that will not be constructed until several years in the future. RSMeans data can be used to estimate the current cost of the project, and the future value formula and the historic index table can be used to predict future cost. In this situation, first decide how many years back you would like to go in the historic table to serve as a basis for cost change per year. Five to ten years will usually suffice. Average the percent cost increase from year to year over that time span, and then take an average of those percent changes. Then use the future value formula:

Future Value = Present Value times average annual increase raised to the n years power:

$FV = PV(1 + i)^n$ where,

FV = Future Value,

PV = Present Value,

i = average % rate of change over n years, and

n = number of years.

The number of years should reflect the time in years from the effective date of the data to the midpoint of construction.

Example: Using January 1, 2011 RSMeans data, calculate the cost to construct a 25,000 SF three story office building if the cost escalation has averaged 3.5% per year. The midpoint of construction is estimated to be July 1, 2015.

Solution: Median cost per SF in 2011 is $120.

$120/SF × 25,000 SF = $3,000,000.

$FV = \$3,000,000 \times (1 + .035)^{4.5}$.

FV = $3,502,295; use $3,500,000.

Chapter 9 Exercises

1. A developer will construct a 60,000 SF supermarket in San Juan, Puerto Rico. How much should he expect to spend?
2. Calculate the cost of a 4-story dormitory at UNH (near Concord, NH). This dorm will serve 250 students.
3. What should the estimator budget for electrical work in a 5,000 SF fire station in Waco, Texas?
4. A developer is planning to build a 250-car parking garage adjacent to a new Nordstrom's in Old Town, Arlington, Virginia. What should the budget be?
5. How big of a gym can be built in Nashville, Tennessee, with a budget of $5,000,000? Assume average construction.
6. A builder constructed an office building in Hartford, Connecticut, in 2005 at a cost of $34,000,000. He plans to construct a similar building nearby in 2011. What is the probable cost?
7. A building constructed in Boston, Massachusetts, in 2011 cost $7,250,000. How much would it cost to build the same building in Albuquerque, New Mexico, in 2011? Calculate using data from the location factor table and from the Historical Cost Index table for 2011.
8. A developer spent $15,650,000 to construct an office building in Tampa, Florida, in 2004. He will construct a similar building in New Orleans, Louisianna, in 2011. What should he budget?
9. What should be budgeted for a college laboratory (40,000 SF) to be built in Portland, Oregon, in 2018, assuming an average annual cost escalation of 4%?
10. What would the budget be for a 1,000,000 SF teaching hospital to be built in New Orleans, Louisianna, if the midpoint of construction is anticipated to be July 2014 (average annual cost escalation at 2.5%)?

Reference Section

All the reference information is in one section, making it easy to find what you need to know . . . and easy to use the book on a daily basis. This section is visually identified by a vertical gray bar on the page edges.

In this Reference Section, we've included Equipment Rental Costs, a listing of rental and operating costs; Crew Listings, a full listing of all crews and equipment, and their costs; Historical Cost Indexes for cost comparisons over time; Location Factors for adjusting costs to the region you are in; Reference Tables, where you will find explanations, estimating information and procedures, or technical data; Square Foot Costs that allow you to make a rough estimate for the overall cost of a project; and an explanation of all the Abbreviations in the book.

Table of Contents

Estimating Tips

- This section contains the average costs to rent and operate hundreds of pieces of construction equipment. This is useful information when estimating the time and material requirements of any particular operation in order to establish a unit or total cost. Equipment costs include not only rental, but also operating costs for equipment under normal use.

Rental Costs

- Equipment rental rates are obtained from industry sources throughout North America-contractors, suppliers, dealers, manufacturers, and distributors.
- Rental rates vary throughout the country, with larger cities generally having lower rates. Lease plans for new equipment are available for periods in excess of six months, with a percentage of payments applying toward purchase.
- Monthly rental rates vary from 2% to 5% of the purchase price of the equipment depending on the anticipated life of the equipment and its wearing parts.
- Weekly rental rates are about 1/3 the monthly rates, and daily rental rates are about 1/3 the weekly rate.
- Rental rates can also be treated as reimbursement costs for contractor-owned equipment. Owned equipment costs include depreciation, loan payments, interest, taxes, insurance, storage, and major repairs.

Operating Costs

- The operating costs include parts and labor for routine servicing, such as repair and replacement of pumps, filters and worn lines. Normal operating expendables, such as fuel, lubricants, tires and electricity (where applicable), are also included.
- Extraordinary operating expendables with highly variable wear patterns, such as diamond bits and blades, are excluded. These costs can be found as material costs in the Unit Price section.
- The hourly operating costs listed do not include the operator's wages.

Equipment Cost/Day

- Any power equipment required by a crew is shown in the Crew Listings with a daily cost.
- The daily cost of equipment needed by a crew is based on dividing the weekly rental rate by 5 (number of working days in the week), and then adding the hourly operating cost times 8 (the number of hours in a day). This "Equipment Cost/Day" is shown in the far right column of the Equipment Rental pages.
- If equipment is needed for only one or two days, it is best to develop your own cost by including components for daily rent and hourly operating cost. This is important when the listed Crew for a task does not contain the equipment needed, such as a crane for lifting mechanical heating/cooling equipment up onto a roof.
- If the quantity of work is less than the crew's Daily Output shown for a Unit Price line item that includes a bare unit equipment cost, it is recommended to estimate one day's rental cost and operating cost for equipment shown in the Crew Listing for that line item.

Mobilization/ Demobilization

- The cost to move construction equipment from an equipment yard or rental company to the jobsite and back again is not included in equipment rental costs listed in the Reference section, nor in the bare equipment cost of any Unit Price line item, nor in any equipment costs shown in the Crew listings.
- Mobilization (to the site) and demobilization (from the site) costs can be found in the Unit Price section.
- If a piece of equipment is already at the jobsite, it is not appropriate to utilize mobil./ demob. costs again in an estimate.

01 54 | Construction Aids

01 54 33 | Equipment Rental

		01 54 33 Equipment Rental	Unit	Hourly Oper. Cost	Rent per Day	Rent per Week	Rent per Month	Equipment Cost/Day	
10	0010	**CONCRETE EQUIPMENT RENTAL** without operators R015433-10							10
	0200	Bucket, concrete lightweight, 1/2 C.Y.	Ea.	.75	23	69	207	19.80	
	0300	1 C.Y.		.80	28	84	252	23.20	
	0400	1-1/2 C.Y.		1.00	38.50	115	345	31	
	0500	2 C.Y.		1.10	45	135	405	35.80	
	0580	8 C.Y.		5.70	265	795	2,375	204.60	
	0600	Cart, concrete, self-propelled, operator walking, 10 C.F.		2.50	58.50	175	525	55	
	0700	Operator riding, 18 C.F.		4.20	96.50	290	870	91.60	
	0800	Conveyer for concrete, portable, gas, 16" wide, 26' long		9.20	127	380	1,150	149.60	
	0900	46' long		9.55	153	460	1,375	168.40	
	1000	56' long		9.70	162	485	1,450	174.60	
	1100	Core drill, electric, 2-1/2 H.P., 1" to 8" bit diameter		1.40	59.50	179	535	47	
	1150	11 H.P., 8" to 18" cores		5.10	115	345	1,025	109.80	
	1200	Finisher, concrete floor, gas, riding trowel, 96" wide		8.55	148	445	1,325	157.40	
	1300	Gas, walk-behind, 3 blade, 36" trowel		1.40	20.50	62	186	23.60	
	1400	4 blade, 48" trowel		2.70	28	84	252	38.40	
	1500	Float, hand-operated (Bull float) 48" wide		.08	13.65	41	123	8.85	
	1570	Curb builder, 14 H.P., gas, single screw		11.95	253	760	2,275	247.60	
	1590	Double screw		12.60	293	880	2,650	276.80	
	1600	Floor grinder, concrete and terrazzo, electric, 22" path		1.62	102	305	915	73.95	
	1700	Edger, concrete, electric, 7" path		.90	51.50	155	465	38.20	
	1750	Vacuum pick-up system for floor grinders, wet/dry		1.35	81.50	245	735	59.80	
	1800	Mixer, powered, mortar and concrete, gas, 6 C.F., 18 H.P.		6.45	120	360	1,075	123.60	
	1900	10 C.F., 25 H.P.		7.90	147	440	1,325	151.20	
	2000	16 C.F.		8.20	170	510	1,525	167.60	
	2100	Concrete, stationary, tilt drum, 2 C.Y.		6.45	238	715	2,150	194.60	
	2120	Pump, concrete, truck mounted 4" line 80' boom		23.25	925	2,775	8,325	741	
	2140	5" line, 110' boom		30.70	1,250	3,770	11,300	999.60	
	2160	Mud jack, 50 C.F. per hr.		6.10	128	385	1,150	125.80	
	2180	225 C.F. per hr		8.15	147	440	1,325	153.20	
	2190	Shotcrete pump rig, 12 C.Y./hr.		14.15	227	680	2,050	249.20	
	2200	35 C.Y./hr.		16.75	243	730	2,200	280	
	2600	Saw, concrete, manual, gas, 18 H.P.		4.65	43.50	130	390	63.20	
	2650	Self-propelled, gas, 30 H.P.		8.90	105	315	945	134.20	
	2700	Vibrators, concrete, electric, 60 cycle, 2 H.P.		.31	9	27	81	7.90	
	2800	3 H.P.		.43	12	36	108	10.65	
	2900	Gas engine, 5 H.P.		1.30	16.35	49	147	20.20	
	3000	8 H.P.		1.75	15.35	46	138	23.20	
	3050	Vibrating screed, gas engine, 8 H.P.		2.81	71.50	215	645	65.50	
	3120	Concrete transit mixer, 6 x 4, 250 H.P., 8 C.Y., rear discharge		43.95	590	1,765	5,300	704.60	
	3200	Front discharge		51.35	725	2,175	6,525	845.80	
	3300	6 x 6, 285 H.P., 12 C.Y., rear discharge		50.45	685	2,050	6,150	813.60	
	3400	Front discharge	▼	52.65	735	2,200	6,600	861.20	
20	0010	**EARTHWORK EQUIPMENT RENTAL** without operators R015433-10							20
	0040	Aggregate spreader, push type 8' to 12' wide	Ea.	2.25	26	78	234	33.60	
	0045	Tailgate type, 8' wide		2.20	33.50	100	300	37.60	
	0055	Earth auger, truck-mounted, for fence & sign posts, utility poles		11.60	515	1,545	4,625	401.80	
	0060	For borings and monitoring wells		38.45	665	1,995	5,975	706.60	
	0070	Portable, trailer mounted		2.45	32.50	98	294	39.20	
	0075	Truck-mounted, for caissons, water wells		83.90	3,025	9,075	27,200	2,486	
	0080	Horizontal boring machine, 12" to 36" diameter, 45 H.P.		20.85	200	600	1,800	286.80	
	0090	12" to 48" diameter, 65 H.P.		28.90	350	1,055	3,175	442.20	
	0095	Auger, for fence posts, gas engine, hand held		.45	6	18	54	7.20	
	0100	Excavator, diesel hydraulic, crawler mounted, 1/2 C.Y. cap.		16.70	385	1,155	3,475	364.60	
	0120	5/8 C.Y. capacity		24.50	555	1,670	5,000	530	
	0140	3/4 C.Y. capacity		30.55	660	1,980	5,950	640.40	
	0150	1 C.Y. capacity	▼	36.35	730	2,195	6,575	729.80	

01 54 | Construction Aids

01 54 33 | Equipment Rental

			UNIT	HOURLY OPER. COST	RENT PER DAY	RENT PER WEEK	RENT PER MONTH	EQUIPMENT COST/DAY	
20	0200	1-1/2 C.Y. capacity	Ea.	44.55	985	2,955	8,875	947.40	20
	0300	2 C.Y. capacity		58.25	1,275	3,790	11,400	1,224	
	0320	2-1/2 C.Y. capacity		77.80	1,650	4,955	14,900	1,613	
	0325	3-1/2 C.Y. capacity		104.95	2,225	6,655	20,000	2,171	
	0330	4-1/2 C.Y. capacity		128.05	2,750	8,225	24,700	2,669	
	0335	6 C.Y. capacity		158.65	3,050	9,160	27,500	3,101	
	0340	7 C.Y. capacity		160.90	3,150	9,460	28,400	3,179	
	0342	Excavator attachments, bucket thumbs		3.00	245	735	2,200	171	
	0345	Grapples		2.70	213	640	1,925	149.60	
	0347	Hydraulic hammer for boom mounting, 5000 ft lbs		12.65	420	1,255	3,775	352.20	
	0349	11,000 ft lbs		21.50	750	2,250	6,750	622	
	0350	Gradall type, truck mounted, 3 ton @ 15' radius, 5/8 C.Y.		36.80	940	2,815	8,450	857.40	
	0370	1 C.Y. capacity		41.00	1,125	3,365	10,100	1,001	
	0400	Backhoe-loader, 40 to 45 H.P., 5/8 C.Y. capacity		11.20	223	670	2,000	223.60	
	0450	45 H.P. to 60 H.P., 3/4 C.Y. capacity		16.30	315	950	2,850	320.40	
	0460	80 H.P., 1-1/4 C.Y. capacity		17.80	345	1,035	3,100	349.40	
	0470	112 H.P., 1-1/2 C.Y. capacity		30.25	610	1,830	5,500	608	
	0482	Backhoe-loader attachment, compactor, 20,000 lb.		5.40	145	435	1,300	130.20	
	0485	Hydraulic hammer, 750 ft lbs		3.10	103	310	930	86.80	
	0486	Hydraulic hammer, 1200 ft lbs		5.50	190	570	1,700	158	
	0500	Brush chipper, gas engine, 6" cutter head, 35 H.P.		8.10	110	330	990	130.80	
	0550	12" cutter head, 130 H.P.		13.25	185	555	1,675	217	
	0600	15" cutter head, 165 H.P.		19.45	222	665	2,000	288.60	
	0750	Bucket, clamshell, general purpose, 3/8 C.Y.		1.20	38.50	115	345	32.60	
	0800	1/2 C.Y.		1.30	45	135	405	37.40	
	0850	3/4 C.Y.		1.45	56.50	170	510	45.60	
	0900	1 C.Y.		1.50	60	180	540	48	
	0950	1-1/2 C.Y.		2.40	81.50	245	735	68.20	
	1000	2 C.Y.		2.50	91.50	275	825	75	
	1010	Bucket, dragline, medium duty, 1/2 C.Y.		.70	24.50	73	219	20.20	
	1020	3/4 C.Y.		.70	25.50	77	231	21	
	1030	1 C.Y.		.75	27	81	243	22.20	
	1040	1-1/2 C.Y.		1.15	41.50	125	375	34.20	
	1050	2 C.Y.		1.20	46.50	140	420	37.60	
	1070	3 C.Y.		1.85	63.50	190	570	52.80	
	1200	Compactor, manually guided 2-drum vibratory smooth roller, 7.5 H.P.		5.85	167	500	1,500	146.80	
	1250	Rammer/tamper, gas, 8"		2.20	43.50	130	390	43.60	
	1260	15"		2.45	50	150	450	49.60	
	1300	Vibratory plate, gas, 18" plate, 3000 lb. blow		2.10	23.50	70	210	30.80	
	1350	21" plate, 5000 lb. blow		2.60	30.50	91	273	39	
	1370	Curb builder/extruder, 14 H.P., gas, single screw		11.95	253	760	2,275	247.60	
	1390	Double screw		12.60	293	880	2,650	276.80	
	1500	Disc harrow attachment, for tractor		.41	68.50	205	615	44.30	
	1810	Feller buncher, shearing & accumulating trees, 100 H.P.		27.75	570	1,710	5,125	564	
	1860	Grader, self-propelled, 25,000 lb.		22.90	470	1,405	4,225	464.20	
	1910	30,000 lb.		26.35	555	1,660	4,975	542.80	
	1920	40,000 lb.		43.05	1,050	3,155	9,475	975.40	
	1930	55,000 lb.		55.95	1,425	4,275	12,800	1,303	
	1950	Hammer, pavement breaker, self-propelled, diesel, 1000 to 1250 lb.		22.90	360	1,080	3,250	399.20	
	2000	1300 to 1500 lb.		34.35	720	2,160	6,475	706.80	
	2050	Pile driving hammer, steam or air, 4150 ft lbs @ 225 bpm		9.80	500	1,495	4,475	377.40	
	2100	8750 ft lbs @ 145 bpm		11.80	700	2,095	6,275	513.40	
	2150	15,000 ft lbs @ 60 bpm		13.40	845	2,535	7,600	614.20	
	2200	24,450 ft lbs @ 111 bpm		14.40	940	2,820	8,450	679.20	
	2250	Leads, 60' high for pile driving hammers up to 20,000 ft lbs		3.10	82	246	740	74	
	2300	90' high for hammers over 20,000 ft lbs		4.70	149	448	1,350	127.20	
	2350	Diesel type hammer, 22,400 ft lbs		24.80	655	1,970	5,900	592.40	
	2400	41,300 ft lbs		31.25	710	2,135	6,400	677	

01 54 | Construction Aids

01 54 33 | Equipment Rental			UNIT	HOURLY OPER. COST	RENT PER DAY	RENT PER WEEK	RENT PER MONTH	EQUIPMENT COST/DAY	
20	2450	141,000 ft lbs	Ea.	47.65	1,225	3,660	11,000	1,113	20
	2500	Vib. elec. hammer/extractor, 200 kW diesel generator, 34 H.P.		36.95	680	2,045	6,125	704.60	
	2550	80 H.P.		64.95	1,000	3,025	9,075	1,125	
	2600	150 H.P.		120.90	1,950	5,865	17,600	2,140	
	2700	Extractor, steam or air, 700 ft lbs		17.75	520	1,560	4,675	454	
	2750	1000 ft lbs		20.20	635	1,910	5,725	543.60	
	2800	Log chipper, up to 22" diameter, 600 H.P.		35.35	455	1,360	4,075	554.80	
	2850	Logger, for skidding & stacking logs, 150 H.P.		41.75	850	2,550	7,650	844	
	2860	Mulcher, diesel powered, trailer mounted		15.60	223	670	2,000	258.80	
	2900	Rake, spring tooth, with tractor		12.66	294	882	2,650	277.70	
	3000	Roller, vibratory, tandem, smooth drum, 20 H.P.		7.40	143	430	1,300	145.20	
	3050	35 H.P.		9.40	248	745	2,225	224.20	
	3100	Towed type vibratory compactor, smooth drum, 50 H.P.		22.45	360	1,075	3,225	394.60	
	3150	Sheepsfoot, 50 H.P.		24.75	420	1,255	3,775	449	
	3170	Landfill compactor, 220 H.P.		68.75	1,425	4,290	12,900	1,408	
	3200	Pneumatic tire roller, 80 H.P.		12.65	345	1,040	3,125	309.20	
	3250	120 H.P.		19.15	600	1,800	5,400	513.20	
	3300	Sheepsfoot vibratory roller, 240 H.P.		55.10	1,100	3,265	9,800	1,094	
	3320	340 H.P.		73.05	1,675	5,025	15,100	1,589	
	3350	Smooth drum vibratory roller, 75 H.P.		20.60	555	1,670	5,000	498.80	
	3400	125 H.P.		25.60	655	1,960	5,875	596.80	
	3410	Rotary mower, brush, 60", with tractor		16.60	268	805	2,425	293.80	
	3420	Rototiller, walk-behind, gas, 5 H.P.		2.23	67.50	203	610	58.45	
	3422	8 H.P.		3.47	103	310	930	89.75	
	3440	Scrapers, towed type, 7 C.Y. capacity		5.25	110	330	990	108	
	3450	10 C.Y. capacity		6.00	153	460	1,375	140	
	3500	15 C.Y. capacity		6.45	178	535	1,600	158.60	
	3525	Self-propelled, single engine, 14 C.Y. capacity		77.50	1,475	4,460	13,400	1,512	
	3550	Dual engine, 21 C.Y. capacity		107.80	1,675	5,050	15,200	1,872	
	3600	31 C.Y. capacity		152.05	2,600	7,805	23,400	2,777	
	3640	44 C.Y. capacity		193.60	3,600	10,780	32,300	3,705	
	3650	Elevating type, single engine, 11 C.Y. capacity		54.35	1,100	3,270	9,800	1,089	
	3700	22 C.Y. capacity		99.85	2,150	6,440	19,300	2,087	
	3710	Screening plant 110 H.P. w/5' x 10' screen		30.50	420	1,260	3,775	496	
	3720	5' x 16' screen		32.65	520	1,560	4,675	573.20	
	3850	Shovel, crawler-mounted, front-loading, 7 C.Y. capacity		168.30	2,950	8,815	26,400	3,109	
	3855	12 C.Y. capacity		225.05	3,750	11,230	33,700	4,046	
	3860	Shovel/backhoe bucket, 1/2 C.Y.		2.25	63.50	190	570	56	
	3870	3/4 C.Y.		2.30	71.50	215	645	61.40	
	3880	1 C.Y.		2.40	80	240	720	67.20	
	3890	1-1/2 C.Y.		2.55	95	285	855	77.40	
	3910	3 C.Y.		2.85	130	390	1,175	100.80	
	3950	Stump chipper, 18" deep, 30 H.P.		7.17	187	560	1,675	169.35	
	4110	Dozer, crawler, torque converter, diesel 80 H.P.		21.70	400	1,195	3,575	412.60	
	4150	105 H.P.		25.80	500	1,495	4,475	505.40	
	4200	140 H.P.		37.35	800	2,400	7,200	778.80	
	4260	200 H.P.		50.75	1,075	3,225	9,675	1,051	
	4310	300 H.P.		71.80	1,725	5,165	15,500	1,607	
	4360	410 H.P.		97.20	2,200	6,635	19,900	2,105	
	4370	500 H.P.		130.15	3,000	9,020	27,100	2,845	
	4380	700 H.P.		191.35	4,375	13,095	39,300	4,150	
	4400	Loader, crawler, torque conv., diesel, 1-1/2 C.Y., 80 H.P.		22.70	475	1,425	4,275	466.60	
	4450	1-1/2 to 1-3/4 C.Y., 95 H.P.		25.75	600	1,795	5,375	565	
	4510	1-3/4 to 2-1/4 C.Y., 130 H.P.		33.70	840	2,520	7,550	773.60	
	4530	2-1/2 to 3-1/4 C.Y., 190 H.P.		46.40	1,050	3,185	9,550	1,008	
	4560	3-1/2 to 5 C.Y., 275 H.P.		64.35	1,525	4,585	13,800	1,432	
	4610	Front end loader, 4WD, articulated frame, 1 to 1-1/4 C.Y., 70 H.P.		13.35	215	645	1,925	235.80	
	4620	1-1/2 to 1-3/4 C.Y., 95 H.P.	↓	17.30	300	905	2,725	319.40	

01 54 | Construction Aids

	01 54 33 | Equipment Rental		UNIT	HOURLY OPER. COST	RENT PER DAY	RENT PER WEEK	RENT PER MONTH	EQUIPMENT COST/DAY	
20	4650	1-3/4 to 2 C.Y., 130 H.P.	Ea.	19.90	370	1,105	3,325	380.20	20
	4710	2-1/2 to 3-1/2 C.Y., 145 H.P.		21.20	380	1,140	3,425	397.60	
	4730	3 to 4-1/2 C.Y., 185 H.P.		28.95	610	1,835	5,500	598.60	
	4760	5-1/4 to 5-3/4 C.Y., 270 H.P.		47.05	930	2,785	8,350	933.40	
	4810	7 to 9 C.Y., 475 H.P.		83.30	1,900	5,725	17,200	1,811	
	4870	9 - 11 C.Y., 620 H.P.		108.70	2,675	8,030	24,100	2,476	
	4880	Skid steer loader, wheeled, 10 C.F., 30 H.P. gas		7.85	148	445	1,325	151.80	
	4890	1 C.Y., 78 H.P., diesel		13.90	250	750	2,250	261.20	
	4892	Skid-steer attachment, auger		.53	88	264	790	57.05	
	4893	Backhoe		.71	119	356	1,075	76.90	
	4894	Broom		.74	123	370	1,100	79.90	
	4895	Forks		.24	40.50	122	365	26.30	
	4896	Grapple		.55	91.50	274	820	59.20	
	4897	Concrete hammer		1.07	178	534	1,600	115.35	
	4898	Tree spade		.78	129	388	1,175	83.85	
	4899	Trencher		.76	127	380	1,150	82.10	
	4900	Trencher, chain, boom type, gas, operator walking, 12 H.P.		3.55	48.50	145	435	57.40	
	4910	Operator riding, 40 H.P.		14.25	305	915	2,750	297	
	5000	Wheel type, diesel, 4' deep, 12" wide		61.50	850	2,545	7,625	1,001	
	5100	6' deep, 20" wide		77.45	1,975	5,910	17,700	1,802	
	5150	Chain type, diesel, 5' deep, 8" wide		27.10	600	1,805	5,425	577.80	
	5200	Diesel, 8' deep, 16" wide		114.45	2,650	7,950	23,900	2,506	
	5202	Rock trencher, wheel type, 6" wide x 18" deep		17.35	325	970	2,900	332.80	
	5206	Chain type, 18" wide x 7' deep		95.95	2,950	8,850	26,600	2,538	
	5210	Tree spade, self-propelled		13.73	267	800	2,400	269.85	
	5250	Truck, dump, 2-axle, 12 ton, 8 C.Y. payload, 220 H.P.		21.80	235	705	2,125	315.40	
	5300	Three axle dump, 16 ton, 12 CY payload, 400 H.P.		38.00	335	1,010	3,025	506	
	5310	Four axle dump, 25 ton, 18 C.Y. payload, 450 H.P.		45.60	495	1,480	4,450	660.80	
	5350	Dump trailer only, rear dump, 16-1/2 C.Y.		5.10	142	425	1,275	125.80	
	5400	20 C.Y.		5.50	160	480	1,450	140	
	5450	Flatbed, single axle, 1-1/2 ton rating		16.05	71.50	215	645	171.40	
	5500	3 ton rating		19.50	102	305	915	217	
	5550	Off highway rear dump, 25 ton capacity		57.25	1,350	4,030	12,100	1,264	
	5600	35 ton capacity		52.80	1,200	3,565	10,700	1,135	
	5610	50 ton capacity		71.20	1,625	4,895	14,700	1,549	
	5620	65 ton capacity		72.90	1,625	4,840	14,500	1,551	
	5630	100 ton capacity		93.85	2,100	6,290	18,900	2,009	
	6000	Vibratory plow, 25 H.P., walking	↓	5.95	63.50	190	570	85.60	
40	0010	**GENERAL EQUIPMENT RENTAL** without operators	R015433 -10						40
	0150	Aerial lift, scissor type, to 15' high, 1000 lb. cap., electric	Ea.	2.90	56.50	170	510	57.20	
	0160	To 25' high, 2000 lb. capacity		3.30	71.50	215	645	69.40	
	0170	Telescoping boom to 40' high, 500 lb. capacity, gas		15.85	320	955	2,875	317.80	
	0180	To 45' high, 500 lb. capacity		16.95	370	1,105	3,325	356.60	
	0190	To 60' high, 600 lb. capacity		19.20	480	1,445	4,325	442.60	
	0195	Air compressor, portable, 6.5 CFM, electric		.33	14	42	126	11.05	
	0196	Gasoline		.80	21	63	189	19	
	0200	Towed type, gas engine, 60 CFM		10.15	48.50	145	435	110.20	
	0300	160 CFM		11.75	51.50	155	465	125	
	0400	Diesel engine, rotary screw, 250 CFM		12.35	107	320	960	162.80	
	0500	365 CFM		16.40	132	395	1,175	210.20	
	0550	450 CFM		20.70	165	495	1,475	264.60	
	0600	600 CFM		35.95	228	685	2,050	424.60	
	0700	750 CFM	↓	36.10	238	715	2,150	431.80	
	0800	For silenced models, small sizes, add to rent		3%	5%	5%	5%		
	0900	Large sizes, add to rent		5%	7%	7%	7%		
	0930	Air tools, breaker, pavement, 60 lb.	Ea.	.50	9.65	29	87	9.80	
	0940	80 lb.		.50	10.35	31	93	10.20	
	0950	Drills, hand (jackhammer) 65 lb.	↓	.55	17.35	52	156	14.80	

Reference Section

01 54 | Construction Aids

01 54 33 | Equipment Rental

		01 54 33 | Equipment Rental	UNIT	HOURLY OPER. COST	RENT PER DAY	RENT PER WEEK	RENT PER MONTH	EQUIPMENT COST/DAY	
40	0960	Track or wagon, swing boom, 4" drifter	Ea.	49.45	755	2,270	6,800	849.60	40
	0970	5" drifter		59.40	815	2,450	7,350	965.20	
	0975	Track mounted quarry drill, 6" diameter drill		82.30	1,250	3,725	11,200	1,403	
	0980	Dust control per drill		.58	18.65	56	168	15.85	
	0990	Hammer, chipping, 12 lb.		.50	24.50	73	219	18.60	
	1000	Hose, air with couplings, 50' long, 3/4" diameter		.03	4.67	14	42	3.05	
	1100	1" diameter		.04	6.35	19	57	4.10	
	1200	1-1/2" diameter		.06	9.65	29	87	6.30	
	1300	2" diameter		.07	12	36	108	7.75	
	1400	2-1/2" diameter		.11	19	57	171	12.30	
	1410	3" diameter		.14	23	69	207	14.90	
	1450	Drill, steel, 7/8" x 2'		.05	8.65	26	78	5.60	
	1460	7/8" x 6'		.05	9	27	81	5.80	
	1520	Moil points		.02	4	12	36	2.55	
	1525	Pneumatic nailer w/accessories		.49	32.50	98	294	23.50	
	1530	Sheeting driver for 60 lb. breaker		.04	6	18	54	3.90	
	1540	For 90 lb. breaker		.12	8	24	72	5.75	
	1550	Spade, 25 lb.		.45	7.35	22	66	8	
	1560	Tamper, single, 35 lb.		.55	36.50	109	325	26.20	
	1570	Triple, 140 lb.		.82	54.50	164	490	39.35	
	1580	Wrenches, impact, air powered, up to 3/4" bolt		.40	10	30	90	9.20	
	1590	Up to 1-1/4" bolt		.50	23.50	71	213	18.20	
	1600	Barricades, barrels, reflectorized, 1 to 50 barrels		.03	4.60	13.80	41.50	3	
	1610	100 to 200 barrels		.02	3.53	10.60	32	2.30	
	1620	Barrels with flashers, 1 to 50 barrels		.03	5.25	15.80	47.50	3.40	
	1630	100 to 200 barrels		.03	4.20	12.60	38	2.75	
	1640	Barrels with steady burn type C lights		.04	7	21	63	4.50	
	1650	Illuminated board, trailer mounted, with generator		.75	122	365	1,100	79	
	1670	Portable barricade, stock, with flashers, 1 to 6 units		.03	5.25	15.80	47.50	3.40	
	1680	25 to 50 units		.03	4.90	14.70	44	3.20	
	1685	Butt fusion machine, wheeled, 1.5 HP electric, 2" - 8" diam. pipe		2.56	167	500	1,500	120.50	
	1690	Tracked, 20 HP diesel, 4"-12" diam. pipe		10.43	490	1,465	4,400	376.45	
	1695	83 HP diesel, 8" - 24" diam. pipe		27.52	975	2,930	8,800	806.15	
	1700	Carts, brick, hand powered, 1000 lb. capacity		.44	72.50	218	655	47.10	
	1800	Gas engine, 1500 lb., 7-1/2' lift		4.04	115	345	1,025	101.30	
	1822	Dehumidifier, medium, 6 lb/hr, 150 CFM		.93	60	180	540	43.45	
	1824	Large, 18 lb/hr, 600 CFM		1.89	122	366	1,100	88.30	
	1830	Distributor, asphalt, trailer mounted, 2000 gal., 38 H.P. diesel		9.60	340	1,020	3,050	280.80	
	1840	3000 gal., 38 H.P. diesel		11.00	370	1,105	3,325	309	
	1850	Drill, rotary hammer, electric		.82	26.50	80	240	22.55	
	1860	Carbide bit, 1-1/2" diameter, add to electric rotary hammer		.02	6.25	18.80	56.50	3.90	
	1865	Rotary, crawler, 250 H.P.		122.20	2,050	6,125	18,400	2,203	
	1870	Emulsion sprayer, 65 gal., 5 H.P. gas engine		2.89	99	297	890	82.50	
	1880	200 gal., 5 H.P. engine		6.85	165	495	1,475	153.80	
	1900	Floor auto-scrubbing machine, walk-behind, 28" path		3.26	213	640	1,925	154.10	
	1930	Floodlight, mercury vapor, or quartz, on tripod, 1000 watt		.27	14	42	126	10.55	
	1940	2000 watt		.50	27	81	243	20.20	
	1950	Floodlights, trailer mounted with generator, 1 - 300 watt light		3.10	73.50	220	660	68.80	
	1960	2 - 1000 watt lights		3.95	98.50	295	885	90.60	
	2000	4 - 300 watt lights		3.75	93.50	280	840	86	
	2005	Foam spray rig, incl. box trailer, compressor, generator, proportioner		23.78	495	1,490	4,475	488.25	
	2020	Forklift, straight mast, 12' lift, 5000 lb., 2 wheel drive, gas		19.80	202	605	1,825	279.40	
	2040	21' lift, 5000 lb., 4 wheel drive, diesel		16.20	248	745	2,225	278.60	
	2050	For rough terrain, 42' lift, 35' reach, 9000 lb., 110 H.P.		22.30	450	1,345	4,025	447.40	
	2060	For plant, 4 ton capacity, 80 H.P., 2 wheel drive, gas		12.05	102	305	915	157.40	
	2080	10 ton capacity, 120 H.P., 2 wheel drive, diesel		17.70	177	530	1,600	247.60	
	2100	Generator, electric, gas engine, 1.5 kW to 3 kW		2.80	11.35	34	102	29.20	
	2200	5 kW	↓	3.55	15.35	46	138	37.60	

01 54 | Construction Aids

	01 54 33 | Equipment Rental		UNIT	HOURLY OPER. COST	RENT PER DAY	RENT PER WEEK	RENT PER MONTH	EQUIPMENT COST/DAY	
40	2300	10 kW	Ea.	6.65	38.50	115	345	76.20	40
	2400	25 kW		7.80	88.50	265	795	115.40	
	2500	Diesel engine, 20 kW		9.00	75	225	675	117	
	2600	50 kW		15.85	105	315	945	189.80	
	2700	100 kW		30.25	145	435	1,300	329	
	2800	250 kW		59.65	253	760	2,275	629.20	
	2850	Hammer, hydraulic, for mounting on boom, to 500 ft lbs		2.40	76.50	230	690	65.20	
	2860	1000 ft lbs		4.05	127	380	1,150	108.40	
	2900	Heaters, space, oil or electric, 50 MBH		1.84	8	24	72	19.50	
	3000	100 MBH		3.31	10.65	32	96	32.90	
	3100	300 MBH		10.62	36.50	110	330	106.95	
	3150	500 MBH		21.42	48.50	145	435	200.35	
	3200	Hose, water, suction with coupling, 20' long, 2" diameter		.02	3	9	27	1.95	
	3210	3" diameter		.03	4.67	14	42	3.05	
	3220	4" diameter		.03	5	15	45	3.25	
	3230	6" diameter		.11	18.35	55	165	11.90	
	3240	8" diameter		.20	33.50	100	300	21.60	
	3250	Discharge hose with coupling, 50' long, 2" diameter		.01	1.33	4	12	.90	
	3260	3" diameter		.02	2.67	8	24	1.75	
	3270	4" diameter		.02	3.67	11	33	2.35	
	3280	6" diameter		.06	9.65	29	87	6.30	
	3290	8" diameter		.20	33.50	100	300	21.60	
	3295	Insulation blower		.20	6	18	54	5.20	
	3300	Ladders, extension type, 16' to 36' long		.15	25.50	76	228	16.40	
	3400	40' to 60' long		.22	36	108	325	23.35	
	3405	Lance for cutting concrete		2.17	104	313	940	79.95	
	3407	Lawn mower, rotary, 22", 5 H.P.		2.14	63.50	190	570	55.10	
	3408	48" self propelled		2.88	94.50	283	850	79.65	
	3410	Level, electronic, automatic, with tripod and leveling rod		1.45	96.50	290	870	69.60	
	3430	Laser type, for pipe and sewer line and grade		.76	50.50	152	455	36.50	
	3440	Rotating beam for interior control		1.16	77.50	232	695	55.70	
	3460	Builder's optical transit, with tripod and rod		.09	14.65	44	132	9.50	
	3500	Light towers, towable, with diesel generator, 2000 watt		3.75	93.50	280	840	86	
	3600	4000 watt		3.95	98.50	295	885	90.60	
	3700	Mixer, powered, plaster and mortar, 6 C.F., 7 H.P.		1.85	30	90	270	32.80	
	3800	10 C.F., 9 H.P.		2.00	31.50	95	285	35	
	3850	Nailer, pneumatic		.49	32.50	98	294	23.50	
	3900	Paint sprayers complete, 8 CFM		.73	48.50	145	435	34.85	
	4000	17 CFM		1.35	89.50	269	805	64.60	
	4020	Pavers, bituminous, rubber tires, 8' wide, 50 H.P., diesel		38.00	1,050	3,165	9,500	937	
	4030	10' wide, 150 H.P.		80.45	1,700	5,120	15,400	1,668	
	4050	Crawler, 8' wide, 100 H.P., diesel		78.80	1,800	5,415	16,200	1,713	
	4060	10' wide, 150 H.P.		90.60	2,125	6,405	19,200	2,006	
	4070	Concrete paver, 12' to 24' wide, 250 H.P.		78.20	1,625	4,875	14,600	1,601	
	4080	Placer-spreader-trimmer, 24' wide, 300 H.P.		111.00	2,650	7,975	23,900	2,483	
	4100	Pump, centrifugal gas pump, 1-1/2" diam., 65 GPM		3.50	48.50	145	435	57	
	4200	2" diameter, 130 GPM		4.50	53.50	160	480	68	
	4300	3" diameter, 250 GPM		4.75	55	165	495	71	
	4400	6" diameter, 1500 GPM		22.95	180	540	1,625	291.60	
	4500	Submersible electric pump, 1-1/4" diameter, 55 GPM		.31	16.35	49	147	12.30	
	4600	1-1/2" diameter, 83 GPM		.35	18.65	56	168	14	
	4700	2" diameter, 120 GPM		1.30	23.50	70	210	24.40	
	4800	3" diameter, 300 GPM		2.20	41.50	125	375	42.60	
	4900	4" diameter, 560 GPM		9.50	163	490	1,475	174	
	5000	6" diameter, 1590 GPM		13.95	220	660	1,975	243.60	
	5100	Diaphragm pump, gas, single, 1-1/2" diameter		1.15	49.50	148	445	38.80	
	5200	2" diameter		3.70	61.50	185	555	66.60	
	5300	3" diameter	↓	3.70	61.50	185	555	66.60	

01 54 | Construction Aids

01 54 33 | Equipment Rental		UNIT	HOURLY OPER. COST	RENT PER DAY	RENT PER WEEK	RENT PER MONTH	EQUIPMENT COST/DAY
40 5400	Double, 4" diameter	Ea.	5.55	108	325	975	109.40
5450	Pressure washer 5 GPM, 3000 psi		3.85	46.50	140	420	58.80
5500	Trash pump, self-priming, gas, 2" diameter		3.70	21	63	189	42.20
5600	Diesel, 4" diameter		9.10	58.50	175	525	107.80
5650	Diesel, 6" diameter		29.10	130	390	1,175	310.80
5655	Grout Pump		14.75	100	300	900	178
5700	Salamanders, L.P. gas fired, 100,000 Btu		3.35	16.35	49	147	36.60
5705	50,000 Btu		2.50	10.35	31	93	26.20
5720	Sandblaster, portable, open top, 3 C.F. capacity		.55	27	81	243	20.60
5730	6 C.F. capacity		.90	40	120	360	31.20
5740	Accessories for above		.12	20.50	61	183	13.15
5750	Sander, floor		.60	20	60	180	16.80
5760	Edger		.61	24.50	74	222	19.70
5800	Saw, chain, gas engine, 18" long		1.75	19.65	59	177	25.80
5900	Hydraulic powered, 36" long		.70	60	180	540	41.60
5950	60" long		.70	61.50	185	555	42.60
6000	Masonry, table mounted, 14" diameter, 5 H.P.		1.25	56.50	170	510	44
6050	Portable cut-off, 8 H.P.		1.90	33	99	297	35
6100	Circular, hand held, electric, 7-1/4" diameter		.14	4.67	14	42	3.90
6200	12" diameter		.21	7.65	23	69	6.30
6250	Wall saw, w/hydraulic power, 10 H.P.		6.50	61.50	185	555	89
6275	Shot blaster, walk-behind, 20" wide		4.65	298	895	2,675	216.20
6280	Sidewalk broom, walk-behind		2.09	60.50	182	545	53.10
6300	Steam cleaner, 100 gallons per hour		3.15	60.50	182	545	61.60
6310	200 gallons per hour		4.25	73.50	220	660	78
6340	Tar Kettle/Pot, 400 gallons		4.97	81.50	245	735	88.75
6350	Torch, cutting, acetylene-oxygen, 150' hose		.50	23.50	70	210	18
6360	Hourly operating cost includes tips and gas		10.35				82.80
6410	Toilet, portable chemical		.12	19.65	59	177	12.75
6420	Recycle flush type		.14	23.50	71	213	15.30
6430	Toilet, fresh water flush, garden hose,		.17	27.50	83	249	17.95
6440	Hoisted, non-flush, for high rise		.14	23.50	70	210	15.10
6450	Toilet, trailers, minimum		.24	40.50	122	365	26.30
6460	Maximum		.74	124	372	1,125	80.30
6465	Tractor, farm with attachment		15.50	305	910	2,725	306
6500	Trailers, platform, flush deck, 2 axle, 25 ton capacity		5.15	117	350	1,050	111.20
6600	40 ton capacity		6.65	162	485	1,450	150.20
6700	3 axle, 50 ton capacity		7.20	180	540	1,625	165.60
6800	75 ton capacity		8.95	237	710	2,125	213.60
6810	Trailer mounted cable reel for high voltage line work		5.04	240	720	2,150	184.30
6820	Trailer mounted cable tensioning rig		10.01	475	1,430	4,300	366.10
6830	Cable pulling rig		67.52	2,675	8,020	24,100	2,144
6900	Water tank trailer, engine driven discharge, 5000 gallons		6.65	155	465	1,400	146.20
6925	10,000 gallons		9.15	218	655	1,975	204.20
6950	Water truck, off highway, 6000 gallons		64.15	860	2,585	7,750	1,030
7010	Tram car for high voltage line work, powered, 2 conductor		4.39	130	391	1,175	113.30
7020	Transit (builder's level) with tripod		.09	14.65	44	132	9.50
7030	Trench box, 3000 lbs. 6' x 8'		.56	93	279	835	60.30
7040	7200 lbs. 6' x 20'		1.05	175	525	1,575	113.40
7050	8000 lbs., 8' x 16'		1.05	175	525	1,575	113.40
7060	9500 lbs., 8' x 20'		1.19	199	596	1,800	128.70
7065	11,000 lbs., 8' x 24'		1.25	209	627	1,875	135.40
7070	12,000 lbs., 10' x 20'		1.36	227	680	2,050	146.90
7100	Truck, pickup, 3/4 ton, 2 wheel drive		8.70	58.50	175	525	104.60
7200	4 wheel drive		8.95	75	225	675	116.60
7250	Crew carrier, 9 passenger		12.30	90	270	810	152.40
7290	Flat bed truck, 20,000 lbs. GVW		13.75	133	400	1,200	190
7300	Tractor, 4 x 2, 220 H.P.	↓	19.40	207	620	1,850	279.20

01 54 | Construction Aids

01 54 33 | Equipment Rental

			UNIT	HOURLY OPER. COST	RENT PER DAY	RENT PER WEEK	RENT PER MONTH	EQUIPMENT COST/DAY	
40	7410	330 H.P.	Ea.	28.45	283	850	2,550	397.60	40
	7500	6 x 4, 380 H.P.		32.55	330	995	2,975	459.40	
	7600	450 H.P.		39.85	400	1,200	3,600	558.80	
	7610	Tractor, with A frame, boom and winch, 225 H.P.		22.10	283	850	2,550	346.80	
	7620	Vacuum truck, hazardous material, 2500 gallons		11.35	315	945	2,825	279.80	
	7625	5,000 gallons		13.89	440	1,320	3,950	375.10	
	7650	Vacuum, HEPA, 16 gallon, wet/dry		.36	18	54	162	13.70	
	7655	55 gallon, wet/dry		.59	27	81	243	20.90	
	7660	Water tank, portable		.17	28.50	85.50	257	18.45	
	7690	Sewer/catch basin vacuum, 14 C.Y., 1500 gallons		18.51	660	1,980	5,950	544.10	
	7700	Welder, electric, 200 amp		1.70	17.35	52	156	24	
	7800	300 amp		2.48	21	63	189	32.45	
	7900	Gas engine, 200 amp		10.15	25.50	76	228	96.40	
	8000	300 amp		11.60	27	81	243	109	
	8100	Wheelbarrow, any size		.08	12.65	38	114	8.25	
	8200	Wrecking ball, 4000 lbs.		2.20	73.50	220	660	61.60	
50	0010	**HIGHWAY EQUIPMENT RENTAL** without operators R015433-10							50
	0050	Asphalt batch plant, portable drum mixer, 100 ton/hr	Ea.	68.35	1,475	4,400	13,200	1,427	
	0060	200 ton/hr		76.45	1,550	4,655	14,000	1,543	
	0070	300 ton/hr		88.65	1,825	5,485	16,500	1,806	
	0100	Backhoe attachment, long stick, up to 185 H.P., 10.5' long		.33	22	66	198	15.85	
	0140	Up to 250 H.P., 12' long		.36	23.50	71	213	17.10	
	0180	Over 250 H.P., 15' long		.49	32.50	97	291	23.30	
	0200	Special dipper arm, up to 100 H.P., 32' long		1.00	66.50	199	595	47.80	
	0240	Over 100 H.P., 33' long		1.24	82.50	248	745	59.50	
	0280	Catch basin/sewer cleaning truck, 3 ton, 9 C.Y., 1000 gal		30.85	415	1,250	3,750	496.80	
	0300	Concrete batch plant, portable, electric, 200 C.Y./hr.		18.20	530	1,585	4,750	462.60	
	0520	Grader/dozer attachment, ripper/scarifier, rear mounted, up to 135 H.P.		3.30	68.50	205	615	67.40	
	0540	Up to 180 H.P.		3.90	86.50	260	780	83.20	
	0580	Up to 250 H.P.		4.30	100	300	900	94.40	
	0700	Pvmt. removal bucket, for hyd. excavator, up to 90 H.P.		1.75	51.50	155	465	45	
	0740	Up to 200 H.P.		1.95	73.50	220	660	59.60	
	0780	Over 200 H.P.		2.10	85	255	765	67.80	
	0900	Aggregate spreader, self-propelled, 187 H.P.		40.05	715	2,140	6,425	748.40	
	1000	Chemical spreader, 3 C.Y.		2.75	45	135	405	49	
	1900	Hammermill, traveling, 250 H.P.		67.55	1,925	5,760	17,300	1,692	
	2000	Horizontal borer, 3" diameter, 13 H.P. gas driven		5.25	56.50	170	510	76	
	2150	Horizontal directional drill, 20,000 lb. thrust, 78 H.P. diesel		26.20	690	2,065	6,200	622.60	
	2160	30,000 lb. thrust, 115 H.P.		32.30	1,050	3,150	9,450	888.40	
	2170	50,000 lb. thrust, 170 H.P.		46.00	1,350	4,030	12,100	1,174	
	2190	Mud trailer for HDD, 1500 gallons, 175 H.P., gas		21.10	162	485	1,450	265.80	
	2200	Hydromulcher, diesel, 3000 gallon, for truck mounting		15.00	263	790	2,375	278	
	2300	Gas, 600 gallon		6.40	103	310	930	113.20	
	2400	Joint & crack cleaner, walk behind, 25 H.P.		2.75	51.50	155	465	53	
	2500	Filler, trailer mounted, 400 gallons, 20 H.P.		7.70	222	665	2,000	194.60	
	3000	Paint striper, self-propelled, double line, 30 H.P.		6.10	165	495	1,475	147.80	
	3200	Post drivers, 6" I-Beam frame, for truck mounting		11.45	440	1,325	3,975	356.60	
	3400	Road sweeper, self-propelled, 8' wide, 90 H.P.		31.10	595	1,780	5,350	604.80	
	3450	Road sweeper, vacuum assisted, 4 C.Y., 220 gallons		51.45	650	1,945	5,825	800.60	
	4000	Road mixer, self-propelled, 130 H.P.		41.45	785	2,350	7,050	801.60	
	4100	310 H.P.		71.70	2,200	6,580	19,700	1,890	
	4220	Cold mix paver, incl. pug mill and bitumen tank, 165 H.P.		90.60	2,150	6,455	19,400	2,016	
	4250	Paver, asphalt, wheel or crawler, 130 H.P., diesel		89.10	2,050	6,180	18,500	1,949	
	4300	Paver, road widener, gas 1' to 6', 67 H.P.		42.15	885	2,650	7,950	867.20	
	4400	Diesel, 2' to 14', 88 H.P.		53.00	1,075	3,215	9,650	1,067	
	4600	Slipform pavers, curb and gutter, 2 track, 75 H.P.		35.60	765	2,300	6,900	744.80	
	4700	4 track, 165 H.P.		44.65	820	2,455	7,375	848.20	
	4800	Median barrier, 215 H.P.		45.40	850	2,550	7,650	873.20	

01 54 | Construction Aids

	01 54 33 | Equipment Rental		UNIT	HOURLY OPER. COST	RENT PER DAY	RENT PER WEEK	RENT PER MONTH	EQUIPMENT COST/DAY	
50	4901	Trailer, low bed, 75 ton capacity	Ea.	9.65	237	710	2,125	219.20	50
	5000	Road planer, walk behind, 10" cutting width, 10 H.P.		2.70	33.50	100	300	41.60	
	5100	Self-propelled, 12" cutting width, 64 H.P.		7.00	115	345	1,025	125	
	5120	Traffic line remover, metal ball blaster, truck mounted, 115 H.P.		47.15	765	2,300	6,900	837.20	
	5140	Grinder, truck mounted, 115 H.P.		50.10	835	2,505	7,525	901.80	
	5160	Walk-behind, 11 H.P.		3.05	55	165	495	57.40	
	5200	Pavement profiler, 4' to 6' wide, 450 H.P.		210.55	3,425	10,285	30,900	3,741	
	5300	8' to 10' wide, 750 H.P.		332.90	4,950	14,815	44,400	5,626	
	5400	Roadway plate, steel, 1" x 8' x 20'		.07	12	36	108	7.75	
	5600	Stabilizer, self-propelled, 150 H.P.		39.35	615	1,850	5,550	684.80	
	5700	310 H.P.		66.85	1,325	3,955	11,900	1,326	
	5800	Striper, thermal, truck mounted 120 gallon paint, 150 H.P.		42.85	520	1,555	4,675	653.80	
	6000	Tar kettle, 330 gallon, trailer mounted		4.58	63.50	190	570	74.65	
	7000	Tunnel locomotive, diesel, 8 to 12 ton		27.45	585	1,755	5,275	570.60	
	7005	Electric, 10 ton		24.75	670	2,005	6,025	599	
	7010	Muck cars, 1/2 C.Y. capacity		1.90	23.50	71	213	29.40	
	7020	1 C.Y. capacity		2.15	32.50	98	294	36.80	
	7030	2 C.Y. capacity		2.25	36.50	110	330	40	
	7040	Side dump, 2 C.Y. capacity		2.45	45	135	405	46.60	
	7050	3 C.Y. capacity		3.30	51.50	155	465	57.40	
	7060	5 C.Y. capacity		4.70	65	195	585	76.60	
	7100	Ventilating blower for tunnel, 7-1/2 H.P.		1.02	51.50	155	465	39.15	
	7110	10 H.P.		1.11	51.50	155	465	39.90	
	7120	20 H.P.		1.67	67.50	202	605	53.75	
	7140	40 H.P.		2.77	96.50	290	870	80.15	
	7160	60 H.P.		4.26	152	455	1,375	125.10	
	7175	75 H.P.		5.51	202	607	1,825	165.50	
	7180	200 H.P.		11.15	305	910	2,725	271.20	
	7800	Windrow loader, elevating	↓	49.40	1,350	4,075	12,200	1,210	
60	0010	**LIFTING AND HOISTING EQUIPMENT RENTAL** without operators	R015433 -10						60
	0120	Aerial lift truck, 2 person, to 80'	Ea.	24.90	730	2,185	6,550	636.20	
	0140	Boom work platform, 40' snorkel	R015433 -15	14.25	278	835	2,500	281	
	0150	Crane, flatbed mounted, 3 ton capacity		13.45	197	590	1,775	225.60	
	0200	Crane, climbing, 106' jib, 6000 lb. capacity, 410 fpm	R312316 -45	29.50	1,525	4,580	13,700	1,152	
	0300	101' jib, 10,250 lb. capacity, 270 fpm		35.60	1,925	5,800	17,400	1,445	
	0500	Tower, static, 130' high, 106' jib, 6200 lb. capacity at 400 fpm		33.05	1,775	5,290	15,900	1,322	
	0600	Crawler mounted, lattice boom, 1/2 C.Y., 15 tons at 12' radius		30.74	685	2,060	6,175	657.90	
	0700	3/4 C.Y., 20 tons at 12' radius		40.99	855	2,570	7,700	841.90	
	0800	1 C.Y., 25 tons at 12' radius		54.65	1,150	3,430	10,300	1,123	
	0900	1-1/2 C.Y., 40 tons at 12' radius		54.65	1,150	3,455	10,400	1,128	
	1000	2 C.Y., 50 tons at 12' radius		58.55	1,350	4,055	12,200	1,279	
	1100	3 C.Y., 75 tons at 12' radius		63.05	1,575	4,710	14,100	1,446	
	1200	100 ton capacity, 60' boom		71.50	1,800	5,405	16,200	1,653	
	1300	165 ton capacity, 60' boom		89.20	2,100	6,325	19,000	1,979	
	1400	200 ton capacity, 70' boom		111.15	2,675	7,995	24,000	2,488	
	1500	350 ton capacity, 80' boom		157.20	4,050	12,160	36,500	3,690	
	1600	Truck mounted, lattice boom, 6 x 4, 20 tons at 10' radius		37.39	1,225	3,680	11,000	1,035	
	1700	25 tons at 10' radius		40.40	1,325	4,000	12,000	1,123	
	1800	8 x 4, 30 tons at 10' radius		43.77	1,425	4,260	12,800	1,202	
	1900	40 tons at 12' radius		46.65	1,475	4,450	13,400	1,263	
	2000	60 tons at 15' radius		52.35	1,575	4,710	14,100	1,361	
	2050	82 tons at 15' radius		58.46	1,675	5,030	15,100	1,474	
	2100	90 tons at 15' radius		65.56	1,825	5,490	16,500	1,622	
	2200	115 tons at 15' radius		73.91	2,050	6,130	18,400	1,817	
	2300	150 tons at 18' radius		76.90	2,150	6,455	19,400	1,906	
	2350	165 tons at 18' radius		86.63	2,275	6,840	20,500	2,061	
	2400	Truck mounted, hydraulic, 12 ton capacity	↓	41.45	540	1,620	4,850	655.60	

01 54 | Construction Aids

	01 54 33 | Equipment Rental		UNIT	HOURLY OPER. COST	RENT PER DAY	RENT PER WEEK	RENT PER MONTH	EQUIPMENT COST/DAY	
60	2500	25 ton capacity	Ea.	43.70	665	1,995	5,975	748.60	60
	2550	33 ton capacity		44.25	680	2,045	6,125	763	
	2560	40 ton capacity		57.20	800	2,405	7,225	938.60	
	2600	55 ton capacity		75.80	1,000	3,010	9,025	1,208	
	2700	80 ton capacity		97.75	1,525	4,555	13,700	1,693	
	2720	100 ton capacity		91.65	1,525	4,585	13,800	1,650	
	2740	120 ton capacity		106.15	1,625	4,910	14,700	1,831	
	2760	150 ton capacity		124.30	2,225	6,710	20,100	2,336	
	2800	Self-propelled, 4 x 4, with telescoping boom, 5 ton		14.60	235	705	2,125	257.80	
	2900	12-1/2 ton capacity		25.80	380	1,140	3,425	434.40	
	3000	15 ton capacity		26.45	405	1,210	3,625	453.60	
	3050	20 ton capacity		28.25	475	1,425	4,275	511	
	3100	25 ton capacity		29.80	535	1,610	4,825	560.40	
	3150	40 ton capacity		36.95	605	1,815	5,450	658.60	
	3200	Derricks, guy, 20 ton capacity, 60' boom, 75' mast		23.35	375	1,122	3,375	411.20	
	3300	100' boom, 115' mast		36.76	645	1,930	5,800	680.10	
	3400	Stiffleg, 20 ton capacity, 70' boom, 37' mast		25.72	485	1,460	4,375	497.75	
	3500	100' boom, 47' mast		39.56	775	2,330	7,000	782.50	
	3550	Helicopter, small, lift to 1250 lbs. maximum, w/pilot		95.92	3,025	9,060	27,200	2,579	
	3600	Hoists, chain type, overhead, manual, 3/4 ton		.15	1	3	9	1.80	
	3900	10 ton		.70	7	21	63	9.80	
	4000	Hoist and tower, 5000 lb. cap., portable electric, 40' high		3.89	215	646	1,950	160.30	
	4100	For each added 10' section, add		.10	17	51	153	11	
	4200	Hoist and single tubular tower, 5000 lb. electric, 100' high		5.34	300	902	2,700	223.10	
	4300	For each added 6'-6" section, add		.17	29	87	261	18.75	
	4400	Hoist and double tubular tower, 5000 lb., 100' high		5.79	330	993	2,975	244.90	
	4500	For each added 6'-6" section, add		.19	32.50	97	291	20.90	
	4550	Hoist and tower, mast type, 6000 lb., 100' high		6.14	345	1,030	3,100	255.10	
	4570	For each added 10' section, add		.12	20.50	61	183	13.15	
	4600	Hoist and tower, personnel, electric, 2000 lb., 100' @ 125 fpm		14.36	915	2,740	8,225	662.90	
	4700	3000 lb., 100' @ 200 fpm		16.38	1,025	3,110	9,325	753.05	
	4800	3000 lb., 150' @ 300 fpm		18.23	1,150	3,480	10,400	841.85	
	4900	4000 lb., 100' @ 300 fpm		18.74	1,175	3,550	10,700	859.90	
	5000	6000 lb., 100' @ 275 fpm		19.92	1,250	3,720	11,200	903.35	
	5100	For added heights up to 500', add	L.F.	.01	1.67	5	15	1.10	
	5200	Jacks, hydraulic, 20 ton	Ea.	.05	2	6	18	1.60	
	5500	100 ton		.35	11.65	35	105	9.80	
	6100	Jacks, hydraulic, climbing w/50' jackrods, control console, 30 ton cap.		1.86	124	371	1,125	89.10	
	6150	For each added 10' jackrod section, add		.05	3.33	10	30	2.40	
	6300	50 ton capacity		2.99	199	597	1,800	143.30	
	6350	For each added 10' jackrod section, add		.06	4	12	36	2.90	
	6500	125 ton capacity		7.85	525	1,570	4,700	376.80	
	6550	For each added 10' jackrod section, add		.54	35.50	107	320	25.70	
	6600	Cable jack, 10 ton capacity with 200' cable		1.56	104	311	935	74.70	
	6650	For each added 50' of cable, add		.18	12	36	108	8.65	
70	0010	**WELLPOINT EQUIPMENT RENTAL** without operators (R015433-10)							70
	0020	Based on 2 months rental							
	0100	Combination jetting & wellpoint pump, 60 H.P. diesel	Ea.	15.74	305	920	2,750	309.90	
	0200	High pressure gas jet pump, 200 H.P., 300 psi	"	36.50	262	786	2,350	449.20	
	0300	Discharge pipe, 8" diameter	L.F.	.01	.50	1.50	4.50	.40	
	0350	12" diameter		.01	.73	2.20	6.60	.50	
	0400	Header pipe, flows up to 150 GPM, 4" diameter		.01	.45	1.35	4.05	.35	
	0500	400 GPM, 6" diameter		.01	.53	1.60	4.80	.40	
	0600	800 GPM, 8" diameter		.01	.73	2.20	6.60	.50	
	0700	1500 GPM, 10" diameter		.01	.77	2.31	6.95	.55	
	0800	2500 GPM, 12" diameter		.02	1.46	4.38	13.15	1.05	
	0900	4500 GPM, 16" diameter		.03	1.87	5.61	16.85	1.35	

01 54 | Construction Aids

		01 54 33 | Equipment Rental	UNIT	HOURLY OPER. COST	RENT PER DAY	RENT PER WEEK	RENT PER MONTH	EQUIPMENT COST/DAY	
70	0950	For quick coupling aluminum and plastic pipe, add	L.F.	.03	1.93	5.80	17.40	1.40	70
	1100	Wellpoint, 25' long, with fittings & riser pipe, 1-1/2" or 2" diameter	Ea.	.06	3.86	11.58	34.50	2.80	
	1200	Wellpoint pump, diesel powered, 4" suction, 20 H.P.		6.81	177	530	1,600	160.50	
	1300	6" suction, 30 H.P.		9.26	219	658	1,975	205.70	
	1400	8" suction, 40 H.P.		12.51	300	902	2,700	280.50	
	1500	10" suction, 75 H.P.		19.00	350	1,054	3,150	362.80	
	1600	12" suction, 100 H.P.		27.26	560	1,680	5,050	554.10	
	1700	12" suction, 175 H.P.	↓	40.08	615	1,850	5,550	690.65	
80	0010	**MARINE EQUIPMENT RENTAL** without operators R015433 -10							80
	0200	Barge, 400 Ton, 30' wide x 90' long	Ea.	16.05	1,150	3,460	10,400	820.40	
	0240	800 Ton, 45' wide x 90' long		19.50	1,400	4,190	12,600	994	
	2000	Tugboat, diesel, 100 H.P.		25.55	215	645	1,925	333.40	
	2040	250 H.P.		49.40	395	1,185	3,550	632.20	
	2080	380 H.P.	↓	108.20	1,175	3,505	10,500	1,567	

Crews

Crew No.	Bare Costs Hr.	Bare Costs Daily	Incl. Subs O&P Hr.	Incl. Subs O&P Daily	Cost Per Labor-Hour Bare Costs	Cost Per Labor-Hour Incl. O&P
Crew A-1	**Hr.**	**Daily**	**Hr.**	**Daily**	**Bare Costs**	**Incl. O&P**
1 Building Laborer	$26.45	$211.60	$44.30	$354.40	$26.45	$44.30
1 Concrete saw, gas manual		63.20		69.52	7.90	8.69
8 L.H., Daily Totals		$274.80		$423.92	$34.35	$52.99
Crew A-1A	**Hr.**	**Daily**	**Hr.**	**Daily**	**Bare Costs**	**Incl. O&P**
1 Skilled Worker	$34.05	$272.40	$57.00	$456.00	$34.05	$57.00
1 Shot Blaster, 20"		216.20		237.82	27.02	29.73
8 L.H., Daily Totals		$488.60		$693.82	$61.08	$86.73
Crew A-1B	**Hr.**	**Daily**	**Hr.**	**Daily**	**Bare Costs**	**Incl. O&P**
1 Building Laborer	$26.45	$211.60	$44.30	$354.40	$26.45	$44.30
1 Concrete Saw		134.20		147.62	16.77	18.45
8 L.H., Daily Totals		$345.80		$502.02	$43.23	$62.75
Crew A-1C	**Hr.**	**Daily**	**Hr.**	**Daily**	**Bare Costs**	**Incl. O&P**
1 Building Laborer	$26.45	$211.60	$44.30	$354.40	$26.45	$44.30
1 Chain Saw, gas, 18"		25.80		28.38	3.23	3.55
8 L.H., Daily Totals		$237.40		$382.78	$29.68	$47.85
Crew A-1D	**Hr.**	**Daily**	**Hr.**	**Daily**	**Bare Costs**	**Incl. O&P**
1 Building Laborer	$26.45	$211.60	$44.30	$354.40	$26.45	$44.30
1 Vibrating plate, gas, 18"		30.80		33.88	3.85	4.24
8 L.H., Daily Totals		$242.40		$388.28	$30.30	$48.53
Crew A-1E	**Hr.**	**Daily**	**Hr.**	**Daily**	**Bare Costs**	**Incl. O&P**
1 Building Laborer	$26.45	$211.60	$44.30	$354.40	$26.45	$44.30
1 Vibratory Plate, gas, 21"		39.00		42.90	4.88	5.36
8 L.H., Daily Totals		$250.60		$397.30	$31.32	$49.66
Crew A-1F	**Hr.**	**Daily**	**Hr.**	**Daily**	**Bare Costs**	**Incl. O&P**
1 Building Laborer	$26.45	$211.60	$44.30	$354.40	$26.45	$44.30
1 Rammer/tamper, gas, 8"		43.60		47.96	5.45	6.00
8 L.H., Daily Totals		$255.20		$402.36	$31.90	$50.30
Crew A-1G	**Hr.**	**Daily**	**Hr.**	**Daily**	**Bare Costs**	**Incl. O&P**
1 Building Laborer	$26.45	$211.60	$44.30	$354.40	$26.45	$44.30
1 Rammer/tamper, gas, 15"		49.60		54.56	6.20	6.82
8 L.H., Daily Totals		$261.20		$408.96	$32.65	$51.12
Crew A-1H	**Hr.**	**Daily**	**Hr.**	**Daily**	**Bare Costs**	**Incl. O&P**
1 Building Laborer	$26.45	$211.60	$44.30	$354.40	$26.45	$44.30
1 Exterior Steam Cleaner		61.60		67.76	7.70	8.47
8 L.H., Daily Totals		$273.20		$422.16	$34.15	$52.77
Crew A-1J	**Hr.**	**Daily**	**Hr.**	**Daily**	**Bare Costs**	**Incl. O&P**
1 Building Laborer	$26.45	$211.60	$44.30	$354.40	$26.45	$44.30
1 Cultivator, Walk-Behind, 5 H.P.		58.45		64.30	7.31	8.04
8 L.H., Daily Totals		$270.05		$418.69	$33.76	$52.34
Crew A-1K	**Hr.**	**Daily**	**Hr.**	**Daily**	**Bare Costs**	**Incl. O&P**
1 Building Laborer	$26.45	$211.60	$44.30	$354.40	$26.45	$44.30
1 Cultivator, Walk-Behind, 8 H.P.		89.75		98.72	11.22	12.34
8 L.H., Daily Totals		$301.35		$453.13	$37.67	$56.64
Crew A-1M	**Hr.**	**Daily**	**Hr.**	**Daily**	**Bare Costs**	**Incl. O&P**
1 Building Laborer	$26.45	$211.60	$44.30	$354.40	$26.45	$44.30
1 Snow Blower, Walk-Behind		53.10		58.41	6.64	7.30
8 L.H., Daily Totals		$264.70		$412.81	$33.09	$51.60

Crew No.	Bare Costs Hr.	Bare Costs Daily	Incl. Subs O&P Hr.	Incl. Subs O&P Daily	Cost Per Labor-Hour Bare Costs	Cost Per Labor-Hour Incl. O&P
Crew A-2	**Hr.**	**Daily**	**Hr.**	**Daily**	**Bare Costs**	**Incl. O&P**
2 Laborers	$26.45	$423.20	$44.30	$708.80	$26.43	$44.13
1 Truck Driver (light)	26.40	211.20	43.80	350.40		
1 Flatbed Truck, Gas, 1.5 Ton		171.40		188.54	7.14	7.86
24 L.H., Daily Totals		$805.80		$1247.74	$33.58	$51.99
Crew A-2A	**Hr.**	**Daily**	**Hr.**	**Daily**	**Bare Costs**	**Incl. O&P**
2 Laborers	$26.45	$423.20	$44.30	$708.80	$26.43	$44.13
1 Truck Driver (light)	26.40	211.20	43.80	350.40		
1 Flatbed Truck, Gas, 1.5 Ton		171.40		188.54		
1 Concrete Saw		134.20		147.62	12.73	14.01
24 L.H., Daily Totals		$940.00		$1395.36	$39.17	$58.14
Crew A-2B	**Hr.**	**Daily**	**Hr.**	**Daily**	**Bare Costs**	**Incl. O&P**
1 Truck Driver (light)	$26.40	$211.20	$43.80	$350.40	$26.40	$43.80
1 Flatbed Truck, Gas, 1.5 Ton		171.40		188.54	21.43	23.57
8 L.H., Daily Totals		$382.60		$538.94	$47.83	$67.37
Crew A-3A	**Hr.**	**Daily**	**Hr.**	**Daily**	**Bare Costs**	**Incl. O&P**
1 Truck Driver (light)	$26.40	$211.20	$43.80	$350.40	$26.40	$43.80
1 Pickup truck, 4 x 4, 3/4 ton		116.60		128.26	14.57	16.03
8 L.H., Daily Totals		$327.80		$478.66	$40.98	$59.83
Crew A-3B	**Hr.**	**Daily**	**Hr.**	**Daily**	**Bare Costs**	**Incl. O&P**
1 Equip. Oper. (med.)	$34.90	$279.20	$57.05	$456.40	$31.05	$51.08
1 Truck Driver (heav.)	27.20	217.60	45.10	360.80		
1 Dump Truck, 12 C.Y., 400 H.P.		506.00		556.60		
1 F.E. Loader, W.M.,2.5 C.Y.		397.60		437.36	56.48	62.12
16 L.H., Daily Totals		$1400.40		$1811.16	$87.53	$113.20
Crew A-3C	**Hr.**	**Daily**	**Hr.**	**Daily**	**Bare Costs**	**Incl. O&P**
1 Equip. Oper. (light)	$33.55	$268.40	$54.80	$438.40	$33.55	$54.80
1 Loader, Skid Steer, 78 H.P.		261.20		287.32	32.65	35.91
8 L.H., Daily Totals		$529.60		$725.72	$66.20	$90.72
Crew A-3D	**Hr.**	**Daily**	**Hr.**	**Daily**	**Bare Costs**	**Incl. O&P**
1 Truck Driver, Light	$26.40	$211.20	$43.80	$350.40	$26.40	$43.80
1 Pickup truck, 4 x 4, 3/4 ton		116.60		128.26		
1 Flatbed Trailer, 25 Ton		111.20		122.32	28.48	31.32
8 L.H., Daily Totals		$439.00		$600.98	$54.88	$75.12
Crew A-3E	**Hr.**	**Daily**	**Hr.**	**Daily**	**Bare Costs**	**Incl. O&P**
1 Equip. Oper. (crane)	$35.80	$286.40	$58.50	$468.00	$31.50	$51.80
1 Truck Driver (heavy)	27.20	217.60	45.10	360.80		
1 Pickup truck, 4 x 4, 3/4 ton		116.60		128.26	7.29	8.02
16 L.H., Daily Totals		$620.60		$957.06	$38.79	$59.82
Crew A-3F	**Hr.**	**Daily**	**Hr.**	**Daily**	**Bare Costs**	**Incl. O&P**
1 Equip. Oper. (crane)	$35.80	$286.40	$58.50	$468.00	$31.50	$51.80
1 Truck Driver (heavy)	27.20	217.60	45.10	360.80		
1 Pickup truck, 4 x 4, 3/4 ton		116.60		128.26		
1 Truck Tractor, 6x4, 380 H.P.		459.40		505.34		
1 Lowbed Trailer, 75 Ton		219.20		241.12	49.70	54.67
16 L.H., Daily Totals		$1299.20		$1703.52	$81.20	$106.47

Crews

Crew No.	Bare Costs Hr.	Bare Costs Daily	Incl. Subs O&P Hr.	Incl. Subs O&P Daily	Cost Per Labor-Hour Bare Costs	Cost Per Labor-Hour Incl. O&P
Crew A-3G	**Hr.**	**Daily**	**Hr.**	**Daily**	**Bare Costs**	**Incl. O&P**
1 Equip. Oper. (crane)	$35.80	$286.40	$58.50	$468.00	$31.50	$51.80
1 Truck Driver (heavy)	27.20	217.60	45.10	360.80		
1 Pickup truck, 4 x 4, 3/4 ton		116.60		128.26		
1 Truck Tractor, 6x4, 450 H.P.		558.80		614.68		
1 Lowbed Trailer, 75 Ton		219.20		241.12	55.91	61.50
16 L.H., Daily Totals		$1398.60		$1812.86	$87.41	$113.30
Crew A-3H	**Hr.**	**Daily**	**Hr.**	**Daily**	**Bare Costs**	**Incl. O&P**
1 Equip. Oper. (crane)	$35.80	$286.40	$58.50	$468.00	$35.80	$58.50
1 Hyd. crane, 12 Ton (daily)		871.60		958.76	108.95	119.85
8 L.H., Daily Totals		$1158.00		$1426.76	$144.75	$178.35
Crew A-3I	**Hr.**	**Daily**	**Hr.**	**Daily**	**Bare Costs**	**Incl. O&P**
1 Equip. Oper. (crane)	$35.80	$286.40	$58.50	$468.00	$35.80	$58.50
1 Hyd. crane, 25 Ton (daily)		1015.00		1116.50	126.88	139.56
8 L.H., Daily Totals		$1301.40		$1584.50	$162.68	$198.06
Crew A-3J	**Hr.**	**Daily**	**Hr.**	**Daily**	**Bare Costs**	**Incl. O&P**
1 Equip. Oper. (crane)	$35.80	$286.40	$58.50	$468.00	$35.80	$58.50
1 Hyd. crane, 40 Ton (daily)		1258.00		1383.80	157.25	172.97
8 L.H., Daily Totals		$1544.40		$1851.80	$193.05	$231.47
Crew A-3K	**Hr.**	**Daily**	**Hr.**	**Daily**	**Bare Costs**	**Incl. O&P**
1 Equip. Oper. (crane)	$35.80	$286.40	$58.50	$468.00	$33.42	$54.63
1 Equip. Oper. Oiler	31.05	248.40	50.75	406.00		
1 Hyd. crane, 55 Ton (daily)		1611.00		1772.10		
1 P/U Truck, 3/4 Ton (daily)		129.60		142.56	108.79	119.67
16 L.H., Daily Totals		$2275.40		$2788.66	$142.21	$174.29
Crew A-3L	**Hr.**	**Daily**	**Hr.**	**Daily**	**Bare Costs**	**Incl. O&P**
1 Equip. Oper. (crane)	$35.80	$286.40	$58.50	$468.00	$33.42	$54.63
1 Equip. Oper. Oiler	31.05	248.40	50.75	406.00		
1 Hyd. crane, 80 Ton (daily)		2302.00		2532.20		
1 P/U Truck, 3/4 Ton (daily)		129.60		142.56	151.97	167.17
16 L.H., Daily Totals		$2966.40		$3548.76	$185.40	$221.80
Crew A-3M	**Hr.**	**Daily**	**Hr.**	**Daily**	**Bare Costs**	**Incl. O&P**
1 Equip. Oper. (crane)	$35.80	$286.40	$58.50	$468.00	$33.42	$54.63
1 Equip. Oper. Oiler	31.05	248.40	50.75	406.00		
1 Hyd. crane, 100 Ton (daily)		2263.00		2489.30		
1 P/U Truck, 3/4 Ton (daily)		129.60		142.56	149.54	164.49
16 L.H., Daily Totals		$2927.40		$3505.86	$182.96	$219.12
Crew A-3N	**Hr.**	**Daily**	**Hr.**	**Daily**	**Bare Costs**	**Incl. O&P**
1 Equip. Oper. (crane)	$35.80	$286.40	$58.50	$468.00	$35.80	$58.50
1 Tower crane (monthly)		987.20		1085.92	123.40	135.74
8 L.H., Daily Totals		$1273.60		$1553.92	$159.20	$194.24
Crew A-3P	**Hr.**	**Daily**	**Hr.**	**Daily**	**Bare Costs**	**Incl. O&P**
1 Equip. Oper., Light	$33.55	$268.40	$54.80	$438.40	$33.55	$54.80
1 A.T. Forklift, 42' lift		447.40		492.14	55.92	61.52
8 L.H., Daily Totals		$715.80		$930.54	$89.47	$116.32
Crew A-4	**Hr.**	**Daily**	**Hr.**	**Daily**	**Bare Costs**	**Incl. O&P**
2 Carpenters	$33.55	$536.80	$56.15	$898.40	$32.25	$53.48
1 Painter, Ordinary	29.65	237.20	48.15	385.20		
24 L.H., Daily Totals		$774.00		$1283.60	$32.25	$53.48

Crew No.	Bare Costs Hr.	Bare Costs Daily	Incl. Subs O&P Hr.	Incl. Subs O&P Daily	Cost Per Labor-Hour Bare Costs	Cost Per Labor-Hour Incl. O&P
Crew A-5	**Hr.**	**Daily**	**Hr.**	**Daily**	**Bare Costs**	**Incl. O&P**
2 Laborers	$26.45	$423.20	$44.30	$708.80	$26.44	$44.24
.25 Truck Driver (light)	26.40	52.80	43.80	87.60		
.25 Flatbed Truck, Gas, 1.5 Ton		42.85		47.13	2.38	2.62
18 L.H., Daily Totals		$518.85		$843.53	$28.82	$46.86
Crew A-6	**Hr.**	**Daily**	**Hr.**	**Daily**	**Bare Costs**	**Incl. O&P**
1 Instrument Man	$34.05	$272.40	$57.00	$456.00	$33.63	$55.75
1 Rodman/Chainman	33.20	265.60	54.50	436.00		
1 Level, electronic		69.60		76.56	4.35	4.79
16 L.H., Daily Totals		$607.60		$968.56	$37.98	$60.53
Crew A-7	**Hr.**	**Daily**	**Hr.**	**Daily**	**Bare Costs**	**Incl. O&P**
1 Chief Of Party	$40.30	$322.40	$67.15	$537.20	$35.85	$59.55
1 Instrument Man	34.05	272.40	57.00	456.00		
1 Rodman/Chainman	33.20	265.60	54.50	436.00		
1 Level, electronic		69.60		76.56	2.90	3.19
24 L.H., Daily Totals		$930.00		$1505.76	$38.75	$62.74
Crew A-8	**Hr.**	**Daily**	**Hr.**	**Daily**	**Bare Costs**	**Incl. O&P**
1 Chief of Party	$40.30	$322.40	$67.15	$537.20	$35.19	$58.29
1 Instrument Man	34.05	272.40	57.00	456.00		
2 Rodmen/Chainmen	33.20	531.20	54.50	872.00		
1 Level, electronic		69.60		76.56	2.17	2.39
32 L.H., Daily Totals		$1195.60		$1941.76	$37.36	$60.68
Crew A-9	**Hr.**	**Daily**	**Hr.**	**Daily**	**Bare Costs**	**Incl. O&P**
1 Asbestos Foreman	$35.20	$281.60	$59.35	$474.80	$34.76	$58.61
7 Asbestos Workers	34.70	1943.20	58.50	3276.00		
64 L.H., Daily Totals		$2224.80		$3750.80	$34.76	$58.61
Crew A-10A	**Hr.**	**Daily**	**Hr.**	**Daily**	**Bare Costs**	**Incl. O&P**
1 Asbestos Foreman	$35.20	$281.60	$59.35	$474.80	$34.87	$58.78
2 Asbestos Workers	34.70	555.20	58.50	936.00		
24 L.H., Daily Totals		$836.80		$1410.80	$34.87	$58.78
Crew A-10B	**Hr.**	**Daily**	**Hr.**	**Daily**	**Bare Costs**	**Incl. O&P**
1 Asbestos Foreman	$35.20	$281.60	$59.35	$474.80	$34.83	$58.71
3 Asbestos Workers	34.70	832.80	58.50	1404.00		
32 L.H., Daily Totals		$1114.40		$1878.80	$34.83	$58.71
Crew A-10C	**Hr.**	**Daily**	**Hr.**	**Daily**	**Bare Costs**	**Incl. O&P**
3 Asbestos Workers	$34.70	$832.80	$58.50	$1404.00	$34.70	$58.50
1 Flatbed Truck, Gas, 1.5 Ton		171.40		188.54	7.14	7.86
24 L.H., Daily Totals		$1004.20		$1592.54	$41.84	$66.36
Crew A-10D	**Hr.**	**Daily**	**Hr.**	**Daily**	**Bare Costs**	**Incl. O&P**
2 Asbestos Workers	$34.70	$555.20	$58.50	$936.00	$34.06	$56.56
1 Equip. Oper. (crane)	35.80	286.40	58.50	468.00		
1 Equip. Oper. Oiler	31.05	248.40	50.75	406.00		
1 Hydraulic Crane, 33 Ton		763.00		839.30	23.84	26.23
32 L.H., Daily Totals		$1853.00		$2649.30	$57.91	$82.79
Crew A-11	**Hr.**	**Daily**	**Hr.**	**Daily**	**Bare Costs**	**Incl. O&P**
1 Asbestos Foreman	$35.20	$281.60	$59.35	$474.80	$34.76	$58.61
7 Asbestos Workers	34.70	1943.20	58.50	3276.00		
2 Chip. Hammers, 12 Lb., Elec.		37.20		40.92	0.58	0.64
64 L.H., Daily Totals		$2262.00		$3791.72	$35.34	$59.25

Crews

Crew No.	Bare Costs		Incl. Subs O&P		Cost Per Labor-Hour	
Crew A-12	**Hr.**	**Daily**	**Hr.**	**Daily**	**Bare Costs**	**Incl. O&P**
1 Asbestos Foreman	$35.20	$281.60	$59.35	$474.80	$34.76	$58.61
7 Asbestos Workers	34.70	1943.20	58.50	3276.00		
1 Trk-mtd vac, 14 CY, 1500 Gal.		544.10		598.51		
1 Flatbed Truck, 20,000 GVW		190.00		209.00	11.47	12.62
64 L.H., Daily Totals		$2958.90		$4558.31	$46.23	$71.22

Crew A-13	**Hr.**	**Daily**	**Hr.**	**Daily**	**Bare Costs**	**Incl. O&P**
1 Equip. Oper. (light)	$33.55	$268.40	$54.80	$438.40	$33.55	$54.80
1 Trk-mtd vac, 14 CY, 1500 Gal.		544.10		598.51		
1 Flatbed Truck, 20,000 GVW		190.00		209.00	91.76	100.94
8 L.H., Daily Totals		$1002.50		$1245.91	$125.31	$155.74

Crew B-1	**Hr.**	**Daily**	**Hr.**	**Daily**	**Bare Costs**	**Incl. O&P**
1 Labor Foreman (outside)	$28.45	$227.60	$47.65	$381.20	$27.12	$45.42
2 Laborers	26.45	423.20	44.30	708.80		
24 L.H., Daily Totals		$650.80		$1090.00	$27.12	$45.42

Crew B-1A	**Hr.**	**Daily**	**Hr.**	**Daily**	**Bare Costs**	**Incl. O&P**
1 Laborer Foreman	$28.45	$227.60	$47.65	$381.20	$27.12	$45.42
2 Laborers	26.45	423.20	44.30	708.80		
2 Cutting Torches		36.00		39.60		
2 Set of Gases		165.60		182.16	8.40	9.24
24 L.H., Daily Totals		$852.40		$1311.76	$35.52	$54.66

Crew B-1B	**Hr.**	**Daily**	**Hr.**	**Daily**	**Bare Costs**	**Incl. O&P**
1 Laborer Foreman	$28.45	$227.60	$47.65	$381.20	$29.29	$48.69
2 Laborers	26.45	423.20	44.30	708.80		
1 Equip. Oper. (crane)	35.80	286.40	58.50	468.00		
2 Cutting Torches		36.00		39.60		
2 Set of Gases		165.60		182.16		
1 Hyd. Crane, 12 Ton		655.60		721.16	26.79	29.47
32 L.H., Daily Totals		$1794.40		$2500.92	$56.08	$78.15

Crew B-2	**Hr.**	**Daily**	**Hr.**	**Daily**	**Bare Costs**	**Incl. O&P**
1 Labor Foreman (outside)	$28.45	$227.60	$47.65	$381.20	$26.85	$44.97
4 Laborers	26.45	846.40	44.30	1417.60		
40 L.H., Daily Totals		$1074.00		$1798.80	$26.85	$44.97

Crew B-3	**Hr.**	**Daily**	**Hr.**	**Daily**	**Bare Costs**	**Incl. O&P**
1 Labor Foreman (outside)	$28.45	$227.60	$47.65	$381.20	$28.44	$47.25
2 Laborers	26.45	423.20	44.30	708.80		
1 Equip. Oper. (med.)	34.90	279.20	57.05	456.40		
2 Truck Drivers (heavy)	27.20	435.20	45.10	721.60		
1 Crawler Loader, 3 C.Y.		1008.00		1108.80		
2 Dump Trucks 12 C.Y., 400 H.P.		1012.00		1113.20	42.08	46.29
48 L.H., Daily Totals		$3385.20		$4490.00	$70.53	$93.54

Crew B-3A	**Hr.**	**Daily**	**Hr.**	**Daily**	**Bare Costs**	**Incl. O&P**
4 Laborers	$26.45	$846.40	$44.30	$1417.60	$28.14	$46.85
1 Equip. Oper. (med.)	34.90	279.20	57.05	456.40		
1 Hyd. Excavator, 1.5 C.Y.		947.40		1042.14	23.68	26.05
40 L.H., Daily Totals		$2073.00		$2916.14	$51.83	$72.90

Crew B-3B	**Hr.**	**Daily**	**Hr.**	**Daily**	**Bare Costs**	**Incl. O&P**
2 Laborers	$26.45	$423.20	$44.30	$708.80	$28.75	$47.69
1 Equip. Oper. (med.)	34.90	279.20	57.05	456.40		
1 Truck Driver (heavy)	27.20	217.60	45.10	360.80		
1 Backhoe Loader, 80 H.P.		349.40		384.34		
1 Dump Truck, 12 C.Y., 400 H.P.		506.00		556.60	26.73	29.40
32 L.H., Daily Totals		$1775.40		$2466.94	$55.48	$77.09

Crew No.	Bare Costs		Incl. Subs O&P		Cost Per Labor-Hour	
Crew B-3C	**Hr.**	**Daily**	**Hr.**	**Daily**	**Bare Costs**	**Incl. O&P**
3 Laborers	$26.45	$634.80	$44.30	$1063.20	$28.56	$47.49
1 Equip. Oper. (med.)	34.90	279.20	57.05	456.40		
1 Crawler Loader, 4 C.Y.		1432.00		1575.20	44.75	49.23
32 L.H., Daily Totals		$2346.00		$3094.80	$73.31	$96.71

Crew B-4	**Hr.**	**Daily**	**Hr.**	**Daily**	**Bare Costs**	**Incl. O&P**
1 Labor Foreman (outside)	$28.45	$227.60	$47.65	$381.20	$26.91	$44.99
4 Laborers	26.45	846.40	44.30	1417.60		
1 Truck Driver (heavy)	27.20	217.60	45.10	360.80		
1 Truck Tractor, 220 H.P.		279.20		307.12		
1 Flatbed Trailer, 40 Ton		150.20		165.22	8.95	9.84
48 L.H., Daily Totals		$1721.00		$2631.94	$35.85	$54.83

Crew B-5	**Hr.**	**Daily**	**Hr.**	**Daily**	**Bare Costs**	**Incl. O&P**
1 Labor Foreman (outside)	$28.45	$227.60	$47.65	$381.20	$28.54	$47.52
3 Laborers	26.45	634.80	44.30	1063.20		
1 Equip. Oper. (med.)	34.90	279.20	57.05	456.40		
1 Air Compressor, 250 cfm		162.80		179.08		
2 Breakers, Pavement, 60 lb.		19.60		21.56		
2 -50' Air Hoses, 1.5"		12.60		13.86		
1 Crawler Loader, 3 C.Y.		1008.00		1108.80	30.07	33.08
40 L.H., Daily Totals		$2344.60		$3224.10	$58.62	$80.60

Crew B-5A	**Hr.**	**Daily**	**Hr.**	**Daily**	**Bare Costs**	**Incl. O&P**
1 Foreman	$28.45	$227.60	$47.65	$381.20	$28.74	$47.71
6 Laborers	26.45	1269.60	44.30	2126.40		
2 Equip. Oper. (med.)	34.90	558.40	57.05	912.80		
1 Equip. Oper. (light)	33.55	268.40	54.80	438.40		
2 Truck Drivers (heavy)	27.20	435.20	45.10	721.60		
1 Air Compressor, 365 cfm		210.20		231.22		
2 Breakers, Pavement, 60 lb.		19.60		21.56		
8 -50' Air Hoses, 1"		32.80		36.08		
2 Dump Trucks, 8 C.Y., 220 H.P.		630.80		693.88	9.31	10.24
96 L.H., Daily Totals		$3652.60		$5563.14	$38.05	$57.95

Crew B-5B	**Hr.**	**Daily**	**Hr.**	**Daily**	**Bare Costs**	**Incl. O&P**
1 Powderman	$34.05	$272.40	$57.00	$456.00	$30.91	$51.07
2 Equip. Oper. (med.)	34.90	558.40	57.05	912.80		
3 Truck Drivers (heavy)	27.20	652.80	45.10	1082.40		
1 F.E. Loader, W.M.,2.5 C.Y.		397.60		437.36		
3 Dump Trucks, 12 C.Y., 400 H.P.		1518.00		1669.80		
1 Air Compressor, 365 cfm		210.20		231.22	44.29	48.72
48 L.H., Daily Totals		$3609.40		$4789.58	$75.20	$99.78

Crew B-5C	**Hr.**	**Daily**	**Hr.**	**Daily**	**Bare Costs**	**Incl. O&P**
3 Laborers	$26.45	$634.80	$44.30	$1063.20	$29.44	$48.67
1 Equip. Oper. (med.)	34.90	279.20	57.05	456.40		
2 Truck Drivers (heav.)	27.20	435.20	45.10	721.60		
1 Equip. Oper. (crane)	35.80	286.40	58.50	468.00		
1 Equip. Oper. Oiler	31.05	248.40	50.75	406.00		
2 Dump Trucks, 12 C.Y., 400 H.P.		1012.00		1113.20		
1 Crawler Loader, 4 C.Y.		1432.00		1575.20		
1 S.P. Crane, 4x4, 25 Ton		560.40		616.44	46.94	51.64
64 L.H., Daily Totals		$4888.40		$6420.04	$76.38	$100.31

Crew B-6	**Hr.**	**Daily**	**Hr.**	**Daily**	**Bare Costs**	**Incl. O&P**
2 Laborers	$26.45	$423.20	$44.30	$708.80	$28.82	$47.80
1 Equip. Oper. (light)	33.55	268.40	54.80	438.40		
1 Backhoe Loader, 48 H.P.		320.40		352.44	13.35	14.69
24 L.H., Daily Totals		$1012.00		$1499.64	$42.17	$62.48

Crews

Crew No.	Bare Costs		Incl. Subs O&P		Cost Per Labor-Hour	
Crew B-6B	**Hr.**	**Daily**	**Hr.**	**Daily**	**Bare Costs**	**Incl. O&P**
2 Labor Foremen (out)	$28.45	$455.20	$47.65	$762.40	$27.12	$45.42
4 Laborers	26.45	846.40	44.30	1417.60		
1 S.P. Crane, 4x4, 5 Ton		257.80		283.58		
1 Flatbed Truck, Gas, 1.5 Ton		171.40		188.54		
1 Butt Fusion Mach., 4"-12" diam.		376.45		414.10	16.78	18.46
48 L.H., Daily Totals		$2107.25		$3066.22	$43.90	$63.88
Crew B-6C	**Hr.**	**Daily**	**Hr.**	**Daily**	**Bare Costs**	**Incl. O&P**
2 Labor Foremen (out)	$28.45	$455.20	$47.65	$762.40	$27.12	$45.42
4 Laborers	26.45	846.40	44.30	1417.60		
1 S.P. Crane, 4x4, 12 Ton		434.40		477.84		
1 Flatbed Truck, Gas, 3 Ton		217.00		238.70		
1 Butt Fusion Mach., 8"-24" diam.		806.15		886.76	30.37	33.40
48 L.H., Daily Totals		$2759.15		$3783.30	$57.48	$78.82
Crew B-7	**Hr.**	**Daily**	**Hr.**	**Daily**	**Bare Costs**	**Incl. O&P**
1 Labor Foreman (outside)	$28.45	$227.60	$47.65	$381.20	$28.19	$46.98
4 Laborers	26.45	846.40	44.30	1417.60		
1 Equip. Oper. (med.)	34.90	279.20	57.05	456.40		
1 Brush Chipper, 12", 130 H.P.		217.00		238.70		
1 Crawler Loader, 3 C.Y.		1008.00		1108.80		
2 Chain Saws, gas, 36" Long		83.20		91.52	27.25	29.98
48 L.H., Daily Totals		$2661.40		$3694.22	$55.45	$76.96
Crew B-7A	**Hr.**	**Daily**	**Hr.**	**Daily**	**Bare Costs**	**Incl. O&P**
2 Laborers	$26.45	$423.20	$44.30	$708.80	$28.82	$47.80
1 Equip. Oper. (light)	33.55	268.40	54.80	438.40		
1 Rake w/Tractor		277.70		305.47		
2 Chain Saws, gas, 18"		51.60		56.76	13.72	15.09
24 L.H., Daily Totals		$1020.90		$1509.43	$42.54	$62.89
Crew B-8	**Hr.**	**Daily**	**Hr.**	**Daily**	**Bare Costs**	**Incl. O&P**
1 Labor Foreman (outside)	$28.45	$227.60	$47.65	$381.20	$29.36	$48.65
2 Laborers	26.45	423.20	44.30	708.80		
2 Equip. Oper. (med.)	34.90	558.40	57.05	912.80		
2 Truck Drivers (heavy)	27.20	435.20	45.10	721.60		
1 Hyd. Crane, 25 Ton		748.60		823.46		
1 Crawler Loader, 3 C.Y.		1008.00		1108.80		
2 Dump Trucks, 12 C.Y., 400 H.P.		1012.00		1113.20	49.44	54.38
56 L.H., Daily Totals		$4413.00		$5769.86	$78.80	$103.03
Crew B-9	**Hr.**	**Daily**	**Hr.**	**Daily**	**Bare Costs**	**Incl. O&P**
1 Labor Foreman (outside)	$28.45	$227.60	$47.65	$381.20	$26.85	$44.97
4 Laborers	26.45	846.40	44.30	1417.60		
1 Air Compressor, 250 cfm		162.80		179.08		
2 Breakers, Pavement, 60 lb.		19.60		21.56		
2 -50' Air Hoses, 1.5"		12.60		13.86	4.88	5.36
40 L.H., Daily Totals		$1269.00		$2013.30	$31.73	$50.33
Crew B-9A	**Hr.**	**Daily**	**Hr.**	**Daily**	**Bare Costs**	**Incl. O&P**
2 Laborers	$26.45	$423.20	$44.30	$708.80	$26.70	$44.57
1 Truck Driver (heavy)	27.20	217.60	45.10	360.80		
1 Water Tank Trailer, 5000 Gal.		146.20		160.82		
1 Truck Tractor, 220 H.P.		279.20		307.12		
2 -50' Discharge Hoses, 3"		3.50		3.85	17.87	19.66
24 L.H., Daily Totals		$1069.70		$1541.39	$44.57	$64.22

Crew No.	Bare Costs		Incl. Subs O&P		Cost Per Labor-Hour	
Crew B-9B	**Hr.**	**Daily**	**Hr.**	**Daily**	**Bare Costs**	**Incl. O&P**
2 Laborers	$26.45	$423.20	$44.30	$708.80	$26.70	$44.57
1 Truck Driver (heavy)	27.20	217.60	45.10	360.80		
2 -50' Discharge Hoses, 3"		3.50		3.85		
1 Water Tank Trailer, 5000 Gal.		146.20		160.82		
1 Truck Tractor, 220 H.P.		279.20		307.12		
1 Pressure Washer		58.80		64.68	20.32	22.35
24 L.H., Daily Totals		$1128.50		$1606.07	$47.02	$66.92
Crew B-9D	**Hr.**	**Daily**	**Hr.**	**Daily**	**Bare Costs**	**Incl. O&P**
1 Labor Foreman (Outside)	$28.45	$227.60	$47.65	$381.20	$26.85	$44.97
4 Common Laborers	26.45	846.40	44.30	1417.60		
1 Air Compressor, 250 cfm		162.80		179.08		
2 -50' Air Hoses, 1.5"		12.60		13.86		
2 Air Powered Tampers		52.40		57.64	5.70	6.26
40 L.H., Daily Totals		$1301.80		$2049.38	$32.55	$51.23
Crew B-10	**Hr.**	**Daily**	**Hr.**	**Daily**	**Bare Costs**	**Incl. O&P**
1 Equip. Oper. (med.)	$34.90	$279.20	$57.05	$456.40	$34.90	$57.05
8 L.H., Daily Totals		$279.20		$456.40	$34.90	$57.05
Crew B-10A	**Hr.**	**Daily**	**Hr.**	**Daily**	**Bare Costs**	**Incl. O&P**
1 Equip. Oper. (med.)	$34.90	$279.20	$57.05	$456.40	$34.90	$57.05
1 Roller, 2-Drum, W.B., 7.5 H.P.		146.80		161.48	18.35	20.18
8 L.H., Daily Totals		$426.00		$617.88	$53.25	$77.23
Crew B-10B	**Hr.**	**Daily**	**Hr.**	**Daily**	**Bare Costs**	**Incl. O&P**
1 Equip. Oper. (med.)	$34.90	$279.20	$57.05	$456.40	$34.90	$57.05
1 Dozer, 200 H.P.		1051.00		1156.10	131.38	144.51
8 L.H., Daily Totals		$1330.20		$1612.50	$166.28	$201.56
Crew B-10C	**Hr.**	**Daily**	**Hr.**	**Daily**	**Bare Costs**	**Incl. O&P**
1 Equip. Oper. (med.)	$34.90	$279.20	$57.05	$456.40	$34.90	$57.05
1 Dozer, 200 H.P.		1051.00		1156.10		
1 Vibratory Roller, Towed, 23 Ton		394.60		434.06	180.70	198.77
8 L.H., Daily Totals		$1724.80		$2046.56	$215.60	$255.82
Crew B-10D	**Hr.**	**Daily**	**Hr.**	**Daily**	**Bare Costs**	**Incl. O&P**
1 Equip. Oper. (med.)	$34.90	$279.20	$57.05	$456.40	$34.90	$57.05
1 Dozer, 200 H.P.		1051.00		1156.10		
1 Sheepsft. Roller, Towed		449.00		493.90	187.50	206.25
8 L.H., Daily Totals		$1779.20		$2106.40	$222.40	$263.30
Crew B-10E	**Hr.**	**Daily**	**Hr.**	**Daily**	**Bare Costs**	**Incl. O&P**
1 Equip. Oper. (med.)	$34.90	$279.20	$57.05	$456.40	$34.90	$57.05
1 Tandem Roller, 5 Ton		145.20		159.72	18.15	19.97
8 L.H., Daily Totals		$424.40		$616.12	$53.05	$77.02
Crew B-10F	**Hr.**	**Daily**	**Hr.**	**Daily**	**Bare Costs**	**Incl. O&P**
1 Equip. Oper. (med.)	$34.90	$279.20	$57.05	$456.40	$34.90	$57.05
1 Tandem Roller, 10 Ton		224.20		246.62	28.02	30.83
8 L.H., Daily Totals		$503.40		$703.02	$62.92	$87.88
Crew B-10G	**Hr.**	**Daily**	**Hr.**	**Daily**	**Bare Costs**	**Incl. O&P**
1 Equip. Oper. (med.)	$34.90	$279.20	$57.05	$456.40	$34.90	$57.05
1 Sheepsft. Roll., 240 H.P.		1094.00		1203.40	136.75	150.43
8 L.H., Daily Totals		$1373.20		$1659.80	$171.65	$207.47

Crews

Crew B-10H	Bare Costs Hr.	Bare Costs Daily	Incl. Subs O&P Hr.	Incl. Subs O&P Daily	Cost Per Labor-Hour Bare Costs	Cost Per Labor-Hour Incl. O&P
1 Equip. Oper. (med.)	$34.90	$279.20	$57.05	$456.40	$34.90	$57.05
1 Diaphragm Water Pump, 2"		66.60		73.26		
1 -20' Suction Hose, 2"		1.95		2.15		
2 -50' Discharge Hoses, 2"		1.80		1.98	8.79	9.67
8 L.H., Daily Totals		$349.55		$533.78	$43.69	$66.72

Crew B-10I	Hr.	Daily	Hr.	Daily	Bare Costs	Incl. O&P
1 Equip. Oper. (med.)	$34.90	$279.20	$57.05	$456.40	$34.90	$57.05
1 Diaphragm Water Pump, 4"		109.40		120.34		
1 -20' Suction Hose, 4"		3.25		3.58		
2 -50' Discharge Hoses, 4"		4.70		5.17	14.67	16.14
8 L.H., Daily Totals		$396.55		$585.49	$49.57	$73.19

Crew B-10J	Hr.	Daily	Hr.	Daily	Bare Costs	Incl. O&P
1 Equip. Oper. (med.)	$34.90	$279.20	$57.05	$456.40	$34.90	$57.05
1 Centrifugal Water Pump, 3"		71.00		78.10		
1 -20' Suction Hose, 3"		3.05		3.36		
2 -50' Discharge Hoses, 3"		3.50		3.85	9.69	10.66
8 L.H., Daily Totals		$356.75		$541.71	$44.59	$67.71

Crew B-10K	Hr.	Daily	Hr.	Daily	Bare Costs	Incl. O&P
1 Equip. Oper. (med.)	$34.90	$279.20	$57.05	$456.40	$34.90	$57.05
1 Centr. Water Pump, 6"		291.60		320.76		
1 -20' Suction Hose, 6"		11.90		13.09		
2 -50' Discharge Hoses, 6"		12.60		13.86	39.51	43.46
8 L.H., Daily Totals		$595.30		$804.11	$74.41	$100.51

Crew B-10L	Hr.	Daily	Hr.	Daily	Bare Costs	Incl. O&P
1 Equip. Oper. (med.)	$34.90	$279.20	$57.05	$456.40	$34.90	$57.05
1 Dozer, 80 H.P.		412.60		453.86	51.58	56.73
8 L.H., Daily Totals		$691.80		$910.26	$86.47	$113.78

Crew B-10M	Hr.	Daily	Hr.	Daily	Bare Costs	Incl. O&P
1 Equip. Oper. (med.)	$34.90	$279.20	$57.05	$456.40	$34.90	$57.05
1 Dozer, 300 H.P.		1607.00		1767.70	200.88	220.96
8 L.H., Daily Totals		$1886.20		$2224.10	$235.78	$278.01

Crew B-10N	Hr.	Daily	Hr.	Daily	Bare Costs	Incl. O&P
1 Equip. Oper. (med.)	$34.90	$279.20	$57.05	$456.40	$34.90	$57.05
1 F.E. Loader, T.M., 1.5 C.Y		466.60		513.26	58.33	64.16
8 L.H., Daily Totals		$745.80		$969.66	$93.22	$121.21

Crew B-10O	Hr.	Daily	Hr.	Daily	Bare Costs	Incl. O&P
1 Equip. Oper. (med.)	$34.90	$279.20	$57.05	$456.40	$34.90	$57.05
1 F.E. Loader, T.M., 2.25 C.Y.		773.60		850.96	96.70	106.37
8 L.H., Daily Totals		$1052.80		$1307.36	$131.60	$163.42

Crew B-10P	Hr.	Daily	Hr.	Daily	Bare Costs	Incl. O&P
1 Equip. Oper. (med.)	$34.90	$279.20	$57.05	$456.40	$34.90	$57.05
1 Crawler Loader, 3 C.Y.		1008.00		1108.80	126.00	138.60
8 L.H., Daily Totals		$1287.20		$1565.20	$160.90	$195.65

Crew B-10Q	Hr.	Daily	Hr.	Daily	Bare Costs	Incl. O&P
1 Equip. Oper. (med.)	$34.90	$279.20	$57.05	$456.40	$34.90	$57.05
1 Crawler Loader, 4 C.Y.		1432.00		1575.20	179.00	196.90
8 L.H., Daily Totals		$1711.20		$2031.60	$213.90	$253.95

Crew B-10R	Hr.	Daily	Hr.	Daily	Bare Costs	Incl. O&P
1 Equip. Oper. (med.)	$34.90	$279.20	$57.05	$456.40	$34.90	$57.05
1 F.E. Loader, W.M., 1 C.Y.		235.80		259.38	29.48	32.42
8 L.H., Daily Totals		$515.00		$715.78	$64.38	$89.47

Crew B-10S	Hr.	Daily	Hr.	Daily	Bare Costs	Incl. O&P
1 Equip. Oper. (med.)	$34.90	$279.20	$57.05	$456.40	$34.90	$57.05
1 F.E. Loader, W.M., 1.5 C.Y.		319.40		351.34	39.92	43.92
8 L.H., Daily Totals		$598.60		$807.74	$74.83	$100.97

Crew B-10T	Hr.	Daily	Hr.	Daily	Bare Costs	Incl. O&P
1 Equip. Oper. (med.)	$34.90	$279.20	$57.05	$456.40	$34.90	$57.05
1 F.E. Loader, W.M.,2.5 C.Y.		397.60		437.36	49.70	54.67
8 L.H., Daily Totals		$676.80		$893.76	$84.60	$111.72

Crew B-10U	Hr.	Daily	Hr.	Daily	Bare Costs	Incl. O&P
1 Equip. Oper. (med.)	$34.90	$279.20	$57.05	$456.40	$34.90	$57.05
1 F.E. Loader, W.M., 5.5 C.Y.		933.40		1026.74	116.68	128.34
8 L.H., Daily Totals		$1212.60		$1483.14	$151.57	$185.39

Crew B-10V	Hr.	Daily	Hr.	Daily	Bare Costs	Incl. O&P
1 Equip. Oper. (med.)	$34.90	$279.20	$57.05	$456.40	$34.90	$57.05
1 Dozer, 700 H.P.		4150.00		4565.00	518.75	570.63
8 L.H., Daily Totals		$4429.20		$5021.40	$553.65	$627.67

Crew B-10W	Hr.	Daily	Hr.	Daily	Bare Costs	Incl. O&P
1 Equip. Oper. (med.)	$34.90	$279.20	$57.05	$456.40	$34.90	$57.05
1 Dozer, 105 H.P.		505.40		555.94	63.17	69.49
8 L.H., Daily Totals		$784.60		$1012.34	$98.08	$126.54

Crew B-10X	Hr.	Daily	Hr.	Daily	Bare Costs	Incl. O&P
1 Equip. Oper. (med.)	$34.90	$279.20	$57.05	$456.40	$34.90	$57.05
1 Dozer, 410 H.P.		2105.00		2315.50	263.13	289.44
8 L.H., Daily Totals		$2384.20		$2771.90	$298.02	$346.49

Crew B-10Y	Hr.	Daily	Hr.	Daily	Bare Costs	Incl. O&P
1 Equip. Oper. (med.)	$34.90	$279.20	$57.05	$456.40	$34.90	$57.05
1 Vibr. Roller, Towed, 12 Ton		498.80		548.68	62.35	68.58
8 L.H., Daily Totals		$778.00		$1005.08	$97.25	$125.64

Crew B-11A	Hr.	Daily	Hr.	Daily	Bare Costs	Incl. O&P
1 Equipment Oper. (med.)	$34.90	$279.20	$57.05	$456.40	$30.68	$50.67
1 Laborer	26.45	211.60	44.30	354.40		
1 Dozer, 200 H.P.		1051.00		1156.10	65.69	72.26
16 L.H., Daily Totals		$1541.80		$1966.90	$96.36	$122.93

Crew B-11B	Hr.	Daily	Hr.	Daily	Bare Costs	Incl. O&P
1 Equipment Oper. (light)	$33.55	$268.40	$54.80	$438.40	$30.00	$49.55
1 Laborer	26.45	211.60	44.30	354.40		
1 Air Powered Tamper		26.20		28.82		
1 Air Compressor, 365 cfm		210.20		231.22		
2 -50' Air Hoses, 1.5"		12.60		13.86	15.56	17.12
16 L.H., Daily Totals		$729.00		$1066.70	$45.56	$66.67

Crew B-11C	Hr.	Daily	Hr.	Daily	Bare Costs	Incl. O&P
1 Equipment Oper. (med.)	$34.90	$279.20	$57.05	$456.40	$30.68	$50.67
1 Laborer	26.45	211.60	44.30	354.40		
1 Backhoe Loader, 48 H.P.		320.40		352.44	20.02	22.03
16 L.H., Daily Totals		$811.20		$1163.24	$50.70	$72.70

Crews

Crew No.	Bare Costs		Incl. Subs O&P		Cost Per Labor-Hour	
Crew B-11K	Hr.	Daily	Hr.	Daily	Bare Costs	Incl. O&P
1 Equipment Oper. (med.)	$34.90	$279.20	$57.05	$456.40	$30.68	$50.67
1 Laborer	26.45	211.60	44.30	354.40		
1 Trencher, Chain Type, 8' D		2506.00		2756.60	156.63	172.29
16 L.H., Daily Totals		$2996.80		$3567.40	$187.30	$222.96
Crew B-11L	Hr.	Daily	Hr.	Daily	Bare Costs	Incl. O&P
1 Equipment Oper. (med.)	$34.90	$279.20	$57.05	$456.40	$30.68	$50.67
1 Laborer	26.45	211.60	44.30	354.40		
1 Grader, 30,000 Lbs.		542.80		597.08	33.92	37.32
16 L.H., Daily Totals		$1033.60		$1407.88	$64.60	$87.99
Crew B-11M	Hr.	Daily	Hr.	Daily	Bare Costs	Incl. O&P
1 Equipment Oper. (med.)	$34.90	$279.20	$57.05	$456.40	$30.68	$50.67
1 Laborer	26.45	211.60	44.30	354.40		
1 Backhoe Loader, 80 H.P.		349.40		384.34	21.84	24.02
16 L.H., Daily Totals		$840.20		$1195.14	$52.51	$74.70
Crew B-11W	Hr.	Daily	Hr.	Daily	Bare Costs	Incl. O&P
1 Equipment Operator (med.)	$34.90	$279.20	$57.05	$456.40	$27.78	$46.03
1 Common Laborer	26.45	211.60	44.30	354.40		
10 Truck Drivers (hvy.)	27.20	2176.00	45.10	3608.00		
1 Dozer, 200 H.P.		1051.00		1156.10		
1 Vibratory Roller, Towed, 23 Ton		394.60		434.06		
10 Dump Trucks, 8 C.Y., 220 H.P.		3154.00		3469.40	47.91	52.70
96 L.H., Daily Totals		$7266.40		$9478.36	$75.69	$98.73
Crew B-11Y	Hr.	Daily	Hr.	Daily	Bare Costs	Incl. O&P
1 Labor Foreman (Outside)	$28.45	$227.60	$47.65	$381.20	$29.49	$48.92
5 Common Laborers	26.45	1058.00	44.30	1772.00		
3 Equipment Operators (med.)	34.90	837.60	57.05	1369.20		
1 Dozer, 80 H.P.		412.60		453.86		
2 Roller, 2-Drum, W.B., 7.5 H.P.		293.60		322.96		
4 Vibratory Plates, gas, 21"		156.00		171.60	11.98	13.17
72 L.H., Daily Totals		$2985.40		$4470.82	$41.46	$62.09
Crew B-12A	Hr.	Daily	Hr.	Daily	Bare Costs	Incl. O&P
1 Equip. Oper. (crane)	$35.80	$286.40	$58.50	$468.00	$31.13	$51.40
1 Laborer	26.45	211.60	44.30	354.40		
1 Hyd. Excavator, 1 C.Y.		729.80		802.78	45.61	50.17
16 L.H., Daily Totals		$1227.80		$1625.18	$76.74	$101.57
Crew B-12B	Hr.	Daily	Hr.	Daily	Bare Costs	Incl. O&P
1 Equip. Oper. (crane)	$35.80	$286.40	$58.50	$468.00	$31.13	$51.40
1 Laborer	26.45	211.60	44.30	354.40		
1 Hyd. Excavator, 1.5 C.Y.		947.40		1042.14	59.21	65.13
16 L.H., Daily Totals		$1445.40		$1864.54	$90.34	$116.53
Crew B-12C	Hr.	Daily	Hr.	Daily	Bare Costs	Incl. O&P
1 Equip. Oper. (crane)	$35.80	$286.40	$58.50	$468.00	$31.13	$51.40
1 Laborer	26.45	211.60	44.30	354.40		
1 Hyd. Excavator, 2 C.Y.		1224.00		1346.40	76.50	84.15
16 L.H., Daily Totals		$1722.00		$2168.80	$107.63	$135.55
Crew B-12D	Hr.	Daily	Hr.	Daily	Bare Costs	Incl. O&P
1 Equip. Oper. (crane)	$35.80	$286.40	$58.50	$468.00	$31.13	$51.40
1 Laborer	26.45	211.60	44.30	354.40		
1 Hyd. Excavator, 3.5 C.Y.		2171.00		2388.10	135.69	149.26
16 L.H., Daily Totals		$2669.00		$3210.50	$166.81	$200.66

Crew No.	Bare Costs		Incl. Subs O&P		Cost Per Labor-Hour	
Crew B-12E	Hr.	Daily	Hr.	Daily	Bare Costs	Incl. O&P
1 Equip. Oper. (crane)	$35.80	$286.40	$58.50	$468.00	$31.13	$51.40
1 Laborer	26.45	211.60	44.30	354.40		
1 Hyd. Excavator, .5 C.Y.		364.60		401.06	22.79	25.07
16 L.H., Daily Totals		$862.60		$1223.46	$53.91	$76.47
Crew B-12F	Hr.	Daily	Hr.	Daily	Bare Costs	Incl. O&P
1 Equip. Oper. (crane)	$35.80	$286.40	$58.50	$468.00	$31.13	$51.40
1 Laborer	26.45	211.60	44.30	354.40		
1 Hyd. Excavator, .75 C.Y.		640.40		704.44	40.02	44.03
16 L.H., Daily Totals		$1138.40		$1526.84	$71.15	$95.43
Crew B-12G	Hr.	Daily	Hr.	Daily	Bare Costs	Incl. O&P
1 Equip. Oper. (crane)	$35.80	$286.40	$58.50	$468.00	$31.13	$51.40
1 Laborer	26.45	211.60	44.30	354.40		
1 Crawler Crane, 15 Ton		657.90		723.69		
1 Clamshell Bucket, .5 C.Y.		37.40		41.14	43.46	47.80
16 L.H., Daily Totals		$1193.30		$1587.23	$74.58	$99.20
Crew B-12H	Hr.	Daily	Hr.	Daily	Bare Costs	Incl. O&P
1 Equip. Oper. (crane)	$35.80	$286.40	$58.50	$468.00	$31.13	$51.40
1 Laborer	26.45	211.60	44.30	354.40		
1 Crawler Crane, 25 Ton		1123.00		1235.30		
1 Clamshell Bucket, 1 C.Y.		48.00		52.80	73.19	80.51
16 L.H., Daily Totals		$1669.00		$2110.50	$104.31	$131.91
Crew B-12I	Hr.	Daily	Hr.	Daily	Bare Costs	Incl. O&P
1 Equip. Oper. (crane)	$35.80	$286.40	$58.50	$468.00	$31.13	$51.40
1 Laborer	26.45	211.60	44.30	354.40		
1 Crawler Crane, 20 Ton		841.90		926.09		
1 Dragline Bucket, .75 C.Y.		21.00		23.10	53.93	59.32
16 L.H., Daily Totals		$1360.90		$1771.59	$85.06	$110.72
Crew B-12J	Hr.	Daily	Hr.	Daily	Bare Costs	Incl. O&P
1 Equip. Oper. (crane)	$35.80	$286.40	$58.50	$468.00	$31.13	$51.40
1 Laborer	26.45	211.60	44.30	354.40		
1 Gradall, 5/8 C.Y.		857.40		943.14	53.59	58.95
16 L.H., Daily Totals		$1355.40		$1765.54	$84.71	$110.35
Crew B-12K	Hr.	Daily	Hr.	Daily	Bare Costs	Incl. O&P
1 Equip. Oper. (crane)	$35.80	$286.40	$58.50	$468.00	$31.13	$51.40
1 Laborer	26.45	211.60	44.30	354.40		
1 Gradall, 3 Ton, 1 C.Y.		1001.00		1101.10	62.56	68.82
16 L.H., Daily Totals		$1499.00		$1923.50	$93.69	$120.22
Crew B-12L	Hr.	Daily	Hr.	Daily	Bare Costs	Incl. O&P
1 Equip. Oper. (crane)	$35.80	$286.40	$58.50	$468.00	$31.13	$51.40
1 Laborer	26.45	211.60	44.30	354.40		
1 Crawler Crane, 15 Ton		657.90		723.69		
1 F.E. Attachment, .5 C.Y.		56.00		61.60	44.62	49.08
16 L.H., Daily Totals		$1211.90		$1607.69	$75.74	$100.48
Crew B-12M	Hr.	Daily	Hr.	Daily	Bare Costs	Incl. O&P
1 Equip. Oper. (crane)	$35.80	$286.40	$58.50	$468.00	$31.13	$51.40
1 Laborer	26.45	211.60	44.30	354.40		
1 Crawler Crane, 20 Ton		841.90		926.09		
1 F.E. Attachment, .75 C.Y.		61.40		67.54	56.46	62.10
16 L.H., Daily Totals		$1401.30		$1816.03	$87.58	$113.50

Crews

Crew No.	Bare Costs		Incl. Subs O&P		Cost Per Labor-Hour	
Crew B-12N	**Hr.**	**Daily**	**Hr.**	**Daily**	**Bare Costs**	**Incl. O&P**
1 Equip. Oper. (crane)	$35.80	$286.40	$58.50	$468.00	$31.13	$51.40
1 Laborer	26.45	211.60	44.30	354.40		
1 Crawler Crane, 25 Ton		1123.00		1235.30		
1 F.E. Attachment, 1 C.Y.		67.20		73.92	74.39	81.83
16 L.H., Daily Totals		$1688.20		$2131.62	$105.51	$133.23

Crew No.	Bare Costs		Incl. Subs O&P		Cost Per Labor-Hour	
Crew B-12O	**Hr.**	**Daily**	**Hr.**	**Daily**	**Bare Costs**	**Incl. O&P**
1 Equip. Oper. (crane)	$35.80	$286.40	$58.50	$468.00	$31.13	$51.40
1 Laborer	26.45	211.60	44.30	354.40		
1 Crawler Crane, 40 Ton		1128.00		1240.80		
1 F.E. Attachment, 1.5 C.Y.		77.40		85.14	75.34	82.87
16 L.H., Daily Totals		$1703.40		$2148.34	$106.46	$134.27

Crew No.	Bare Costs		Incl. Subs O&P		Cost Per Labor-Hour	
Crew B-12P	**Hr.**	**Daily**	**Hr.**	**Daily**	**Bare Costs**	**Incl. O&P**
1 Equip. Oper. (crane)	$35.80	$286.40	$58.50	$468.00	$31.13	$51.40
1 Laborer	26.45	211.60	44.30	354.40		
1 Crawler Crane, 40 Ton		1128.00		1240.80		
1 Dragline Bucket, 1.5 C.Y.		34.20		37.62	72.64	79.90
16 L.H., Daily Totals		$1660.20		$2100.82	$103.76	$131.30

Crew No.	Bare Costs		Incl. Subs O&P		Cost Per Labor-Hour	
Crew B-12Q	**Hr.**	**Daily**	**Hr.**	**Daily**	**Bare Costs**	**Incl. O&P**
1 Equip. Oper. (crane)	$35.80	$286.40	$58.50	$468.00	$31.13	$51.40
1 Laborer	26.45	211.60	44.30	354.40		
1 Hyd. Excavator, 5/8 C.Y.		530.00		583.00	33.13	36.44
16 L.H., Daily Totals		$1028.00		$1405.40	$64.25	$87.84

Crew No.	Bare Costs		Incl. Subs O&P		Cost Per Labor-Hour	
Crew B-12S	**Hr.**	**Daily**	**Hr.**	**Daily**	**Bare Costs**	**Incl. O&P**
1 Equip. Oper. (crane)	$35.80	$286.40	$58.50	$468.00	$31.13	$51.40
1 Laborer	26.45	211.60	44.30	354.40		
1 Hyd. Excavator, 2.5 C.Y.		1613.00		1774.30	100.81	110.89
16 L.H., Daily Totals		$2111.00		$2596.70	$131.94	$162.29

Crew No.	Bare Costs		Incl. Subs O&P		Cost Per Labor-Hour	
Crew B-12T	**Hr.**	**Daily**	**Hr.**	**Daily**	**Bare Costs**	**Incl. O&P**
1 Equip. Oper. (crane)	$35.80	$286.40	$58.50	$468.00	$31.13	$51.40
1 Laborer	26.45	211.60	44.30	354.40		
1 Crawler Crane, 75 Ton		1446.00		1590.60		
1 F.E. Attachment, 3 C.Y.		100.80		110.88	96.67	106.34
16 L.H., Daily Totals		$2044.80		$2523.88	$127.80	$157.74

Crew No.	Bare Costs		Incl. Subs O&P		Cost Per Labor-Hour	
Crew B-12V	**Hr.**	**Daily**	**Hr.**	**Daily**	**Bare Costs**	**Incl. O&P**
1 Equip. Oper. (crane)	$35.80	$286.40	$58.50	$468.00	$31.13	$51.40
1 Laborer	26.45	211.60	44.30	354.40		
1 Crawler Crane, 75 Ton		1446.00		1590.60		
1 Dragline Bucket, 3 C.Y.		52.80		58.08	93.67	103.04
16 L.H., Daily Totals		$1996.80		$2471.08	$124.80	$154.44

Crew No.	Bare Costs		Incl. Subs O&P		Cost Per Labor-Hour	
Crew B-12Y	**Hr.**	**Daily**	**Hr.**	**Daily**	**Bare Costs**	**Incl. O&P**
1 Equip. Oper. (crane)	$35.80	$286.40	$58.50	$468.00	$29.57	$49.03
2 Laborers	26.45	423.20	44.30	708.80		
1 Hyd. Excavator, 3.5 C.Y.		2171.00		2388.10	90.46	99.50
24 L.H., Daily Totals		$2880.60		$3564.90	$120.03	$148.54

Crew No.	Bare Costs		Incl. Subs O&P		Cost Per Labor-Hour	
Crew B-12Z	**Hr.**	**Daily**	**Hr.**	**Daily**	**Bare Costs**	**Incl. O&P**
1 Equip. Oper. (crane)	$35.80	$286.40	$58.50	$468.00	$29.57	$49.03
2 Laborers	26.45	423.20	44.30	708.80		
1 Hyd. Excavator, 2.5 C.Y.		1613.00		1774.30	67.21	73.93
24 L.H., Daily Totals		$2322.60		$2951.10	$96.78	$122.96

Crew No.	Bare Costs		Incl. Subs O&P		Cost Per Labor-Hour	
Crew B-13	**Hr.**	**Daily**	**Hr.**	**Daily**	**Bare Costs**	**Incl. O&P**
1 Labor Foreman (outside)	$28.45	$227.60	$47.65	$381.20	$28.34	$47.23
4 Laborers	26.45	846.40	44.30	1417.60		
1 Equip. Oper. (crane)	35.80	286.40	58.50	468.00		
1 Hyd. Crane, 25 Ton		748.60		823.46	15.60	17.16
48 L.H., Daily Totals		$2109.00		$3090.26	$43.94	$64.38

Crew No.	Bare Costs		Incl. Subs O&P		Cost Per Labor-Hour	
Crew B-13A	**Hr.**	**Daily**	**Hr.**	**Daily**	**Bare Costs**	**Incl. O&P**
1 Foreman	$28.45	$227.60	$47.65	$381.20	$29.36	$48.65
2 Laborers	26.45	423.20	44.30	708.80		
2 Equipment Operators (med.)	34.90	558.40	57.05	912.80		
2 Truck Drivers (heavy)	27.20	435.20	45.10	721.60		
1 Crawler Crane, 75 Ton		1446.00		1590.60		
1 Crawler Loader, 4 C.Y.		1432.00		1575.20		
2 Dump Trucks, 8 C.Y., 220 H.P.		630.80		693.88	62.66	68.92
56 L.H., Daily Totals		$5153.20		$6584.08	$92.02	$117.57

Crew No.	Bare Costs		Incl. Subs O&P		Cost Per Labor-Hour	
Crew B-13B	**Hr.**	**Daily**	**Hr.**	**Daily**	**Bare Costs**	**Incl. O&P**
1 Labor Foreman (outside)	$28.45	$227.60	$47.65	$381.20	$28.73	$47.73
4 Laborers	26.45	846.40	44.30	1417.60		
1 Equip. Oper. (crane)	35.80	286.40	58.50	468.00		
1 Equip. Oper. Oiler	31.05	248.40	50.75	406.00		
1 Hyd. Crane, 55 Ton		1208.00		1328.80	21.57	23.73
56 L.H., Daily Totals		$2816.80		$4001.60	$50.30	$71.46

Crew No.	Bare Costs		Incl. Subs O&P		Cost Per Labor-Hour	
Crew B-13C	**Hr.**	**Daily**	**Hr.**	**Daily**	**Bare Costs**	**Incl. O&P**
1 Labor Foreman (outside)	$28.45	$227.60	$47.65	$381.20	$28.73	$47.73
4 Laborers	26.45	846.40	44.30	1417.60		
1 Equip. Oper. (crane)	35.80	286.40	58.50	468.00		
1 Equip. Oper. Oiler	31.05	248.40	50.75	406.00		
1 Crawler Crane, 100 Ton		1653.00		1818.30	29.52	32.47
56 L.H., Daily Totals		$3261.80		$4491.10	$58.25	$80.20

Crew No.	Bare Costs		Incl. Subs O&P		Cost Per Labor-Hour	
Crew B-13D	**Hr.**	**Daily**	**Hr.**	**Daily**	**Bare Costs**	**Incl. O&P**
1 Laborer	$26.45	$211.60	$44.30	$354.40	$31.13	$51.40
1 Equip. Oper. (crane)	35.80	286.40	58.50	468.00		
1 Hyd. Excavator, 1 C.Y.		729.80		802.78		
1 Trench Box		113.40		124.74	52.70	57.97
16 L.H., Daily Totals		$1341.20		$1749.92	$83.83	$109.37

Crew No.	Bare Costs		Incl. Subs O&P		Cost Per Labor-Hour	
Crew B-13E	**Hr.**	**Daily**	**Hr.**	**Daily**	**Bare Costs**	**Incl. O&P**
1 Laborer	$26.45	$211.60	$44.30	$354.40	$31.13	$51.40
1 Equip. Oper. (crane)	35.80	286.40	58.50	468.00		
1 Hyd. Excavator, 1.5 C.Y.		947.40		1042.14		
1 Trench Box		113.40		124.74	66.30	72.93
16 L.H., Daily Totals		$1558.80		$1989.28	$97.42	$124.33

Crew No.	Bare Costs		Incl. Subs O&P		Cost Per Labor-Hour	
Crew B-13F	**Hr.**	**Daily**	**Hr.**	**Daily**	**Bare Costs**	**Incl. O&P**
1 Laborer	$26.45	$211.60	$44.30	$354.40	$31.13	$51.40
1 Equip. Oper. (crane)	35.80	286.40	58.50	468.00		
1 Hyd. Excavator, 3.5 C.Y.		2171.00		2388.10		
1 Trench Box		113.40		124.74	142.78	157.05
16 L.H., Daily Totals		$2782.40		$3335.24	$173.90	$208.45

Crew No.	Bare Costs		Incl. Subs O&P		Cost Per Labor-Hour	
Crew B-13G	**Hr.**	**Daily**	**Hr.**	**Daily**	**Bare Costs**	**Incl. O&P**
1 Laborer	$26.45	$211.60	$44.30	$354.40	$31.13	$51.40
1 Equip. Oper. (crane)	35.80	286.40	58.50	468.00		
1 Hyd. Excavator, .75 C.Y.		640.40		704.44		
1 Trench Box		113.40		124.74	47.11	51.82
16 L.H., Daily Totals		$1251.80		$1651.58	$78.24	$103.22

Crews

Crew No.	Bare Costs		Incl. Subs O&P		Cost Per Labor-Hour	
Crew B-13H	**Hr.**	**Daily**	**Hr.**	**Daily**	**Bare Costs**	**Incl. O&P**
1 Laborer	$26.45	$211.60	$44.30	$354.40	$31.13	$51.40
1 Equip. Oper. (crane)	35.80	286.40	58.50	468.00		
1 Gradall, 5/8 C.Y.		857.40		943.14		
1 Trench Box		113.40		124.74	60.67	66.74
16 L.H., Daily Totals		$1468.80		$1890.28	$91.80	$118.14
Crew B-13I	**Hr.**	**Daily**	**Hr.**	**Daily**	**Bare Costs**	**Incl. O&P**
1 Laborer	$26.45	$211.60	$44.30	$354.40	$31.13	$51.40
1 Equip. Oper. (crane)	35.80	286.40	58.50	468.00		
1 Gradall, 3 Ton, 1 C.Y.		1001.00		1101.10		
1 Trench Box		113.40		124.74	69.65	76.61
16 L.H., Daily Totals		$1612.40		$2048.24	$100.78	$128.01
Crew B-13J	**Hr.**	**Daily**	**Hr.**	**Daily**	**Bare Costs**	**Incl. O&P**
1 Laborer	$26.45	$211.60	$44.30	$354.40	$31.13	$51.40
1 Equip. Oper. (crane)	35.80	286.40	58.50	468.00		
1 Hyd. Excavator, 2.5 C.Y.		1613.00		1774.30		
1 Trench Box		113.40		124.74	107.90	118.69
16 L.H., Daily Totals		$2224.40		$2721.44	$139.03	$170.09
Crew B-14	**Hr.**	**Daily**	**Hr.**	**Daily**	**Bare Costs**	**Incl. O&P**
1 Labor Foreman (outside)	$28.45	$227.60	$47.65	$381.20	$27.97	$46.61
4 Laborers	26.45	846.40	44.30	1417.60		
1 Equip. Oper. (light)	33.55	268.40	54.80	438.40		
1 Backhoe Loader, 48 H.P.		320.40		352.44	6.67	7.34
48 L.H., Daily Totals		$1662.80		$2589.64	$34.64	$53.95
Crew B-14A	**Hr.**	**Daily**	**Hr.**	**Daily**	**Bare Costs**	**Incl. O&P**
1 Equip. Oper. (crane)	$35.80	$286.40	$58.50	$468.00	$32.68	$53.77
.5 Laborer	26.45	105.80	44.30	177.20		
1 Hyd. Excavator, 4.5 C.Y.		2669.00		2935.90	222.42	244.66
12 L.H., Daily Totals		$3061.20		$3581.10	$255.10	$298.43
Crew B-14B	**Hr.**	**Daily**	**Hr.**	**Daily**	**Bare Costs**	**Incl. O&P**
1 Equip. Oper. (crane)	$35.80	$286.40	$58.50	$468.00	$32.68	$53.77
.5 Laborer	26.45	105.80	44.30	177.20		
1 Hyd. Excavator, 6 C.Y.		3101.00		3411.10	258.42	284.26
12 L.H., Daily Totals		$3493.20		$4056.30	$291.10	$338.02
Crew B-14C	**Hr.**	**Daily**	**Hr.**	**Daily**	**Bare Costs**	**Incl. O&P**
1 Equip. Oper. (crane)	$35.80	$286.40	$58.50	$468.00	$32.68	$53.77
.5 Laborer	26.45	105.80	44.30	177.20		
1 Hyd. Excavator, 7 C.Y.		3179.00		3496.90	264.92	291.41
12 L.H., Daily Totals		$3571.20		$4142.10	$297.60	$345.18
Crew B-14F	**Hr.**	**Daily**	**Hr.**	**Daily**	**Bare Costs**	**Incl. O&P**
1 Equip. Oper. (crane)	$35.80	$286.40	$58.50	$468.00	$32.68	$53.77
.5 Laborer	26.45	105.80	44.30	177.20		
1 Hyd. Shovel, 7 C.Y.		3109.00		3419.90	259.08	284.99
12 L.H., Daily Totals		$3501.20		$4065.10	$291.77	$338.76
Crew B-14G	**Hr.**	**Daily**	**Hr.**	**Daily**	**Bare Costs**	**Incl. O&P**
1 Equip. Oper. (crane)	$35.80	$286.40	$58.50	$468.00	$32.68	$53.77
.5 Laborer	26.45	105.80	44.30	177.20		
1 Hyd. Shovel, 12 C.Y.		4046.00		4450.60	337.17	370.88
12 L.H., Daily Totals		$4438.20		$5095.80	$369.85	$424.65

Crew No.	Bare Costs		Incl. Subs O&P		Cost Per Labor-Hour	
Crew B-14J	**Hr.**	**Daily**	**Hr.**	**Daily**	**Bare Costs**	**Incl. O&P**
1 Equip. Oper. (med.)	$34.90	$279.20	$57.05	$456.40	$32.08	$52.80
.5 Laborer	26.45	105.80	44.30	177.20		
1 F.E. Loader, 8 C.Y.		1811.00		1992.10	150.92	166.01
12 L.H., Daily Totals		$2196.00		$2625.70	$183.00	$218.81
Crew B-14K	**Hr.**	**Daily**	**Hr.**	**Daily**	**Bare Costs**	**Incl. O&P**
1 Equip. Oper. (med.)	$34.90	$279.20	$57.05	$456.40	$32.08	$52.80
.5 Laborer	26.45	105.80	44.30	177.20		
1 F.E. Loader, 10 C.Y.		2476.00		2723.60	206.33	226.97
12 L.H., Daily Totals		$2861.00		$3357.20	$238.42	$279.77
Crew B-15	**Hr.**	**Daily**	**Hr.**	**Daily**	**Bare Costs**	**Incl. O&P**
1 Equipment Oper. (med.)	$34.90	$279.20	$57.05	$456.40	$29.29	$48.40
.5 Laborer	26.45	105.80	44.30	177.20		
2 Truck Drivers (heavy)	27.20	435.20	45.10	721.60		
2 Dump Trucks, 12 C.Y., 400 H.P.		1012.00		1113.20		
1 Dozer, 200 H.P.		1051.00		1156.10	73.68	81.05
28 L.H., Daily Totals		$2883.20		$3624.50	$102.97	$129.45
Crew B-16	**Hr.**	**Daily**	**Hr.**	**Daily**	**Bare Costs**	**Incl. O&P**
1 Labor Foreman (outside)	$28.45	$227.60	$47.65	$381.20	$27.14	$45.34
2 Laborers	26.45	423.20	44.30	708.80		
1 Truck Driver (heavy)	27.20	217.60	45.10	360.80		
1 Dump Truck, 12 C.Y., 400 H.P.		506.00		556.60	15.81	17.39
32 L.H., Daily Totals		$1374.40		$2007.40	$42.95	$62.73
Crew B-17	**Hr.**	**Daily**	**Hr.**	**Daily**	**Bare Costs**	**Incl. O&P**
2 Laborers	$26.45	$423.20	$44.30	$708.80	$28.41	$47.13
1 Equip. Oper. (light)	33.55	268.40	54.80	438.40		
1 Truck Driver (heavy)	27.20	217.60	45.10	360.80		
1 Backhoe Loader, 48 H.P.		320.40		352.44		
1 Dump Truck, 8 C.Y., 220 H.P.		315.40		346.94	19.87	21.86
32 L.H., Daily Totals		$1545.00		$2207.38	$48.28	$68.98
Crew B-17A	**Hr.**	**Daily**	**Hr.**	**Daily**	**Bare Costs**	**Incl. O&P**
2 Laborer Foremen	$28.45	$455.20	$47.65	$762.40	$28.57	$47.84
6 Laborers	26.45	1269.60	44.30	2126.40		
1 Skilled Worker Foreman	36.05	288.40	60.35	482.80		
1 Skilled Worker	34.05	272.40	57.00	456.00		
80 L.H., Daily Totals		$2285.60		$3827.60	$28.57	$47.84
Crew B-18	**Hr.**	**Daily**	**Hr.**	**Daily**	**Bare Costs**	**Incl. O&P**
1 Labor Foreman (outside)	$28.45	$227.60	$47.65	$381.20	$27.12	$45.42
2 Laborers	26.45	423.20	44.30	708.80		
1 Vibratory Plate, gas, 21"		39.00		42.90	1.63	1.79
24 L.H., Daily Totals		$689.80		$1132.90	$28.74	$47.20
Crew B-19	**Hr.**	**Daily**	**Hr.**	**Daily**	**Bare Costs**	**Incl. O&P**
1 Pile Driver Foreman	$34.75	$278.00	$60.55	$484.40	$32.57	$55.96
4 Pile Drivers	32.75	1048.00	57.10	1827.20		
1 Equip. Oper. (crane)	35.80	286.40	58.50	468.00		
1 Building Laborer	26.45	211.60	44.30	354.40		
1 Crawler Crane, 40 Ton		1128.00		1240.80		
1 Lead, 90' high		127.20		139.92		
1 Hammer, Diesel, 22k ft-lb		592.40		651.64	32.99	36.29
56 L.H., Daily Totals		$3671.60		$5166.36	$65.56	$92.26

Crews

Crew No.	Bare Costs		Incl. Subs O&P		Cost Per Labor-Hour	
Crew B-19A	**Hr.**	**Daily**	**Hr.**	**Daily**	**Bare Costs**	**Incl. O&P**
1 Pile Driver Foreman	$34.75	$278.00	$60.55	$484.40	$33.55	$57.09
4 Pile Drivers	32.75	1048.00	57.10	1827.20		
2 Equip. Oper. (crane)	35.80	572.80	58.50	936.00		
1 Equip. Oper. Oiler	31.05	248.40	50.75	406.00		
1 Crawler Crane, 75 Ton		1446.00		1590.60		
1 Lead, 90' high		127.20		139.92		
1 Hammer, Diesel, 41k ft-lb		677.00		744.70	35.16	38.68
64 L.H., Daily Totals		$4397.40		$6128.82	$68.71	$95.76

Crew B-20	**Hr.**	**Daily**	**Hr.**	**Daily**	**Bare Costs**	**Incl. O&P**
1 Labor Foreman (out)	$28.45	$227.60	$47.65	$381.20	$27.12	$45.42
2 Laborers	26.45	423.20	44.30	708.80		
24 L.H., Daily Totals		$650.80		$1090.00	$27.12	$45.42

Crew B-20A	**Hr.**	**Daily**	**Hr.**	**Daily**	**Bare Costs**	**Incl. O&P**
1 Labor Foreman	$28.45	$227.60	$47.65	$381.20	$31.56	$52.11
1 Laborer	26.45	211.60	44.30	354.40		
1 Plumber	39.65	317.20	64.75	518.00		
1 Plumber Apprentice	31.70	253.60	51.75	414.00		
32 L.H., Daily Totals		$1010.00		$1667.60	$31.56	$52.11

Crew B-21	**Hr.**	**Daily**	**Hr.**	**Daily**	**Bare Costs**	**Incl. O&P**
1 Labor Foreman (out)	$28.45	$227.60	$47.65	$381.20	$28.36	$47.29
2 Laborers	26.45	423.20	44.30	708.80		
.5 Equip. Oper. (crane)	35.80	143.20	58.50	234.00		
.5 S.P. Crane, 4x4, 5 Ton		128.90		141.79	4.60	5.06
28 L.H., Daily Totals		$922.90		$1465.79	$32.96	$52.35

Crew B-21A	**Hr.**	**Daily**	**Hr.**	**Daily**	**Bare Costs**	**Incl. O&P**
1 Labor Foreman	$28.45	$227.60	$47.65	$381.20	$32.41	$53.39
1 Laborer	26.45	211.60	44.30	354.40		
1 Plumber	39.65	317.20	64.75	518.00		
1 Plumber Apprentice	31.70	253.60	51.75	414.00		
1 Equip. Oper. (crane)	35.80	286.40	58.50	468.00		
1 S.P. Crane, 4x4, 12 Ton		434.40		477.84	10.86	11.95
40 L.H., Daily Totals		$1730.80		$2613.44	$43.27	$65.34

Crew B-21B	**Hr.**	**Daily**	**Hr.**	**Daily**	**Bare Costs**	**Incl. O&P**
1 Laborer Foreman	$28.45	$227.60	$47.65	$381.20	$28.72	$47.81
3 Laborers	26.45	634.80	44.30	1063.20		
1 Equip. Oper. (crane)	35.80	286.40	58.50	468.00		
1 Hyd. Crane, 12 Ton		655.60		721.16	16.39	18.03
40 L.H., Daily Totals		$1804.40		$2633.56	$45.11	$65.84

Crew B-21C	**Hr.**	**Daily**	**Hr.**	**Daily**	**Bare Costs**	**Incl. O&P**
1 Laborer Foreman	$28.45	$227.60	$47.65	$381.20	$28.73	$47.73
4 Laborers	26.45	846.40	44.30	1417.60		
1 Equip. Oper. (crane)	35.80	286.40	58.50	468.00		
1 Equip. Oper. Oiler	31.05	248.40	50.75	406.00		
2 Cutting Torches		36.00		39.60		
2 Set of Gases		165.60		182.16		
1 Lattice Boom Crane, 90 Ton		1622.00		1784.20	32.56	35.82
56 L.H., Daily Totals		$3432.40		$4678.76	$61.29	$83.55

Crew B-22	**Hr.**	**Daily**	**Hr.**	**Daily**	**Bare Costs**	**Incl. O&P**
1 Labor Foreman (out)	$28.45	$227.60	$47.65	$381.20	$28.85	$48.03
2 Laborers	26.45	423.20	44.30	708.80		
.75 Equip. Oper. (crane)	35.80	214.80	58.50	351.00		
.75 S.P. Crane, 4x4, 5 Ton		193.35		212.69	6.45	7.09
30 L.H., Daily Totals		$1058.95		$1653.68	$35.30	$55.12

Crew No.	Bare Costs		Incl. Subs O&P		Cost Per Labor-Hour	
Crew B-22A	**Hr.**	**Daily**	**Hr.**	**Daily**	**Bare Costs**	**Incl. O&P**
1 Labor Foreman (out)	$28.45	$227.60	$47.65	$381.20	$30.24	$50.35
1 Skilled Worker	34.05	272.40	57.00	456.00		
2 Laborers	26.45	423.20	44.30	708.80		
1 Equipment Oper. (crane)	35.80	286.40	58.50	468.00		
1 S.P. Crane, 4x4, 5 Ton		257.80		283.58		
1 Butt Fusion Mach., 4"-12" diam.		376.45		414.10	15.86	17.44
40 L.H., Daily Totals		$1843.85		$2711.68	$46.10	$67.79

Crew B-22B	**Hr.**	**Daily**	**Hr.**	**Daily**	**Bare Costs**	**Incl. O&P**
1 Labor Foreman (out)	$28.45	$227.60	$47.65	$381.20	$30.24	$50.35
1 Skilled Worker	34.05	272.40	57.00	456.00		
2 Laborers	26.45	423.20	44.30	708.80		
1 Equip. Oper. (crane)	35.80	286.40	58.50	468.00		
1 S.P. Crane, 4x4, 5 Ton		257.80		283.58		
1 Butt Fusion Mach., 8"-24" diam.		806.15		886.76	26.60	29.26
40 L.H., Daily Totals		$2273.55		$3184.34	$56.84	$79.61

Crew B-22C	**Hr.**	**Daily**	**Hr.**	**Daily**	**Bare Costs**	**Incl. O&P**
1 Skilled Worker	$34.05	$272.40	$57.00	$456.00	$30.25	$50.65
1 Laborer	26.45	211.60	44.30	354.40		
1 Butt Fusion Mach., 2"-8" diam.		120.50		132.55	7.53	8.28
16 L.H., Daily Totals		$604.50		$942.95	$37.78	$58.93

Crew B-23	**Hr.**	**Daily**	**Hr.**	**Daily**	**Bare Costs**	**Incl. O&P**
1 Labor Foreman (outside)	$28.45	$227.60	$47.65	$381.20	$26.85	$44.97
4 Laborers	26.45	846.40	44.30	1417.60		
1 Drill Rig, Truck-Mounted		2486.00		2734.60		
1 Flatbed Truck, Gas, 3 Ton		217.00		238.70	67.58	74.33
40 L.H., Daily Totals		$3777.00		$4772.10	$94.42	$119.30

Crew B-23A	**Hr.**	**Daily**	**Hr.**	**Daily**	**Bare Costs**	**Incl. O&P**
1 Labor Foreman (outside)	$28.45	$227.60	$47.65	$381.20	$29.93	$49.67
1 Laborer	26.45	211.60	44.30	354.40		
1 Equip. Operator (med.)	34.90	279.20	57.05	456.40		
1 Drill Rig, Truck-Mounted		2486.00		2734.60		
1 Pickup Truck, 3/4 Ton		104.60		115.06	107.94	118.74
24 L.H., Daily Totals		$3309.00		$4041.66	$137.88	$168.40

Crew B-23B	**Hr.**	**Daily**	**Hr.**	**Daily**	**Bare Costs**	**Incl. O&P**
1 Labor Foreman (outside)	$28.45	$227.60	$47.65	$381.20	$29.93	$49.67
1 Laborer	26.45	211.60	44.30	354.40		
1 Equip. Operator (med.)	34.90	279.20	57.05	456.40		
1 Drill Rig, Truck-Mounted		2486.00		2734.60		
1 Pickup Truck, 3/4 Ton		104.60		115.06		
1 Centr. Water Pump, 6"		291.60		320.76	120.09	132.10
24 L.H., Daily Totals		$3600.60		$4362.42	$150.03	$181.77

Crew B-24	**Hr.**	**Daily**	**Hr.**	**Daily**	**Bare Costs**	**Incl. O&P**
1 Cement Finisher	$31.85	$254.80	$50.85	$406.80	$30.62	$50.43
1 Laborer	26.45	211.60	44.30	354.40		
1 Carpenter	33.55	268.40	56.15	449.20		
24 L.H., Daily Totals		$734.80		$1210.40	$30.62	$50.43

Crew B-25	**Hr.**	**Daily**	**Hr.**	**Daily**	**Bare Costs**	**Incl. O&P**
1 Labor Foreman	$28.45	$227.60	$47.65	$381.20	$28.94	$48.08
7 Laborers	26.45	1481.20	44.30	2480.80		
3 Equip. Oper. (med.)	34.90	837.60	57.05	1369.20		
1 Asphalt Paver, 130 H.P.		1949.00		2143.90		
1 Tandem Roller, 10 Ton		224.20		246.62		
1 Roller, Pneum. Whl, 12 Ton		309.20		340.12	28.21	31.03
88 L.H., Daily Totals		$5028.80		$6961.84	$57.15	$79.11

Crews

Crew No.	Bare Costs		Incl. Subs O&P		Cost Per Labor-Hour	
Crew B-25B	**Hr.**	**Daily**	**Hr.**	**Daily**	**Bare Costs**	**Incl. O&P**
1 Labor Foreman	$28.45	$227.60	$47.65	$381.20	$29.43	$48.83
7 Laborers	26.45	1481.20	44.30	2480.80		
4 Equip. Oper. (med.)	34.90	1116.80	57.05	1825.60		
1 Asphalt Paver, 130 H.P.		1949.00		2143.90		
2 Tandem Rollers, 10 Ton		448.40		493.24		
1 Roller, Pneum. Whl, 12 Ton		309.20		340.12	28.19	31.01
96 L.H., Daily Totals		$5532.20		$7664.86	$57.63	$79.84
Crew B-25C	**Hr.**	**Daily**	**Hr.**	**Daily**	**Bare Costs**	**Incl. O&P**
1 Labor Foreman	$28.45	$227.60	$47.65	$381.20	$29.60	$49.11
3 Laborers	26.45	634.80	44.30	1063.20		
2 Equip. Oper. (med.)	34.90	558.40	57.05	912.80		
1 Asphalt Paver, 130 H.P.		1949.00		2143.90		
1 Tandem Roller, 10 Ton		224.20		246.62	45.27	49.80
48 L.H., Daily Totals		$3594.00		$4747.72	$74.88	$98.91
Crew B-26	**Hr.**	**Daily**	**Hr.**	**Daily**	**Bare Costs**	**Incl. O&P**
1 Labor Foreman (outside)	$28.45	$227.60	$47.65	$381.20	$29.55	$49.23
6 Laborers	26.45	1269.60	44.30	2126.40		
2 Equip. Oper. (med.)	34.90	558.40	57.05	912.80		
1 Rodman (reinf.)	36.30	290.40	63.10	504.80		
1 Cement Finisher	31.85	254.80	50.85	406.80		
1 Grader, 30,000 Lbs.		542.80		597.08		
1 Paving Mach. & Equip.		2483.00		2731.30	34.38	37.82
88 L.H., Daily Totals		$5626.60		$7660.38	$63.94	$87.05
Crew B-26A	**Hr.**	**Daily**	**Hr.**	**Daily**	**Bare Costs**	**Incl. O&P**
1 Labor Foreman (outside)	$28.45	$227.60	$47.65	$381.20	$29.55	$49.23
6 Laborers	26.45	1269.60	44.30	2126.40		
2 Equip. Oper. (med.)	34.90	558.40	57.05	912.80		
1 Rodman (reinf.)	36.30	290.40	63.10	504.80		
1 Cement Finisher	31.85	254.80	50.85	406.80		
1 Grader, 30,000 Lbs.		542.80		597.08		
1 Paving Mach. & Equip.		2483.00		2731.30		
1 Concrete Saw		134.20		147.62	35.91	39.50
88 L.H., Daily Totals		$5760.80		$7808.00	$65.46	$88.73
Crew B-26B	**Hr.**	**Daily**	**Hr.**	**Daily**	**Bare Costs**	**Incl. O&P**
1 Labor Foreman (outside)	$28.45	$227.60	$47.65	$381.20	$30.00	$49.88
6 Laborers	26.45	1269.60	44.30	2126.40		
3 Equip. Oper. (med.)	34.90	837.60	57.05	1369.20		
1 Rodman (reinf.)	36.30	290.40	63.10	504.80		
1 Cement Finisher	31.85	254.80	50.85	406.80		
1 Grader, 30,000 Lbs.		542.80		597.08		
1 Paving Mach. & Equip.		2483.00		2731.30		
1 Concrete Pump, 110' Boom		999.60		1099.56	41.93	46.12
96 L.H., Daily Totals		$6905.40		$9216.34	$71.93	$96.00
Crew B-27	**Hr.**	**Daily**	**Hr.**	**Daily**	**Bare Costs**	**Incl. O&P**
1 Labor Foreman (outside)	$28.45	$227.60	$47.65	$381.20	$26.95	$45.14
3 Laborers	26.45	634.80	44.30	1063.20		
1 Berm Machine		276.80		304.48	8.65	9.52
32 L.H., Daily Totals		$1139.20		$1748.88	$35.60	$54.65
Crew B-28	**Hr.**	**Daily**	**Hr.**	**Daily**	**Bare Costs**	**Incl. O&P**
2 Carpenters	$33.55	$536.80	$56.15	$898.40	$31.18	$52.20
1 Laborer	26.45	211.60	44.30	354.40		
24 L.H., Daily Totals		$748.40		$1252.80	$31.18	$52.20

Crew No.	Bare Costs		Incl. Subs O&P		Cost Per Labor-Hour	
Crew B-29	**Hr.**	**Daily**	**Hr.**	**Daily**	**Bare Costs**	**Incl. O&P**
1 Labor Foreman (outside)	$28.45	$227.60	$47.65	$381.20	$28.34	$47.23
4 Laborers	26.45	846.40	44.30	1417.60		
1 Equip. Oper. (crane)	35.80	286.40	58.50	468.00		
1 Gradall, 5/8 C.Y.		857.40		943.14	17.86	19.65
48 L.H., Daily Totals		$2217.80		$3209.94	$46.20	$66.87
Crew B-30	**Hr.**	**Daily**	**Hr.**	**Daily**	**Bare Costs**	**Incl. O&P**
1 Equip. Oper. (med.)	$34.90	$279.20	$57.05	$456.40	$29.77	$49.08
2 Truck Drivers (heavy)	27.20	435.20	45.10	721.60		
1 Hyd. Excavator, 1.5 C.Y.		947.40		1042.14		
2 Dump Trucks, 12 C.Y., 400 H.P.		1012.00		1113.20	81.64	89.81
24 L.H., Daily Totals		$2673.80		$3333.34	$111.41	$138.89
Crew B-31	**Hr.**	**Daily**	**Hr.**	**Daily**	**Bare Costs**	**Incl. O&P**
1 Labor Foreman (outside)	$28.45	$227.60	$47.65	$381.20	$26.85	$44.97
4 Laborers	26.45	846.40	44.30	1417.60		
1 Air Compressor, 250 cfm		162.80		179.08		
1 Sheeting Driver		5.75		6.33		
2 -50' Air Hoses, 1.5"		12.60		13.86	4.53	4.98
40 L.H., Daily Totals		$1255.15		$1998.07	$31.38	$49.95
Crew B-32	**Hr.**	**Daily**	**Hr.**	**Daily**	**Bare Costs**	**Incl. O&P**
1 Laborer	$26.45	$211.60	$44.30	$354.40	$32.79	$53.86
3 Equip. Oper. (med.)	34.90	837.60	57.05	1369.20		
1 Grader, 30,000 Lbs.		542.80		597.08		
1 Tandem Roller, 10 Ton		224.20		246.62		
1 Dozer, 200 H.P.		1051.00		1156.10	56.81	62.49
32 L.H., Daily Totals		$2867.20		$3723.40	$89.60	$116.36
Crew B-32A	**Hr.**	**Daily**	**Hr.**	**Daily**	**Bare Costs**	**Incl. O&P**
1 Laborer	$26.45	$211.60	$44.30	$354.40	$32.08	$52.80
2 Equip. Oper. (med.)	34.90	558.40	57.05	912.80		
1 Grader, 30,000 Lbs.		542.80		597.08		
1 Roller, Vibratory, 25 Ton		596.80		656.48	47.48	52.23
24 L.H., Daily Totals		$1909.60		$2520.76	$79.57	$105.03
Crew B-32B	**Hr.**	**Daily**	**Hr.**	**Daily**	**Bare Costs**	**Incl. O&P**
1 Laborer	$26.45	$211.60	$44.30	$354.40	$32.08	$52.80
2 Equip. Oper. (med.)	34.90	558.40	57.05	912.80		
1 Dozer, 200 H.P.		1051.00		1156.10		
1 Roller, Vibratory, 25 Ton		596.80		656.48	68.66	75.52
24 L.H., Daily Totals		$2417.80		$3079.78	$100.74	$128.32
Crew B-32C	**Hr.**	**Daily**	**Hr.**	**Daily**	**Bare Costs**	**Incl. O&P**
1 Labor Foreman	$28.45	$227.60	$47.65	$381.20	$31.01	$51.23
2 Laborers	26.45	423.20	44.30	708.80		
3 Equip. Oper. (med.)	34.90	837.60	57.05	1369.20		
1 Grader, 30,000 Lbs.		542.80		597.08		
1 Tandem Roller, 10 Ton		224.20		246.62		
1 Dozer, 200 H.P.		1051.00		1156.10	37.88	41.66
48 L.H., Daily Totals		$3306.40		$4459.00	$68.88	$92.90
Crew B-33A	**Hr.**	**Daily**	**Hr.**	**Daily**	**Bare Costs**	**Incl. O&P**
1 Equip. Oper. (med.)	$34.90	$279.20	$57.05	$456.40	$34.90	$57.05
.25 Equip. Oper. (med.)	34.90	69.80	57.05	114.10		
1 Scraper, Towed, 7 C.Y.		108.00		118.80		
1.250 Dozers, 300 H.P.		2008.75		2209.63	211.68	232.84
10 L.H., Daily Totals		$2465.75		$2898.93	$246.57	$289.89

Crews

Crew No.	Bare Costs		Incl. Subs O&P		Cost Per Labor-Hour	
Crew B-33B	**Hr.**	**Daily**	**Hr.**	**Daily**	**Bare Costs**	**Incl. O&P**
1 Equip. Oper. (med.)	$34.90	$279.20	$57.05	$456.40	$34.90	$57.05
.25 Equip. Oper. (med.)	34.90	69.80	57.05	114.10		
1 Scraper, Towed, 10 C.Y.		140.00		154.00		
1.250 Dozers, 300 H.P.		2008.75		2209.63	214.88	236.36
10 L.H., Daily Totals		$2497.75		$2934.13	$249.78	$293.41

Crew B-33C	**Hr.**	**Daily**	**Hr.**	**Daily**	**Bare Costs**	**Incl. O&P**
1 Equip. Oper. (med.)	$34.90	$279.20	$57.05	$456.40	$34.90	$57.05
.25 Equip. Oper. (med.)	34.90	69.80	57.05	114.10		
1 Scraper, Towed, 15 C.Y.		158.60		174.46		
1.250 Dozers, 300 H.P.		2008.75		2209.63	216.74	238.41
10 L.H., Daily Totals		$2516.35		$2954.59	$251.63	$295.46

Crew B-33D	**Hr.**	**Daily**	**Hr.**	**Daily**	**Bare Costs**	**Incl. O&P**
1 Equip. Oper. (med.)	$34.90	$279.20	$57.05	$456.40	$34.90	$57.05
.25 Equip. Oper. (med.)	34.90	69.80	57.05	114.10		
1 S.P. Scraper, 14 C.Y.		1512.00		1663.20		
.25 Dozer, 300 H.P.		401.75		441.93	191.38	210.51
10 L.H., Daily Totals		$2262.75		$2675.63	$226.28	$267.56

Crew B-33E	**Hr.**	**Daily**	**Hr.**	**Daily**	**Bare Costs**	**Incl. O&P**
1 Equip. Oper. (med.)	$34.90	$279.20	$57.05	$456.40	$34.90	$57.05
.25 Equip. Oper. (med.)	34.90	69.80	57.05	114.10		
1 S.P. Scraper, 21 C.Y.		1872.00		2059.20		
.25 Dozer, 300 H.P.		401.75		441.93	227.38	250.11
10 L.H., Daily Totals		$2622.75		$3071.63	$262.27	$307.16

Crew B-33F	**Hr.**	**Daily**	**Hr.**	**Daily**	**Bare Costs**	**Incl. O&P**
1 Equip. Oper. (med.)	$34.90	$279.20	$57.05	$456.40	$34.90	$57.05
.25 Equip. Oper. (med.)	34.90	69.80	57.05	114.10		
1 Elev. Scraper, 11 C.Y.		1089.00		1197.90		
.25 Dozer, 300 H.P.		401.75		441.93	149.07	163.98
10 L.H., Daily Totals		$1839.75		$2210.32	$183.97	$221.03

Crew B-33G	**Hr.**	**Daily**	**Hr.**	**Daily**	**Bare Costs**	**Incl. O&P**
1 Equip. Oper. (med.)	$34.90	$279.20	$57.05	$456.40	$34.90	$57.05
.25 Equip. Oper. (med.)	34.90	69.80	57.05	114.10		
1 Elev. Scraper, 22 C.Y.		2087.00		2295.70		
.25 Dozer, 300 H.P.		401.75		441.93	248.88	273.76
10 L.H., Daily Totals		$2837.75		$3308.13	$283.77	$330.81

Crew B-33K	**Hr.**	**Daily**	**Hr.**	**Daily**	**Bare Costs**	**Incl. O&P**
1 Equipment Operator (med.)	$34.90	$279.20	$57.05	$456.40	$32.49	$53.41
.25 Equipment Operator (med.)	34.90	69.80	57.05	114.10		
.5 Laborer	26.45	105.80	44.30	177.20		
1 S.P. Scraper, 31 C.Y.		2777.00		3054.70		
.25 Dozer, 410 H.P.		526.25		578.88	235.95	259.54
14 L.H., Daily Totals		$3758.05		$4381.27	$268.43	$312.95

Crew B-34A	**Hr.**	**Daily**	**Hr.**	**Daily**	**Bare Costs**	**Incl. O&P**
1 Truck Driver (heavy)	$27.20	$217.60	$45.10	$360.80	$27.20	$45.10
1 Dump Truck, 8 C.Y., 220 H.P.		315.40		346.94	39.42	43.37
8 L.H., Daily Totals		$533.00		$707.74	$66.63	$88.47

Crew B-34B	**Hr.**	**Daily**	**Hr.**	**Daily**	**Bare Costs**	**Incl. O&P**
1 Truck Driver (heavy)	$27.20	$217.60	$45.10	$360.80	$27.20	$45.10
1 Dump Truck, 12 C.Y., 400 H.P.		506.00		556.60	63.25	69.58
8 L.H., Daily Totals		$723.60		$917.40	$90.45	$114.68

Crew No.	Bare Costs		Incl. Subs O&P		Cost Per Labor-Hour	
Crew B-34C	**Hr.**	**Daily**	**Hr.**	**Daily**	**Bare Costs**	**Incl. O&P**
1 Truck Driver (heavy)	$27.20	$217.60	$45.10	$360.80	$27.20	$45.10
1 Truck Tractor, 6x4, 380 H.P.		459.40		505.34		
1 Dump Trailer, 16.5 C.Y.		125.80		138.38	73.15	80.47
8 L.H., Daily Totals		$802.80		$1004.52	$100.35	$125.57

Crew B-34D	**Hr.**	**Daily**	**Hr.**	**Daily**	**Bare Costs**	**Incl. O&P**
1 Truck Driver (heavy)	$27.20	$217.60	$45.10	$360.80	$27.20	$45.10
1 Truck Tractor, 6x4, 380 H.P.		459.40		505.34		
1 Dump Trailer, 20 C.Y.		140.00		154.00	74.92	82.42
8 L.H., Daily Totals		$817.00		$1020.14	$102.13	$127.52

Crew B-34E	**Hr.**	**Daily**	**Hr.**	**Daily**	**Bare Costs**	**Incl. O&P**
1 Truck Driver (heavy)	$27.20	$217.60	$45.10	$360.80	$27.20	$45.10
1 Dump Truck, Off Hwy., 25 Ton		1264.00		1390.40	158.00	173.80
8 L.H., Daily Totals		$1481.60		$1751.20	$185.20	$218.90

Crew B-34F	**Hr.**	**Daily**	**Hr.**	**Daily**	**Bare Costs**	**Incl. O&P**
1 Truck Driver (heavy)	$27.20	$217.60	$45.10	$360.80	$27.20	$45.10
1 Dump Truck, Off Hwy., 35 Ton		1135.00		1248.50	141.88	156.06
8 L.H., Daily Totals		$1352.60		$1609.30	$169.07	$201.16

Crew B-34G	**Hr.**	**Daily**	**Hr.**	**Daily**	**Bare Costs**	**Incl. O&P**
1 Truck Driver (heavy)	$27.20	$217.60	$45.10	$360.80	$27.20	$45.10
1 Dump Truck, Off Hwy., 50 Ton		1549.00		1703.90	193.63	212.99
8 L.H., Daily Totals		$1766.60		$2064.70	$220.82	$258.09

Crew B-34H	**Hr.**	**Daily**	**Hr.**	**Daily**	**Bare Costs**	**Incl. O&P**
1 Truck Driver (heavy)	$27.20	$217.60	$45.10	$360.80	$27.20	$45.10
1 Dump Truck, Off Hwy., 65 Ton		1551.00		1706.10	193.88	213.26
8 L.H., Daily Totals		$1768.60		$2066.90	$221.07	$258.36

Crew B-34I	**Hr.**	**Daily**	**Hr.**	**Daily**	**Bare Costs**	**Incl. O&P**
1 Truck Driver (heavy)	$27.20	$217.60	$45.10	$360.80	$27.20	$45.10
1 Dump Truck, 18 C.Y., 450 H.P.		660.80		726.88	82.60	90.86
8 L.H., Daily Totals		$878.40		$1087.68	$109.80	$135.96

Crew B-34J	**Hr.**	**Daily**	**Hr.**	**Daily**	**Bare Costs**	**Incl. O&P**
1 Truck Driver (heavy)	$27.20	$217.60	$45.10	$360.80	$27.20	$45.10
1 Dump Truck, Off Hwy., 100 Ton		2009.00		2209.90	251.13	276.24
8 L.H., Daily Totals		$2226.60		$2570.70	$278.32	$321.34

Crew B-34K	**Hr.**	**Daily**	**Hr.**	**Daily**	**Bare Costs**	**Incl. O&P**
1 Truck Driver (heavy)	$27.20	$217.60	$45.10	$360.80	$27.20	$45.10
1 Truck Tractor, 6x4, 450 H.P.		558.80		614.68		
1 Lowbed Trailer, 75 Ton		219.20		241.12	97.25	106.97
8 L.H., Daily Totals		$995.60		$1216.60	$124.45	$152.07

Crew B-34L	**Hr.**	**Daily**	**Hr.**	**Daily**	**Bare Costs**	**Incl. O&P**
1 Equip. Oper. (light)	$33.55	$268.40	$54.80	$438.40	$33.55	$54.80
1 Flatbed Truck, Gas, 1.5 Ton		171.40		188.54	21.43	23.57
8 L.H., Daily Totals		$439.80		$626.94	$54.98	$78.37

Crew B-34N	**Hr.**	**Daily**	**Hr.**	**Daily**	**Bare Costs**	**Incl. O&P**
1 Truck Driver (heavy)	$27.20	$217.60	$45.10	$360.80	$27.20	$45.10
1 Dump Truck, 8 C.Y., 220 H.P.		315.40		346.94		
1 Flatbed Trailer, 40 Ton		150.20		165.22	58.20	64.02
8 L.H., Daily Totals		$683.20		$872.96	$85.40	$109.12

Crews

Crew No.	Bare Costs		Incl. Subs O&P		Cost Per Labor-Hour	
Crew B-34P	**Hr.**	**Daily**	**Hr.**	**Daily**	**Bare Costs**	**Incl. O&P**
1 Pipe Fitter	$40.05	$320.40	$65.40	$523.20	$33.78	$55.42
1 Truck Driver (light)	26.40	211.20	43.80	350.40		
1 Equip. Oper. (med.)	34.90	279.20	57.05	456.40		
1 Flatbed Truck, Gas, 3 Ton		217.00		238.70		
1 Backhoe Loader, 48 H.P.		320.40		352.44	22.39	24.63
24 L.H., Daily Totals		$1348.20		$1921.14	$56.17	$80.05
Crew B-34Q	**Hr.**	**Daily**	**Hr.**	**Daily**	**Bare Costs**	**Incl. O&P**
1 Pipe Fitter	$40.05	$320.40	$65.40	$523.20	$34.08	$55.90
1 Truck Driver (light)	26.40	211.20	43.80	350.40		
1 Eqip. Oper. (crane)	35.80	286.40	58.50	468.00		
1 Flatbed Trailer, 25 Ton		111.20		122.32		
1 Dump Truck, 8 C.Y., 220 H.P.		315.40		346.94		
1 Hyd. Crane, 25 Ton		748.60		823.46	48.97	53.86
24 L.H., Daily Totals		$1993.20		$2634.32	$83.05	$109.76
Crew B-34R	**Hr.**	**Daily**	**Hr.**	**Daily**	**Bare Costs**	**Incl. O&P**
1 Pipe Fitter	$40.05	$320.40	$65.40	$523.20	$34.08	$55.90
1 Truck Driver (light)	26.40	211.20	43.80	350.40		
1 Eqip. Oper. (crane)	35.80	286.40	58.50	468.00		
1 Flatbed Trailer, 25 Ton		111.20		122.32		
1 Dump Truck, 8 C.Y., 220 H.P.		315.40		346.94		
1 Hyd. Crane, 25 Ton		748.60		823.46		
1 Hyd. Excavator, 1 C.Y.		729.80		802.78	79.38	87.31
24 L.H., Daily Totals		$2723.00		$3437.10	$113.46	$143.21
Crew B-34S	**Hr.**	**Daily**	**Hr.**	**Daily**	**Bare Costs**	**Incl. O&P**
2 Pipe Fitters	$40.05	$640.80	$65.40	$1046.40	$35.77	$58.60
1 Truck Driver (heavy)	27.20	217.60	45.10	360.80		
1 Eqip. Oper. (crane)	35.80	286.40	58.50	468.00		
1 Flatbed Trailer, 40 Ton		150.20		165.22		
1 Truck Tractor, 6x4, 380 H.P.		459.40		505.34		
1 Hyd. Crane, 80 Ton		1693.00		1862.30		
1 Hyd. Excavator, 2 C.Y.		1224.00		1346.40	110.21	121.23
32 L.H., Daily Totals		$4671.40		$5754.46	$145.98	$179.83
Crew B-34T	**Hr.**	**Daily**	**Hr.**	**Daily**	**Bare Costs**	**Incl. O&P**
2 Pipe Fitters	$40.05	$640.80	$65.40	$1046.40	$35.77	$58.60
1 Truck Driver (heavy)	27.20	217.60	45.10	360.80		
1 Eqip. Oper. (crane)	35.80	286.40	58.50	468.00		
1 Flatbed Trailer, 40 Ton		150.20		165.22		
1 Truck Tractor, 6x4, 380 H.P.		459.40		505.34		
1 Hyd. Crane, 80 Ton		1693.00		1862.30	71.96	79.15
32 L.H., Daily Totals		$3447.40		$4408.06	$107.73	$137.75
Crew B-35	**Hr.**	**Daily**	**Hr.**	**Daily**	**Bare Costs**	**Incl. O&P**
1 Laborer Foreman (out)	$28.45	$227.60	$47.65	$381.20	$32.88	$54.44
1 Skilled Worker	34.05	272.40	57.00	456.00		
1 Welder (plumber)	39.65	317.20	64.75	518.00		
1 Laborer	26.45	211.60	44.30	354.40		
1 Equip. Oper. (crane)	35.80	286.40	58.50	468.00		
1 Welder, electric, 300 amp		32.45		35.70		
1 Hyd. Excavator, .75 C.Y.		640.40		704.44	16.82	18.50
40 L.H., Daily Totals		$1988.05		$2917.74	$49.70	$72.94

Crew No.	Bare Costs		Incl. Subs O&P		Cost Per Labor-Hour	
Crew B-35A	**Hr.**	**Daily**	**Hr.**	**Daily**	**Bare Costs**	**Incl. O&P**
1 Laborer Foreman (out)	$28.45	$227.60	$47.65	$381.20	$31.70	$52.46
2 Laborers	26.45	423.20	44.30	708.80		
1 Skilled Worker	34.05	272.40	57.00	456.00		
1 Welder (plumber)	39.65	317.20	64.75	518.00		
1 Equip. Oper. (crane)	35.80	286.40	58.50	468.00		
1 Equip. Oper. Oiler	31.05	248.40	50.75	406.00		
1 Welder, gas engine, 300 amp		109.00		119.90		
1 Crawler Crane, 75 Ton		1446.00		1590.60	27.77	30.54
56 L.H., Daily Totals		$3330.20		$4648.50	$59.47	$83.01
Crew B-36	**Hr.**	**Daily**	**Hr.**	**Daily**	**Bare Costs**	**Incl. O&P**
1 Labor Foreman (outside)	$28.45	$227.60	$47.65	$381.20	$30.23	$50.07
2 Laborers	26.45	423.20	44.30	708.80		
2 Equip. Oper. (med.)	34.90	558.40	57.05	912.80		
1 Dozer, 200 H.P.		1051.00		1156.10		
1 Aggregate Spreader		33.60		36.96		
1 Tandem Roller, 10 Ton		224.20		246.62	32.72	35.99
40 L.H., Daily Totals		$2518.00		$3442.48	$62.95	$86.06
Crew B-36A	**Hr.**	**Daily**	**Hr.**	**Daily**	**Bare Costs**	**Incl. O&P**
1 Labor Foreman (outside)	$28.45	$227.60	$47.65	$381.20	$31.56	$52.06
2 Laborers	26.45	423.20	44.30	708.80		
4 Equip. Oper. (med.)	34.90	1116.80	57.05	1825.60		
1 Dozer, 200 H.P.		1051.00		1156.10		
1 Aggregate Spreader		33.60		36.96		
1 Tandem Roller, 10 Ton		224.20		246.62		
1 Roller, Pneum. Whl, 12 Ton		309.20		340.12	28.89	31.78
56 L.H., Daily Totals		$3385.60		$4695.40	$60.46	$83.85
Crew B-36B	**Hr.**	**Daily**	**Hr.**	**Daily**	**Bare Costs**	**Incl. O&P**
1 Labor Foreman (outside)	$28.45	$227.60	$47.65	$381.20	$31.02	$51.19
2 Laborers	26.45	423.20	44.30	708.80		
4 Equip. Oper. (med.)	34.90	1116.80	57.05	1825.60		
1 Truck Driver, Heavy	27.20	217.60	45.10	360.80		
1 Grader, 30,000 Lbs.		542.80		597.08		
1 F.E. Loader, crl, 1.5 C.Y.		565.00		621.50		
1 Dozer, 300 H.P.		1607.00		1767.70		
1 Roller, Vibratory, 25 Ton		596.80		656.48		
1 Truck Tractor, 6x4, 450 H.P.		558.80		614.68		
1 Water Tank Trailer, 5000 Gal.		146.20		160.82	62.76	69.04
64 L.H., Daily Totals		$6001.80		$7694.66	$93.78	$120.23
Crew B-36C	**Hr.**	**Daily**	**Hr.**	**Daily**	**Bare Costs**	**Incl. O&P**
1 Labor Foreman (outside)	$28.45	$227.60	$47.65	$381.20	$32.07	$52.78
3 Equip. Oper. (med.)	34.90	837.60	57.05	1369.20		
1 Truck Driver, Heavy	27.20	217.60	45.10	360.80		
1 Grader, 30,000 Lbs.		542.80		597.08		
1 Dozer, 300 H.P.		1607.00		1767.70		
1 Roller, Vibratory, 25 Ton		596.80		656.48		
1 Truck Tractor, 6x4, 450 H.P.		558.80		614.68		
1 Water Tank Trailer, 5000 Gal.		146.20		160.82	86.29	94.92
40 L.H., Daily Totals		$4734.40		$5907.96	$118.36	$147.70

Crews

Crew No.	Bare Costs		Incl. Subs O&P		Cost Per Labor-Hour	
Crew B-36E	**Hr.**	**Daily**	**Hr.**	**Daily**	**Bare Costs**	**Incl. O&P**
1 Labor Foreman (outside)	$28.45	$227.60	$47.65	$381.20	$32.54	$53.49
4 Equip. Oper. (medium)	34.90	1116.80	57.05	1825.60		
1 Truck Driver, Heavy	27.20	217.60	45.10	360.80		
1 Grader, 30,000 Lbs.		542.80		597.08		
1 Dozer, 300 H.P.		1607.00		1767.70		
1 Roller, Vibratory, 25 Ton		596.80		656.48		
1 Truck Tractor, 6x4, 380 H.P.		459.40		505.34		
1 Dist. Tanker, 3000 Gallon		309.00		339.90	73.23	80.55
48 L.H., Daily Totals		$5077.00		$6434.10	$105.77	$134.04
Crew B-37	**Hr.**	**Daily**	**Hr.**	**Daily**	**Bare Costs**	**Incl. O&P**
1 Labor Foreman (outside)	$28.45	$227.60	$47.65	$381.20	$27.97	$46.61
4 Laborers	26.45	846.40	44.30	1417.60		
1 Equip. Oper. (light)	33.55	268.40	54.80	438.40		
1 Tandem Roller, 5 Ton		145.20		159.72	3.02	3.33
48 L.H., Daily Totals		$1487.60		$2396.92	$30.99	$49.94
Crew B-37A	**Hr.**	**Daily**	**Hr.**	**Daily**	**Bare Costs**	**Incl. O&P**
2 Laborers	$26.45	$423.20	$44.30	$708.80	$26.43	$44.13
1 Truck Driver (light)	26.40	211.20	43.80	350.40		
1 Flatbed Truck, Gas, 1.5 Ton		171.40		188.54		
1 Tar Kettle, T.M.		74.65		82.11	10.25	11.28
24 L.H., Daily Totals		$880.45		$1329.86	$36.69	$55.41
Crew B-37B	**Hr.**	**Daily**	**Hr.**	**Daily**	**Bare Costs**	**Incl. O&P**
3 Laborers	$26.45	$634.80	$44.30	$1063.20	$26.44	$44.17
1 Truck Driver (light)	26.40	211.20	43.80	350.40		
1 Flatbed Truck, Gas, 1.5 Ton		171.40		188.54		
1 Tar Kettle, T.M.		74.65		82.11	7.69	8.46
32 L.H., Daily Totals		$1092.05		$1684.26	$34.13	$52.63
Crew B-37C	**Hr.**	**Daily**	**Hr.**	**Daily**	**Bare Costs**	**Incl. O&P**
2 Laborers	$26.45	$423.20	$44.30	$708.80	$26.43	$44.05
2 Truck Drivers (light)	26.40	422.40	43.80	700.80		
2 Flatbed Truck, Gas, 1.5 Ton		342.80		377.08		
1 Tar Kettle, T.M.		74.65		82.11	13.05	14.35
32 L.H., Daily Totals		$1263.05		$1868.80	$39.47	$58.40
Crew B-37D	**Hr.**	**Daily**	**Hr.**	**Daily**	**Bare Costs**	**Incl. O&P**
1 Laborer	$26.45	$211.60	$44.30	$354.40	$26.43	$44.05
1 Truck Driver (light)	26.40	211.20	43.80	350.40		
1 Pickup Truck, 3/4 Ton		104.60		115.06	6.54	7.19
16 L.H., Daily Totals		$527.40		$819.86	$32.96	$51.24
Crew B-37E	**Hr.**	**Daily**	**Hr.**	**Daily**	**Bare Costs**	**Incl. O&P**
3 Laborers	$26.45	$634.80	$44.30	$1063.20	$28.66	$47.48
1 Equip. Oper. (light)	33.55	268.40	54.80	438.40		
1 Equip. Oper. (medium)	34.90	279.20	57.05	456.40		
2 Truck Drivers (light)	26.40	422.40	43.80	700.80		
4 Barrel w/ Flasher		13.60		14.96		
1 Concrete Saw		134.20		147.62		
1 Rotary Hammer Drill		22.55		24.81		
1 Hammer Drill Bit		3.90		4.29		
1 Loader, Skid Steer, 30 H.P., gas		151.80		166.98		
1 Conc. Hammer Attach.		115.35		126.89		
1 Vibrating plate, gas, 18"		30.80		33.88		
2 Flatbed Truck, Gas, 1.5 Ton		342.80		377.08	14.55	16.01
56 L.H., Daily Totals		$2419.80		$3555.30	$43.21	$63.49

Crew No.	Bare Costs		Incl. Subs O&P		Cost Per Labor-Hour	
Crew B-37F	**Hr.**	**Daily**	**Hr.**	**Daily**	**Bare Costs**	**Incl. O&P**
3 Laborers	$26.45	$634.80	$44.30	$1063.20	$26.44	$44.17
1 Truck Driver (light)	26.40	211.20	43.80	350.40		
4 Barrel w/ Flasher		13.60		14.96		
1 Concrete Mixer, 10 C.F.		151.20		166.32		
1 Air Compressor, 60 cfm		110.20		121.22		
1 -50' Air Hose, 3/4"		3.05		3.36		
1 Spade (chipper)		8.00		8.80		
1 Flatbed Truck, Gas, 1.5 Ton		171.40		188.54	14.30	15.72
32 L.H., Daily Totals		$1303.45		$1916.80	$40.73	$59.90
Crew B-38	**Hr.**	**Daily**	**Hr.**	**Daily**	**Bare Costs**	**Incl. O&P**
2 Laborers	$26.45	$423.20	$44.30	$708.80	$28.82	$47.80
1 Equip. Oper. (light)	33.55	268.40	54.80	438.40		
1 Backhoe Loader, 48 H.P.		320.40		352.44		
1 Hyd.Hammer, (1200 lb.)		158.00		173.80	19.93	21.93
24 L.H., Daily Totals		$1170.00		$1673.44	$48.75	$69.73
Crew B-39	**Hr.**	**Daily**	**Hr.**	**Daily**	**Bare Costs**	**Incl. O&P**
1 Labor Foreman (outside)	$28.45	$227.60	$47.65	$381.20	$26.78	$44.86
5 Laborers	26.45	1058.00	44.30	1772.00		
1 Air Compressor, 250 cfm		162.80		179.08		
2 Breakers, Pavement, 60 lb.		19.60		21.56		
2 -50' Air Hoses, 1.5"		12.60		13.86	4.06	4.47
48 L.H., Daily Totals		$1480.60		$2367.70	$30.85	$49.33
Crew B-40	**Hr.**	**Daily**	**Hr.**	**Daily**	**Bare Costs**	**Incl. O&P**
1 Pile Driver Foreman (out)	$34.75	$278.00	$60.55	$484.40	$32.57	$55.96
4 Pile Drivers	32.75	1048.00	57.10	1827.20		
1 Building Laborer	26.45	211.60	44.30	354.40		
1 Equip. Oper. (crane)	35.80	286.40	58.50	468.00		
1 Crawler Crane, 40 Ton		1128.00		1240.80		
1 Vibratory Hammer & Gen.		2140.00		2354.00	58.36	64.19
56 L.H., Daily Totals		$5092.00		$6728.80	$90.93	$120.16
Crew B-40B	**Hr.**	**Daily**	**Hr.**	**Daily**	**Bare Costs**	**Incl. O&P**
1 Laborer Foreman	$28.45	$227.60	$47.65	$381.20	$29.11	$48.30
3 Laborers	26.45	634.80	44.30	1063.20		
1 Equip. Oper. (crane)	35.80	286.40	58.50	468.00		
1 Equip. Oper. Oiler	31.05	248.40	50.75	406.00		
1 Lattice Boom Crane, 40 Ton		1263.00		1389.30	26.31	28.94
48 L.H., Daily Totals		$2660.20		$3707.70	$55.42	$77.24
Crew B-41	**Hr.**	**Daily**	**Hr.**	**Daily**	**Bare Costs**	**Incl. O&P**
1 Labor Foreman (outside)	$28.45	$227.60	$47.65	$381.20	$27.45	$45.85
4 Laborers	26.45	846.40	44.30	1417.60		
.25 Equip. Oper. (crane)	35.80	71.60	58.50	117.00		
.25 Equip. Oper. Oiler	31.05	62.10	50.75	101.50		
.25 Crawler Crane, 40 Ton		282.00		310.20	6.41	7.05
44 L.H., Daily Totals		$1489.70		$2327.50	$33.86	$52.90
Crew B-42	**Hr.**	**Daily**	**Hr.**	**Daily**	**Bare Costs**	**Incl. O&P**
1 Labor Foreman (outside)	$28.45	$227.60	$47.65	$381.20	$29.96	$49.73
4 Laborers	26.45	846.40	44.30	1417.60		
1 Equip. Oper. (crane)	35.80	286.40	58.50	468.00		
1 Welder	39.65	317.20	64.75	518.00		
1 Hyd. Crane, 25 Ton		748.60		823.46		
1 Welder, gas engine, 300 amp		109.00		119.90		
1 Horz. Boring Csg. Mch.		442.20		486.42	23.21	25.53
56 L.H., Daily Totals		$2977.40		$4214.58	$53.17	$75.26

Crews

Crew No.	Bare Costs		Incl. Subs O&P		Cost Per Labor-Hour	
Crew B-43	**Hr.**	**Daily**	**Hr.**	**Daily**	**Bare Costs**	**Incl. O&P**
1 Labor Foreman (outside)	$28.45	$227.60	$47.65	$381.20	$26.85	$44.97
4 Laborers	26.45	846.40	44.30	1417.60		
1 Drill Rig, Truck-Mounted		2486.00		2734.60	62.15	68.36
40 L.H., Daily Totals		$3560.00		$4533.40	$89.00	$113.33
Crew B-44	**Hr.**	**Daily**	**Hr.**	**Daily**	**Bare Costs**	**Incl. O&P**
1 Pile Driver Foreman	$34.75	$278.00	$60.55	$484.40	$31.81	$54.51
4 Pile Drivers	32.75	1048.00	57.10	1827.20		
1 Equip. Oper. (crane)	35.80	286.40	58.50	468.00		
2 Laborers	26.45	423.20	44.30	708.80		
1 Crawler Crane, 40 Ton		1128.00		1240.80		
1 Lead, 60' high		74.00		81.40		
1 Hammer, diesel, 15K ft.-lbs.		614.20		675.62	28.38	31.22
64 L.H., Daily Totals		$3851.80		$5486.22	$60.18	$85.72
Crew B-45	**Hr.**	**Daily**	**Hr.**	**Daily**	**Bare Costs**	**Incl. O&P**
1 Building Laborer	$26.45	$211.60	$44.30	$354.40	$26.82	$44.70
1 Truck Driver (heavy)	27.20	217.60	45.10	360.80		
1 Dist. Tanker, 3000 Gallon		309.00		339.90	19.31	21.24
16 L.H., Daily Totals		$738.20		$1055.10	$46.14	$65.94
Crew B-46	**Hr.**	**Daily**	**Hr.**	**Daily**	**Bare Costs**	**Incl. O&P**
1 Pile Driver Foreman	$34.75	$278.00	$60.55	$484.40	$29.93	$51.27
2 Pile Drivers	32.75	524.00	57.10	913.60		
3 Laborers	26.45	634.80	44.30	1063.20		
1 Chain Saw, gas, 36" Long		41.60		45.76	0.87	0.95
48 L.H., Daily Totals		$1478.40		$2506.96	$30.80	$52.23
Crew B-47	**Hr.**	**Daily**	**Hr.**	**Daily**	**Bare Costs**	**Incl. O&P**
1 Blast Foreman	$28.45	$227.60	$47.65	$381.20	$27.45	$45.98
1 Driller	26.45	211.60	44.30	354.40		
1 Air Track Drill, 4"		849.60		934.56		
1 Air Compressor, 600 cfm		424.60		467.06		
2 -50' Air Hoses, 3"		29.80		32.78	81.50	89.65
16 L.H., Daily Totals		$1743.20		$2170.00	$108.95	$135.63
Crew B-47A	**Hr.**	**Daily**	**Hr.**	**Daily**	**Bare Costs**	**Incl. O&P**
1 Drilling Foreman	$28.45	$227.60	$47.65	$381.20	$31.77	$52.30
1 Equip. Oper. (heavy)	35.80	286.40	58.50	468.00		
1 Oiler	31.05	248.40	50.75	406.00		
1 Air Track Drill, 5"		965.20		1061.72	40.22	44.24
24 L.H., Daily Totals		$1727.60		$2316.92	$71.98	$96.54
Crew B-47C	**Hr.**	**Daily**	**Hr.**	**Daily**	**Bare Costs**	**Incl. O&P**
1 Laborer	$26.45	$211.60	$44.30	$354.40	$30.00	$49.55
1 Equip. Oper. (light)	33.55	268.40	54.80	438.40		
1 Air Compressor, 750 cfm		431.80		474.98		
2 -50' Air Hoses, 3"		29.80		32.78		
1 Air Track Drill, 4"		849.60		934.56	81.95	90.14
16 L.H., Daily Totals		$1791.20		$2235.12	$111.95	$139.69
Crew B-47E	**Hr.**	**Daily**	**Hr.**	**Daily**	**Bare Costs**	**Incl. O&P**
1 Laborer Foreman	$28.45	$227.60	$47.65	$381.20	$26.95	$45.14
3 Laborers	26.45	634.80	44.30	1063.20		
1 Flatbed Truck, Gas, 3 Ton		217.00		238.70	6.78	7.46
32 L.H., Daily Totals		$1079.40		$1683.10	$33.73	$52.60

Crew No.	Bare Costs		Incl. Subs O&P		Cost Per Labor-Hour	
Crew B-47G	**Hr.**	**Daily**	**Hr.**	**Daily**	**Bare Costs**	**Incl. O&P**
1 Laborer Foreman	$28.45	$227.60	$47.65	$381.20	$27.12	$45.42
2 Laborers	26.45	423.20	44.30	708.80		
1 Air Track Drill, 4"		849.60		934.56		
1 Air Compressor, 600 cfm		424.60		467.06		
2 -50' Air Hoses, 3"		29.80		32.78		
1 Gunnite Pump rig		178.00		195.80	61.75	67.92
24 L.H., Daily Totals		$2132.80		$2720.20	$88.87	$113.34
Crew B-47H	**Hr.**	**Daily**	**Hr.**	**Daily**	**Bare Costs**	**Incl. O&P**
1 Skilled Worker Foreman	$36.05	$288.40	$60.35	$482.80	$34.55	$57.84
3 Skilled Workers	34.05	817.20	57.00	1368.00		
1 Flatbed Truck, Gas, 3 Ton		217.00		238.70	6.78	7.46
32 L.H., Daily Totals		$1322.60		$2089.50	$41.33	$65.30
Crew B-48	**Hr.**	**Daily**	**Hr.**	**Daily**	**Bare Costs**	**Incl. O&P**
1 Labor Foreman (outside)	$28.45	$227.60	$47.65	$381.20	$28.34	$47.23
4 Laborers	26.45	846.40	44.30	1417.60		
1 Equip. Oper. (crane)	35.80	286.40	58.50	468.00		
1 Centr. Water Pump, 6"		291.60		320.76		
1 -20' Suction Hose, 6"		11.90		13.09		
1 -50' Discharge Hose, 6"		6.30		6.93		
1 Drill Rig, Truck-Mounted		2486.00		2734.60	58.25	64.07
48 L.H., Daily Totals		$4156.20		$5342.18	$86.59	$111.30
Crew B-49	**Hr.**	**Daily**	**Hr.**	**Daily**	**Bare Costs**	**Incl. O&P**
1 Labor Foreman (outside)	$28.45	$227.60	$47.65	$381.20	$29.11	$49.09
5 Laborers	26.45	1058.00	44.30	1772.00		
1 Equip. Oper. (crane)	35.80	286.40	58.50	468.00		
2 Pile Drivers	32.75	524.00	57.10	913.60		
1 Hyd. Crane, 25 Ton		748.60		823.46		
1 Centr. Water Pump, 6"		291.60		320.76		
1 -20' Suction Hose, 6"		11.90		13.09		
1 -50' Discharge Hose, 6"		6.30		6.93		
1 Drill Rig, Truck-Mounted		2486.00		2734.60	49.23	54.15
72 L.H., Daily Totals		$5640.40		$7433.64	$78.34	$103.25
Crew B-50	**Hr.**	**Daily**	**Hr.**	**Daily**	**Bare Costs**	**Incl. O&P**
1 Pile Driver Foreman	$34.75	$278.00	$60.55	$484.40	$30.72	$52.55
6 Pile Drivers	32.75	1572.00	57.10	2740.80		
1 Equip. Oper. (crane)	35.80	286.40	58.50	468.00		
5 Laborers	26.45	1058.00	44.30	1772.00		
1 Crawler Crane, 40 Ton		1128.00		1240.80		
1 Lead, 60' high		74.00		81.40		
1 Hammer, diesel, 15K ft.-lbs.		614.20		675.62		
1 Air Compressor, 600 cfm		424.60		467.06		
2 -50' Air Hoses, 3"		29.80		32.78		
1 Chain Saw, gas, 36" Long		41.60		45.76	22.23	24.46
104 L.H., Daily Totals		$5506.60		$8008.62	$52.95	$77.01
Crew B-51	**Hr.**	**Daily**	**Hr.**	**Daily**	**Bare Costs**	**Incl. O&P**
1 Labor Foreman (outside)	$28.45	$227.60	$47.65	$381.20	$26.77	$44.77
4 Laborers	26.45	846.40	44.30	1417.60		
1 Truck Driver (light)	26.40	211.20	43.80	350.40		
1 Flatbed Truck, Gas, 1.5 Ton		171.40		188.54	3.57	3.93
48 L.H., Daily Totals		$1456.60		$2337.74	$30.35	$48.70

Crews

Crew No.	Bare Costs		Incl. Subs O&P		Cost Per Labor-Hour	
Crew B-52	**Hr.**	**Daily**	**Hr.**	**Daily**	**Bare Costs**	**Incl. O&P**
1 Labor Foreman	$28.45	$227.60	$47.65	$381.20	$29.06	$48.73
1 Carpenter	33.55	268.40	56.15	449.20		
4 Laborers	26.45	846.40	44.30	1417.60		
.5 Rodman (reinf.)	36.30	145.20	63.10	252.40		
.5 Equip. Oper. (med.)	34.90	139.60	57.05	228.20		
.5 Crawler Loader, 3 C.Y.		504.00		554.40	9.00	9.90
56 L.H., Daily Totals		$2131.20		$3283.00	$38.06	$58.63

Crew B-53	**Hr.**	**Daily**	**Hr.**	**Daily**	**Bare Costs**	**Incl. O&P**
1 Building Laborer	$26.45	$211.60	$44.30	$354.40	$26.45	$44.30
1 Trencher, Chain, 12 H.P.		57.40		63.14	7.17	7.89
8 L.H., Daily Totals		$269.00		$417.54	$33.63	$52.19

Crew B-54	**Hr.**	**Daily**	**Hr.**	**Daily**	**Bare Costs**	**Incl. O&P**
1 Equip. Oper. (light)	$33.55	$268.40	$54.80	$438.40	$33.55	$54.80
1 Trencher, Chain, 40 H.P.		297.00		326.70	37.13	40.84
8 L.H., Daily Totals		$565.40		$765.10	$70.67	$95.64

Crew B-54A	**Hr.**	**Daily**	**Hr.**	**Daily**	**Bare Costs**	**Incl. O&P**
.17 Labor Foreman (outside)	$28.45	$38.69	$47.65	$64.80	$33.96	$55.68
1 Equipment Operator (med.)	34.90	279.20	57.05	456.40		
1 Wheel Trencher, 67 H.P.		1001.00		1101.10	106.94	117.64
9.36 L.H., Daily Totals		$1318.89		$1622.30	$140.91	$173.32

Crew B-54B	**Hr.**	**Daily**	**Hr.**	**Daily**	**Bare Costs**	**Incl. O&P**
.25 Labor Foreman (outside)	$28.45	$56.90	$47.65	$95.30	$33.61	$55.17
1 Equipment Operator (med.)	34.90	279.20	57.05	456.40		
1 Wheel Trencher, 150 H.P.		1802.00		1982.20	180.20	198.22
10 L.H., Daily Totals		$2138.10		$2533.90	$213.81	$253.39

Crew B-54D	**Hr.**	**Daily**	**Hr.**	**Daily**	**Bare Costs**	**Incl. O&P**
1 Laborer	$26.45	$211.60	$44.30	$354.40	$30.68	$50.67
1 Equipment Operator (med.)	34.90	279.20	57.05	456.40		
1 Rock trencher, 6" width		332.80		366.08	20.80	22.88
16 L.H., Daily Totals		$823.60		$1176.88	$51.48	$73.56

Crew B-54E	**Hr.**	**Daily**	**Hr.**	**Daily**	**Bare Costs**	**Incl. O&P**
1 Laborer	$26.45	$211.60	$44.30	$354.40	$30.68	$50.67
1 Equipment Operator (med.)	34.90	279.20	57.05	456.40		
1 Rock trencher, 18" width		2538.00		2791.80	158.63	174.49
16 L.H., Daily Totals		$3028.80		$3602.60	$189.30	$225.16

Crew B-55	**Hr.**	**Daily**	**Hr.**	**Daily**	**Bare Costs**	**Incl. O&P**
1 Laborer	$26.45	$211.60	$44.30	$354.40	$26.43	$44.05
1 Truck Driver (light)	26.40	211.20	43.80	350.40		
1 Truck-mounted earth auger		706.60		777.26		
1 Flatbed Truck, Gas, 3 Ton		217.00		238.70	57.73	63.50
16 L.H., Daily Totals		$1346.40		$1720.76	$84.15	$107.55

Crew B-56	**Hr.**	**Daily**	**Hr.**	**Daily**	**Bare Costs**	**Incl. O&P**
2 Laborers	$26.45	$423.20	$44.30	$708.80	$26.45	$44.30
1 Air Track Drill, 4"		849.60		934.56		
1 Air Compressor, 600 cfm		424.60		467.06		
1 -50' Air Hose, 3"		14.90		16.39	80.57	88.63
16 L.H., Daily Totals		$1712.30		$2126.81	$107.02	$132.93

Crew B-57	**Hr.**	**Daily**	**Hr.**	**Daily**	**Bare Costs**	**Incl. O&P**
1 Labor Foreman (outside)	$28.45	$227.60	$47.65	$381.20	$28.72	$47.81
3 Laborers	26.45	634.80	44.30	1063.20		
1 Equip. Oper. (crane)	35.80	286.40	58.50	468.00		
1 Barge, 400 Ton		820.40		902.44		
1 Crawler Crane, 25 Ton		1123.00		1235.30		
1 Clamshell Bucket, 1 C.Y.		48.00		52.80		
1 Centr. Water Pump, 6"		291.60		320.76		
1 -20' Suction Hose, 6"		11.90		13.09		
20 -50' Discharge Hoses, 6"		126.00		138.60	60.52	66.57
40 L.H., Daily Totals		$3569.70		$4575.39	$89.24	$114.38

Crew B-58	**Hr.**	**Daily**	**Hr.**	**Daily**	**Bare Costs**	**Incl. O&P**
2 Laborers	$26.45	$423.20	$44.30	$708.80	$28.82	$47.80
1 Equip. Oper. (light)	33.55	268.40	54.80	438.40		
1 Backhoe Loader, 48 H.P.		320.40		352.44		
1 Small Helicopter, w/pilot		2579.00		2836.90	120.81	132.89
24 L.H., Daily Totals		$3591.00		$4336.54	$149.63	$180.69

Crew B-59	**Hr.**	**Daily**	**Hr.**	**Daily**	**Bare Costs**	**Incl. O&P**
1 Truck Driver (heavy)	$27.20	$217.60	$45.10	$360.80	$27.20	$45.10
1 Truck Tractor, 220 H.P.		279.20		307.12		
1 Water Tank Trailer, 5000 Gal.		146.20		160.82	53.17	58.49
8 L.H., Daily Totals		$643.00		$828.74	$80.38	$103.59

Crew B-60	**Hr.**	**Daily**	**Hr.**	**Daily**	**Bare Costs**	**Incl. O&P**
1 Labor Foreman (outside)	$28.45	$227.60	$47.65	$381.20	$29.52	$48.98
3 Laborers	26.45	634.80	44.30	1063.20		
1 Equip. Oper. (crane)	35.80	286.40	58.50	468.00		
1 Equip. Oper. (light)	33.55	268.40	54.80	438.40		
1 Crawler Crane, 40 Ton		1128.00		1240.80		
1 Lead, 60' high		74.00		81.40		
1 Hammer, diesel, 15K ft.-lbs.		614.20		675.62		
1 Backhoe Loader, 48 H.P.		320.40		352.44	44.51	48.96
48 L.H., Daily Totals		$3553.80		$4701.06	$74.04	$97.94

Crew B-61	**Hr.**	**Daily**	**Hr.**	**Daily**	**Bare Costs**	**Incl. O&P**
1 Labor Foreman (outside)	$28.45	$227.60	$47.65	$381.20	$26.85	$44.97
4 Laborers	26.45	846.40	44.30	1417.60		
1 Cement Mixer, 2 C.Y.		194.60		214.06		
1 Air Compressor, 160 cfm		125.00		137.50	7.99	8.79
40 L.H., Daily Totals		$1393.60		$2150.36	$34.84	$53.76

Crew B-62	**Hr.**	**Daily**	**Hr.**	**Daily**	**Bare Costs**	**Incl. O&P**
2 Laborers	$26.45	$423.20	$44.30	$708.80	$28.82	$47.80
1 Equip. Oper. (light)	33.55	268.40	54.80	438.40		
1 Loader, Skid Steer, 30 H.P., gas		151.80		166.98	6.33	6.96
24 L.H., Daily Totals		$843.40		$1314.18	$35.14	$54.76

Crew B-63	**Hr.**	**Daily**	**Hr.**	**Daily**	**Bare Costs**	**Incl. O&P**
5 Laborers	$26.45	$1058.00	$44.30	$1772.00	$26.45	$44.30
1 Loader, Skid Steer, 30 H.P., gas		151.80		166.98	3.79	4.17
40 L.H., Daily Totals		$1209.80		$1938.98	$30.25	$48.47

Crew B-64	**Hr.**	**Daily**	**Hr.**	**Daily**	**Bare Costs**	**Incl. O&P**
1 Laborer	$26.45	$211.60	$44.30	$354.40	$26.43	$44.05
1 Truck Driver (light)	26.40	211.20	43.80	350.40		
1 Power Mulcher (small)		130.80		143.88		
1 Flatbed Truck, Gas, 1.5 Ton		171.40		188.54	18.89	20.78
16 L.H., Daily Totals		$725.00		$1037.22	$45.31	$64.83

Crews

Crew No.	Bare Costs		Incl. Subs O&P		Cost Per Labor-Hour	
Crew B-65	**Hr.**	**Daily**	**Hr.**	**Daily**	**Bare Costs**	**Incl. O&P**
1 Laborer	$26.45	$211.60	$44.30	$354.40	$26.43	$44.05
1 Truck Driver (light)	26.40	211.20	43.80	350.40		
1 Power Mulcher (large)		258.80		284.68		
1 Flatbed Truck, Gas, 1.5 Ton		171.40		188.54	26.89	29.58
16 L.H., Daily Totals		$853.00		$1178.02	$53.31	$73.63

Crew B-66	**Hr.**	**Daily**	**Hr.**	**Daily**	**Bare Costs**	**Incl. O&P**
1 Equip. Oper. (light)	$33.55	$268.40	$54.80	$438.40	$33.55	$54.80
1 Loader-Backhoe		223.60		245.96	27.95	30.75
8 L.H., Daily Totals		$492.00		$684.36	$61.50	$85.55

Crew B-67	**Hr.**	**Daily**	**Hr.**	**Daily**	**Bare Costs**	**Incl. O&P**
1 Millwright	$35.35	$282.80	$56.45	$451.60	$34.45	$55.63
1 Equip. Oper. (light)	33.55	268.40	54.80	438.40		
1 Forklift, R/T, 4,000 Lb.		278.60		306.46	17.41	19.15
16 L.H., Daily Totals		$829.80		$1196.46	$51.86	$74.78

Crew B-68	**Hr.**	**Daily**	**Hr.**	**Daily**	**Bare Costs**	**Incl. O&P**
2 Millwrights	$35.35	$565.60	$56.45	$903.20	$34.75	$55.90
1 Equip. Oper. (light)	33.55	268.40	54.80	438.40		
1 Forklift, R/T, 4,000 Lb.		278.60		306.46	11.61	12.77
24 L.H., Daily Totals		$1112.60		$1648.06	$46.36	$68.67

Crew B-69	**Hr.**	**Daily**	**Hr.**	**Daily**	**Bare Costs**	**Incl. O&P**
1 Labor Foreman (outside)	$28.45	$227.60	$47.65	$381.20	$29.11	$48.30
3 Laborers	26.45	634.80	44.30	1063.20		
1 Equip Oper. (crane)	35.80	286.40	58.50	468.00		
1 Equip Oper. Oiler	31.05	248.40	50.75	406.00		
1 Hyd. Crane, 80 Ton		1693.00		1862.30	35.27	38.80
48 L.H., Daily Totals		$3090.20		$4180.70	$64.38	$87.10

Crew B-69A	**Hr.**	**Daily**	**Hr.**	**Daily**	**Bare Costs**	**Incl. O&P**
1 Labor Foreman	$28.45	$227.60	$47.65	$381.20	$29.09	$48.08
3 Laborers	26.45	634.80	44.30	1063.20		
1 Equip. Oper. (med.)	34.90	279.20	57.05	456.40		
1 Concrete Finisher	31.85	254.80	50.85	406.80		
1 Curb/Gutter Paver, 2-Track		744.80		819.28	15.52	17.07
48 L.H., Daily Totals		$2141.20		$3126.88	$44.61	$65.14

Crew B-69B	**Hr.**	**Daily**	**Hr.**	**Daily**	**Bare Costs**	**Incl. O&P**
1 Labor Foreman	$28.45	$227.60	$47.65	$381.20	$29.09	$48.08
3 Laborers	26.45	634.80	44.30	1063.20		
1 Equip. Oper. (med.)	34.90	279.20	57.05	456.40		
1 Cement Finisher	31.85	254.80	50.85	406.80		
1 Curb/Gutter Paver, 4-Track		848.20		933.02	17.67	19.44
48 L.H., Daily Totals		$2244.60		$3240.62	$46.76	$67.51

Crew B-70	**Hr.**	**Daily**	**Hr.**	**Daily**	**Bare Costs**	**Incl. O&P**
1 Labor Foreman (outside)	$28.45	$227.60	$47.65	$381.20	$30.36	$50.24
3 Laborers	26.45	634.80	44.30	1063.20		
3 Equip. Oper. (med.)	34.90	837.60	57.05	1369.20		
1 Grader, 30,000 Lbs.		542.80		597.08		
1 Ripper, beam & 1 shank		83.20		91.52		
1 Road Sweeper, S.P., 8' wide		604.80		665.28		
1 F.E. Loader, W.M., 1.5 C.Y.		319.40		351.34	27.68	30.45
56 L.H., Daily Totals		$3250.20		$4518.82	$58.04	$80.69

Crew No.	Bare Costs		Incl. Subs O&P		Cost Per Labor-Hour	
Crew B-71	**Hr.**	**Daily**	**Hr.**	**Daily**	**Bare Costs**	**Incl. O&P**
1 Labor Foreman (outside)	$28.45	$227.60	$47.65	$381.20	$30.36	$50.24
3 Laborers	26.45	634.80	44.30	1063.20		
3 Equip. Oper. (med.)	34.90	837.60	57.05	1369.20		
1 Pvmt. Profiler, 750 H.P.		5626.00		6188.60		
1 Road Sweeper, S.P., 8' wide		604.80		665.28		
1 F.E. Loader, W.M., 1.5 C.Y.		319.40		351.34	116.97	128.66
56 L.H., Daily Totals		$8250.20		$10018.82	$147.32	$178.91

Crew B-72	**Hr.**	**Daily**	**Hr.**	**Daily**	**Bare Costs**	**Incl. O&P**
1 Labor Foreman (outside)	$28.45	$227.60	$47.65	$381.20	$30.93	$51.09
3 Laborers	26.45	634.80	44.30	1063.20		
4 Equip. Oper. (med.)	34.90	1116.80	57.05	1825.60		
1 Pvmt. Profiler, 750 H.P.		5626.00		6188.60		
1 Hammermill, 250 H.P.		1692.00		1861.20		
1 Windrow Loader		1210.00		1331.00		
1 Mix Paver 165 H.P.		2016.00		2217.60		
1 Roller, Pneum. Whl, 12 Ton		309.20		340.12	169.58	186.54
64 L.H., Daily Totals		$12832.40		$15208.52	$200.51	$237.63

Crew B-73	**Hr.**	**Daily**	**Hr.**	**Daily**	**Bare Costs**	**Incl. O&P**
1 Labor Foreman (outside)	$28.45	$227.60	$47.65	$381.20	$31.98	$52.69
2 Laborers	26.45	423.20	44.30	708.80		
5 Equip. Oper. (med.)	34.90	1396.00	57.05	2282.00		
1 Road Mixer, 310 H.P.		1890.00		2079.00		
1 Tandem Roller, 10 Ton		224.20		246.62		
1 Hammermill, 250 H.P.		1692.00		1861.20		
1 Grader, 30,000 Lbs.		542.80		597.08		
.5 F.E. Loader, W.M., 1.5 C.Y.		159.70		175.67		
.5 Truck Tractor, 220 H.P.		139.60		153.56		
.5 Water Tank Trailer, 5000 Gal.		73.10		80.41	73.77	81.15
64 L.H., Daily Totals		$6768.20		$8565.54	$105.75	$133.84

Crew B-74	**Hr.**	**Daily**	**Hr.**	**Daily**	**Bare Costs**	**Incl. O&P**
1 Labor Foreman (outside)	$28.45	$227.60	$47.65	$381.20	$31.11	$51.29
1 Laborer	26.45	211.60	44.30	354.40		
4 Equip. Oper. (med.)	34.90	1116.80	57.05	1825.60		
2 Truck Drivers (heavy)	27.20	435.20	45.10	721.60		
1 Grader, 30,000 Lbs.		542.80		597.08		
1 Ripper, beam & 1 shank		83.20		91.52		
2 Stabilizers, 310 H.P.		2652.00		2917.20		
1 Flatbed Truck, Gas, 3 Ton		217.00		238.70		
1 Chem. Spreader, Towed		49.00		53.90		
1 Roller, Vibratory, 25 Ton		596.80		656.48		
1 Water Tank Trailer, 5000 Gal.		146.20		160.82		
1 Truck Tractor, 220 H.P.		279.20		307.12	71.35	78.48
64 L.H., Daily Totals		$6557.40		$8305.62	$102.46	$129.78

Crew B-75	**Hr.**	**Daily**	**Hr.**	**Daily**	**Bare Costs**	**Incl. O&P**
1 Labor Foreman (outside)	$28.45	$227.60	$47.65	$381.20	$31.67	$52.18
1 Laborer	26.45	211.60	44.30	354.40		
4 Equip. Oper. (med.)	34.90	1116.80	57.05	1825.60		
1 Truck Driver (heavy)	27.20	217.60	45.10	360.80		
1 Grader, 30,000 Lbs.		542.80		597.08		
1 Ripper, beam & 1 shank		83.20		91.52		
2 Stabilizers, 310 H.P.		2652.00		2917.20		
1 Dist. Tanker, 3000 Gallon		309.00		339.90		
1 Truck Tractor, 6x4, 380 H.P.		459.40		505.34		
1 Roller, Vibratory, 25 Ton		596.80		656.48	82.91	91.21
56 L.H., Daily Totals		$6416.80		$8029.52	$114.59	$143.38

Crews

Crew No.	Bare Costs		Incl. Subs O&P		Cost Per Labor-Hour	
Crew B-76	**Hr.**	**Daily**	**Hr.**	**Daily**	**Bare Costs**	**Incl. O&P**
1 Dock Builder Foreman	$34.75	$278.00	$60.55	$484.40	$33.46	$57.09
5 Dock Builders	32.75	1310.00	57.10	2284.00		
2 Equip. Oper. (crane)	35.80	572.80	58.50	936.00		
1 Equip. Oper. Oiler	31.05	248.40	50.75	406.00		
1 Crawler Crane, 50 Ton		1279.00		1406.90		
1 Barge, 400 Ton		820.40		902.44		
1 Hammer, diesel, 15K ft.-lbs.		614.20		675.62		
1 Lead, 60' high		74.00		81.40		
1 Air Compressor, 600 cfm		424.60		467.06		
2 -50' Air Hoses, 3"		29.80		32.78	45.03	49.53
72 L.H., Daily Totals		$5651.20		$7676.60	$78.49	$106.62

Crew No.	Bare Costs		Incl. Subs O&P		Cost Per Labor-Hour	
Crew B-76A	**Hr.**	**Daily**	**Hr.**	**Daily**	**Bare Costs**	**Incl. O&P**
1 Laborer Foreman	$28.45	$227.60	$47.65	$381.20	$28.44	$47.30
5 Laborers	26.45	1058.00	44.30	1772.00		
1 Equip. Oper. (crane)	35.80	286.40	58.50	468.00		
1 Equip. Oper. Oiler	31.05	248.40	50.75	406.00		
1 Crawler Crane, 50 Ton		1279.00		1406.90		
1 Barge, 400 Ton		820.40		902.44	32.80	36.08
64 L.H., Daily Totals		$3919.80		$5336.54	$61.25	$83.38

Crew No.	Bare Costs		Incl. Subs O&P		Cost Per Labor-Hour	
Crew B-77	**Hr.**	**Daily**	**Hr.**	**Daily**	**Bare Costs**	**Incl. O&P**
1 Labor Foreman	$28.45	$227.60	$47.65	$381.20	$26.84	$44.87
3 Laborers	26.45	634.80	44.30	1063.20		
1 Truck Driver (light)	26.40	211.20	43.80	350.40		
1 Crack Cleaner, 25 H.P.		53.00		58.30		
1 Crack Filler, Trailer Mtd.		194.60		214.06		
1 Flatbed Truck, Gas, 3 Ton		217.00		238.70	11.62	12.78
40 L.H., Daily Totals		$1538.20		$2305.86	$38.45	$57.65

Crew No.	Bare Costs		Incl. Subs O&P		Cost Per Labor-Hour	
Crew B-78	**Hr.**	**Daily**	**Hr.**	**Daily**	**Bare Costs**	**Incl. O&P**
1 Labor Foreman	$28.45	$227.60	$47.65	$381.20	$26.85	$44.97
4 Laborers	26.45	846.40	44.30	1417.60		
1 Paint Striper, S.P.		147.80		162.58		
1 Flatbed Truck, Gas, 3 Ton		217.00		238.70		
1 Pickup Truck, 3/4 Ton		104.60		115.06	11.73	12.91
40 L.H., Daily Totals		$1543.40		$2315.14	$38.59	$57.88

Crew No.	Bare Costs		Incl. Subs O&P		Cost Per Labor-Hour	
Crew B-78B	**Hr.**	**Daily**	**Hr.**	**Daily**	**Bare Costs**	**Incl. O&P**
2 Laborers	$26.45	$423.20	$44.30	$708.80	$27.24	$45.47
.25 Equip. Oper. (light)	33.55	67.10	54.80	109.60		
1 Pickup Truck, 3/4 Ton		104.60		115.06		
1 Line Rem., 11 H.P., walk behind		57.40		63.14		
.25 Road Sweeper, S.P., 8' wide		151.20		166.32	17.40	19.14
18 L.H., Daily Totals		$803.50		$1162.92	$44.64	$64.61

Crew No.	Bare Costs		Incl. Subs O&P		Cost Per Labor-Hour	
Crew B-79	**Hr.**	**Daily**	**Hr.**	**Daily**	**Bare Costs**	**Incl. O&P**
1 Labor Foreman	$28.45	$227.60	$47.65	$381.20	$26.95	$45.14
3 Laborers	26.45	634.80	44.30	1063.20		
1 Thermo. Striper, T.M.		653.80		719.18		
1 Flatbed Truck, Gas, 3 Ton		217.00		238.70		
2 Pickup Trucks, 3/4 Ton		209.20		230.12	33.75	37.13
32 L.H., Daily Totals		$1942.40		$2632.40	$60.70	$82.26

Crew No.	Bare Costs		Incl. Subs O&P		Cost Per Labor-Hour	
Crew B-80	**Hr.**	**Daily**	**Hr.**	**Daily**	**Bare Costs**	**Incl. O&P**
1 Labor Foreman	$28.45	$227.60	$47.65	$381.20	$27.12	$45.42
2 Laborers	26.45	423.20	44.30	708.80		
1 Flatbed Truck, Gas, 3 Ton		217.00		238.70		
1 Earth Auger, Truck-Mtd.		401.80		441.98	25.78	28.36
24 L.H., Daily Totals		$1269.60		$1770.68	$52.90	$73.78

Crew No.	Bare Costs		Incl. Subs O&P		Cost Per Labor-Hour	
Crew B-80A	**Hr.**	**Daily**	**Hr.**	**Daily**	**Bare Costs**	**Incl. O&P**
3 Laborers	$26.45	$634.80	$44.30	$1063.20	$26.45	$44.30
1 Flatbed Truck, Gas, 3 Ton		217.00		238.70	9.04	9.95
24 L.H., Daily Totals		$851.80		$1301.90	$35.49	$54.25

Crew No.	Bare Costs		Incl. Subs O&P		Cost Per Labor-Hour	
Crew B-80B	**Hr.**	**Daily**	**Hr.**	**Daily**	**Bare Costs**	**Incl. O&P**
3 Laborers	$26.45	$634.80	$44.30	$1063.20	$28.23	$46.92
1 Equip. Oper. (light)	33.55	268.40	54.80	438.40		
1 Crane, Flatbed Mounted, 3 Ton		225.60		248.16	7.05	7.75
32 L.H., Daily Totals		$1128.80		$1749.76	$35.27	$54.68

Crew No.	Bare Costs		Incl. Subs O&P		Cost Per Labor-Hour	
Crew B-80C	**Hr.**	**Daily**	**Hr.**	**Daily**	**Bare Costs**	**Incl. O&P**
2 Laborers	$26.45	$423.20	$44.30	$708.80	$26.43	$44.13
1 Truck Driver (light)	26.40	211.20	43.80	350.40		
1 Flatbed Truck, Gas, 1.5 Ton		171.40		188.54		
1 Manual fence post auger, gas		7.20		7.92	7.44	8.19
24 L.H., Daily Totals		$813.00		$1255.66	$33.88	$52.32

Crew No.	Bare Costs		Incl. Subs O&P		Cost Per Labor-Hour	
Crew B-81	**Hr.**	**Daily**	**Hr.**	**Daily**	**Bare Costs**	**Incl. O&P**
1 Laborer	$26.45	$211.60	$44.30	$354.40	$26.82	$44.70
1 Truck Driver (heavy)	27.20	217.60	45.10	360.80		
1 Hydromulcher, T.M., 3000 Gal.		278.00		305.80		
1 Truck Tractor, 220 H.P.		279.20		307.12	34.83	38.31
16 L.H., Daily Totals		$986.40		$1328.12	$61.65	$83.01

Crew No.	Bare Costs		Incl. Subs O&P		Cost Per Labor-Hour	
Crew B-81A	**Hr.**	**Daily**	**Hr.**	**Daily**	**Bare Costs**	**Incl. O&P**
1 Laborer	$26.45	$211.60	$44.30	$354.40	$26.43	$44.05
1 Truck Driver (light)	26.40	211.20	43.80	350.40		
1 Hydromulcher, T.M., 600 Gal.		113.20		124.52		
1 Flatbed Truck, Gas, 3 Ton		217.00		238.70	20.64	22.70
16 L.H., Daily Totals		$753.00		$1068.02	$47.06	$66.75

Crew No.	Bare Costs		Incl. Subs O&P		Cost Per Labor-Hour	
Crew B-82	**Hr.**	**Daily**	**Hr.**	**Daily**	**Bare Costs**	**Incl. O&P**
1 Laborer	$26.45	$211.60	$44.30	$354.40	$30.00	$49.55
1 Equip. Oper. (light)	33.55	268.40	54.80	438.40		
1 Horiz. Borer, 6 H.P.		76.00		83.60	4.75	5.22
16 L.H., Daily Totals		$556.00		$876.40	$34.75	$54.77

Crew No.	Bare Costs		Incl. Subs O&P		Cost Per Labor-Hour	
Crew B-82A	**Hr.**	**Daily**	**Hr.**	**Daily**	**Bare Costs**	**Incl. O&P**
1 Laborer	$26.45	$211.60	$44.30	$354.40	$30.00	$49.55
1 Equip. Oper. (light)	33.55	268.40	54.80	438.40		
1 Flatbed Truck, Gas, 3 Ton		217.00		238.70		
1 Flatbed Trailer, 25 Ton		111.20		122.32		
1 Horiz. Dir. Drill, 20k lb. thrust		622.60		684.86	59.42	65.37
16 L.H., Daily Totals		$1430.80		$1838.68	$89.42	$114.92

Crew No.	Bare Costs		Incl. Subs O&P		Cost Per Labor-Hour	
Crew B-82B	**Hr.**	**Daily**	**Hr.**	**Daily**	**Bare Costs**	**Incl. O&P**
2 Laborers	$26.45	$423.20	$44.30	$708.80	$28.82	$47.80
1 Equip. Oper. (light)	33.55	268.40	54.80	438.40		
1 Flatbed Truck, Gas, 3 Ton		217.00		238.70		
1 Flatbed Trailer, 25 Ton		111.20		122.32		
1 Horiz. Dir. Drill, 30k lb. thrust		888.40		977.24	50.69	55.76
24 L.H., Daily Totals		$1908.20		$2485.46	$79.51	$103.56

Crew No.	Bare Costs		Incl. Subs O&P		Cost Per Labor-Hour	
Crew B-82C	**Hr.**	**Daily**	**Hr.**	**Daily**	**Bare Costs**	**Incl. O&P**
2 Laborers	$26.45	$423.20	$44.30	$708.80	$28.82	$47.80
1 Equip. Oper. (light)	33.55	268.40	54.80	438.40		
1 Flatbed Truck, Gas, 3 Ton		217.00		238.70		
1 Flatbed Trailer, 25 Ton		111.20		122.32		
1 Horiz. Dir. Drill, 50k lb. thrust		1174.00		1291.40	62.59	68.85
24 L.H., Daily Totals		$2193.80		$2799.62	$91.41	$116.65

Crews

Crew No.	Bare Costs		Incl. Subs O&P		Cost Per Labor-Hour	
Crew B-82D	**Hr.**	**Daily**	**Hr.**	**Daily**	**Bare Costs**	**Incl. O&P**
1 Equip. Oper. (light)	$33.55	$268.40	$54.80	$438.40	$33.55	$54.80
1 Mud Trailer for HDD, 1500 gallon		265.80		292.38	33.23	36.55
8 L.H., Daily Totals		$534.20		$730.78	$66.78	$91.35

Crew No.	Bare Costs		Incl. Subs O&P		Cost Per Labor-Hour	
Crew B-83	**Hr.**	**Daily**	**Hr.**	**Daily**	**Bare Costs**	**Incl. O&P**
1 Tugboat Captain	$34.90	$279.20	$57.05	$456.40	$30.68	$50.67
1 Tugboat Hand	26.45	211.60	44.30	354.40		
1 Tugboat, 250 H.P.		632.20		695.42	39.51	43.46
16 L.H., Daily Totals		$1123.00		$1506.22	$70.19	$94.14

Crew No.	Bare Costs		Incl. Subs O&P		Cost Per Labor-Hour	
Crew B-84	**Hr.**	**Daily**	**Hr.**	**Daily**	**Bare Costs**	**Incl. O&P**
1 Equip. Oper. (med.)	$34.90	$279.20	$57.05	$456.40	$34.90	$57.05
1 Rotary Mower/Tractor		293.80		323.18	36.73	40.40
8 L.H., Daily Totals		$573.00		$779.58	$71.63	$97.45

Crew No.	Bare Costs		Incl. Subs O&P		Cost Per Labor-Hour	
Crew B-85	**Hr.**	**Daily**	**Hr.**	**Daily**	**Bare Costs**	**Incl. O&P**
3 Laborers	$26.45	$634.80	$44.30	$1063.20	$28.29	$47.01
1 Equip. Oper. (med.)	34.90	279.20	57.05	456.40		
1 Truck Driver (heavy)	27.20	217.60	45.10	360.80		
1 Aerial Lift Truck, 80'		636.20		699.82		
1 Brush Chipper, 12", 130 H.P.		217.00		238.70		
1 Pruning Saw, Rotary		6.30		6.93	21.49	23.64
40 L.H., Daily Totals		$1991.10		$2825.85	$49.78	$70.65

Crew No.	Bare Costs		Incl. Subs O&P		Cost Per Labor-Hour	
Crew B-86	**Hr.**	**Daily**	**Hr.**	**Daily**	**Bare Costs**	**Incl. O&P**
1 Equip. Oper. (med.)	$34.90	$279.20	$57.05	$456.40	$34.90	$57.05
1 Stump Chipper, S.P.		169.35		186.29	21.17	23.29
8 L.H., Daily Totals		$448.55		$642.68	$56.07	$80.34

Crew No.	Bare Costs		Incl. Subs O&P		Cost Per Labor-Hour	
Crew B-86A	**Hr.**	**Daily**	**Hr.**	**Daily**	**Bare Costs**	**Incl. O&P**
1 Equip. Oper. (med.)	$34.90	$279.20	$57.05	$456.40	$34.90	$57.05
1 Grader, 30,000 Lbs.		542.80		597.08	67.85	74.64
8 L.H., Daily Totals		$822.00		$1053.48	$102.75	$131.69

Crew No.	Bare Costs		Incl. Subs O&P		Cost Per Labor-Hour	
Crew B-86B	**Hr.**	**Daily**	**Hr.**	**Daily**	**Bare Costs**	**Incl. O&P**
1 Equip. Oper. (med.)	$34.90	$279.20	$57.05	$456.40	$34.90	$57.05
1 Dozer, 200 H.P.		1051.00		1156.10	131.38	144.51
8 L.H., Daily Totals		$1330.20		$1612.50	$166.28	$201.56

Crew No.	Bare Costs		Incl. Subs O&P		Cost Per Labor-Hour	
Crew B-87	**Hr.**	**Daily**	**Hr.**	**Daily**	**Bare Costs**	**Incl. O&P**
1 Laborer	$26.45	$211.60	$44.30	$354.40	$33.21	$54.50
4 Equip. Oper. (med.)	34.90	1116.80	57.05	1825.60		
2 Feller Bunchers, 100 H.P.		1128.00		1240.80		
1 Log Chipper, 22" Tree		554.80		610.28		
1 Dozer, 105 H.P.		505.40		555.94		
1 Chain Saw, gas, 36" Long		41.60		45.76	55.74	61.32
40 L.H., Daily Totals		$3558.20		$4632.78	$88.95	$115.82

Crew No.	Bare Costs		Incl. Subs O&P		Cost Per Labor-Hour	
Crew B-88	**Hr.**	**Daily**	**Hr.**	**Daily**	**Bare Costs**	**Incl. O&P**
1 Laborer	$26.45	$211.60	$44.30	$354.40	$33.69	$55.23
6 Equip. Oper. (med.)	34.90	1675.20	57.05	2738.40		
2 Feller Bunchers, 100 H.P.		1128.00		1240.80		
1 Log Chipper, 22" Tree		554.80		610.28		
2 Log Skidders, 50 H.P.		1688.00		1856.80		
1 Dozer, 105 H.P.		505.40		555.94		
1 Chain Saw, gas, 36" Long		41.60		45.76	69.96	76.96
56 L.H., Daily Totals		$5804.60		$7402.38	$103.65	$132.19

Crew No.	Bare Costs		Incl. Subs O&P		Cost Per Labor-Hour	
Crew B-89	**Hr.**	**Daily**	**Hr.**	**Daily**	**Bare Costs**	**Incl. O&P**
1 Skilled Worker	$34.05	$272.40	$57.00	$456.00	$30.25	$50.65
1 Building Laborer	26.45	211.60	44.30	354.40		
1 Flatbed Truck, Gas, 3 Ton		217.00		238.70		
1 Concrete Saw		134.20		147.62		
1 Water Tank, 65 Gal.		18.45		20.30	23.10	25.41
16 L.H., Daily Totals		$853.65		$1217.02	$53.35	$76.06

Crew No.	Bare Costs		Incl. Subs O&P		Cost Per Labor-Hour	
Crew B-89A	**Hr.**	**Daily**	**Hr.**	**Daily**	**Bare Costs**	**Incl. O&P**
1 Skilled Worker	$34.05	$272.40	$57.00	$456.00	$30.25	$50.65
1 Laborer	26.45	211.60	44.30	354.40		
1 Core Drill (large)		109.80		120.78	6.86	7.55
16 L.H., Daily Totals		$593.80		$931.18	$37.11	$58.20

Crew No.	Bare Costs		Incl. Subs O&P		Cost Per Labor-Hour	
Crew B-89B	**Hr.**	**Daily**	**Hr.**	**Daily**	**Bare Costs**	**Incl. O&P**
1 Equip. Oper. (light)	$33.55	$268.40	$54.80	$438.40	$29.98	$49.30
1 Truck Driver, Light	26.40	211.20	43.80	350.40		
1 Wall Saw, Hydraulic, 10 H.P.		89.00		97.90		
1 Generator, Diesel, 100 kW		329.00		361.90		
1 Water Tank, 65 Gal.		18.45		20.30		
1 Flatbed Truck, Gas, 3 Ton		217.00		238.70	40.84	44.92
16 L.H., Daily Totals		$1133.05		$1507.60	$70.82	$94.22

Crew No.	Bare Costs		Incl. Subs O&P		Cost Per Labor-Hour	
Crew B-90	**Hr.**	**Daily**	**Hr.**	**Daily**	**Bare Costs**	**Incl. O&P**
1 Labor Foreman (outside)	$28.45	$227.60	$47.65	$381.20	$28.66	$47.54
3 Laborers	26.45	634.80	44.30	1063.20		
2 Equip. Oper. (light)	33.55	536.80	54.80	876.80		
2 Truck Drivers (heavy)	27.20	435.20	45.10	721.60		
1 Road Mixer, 310 H.P.		1890.00		2079.00		
1 Dist. Truck, 2000 Gal.		280.80		308.88	33.92	37.31
64 L.H., Daily Totals		$4005.20		$5430.68	$62.58	$84.85

Crew No.	Bare Costs		Incl. Subs O&P		Cost Per Labor-Hour	
Crew B-90A	**Hr.**	**Daily**	**Hr.**	**Daily**	**Bare Costs**	**Incl. O&P**
1 Labor Foreman	$28.45	$227.60	$47.65	$381.20	$31.56	$52.06
2 Laborers	26.45	423.20	44.30	708.80		
4 Equip. Oper. (med.)	34.90	1116.80	57.05	1825.60		
2 Graders, 30,000 Lbs.		1085.60		1194.16		
1 Tandem Roller, 10 Ton		224.20		246.62		
1 Roller, Pneum. Whl, 12 Ton		309.20		340.12	28.91	31.80
56 L.H., Daily Totals		$3386.60		$4696.50	$60.48	$83.87

Crew No.	Bare Costs		Incl. Subs O&P		Cost Per Labor-Hour	
Crew B-90B	**Hr.**	**Daily**	**Hr.**	**Daily**	**Bare Costs**	**Incl. O&P**
1 Labor Foreman	$28.45	$227.60	$47.65	$381.20	$31.01	$51.23
2 Laborers	26.45	423.20	44.30	708.80		
3 Equip. Oper. (med.)	34.90	837.60	57.05	1369.20		
1 Tandem Roller, 10 Ton		224.20		246.62		
1 Roller, Pneum. Whl, 12 Ton		309.20		340.12		
1 Road Mixer, 310 H.P.		1890.00		2079.00	50.49	55.54
48 L.H., Daily Totals		$3911.80		$5124.94	$81.50	$106.77

Crew No.	Bare Costs		Incl. Subs O&P		Cost Per Labor-Hour	
Crew B-91	**Hr.**	**Daily**	**Hr.**	**Daily**	**Bare Costs**	**Incl. O&P**
1 Labor Foreman (outside)	$28.45	$227.60	$47.65	$381.20	$31.02	$51.19
2 Laborers	26.45	423.20	44.30	708.80		
4 Equip. Oper. (med.)	34.90	1116.80	57.05	1825.60		
1 Truck Driver (heavy)	27.20	217.60	45.10	360.80		
1 Dist. Tanker, 3000 Gallon		309.00		339.90		
1 Truck Tractor, 6x4, 380 H.P.		459.40		505.34		
1 Aggreg. Spreader, S.P.		748.40		823.24		
1 Roller, Pneum. Whl, 12 Ton		309.20		340.12		
1 Tandem Roller, 10 Ton		224.20		246.62	32.03	35.24
64 L.H., Daily Totals		$4035.40		$5531.62	$63.05	$86.43

Crews

Crew No.	Bare Costs		Incl. Subs O&P		Cost Per Labor-Hour	

Crew B-92	Hr.	Daily	Hr.	Daily	Bare Costs	Incl. O&P
1 Labor Foreman (outside)	$28.45	$227.60	$47.65	$381.20	$26.95	$45.14
3 Laborers	26.45	634.80	44.30	1063.20		
1 Crack Cleaner, 25 H.P.		53.00		58.30		
1 Air Compressor, 60 cfm		110.20		121.22		
1 Tar Kettle, T.M.		74.65		82.11		
1 Flatbed Truck, Gas, 3 Ton		217.00		238.70	14.21	15.64
32 L.H., Daily Totals		$1317.25		$1944.73	$41.16	$60.77

Crew B-93	Hr.	Daily	Hr.	Daily	Bare Costs	Incl. O&P
1 Equip. Oper. (med.)	$34.90	$279.20	$57.05	$456.40	$34.90	$57.05
1 Feller Buncher, 100 H.P.		564.00		620.40	70.50	77.55
8 L.H., Daily Totals		$843.20		$1076.80	$105.40	$134.60

Crew B-94A	Hr.	Daily	Hr.	Daily	Bare Costs	Incl. O&P
1 Laborer	$26.45	$211.60	$44.30	$354.40	$26.45	$44.30
1 Diaphragm Water Pump, 2"		66.60		73.26		
1 -20' Suction Hose, 2"		1.95		2.15		
2 -50' Discharge Hoses, 2"		1.80		1.98	8.79	9.67
8 L.H., Daily Totals		$281.95		$431.79	$35.24	$53.97

Crew B-94B	Hr.	Daily	Hr.	Daily	Bare Costs	Incl. O&P
1 Laborer	$26.45	$211.60	$44.30	$354.40	$26.45	$44.30
1 Diaphragm Water Pump, 4"		109.40		120.34		
1 -20' Suction Hose, 4"		3.25		3.58		
2 -50' Discharge Hoses, 4"		4.70		5.17	14.67	16.14
8 L.H., Daily Totals		$328.95		$483.49	$41.12	$60.44

Crew B-94C	Hr.	Daily	Hr.	Daily	Bare Costs	Incl. O&P
1 Laborer	$26.45	$211.60	$44.30	$354.40	$26.45	$44.30
1 Centrifugal Water Pump, 3"		71.00		78.10		
1 -20' Suction Hose, 3"		3.05		3.36		
2 -50' Discharge Hoses, 3"		3.50		3.85	9.69	10.66
8 L.H., Daily Totals		$289.15		$439.70	$36.14	$54.96

Crew B-94D	Hr.	Daily	Hr.	Daily	Bare Costs	Incl. O&P
1 Laborer	$26.45	$211.60	$44.30	$354.40	$26.45	$44.30
1 Centr. Water Pump, 6"		291.60		320.76		
1 -20' Suction Hose, 6"		11.90		13.09		
2 -50' Discharge Hoses, 6"		12.60		13.86	39.51	43.46
8 L.H., Daily Totals		$527.70		$702.11	$65.96	$87.76

Crew C-1	Hr.	Daily	Hr.	Daily	Bare Costs	Incl. O&P
2 Carpenters	$33.55	$536.80	$56.15	$898.40	$29.73	$49.73
1 Carpenter Helper	25.35	202.80	42.30	338.40		
1 Laborer	26.45	211.60	44.30	354.40		
32 L.H., Daily Totals		$951.20		$1591.20	$29.73	$49.73

Crew C-2	Hr.	Daily	Hr.	Daily	Bare Costs	Incl. O&P
1 Carpenter Foreman (out)	$35.55	$284.40	$59.50	$476.00	$29.97	$50.12
2 Carpenters	33.55	536.80	56.15	898.40		
2 Carpenter Helpers	25.35	405.60	42.30	676.80		
1 Laborer	26.45	211.60	44.30	354.40		
48 L.H., Daily Totals		$1438.40		$2405.60	$29.97	$50.12

Crew C-2A	Hr.	Daily	Hr.	Daily	Bare Costs	Incl. O&P
1 Carpenter Foreman (out)	$35.55	$284.40	$59.50	$476.00	$32.42	$53.85
3 Carpenters	33.55	805.20	56.15	1347.60		
1 Cement Finisher	31.85	254.80	50.85	406.80		
1 Laborer	26.45	211.60	44.30	354.40		
48 L.H., Daily Totals		$1556.00		$2584.80	$32.42	$53.85

Crew No.	Bare Costs		Incl. Subs O&P		Cost Per Labor-Hour	

Crew C-3	Hr.	Daily	Hr.	Daily	Bare Costs	Incl. O&P
1 Rodman Foreman	$38.30	$306.40	$66.55	$532.40	$32.51	$55.44
3 Rodmen (reinf.)	36.30	871.20	63.10	1514.40		
1 Equip. Oper. (light)	33.55	268.40	54.80	438.40		
3 Laborers	26.45	634.80	44.30	1063.20		
3 Stressing Equipment		29.40		32.34		
.5 Grouting Equipment		76.60		84.26	1.66	1.82
64 L.H., Daily Totals		$2186.80		$3665.00	$34.17	$57.27

Crew C-4	Hr.	Daily	Hr.	Daily	Bare Costs	Incl. O&P
1 Rodman Foreman	$38.30	$306.40	$66.55	$532.40	$34.34	$59.26
2 Rodmen (reinf.)	36.30	580.80	63.10	1009.60		
1 Building Laborer	26.45	211.60	44.30	354.40		
3 Stressing Equipment		29.40		32.34	0.92	1.01
32 L.H., Daily Totals		$1128.20		$1928.74	$35.26	$60.27

Crew C-4A	Hr.	Daily	Hr.	Daily	Bare Costs	Incl. O&P
2 Rodmen (reinf.)	$36.30	$580.80	$63.10	$1009.60	$36.30	$63.10
4 Stressing Equipment		39.20		43.12	2.45	2.69
16 L.H., Daily Totals		$620.00		$1052.72	$38.75	$65.80

Crew C-5	Hr.	Daily	Hr.	Daily	Bare Costs	Incl. O&P
1 Rodman Foreman	$38.30	$306.40	$66.55	$532.40	$33.27	$56.64
2 Rodmen (reinf.)	36.30	580.80	63.10	1009.60		
1 Equip. Oper. (crane)	35.80	286.40	58.50	468.00		
2 Building Laborers	26.45	423.20	44.30	708.80		
1 Hyd. Crane, 25 Ton		748.60		823.46	15.60	17.16
48 L.H., Daily Totals		$2345.40		$3542.26	$48.86	$73.80

Crew C-6	Hr.	Daily	Hr.	Daily	Bare Costs	Incl. O&P
1 Labor Foreman (outside)	$28.45	$227.60	$47.65	$381.20	$27.68	$45.95
4 Laborers	26.45	846.40	44.30	1417.60		
1 Cement Finisher	31.85	254.80	50.85	406.80		
2 Gas Engine Vibrators		46.40		51.04	0.97	1.06
48 L.H., Daily Totals		$1375.20		$2256.64	$28.65	$47.01

Crew C-7	Hr.	Daily	Hr.	Daily	Bare Costs	Incl. O&P
1 Labor Foreman (outside)	$28.45	$227.60	$47.65	$381.20	$28.72	$47.53
5 Laborers	26.45	1058.00	44.30	1772.00		
1 Cement Finisher	31.85	254.80	50.85	406.80		
1 Equip. Oper. (med.)	34.90	279.20	57.05	456.40		
1 Equip. Oper. (oiler)	31.05	248.40	50.75	406.00		
2 Gas Engine Vibrators		46.40		51.04		
1 Concrete Bucket, 1 C.Y.		23.20		25.52		
1 Hyd. Crane, 55 Ton		1208.00		1328.80	17.74	19.52
72 L.H., Daily Totals		$3345.60		$4827.76	$46.47	$67.05

Crew C-8	Hr.	Daily	Hr.	Daily	Bare Costs	Incl. O&P
1 Labor Foreman (outside)	$28.45	$227.60	$47.65	$381.20	$29.49	$48.47
3 Laborers	26.45	634.80	44.30	1063.20		
2 Cement Finishers	31.85	509.60	50.85	813.60		
1 Equip. Oper. (med.)	34.90	279.20	57.05	456.40		
1 Concrete Pump (small)		741.00		815.10	13.23	14.56
56 L.H., Daily Totals		$2392.20		$3529.50	$42.72	$63.03

Crew C-8A	Hr.	Daily	Hr.	Daily	Bare Costs	Incl. O&P
1 Labor Foreman (outside)	$28.45	$227.60	$47.65	$381.20	$28.58	$47.04
3 Laborers	26.45	634.80	44.30	1063.20		
2 Cement Finishers	31.85	509.60	50.85	813.60		
48 L.H., Daily Totals		$1372.00		$2258.00	$28.58	$47.04

Crews

Crew No.	Bare Costs		Incl. Subs O&P		Cost Per Labor-Hour	
Crew C-8B	**Hr.**	**Daily**	**Hr.**	**Daily**	**Bare Costs**	**Incl. O&P**
1 Labor Foreman (outside)	$28.45	$227.60	$47.65	$381.20	$28.54	$47.52
3 Laborers	26.45	634.80	44.30	1063.20		
1 Equip. Oper. (med.)	34.90	279.20	57.05	456.40		
1 Vibrating Power Screed		65.50		72.05		
1 Roller, Vibratory, 25 Ton		596.80		656.48		
1 Dozer, 200 H.P.		1051.00		1156.10	42.83	47.12
40 L.H., Daily Totals		$2854.90		$3785.43	$71.37	$94.64

Crew No.	Hr.	Daily	Hr.	Daily	Bare Costs	Incl. O&P
Crew C-8C						
1 Labor Foreman (outside)	$28.45	$227.60	$47.65	$381.20	$29.09	$48.08
3 Laborers	26.45	634.80	44.30	1063.20		
1 Cement Finisher	31.85	254.80	50.85	406.80		
1 Equip. Oper. (med.)	34.90	279.20	57.05	456.40		
1 Shotcrete Rig, 12 C.Y./hr		249.20		274.12		
1 Air Compressor, 160 cfm		125.00		137.50		
4 -50' Air Hoses, 1"		16.40		18.04		
4 -50' Air Hoses, 2"		31.00		34.10	8.78	9.66
48 L.H., Daily Totals		$1818.00		$2771.36	$37.88	$57.74

Crew No.	Hr.	Daily	Hr.	Daily	Bare Costs	Incl. O&P
Crew C-8D						
1 Labor Foreman (outside)	$28.45	$227.60	$47.65	$381.20	$30.07	$49.40
1 Laborer	26.45	211.60	44.30	354.40		
1 Cement Finisher	31.85	254.80	50.85	406.80		
1 Equipment Oper. (light)	33.55	268.40	54.80	438.40		
1 Air Compressor, 250 cfm		162.80		179.08		
2 -50' Air Hoses, 1"		8.20		9.02	5.34	5.88
32 L.H., Daily Totals		$1133.40		$1768.90	$35.42	$55.28

Crew No.	Hr.	Daily	Hr.	Daily	Bare Costs	Incl. O&P
Crew C-8E						
1 Labor Foreman (outside)	$28.45	$227.60	$47.65	$381.20	$28.87	$47.70
3 Laborers	26.45	634.80	44.30	1063.20		
1 Cement Finisher	31.85	254.80	50.85	406.80		
1 Equipment Oper. (light)	33.55	268.40	54.80	438.40		
1 Shotcrete Rig, 35 C.Y./hr		280.00		308.00		
1 Air Compressor, 250 cfm		162.80		179.08		
4 -50' Air Hoses, 1"		16.40		18.04		
4 -50' Air Hoses, 2"		31.00		34.10	10.21	11.23
48 L.H., Daily Totals		$1875.80		$2828.82	$39.08	$58.93

Crew No.	Hr.	Daily	Hr.	Daily	Bare Costs	Incl. O&P
Crew C-10						
1 Laborer	$26.45	$211.60	$44.30	$354.40	$30.05	$48.67
2 Cement Finishers	31.85	509.60	50.85	813.60		
24 L.H., Daily Totals		$721.20		$1168.00	$30.05	$48.67

Crew No.	Hr.	Daily	Hr.	Daily	Bare Costs	Incl. O&P
Crew C-10B						
3 Laborers	$26.45	$634.80	$44.30	$1063.20	$28.61	$46.92
2 Cement Finishers	31.85	509.60	50.85	813.60		
1 Concrete Mixer, 10 C.F.		151.20		166.32		
2 Trowels, 48" Walk-Behind		76.80		84.48	5.70	6.27
40 L.H., Daily Totals		$1372.40		$2127.60	$34.31	$53.19

Crew No.	Hr.	Daily	Hr.	Daily	Bare Costs	Incl. O&P
Crew C-10C						
1 Laborer	$26.45	$211.60	$44.30	$354.40	$30.05	$48.67
2 Cement Finishers	31.85	509.60	50.85	813.60		
1 Trowel, 48" Walk-Behind		38.40		42.24	1.60	1.76
24 L.H., Daily Totals		$759.60		$1210.24	$31.65	$50.43

Crew No.	Bare Costs		Incl. Subs O&P		Cost Per Labor-Hour	
Crew C-10D	**Hr.**	**Daily**	**Hr.**	**Daily**	**Bare Costs**	**Incl. O&P**
1 Laborer	$26.45	$211.60	$44.30	$354.40	$30.05	$48.67
2 Cement Finishers	31.85	509.60	50.85	813.60		
1 Vibrating Power Screed		65.50		72.05		
1 Trowel, 48" Walk-Behind		38.40		42.24	4.33	4.76
24 L.H., Daily Totals		$825.10		$1282.29	$34.38	$53.43

Crew No.	Hr.	Daily	Hr.	Daily	Bare Costs	Incl. O&P
Crew C-10E						
1 Laborer	$26.45	$211.60	$44.30	$354.40	$30.05	$48.67
2 Cement Finishers	31.85	509.60	50.85	813.60		
1 Vibrating Power Screed		65.50		72.05		
1 Cement Trowel, 96" Ride-On		157.40		173.14	9.29	10.22
24 L.H., Daily Totals		$944.10		$1413.19	$39.34	$58.88

Crew No.	Hr.	Daily	Hr.	Daily	Bare Costs	Incl. O&P
Crew C-11						
1 Skilled Worker Foreman	$36.05	$288.40	$60.35	$482.80	$34.59	$57.69
5 Skilled Workers	34.05	1362.00	57.00	2280.00		
1 Equip. Oper. (crane)	35.80	286.40	58.50	468.00		
1 Lattice Boom Crane, 150 Ton		1906.00		2096.60	34.04	37.44
56 L.H., Daily Totals		$3842.80		$5327.40	$68.62	$95.13

Crew No.	Hr.	Daily	Hr.	Daily	Bare Costs	Incl. O&P
Crew C-12						
1 Carpenter Foreman (out)	$35.55	$284.40	$59.50	$476.00	$33.08	$55.13
3 Carpenters	33.55	805.20	56.15	1347.60		
1 Laborer	26.45	211.60	44.30	354.40		
1 Equip. Oper. (crane)	35.80	286.40	58.50	468.00		
1 Hyd. Crane, 12 Ton		655.60		721.16	13.66	15.02
48 L.H., Daily Totals		$2243.20		$3367.16	$46.73	$70.15

Crew No.	Hr.	Daily	Hr.	Daily	Bare Costs	Incl. O&P
Crew C-13						
2 Struc. Steel Workers	$36.40	$582.40	$69.40	$1110.40	$35.45	$64.98
1 Carpenter	33.55	268.40	56.15	449.20		
1 Welder, gas engine, 300 amp		109.00		119.90	4.54	5.00
24 L.H., Daily Totals		$959.80		$1679.50	$39.99	$69.98

Crew No.	Hr.	Daily	Hr.	Daily	Bare Costs	Incl. O&P
Crew C-14						
1 Carpenter Foreman (out)	$35.55	$284.40	$59.50	$476.00	$30.32	$50.63
3 Carpenters	33.55	805.20	56.15	1347.60		
2 Carpenter Helpers	25.35	405.60	42.30	676.80		
4 Laborers	26.45	846.40	44.30	1417.60		
2 Rodmen (reinf.)	36.30	580.80	63.10	1009.60		
2 Rodman Helpers	25.35	405.60	42.30	676.80		
2 Cement Finishers	31.85	509.60	50.85	813.60		
1 Equip. Oper. (crane)	35.80	286.40	58.50	468.00		
1 Hyd. Crane, 80 Ton		1693.00		1862.30	12.45	13.69
136 L.H., Daily Totals		$5817.00		$8748.30	$42.77	$64.33

Crew No.	Hr.	Daily	Hr.	Daily	Bare Costs	Incl. O&P
Crew C-14A						
1 Carpenter Foreman (out)	$35.55	$284.40	$59.50	$476.00	$33.49	$56.27
16 Carpenters	33.55	4294.40	56.15	7187.20		
4 Rodmen (reinf.)	36.30	1161.60	63.10	2019.20		
2 Laborers	26.45	423.20	44.30	708.80		
1 Cement Finisher	31.85	254.80	50.85	406.80		
1 Equip. Oper. (med.)	34.90	279.20	57.05	456.40		
1 Gas Engine Vibrator		23.20		25.52		
1 Concrete Pump (small)		741.00		815.10	3.82	4.20
200 L.H., Daily Totals		$7461.80		$12095.02	$37.31	$60.48

Crews

Crew No.	Bare Costs		Incl. Subs O&P		Cost Per Labor-Hour	
Crew C-14B	**Hr.**	**Daily**	**Hr.**	**Daily**	**Bare Costs**	**Incl. O&P**
1 Carpenter Foreman (out)	$35.55	$284.40	$59.50	$476.00	$33.42	$56.06
16 Carpenters	33.55	4294.40	56.15	7187.20		
4 Rodmen (reinf.)	36.30	1161.60	63.10	2019.20		
2 Laborers	26.45	423.20	44.30	708.80		
2 Cement Finishers	31.85	509.60	50.85	813.60		
1 Equip. Oper. (med.)	34.90	279.20	57.05	456.40		
1 Gas Engine Vibrator		23.20		25.52		
1 Concrete Pump (small)		741.00		815.10	3.67	4.04
208 L.H., Daily Totals		$7716.60		$12501.82	$37.10	$60.10

Crew No.	Bare Costs		Incl. Subs O&P		Cost Per Labor-Hour	
Crew C-14C	**Hr.**	**Daily**	**Hr.**	**Daily**	**Bare Costs**	**Incl. O&P**
1 Carpenter Foreman (out)	$35.55	$284.40	$59.50	$476.00	$31.94	$53.62
6 Carpenters	33.55	1610.40	56.15	2695.20		
2 Rodmen (reinf.)	36.30	580.80	63.10	1009.60		
4 Laborers	26.45	846.40	44.30	1417.60		
1 Cement Finisher	31.85	254.80	50.85	406.80		
1 Gas Engine Vibrator		23.20		25.52	0.21	0.23
112 L.H., Daily Totals		$3600.00		$6030.72	$32.14	$53.85

Crew No.	Bare Costs		Incl. Subs O&P		Cost Per Labor-Hour	
Crew C-14D	**Hr.**	**Daily**	**Hr.**	**Daily**	**Bare Costs**	**Incl. O&P**
1 Carpenter Foreman (out)	$35.55	$284.40	$59.50	$476.00	$33.27	$55.72
18 Carpenters	33.55	4831.20	56.15	8085.60		
2 Rodmen (reinf.)	36.30	580.80	63.10	1009.60		
2 Laborers	26.45	423.20	44.30	708.80		
1 Cement Finisher	31.85	254.80	50.85	406.80		
1 Equip. Oper. (med.)	34.90	279.20	57.05	456.40		
1 Gas Engine Vibrator		23.20		25.52		
1 Concrete Pump (small)		741.00		815.10	3.82	4.20
200 L.H., Daily Totals		$7417.80		$11983.82	$37.09	$59.92

Crew No.	Bare Costs		Incl. Subs O&P		Cost Per Labor-Hour	
Crew C-14E	**Hr.**	**Daily**	**Hr.**	**Daily**	**Bare Costs**	**Incl. O&P**
1 Carpenter Foreman (out)	$35.55	$284.40	$59.50	$476.00	$32.64	$55.27
2 Carpenters	33.55	536.80	56.15	898.40		
4 Rodmen (reinf.)	36.30	1161.60	63.10	2019.20		
3 Laborers	26.45	634.80	44.30	1063.20		
1 Cement Finisher	31.85	254.80	50.85	406.80		
1 Gas Engine Vibrator		23.20		25.52	0.26	0.29
88 L.H., Daily Totals		$2895.60		$4889.12	$32.90	$55.56

Crew No.	Bare Costs		Incl. Subs O&P		Cost Per Labor-Hour	
Crew C-14F	**Hr.**	**Daily**	**Hr.**	**Daily**	**Bare Costs**	**Incl. O&P**
1 Laborer Foreman (out)	$28.45	$227.60	$47.65	$381.20	$30.27	$49.04
2 Laborers	26.45	423.20	44.30	708.80		
6 Cement Finishers	31.85	1528.80	50.85	2440.80		
1 Gas Engine Vibrator		23.20		25.52	0.32	0.35
72 L.H., Daily Totals		$2202.80		$3556.32	$30.59	$49.39

Crew No.	Bare Costs		Incl. Subs O&P		Cost Per Labor-Hour	
Crew C-14G	**Hr.**	**Daily**	**Hr.**	**Daily**	**Bare Costs**	**Incl. O&P**
1 Laborer Foreman (out)	$28.45	$227.60	$47.65	$381.20	$29.82	$48.52
2 Laborers	26.45	423.20	44.30	708.80		
4 Cement Finishers	31.85	1019.20	50.85	1627.20		
1 Gas Engine Vibrator		23.20		25.52	0.41	0.46
56 L.H., Daily Totals		$1693.20		$2742.72	$30.24	$48.98

Crew No.	Bare Costs		Incl. Subs O&P		Cost Per Labor-Hour	
Crew C-14H	**Hr.**	**Daily**	**Hr.**	**Daily**	**Bare Costs**	**Incl. O&P**
1 Carpenter Foreman (out)	$35.55	$284.40	$59.50	$476.00	$32.88	$55.01
2 Carpenters	33.55	536.80	56.15	898.40		
1 Rodman (reinf.)	36.30	290.40	63.10	504.80		
1 Laborer	26.45	211.60	44.30	354.40		
1 Cement Finisher	31.85	254.80	50.85	406.80		
1 Gas Engine Vibrator		23.20		25.52	0.48	0.53
48 L.H., Daily Totals		$1601.20		$2665.92	$33.36	$55.54

Crew No.	Bare Costs		Incl. Subs O&P		Cost Per Labor-Hour	
Crew C-14L	**Hr.**	**Daily**	**Hr.**	**Daily**	**Bare Costs**	**Incl. O&P**
1 Carpenter Foreman (out)	$35.55	$284.40	$59.50	$476.00	$31.21	$52.04
6 Carpenters	33.55	1610.40	56.15	2695.20		
4 Laborers	26.45	846.40	44.30	1417.60		
1 Cement Finisher	31.85	254.80	50.85	406.80		
1 Gas Engine Vibrator		23.20		25.52	0.24	0.27
96 L.H., Daily Totals		$3019.20		$5021.12	$31.45	$52.30

Crew No.	Bare Costs		Incl. Subs O&P		Cost Per Labor-Hour	
Crew C-15	**Hr.**	**Daily**	**Hr.**	**Daily**	**Bare Costs**	**Incl. O&P**
1 Carpenter Foreman (out)	$35.55	$284.40	$59.50	$476.00	$31.33	$52.17
2 Carpenters	33.55	536.80	56.15	898.40		
3 Laborers	26.45	634.80	44.30	1063.20		
2 Cement Finishers	31.85	509.60	50.85	813.60		
1 Rodman (reinf.)	36.30	290.40	63.10	504.80		
72 L.H., Daily Totals		$2256.00		$3756.00	$31.33	$52.17

Crew No.	Bare Costs		Incl. Subs O&P		Cost Per Labor-Hour	
Crew C-16	**Hr.**	**Daily**	**Hr.**	**Daily**	**Bare Costs**	**Incl. O&P**
1 Labor Foreman (outside)	$28.45	$227.60	$47.65	$381.20	$29.49	$48.47
3 Laborers	26.45	634.80	44.30	1063.20		
2 Cement Finishers	31.85	509.60	50.85	813.60		
1 Equip. Oper. (med.)	34.90	279.20	57.05	456.40		
1 Gunite Pump Rig		178.00		195.80		
2 -50' Air Hoses, 3/4"		6.10		6.71		
2 -50' Air Hoses, 2"		15.50		17.05	3.56	3.92
56 L.H., Daily Totals		$1850.80		$2933.96	$33.05	$52.39

Crew No.	Bare Costs		Incl. Subs O&P		Cost Per Labor-Hour	
Crew C-17	**Hr.**	**Daily**	**Hr.**	**Daily**	**Bare Costs**	**Incl. O&P**
2 Skilled Worker Foremen	$36.05	$576.80	$60.35	$965.60	$34.45	$57.67
8 Skilled Workers	34.05	2179.20	57.00	3648.00		
80 L.H., Daily Totals		$2756.00		$4613.60	$34.45	$57.67

Crew No.	Bare Costs		Incl. Subs O&P		Cost Per Labor-Hour	
Crew C-17A	**Hr.**	**Daily**	**Hr.**	**Daily**	**Bare Costs**	**Incl. O&P**
2 Skilled Worker Foremen	$36.05	$576.80	$60.35	$965.60	$34.47	$57.68
8 Skilled Workers	34.05	2179.20	57.00	3648.00		
.125 Equip. Oper. (crane)	35.80	35.80	58.50	58.50		
.125 Hyd. Crane, 80 Ton		211.63		232.79	2.61	2.87
81 L.H., Daily Totals		$3003.43		$4904.89	$37.08	$60.55

Crew No.	Bare Costs		Incl. Subs O&P		Cost Per Labor-Hour	
Crew C-17B	**Hr.**	**Daily**	**Hr.**	**Daily**	**Bare Costs**	**Incl. O&P**
2 Skilled Worker Foremen	$36.05	$576.80	$60.35	$965.60	$34.48	$57.69
8 Skilled Workers	34.05	2179.20	57.00	3648.00		
.25 Equip. Oper. (crane)	35.80	71.60	58.50	117.00		
.25 Hyd. Crane, 80 Ton		423.25		465.57		
.25 Trowel, 48" Walk-Behind		9.60		10.56	5.28	5.81
82 L.H., Daily Totals		$3260.45		$5206.73	$39.76	$63.50

Crew No.	Bare Costs		Incl. Subs O&P		Cost Per Labor-Hour	
Crew C-17C	**Hr.**	**Daily**	**Hr.**	**Daily**	**Bare Costs**	**Incl. O&P**
2 Skilled Worker Foremen	$36.05	$576.80	$60.35	$965.60	$34.50	$57.70
8 Skilled Workers	34.05	2179.20	57.00	3648.00		
.375 Equip. Oper. (crane)	35.80	107.40	58.50	175.50		
.375 Hyd. Crane, 80 Ton		634.88		698.36	7.65	8.41
83 L.H., Daily Totals		$3498.28		$5487.46	$42.15	$66.11

Crew No.	Bare Costs		Incl. Subs O&P		Cost Per Labor-Hour	
Crew C-17D	**Hr.**	**Daily**	**Hr.**	**Daily**	**Bare Costs**	**Incl. O&P**
2 Skilled Worker Foremen	$36.05	$576.80	$60.35	$965.60	$34.51	$57.71
8 Skilled Workers	34.05	2179.20	57.00	3648.00		
.5 Equip. Oper. (crane)	35.80	143.20	58.50	234.00		
.5 Hyd. Crane, 80 Ton		846.50		931.15	10.08	11.09
84 L.H., Daily Totals		$3745.70		$5778.75	$44.59	$68.79

Crews

Crew No.	Bare Costs		Incl. Subs O&P		Cost Per Labor-Hour	
Crew C-17E	**Hr.**	**Daily**	**Hr.**	**Daily**	**Bare Costs**	**Incl. O&P**
2 Skilled Worker Foremen	$36.05	$576.80	$60.35	$965.60	$34.45	$57.67
8 Skilled Workers	34.05	2179.20	57.00	3648.00		
1 Hyd. Jack with Rods		89.10		98.01	1.11	1.23
80 L.H., Daily Totals		$2845.10		$4711.61	$35.56	$58.90
Crew C-18	**Hr.**	**Daily**	**Hr.**	**Daily**	**Bare Costs**	**Incl. O&P**
.125 Labor Foreman (out)	$28.45	$28.45	$47.65	$47.65	$26.67	$44.67
1 Laborer	26.45	211.60	44.30	354.40		
1 Concrete Cart, 10 C.F.		55.00		60.50	6.11	6.72
9 L.H., Daily Totals		$295.05		$462.55	$32.78	$51.39
Crew C-19	**Hr.**	**Daily**	**Hr.**	**Daily**	**Bare Costs**	**Incl. O&P**
.125 Labor Foreman (out)	$28.45	$28.45	$47.65	$47.65	$26.67	$44.67
1 Laborer	26.45	211.60	44.30	354.40		
1 Concrete Cart, 18 C.F.		91.60		100.76	10.18	11.20
9 L.H., Daily Totals		$331.65		$502.81	$36.85	$55.87
Crew C-20	**Hr.**	**Daily**	**Hr.**	**Daily**	**Bare Costs**	**Incl. O&P**
1 Labor Foreman (outside)	$28.45	$227.60	$47.65	$381.20	$28.43	$47.13
5 Laborers	26.45	1058.00	44.30	1772.00		
1 Cement Finisher	31.85	254.80	50.85	406.80		
1 Equip. Oper. (med.)	34.90	279.20	57.05	456.40		
2 Gas Engine Vibrators		46.40		51.04		
1 Concrete Pump (small)		741.00		815.10	12.30	13.53
64 L.H., Daily Totals		$2607.00		$3882.54	$40.73	$60.66
Crew C-21	**Hr.**	**Daily**	**Hr.**	**Daily**	**Bare Costs**	**Incl. O&P**
1 Labor Foreman (outside)	$28.45	$227.60	$47.65	$381.20	$28.43	$47.13
5 Laborers	26.45	1058.00	44.30	1772.00		
1 Cement Finisher	31.85	254.80	50.85	406.80		
1 Equip. Oper. (med.)	34.90	279.20	57.05	456.40		
2 Gas Engine Vibrators		46.40		51.04		
1 Concrete Conveyer		174.60		192.06	3.45	3.80
64 L.H., Daily Totals		$2040.60		$3259.50	$31.88	$50.93
Crew C-22	**Hr.**	**Daily**	**Hr.**	**Daily**	**Bare Costs**	**Incl. O&P**
1 Rodman Foreman	$38.30	$306.40	$66.55	$532.40	$36.54	$63.35
4 Rodmen (reinf.)	36.30	1161.60	63.10	2019.20		
.125 Equip. Oper. (crane)	35.80	35.80	58.50	58.50		
.125 Equip. Oper. Oiler	31.05	31.05	50.75	50.75		
.125 Hyd. Crane, 25 Ton		93.58		102.93	2.23	2.45
42 L.H., Daily Totals		$1628.43		$2763.78	$38.77	$65.80
Crew C-23	**Hr.**	**Daily**	**Hr.**	**Daily**	**Bare Costs**	**Incl. O&P**
2 Skilled Worker Foremen	$36.05	$576.80	$60.35	$965.60	$34.33	$57.20
6 Skilled Workers	34.05	1634.40	57.00	2736.00		
1 Equip. Oper. (crane)	35.80	286.40	58.50	468.00		
1 Equip. Oper. Oiler	31.05	248.40	50.75	406.00		
1 Lattice Boom Crane, 90 Ton		1622.00		1784.20	20.27	22.30
80 L.H., Daily Totals		$4368.00		$6359.80	$54.60	$79.50
Crew C-24	**Hr.**	**Daily**	**Hr.**	**Daily**	**Bare Costs**	**Incl. O&P**
2 Skilled Worker Foremen	$36.05	$576.80	$60.35	$965.60	$34.33	$57.20
6 Skilled Workers	34.05	1634.40	57.00	2736.00		
1 Equip. Oper. (crane)	35.80	286.40	58.50	468.00		
1 Equip. Oper. Oiler	31.05	248.40	50.75	406.00		
1 Lattice Boom Crane, 150 Ton		1906.00		2096.60	23.82	26.21
80 L.H., Daily Totals		$4652.00		$6672.20	$58.15	$83.40

Crew No.	Bare Costs		Incl. Subs O&P		Cost Per Labor-Hour	
Crew C-25	**Hr.**	**Daily**	**Hr.**	**Daily**	**Bare Costs**	**Incl. O&P**
2 Rodmen (reinf.)	$36.30	$580.80	$63.10	$1009.60	$28.60	$50.40
2 Rodmen Helpers	20.90	334.40	37.70	603.20		
32 L.H., Daily Totals		$915.20		$1612.80	$28.60	$50.40
Crew C-27	**Hr.**	**Daily**	**Hr.**	**Daily**	**Bare Costs**	**Incl. O&P**
2 Cement Finishers	$31.85	$509.60	$50.85	$813.60	$31.85	$50.85
1 Concrete Saw		134.20		147.62	8.39	9.23
16 L.H., Daily Totals		$643.80		$961.22	$40.24	$60.08
Crew C-28	**Hr.**	**Daily**	**Hr.**	**Daily**	**Bare Costs**	**Incl. O&P**
1 Cement Finisher	$31.85	$254.80	$50.85	$406.80	$31.85	$50.85
1 Portable Air Compressor, Gas		19.00		20.90	2.38	2.61
8 L.H., Daily Totals		$273.80		$427.70	$34.23	$53.46
Crew C-29	**Hr.**	**Daily**	**Hr.**	**Daily**	**Bare Costs**	**Incl. O&P**
1 Laborer	$26.45	$211.60	$44.30	$354.40	$26.45	$44.30
1 Pressure Washer		58.80		64.68	7.35	8.09
8 L.H., Daily Totals		$270.40		$419.08	$33.80	$52.38
Crew D-1	**Hr.**	**Daily**	**Hr.**	**Daily**	**Bare Costs**	**Incl. O&P**
1 Bricklayer	$33.65	$269.20	$55.20	$441.60	$30.40	$49.88
1 Bricklayer Helper	27.15	217.20	44.55	356.40		
16 L.H., Daily Totals		$486.40		$798.00	$30.40	$49.88
Crew D-2	**Hr.**	**Daily**	**Hr.**	**Daily**	**Bare Costs**	**Incl. O&P**
3 Bricklayers	$33.65	$807.60	$55.20	$1324.80	$31.05	$50.94
2 Bricklayer Helpers	27.15	434.40	44.55	712.80		
40 L.H., Daily Totals		$1242.00		$2037.60	$31.05	$50.94
Crew D-3	**Hr.**	**Daily**	**Hr.**	**Daily**	**Bare Costs**	**Incl. O&P**
3 Bricklayers	$33.65	$807.60	$55.20	$1324.80	$31.17	$51.19
2 Bricklayer Helpers	27.15	434.40	44.55	712.80		
.25 Carpenter	33.55	67.10	56.15	112.30		
42 L.H., Daily Totals		$1309.10		$2149.90	$31.17	$51.19
Crew D-4	**Hr.**	**Daily**	**Hr.**	**Daily**	**Bare Costs**	**Incl. O&P**
1 Bricklayer	$33.65	$269.20	$55.20	$441.60	$28.31	$46.63
3 Bricklayer Helpers	27.15	651.60	44.55	1069.20		
1 Building Laborer	26.45	211.60	44.30	354.40		
1 Grout Pump, 50 C.F./hr.		125.80		138.38	3.15	3.46
40 L.H., Daily Totals		$1258.20		$2003.58	$31.45	$50.09
Crew D-5	**Hr.**	**Daily**	**Hr.**	**Daily**	**Bare Costs**	**Incl. O&P**
1 Block Mason Helper	$27.15	$217.20	$44.55	$356.40	$27.15	$44.55
8 L.H., Daily Totals		$217.20		$356.40	$27.15	$44.55
Crew D-6	**Hr.**	**Daily**	**Hr.**	**Daily**	**Bare Costs**	**Incl. O&P**
3 Bricklayers	$33.65	$807.60	$55.20	$1324.80	$30.40	$49.88
3 Bricklayer Helpers	27.15	651.60	44.55	1069.20		
48 L.H., Daily Totals		$1459.20		$2394.00	$30.40	$49.88
Crew D-7	**Hr.**	**Daily**	**Hr.**	**Daily**	**Bare Costs**	**Incl. O&P**
1 Tile Layer	$31.55	$252.40	$50.40	$403.20	$28.20	$45.05
1 Tile Layer Helper	24.85	198.80	39.70	317.60		
16 L.H., Daily Totals		$451.20		$720.80	$28.20	$45.05

Crews

Crew No.	Bare Costs		Incl. Subs O&P		Cost Per Labor-Hour	
Crew D-8	**Hr.**	**Daily**	**Hr.**	**Daily**	**Bare Costs**	**Incl. O&P**
3 Bricklayers	$33.65	$807.60	$55.20	$1324.80	$31.05	$50.94
2 Bricklayer Helpers	27.15	434.40	44.55	712.80		
40 L.H., Daily Totals		$1242.00		$2037.60	$31.05	$50.94
Crew D-9	**Hr.**	**Daily**	**Hr.**	**Daily**	**Bare Costs**	**Incl. O&P**
3 Bricklayers	$33.65	$807.60	$55.20	$1324.80	$30.40	$49.88
3 Bricklayer Helpers	27.15	651.60	44.55	1069.20		
48 L.H., Daily Totals		$1459.20		$2394.00	$30.40	$49.88
Crew D-10	**Hr.**	**Daily**	**Hr.**	**Daily**	**Bare Costs**	**Incl. O&P**
1 Bricklayer Foreman	$35.65	$285.20	$58.45	$467.60	$33.06	$54.17
1 Bricklayer	33.65	269.20	55.20	441.60		
1 Bricklayer Helper	27.15	217.20	44.55	356.40		
1 Equip. Oper. (crane)	35.80	286.40	58.50	468.00		
1 S.P. Crane, 4x4, 12 Ton		434.40		477.84	13.57	14.93
32 L.H., Daily Totals		$1492.40		$2211.44	$46.64	$69.11
Crew D-11	**Hr.**	**Daily**	**Hr.**	**Daily**	**Bare Costs**	**Incl. O&P**
2 Bricklayers	$33.65	$538.40	$55.20	$883.20	$31.48	$51.65
1 Bricklayer Helper	27.15	217.20	44.55	356.40		
24 L.H., Daily Totals		$755.60		$1239.60	$31.48	$51.65
Crew D-12	**Hr.**	**Daily**	**Hr.**	**Daily**	**Bare Costs**	**Incl. O&P**
2 Bricklayers	$33.65	$538.40	$55.20	$883.20	$30.40	$49.88
2 Bricklayer Helpers	27.15	434.40	44.55	712.80		
32 L.H., Daily Totals		$972.80		$1596.00	$30.40	$49.88
Crew D-13	**Hr.**	**Daily**	**Hr.**	**Daily**	**Bare Costs**	**Incl. O&P**
1 Bricklayer Foreman	$35.65	$285.20	$58.45	$467.60	$32.17	$52.74
2 Bricklayers	33.65	538.40	55.20	883.20		
2 Bricklayer Helpers	27.15	434.40	44.55	712.80		
1 Equip. Oper. (crane)	35.80	286.40	58.50	468.00		
1 S.P. Crane, 4x4, 12 Ton		434.40		477.84	9.05	9.96
48 L.H., Daily Totals		$1978.80		$3009.44	$41.23	$62.70
Crew E-1	**Hr.**	**Daily**	**Hr.**	**Daily**	**Bare Costs**	**Incl. O&P**
2 Struc. Steel Workers	$36.40	$582.40	$69.40	$1110.40	$36.40	$69.40
1 Welder, gas engine, 300 amp		109.00		119.90	6.81	7.49
16 L.H., Daily Totals		$691.40		$1230.30	$43.21	$76.89
Crew E-2	**Hr.**	**Daily**	**Hr.**	**Daily**	**Bare Costs**	**Incl. O&P**
1 Struc. Steel Foreman	$38.40	$307.20	$73.20	$585.60	$36.63	$68.22
4 Struc. Steel Workers	36.40	1164.80	69.40	2220.80		
1 Equip. Oper. (crane)	35.80	286.40	58.50	468.00		
1 Lattice Boom Crane, 90 Ton		1622.00		1784.20	33.79	37.17
48 L.H., Daily Totals		$3380.40		$5058.60	$70.42	$105.39
Crew E-3	**Hr.**	**Daily**	**Hr.**	**Daily**	**Bare Costs**	**Incl. O&P**
1 Struc. Steel Foreman	$38.40	$307.20	$73.20	$585.60	$37.07	$70.67
2 Struc. Steel Workers	36.40	582.40	69.40	1110.40		
1 Welder, gas engine, 300 amp		109.00		119.90	4.54	5.00
24 L.H., Daily Totals		$998.60		$1815.90	$41.61	$75.66
Crew E-4	**Hr.**	**Daily**	**Hr.**	**Daily**	**Bare Costs**	**Incl. O&P**
1 Struc. Steel Foreman	$38.40	$307.20	$73.20	$585.60	$36.90	$70.35
3 Struc. Steel Workers	36.40	873.60	69.40	1665.60		
1 Welder, gas engine, 300 amp		109.00		119.90	3.41	3.75
32 L.H., Daily Totals		$1289.80		$2371.10	$40.31	$74.10

Crew No.	Bare Costs		Incl. Subs O&P		Cost Per Labor-Hour	
Crew E-5	**Hr.**	**Daily**	**Hr.**	**Daily**	**Bare Costs**	**Incl. O&P**
1 Struc. Steel Foreman	$38.40	$307.20	$73.20	$585.60	$36.56	$68.61
7 Struc. Steel Workers	36.40	2038.40	69.40	3886.40		
1 Equip. Oper. (crane)	35.80	286.40	58.50	468.00		
1 Lattice Boom Crane, 90 Ton		1622.00		1784.20		
1 Welder, gas engine, 300 amp		109.00		119.90	24.04	26.45
72 L.H., Daily Totals		$4363.00		$6844.10	$60.60	$95.06
Crew E-6	**Hr.**	**Daily**	**Hr.**	**Daily**	**Bare Costs**	**Incl. O&P**
1 Struc. Steel Foreman	$38.40	$307.20	$73.20	$585.60	$36.30	$67.95
12 Struc. Steel Workers	36.40	3494.40	69.40	6662.40		
1 Equip. Oper. (crane)	35.80	286.40	58.50	468.00		
1 Equip. Oper. (light)	33.55	268.40	54.80	438.40		
1 Lattice Boom Crane, 90 Ton		1622.00		1784.20		
1 Welder, gas engine, 300 amp		109.00		119.90		
1 Air Compressor, 160 cfm		125.00		137.50		
2 Impact Wrenches		36.40		40.04	15.77	17.35
120 L.H., Daily Totals		$6248.80		$10236.04	$52.07	$85.30
Crew E-7	**Hr.**	**Daily**	**Hr.**	**Daily**	**Bare Costs**	**Incl. O&P**
1 Struc. Steel Foreman	$38.40	$307.20	$73.20	$585.60	$36.56	$68.61
7 Struc. Steel Workers	36.40	2038.40	69.40	3886.40		
1 Equip. Oper. (crane)	35.80	286.40	58.50	468.00		
1 Lattice Boom Crane, 90 Ton		1622.00		1784.20		
2 Welders, gas engine, 300 amp		218.00		239.80	25.56	28.11
72 L.H., Daily Totals		$4472.00		$6964.00	$62.11	$96.72
Crew E-8	**Hr.**	**Daily**	**Hr.**	**Daily**	**Bare Costs**	**Incl. O&P**
1 Struc. Steel Foreman	$38.40	$307.20	$73.20	$585.60	$36.53	$68.75
9 Struc. Steel Workers	36.40	2620.80	69.40	4996.80		
1 Equip. Oper. (crane)	35.80	286.40	58.50	468.00		
1 Lattice Boom Crane, 90 Ton		1622.00		1784.20		
4 Welders, gas engine, 300 amp		436.00		479.60	23.39	25.73
88 L.H., Daily Totals		$5272.40		$8314.20	$59.91	$94.48
Crew E-9	**Hr.**	**Daily**	**Hr.**	**Daily**	**Bare Costs**	**Incl. O&P**
2 Struc. Steel Foremen	$38.40	$614.40	$73.20	$1171.20	$36.23	$67.35
5 Struc. Steel Workers	36.40	1456.00	69.40	2776.00		
1 Welder Foreman	38.40	307.20	73.20	585.60		
5 Welders	36.40	1456.00	69.40	2776.00		
1 Equip. Oper. (crane)	35.80	286.40	58.50	468.00		
1 Equip. Oper. Oiler	31.05	248.40	50.75	406.00		
1 Equip. Oper. (light)	33.55	268.40	54.80	438.40		
1 Lattice Boom Crane, 90 Ton		1622.00		1784.20		
5 Welders, gas engine, 300 amp		545.00		599.50	16.93	18.62
128 L.H., Daily Totals		$6803.80		$11004.90	$53.15	$85.98
Crew E-10	**Hr.**	**Daily**	**Hr.**	**Daily**	**Bare Costs**	**Incl. O&P**
1 Struc. Steel Foreman	$38.40	$307.20	$73.20	$585.60	$37.07	$70.67
2 Struc. Steel Workers	36.40	582.40	69.40	1110.40		
1 Welder, gas engine, 300 amp		109.00		119.90		
1 Flatbed Truck, Gas, 3 Ton		217.00		238.70	13.58	14.94
24 L.H., Daily Totals		$1215.60		$2054.60	$50.65	$85.61
Crew E-11	**Hr.**	**Daily**	**Hr.**	**Daily**	**Bare Costs**	**Incl. O&P**
2 Painters, Struc. Steel	$30.55	$488.80	$58.95	$943.20	$30.27	$54.25
1 Building Laborer	26.45	211.60	44.30	354.40		
1 Equip. Oper. (light)	33.55	268.40	54.80	438.40		
1 Air Compressor, 250 cfm		162.80		179.08		
1 Sandblaster, portable, 3 C.F.		20.60		22.66		
1 Set Sand Blasting Accessories		13.15		14.47	6.14	6.76
32 L.H., Daily Totals		$1165.35		$1952.20	$36.42	$61.01

Crews

Crew No.	Bare Costs		Incl. Subs O&P		Cost Per Labor-Hour	
Crew E-12	**Hr.**	**Daily**	**Hr.**	**Daily**	**Bare Costs**	**Incl. O&P**
1 Welder Foreman	$38.40	$307.20	$73.20	$585.60	$35.98	$64.00
1 Equip. Oper. (light)	33.55	268.40	54.80	438.40		
1 Welder, gas engine, 300 amp		109.00		119.90	6.81	7.49
16 L.H., Daily Totals		$684.60		$1143.90	$42.79	$71.49

Crew E-13	Hr.	Daily	Hr.	Daily	Bare Costs	Incl. O&P
1 Welder Foreman	$38.40	$307.20	$73.20	$585.60	$36.78	$67.07
.5 Equip. Oper. (light)	33.55	134.20	54.80	219.20		
1 Welder, gas engine, 300 amp		109.00		119.90	9.08	9.99
12 L.H., Daily Totals		$550.40		$924.70	$45.87	$77.06

Crew E-14	Hr.	Daily	Hr.	Daily	Bare Costs	Incl. O&P
1 Struc. Steel Worker	$36.40	$291.20	$69.40	$555.20	$36.40	$69.40
1 Welder, gas engine, 300 amp		109.00		119.90	13.63	14.99
8 L.H., Daily Totals		$400.20		$675.10	$50.02	$84.39

Crew E-16	Hr.	Daily	Hr.	Daily	Bare Costs	Incl. O&P
1 Welder Foreman	$38.40	$307.20	$73.20	$585.60	$37.40	$71.30
1 Welder	36.40	291.20	69.40	555.20		
1 Welder, gas engine, 300 amp		109.00		119.90	6.81	7.49
16 L.H., Daily Totals		$707.40		$1260.70	$44.21	$78.79

Crew E-17	Hr.	Daily	Hr.	Daily	Bare Costs	Incl. O&P
1 Structural Steel Foreman	$38.40	$307.20	$73.20	$585.60	$37.40	$71.30
1 Structural Steel Worker	36.40	291.20	69.40	555.20		
16 L.H., Daily Totals		$598.40		$1140.80	$37.40	$71.30

Crew E-18	Hr.	Daily	Hr.	Daily	Bare Costs	Incl. O&P
1 Structural Steel Foreman	$38.40	$307.20	$73.20	$585.60	$36.50	$67.69
3 Structural Steel Workers	36.40	873.60	69.40	1665.60		
1 Equipment Operator (med.)	34.90	279.20	57.05	456.40		
1 Lattice Boom Crane, 20 Ton		1035.00		1138.50	25.88	28.46
40 L.H., Daily Totals		$2495.00		$3846.10	$62.38	$96.15

Crew E-19	Hr.	Daily	Hr.	Daily	Bare Costs	Incl. O&P
1 Structural Steel Worker	$36.40	$291.20	$69.40	$555.20	$36.12	$65.80
1 Structural Steel Foreman	38.40	307.20	73.20	585.60		
1 Equip. Oper. (light)	33.55	268.40	54.80	438.40		
1 Lattice Boom Crane, 20 Ton		1035.00		1138.50	43.13	47.44
24 L.H., Daily Totals		$1901.80		$2717.70	$79.24	$113.24

Crew E-20	Hr.	Daily	Hr.	Daily	Bare Costs	Incl. O&P
1 Structural Steel Foreman	$38.40	$307.20	$73.20	$585.60	$35.91	$66.18
5 Structural Steel Workers	36.40	1456.00	69.40	2776.00		
1 Equip. Oper. (crane)	35.80	286.40	58.50	468.00		
1 Oiler	31.05	248.40	50.75	406.00		
1 Lattice Boom Crane, 40 Ton		1263.00		1389.30	19.73	21.71
64 L.H., Daily Totals		$3561.00		$5624.90	$55.64	$87.89

Crew E-22	Hr.	Daily	Hr.	Daily	Bare Costs	Incl. O&P
1 Skilled Worker Foreman	$36.05	$288.40	$60.35	$482.80	$34.72	$58.12
2 Skilled Workers	34.05	544.80	57.00	912.00		
24 L.H., Daily Totals		$833.20		$1394.80	$34.72	$58.12

Crew E-24	Hr.	Daily	Hr.	Daily	Bare Costs	Incl. O&P
3 Structural Steel Workers	$36.40	$873.60	$69.40	$1665.60	$36.02	$66.31
1 Equipment Operator (med.)	34.90	279.20	57.05	456.40		
1 Hyd. Crane, 25 Ton		748.60		823.46	23.39	25.73
32 L.H., Daily Totals		$1901.40		$2945.46	$59.42	$92.05

Crew No.	Bare Costs		Incl. Subs O&P		Cost Per Labor-Hour	
Crew E-25	**Hr.**	**Daily**	**Hr.**	**Daily**	**Bare Costs**	**Incl. O&P**
1 Welder	$36.40	$291.20	$69.40	$555.20	$36.40	$69.40
1 Cutting Torch		18.00		19.80		
1 Set of Gases		82.80		91.08	12.60	13.86
8 L.H., Daily Totals		$392.00		$666.08	$49.00	$83.26

Crew F-3	Hr.	Daily	Hr.	Daily	Bare Costs	Incl. O&P
2 Carpenters	$33.55	$536.80	$56.15	$898.40	$30.72	$51.08
2 Carpenter Helpers	25.35	405.60	42.30	676.80		
1 Equip. Oper. (crane)	35.80	286.40	58.50	468.00		
1 Hyd. Crane, 12 Ton		655.60		721.16	16.39	18.03
40 L.H., Daily Totals		$1884.40		$2764.36	$47.11	$69.11

Crew F-4	Hr.	Daily	Hr.	Daily	Bare Costs	Incl. O&P
2 Carpenters	$33.55	$536.80	$56.15	$898.40	$30.72	$51.08
2 Carpenter Helpers	25.35	405.60	42.30	676.80		
1 Equip. Oper. (crane)	35.80	286.40	58.50	468.00		
1 Hyd. Crane, 55 Ton		1208.00		1328.80	30.20	33.22
40 L.H., Daily Totals		$2436.80		$3372.00	$60.92	$84.30

Crew F-5	Hr.	Daily	Hr.	Daily	Bare Costs	Incl. O&P
2 Carpenters	$33.55	$536.80	$56.15	$898.40	$29.45	$49.23
2 Carpenter Helpers	25.35	405.60	42.30	676.80		
32 L.H., Daily Totals		$942.40		$1575.20	$29.45	$49.23

Crew F-6	Hr.	Daily	Hr.	Daily	Bare Costs	Incl. O&P
2 Carpenters	$33.55	$536.80	$56.15	$898.40	$31.16	$51.88
2 Building Laborers	26.45	423.20	44.30	708.80		
1 Equip. Oper. (crane)	35.80	286.40	58.50	468.00		
1 Hyd. Crane, 12 Ton		655.60		721.16	16.39	18.03
40 L.H., Daily Totals		$1902.00		$2796.36	$47.55	$69.91

Crew F-7	Hr.	Daily	Hr.	Daily	Bare Costs	Incl. O&P
2 Carpenters	$33.55	$536.80	$56.15	$898.40	$30.00	$50.23
2 Building Laborers	26.45	423.20	44.30	708.80		
32 L.H., Daily Totals		$960.00		$1607.20	$30.00	$50.23

Crew G-1	Hr.	Daily	Hr.	Daily	Bare Costs	Incl. O&P
1 Roofer Foreman	$30.15	$241.20	$54.40	$435.20	$26.36	$47.57
4 Roofers, Composition	28.15	900.80	50.80	1625.60		
2 Roofer Helpers	20.90	334.40	37.70	603.20		
1 Application Equipment		168.40		185.24		
1 Tar Kettle/Pot		88.75		97.63		
1 Crew Truck		152.40		167.64	7.31	8.04
56 L.H., Daily Totals		$1885.95		$3114.51	$33.68	$55.62

Crew G-2	Hr.	Daily	Hr.	Daily	Bare Costs	Incl. O&P
1 Plasterer	$30.70	$245.60	$49.95	$399.60	$28.17	$46.25
1 Plasterer Helper	27.35	218.80	44.50	356.00		
1 Building Laborer	26.45	211.60	44.30	354.40		
1 Grout Pump, 50 C.F./hr		125.80		138.38	5.24	5.77
24 L.H., Daily Totals		$801.80		$1248.38	$33.41	$52.02

Crew G-2A	Hr.	Daily	Hr.	Daily	Bare Costs	Incl. O&P
1 Roofer, composition	$28.15	$225.20	$50.80	$406.40	$25.17	$44.27
1 Roofer Helper	20.90	167.20	37.70	301.60		
1 Building Laborer	26.45	211.60	44.30	354.40		
1 Foam spray rig, trailer-mtd.		488.25		537.08		
1 Pickup Truck, 3/4 Ton		104.60		115.06	24.70	27.17
24 L.H., Daily Totals		$1196.85		$1714.54	$49.87	$71.44

Crews

Crew No.	Bare Costs Hr.	Bare Costs Daily	Incl. Subs O&P Hr.	Incl. Subs O&P Daily	Cost Per Labor-Hour Bare Costs	Cost Per Labor-Hour Incl. O&P
Crew G-3	**Hr.**	**Daily**	**Hr.**	**Daily**	**Bare Costs**	**Incl. O&P**
2 Sheet Metal Workers	$38.20	$611.20	$63.10	$1009.60	$32.33	$53.70
2 Building Laborers	26.45	423.20	44.30	708.80		
32 L.H., Daily Totals		$1034.40		$1718.40	$32.33	$53.70

Crew G-4	Hr.	Daily	Hr.	Daily	Bare Costs	Incl. O&P
1 Labor Foreman (outside)	$28.45	$227.60	$47.65	$381.20	$27.12	$45.42
2 Building Laborers	26.45	423.20	44.30	708.80		
1 Flatbed Truck, Gas, 1.5 Ton		171.40		188.54		
1 Air Compressor, 160 cfm		125.00		137.50	12.35	13.59
24 L.H., Daily Totals		$947.20		$1416.04	$39.47	$59.00

Crew G-5	Hr.	Daily	Hr.	Daily	Bare Costs	Incl. O&P
1 Roofer Foreman	$30.15	$241.20	$54.40	$435.20	$25.65	$46.28
2 Roofers, Composition	28.15	450.40	50.80	812.80		
2 Roofer Helpers	20.90	334.40	37.70	603.20		
1 Application Equipment		168.40		185.24	4.21	4.63
40 L.H., Daily Totals		$1194.40		$2036.44	$29.86	$50.91

Crew G-6A	Hr.	Daily	Hr.	Daily	Bare Costs	Incl. O&P
2 Roofers Composition	$28.15	$450.40	$50.80	$812.80	$28.15	$50.80
1 Small Compressor, Electric		11.05		12.15		
2 Pneumatic Nailers		47.00		51.70	3.63	3.99
16 L.H., Daily Totals		$508.45		$876.65	$31.78	$54.79

Crew G-7	Hr.	Daily	Hr.	Daily	Bare Costs	Incl. O&P
1 Carpenter	$33.55	$268.40	$56.15	$449.20	$33.55	$56.15
1 Small Compressor, Electric		11.05		12.15		
1 Pneumatic Nailer		23.50		25.85	4.32	4.75
8 L.H., Daily Totals		$302.95		$487.20	$37.87	$60.90

Crew H-1	Hr.	Daily	Hr.	Daily	Bare Costs	Incl. O&P
2 Glaziers	$33.20	$531.20	$54.50	$872.00	$34.80	$61.95
2 Struc. Steel Workers	36.40	582.40	69.40	1110.40		
32 L.H., Daily Totals		$1113.60		$1982.40	$34.80	$61.95

Crew H-2	Hr.	Daily	Hr.	Daily	Bare Costs	Incl. O&P
2 Glaziers	$33.20	$531.20	$54.50	$872.00	$30.95	$51.10
1 Building Laborer	26.45	211.60	44.30	354.40		
24 L.H., Daily Totals		$742.80		$1226.40	$30.95	$51.10

Crew H-3	Hr.	Daily	Hr.	Daily	Bare Costs	Incl. O&P
1 Glazier	$33.20	$265.60	$54.50	$436.00	$29.27	$48.40
1 Helper	25.35	202.80	42.30	338.40		
16 L.H., Daily Totals		$468.40		$774.40	$29.27	$48.40

Crew H-4	Hr.	Daily	Hr.	Daily	Bare Costs	Incl. O&P
1 Carpenter	$33.55	$268.40	$56.15	$449.20	$31.61	$52.43
1 Carpenter Helper	25.35	202.80	42.30	338.40		
.5 Electrician	40.25	161.00	65.25	261.00		
20 L.H., Daily Totals		$632.20		$1048.60	$31.61	$52.43

Crew J-1	Hr.	Daily	Hr.	Daily	Bare Costs	Incl. O&P
3 Plasterers	$30.70	$736.80	$49.95	$1198.80	$29.36	$47.77
2 Plasterer Helpers	27.35	437.60	44.50	712.00		
1 Mixing Machine, 6 C.F.		123.60		135.96	3.09	3.40
40 L.H., Daily Totals		$1298.00		$2046.76	$32.45	$51.17

Crew J-2	Hr.	Daily	Hr.	Daily	Bare Costs	Incl. O&P
3 Plasterers	$30.70	$736.80	$49.95	$1198.80	$29.48	$47.82
2 Plasterer Helpers	27.35	437.60	44.50	712.00		
1 Lather	30.05	240.40	48.05	384.40		
1 Mixing Machine, 6 C.F.		123.60		135.96	2.58	2.83
48 L.H., Daily Totals		$1538.40		$2431.16	$32.05	$50.65

Crew J-3	Hr.	Daily	Hr.	Daily	Bare Costs	Incl. O&P
1 Terrazzo Worker	$31.30	$250.40	$50.00	$400.00	$28.70	$45.85
1 Terrazzo Helper	26.10	208.80	41.70	333.60		
1 Floor Grinder, 22" path		73.95		81.34		
1 Terrazzo Mixer		167.60		184.36	15.10	16.61
16 L.H., Daily Totals		$700.75		$999.30	$43.80	$62.46

Crew J-4	Hr.	Daily	Hr.	Daily	Bare Costs	Incl. O&P
2 Cement Finishers	$31.85	$509.60	$50.85	$813.60	$30.05	$48.67
1 Laborer	26.45	211.60	44.30	354.40		
1 Floor Grinder, 22" path		73.95		81.34		
1 Floor Edger, 7" path		38.20		42.02		
1 Vacuum Pick-Up System		59.80		65.78	7.16	7.88
24 L.H., Daily Totals		$893.15		$1357.15	$37.21	$56.55

Crew J-4A	Hr.	Daily	Hr.	Daily	Bare Costs	Incl. O&P
2 Cement Finishers	$31.85	$509.60	$50.85	$813.60	$29.15	$47.58
2 Laborers	26.45	423.20	44.30	708.80		
1 Floor Grinder, 22" path		73.95		81.34		
1 Floor Edger, 7" path		38.20		42.02		
1 Vacuum Pick-Up System		59.80		65.78		
1 Floor Auto Scrubber		154.10		169.51	10.19	11.21
32 L.H., Daily Totals		$1258.85		$1881.06	$39.34	$58.78

Crew J-4B	Hr.	Daily	Hr.	Daily	Bare Costs	Incl. O&P
1 Laborer	$26.45	$211.60	$44.30	$354.40	$26.45	$44.30
1 Floor Auto Scrubber		154.10		169.51	19.26	21.19
8 L.H., Daily Totals		$365.70		$523.91	$45.71	$65.49

Crew K-1	Hr.	Daily	Hr.	Daily	Bare Costs	Incl. O&P
1 Carpenter	$33.55	$268.40	$56.15	$449.20	$29.98	$49.98
1 Truck Driver (light)	26.40	211.20	43.80	350.40		
1 Flatbed Truck, Gas, 3 Ton		217.00		238.70	13.56	14.92
16 L.H., Daily Totals		$696.60		$1038.30	$43.54	$64.89

Crew K-2	Hr.	Daily	Hr.	Daily	Bare Costs	Incl. O&P
1 Struc. Steel Foreman	$38.40	$307.20	$73.20	$585.60	$33.73	$62.13
1 Struc. Steel Worker	36.40	291.20	69.40	555.20		
1 Truck Driver (light)	26.40	211.20	43.80	350.40		
1 Flatbed Truck, Gas, 3 Ton		217.00		238.70	9.04	9.95
24 L.H., Daily Totals		$1026.60		$1729.90	$42.77	$72.08

Crew L-1	Hr.	Daily	Hr.	Daily	Bare Costs	Incl. O&P
.25 Electrician	$40.25	$80.50	$65.25	$130.50	$39.77	$64.85
1 Plumber	39.65	317.20	64.75	518.00		
10 L.H., Daily Totals		$397.70		$648.50	$39.77	$64.85

Crew L-2	Hr.	Daily	Hr.	Daily	Bare Costs	Incl. O&P
1 Carpenter	$33.55	$268.40	$56.15	$449.20	$29.45	$49.23
1 Carpenter Helper	25.35	202.80	42.30	338.40		
16 L.H., Daily Totals		$471.20		$787.60	$29.45	$49.23

Crews

Crew No.	Bare Costs		Incl. Subs O&P		Cost Per Labor-Hour	
Crew L-3	**Hr.**	**Daily**	**Hr.**	**Daily**	**Bare Costs**	**Incl. O&P**
1 Carpenter	$33.55	$268.40	$56.15	$449.20	$34.89	$57.97
.25 Electrician	40.25	80.50	65.25	130.50		
10 L.H., Daily Totals		$348.90		$579.70	$34.89	$57.97
Crew L-3A	**Hr.**	**Daily**	**Hr.**	**Daily**	**Bare Costs**	**Incl. O&P**
1 Carpenter Foreman (outside)	$35.55	$284.40	$59.50	$476.00	$36.43	$60.70
.5 Sheet Metal Worker	38.20	152.80	63.10	252.40		
12 L.H., Daily Totals		$437.20		$728.40	$36.43	$60.70
Crew L-4	**Hr.**	**Daily**	**Hr.**	**Daily**	**Bare Costs**	**Incl. O&P**
1 Skilled Worker	$34.05	$272.40	$57.00	$456.00	$29.70	$49.65
1 Helper	25.35	202.80	42.30	338.40		
16 L.H., Daily Totals		$475.20		$794.40	$29.70	$49.65
Crew L-5	**Hr.**	**Daily**	**Hr.**	**Daily**	**Bare Costs**	**Incl. O&P**
1 Struc. Steel Foreman	$38.40	$307.20	$73.20	$585.60	$36.60	$68.39
5 Struc. Steel Workers	36.40	1456.00	69.40	2776.00		
1 Equip. Oper. (crane)	35.80	286.40	58.50	468.00		
1 Hyd. Crane, 25 Ton		748.60		823.46	13.37	14.70
56 L.H., Daily Totals		$2798.20		$4653.06	$49.97	$83.09
Crew L-5A	**Hr.**	**Daily**	**Hr.**	**Daily**	**Bare Costs**	**Incl. O&P**
1 Structural Steel Foreman	$38.40	$307.20	$73.20	$585.60	$36.75	$67.63
2 Structural Steel Workers	36.40	582.40	69.40	1110.40		
1 Equip. Oper. (crane)	35.80	286.40	58.50	468.00		
1 S.P. Crane, 4x4, 25 Ton		560.40		616.44	17.51	19.26
32 L.H., Daily Totals		$1736.40		$2780.44	$54.26	$86.89
Crew L-5B	**Hr.**	**Daily**	**Hr.**	**Daily**	**Bare Costs**	**Incl. O&P**
1 Structural Steel Foreman	$38.40	$307.20	$73.20	$585.60	$37.12	$64.01
2 Structural Steel Workers	36.40	582.40	69.40	1110.40		
2 Electricians	40.25	644.00	65.25	1044.00		
2 Steamfitters/Pipefitters	40.05	640.80	65.40	1046.40		
1 Equip. Oper. (crane)	35.80	286.40	58.50	468.00		
1 Common Building Laborer	26.45	211.60	44.30	354.40		
1 Hyd. Crane, 80 Ton		1693.00		1862.30	23.51	25.87
72 L.H., Daily Totals		$4365.40		$6471.10	$60.63	$89.88
Crew L-6	**Hr.**	**Daily**	**Hr.**	**Daily**	**Bare Costs**	**Incl. O&P**
1 Plumber	$39.65	$317.20	$64.75	$518.00	$39.85	$64.92
.5 Electrician	40.25	161.00	65.25	261.00		
12 L.H., Daily Totals		$478.20		$779.00	$39.85	$64.92
Crew L-7	**Hr.**	**Daily**	**Hr.**	**Daily**	**Bare Costs**	**Incl. O&P**
1 Carpenter	$33.55	$268.40	$56.15	$449.20	$29.02	$48.33
2 Carpenter Helpers	25.35	405.60	42.30	676.80		
.25 Electrician	40.25	80.50	65.25	130.50		
26 L.H., Daily Totals		$754.50		$1256.50	$29.02	$48.33
Crew L-8	**Hr.**	**Daily**	**Hr.**	**Daily**	**Bare Costs**	**Incl. O&P**
1 Carpenter	$33.55	$268.40	$56.15	$449.20	$31.49	$52.33
1 Carpenter Helper	25.35	202.80	42.30	338.40		
.5 Plumber	39.65	158.60	64.75	259.00		
20 L.H., Daily Totals		$629.80		$1046.60	$31.49	$52.33

Crew No.	Bare Costs		Incl. Subs O&P		Cost Per Labor-Hour	
Crew L-9	**Hr.**	**Daily**	**Hr.**	**Daily**	**Bare Costs**	**Incl. O&P**
1 Skilled Worker Foreman	$36.05	$288.40	$60.35	$482.80	$31.32	$52.13
1 Skilled Worker	34.05	272.40	57.00	456.00		
2 Helpers	25.35	405.60	42.30	676.80		
.5 Electrician	40.25	161.00	65.25	261.00		
36 L.H., Daily Totals		$1127.40		$1876.60	$31.32	$52.13
Crew L-10	**Hr.**	**Daily**	**Hr.**	**Daily**	**Bare Costs**	**Incl. O&P**
1 Structural Steel Foreman	$38.40	$307.20	$73.20	$585.60	$36.87	$67.03
1 Structural Steel Worker	36.40	291.20	69.40	555.20		
1 Equip. Oper. (crane)	35.80	286.40	58.50	468.00		
1 Hyd. Crane, 12 Ton		655.60		721.16	27.32	30.05
24 L.H., Daily Totals		$1540.40		$2329.96	$64.18	$97.08
Crew L-11	**Hr.**	**Daily**	**Hr.**	**Daily**	**Bare Costs**	**Incl. O&P**
2 Wreckers	$26.45	$423.20	$48.20	$771.20	$30.56	$52.42
1 Equip. Oper. (crane)	35.80	286.40	58.50	468.00		
1 Equip. Oper. (light)	33.55	268.40	54.80	438.40		
1 Hyd. Excavator, 2.5 C.Y.		1613.00		1774.30		
1 Loader, Skid Steer, 78 H.P.		261.20		287.32	58.57	64.43
32 L.H., Daily Totals		$2852.20		$3739.22	$89.13	$116.85
Crew M-1	**Hr.**	**Daily**	**Hr.**	**Daily**	**Bare Costs**	**Incl. O&P**
3 Elevator Constructors	$50.55	$1213.20	$82.10	$1970.40	$48.02	$78.00
1 Elevator Apprentice	40.45	323.60	65.70	525.60		
5 Hand Tools		49.00		53.90	1.53	1.68
32 L.H., Daily Totals		$1585.80		$2549.90	$49.56	$79.68
Crew M-3	**Hr.**	**Daily**	**Hr.**	**Daily**	**Bare Costs**	**Incl. O&P**
1 Electrician Foreman (out)	$42.25	$338.00	$68.50	$548.00	$39.63	$64.67
1 Common Laborer	26.45	211.60	44.30	354.40		
.25 Equipment Operator, Med.	34.90	69.80	57.05	114.10		
1 Elevator Constructor	50.55	404.40	82.10	656.80		
1 Elevator Apprentice	40.45	323.60	65.70	525.60		
.25 S.P. Crane, 4x4, 20 Ton		127.75		140.53	3.76	4.13
34 L.H., Daily Totals		$1475.15		$2339.43	$43.39	$68.81
Crew M-4	**Hr.**	**Daily**	**Hr.**	**Daily**	**Bare Costs**	**Incl. O&P**
1 Electrician Foreman (out)	$42.25	$338.00	$68.50	$548.00	$39.20	$63.98
1 Common Laborer	26.45	211.60	44.30	354.40		
.25 Equipment Operator, Crane	35.80	71.60	58.50	117.00		
.25 Equipment Operator, Oiler	31.05	62.10	50.75	101.50		
1 Elevator Constructor	50.55	404.40	82.10	656.80		
1 Elevator Apprentice	40.45	323.60	65.70	525.60		
.25 S.P. Crane, 4x4, 40 Ton		164.65		181.12	4.57	5.03
36 L.H., Daily Totals		$1575.95		$2484.42	$43.78	$69.01
Crew Q-1	**Hr.**	**Daily**	**Hr.**	**Daily**	**Bare Costs**	**Incl. O&P**
1 Plumber	$39.65	$317.20	$64.75	$518.00	$35.67	$58.25
1 Plumber Apprentice	31.70	253.60	51.75	414.00		
16 L.H., Daily Totals		$570.80		$932.00	$35.67	$58.25
Crew Q-1A	**Hr.**	**Daily**	**Hr.**	**Daily**	**Bare Costs**	**Incl. O&P**
.25 Plumber Foreman (out)	$41.65	$83.30	$68.00	$136.00	$40.05	$65.40
1 Plumber	39.65	317.20	64.75	518.00		
10 L.H., Daily Totals		$400.50		$654.00	$40.05	$65.40

Crews

Crew No.	Bare Costs		Incl. Subs O&P		Cost Per Labor-Hour	
Crew Q-1C	**Hr.**	**Daily**	**Hr.**	**Daily**	**Bare Costs**	**Incl. O&P**
1 Plumber	$39.65	$317.20	$64.75	$518.00	$35.42	$57.85
1 Plumber Apprentice	31.70	253.60	51.75	414.00		
1 Equip. Oper. (medium)	34.90	279.20	57.05	456.40		
1 Trencher, Chain Type, 8' D		2506.00		2756.60	104.42	114.86
24 L.H., Daily Totals		$3356.00		$4145.00	$139.83	$172.71

Crew Q-2	Hr.	Daily	Hr.	Daily	Bare Costs	Incl. O&P
1 Plumber	$39.65	$317.20	$64.75	$518.00	$34.35	$56.08
2 Plumber Apprentices	31.70	507.20	51.75	828.00		
24 L.H., Daily Totals		$824.40		$1346.00	$34.35	$56.08

Crew Q-3	Hr.	Daily	Hr.	Daily	Bare Costs	Incl. O&P
2 Plumbers	$39.65	$634.40	$64.75	$1036.00	$35.67	$58.25
2 Plumber Apprentices	31.70	507.20	51.75	828.00		
32 L.H., Daily Totals		$1141.60		$1864.00	$35.67	$58.25

Crew Q-4	Hr.	Daily	Hr.	Daily	Bare Costs	Incl. O&P
2 Plumbers	$39.65	$634.40	$64.75	$1036.00	$37.66	$61.50
1 Welder (plumber)	39.65	317.20	64.75	518.00		
1 Plumber Apprentice	31.70	253.60	51.75	414.00		
1 Welder, electric, 300 amp		32.45		35.70	1.01	1.12
32 L.H., Daily Totals		$1237.65		$2003.69	$38.68	$62.62

Crew Q-5	Hr.	Daily	Hr.	Daily	Bare Costs	Incl. O&P
1 Steamfitter	$40.05	$320.40	$65.40	$523.20	$36.05	$58.88
1 Steamfitter Apprentice	32.05	256.40	52.35	418.80		
16 L.H., Daily Totals		$576.80		$942.00	$36.05	$58.88

Crew Q-6	Hr.	Daily	Hr.	Daily	Bare Costs	Incl. O&P
1 Steamfitter	$40.05	$320.40	$65.40	$523.20	$34.72	$56.70
2 Steamfitter Apprentices	32.05	512.80	52.35	837.60		
24 L.H., Daily Totals		$833.20		$1360.80	$34.72	$56.70

Crew Q-7	Hr.	Daily	Hr.	Daily	Bare Costs	Incl. O&P
2 Steamfitters	$40.05	$640.80	$65.40	$1046.40	$36.05	$58.88
2 Steamfitter Apprentices	32.05	512.80	52.35	837.60		
32 L.H., Daily Totals		$1153.60		$1884.00	$36.05	$58.88

Crew Q-8	Hr.	Daily	Hr.	Daily	Bare Costs	Incl. O&P
2 Steamfitters	$40.05	$640.80	$65.40	$1046.40	$38.05	$62.14
1 Welder (steamfitter)	40.05	320.40	65.40	523.20		
1 Steamfitter Apprentice	32.05	256.40	52.35	418.80		
1 Welder, electric, 300 amp		32.45		35.70	1.01	1.12
32 L.H., Daily Totals		$1250.05		$2024.10	$39.06	$63.25

Crew Q-9	Hr.	Daily	Hr.	Daily	Bare Costs	Incl. O&P
1 Sheet Metal Worker	$38.20	$305.60	$63.10	$504.80	$34.38	$56.77
1 Sheet Metal Apprentice	30.55	244.40	50.45	403.60		
16 L.H., Daily Totals		$550.00		$908.40	$34.38	$56.77

Crew Q-10	Hr.	Daily	Hr.	Daily	Bare Costs	Incl. O&P
2 Sheet Metal Workers	$38.20	$611.20	$63.10	$1009.60	$35.65	$58.88
1 Sheet Metal Apprentice	30.55	244.40	50.45	403.60		
24 L.H., Daily Totals		$855.60		$1413.20	$35.65	$58.88

Crew Q-11	Hr.	Daily	Hr.	Daily	Bare Costs	Incl. O&P
2 Sheet Metal Workers	$38.20	$611.20	$63.10	$1009.60	$34.38	$56.77
2 Sheet Metal Apprentices	30.55	488.80	50.45	807.20		
32 L.H., Daily Totals		$1100.00		$1816.80	$34.38	$56.77

Crew No.	Bare Costs		Incl. Subs O&P		Cost Per Labor-Hour	
Crew Q-12	**Hr.**	**Daily**	**Hr.**	**Daily**	**Bare Costs**	**Incl. O&P**
1 Sprinkler Installer	$39.30	$314.40	$64.20	$513.60	$35.38	$57.80
1 Sprinkler Apprentice	31.45	251.60	51.40	411.20		
16 L.H., Daily Totals		$566.00		$924.80	$35.38	$57.80

Crew Q-13	Hr.	Daily	Hr.	Daily	Bare Costs	Incl. O&P
2 Sprinkler Installers	$39.30	$628.80	$64.20	$1027.20	$35.38	$57.80
2 Sprinkler Apprentices	31.45	503.20	51.40	822.40		
32 L.H., Daily Totals		$1132.00		$1849.60	$35.38	$57.80

Crew Q-14	Hr.	Daily	Hr.	Daily	Bare Costs	Incl. O&P
1 Asbestos Worker	$34.70	$277.60	$58.50	$468.00	$31.23	$52.65
1 Asbestos Apprentice	27.75	222.00	46.80	374.40		
16 L.H., Daily Totals		$499.60		$842.40	$31.23	$52.65

Crew Q-15	Hr.	Daily	Hr.	Daily	Bare Costs	Incl. O&P
1 Plumber	$39.65	$317.20	$64.75	$518.00	$35.67	$58.25
1 Plumber Apprentice	31.70	253.60	51.75	414.00		
1 Welder, electric, 300 amp		32.45		35.70	2.03	2.23
16 L.H., Daily Totals		$603.25		$967.70	$37.70	$60.48

Crew Q-16	Hr.	Daily	Hr.	Daily	Bare Costs	Incl. O&P
2 Plumbers	$39.65	$634.40	$64.75	$1036.00	$37.00	$60.42
1 Plumber Apprentice	31.70	253.60	51.75	414.00		
1 Welder, electric, 300 amp		32.45		35.70	1.35	1.49
24 L.H., Daily Totals		$920.45		$1485.69	$38.35	$61.90

Crew Q-17	Hr.	Daily	Hr.	Daily	Bare Costs	Incl. O&P
1 Steamfitter	$40.05	$320.40	$65.40	$523.20	$36.05	$58.88
1 Steamfitter Apprentice	32.05	256.40	52.35	418.80		
1 Welder, electric, 300 amp		32.45		35.70	2.03	2.23
16 L.H., Daily Totals		$609.25		$977.70	$38.08	$61.11

Crew Q-17A	Hr.	Daily	Hr.	Daily	Bare Costs	Incl. O&P
1 Steamfitter	$40.05	$320.40	$65.40	$523.20	$35.97	$58.75
1 Steamfitter Apprentice	32.05	256.40	52.35	418.80		
1 Equip. Oper. (crane)	35.80	286.40	58.50	468.00		
1 Hyd. Crane, 12 Ton		655.60		721.16		
1 Welder, electric, 300 amp		32.45		35.70	28.67	31.54
24 L.H., Daily Totals		$1551.25		$2166.86	$64.64	$90.29

Crew Q-18	Hr.	Daily	Hr.	Daily	Bare Costs	Incl. O&P
2 Steamfitters	$40.05	$640.80	$65.40	$1046.40	$37.38	$61.05
1 Steamfitter Apprentice	32.05	256.40	52.35	418.80		
1 Welder, electric, 300 amp		32.45		35.70	1.35	1.49
24 L.H., Daily Totals		$929.65		$1500.90	$38.74	$62.54

Crew Q-19	Hr.	Daily	Hr.	Daily	Bare Costs	Incl. O&P
1 Steamfitter	$40.05	$320.40	$65.40	$523.20	$37.45	$61.00
1 Steamfitter Apprentice	32.05	256.40	52.35	418.80		
1 Electrician	40.25	322.00	65.25	522.00		
24 L.H., Daily Totals		$898.80		$1464.00	$37.45	$61.00

Crew Q-20	Hr.	Daily	Hr.	Daily	Bare Costs	Incl. O&P
1 Sheet Metal Worker	$38.20	$305.60	$63.10	$504.80	$35.55	$58.47
1 Sheet Metal Apprentice	30.55	244.40	50.45	403.60		
.5 Electrician	40.25	161.00	65.25	261.00		
20 L.H., Daily Totals		$711.00		$1169.40	$35.55	$58.47

Crews

Crew No.	Bare Costs Hr.	Bare Costs Daily	Incl. Subs O&P Hr.	Incl. Subs O&P Daily	Cost Per Labor-Hour Bare Costs	Cost Per Labor-Hour Incl. O&P
Crew Q-21	**Hr.**	**Daily**	**Hr.**	**Daily**	**Bare Costs**	**Incl. O&P**
2 Steamfitters	$40.05	$640.80	$65.40	$1046.40	$38.10	$62.10
1 Steamfitter Apprentice	32.05	256.40	52.35	418.80		
1 Electrician	40.25	322.00	65.25	522.00		
32 L.H., Daily Totals		$1219.20		$1987.20	$38.10	$62.10
Crew Q-22	**Hr.**	**Daily**	**Hr.**	**Daily**	**Bare Costs**	**Incl. O&P**
1 Plumber	$39.65	$317.20	$64.75	$518.00	$35.67	$58.25
1 Plumber Apprentice	31.70	253.60	51.75	414.00		
1 Hyd. Crane, 12 Ton		655.60		721.16	40.98	45.07
16 L.H., Daily Totals		$1226.40		$1653.16	$76.65	$103.32
Crew Q-22A	**Hr.**	**Daily**	**Hr.**	**Daily**	**Bare Costs**	**Incl. O&P**
1 Plumber	$39.65	$317.20	$64.75	$518.00	$33.40	$54.83
1 Plumber Apprentice	31.70	253.60	51.75	414.00		
1 Laborer	26.45	211.60	44.30	354.40		
1 Equip. Oper. (crane)	35.80	286.40	58.50	468.00		
1 Hyd. Crane, 12 Ton		655.60		721.16	20.49	22.54
32 L.H., Daily Totals		$1724.40		$2475.56	$53.89	$77.36
Crew Q-23	**Hr.**	**Daily**	**Hr.**	**Daily**	**Bare Costs**	**Incl. O&P**
1 Plumber Foreman	$41.65	$333.20	$68.00	$544.00	$38.73	$63.27
1 Plumber	39.65	317.20	64.75	518.00		
1 Equip. Oper. (med.)	34.90	279.20	57.05	456.40		
1 Lattice Boom Crane, 20 Ton		1035.00		1138.50	43.13	47.44
24 L.H., Daily Totals		$1964.60		$2656.90	$81.86	$110.70
Crew R-1	**Hr.**	**Daily**	**Hr.**	**Daily**	**Bare Costs**	**Incl. O&P**
1 Electrician Foreman	$40.75	$326.00	$66.05	$528.40	$35.37	$57.73
3 Electricians	40.25	966.00	65.25	1566.00		
2 Helpers	25.35	405.60	42.30	676.80		
48 L.H., Daily Totals		$1697.60		$2771.20	$35.37	$57.73
Crew R-1A	**Hr.**	**Daily**	**Hr.**	**Daily**	**Bare Costs**	**Incl. O&P**
1 Electrician	$40.25	$322.00	$65.25	$522.00	$32.80	$53.77
1 Helper	25.35	202.80	42.30	338.40		
16 L.H., Daily Totals		$524.80		$860.40	$32.80	$53.77
Crew R-2	**Hr.**	**Daily**	**Hr.**	**Daily**	**Bare Costs**	**Incl. O&P**
1 Electrician Foreman	$40.75	$326.00	$66.05	$528.40	$35.43	$57.84
3 Electricians	40.25	966.00	65.25	1566.00		
2 Helpers	25.35	405.60	42.30	676.80		
1 Equip. Oper. (crane)	35.80	286.40	58.50	468.00		
1 S.P. Crane, 4x4, 5 Ton		257.80		283.58	4.60	5.06
56 L.H., Daily Totals		$2241.80		$3522.78	$40.03	$62.91
Crew R-3	**Hr.**	**Daily**	**Hr.**	**Daily**	**Bare Costs**	**Incl. O&P**
1 Electrician Foreman	$40.75	$326.00	$66.05	$528.40	$39.56	$64.22
1 Electrician	40.25	322.00	65.25	522.00		
.5 Equip. Oper. (crane)	35.80	143.20	58.50	234.00		
.5 S.P. Crane, 4x4, 5 Ton		128.90		141.79	6.45	7.09
20 L.H., Daily Totals		$920.10		$1426.19	$46.01	$71.31
Crew R-4	**Hr.**	**Daily**	**Hr.**	**Daily**	**Bare Costs**	**Incl. O&P**
1 Struc. Steel Foreman	$38.40	$307.20	$73.20	$585.60	$37.57	$69.33
3 Struc. Steel Workers	36.40	873.60	69.40	1665.60		
1 Electrician	40.25	322.00	65.25	522.00		
1 Welder, gas engine, 300 amp		109.00		119.90	2.73	3.00
40 L.H., Daily Totals		$1611.80		$2893.10	$40.30	$72.33

Crew No.	Bare Costs Hr.	Bare Costs Daily	Incl. Subs O&P Hr.	Incl. Subs O&P Daily	Cost Per Labor-Hour Bare Costs	Cost Per Labor-Hour Incl. O&P
Crew R-5	**Hr.**	**Daily**	**Hr.**	**Daily**	**Bare Costs**	**Incl. O&P**
1 Electrician Foreman	$40.75	$326.00	$66.05	$528.40	$34.88	$56.98
4 Electrician Linemen	40.25	1288.00	65.25	2088.00		
2 Electrician Operators	40.25	644.00	65.25	1044.00		
4 Electrician Groundmen	25.35	811.20	42.30	1353.60		
1 Crew Truck		152.40		167.64		
1 Flatbed Truck, 20,000 GVW		190.00		209.00		
1 Pickup Truck, 3/4 Ton		104.60		115.06		
.2 Hyd. Crane, 55 Ton		241.60		265.76		
.2 Hyd. Crane, 12 Ton		131.12		144.23		
.2 Earth Auger, Truck-Mtd.		80.36		88.40		
1 Tractor w/Winch		346.80		381.48	14.17	15.59
88 L.H., Daily Totals		$4316.08		$6385.57	$49.05	$72.56
Crew R-6	**Hr.**	**Daily**	**Hr.**	**Daily**	**Bare Costs**	**Incl. O&P**
1 Electrician Foreman	$40.75	$326.00	$66.05	$528.40	$34.88	$56.98
4 Electrician Linemen	40.25	1288.00	65.25	2088.00		
2 Electrician Operators	40.25	644.00	65.25	1044.00		
4 Electrician Groundmen	25.35	811.20	42.30	1353.60		
1 Crew Truck		152.40		167.64		
1 Flatbed Truck, 20,000 GVW		190.00		209.00		
1 Pickup Truck, 3/4 Ton		104.60		115.06		
.2 Hyd. Crane, 55 Ton		241.60		265.76		
.2 Hyd. Crane, 12 Ton		131.12		144.23		
.2 Earth Auger, Truck-Mtd.		80.36		88.40		
1 Tractor w/Winch		346.80		381.48		
3 Cable Trailers		552.90		608.19		
.5 Tensioning Rig		183.05		201.35		
.5 Cable Pulling Rig		1072.00		1179.20	34.71	38.19
88 L.H., Daily Totals		$6124.03		$8374.31	$69.59	$95.16
Crew R-7	**Hr.**	**Daily**	**Hr.**	**Daily**	**Bare Costs**	**Incl. O&P**
1 Electrician Foreman	$40.75	$326.00	$66.05	$528.40	$27.92	$46.26
5 Electrician Groundmen	25.35	1014.00	42.30	1692.00		
1 Crew Truck		152.40		167.64	3.17	3.49
48 L.H., Daily Totals		$1492.40		$2388.04	$31.09	$49.75
Crew R-8	**Hr.**	**Daily**	**Hr.**	**Daily**	**Bare Costs**	**Incl. O&P**
1 Electrician Foreman	$40.75	$326.00	$66.05	$528.40	$35.37	$57.73
3 Electrician Linemen	40.25	966.00	65.25	1566.00		
2 Electrician Groundmen	25.35	405.60	42.30	676.80		
1 Pickup Truck, 3/4 Ton		104.60		115.06		
1 Crew Truck		152.40		167.64	5.35	5.89
48 L.H., Daily Totals		$1954.60		$3053.90	$40.72	$63.62
Crew R-9	**Hr.**	**Daily**	**Hr.**	**Daily**	**Bare Costs**	**Incl. O&P**
1 Electrician Foreman	$40.75	$326.00	$66.05	$528.40	$32.86	$53.88
1 Electrician Lineman	40.25	322.00	65.25	522.00		
2 Electrician Operators	40.25	644.00	65.25	1044.00		
4 Electrician Groundmen	25.35	811.20	42.30	1353.60		
1 Pickup Truck, 3/4 Ton		104.60		115.06		
1 Crew Truck		152.40		167.64	4.02	4.42
64 L.H., Daily Totals		$2360.20		$3730.70	$36.88	$58.29
Crew R-10	**Hr.**	**Daily**	**Hr.**	**Daily**	**Bare Costs**	**Incl. O&P**
1 Electrician Foreman	$40.75	$326.00	$66.05	$528.40	$37.85	$61.56
4 Electrician Linemen	40.25	1288.00	65.25	2088.00		
1 Electrician Groundman	25.35	202.80	42.30	338.40		
1 Crew Truck		152.40		167.64		
3 Tram Cars		339.90		373.89	10.26	11.28
48 L.H., Daily Totals		$2309.10		$3496.33	$48.11	$72.84

Crews

Crew R-11	Bare Costs Hr.	Bare Costs Daily	Incl. Subs O&P Hr.	Incl. Subs O&P Daily	Cost Per Labor-Hour Bare Costs	Cost Per Labor-Hour Incl. O&P
1 Electrician Foreman	$40.75	$326.00	$66.05	$528.40	$37.71	$61.41
4 Electricians	40.25	1288.00	65.25	2088.00		
1 Equip. Oper. (crane)	35.80	286.40	58.50	468.00		
1 Common Laborer	26.45	211.60	44.30	354.40		
1 Crew Truck		152.40		167.64		
1 Hyd. Crane, 12 Ton		655.60		721.16	14.43	15.87
56 L.H., Daily Totals		$2920.00		$4327.60	$52.14	$77.28

Crew R-12	Hr.	Daily	Hr.	Daily	Bare Costs	Incl. O&P
1 Carpenter Foreman	$34.05	$272.40	$57.00	$456.00	$31.40	$53.20
4 Carpenters	33.55	1073.60	56.15	1796.80		
4 Common Laborers	26.45	846.40	44.30	1417.60		
1 Equip. Oper. (med.)	34.90	279.20	57.05	456.40		
1 Steel Worker	36.40	291.20	69.40	555.20		
1 Dozer, 200 H.P.		1051.00		1156.10		
1 Pickup Truck, 3/4 Ton		104.60		115.06	13.13	14.45
88 L.H., Daily Totals		$3918.40		$5953.16	$44.53	$67.65

Crew R-15	Hr.	Daily	Hr.	Daily	Bare Costs	Incl. O&P
1 Electrician Foreman	$40.75	$326.00	$66.05	$528.40	$39.22	$63.64
4 Electricians	40.25	1288.00	65.25	2088.00		
1 Equipment Oper. (light)	33.55	268.40	54.80	438.40		
1 Aerial Lift Truck		317.80		349.58	6.62	7.28
48 L.H., Daily Totals		$2200.20		$3404.38	$45.84	$70.92

Crew R-15A	Hr.	Daily	Hr.	Daily	Bare Costs	Incl. O&P
1 Electrician Foreman	$40.75	$326.00	$66.05	$528.40	$34.62	$56.66
2 Electricians	40.25	644.00	65.25	1044.00		
2 Common Laborers	26.45	423.20	44.30	708.80		
1 Equipment Operator	33.55	268.40	54.80	438.40		
1 Aerial Lift Truck		317.80		349.58	6.62	7.28
48 L.H., Daily Totals		$1979.40		$3069.18	$41.24	$63.94

Crew R-18	Hr.	Daily	Hr.	Daily	Bare Costs	Incl. O&P
.25 Electrician Foreman	$40.75	$81.50	$66.05	$132.10	$31.12	$51.19
1 Electrician	40.25	322.00	65.25	522.00		
2 Helpers	25.35	405.60	42.30	676.80		
26 L.H., Daily Totals		$809.10		$1330.90	$31.12	$51.19

Crew R-19	Hr.	Daily	Hr.	Daily	Bare Costs	Incl. O&P
.5 Electrician Foreman	$40.75	$163.00	$66.05	$264.20	$40.35	$65.41
2 Electricians	40.25	644.00	65.25	1044.00		
20 L.H., Daily Totals		$807.00		$1308.20	$40.35	$65.41

Crew R-21	Hr.	Daily	Hr.	Daily	Bare Costs	Incl. O&P
1 Electrician Foreman	$40.75	$326.00	$66.05	$528.40	$40.24	$65.25
3 Electricians	40.25	966.00	65.25	1566.00		
.1 Equip. Oper. (med.)	34.90	27.92	57.05	45.64		
.1 S.P. Crane, 4x4, 25 Ton		56.04		61.64	1.71	1.88
32.8 L.H., Daily Totals		$1375.96		$2201.68	$41.95	$67.12

Crew R-22	Hr.	Daily	Hr.	Daily	Bare Costs	Incl. O&P
.66 Electrician Foreman	$40.75	$215.16	$66.05	$348.74	$33.93	$55.51
2 Helpers	25.35	405.60	42.30	676.80		
2 Electricians	40.25	644.00	65.25	1044.00		
37.28 L.H., Daily Totals		$1264.76		$2069.54	$33.93	$55.51

Crew R-30	Bare Costs Hr.	Bare Costs Daily	Incl. Subs O&P Hr.	Incl. Subs O&P Daily	Cost Per Labor-Hour Bare Costs	Cost Per Labor-Hour Incl. O&P
.25 Electrician Foreman (out)	$42.25	$84.50	$68.50	$137.00	$31.91	$52.61
1 Electrician	40.25	322.00	65.25	522.00		
2 Laborers, (Semi-Skilled)	26.45	423.20	44.30	708.80		
26 L.H., Daily Totals		$829.70		$1367.80	$31.91	$52.61

Historical Cost Indexes

The following tables are the estimated Historical Cost Indexes based on a 30-city national average with a base of 100 on January 1, 1993.

The indexes may be used to:

1. Estimate and compare construction costs for different years in the same city.
2. Estimate and compare construction costs in different cities for the same year.
3. Estimate and compare construction costs in different cities for different years.
4. Compare construction trends in any city with the national average.

EXAMPLES

1. Estimate and compare construction costs for different years in the same city.

A. To estimate the construction cost of a building in Lexington, KY in 1970, knowing that it cost $915,000 in 2011.

Index Lexington, KY in 1970 = 26.9
Index Lexington, KY in 2011 = 160.8

$$\frac{\text{Index 1970}}{\text{Index 2011}} \times \text{Cost 2011} = \text{Cost 1970}$$

$$\frac{26.9}{160.8} \times \$915{,}000 = \$153{,}000$$

Construction Cost in Lexington, KY in 1970 = $153,000

B. To estimate the current construction cost of a building in Boston, MA that was built in 1980 for $900,000.

Index Boston, MA in 1980 = 64.0
Index Boston, MA in 2011 = 217.8

$$\frac{\text{Index 2011}}{\text{Index 1980}} \times \text{Cost 1980} = \text{Cost 2011}$$

$$\frac{217.8}{64.0} \times \$900{,}000 = \$3{,}063{,}000$$

Construction Cost in Boston in 2011 = $3,063,000

2. Estimate and compare construction costs in different cities for the same year.

To compare the construction cost of a building in Topeka, KS in 2011 with the known cost of $800,000 in Baltimore, MD in 2011

Index Topeka, KS in 2011 = 155.3
Index Baltimore, MD in 2011 = 171.3

$$\frac{\text{Index Topeka}}{\text{Index Baltimore}} \times \text{Cost Baltimore} = \text{Cost Topeka}$$

$$\frac{155.3}{171.3} \times \$800{,}000 = \$725{,}500$$

Construction Cost in Topeka in 2011 = $725,500

3. Estimate and compare construction costs in different cities for different years.

To compare the construction cost of a building in Detroit, MI in 2011 with the known construction cost of $5,000,000 for the same building in San Francisco, CA in 1980.

Index Detroit, MI in 2011 = 190.1
Index San Francisco, CA in 1980 = 75.2

$$\frac{\text{Index Detroit 2011}}{\text{Index San Francisco 1980}} \times \text{Cost San Francisco 1980} = \text{Cost Detroit 2011}$$

$$\frac{190.1}{75.2} \times \$5{,}000{,}000 = \$12{,}639{,}500$$

Construction Cost in Detroit in 2011 = $12,639,500

4. Compare construction trends in any city with the national average.

To compare the construction cost in Las Vegas, NV from 1975 to 2011 with the increase in the National Average during the same time period.

Index Las Vegas, NV for 1975 = 42.8 For 2011 = 195.9
Index 30 City Average for 1975 = 43.7 For 2011 = 185.0

A. National Average escalation From 1975 to 2011 $= \frac{\text{Index — 30 City 2011}}{\text{Index — 30 City 1975}} = \frac{185.0}{43.7}$

National Average escalation From 1975 to 2011 = 4.23 or increased by 323%

B. Escalation for Las Vegas, NV From 1975 to 2011 $= \frac{\text{Index Las Vegas, NV 2011}}{\text{Index Las Vegas, NV 1975}} = \frac{195.9}{42.8}$

Las Vegas escalation From 1975 to 2011 = 4.58 or increased by 358%

Conclusion: Construction costs in Las Vegas are higher than National average costs and increased at a greater rate from 1975 to 2011 than the National Average.

Historical Cost Indexes

Year	National	Alabama					Alaska	Arizona		Arkansas		California				
	30 City Average	Birmingham	Huntsville	Mobile	Montgomery	Tuscaloosa	Anchorage	Phoenix	Tuscon	Fort Smith	Little Rock	Anaheim	Bakersfield	Fresno	Los Angeles	Oxnard
Jan 2011	185.0E	161.8E	155.1E	155.3E	146.4E	146.9E	224.8E	163.6E	158.7E	151.7E	153.7E	196.3E	195.0E	199.3E	198.3E	197.0E
2010	181.6	159.6	152.9	153.1	144.4	144.9	218.3	160.7	157.3	150.3	152.5	192.8	190.8	195.0	194.9	192.8
2009	182.5	162.7	157.2	155.6	148.8	149.4	222.9	161.3	156.8	149.2	156.3	194.5	192.3	195.0	196.6	194.5
2008	171.0	150.3	146.9	143.7	138.4	138.8	210.8	152.2	148.4	138.6	145.4	182.7	179.8	183.2	184.7	182.6
2007	165.0	146.9	143.3	140.3	134.8	135.4	206.8	147.7	143.1	134.8	141.7	175.1	173.7	176.7	177.5	176.2
2006	156.2	135.7	133.6	126.7	124.0	122.2	196.4	137.9	134.8	123.2	127.2	166.8	164.9	169.8	167.3	167.4
2005	146.7	127.9	125.4	119.3	116.6	114.6	185.6	128.5	124.1	115.4	119.4	156.5	153.0	157.9	157.1	156.4
2004	132.8	115.9	112.9	107.0	104.9	102.9	167.0	116.5	113.6	103.8	108.6	142.1	139.5	143.8	142.0	140.9
2003	129.7	113.1	110.7	104.8	102.6	100.8	163.5	113.9	110.6	101.8	105.8	139.4	137.2	142.1	139.6	138.7
2002	126.7	110.0	103.4	103.6	101.7	99.7	159.5	113.3	110.2	100.4	102.1	136.5	133.4	136.0	136.4	136.3
2001	122.2	106.0	100.5	100.6	98.3	95.9	152.6	109.0	106.4	97.3	98.7	132.5	129.3	132.8	132.4	132.4
2000	118.9	104.1	98.9	99.1	94.2	94.6	148.3	106.9	104.9	94.4	95.7	129.4	125.2	129.4	129.9	129.7
1999	116.6	101.2	97.4	97.7	92.5	92.8	145.9	105.5	103.3	93.1	94.4	127.9	123.6	126.9	128.7	128.2
1998	113.6	96.2	94.0	94.8	90.3	89.8	143.8	102.1	101.0	90.6	91.4	125.2	120.3	123.9	125.8	124.7
1997	111.5	94.6	92.4	93.3	88.8	88.3	142.0	101.8	100.7	89.3	90.1	124.0	119.1	122.3	124.6	123.5
1996	108.9	90.9	91.4	92.3	87.5	86.5	140.4	98.3	97.4	86.5	86.7	121.7	117.2	120.2	122.4	121.5
1995	105.6	87.8	88.0	88.8	84.5	83.5	138.0	96.1	95.5	85.0	85.6	120.1	115.7	117.5	120.9	120.0
1994	103.0	85.9	86.1	86.8	82.6	81.7	134.0	93.7	93.1	82.4	83.9	118.1	113.6	114.5	119.0	117.5
1993	100.0	82.8	82.7	86.1	82.1	78.7	132.0	90.9	90.9	80.9	82.4	115.0	111.3	112.9	115.6	115.4
1992	97.9	81.7	81.5	84.9	80.9	77.6	128.6	88.8	89.4	79.7	81.2	113.5	108.1	110.3	113.7	113.7
1991	95.7	80.5	78.9	83.9	79.8	76.6	127.4	88.1	88.5	78.3	80.0	111.0	105.8	108.2	110.9	111.4
1990	93.2	79.4	77.6	82.7	78.4	75.2	125.8	86.4	87.0	77.1	77.9	107.7	102.7	103.3	107.5	107.4
1985	81.8	71.1	70.5	72.7	70.9	67.9	116.0	78.1	77.7	69.6	71.0	95.4	92.2	92.6	94.6	96.6
1980	60.7	55.2	54.1	56.8	56.8	54.0	91.4	63.7	62.4	53.1	55.8	68.7	69.4	68.7	67.4	69.9
1975	43.7	40.0	40.9	41.8	40.1	37.8	57.3	44.5	45.2	39.0	38.7	47.6	46.6	47.7	48.3	47.0
1970	27.8	24.1	25.1	25.8	25.4	24.2	43.0	27.2	28.5	24.3	22.3	31.0	30.8	31.1	29.0	31.0
1965	21.5	19.6	19.3	19.6	19.5	18.7	34.9	21.8	22.0	18.7	18.5	23.9	23.7	24.0	22.7	23.9
1960	19.5	17.9	17.5	17.8	17.7	16.9	31.7	19.9	20.0	17.0	16.8	21.7	21.5	21.8	20.6	21.7
1955	16.3	14.8	14.7	14.9	14.9	14.2	26.6	16.7	16.7	14.3	14.5	18.2	18.1	18.3	17.3	18.2
1950	13.5	12.2	12.1	12.3	12.3	11.7	21.9	13.8	13.8	11.8	11.6	15.1	15.0	15.1	14.3	15.1
1945	8.6	7.8	7.7	7.8	7.8	7.5	14.0	8.8	8.8	7.5	7.4	9.6	9.5	9.6	9.1	9.6
▼ 1940	6.6	6.0	6.0	6.1	6.0	5.8	10.8	6.8	6.8	5.8	5.7	7.4	7.4	7.4	7.0	7.4

Year	National	California							Colorado			Connecticut				
	30 City Average	Riverside	Sacramento	San Diego	San Francisco	Santa Barbara	Stockton	Vallejo	Colorado Springs	Denver	Pueblo	Bridge-Port	Bristol	Hartford	New Britain	New Haven
Jan 2011	185.0E	195.9E	202.4E	192.3E	228.5E	196.5E	201.2E	210.9E	171.7E	173.9E	169.4E	200.8E	201.1E	201.8E	200.8E	202.6E
2010	181.6	192.7	196.7	188.2	223.0	192.9	195.9	204.6	170.5	172.6	168.1	198.8	199.1	199.8	198.8	200.7
2009	182.5	193.1	198.1	191.6	224.9	193.6	194.5	202.7	168.4	172.0	166.7	199.0	197.7	198.7	197.4	199.3
2008	171.0	181.3	186.0	179.9	210.6	181.5	183.2	191.6	158.8	161.9	157.9	185.8	184.8	185.7	184.5	186.1
2007	165.0	174.7	179.1	173.6	201.1	175.2	178.2	185.9	152.6	155.9	152.3	179.6	178.3	179.6	178.0	179.5
2006	156.2	166.0	172.2	164.0	191.2	166.4	171.0	177.9	146.1	149.2	145.0	169.5	168.0	169.5	167.7	169.8
2005	146.7	155.4	161.1	153.8	179.7	155.7	160.0	167.2	138.3	141.1	136.4	160.8	159.4	159.6	159.1	160.8
2004	132.8	141.0	147.2	139.0	163.6	141.4	144.2	148.6	125.6	127.2	123.5	143.6	142.9	142.4	142.7	144.3
2003	129.7	138.7	144.9	136.6	162.1	139.2	141.8	146.4	123.1	123.7	120.7	141.3	140.2	140.3	140.0	141.6
2002	126.7	135.3	138.4	134.2	157.9	135.9	136.9	143.8	118.7	121.3	116.7	133.8	133.6	133.1	133.3	134.8
2001	122.2	131.4	135.5	129.7	151.8	131.5	134.1	140.2	113.3	117.2	113.1	128.6	128.5	128.5	128.3	128.6
2000	118.9	128.0	131.5	127.1	146.9	128.7	130.7	137.1	109.8	111.8	108.8	122.7	122.6	122.9	122.4	122.7
1999	116.6	126.5	129.4	124.9	145.1	127.2	127.9	135.3	107.0	109.1	107.1	121.1	121.3	121.2	121.1	121.3
1998	113.6	123.7	125.9	121.3	141.9	123.7	124.7	132.5	103.3	106.5	103.7	119.1	119.4	120.0	119.7	120.0
1997	111.5	122.6	124.7	120.3	139.2	122.4	123.3	130.6	101.1	104.4	102.0	119.2	119.5	119.9	119.7	120.0
1996	108.9	120.6	122.4	118.4	136.8	120.5	121.4	128.1	98.5	101.4	99.7	117.4	117.6	117.9	117.8	118.1
1995	105.6	119.2	119.5	115.4	133.8	119.0	119.0	122.5	96.1	98.9	96.8	116.0	116.5	116.9	116.3	116.5
1994	103.0	116.6	114.8	113.4	131.4	116.4	115.5	119.7	94.2	95.9	94.6	114.3	115.0	115.3	114.7	114.8
1993	100.0	114.3	112.5	111.3	129.6	114.3	115.2	117.5	92.1	93.8	92.6	108.7	107.8	108.4	106.3	106.1
1992	97.9	111.8	110.8	109.5	127.9	112.2	112.7	115.2	90.6	91.5	90.9	106.7	106.0	106.6	104.5	104.4
1991	95.7	109.4	108.5	107.7	125.8	110.0	111.0	113.4	88.9	90.4	89.4	97.6	97.3	98.0	97.3	97.9
1990	93.2	107.0	104.9	105.6	121.8	106.4	105.4	111.4	86.9	88.8	88.8	96.3	95.9	96.6	95.9	96.5
1985	81.8	94.8	92.2	94.2	106.2	93.2	93.7	97.0	79.6	81.0	80.4	86.1	86.2	87.2	86.1	86.3
1980	60.7	68.7	71.3	68.1	75.2	71.1	71.2	71.9	60.7	60.9	59.5	61.5	60.7	61.9	60.7	61.4
1975	43.7	47.3	49.1	47.7	49.8	46.1	47.6	46.5	43.1	42.7	42.5	45.2	45.5	46.0	45.2	46.0
1970	27.8	31.0	32.1	30.7	31.6	31.3	31.8	31.8	27.6	26.1	27.3	29.2	28.2	29.6	28.2	29.3
1965	21.5	23.9	24.8	24.0	23.7	24.1	24.5	24.5	21.3	20.9	21.0	22.4	21.7	22.6	21.7	23.2
1960	19.5	21.7	22.5	21.7	21.5	21.9	22.3	22.2	19.3	19.0	19.1	20.0	19.8	20.0	19.8	20.0
1955	16.3	18.2	18.9	18.2	18.0	18.4	18.7	18.6	16.2	15.9	16.0	16.8	16.6	16.8	16.6	16.8
1950	13.5	15.0	15.6	15.0	14.9	15.2	15.4	15.4	13.4	13.2	13.2	13.9	13.7	13.8	13.7	13.9
1945	8.6	9.6	10.0	9.6	9.5	9.7	9.9	9.8	8.5	8.4	8.4	8.8	8.7	8.8	8.7	8.9
▼ 1940	6.6	7.4	7.7	7.4	7.3	7.5	7.6	7.6	6.6	6.5	6.5	6.8	6.8	6.8	6.8	6.8

Historical Cost Indexes

Year	National 30 City Average	Connecticut Norwalk	Connecticut Stamford	Connecticut Waterbury	Delaware Wilmington	D.C. Washington	Florida Fort Lauderdale	Florida Jacksonville	Florida Miami	Florida Orlando	Florida Tallahassee	Florida Tampa	Georgia Albany	Georgia Atlanta	Georgia Columbus	Georgia Macon
Jan 2011	185.0E	200.3E	205.9E	201.4E	191.6E	181.2E	162.9E	157.7E	165.9E	164.9E	150.3E	170.4E	150.3E	162.9E	152.9E	151.5E
2010	181.6	198.3	203.8	199.4	187.7	179.0	160.3	155.4	163.3	162.5	148.1	167.5	148.3	160.1	150.9	149.4
2009	182.5	198.8	204.3	198.4	188.9	181.4	160.7	152.0	165.2	163.9	145.1	165.1	152.0	164.5	155.6	153.5
2008	171.0	185.2	189.2	185.3	177.5	170.4	149.6	142.5	152.9	153.0	135.3	155.6	140.4	153.2	143.9	142.5
2007	165.0	178.6	183.3	179.1	173.4	163.1	145.6	139.0	148.9	148.1	131.7	152.0	135.8	148.0	140.1	138.1
2006	156.2	169.7	173.7	169.1	158.3	153.0	136.0	126.5	136.7	134.3	118.4	136.6	123.6	139.9	126.3	124.3
2005	146.7	160.9	164.1	160.5	149.6	142.7	126.0	119.2	127.1	126.0	110.7	127.7	115.7	131.7	118.9	116.1
2004	132.8	143.6	146.5	143.2	136.1	126.8	114.8	107.8	115.6	113.2	100.6	116.6	102.8	119.6	102.0	105.0
2003	129.7	140.6	145.1	140.8	133.0	124.6	109.0	105.6	110.5	107.8	98.0	103.5	100.5	115.8	99.1	102.7
2002	126.7	133.5	136.2	133.9	129.4	120.3	107.1	104.6	107.4	106.7	97.3	102.8	99.7	113.0	98.3	101.9
2001	122.2	128.2	131.5	128.8	124.8	115.9	104.3	100.8	104.6	103.8	94.8	100.3	96.4	109.1	95.5	99.0
2000	118.9	121.5	126.4	123.2	117.4	113.8	102.4	99.0	101.8	101.2	93.6	98.9	94.9	106.1	94.1	97.0
1999	116.6	120.0	122.0	121.2	116.6	111.5	101.4	98.0	100.8	100.1	92.5	97.9	93.5	102.9	92.8	95.8
1998	113.6	118.8	120.8	120.1	112.3	109.6	99.4	96.2	99.0	98.4	90.9	96.3	91.4	100.8	90.4	93.3
1997	111.5	118.9	120.9	120.2	110.7	106.4	98.6	95.4	98.2	97.2	89.9	95.5	89.9	98.5	88.7	91.6
1996	108.9	117.0	119.0	118.4	108.6	105.4	95.9	92.6	95.9	95.1	88.0	93.6	86.6	94.3	83.9	88.3
1995	105.6	115.5	117.9	117.3	106.1	102.3	94.0	90.9	93.7	93.5	86.5	92.2	85.0	92.0	82.6	86.8
1994	103.0	113.9	116.4	115.8	105.0	99.6	92.2	88.9	91.8	91.5	84.5	90.2	82.3	89.6	80.6	83.7
1993	100.0	108.8	110.6	104.8	101.5	96.3	87.4	86.1	87.1	88.5	82.1	87.7	79.5	85.7	77.8	80.9
1992	97.9	107.2	109.0	103.1	100.3	94.7	85.7	84.0	85.3	87.1	80.8	86.2	78.2	84.3	76.5	79.6
1991	95.7	100.6	103.2	96.5	94.5	92.9	85.1	82.8	85.2	85.5	79.7	86.3	76.3	82.6	75.4	78.4
1990	93.2	96.3	98.9	95.1	92.5	90.4	83.9	81.1	84.0	82.9	78.4	85.0	75.0	80.4	74.0	76.9
1985	81.8	85.3	86.8	85.6	81.1	78.8	76.7	73.4	78.3	73.9	70.6	77.3	66.9	70.3	66.1	68.1
1980	60.7	60.7	60.9	62.3	58.5	59.6	55.3	55.8	56.5	56.7	53.5	57.2	51.7	54.0	51.1	51.3
1975	43.7	44.7	45.0	46.3	42.9	43.7	42.1	40.3	43.2	41.5	38.1	41.3	37.5	38.4	36.2	36.5
1970	27.8	28.1	28.2	28.9	27.0	26.3	25.7	22.8	27.0	26.2	24.5	24.2	23.8	25.2	22.8	23.4
1965	21.5	21.6	21.7	22.2	20.9	21.8	19.8	17.4	19.3	20.2	18.9	18.6	18.3	19.8	17.6	18.0
1960	19.5	19.7	19.7	20.2	18.9	19.4	18.0	15.8	17.6	18.3	17.2	16.9	16.7	17.1	16.0	16.4
1955	16.3	16.5	16.5	17.0	15.9	16.3	15.1	13.2	14.7	15.4	14.4	14.1	14.0	14.4	13.4	13.7
1950	13.5	13.6	13.7	14.0	13.1	13.4	12.5	11.0	12.2	12.7	11.9	11.7	11.5	11.9	11.0	11.3
1945	8.6	8.7	8.7	8.9	8.4	8.6	7.9	7.0	7.8	8.1	7.6	7.5	7.4	7.6	7.0	7.2
▼ 1940	6.6	6.7	6.7	6.9	6.4	6.6	6.1	5.4	6.0	6.2	5.8	5.7	5.7	5.8	5.5	5.6

Year	National 30 City Average	Georgia Savannah	Hawaii Honolulu	Idaho Boise	Idaho Pocatello	Illinois Chicago	Illinois Decatur	Illinois Joliet	Illinois Peoria	Illinois Rockford	Illinois Springfield	Indiana Anderson	Indiana Evansville	Indiana Fort Wayne	Indiana Gary	Indiana Indianapolis
Jan 2011	185.0E	150.2E	217.4E	162.4E	163.1E	216.8E	186.3E	215.0E	190.2E	201.8E	187.6E	165.9E	169.2E	163.6E	188.4E	171.5E
2010	181.6	147.9	214.6	161.7	162.4	210.6	181.2	208.3	186.2	197.6	182.2	162.8	165.9	160.0	183.7	168.1
2009	182.5	152.8	218.6	162.6	163.4	209.0	182.2	205.9	186.2	196.5	184.0	164.9	166.9	163.0	185.2	170.3
2008	171.0	141.1	205.5	153.2	152.8	195.7	168.7	185.5	172.2	180.2	168.5	152.4	155.8	151.0	169.0	159.2
2007	165.0	136.9	200.4	148.0	148.1	186.5	160.5	177.7	166.0	175.0	159.2	149.0	152.6	147.7	165.4	154.1
2006	156.2	124.0	191.9	141.6	141.2	177.8	153.8	168.7	155.2	163.3	152.8	139.8	144.4	139.1	154.5	144.6
2005	146.7	116.9	181.2	133.8	132.5	164.4	145.0	158.3	146.5	151.2	145.4	131.4	136.2	131.0	146.4	137.4
2004	132.8	105.6	161.2	122.0	120.5	149.3	129.3	145.1	133.9	138.1	130.3	120.9	123.5	120.1	132.2	125.0
2003	129.7	103.0	159.4	120.2	118.7	146.2	127.4	143.5	132.4	137.3	128.4	119.7	121.4	118.2	131.4	122.5
2002	126.7	102.0	157.2	118.3	117.1	141.2	123.8	138.5	129.6	132.9	125.3	117.5	119.0	116.4	129.1	120.5
2001	122.2	99.0	150.0	114.3	113.4	135.8	120.1	133.7	124.3	127.8	119.8	113.4	115.6	112.1	123.4	116.4
2000	118.9	97.5	144.8	112.9	112.1	131.2	115.1	124.6	119.0	122.2	116.2	109.8	111.5	108.4	117.8	113.2
1999	116.6	96.0	143.0	110.2	109.6	129.6	113.0	122.6	116.4	120.7	113.8	107.1	109.3	106.8	112.8	110.6
1998	113.6	93.7	140.4	107.4	107.0	125.2	110.1	119.8	113.8	115.5	111.1	105.0	107.2	104.5	111.1	108.0
1997	111.5	92.0	139.8	104.6	104.7	121.3	107.8	117.5	111.5	113.1	108.9	101.9	104.4	101.8	110.3	105.2
1996	108.9	88.6	134.5	102.2	102.1	118.8	106.6	116.2	109.3	111.5	106.5	100.0	102.1	99.9	107.5	102.7
1995	105.6	87.4	130.3	99.5	98.2	114.2	98.5	110.5	102.3	103.6	98.1	96.4	97.2	95.0	100.7	100.1
1994	103.0	85.3	124.0	94.8	95.0	111.3	97.3	108.9	100.9	102.2	97.0	93.6	95.8	93.6	99.1	97.1
1993	100.0	82.0	122.0	92.2	92.1	107.6	95.6	106.8	98.9	99.6	95.2	91.2	94.3	91.5	96.7	93.9
1992	97.9	80.8	120.0	91.0	91.0	104.3	94.4	104.2	97.3	98.2	94.0	89.5	92.9	89.9	95.0	91.5
1991	95.7	79.5	106.1	89.5	89.4	100.9	92.3	100.0	95.9	95.8	91.5	87.8	91.4	88.3	93.3	89.1
1990	93.2	77.9	104.7	88.2	88.1	98.4	90.9	98.4	93.7	94.0	90.1	84.6	89.3	83.4	88.4	87.1
1985	81.8	68.9	94.7	78.0	78.0	82.4	81.9	83.4	83.7	83.0	81.5	75.2	79.8	75.0	77.8	77.1
1980	60.7	52.2	68.9	60.3	59.5	62.8	62.3	63.4	64.5	61.6	61.1	56.5	59.0	56.7	59.8	57.9
1975	43.7	36.9	44.6	40.8	40.5	45.7	43.1	44.5	44.7	42.7	42.4	39.5	41.7	39.9	41.9	40.6
1970	27.8	21.0	30.4	26.7	26.6	29.1	28.0	28.6	29.0	27.5	27.6	25.4	26.4	25.5	27.1	26.2
1965	21.5	16.4	21.8	20.6	20.5	22.7	21.5	22.1	22.4	21.2	21.3	19.5	20.4	19.7	20.8	20.7
1960	19.5	14.9	19.8	18.7	18.6	20.2	19.6	20.0	20.3	19.2	19.3	17.7	18.7	17.9	18.9	18.4
1955	16.3	12.5	16.6	15.7	15.6	16.9	16.4	16.8	17.0	16.1	16.2	14.9	15.7	15.0	15.9	15.5
1950	13.5	10.3	13.7	13.0	12.9	14.0	13.6	13.9	14.0	13.3	13.4	12.3	12.9	12.4	13.1	12.8
1945	8.6	6.6	8.8	8.3	8.2	8.9	8.6	8.9	9.0	8.5	8.6	7.8	8.3	7.9	8.4	8.1
▼ 1940	6.6	5.1	6.8	6.4	6.3	6.9	6.7	6.8	6.9	6.5	6.6	6.0	6.4	6.1	6.5	6.3

Historical Cost Indexes

Year	National	Indiana			Iowa					Kansas		Kentucky		Louisiana		
	30 City Average	Muncie	South Bend	Terre Haute	Cedar Rapids	Davenport	Des Moines	Sioux City	Waterloo	Topeka	Wichita	Lexington	Louisville	Baton Rouge	Lake Charles	New Orleans
Jan 2011	185.0E	166.6E	166.3E	170.6E	173.6E	177.7E	172.5E	161.9E	162.4E	155.3E	154.3E	160.8E	168.9E	156.4E	156.0E	161.6E
2010	181.6	163.1	163.5	167.8	170.4	175.6	170.2	159.0	160.2	152.8	151.3	158.7	166.9	153.7	153.4	160.3
2009	182.5	164.5	167.1	168.0	165.9	172.0	162.1	156.3	147.6	154.5	152.4	160.7	167.5	156.9	154.4	161.9
2008	171.0	153.0	153.1	155.9	157.3	163.4	152.4	147.8	139.2	144.9	143.1	152.0	156.8	144.0	141.5	149.3
2007	165.0	149.7	149.5	152.7	152.4	157.6	148.9	144.0	135.0	141.0	138.7	147.6	150.9	139.4	138.0	145.2
2006	156.2	141.2	141.3	144.0	145.8	151.1	142.9	137.3	128.4	134.7	132.9	129.0	143.0	129.4	129.8	135.8
2005	146.7	131.0	133.8	135.2	135.7	142.1	133.5	129.4	120.0	125.9	125.6	121.5	135.1	120.8	117.7	126.2
2004	132.8	120.0	120.6	123.0	122.7	128.6	121.6	116.2	108.3	112.6	113.4	110.1	120.3	105.6	109.0	115.3
2003	129.7	118.7	119.2	121.7	120.9	126.4	120.2	114.6	105.9	109.4	111.3	107.8	118.7	103.3	106.1	112.4
2002	126.7	116.7	117.1	119.9	116.0	122.1	116.7	111.4	103.7	107.8	109.6	106.2	116.4	102.5	105.0	110.6
2001	122.2	112.8	111.6	115.5	112.5	117.5	113.2	107.7	100.6	104.3	105.2	103.3	112.7	99.4	101.3	104.6
2000	118.9	109.1	106.9	110.8	108.8	112.6	108.9	99.0	98.1	101.0	101.1	101.4	109.3	97.8	99.7	102.1
1999	116.6	105.7	103.7	107.9	104.1	109.2	107.6	96.7	96.4	98.7	99.3	99.7	106.6	96.1	97.6	99.5
1998	113.6	103.6	102.3	106.0	102.5	106.8	103.8	95.1	94.8	97.4	97.4	97.4	101.5	94.7	96.1	97.7
1997	111.5	101.3	101.2	104.6	101.2	105.6	102.5	93.8	93.5	96.3	96.2	96.1	99.9	93.3	96.8	96.2
1996	108.9	99.4	99.1	102.1	99.1	102.0	99.1	90.9	90.7	93.6	93.8	94.4	98.3	91.5	94.9	94.3
1995	105.6	95.6	94.6	97.2	95.4	96.6	94.4	88.2	88.9	91.1	91.0	91.6	94.6	89.6	92.7	91.6
1994	103.0	93.2	93.0	95.8	93.0	92.3	92.4	85.9	86.8	89.4	88.1	89.6	92.2	87.7	89.9	89.5
1993	100.0	91.0	90.7	94.4	91.1	90.5	90.7	84.1	85.2	87.4	86.2	87.3	89.4	86.4	88.5	87.8
1992	97.9	89.4	89.3	93.1	89.8	89.3	89.4	82.9	83.8	86.2	85.0	85.8	88.0	85.2	87.3	86.6
1991	95.7	87.7	87.5	91.7	88.5	88.0	88.4	81.8	81.6	84.4	83.9	84.1	84.4	83.2	85.3	85.8
1990	93.2	83.9	85.1	89.1	87.1	86.5	86.9	80.4	80.3	83.2	82.6	82.8	82.6	82.0	84.1	84.5
1985	81.8	75.2	76.0	79.2	75.9	77.6	77.1	72.1	72.3	75.0	74.7	75.5	74.5	74.9	78.5	78.2
1980	60.7	56.1	58.3	59.7	62.7	59.6	61.8	59.3	57.7	58.9	58.0	59.3	59.8	59.1	60.0	57.2
1975	43.7	39.2	40.3	41.9	43.1	41.0	43.1	41.0	40.0	42.0	43.1	42.7	42.5	40.4	40.6	41.5
1970	27.8	25.3	26.2	27.0	28.2	26.9	27.6	26.6	25.9	27.0	25.5	26.9	25.9	25.0	26.7	27.2
1965	21.5	19.5	20.2	20.9	21.7	20.8	21.7	20.5	20.0	20.8	19.6	20.7	20.3	19.4	20.6	20.4
1960	19.5	17.8	18.3	18.9	19.7	18.8	19.5	18.6	18.2	18.9	17.8	18.8	18.4	17.6	18.7	18.5
1955	16.3	14.9	15.4	15.9	16.6	15.8	16.3	15.6	15.2	15.8	15.0	15.8	15.4	14.8	15.7	15.6
1950	13.5	12.3	12.7	13.1	13.7	13.1	13.5	12.9	12.6	13.1	12.4	13.0	12.8	12.2	13.0	12.8
1945	8.6	7.8	8.1	8.4	8.7	8.3	8.6	8.2	8.0	8.3	7.9	8.3	8.1	7.8	8.3	8.2
1940	6.6	6.0	6.3	6.5	6.7	6.4	6.6	6.3	6.2	6.4	6.1	6.4	6.3	6.0	6.4	6.3

Year	National	Louisiana	Maine		Maryland	Massachusetts									Michigan	
	30 City Average	Shreveport	Lewiston	Portland	Baltimore	Boston	Brockton	Fall River	Lawrence	Lowell	New Bedford	Pittsfield	Springfield	Worcester	Ann Arbor	Dearborn
Jan 2011	185.0E	146.4E	165.7E	167.4E	171.3E	217.8E	206.2E	204.2E	208.6E	208.3E	203.4E	189.0E	192.9E	203.8E	186.4E	189.1E
2010	181.6	145.1	162.1	163.7	168.3	214.3	202.7	200.7	204.9	204.7	200.0	185.9	189.4	201.1	183.3	186.8
2009	182.5	147.6	159.4	161.8	169.1	211.8	199.2	197.6	201.9	201.9	196.9	185.1	187.1	198.2	179.2	186.9
2008	171.0	135.9	149.1	150.9	157.7	198.6	186.4	185.1	188.7	189.1	184.5	171.3	173.7	184.5	169.6	176.4
2007	165.0	132.4	146.4	148.2	152.5	191.8	180.4	179.7	182.0	181.7	179.1	166.8	169.1	180.4	166.8	172.5
2006	156.2	125.2	140.2	139.8	144.9	180.4	171.6	170.4	172.3	173.5	170.4	157.3	159.7	171.6	161.1	165.5
2005	146.7	117.4	131.7	131.4	135.8	169.6	161.3	159.6	162.2	162.3	159.6	147.6	150.7	158.8	147.5	155.8
2004	132.8	105.7	119.8	119.4	121.4	154.1	144.4	143.6	146.4	146.7	143.6	131.9	136.5	143.2	135.9	142.1
2003	129.7	103.9	117.7	117.3	118.1	150.2	139.6	140.6	143.3	143.6	140.5	128.9	133.8	139.0	134.1	139.0
2002	126.7	102.1	117.5	117.1	115.6	145.6	136.0	135.0	136.9	137.3	134.9	124.7	128.3	134.9	131.4	134.3
2001	122.2	98.3	114.6	114.3	111.7	140.9	132.1	131.4	133.1	132.9	131.3	120.3	124.5	130.1	126.8	129.8
2000	118.9	96.2	105.7	105.4	107.7	138.9	129.4	128.4	129.7	130.3	128.3	116.7	121.0	127.3	124.3	125.8
1999	116.6	94.8	105.0	104.7	106.4	136.2	126.7	126.3	127.3	127.4	126.2	115.0	118.8	123.8	117.5	122.7
1998	113.6	92.0	102.8	102.5	104.1	132.8	125.0	124.7	125.0	125.3	124.6	114.2	117.1	122.6	116.0	119.7
1997	111.5	90.1	101.7	101.4	102.2	132.1	124.4	124.7	125.1	125.6	124.7	114.6	117.4	122.9	113.5	117.9
1996	108.9	88.5	99.5	99.2	99.6	128.7	121.6	122.2	121.7	122.1	122.2	112.7	115.0	120.2	112.9	116.2
1995	105.6	86.7	96.8	96.5	96.1	128.6	119.6	117.7	120.5	119.6	117.2	110.7	112.7	114.8	106.1	110.5
1994	103.0	85.0	95.2	94.9	94.4	124.9	114.2	113.7	116.8	116.9	111.6	109.2	111.1	112.9	105.0	109.0
1993	100.0	83.6	93.0	93.0	93.1	121.1	111.7	111.5	114.9	114.3	109.4	107.1	108.9	110.3	102.8	105.8
1992	97.9	82.3	91.7	91.7	90.9	118.0	110.0	109.8	110.9	110.2	107.7	105.7	107.2	108.6	101.5	103.2
1991	95.7	81.6	89.8	89.9	89.1	115.2	105.9	104.7	107.3	105.5	105.1	100.6	103.2	105.7	95.1	98.3
1990	93.2	80.3	88.5	88.6	85.6	110.9	103.7	102.9	105.2	102.8	102.6	98.7	101.4	103.2	93.2	96.5
1985	81.8	73.6	76.7	77.0	72.7	92.8	88.8	88.7	89.4	88.2	88.6	85.0	85.6	86.7	80.1	82.7
1980	60.7	58.7	57.3	58.5	53.6	64.0	63.7	64.1	63.4	62.7	63.1	61.8	62.0	62.3	62.9	64.0
1975	43.7	40.5	41.9	42.1	39.8	46.6	45.8	45.7	46.2	45.7	46.1	45.5	45.8	46.0	44.1	44.9
1970	27.8	26.4	26.5	25.8	25.1	29.2	29.1	29.2	29.1	28.9	29.0	28.6	28.5	28.7	28.5	28.9
1965	21.5	20.3	20.4	19.4	20.2	23.0	22.5	22.5	22.5	22.2	22.4	22.0	22.4	22.1	22.0	22.3
1960	19.5	18.5	18.6	17.6	17.5	20.5	20.4	20.4	20.4	20.2	20.3	20.0	20.1	20.1	20.0	20.2
1955	16.3	15.5	15.6	14.7	14.7	17.2	17.1	17.1	17.1	16.9	17.1	16.8	16.9	16.8	16.7	16.9
1950	13.5	12.8	12.9	12.2	12.1	14.2	14.1	14.1	14.1	14.0	14.1	13.8	14.0	13.9	13.8	14.0
1945	8.6	8.1	8.2	7.8	7.7	9.1	9.0	9.0	9.0	8.9	9.0	8.9	8.9	8.9	8.8	8.9
1940	6.6	6.3	6.3	6.0	6.0	7.0	6.9	7.0	7.0	6.9	6.9	6.8	6.9	6.8	6.8	6.9

Historical Cost Indexes

Year	National 30 City Average	Michigan: Detroit	Flint	Grand Rapids	Kala-mazoo	Lansing	Sagi-naw	Minnesota: Duluth	Minne-apolis	Roches-ter	Mississippi: Biloxi	Jackson	Missouri: Kansas City	St. Joseph	St. Louis	Spring-field
Jan 2011	185.0E	190.1E	178.2E	163.0E	171.4E	177.7E	174.4E	195.7E	208.0E	193.6E	150.1E	153.0E	189.9E	177.5E	189.8E	167.3E
2010	181.6	187.5	176.2	160.6	167.3	175.8	171.9	193.1	203.8	188.9	148.0	151.0	186.1	174.2	185.9	164.3
2009	182.5	187.8	176.0	156.7	166.3	173.7	168.6	191.8	203.1	188.0	152.8	156.6	185.6	172.3	187.2	163.7
2008	171.0	177.0	164.4	140.2	155.7	161.6	158.7	177.4	190.6	175.0	141.8	147.1	175.5	164.2	176.2	152.9
2007	165.0	172.9	161.7	137.2	152.6	158.7	156.1	174.2	184.5	171.4	137.7	131.2	169.0	157.8	170.6	145.3
2006	156.2	165.8	153.8	131.1	146.0	152.5	150.6	166.2	173.9	161.9	123.3	117.1	162.0	152.4	159.3	139.4
2005	146.7	156.2	145.1	123.1	134.9	142.0	140.6	157.3	164.6	152.8	116.0	109.9	151.3	143.1	149.4	131.7
2004	132.8	141.9	129.6	112.8	122.5	129.7	127.6	139.1	150.1	137.0	105.3	99.3	135.1	126.9	135.5	116.3
2003	129.7	138.7	127.9	110.9	120.8	127.7	126.0	136.7	146.5	134.5	101.4	96.7	131.9	124.8	133.3	113.9
2002	126.7	134.2	126.9	107.4	119.9	125.3	124.4	131.9	139.5	130.0	101.0	96.3	128.4	122.7	129.6	112.5
2001	122.2	129.4	122.4	104.3	115.2	120.1	119.7	131.4	136.1	124.9	98.7	93.9	121.8	114.7	125.5	106.7
2000	118.9	125.3	119.8	102.7	111.6	117.6	115.9	124.1	131.1	120.1	97.2	92.9	118.2	111.9	122.4	104.3
1999	116.6	122.6	113.7	100.9	106.1	111.5	110.1	120.3	126.5	117.1	95.7	91.8	114.9	106.1	119.8	101.1
1998	113.6	119.5	112.3	99.7	104.9	110.1	108.7	117.7	124.6	115.5	92.1	89.5	108.2	104.2	115.9	98.9
1997	111.5	117.6	110.8	98.9	104.2	108.7	107.5	115.9	121.9	114.1	90.9	88.3	106.4	102.4	113.2	97.3
1996	108.9	116.0	109.7	93.9	103.5	107.8	106.9	115.0	120.4	113.5	87.1	85.4	103.2	99.9	110.1	94.6
1995	105.6	110.1	104.1	91.1	96.4	98.3	102.3	100.3	111.9	102.5	84.2	83.5	99.7	96.1	106.3	89.5
1994	103.0	108.5	102.9	89.7	95.0	97.1	101.1	99.1	109.3	101.1	82.4	81.7	97.3	92.8	103.2	88.2
1993	100.0	105.4	100.7	87.7	92.7	95.3	98.8	99.8	106.7	99.9	80.0	79.4	94.5	90.9	99.9	86.2
1992	97.9	102.9	99.4	86.4	91.4	94.0	97.4	98.5	105.3	98.6	78.8	78.2	92.9	89.5	98.6	84.9
1991	95.7	97.6	93.6	85.3	90.0	92.1	90.3	96.0	102.9	97.3	77.6	76.9	90.9	88.1	96.4	83.9
1990	93.2	96.0	91.7	84.0	87.4	90.2	88.9	94.3	100.5	95.6	76.4	75.6	89.5	86.8	94.0	82.6
1985	81.8	81.6	80.7	75.8	79.4	80.0	80.5	85.2	87.9	86.5	69.4	68.5	78.2	77.3	80.8	73.2
1980	60.7	62.5	63.1	59.8	61.7	60.3	63.3	63.4	64.1	64.8	54.3	54.3	59.1	60.4	59.7	56.7
1975	43.7	45.8	44.9	41.8	43.7	44.1	44.8	45.2	46.0	45.3	37.9	37.7	42.7	45.4	44.8	41.1
1970	27.8	29.7	28.5	26.7	28.1	27.8	28.7	28.9	29.5	28.9	24.4	21.3	25.3	28.3	28.4	26.0
1965	21.5	22.1	21.9	20.6	21.7	21.5	22.2	22.3	23.4	22.3	18.8	16.4	20.6	21.8	21.8	20.0
1960	19.5	20.1	20.0	18.7	19.7	19.5	20.1	20.3	20.5	20.2	17.1	14.9	19.0	19.8	19.5	18.2
1955	16.3	16.8	16.7	15.7	16.5	16.3	16.9	17.0	17.2	17.0	14.3	12.5	16.0	16.6	16.3	15.2
1950	13.5	13.9	13.8	13.0	13.6	13.5	13.9	14.0	14.2	14.0	11.8	10.3	13.2	13.7	13.5	12.6
1945	8.6	8.9	8.8	8.3	8.7	8.6	8.9	8.9	9.0	8.9	7.6	6.6	8.4	8.8	8.6	8.0
▼ 1940	6.6	6.8	6.8	6.4	6.7	6.6	6.9	6.9	7.0	6.9	5.8	5.1	6.5	6.7	6.7	6.2

Year	National 30 City Average	Montana: Billings	Great Falls	Nebraska: Lincoln	Omaha	Nevada: Las Vegas	Reno	New Hampshire: Man-chester	Nashua	New Jersey: Camden	Jersey City	Newark	Pater-son	Trenton	NM: Albu-querque	NY: Albany
Jan 2011	185.0E	168.4E	170.4E	161.9E	168.8E	195.9E	176.8E	174.3E	173.7E	204.1E	205.2E	208.5E	207.5E	203.1E	163.4E	180.4E
2010	181.6	166.5	168.6	159.3	165.8	193.7	175.6	172.5	172.0	201.2	203.3	206.3	205.3	200.8	162.5	177.8
2009	182.5	165.3	166.0	161.7	165.0	191.5	176.5	174.0	173.5	196.8	200.5	203.6	202.0	198.6	163.0	178.1
2008	171.0	153.1	154.6	152.0	154.8	176.2	167.0	162.5	161.9	184.0	187.3	189.6	188.7	185.6	152.6	166.0
2007	165.0	147.7	147.8	147.7	150.3	166.7	161.4	159.3	158.7	179.3	183.3	185.2	184.6	181.7	146.7	159.4
2006	156.2	140.5	140.8	132.4	140.7	160.0	154.5	146.7	146.4	167.7	171.0	173.6	172.2	169.9	140.1	151.6
2005	146.7	131.9	132.4	124.1	132.8	149.5	145.4	136.8	136.5	159.0	162.4	163.9	163.0	161.5	130.4	142.2
2004	132.8	118.2	117.9	112.1	119.4	137.1	130.5	124.2	123.9	143.4	145.4	147.2	146.6	146.0	118.3	128.6
2003	129.7	115.8	115.8	110.2	117.3	133.8	128.3	122.4	122.2	142.0	144.3	145.7	145.0	143.0	116.4	126.8
2002	126.7	114.6	114.5	108.2	115.0	131.9	126.4	119.9	119.6	135.9	138.5	140.2	140.1	137.6	114.6	122.6
2001	122.2	117.5	117.7	101.3	111.6	127.8	122.5	116.2	116.2	133.4	136.7	136.9	136.8	136.0	111.4	119.2
2000	118.9	113.7	113.9	98.8	107.0	125.8	118.2	111.9	111.9	128.4	130.5	132.6	132.4	130.6	109.0	116.5
1999	116.6	112.1	112.7	96.6	104.8	121.9	114.4	109.6	109.6	125.3	128.8	131.6	129.5	129.6	106.7	114.6
1998	113.6	109.7	109.1	94.9	101.2	118.1	111.4	110.1	110.0	124.3	127.7	128.9	128.5	127.9	103.8	113.0
1997	111.5	107.4	107.6	93.3	99.5	114.6	109.8	108.6	108.6	121.9	125.1	126.4	126.4	125.2	100.8	110.0
1996	108.9	108.0	107.5	91.4	97.4	111.8	108.9	106.5	106.4	119.1	122.5	122.5	123.8	121.8	98.6	108.3
1995	105.6	104.7	104.9	85.6	93.4	108.5	105.0	100.9	100.8	107.0	112.2	111.9	112.1	111.2	96.3	103.6
1994	103.0	100.2	100.7	84.3	91.0	105.3	102.2	97.8	97.7	105.4	110.6	110.1	110.6	108.2	93.4	102.4
1993	100.0	97.9	97.8	82.3	88.7	102.8	99.9	95.4	95.4	103.7	109.0	108.2	109.0	106.4	90.1	99.8
1992	97.9	95.5	96.4	81.1	87.4	99.4	98.4	90.4	90.4	102.0	107.5	107.0	107.8	103.9	87.5	98.5
1991	95.7	94.2	95.1	80.1	86.4	97.5	95.8	87.8	87.8	95.0	98.5	95.6	100.0	96.6	86.2	96.1
1990	93.2	92.9	93.8	78.8	85.0	96.3	94.5	86.3	86.3	93.0	93.5	93.8	97.2	94.5	84.9	93.2
1985	81.8	83.9	84.3	71.5	77.5	87.6	85.0	78.1	78.1	81.3	83.3	83.5	84.5	82.8	76.7	79.5
1980	60.7	63.9	64.6	58.5	63.5	64.6	63.0	56.6	56.0	58.6	60.6	60.1	60.0	58.9	59.0	59.5
1975	43.7	43.1	43.8	40.9	43.2	42.8	41.9	41.3	40.8	42.3	43.4	44.2	43.9	43.8	40.3	43.9
1970	27.8	28.5	28.9	26.4	26.8	29.4	28.0	26.2	25.6	27.2	27.8	29.0	27.8	27.4	26.4	28.3
1965	21.5	22.0	22.3	20.3	20.6	22.4	21.6	20.6	19.7	20.9	21.4	23.8	21.4	21.3	20.6	22.3
1960	19.5	20.0	20.3	18.5	18.7	20.2	19.6	18.0	17.9	19.0	19.4	19.4	19.4	19.2	18.5	19.3
1955	16.3	16.7	17.0	15.5	15.7	16.9	16.4	15.1	15.0	16.0	16.3	16.3	16.3	16.1	15.6	16.2
1950	13.5	13.9	14.0	12.8	13.0	14.0	13.6	12.5	12.4	13.2	13.5	13.5	13.5	13.3	12.9	13.4
1945	8.6	8.8	9.0	8.1	8.3	8.9	8.7	8.0	7.9	8.4	8.6	8.6	8.6	8.5	8.2	8.5
▼ 1940	6.6	6.8	6.9	6.3	6.4	6.9	6.7	6.2	6.1	6.5	6.6	6.6	6.6	6.6	6.3	6.6

Historical Cost Indexes

Year	National 30 City Average	New York: Binghamton	Buffalo	New York	Rochester	Schenectady	Syracuse	Utica	Yonkers	North Carolina: Charlotte	Durham	Greensboro	Raleigh	Winston-Salem	N. Dakota: Fargo	Ohio: Akron
Jan 2011	185.0E	175.7E	187.9E	246.0E	181.7E	181.1E	179.0E	172.8E	218.4E	141.7E	142.8E	141.2E	141.8E	140.0E	157.9E	178.6E
2010	181.6	173.8	184.0	241.4	179.6	179.0	176.7	170.5	217.1	140.5	141.6	140.0	140.6	138.8	156.2	174.5
2009	182.5	172.9	184.4	239.9	179.6	178.8	176.9	171.3	217.5	145.1	145.8	144.0	145.1	142.7	154.3	175.7
2008	171.0	160.7	173.9	226.8	168.0	167.2	165.9	161.6	202.1	135.8	136.6	134.6	135.3	133.8	145.0	165.4
2007	165.0	155.7	168.6	215.2	163.5	159.8	159.8	155.0	196.5	132.8	133.7	131.7	132.1	131.0	139.8	159.0
2006	156.2	147.5	159.2	204.5	155.2	151.1	150.9	146.4	186.0	125.1	124.6	123.8	124.2	123.0	133.2	152.8
2005	146.7	137.5	149.4	194.0	147.3	142.2	141.9	136.9	176.1	110.5	112.1	112.1	112.1	111.1	125.3	144.4
2004	132.8	123.5	136.1	177.7	132.0	128.4	127.3	124.4	161.8	98.9	100.0	100.1	100.3	99.4	113.0	131.9
2003	129.7	121.8	132.8	173.4	130.3	126.7	124.7	121.7	160.0	96.2	97.5	97.5	97.8	96.8	110.6	129.6
2002	126.7	119.0	128.5	170.1	127.1	123.0	121.8	118.4	154.8	94.8	96.1	96.1	96.3	95.5	106.3	127.2
2001	122.2	116.0	125.2	164.4	123.1	120.1	118.6	115.3	151.4	91.5	92.9	92.9	93.3	92.3	103.1	123.5
2000	118.9	112.4	122.3	159.2	120.0	117.5	115.1	112.4	144.8	90.3	91.6	91.6	91.9	91.0	97.9	117.8
1999	116.6	108.6	120.2	155.9	116.8	114.7	113.7	108.5	140.6	89.3	90.2	90.3	90.5	90.1	96.8	116.0
1998	113.6	109.0	119.1	154.4	117.2	114.2	113.6	108.6	141.4	88.2	89.1	89.2	89.4	89.0	95.2	113.1
1997	111.5	107.2	115.7	150.3	115.2	111.2	110.6	107.0	138.8	86.8	87.7	87.8	87.9	87.6	93.5	110.6
1996	108.9	105.5	114.0	148.0	113.3	109.5	108.2	105.2	137.2	85.0	85.9	86.0	86.1	85.8	91.7	107.9
1995	105.6	99.3	110.1	140.7	106.6	104.6	104.0	97.7	129.4	81.8	82.6	82.6	82.7	82.6	88.4	103.4
1994	103.0	98.1	107.2	137.1	105.2	103.5	102.3	96.5	128.1	80.3	81.1	81.0	81.2	81.1	87.0	102.2
1993	100.0	95.7	102.2	133.3	102.0	100.8	99.6	94.1	126.3	78.2	78.9	78.9	78.9	78.8	85.8	100.3
1992	97.9	93.7	100.1	128.2	99.3	99.3	98.1	90.5	123.9	77.1	77.7	77.8	77.8	77.6	83.1	98.0
1991	95.7	89.5	96.8	124.4	96.0	96.7	95.5	88.7	121.5	76.0	76.6	76.8	76.7	76.6	82.6	96.8
1990	93.2	87.0	94.1	118.1	94.6	93.9	91.0	85.6	111.4	74.8	75.4	75.5	75.5	75.3	81.3	94.6
1985	81.8	77.5	83.2	94.9	81.6	80.1	81.0	77.0	92.8	66.9	67.6	67.7	67.6	67.5	73.4	86.8
1980	60.7	58.0	60.6	66.0	60.5	60.3	61.6	58.5	65.9	51.1	52.2	52.5	51.7	50.8	57.4	62.3
1975	43.7	42.5	45.1	49.5	44.8	43.7	44.8	42.7	47.1	36.1	37.0	37.0	37.3	36.1	39.3	44.7
1970	27.8	27.0	28.9	32.8	29.2	27.8	28.5	26.9	30.0	20.9	23.7	23.6	23.4	23.0	25.8	28.3
1965	21.5	20.8	22.2	25.5	22.8	21.4	21.9	20.7	23.1	16.0	18.2	18.2	18.0	17.7	19.9	21.8
1960	19.5	18.9	19.9	21.5	19.7	19.5	19.9	18.8	21.0	14.4	16.6	16.6	16.4	16.1	18.1	19.9
1955	16.3	15.8	16.7	18.1	16.5	16.3	16.7	15.8	17.6	12.1	13.9	13.9	13.7	13.5	15.2	16.6
1950	13.5	13.1	13.8	14.9	13.6	13.5	13.8	13.0	14.5	10.0	11.5	11.5	11.4	11.2	12.5	13.7
1945	8.6	8.3	8.8	9.5	8.7	8.6	8.8	8.3	9.3	6.4	7.3	7.3	7.2	7.1	8.0	8.8
▼ 1940	6.6	6.4	6.8	7.4	6.7	6.7	6.8	6.4	7.2	4.9	5.6	5.6	5.6	5.5	6.2	6.8

Year	National 30 City Average	Ohio: Canton	Cincinnati	Cleveland	Columbus	Dayton	Lorain	Springfield	Toledo	Youngstown	Oklahoma: Lawton	Oklahoma City	Tulsa	Oregon: Eugene	Portland	PA: Allentown
Jan 2011	185.0E	169.8E	169.6E	184.4E	174.8E	167.1E	176.7E	167.3E	179.7E	173.7E	151.3E	151.3E	144.5E	183.4E	184.8E	190.3E
2010	181.6	167.5	166.5	180.2	170.3	163.4	173.0	164.6	176.5	171.7	149.5	149.4	143.1	181.5	182.9	187.9
2009	182.5	168.4	168.1	181.6	171.2	165.3	174.0	166.0	178.5	172.8	152.5	153.3	146.8	181.3	183.6	188.3
2008	171.0	158.4	159.0	170.6	160.2	157.5	163.9	157.5	168.2	161.5	141.3	140.6	136.2	172.3	174.7	175.7
2007	165.0	153.2	152.4	164.9	155.1	149.4	158.1	150.3	161.6	156.7	136.2	135.6	132.5	168.5	169.5	169.4
2006	156.2	147.0	144.9	156.9	146.9	142.9	151.8	143.4	154.5	150.3	129.5	129.5	125.7	160.5	161.7	160.1
2005	146.7	137.7	136.8	147.9	138.0	134.5	143.3	134.8	145.5	141.1	121.6	121.6	117.8	150.9	152.2	150.2
2004	132.8	125.6	124.3	135.6	126.2	121.1	131.5	122.0	132.8	129.3	109.3	109.4	107.3	136.8	137.8	133.9
2003	129.7	123.7	121.3	132.7	123.7	119.4	126.1	120.0	130.4	126.2	107.5	107.8	105.0	134.2	135.9	130.3
2002	126.7	120.9	119.2	130.0	120.9	117.5	124.3	118.0	128.4	123.6	106.3	106.0	104.1	132.2	133.9	128.1
2001	122.2	117.7	115.9	125.9	117.1	114.0	120.4	114.7	123.8	119.9	102.0	102.1	99.7	129.8	131.2	123.5
2000	118.9	112.7	110.1	121.3	112.5	109.0	115.0	109.2	115.7	114.1	98.7	98.9	97.5	126.3	127.4	119.9
1999	116.6	110.6	107.9	118.7	109.5	107.1	112.0	106.4	113.6	111.9	97.5	97.4	96.2	120.9	124.3	117.8
1998	113.6	108.5	105.0	114.8	106.8	104.5	109.4	104.1	111.2	109.0	94.3	94.5	94.5	120.0	122.2	115.7
1997	111.5	106.4	102.9	112.9	104.7	102.6	107.4	102.2	108.5	107.2	92.9	93.2	93.0	118.1	119.6	113.9
1996	108.9	103.2	100.2	110.1	101.1	98.8	103.8	98.4	105.1	104.1	90.9	91.3	91.2	114.4	116.1	112.0
1995	105.6	98.8	97.1	106.4	99.1	94.6	97.1	92.0	100.6	100.1	85.3	88.0	89.0	112.2	114.3	108.3
1994	103.0	97.7	95.0	104.8	95.4	93.0	96.0	90.1	99.3	98.9	84.0	86.5	87.6	107.6	109.2	106.4
1993	100.0	95.9	92.3	101.9	93.9	90.6	93.9	87.7	98.3	97.2	81.3	83.9	84.8	107.2	108.8	103.6
1992	97.9	94.5	90.6	98.7	92.6	89.2	92.2	86.4	97.1	96.0	80.1	82.7	83.4	101.1	102.6	101.6
1991	95.7	93.4	88.6	97.2	90.6	87.7	91.1	84.9	94.2	92.7	78.3	80.7	81.4	99.5	101.1	98.9
1990	93.2	92.1	86.7	95.1	88.1	85.9	89.4	83.4	92.9	91.8	77.1	79.9	80.0	98.0	99.4	95.0
1985	81.8	83.7	78.2	86.2	77.7	76.3	80.7	74.3	84.7	82.7	71.2	73.9	73.8	88.7	90.1	81.9
1980	60.7	60.6	59.8	61.0	58.6	56.6	60.8	56.2	64.0	61.5	52.2	55.5	57.2	68.9	68.4	58.7
1975	43.7	44.0	43.7	44.6	42.2	40.5	42.7	40.0	44.9	44.5	38.6	38.6	39.2	44.6	44.9	42.3
1970	27.8	27.8	28.2	29.6	26.7	27.0	27.5	25.4	29.1	29.6	24.1	23.4	24.1	30.4	30.0	27.3
1965	21.5	21.5	20.7	21.1	20.2	20.0	21.1	19.6	21.3	21.0	18.5	17.8	20.6	23.4	22.6	21.0
1960	19.5	19.5	19.3	19.6	18.7	18.1	19.2	17.8	19.4	19.1	16.9	16.1	18.1	21.2	21.1	19.1
1955	16.3	16.3	16.2	16.5	15.7	15.2	16.1	14.9	16.2	16.0	14.1	13.5	15.1	17.8	17.7	16.0
1950	13.5	13.5	13.3	13.6	13.0	12.5	13.3	12.3	13.4	13.2	11.7	11.2	12.5	14.7	14.6	13.2
1945	8.6	8.6	8.5	8.7	8.3	8.0	8.5	7.8	8.6	8.4	7.4	7.1	8.0	9.4	9.3	8.4
▼ 1940	6.6	6.7	6.5	6.7	6.4	6.2	6.5	6.1	6.6	6.5	5.7	5.5	6.1	7.2	7.2	6.5

Historical Cost Indexes

Year	National 30 City Average	Pennsylvania: Erie	Pennsylvania: Harrisburg	Pennsylvania: Philadelphia	Pennsylvania: Pittsburgh	Pennsylvania: Reading	Pennsylvania: Scranton	RI: Providence	South Carolina: Charleston	South Carolina: Columbia	South Dakota: Rapid City	South Dakota: Sioux Falls	Tennessee: Chattanooga	Tennessee: Knoxville	Tennessee: Memphis	Tennessee: Nashville
Jan 2011	185.0E	174.3E	177.8E	211.6E	186.4E	182.1E	183.0E	194.8E	148.6E	141.3E	149.0E	150.7E	155.7E	147.7E	157.0E	159.0E
2010	181.6	171.0	175.2	209.3	182.3	179.8	180.8	192.8	147.4	140.2	147.3	149.0	152.4	144.7	154.2	156.7
2009	182.5	171.9	176.5	209.5	181.7	180.7	181.3	192.9	150.8	144.3	148.2	151.1	155.7	148.0	157.2	159.8
2008	171.0	161.0	165.7	196.2	169.2	169.3	167.9	178.2	142.1	134.8	138.4	141.3	136.7	133.4	146.0	147.5
2007	165.0	156.0	159.4	188.4	163.2	164.9	163.1	174.4	139.3	131.7	129.5	133.0	133.7	130.6	143.1	144.4
2006	156.2	148.7	150.7	177.8	155.5	155.0	153.9	163.6	122.4	118.7	122.8	126.6	127.0	123.9	136.5	135.5
2005	146.7	140.6	141.5	166.4	146.6	145.5	142.9	155.6	110.2	109.6	115.2	118.5	116.0	115.1	128.6	127.9
2004	132.8	127.0	126.2	148.4	133.1	128.7	129.0	138.8	98.9	98.4	103.9	107.3	105.0	104.5	115.6	115.8
2003	129.7	124.5	123.8	145.5	130.5	126.3	126.0	135.9	96.1	95.7	101.3	104.0	102.8	102.3	111.1	110.9
2002	126.7	122.1	121.1	142.0	127.8	123.4	124.2	131.1	94.8	93.9	100.2	103.2	102.2	101.2	107.6	109.2
2001	122.2	118.9	118.4	136.9	124.1	120.4	121.1	127.8	92.1	91.3	96.8	99.7	98.2	96.5	102.6	104.3
2000	118.9	114.6	114.0	132.1	120.9	115.9	117.7	122.8	89.9	89.1	94.2	98.0	96.9	95.0	100.8	100.8
1999	116.6	113.8	112.8	129.8	119.6	114.8	116.3	121.6	89.0	88.1	92.4	95.8	95.9	93.4	99.7	98.6
1998	113.6	109.8	110.5	126.6	117.1	112.5	113.7	120.3	88.0	87.2	90.7	93.6	94.9	92.4	97.2	96.8
1997	111.5	108.5	108.8	123.3	113.9	110.8	112.3	118.9	86.5	85.6	89.3	92.1	93.6	90.8	96.1	94.9
1996	108.9	106.8	105.7	120.3	110.8	108.7	109.9	117.2	84.9	84.1	87.2	90.0	91.6	88.8	94.0	91.9
1995	105.6	99.8	100.6	117.1	106.3	103.6	103.8	111.1	82.7	82.2	84.0	84.7	89.2	86.0	91.2	87.6
1994	103.0	98.2	99.4	115.2	103.7	102.2	102.6	109.6	81.2	80.6	82.7	83.1	87.4	84.1	89.0	84.8
1993	100.0	94.9	96.6	107.4	99.0	98.6	99.8	108.2	78.4	77.9	81.2	81.9	85.1	82.3	86.8	81.9
1992	97.9	92.1	94.8	105.3	96.5	96.7	97.8	106.9	77.1	76.8	80.0	80.7	83.6	77.7	85.4	80.7
1991	95.7	90.5	92.5	101.7	93.2	94.1	94.8	96.1	75.0	75.8	78.7	79.7	81.9	76.6	83.0	79.1
1990	93.2	88.5	89.5	98.5	91.2	90.8	91.3	94.1	73.7	74.5	77.2	78.4	79.9	75.1	81.3	77.1
1985	81.8	79.6	77.2	82.2	81.5	77.7	80.0	83.0	65.9	66.9	69.7	71.2	72.5	67.7	74.3	66.7
1980	60.7	59.4	56.9	58.7	61.3	57.7	58.4	59.2	50.6	51.1	54.8	57.1	55.0	51.6	55.9	53.1
1975	43.7	43.4	42.2	44.5	44.5	43.6	41.9	42.7	35.0	36.4	37.7	39.6	39.1	37.3	40.7	37.0
1970	27.8	27.5	25.8	27.5	28.7	27.1	27.0	27.3	22.8	22.9	24.8	25.8	24.9	22.0	23.0	22.8
1965	21.5	21.1	20.1	21.7	22.4	20.9	20.8	21.9	17.6	17.6	19.1	19.9	19.2	17.1	18.3	17.4
1960	19.5	19.1	18.6	19.4	19.7	19.0	18.9	19.1	16.0	16.0	17.3	18.1	17.4	15.5	16.6	15.8
1955	16.3	16.1	15.6	16.3	16.5	15.9	15.9	16.0	13.4	13.4	14.5	15.1	14.6	13.0	13.9	13.3
1950	13.5	13.3	12.9	13.5	13.6	13.2	13.1	13.2	11.1	11.1	12.0	12.5	12.1	10.7	11.5	10.9
1945	8.6	8.5	8.2	8.6	8.7	8.4	8.3	8.4	7.1	7.1	7.7	8.0	7.7	6.8	7.3	7.0
▼ 1940	6.6	6.5	6.3	6.6	6.7	6.4	6.5	6.5	5.4	5.5	5.9	6.2	6.0	5.3	5.6	5.4

Year	National 30 City Average	Texas: Abilene	Texas: Amarillo	Texas: Austin	Texas: Beaumont	Texas: Corpus Christi	Texas: Dallas	Texas: El Paso	Texas: Fort Worth	Texas: Houston	Texas: Lubbock	Texas: Odessa	Texas: San Antonio	Texas: Waco	Texas: Wichita Falls	Utah: Ogden
Jan 2011	185.0E	145.5E	151.5E	146.3E	152.4E	143.5E	158.1E	142.1E	152.5E	160.5E	149.5E	142.1E	151.0E	147.0E	146.3E	158.5E
2010	181.6	144.2	150.1	144.7	150.5	142.1	155.1	140.8	149.6	157.3	148.1	140.7	147.9	145.8	145.0	157.5
2009	182.5	143.5	147.3	146.4	149.8	141.7	155.5	141.1	150.1	160.9	144.6	139.1	150.8	146.7	146.1	155.1
2008	171.0	132.4	137.0	137.5	140.3	133.2	144.1	130.7	138.3	149.1	134.7	128.9	141.0	136.1	136.1	144.8
2007	165.0	128.5	132.2	131.6	137.1	128.8	138.6	126.2	134.8	146.0	130.2	125.0	136.4	132.1	132.1	140.0
2006	156.2	121.6	125.7	125.5	130.1	122.0	131.1	120.3	127.6	138.2	123.2	118.3	129.8	125.2	125.3	133.4
2005	146.7	113.8	117.3	117.9	121.3	114.4	123.7	112.5	119.4	129.0	115.5	110.5	121.3	116.1	117.3	126.0
2004	132.8	102.9	106.3	105.7	108.5	102.9	112.0	101.5	108.4	115.9	104.7	99.9	108.4	104.8	105.1	115.2
2003	129.7	100.5	104.4	103.9	106.1	100.4	109.3	99.5	105.5	113.4	102.3	97.6	104.8	102.7	103.0	112.7
2002	126.7	99.9	101.9	102.4	104.5	98.9	107.9	98.7	104.6	111.5	100.5	95.9	103.7	100.5	101.4	110.8
2001	122.2	93.4	98.4	99.8	102.3	96.6	103.8	95.5	100.9	107.8	97.6	93.4	100.5	97.8	97.3	107.7
2000	118.9	93.4	98.1	99.1	101.6	96.9	102.7	92.4	99.9	106.0	97.6	93.4	99.4	97.3	96.9	104.6
1999	116.6	91.8	94.5	96.0	99.7	94.0	101.0	90.7	97.6	104.6	96.0	92.1	98.0	94.8	95.5	103.3
1998	113.6	89.8	92.7	94.2	97.9	91.8	97.9	88.4	94.5	101.3	93.3	90.1	94.8	92.7	92.9	98.5
1997	111.5	88.4	91.3	92.8	96.8	90.3	96.1	87.0	93.3	100.1	91.9	88.8	93.4	91.4	91.5	96.1
1996	108.9	86.8	89.6	90.9	95.3	88.5	94.1	86.7	91.5	97.9	90.2	87.1	92.3	89.7	89.9	94.1
1995	105.6	85.2	87.4	89.3	93.7	87.4	91.4	85.2	89.5	95.9	88.4	85.6	88.9	86.4	86.8	92.2
1994	103.0	83.3	85.5	87.0	91.8	84.6	89.6	82.2	87.4	93.4	87.0	83.9	87.0	84.9	85.4	89.4
1993	100.0	81.4	83.4	84.9	90.0	82.8	87.8	80.2	85.5	91.1	84.8	81.9	85.0	83.0	83.5	87.1
1992	97.9	80.3	82.3	83.8	88.9	81.6	86.2	79.0	84.1	89.8	83.6	80.7	83.9	81.8	82.4	85.1
1991	95.7	79.2	81.1	82.8	87.8	80.6	85.9	78.0	83.3	87.9	82.7	79.8	83.3	80.6	81.5	84.1
1990	93.2	78.0	80.1	81.3	86.5	79.3	84.5	76.7	82.1	85.4	81.5	78.6	80.7	79.6	80.3	83.4
1985	81.8	71.1	72.5	74.5	79.3	72.3	77.6	69.4	75.1	79.6	74.0	71.2	73.9	71.7	73.3	75.2
1980	60.7	53.4	55.2	54.5	57.6	54.5	57.9	53.1	57.0	59.4	55.6	57.2	55.0	54.9	55.4	62.2
1975	43.7	37.6	39.0	39.0	39.6	38.1	40.7	38.0	40.4	41.2	38.9	37.9	39.0	38.6	38.0	40.0
1970	27.8	24.5	24.9	24.9	25.7	24.5	25.5	23.7	25.9	25.4	25.1	24.6	23.3	24.8	24.5	26.8
1965	21.5	18.9	19.2	19.2	19.9	18.9	19.9	19.0	19.9	20.0	19.4	19.0	18.5	19.2	18.9	20.6
1960	19.5	17.1	17.4	17.4	18.1	17.1	18.2	17.0	18.1	18.2	17.6	17.3	16.8	17.4	17.2	18.8
1955	16.3	14.4	14.6	14.6	15.1	14.4	15.3	14.3	15.2	15.2	14.8	14.5	14.1	14.6	14.4	15.7
1950	13.5	11.9	12.1	12.1	12.5	11.9	12.6	11.8	12.5	12.6	12.2	12.0	11.6	12.1	11.9	13.0
1945	8.6	7.6	7.7	7.7	8.0	7.6	8.0	7.5	8.0	8.0	7.8	7.6	7.4	7.7	7.6	8.3
▼ 1940	6.6	5.9	5.9	5.9	6.1	5.8	6.2	5.8	6.2	6.2	6.0	5.9	5.7	5.9	5.8	6.4

Historical Cost Indexes

Year	National 30 City Average	Utah: Salt Lake City	Vermont: Burlington	Vermont: Rutland	Virginia: Alexandria	Virginia: Newport News	Virginia: Norfolk	Virginia: Richmond	Virginia: Roanoke	Washington: Seattle	Washington: Spokane	Washington: Tacoma	West Virginia: Charleston	West Virginia: Huntington	Wisconsin: Green Bay	Wisconsin: Kenosha
Jan 2011	185.0E	161.8E	157.9E	156.2E	173.2E	158.2E	159.7E	157.7E	153.2E	194.3E	173.0E	187.6E	175.7E	178.2E	181.2E	187.3E
2010	181.6	160.9	156.3	154.6	170.9	156.7	158.0	156.6	151.8	191.8	171.5	186.0	171.6	176.1	179.0	185.1
2009	182.5	160.4	158.3	156.7	172.5	160.0	161.9	160.7	155.6	188.5	173.0	186.3	174.6	177.4	175.5	182.5
2008	171.0	149.7	147.2	145.8	161.8	150.8	150.8	150.9	145.9	176.9	162.4	174.9	162.8	165.7	165.6	172.2
2007	165.0	144.6	144.4	143.2	155.0	146.2	146.1	147.3	142.9	171.4	156.7	168.7	158.7	160.1	159.4	164.8
2006	156.2	137.7	131.7	130.7	146.2	133.7	134.6	134.8	129.7	162.9	150.1	160.7	149.4	151.4	152.7	157.1
2005	146.7	129.4	124.6	123.8	136.5	123.5	124.4	125.4	112.2	153.9	141.9	151.5	140.8	141.0	144.4	148.3
2004	132.8	117.8	113.0	112.3	121.5	108.9	110.2	110.9	99.7	138.0	127.6	134.5	124.8	125.6	128.9	132.4
2003	129.7	116.0	110.7	110.1	119.5	104.8	106.2	108.6	97.0	134.9	125.9	133.0	123.3	123.6	127.4	130.9
2002	126.7	113.7	109.0	108.4	115.1	102.9	104.1	106.6	95.2	132.7	123.9	131.4	121.2	120.9	123.0	127.3
2001	122.2	109.1	105.7	105.2	110.8	99.8	100.3	102.9	92.1	127.9	120.3	125.7	114.6	117.5	119.1	123.6
2000	118.9	106.5	98.9	98.3	108.1	96.5	97.6	100.2	90.7	124.6	118.3	122.9	111.5	114.4	114.6	119.1
1999	116.6	104.5	98.2	97.7	106.1	95.6	96.5	98.8	89.8	123.3	116.7	121.6	110.6	113.4	112.1	115.8
1998	113.6	99.5	97.8	97.3	104.1	93.7	93.9	97.0	88.3	119.4	114.3	118.3	106.7	109.0	109.5	112.9
1997	111.5	97.2	96.6	96.3	101.2	91.6	91.7	92.9	86.9	118.1	111.7	117.2	105.3	107.7	105.6	109.1
1996	108.9	94.9	95.1	94.8	99.7	90.2	90.4	91.6	85.5	115.2	109.2	114.3	103.1	104.8	103.8	106.4
1995	105.6	93.1	91.1	90.8	96.3	86.0	86.4	87.8	82.8	113.7	107.4	112.8	95.8	97.2	97.6	97.9
1994	103.0	90.2	89.5	89.3	93.9	84.6	84.8	86.3	81.4	109.9	104.0	108.3	94.3	95.3	96.3	96.2
1993	100.0	87.9	87.6	87.6	91.6	82.9	83.0	84.3	79.5	107.3	103.9	106.7	92.6	93.5	94.0	94.3
1992	97.9	86.0	86.1	86.1	90.1	81.0	81.6	82.0	78.3	105.1	101.4	103.7	91.4	92.3	92.0	92.1
1991	95.7	84.9	84.2	84.2	88.2	77.6	77.9	79.8	77.3	102.2	100.0	102.2	89.7	88.6	88.6	89.8
1990	93.2	84.3	83.0	82.9	86.1	76.3	76.7	77.6	76.1	100.1	98.5	100.5	86.1	86.8	86.7	87.8
1985	81.8	75.9	74.8	74.9	75.1	68.7	68.8	69.5	67.2	88.3	89.0	91.2	77.7	77.7	76.7	77.4
1980	60.7	57.0	55.3	58.3	57.3	52.5	52.4	54.3	51.3	67.9	66.3	66.7	57.7	58.3	58.6	58.3
1975	43.7	40.1	41.8	43.9	41.7	37.2	36.9	37.1	37.1	44.9	44.4	44.5	41.0	40.0	40.9	40.5
1970	27.8	26.1	25.4	26.8	26.2	23.9	21.5	22.0	23.7	28.8	29.3	29.6	26.1	25.8	26.4	26.5
1965	21.5	20.0	19.8	20.6	20.2	18.4	17.1	17.2	18.3	22.4	22.5	22.8	20.1	19.9	20.3	20.4
1960	19.5	18.4	18.0	18.8	18.4	16.7	15.4	15.6	16.6	20.4	20.8	20.8	18.3	18.1	18.4	18.6
1955	16.3	15.4	15.1	15.7	15.4	14.0	12.9	13.1	13.9	17.1	17.4	17.4	15.4	15.2	15.5	15.6
1950	13.5	12.7	12.4	13.0	12.7	11.6	10.7	10.8	11.5	14.1	14.4	14.4	12.7	12.5	12.8	12.9
1945	8.6	8.1	7.9	8.3	8.1	7.4	6.8	6.9	7.3	9.0	9.2	9.2	8.1	8.0	8.1	8.2
1940	6.6	6.3	6.1	6.4	6.2	5.7	5.3	5.3	5.7	7.0	7.1	7.1	6.2	6.2	6.3	6.3

Year	National 30 City Average	Wisconsin: Madison	Wisconsin: Milwaukee	Wisconsin: Racine	Wyoming: Cheyenne	Canada: Calgary	Canada: Edmonton	Canada: Hamilton	Canada: London	Canada: Montreal	Canada: Ottawa	Canada: Quebec	Canada: Toronto	Canada: Vancouver	Canada: Winnipeg	
Jan 2011	185.0E	183.2E	191.4E	186.7E	155.4E	212.3E	212.5E	204.3E	200.2E	202.1E	202.1E	200.5E	209.8E	207.0E	192.5E	
2010	181.6	181.3	187.1	184.4	154.4	200.6	200.7	197.7	193.7	193.9	195.6	192.7	200.7	192.7	182.9	
2009	182.5	180.0	187.3	182.8	155.2	205.4	206.6	203.0	198.9	198.8	200.1	197.4	205.6	199.6	188.5	
2008	171.0	168.4	176.3	173.0	146.5	190.2	191.0	194.3	188.6	186.7	188.0	186.8	194.6	184.9	174.4	
2007	165.0	160.7	168.9	164.7	141.3	183.1	184.4	186.8	182.4	181.7	182.4	182.0	187.7	180.6	170.1	
2006	156.2	152.8	158.8	157.2	128.0	163.6	164.8	169.0	164.9	158.6	163.7	159.0	170.6	169.3	155.9	
2005	146.7	145.2	148.4	147.9	118.9	154.4	155.7	160.0	156.2	149.2	155.1	150.1	162.6	159.4	146.9	
2004	132.8	129.7	134.2	132.7	104.7	138.8	139.4	142.6	140.4	134.5	141.0	135.7	146.0	141.7	129.7	
2003	129.7	128.4	131.1	131.5	102.6	133.6	134.2	139.1	137.0	130.2	137.4	131.4	142.3	137.0	124.4	
2002	126.7	124.5	128.2	126.5	101.7	122.5	122.2	136.3	134.1	127.2	134.9	128.4	139.8	134.6	121.4	
2001	122.2	120.6	123.9	123.3	99.0	117.5	117.4	131.5	129.1	124.4	130.2	125.5	134.7	130.2	117.2	
2000	118.9	116.6	120.5	118.7	98.1	115.9	115.8	130.0	127.5	122.8	128.8	124.1	133.1	128.4	115.6	
1999	116.6	115.9	117.4	115.5	96.9	115.3	115.2	128.1	125.6	120.8	126.8	121.9	131.2	127.1	115.2	
1998	113.6	110.8	113.4	112.7	95.4	112.5	112.4	126.3	123.9	119.0	124.7	119.6	128.5	123.8	113.7	
1997	111.5	106.1	110.1	109.3	93.2	110.7	110.6	124.4	121.9	114.6	122.8	115.4	125.7	121.9	111.4	
1996	108.9	104.4	107.1	106.5	91.1	109.1	109.0	122.6	120.2	112.9	121.0	113.6	123.9	119.0	109.7	
1995	105.6	96.5	103.9	97.8	87.6	107.4	107.4	119.9	117.5	110.8	118.2	111.5	121.6	116.2	107.6	
1994	103.0	94.5	100.6	96.1	85.4	106.7	106.6	116.5	114.2	109.5	115.0	110.2	117.8	115.1	105.5	
1993	100.0	91.3	96.7	93.8	82.9	104.7	104.5	113.7	111.9	106.9	112.0	107.0	114.9	109.2	102.8	
1992	97.9	89.2	93.9	91.7	81.7	103.4	103.2	112.4	110.7	104.2	110.8	103.6	113.7	108.0	101.6	
1991	95.7	86.2	91.6	89.3	80.3	102.1	102.0	108.2	106.7	101.8	106.9	100.4	109.0	106.8	98.6	
1990	93.2	84.3	88.9	87.3	79.1	98.0	97.1	103.8	101.4	99.0	102.7	96.8	104.6	103.2	95.1	
1985	81.8	74.3	77.4	77.0	72.3	90.2	89.1	85.8	84.9	82.4	83.9	79.9	86.5	89.8	83.4	
1980	60.7	56.8	58.8	58.1	56.9	64.9	63.3	63.7	61.9	59.2	60.9	59.3	60.9	65.0	61.7	
1975	43.7	40.7	43.3	40.7	40.6	42.2	41.6	42.9	41.6	39.7	41.5	39.0	42.2	42.4	39.2	
1970	27.8	26.5	29.4	26.5	26.0	28.9	28.6	28.5	27.8	25.6	27.6	26.0	25.6	26.0	23.1	
1965	21.5	20.6	21.8	20.4	20.0	22.3	22.0	22.0	21.4	18.7	21.2	20.1	19.4	20.5	17.5	
1960	19.5	18.1	19.0	18.6	18.2	20.2	20.0	20.0	19.5	17.0	19.3	18.2	17.6	18.6	15.8	
1955	16.3	15.2	15.9	15.6	15.2	17.0	16.8	16.7	16.3	14.3	16.2	15.3	14.8	15.5	13.3	
1950	13.5	12.5	13.2	12.9	12.6	14.0	13.9	13.8	13.5	11.8	13.4	12.6	12.2	12.8	10.9	
1945	8.6	8.0	8.4	8.2	8.0	9.0	8.8	8.8	8.6	7.5	8.5	8.0	7.8	8.2	7.0	
1940	6.6	6.2	6.5	6.3	6.2	6.9	6.8	6.8	6.6	5.8	6.6	6.2	6.0	6.3	5.4	

Location Factors

Costs shown in RSMeans cost data publications are based on national averages for materials and installation. To adjust these costs to a specific location, simply multiply the base cost by the factor and divide by 100 for that city. The data is arranged alphabetically by state and postal zip code numbers. For a city not listed, use the factor for a nearby city with similar economic characteristics.

STATE/ZIP	CITY	MAT.	INST.	TOTAL
ALABAMA				
350-352	Birmingham	97.0	75.6	87.4
354	Tuscaloosa	95.8	59.2	79.4
355	Jasper	96.0	58.2	79.1
356	Decatur	95.8	60.8	80.1
357-358	Huntsville	95.8	69.0	83.8
359	Gadsden	95.6	58.2	78.9
360-361	Montgomery	96.8	57.2	79.1
362	Anniston	95.1	65.5	81.9
363	Dothan	95.8	52.6	76.5
364	Evergreen	95.3	54.1	76.9
365-366	Mobile	96.8	67.9	83.9
367	Selma	95.5	52.5	76.3
368	Phenix City	96.2	56.3	78.4
369	Butler	95.7	52.6	76.4
ALASKA				
995-996	Anchorage	127.2	114.5	121.5
997	Fairbanks	124.4	115.0	120.2
998	Juneau	122.8	112.6	118.3
999	Ketchikan	136.9	112.6	126.0
ARIZONA				
850,853	Phoenix	98.6	75.9	88.5
851,852	Mesa/Tempe	98.2	72.8	86.8
855	Globe	98.5	62.0	82.2
856-857	Tucson	96.9	72.0	85.8
859	Show Low	98.6	63.5	82.9
860	Flagstaff	100.7	74.4	88.9
863	Prescott	98.4	63.0	82.6
864	Kingman	96.7	74.2	86.7
865	Chambers	96.8	63.5	81.9
ARKANSAS				
716	Pine Bluff	96.7	64.8	82.4
717	Camden	94.2	50.7	74.8
718	Texarkana	95.9	51.9	76.3
719	Hot Springs	93.4	53.6	75.6
720-722	Little Rock	97.0	65.9	83.1
723	West Memphis	95.9	63.7	81.5
724	Jonesboro	96.7	61.6	81.0
725	Batesville	94.3	56.9	77.6
726	Harrison	95.7	53.0	76.6
727	Fayetteville	92.9	54.9	75.9
728	Russellville	94.2	56.1	77.2
729	Fort Smith	97.0	63.4	82.0
CALIFORNIA				
900-902	Los Angeles	100.6	115.4	107.2
903-905	Inglewood	96.1	112.6	103.5
906-908	Long Beach	97.8	112.6	104.5
910-912	Pasadena	96.5	112.7	103.7
913-916	Van Nuys	99.7	112.7	105.5
917-918	Alhambra	98.6	112.7	104.9
919-921	San Diego	100.6	108.2	104.0
922	Palm Springs	97.5	111.5	103.8
923-924	San Bernardino	95.2	111.2	102.3
925	Riverside	100.0	113.3	105.9
926-927	Santa Ana	97.2	111.4	103.6
928	Anaheim	99.9	113.7	106.1
930	Oxnard	101.4	113.1	106.6
931	Santa Barbara	100.6	113.1	106.2
932-933	Bakersfield	101.4	110.4	105.4
934	San Luis Obispo	101.2	109.8	105.0
935	Mojave	98.2	107.7	102.4
936-938	Fresno	101.7	115.3	107.8
939	Salinas	102.0	123.0	111.4
940-941	San Francisco	109.6	140.8	123.6
942,956-958	Sacramento	103.9	117.2	109.8
943	Palo Alto	102.3	129.4	114.4
944	San Mateo	105.1	131.4	116.9
945	Vallejo	102.9	127.8	114.0
946	Oakland	107.5	133.7	119.2
947	Berkeley	106.8	133.6	118.8
948	Richmond	106.1	128.9	116.3
949	San Rafael	107.4	130.3	117.6
950	Santa Cruz	106.4	123.1	113.9

STATE/ZIP	CITY	MAT.	INST.	TOTAL
CALIFORNIA (CONT'D)				
951	San Jose	104.6	134.3	117.9
952	Stockton	101.9	117.3	108.8
953	Modesto	101.9	116.6	108.5
954	Santa Rosa	101.7	134.9	116.5
955	Eureka	103.1	115.2	108.5
959	Marysville	102.4	113.8	107.5
960	Redding	104.1	113.8	108.4
961	Susanville	103.1	113.8	107.9
COLORADO				
800-802	Denver	101.4	84.9	94.0
803	Boulder	98.5	83.7	91.8
804	Golden	100.6	82.5	92.5
805	Fort Collins	102.3	77.1	91.0
806	Greeley	99.6	73.6	88.0
807	Fort Morgan	99.1	82.3	91.6
808-809	Colorado Springs	101.3	82.4	92.8
810	Pueblo	100.8	80.2	91.6
811	Alamosa	102.1	76.1	90.5
812	Salida	102.0	77.8	91.2
813	Durango	102.6	75.3	90.4
814	Montrose	101.2	74.7	89.4
815	Grand Junction	104.7	75.2	91.5
816	Glenwood Springs	102.1	78.0	91.3
CONNECTICUT				
060	New Britain	101.0	117.8	108.5
061	Hartford	102.1	117.8	109.1
062	Willimantic	101.6	117.8	108.9
063	New London	97.7	117.7	106.6
064	Meriden	99.7	117.7	107.8
065	New Haven	102.9	117.7	109.5
066	Bridgeport	102.2	116.4	108.6
067	Waterbury	101.7	117.7	108.9
068	Norwalk	101.6	116.5	108.3
069	Stamford	101.8	123.0	111.3
D.C.				
200-205	Washington	102.8	91.9	97.9
DELAWARE				
197	Newark	99.3	108.6	103.4
198	Wilmington	99.6	108.6	103.6
199	Dover	99.9	108.6	103.8
FLORIDA				
320,322	Jacksonville	97.8	69.7	85.2
321	Daytona Beach	98.0	76.4	88.3
323	Tallahassee	98.6	59.7	81.2
324	Panama City	99.2	57.5	80.6
325	Pensacola	101.9	65.1	85.4
326,344	Gainesville	99.5	70.6	86.6
327-328,347	Orlando	101.2	74.2	89.1
329	Melbourne	100.4	79.6	91.1
330-332,340	Miami	99.1	78.1	89.7
333	Fort Lauderdale	97.3	76.7	88.1
334,349	West Palm Beach	96.0	75.7	86.9
335-336,346	Tampa	99.0	83.6	92.1
337	St. Petersburg	101.3	66.1	85.6
338	Lakeland	98.0	83.5	91.5
339,341	Fort Myers	97.3	75.4	87.5
342	Sarasota	99.2	77.3	89.4
GEORGIA				
300-303,399	Atlanta	97.2	76.8	88.1
304	Statesboro	96.8	54.1	77.7
305	Gainesville	95.5	64.7	81.7
306	Athens	94.9	64.9	81.5
307	Dalton	96.8	60.5	80.6
308-309	Augusta	95.9	65.5	82.3
310-312	Macon	95.6	65.0	81.9
313-314	Savannah	97.7	60.8	81.2
315	Waycross	96.7	61.8	81.1
316	Valdosta	96.7	61.2	80.9
317,398	Albany	96.9	62.0	81.3
318-319	Columbus	96.8	65.3	82.7

Location Factors

STATE/ZIP	CITY	MAT.	INST.	TOTAL
HAWAII				
967	Hilo	111.9	120.6	115.8
968	Honolulu	115.1	120.6	117.5
STATES & POSS.				
969	Guam	133.3	61.7	101.3
IDAHO				
832	Pocatello	100.4	73.0	88.2
833	Twin Falls	101.7	53.1	80.0
834	Idaho Falls	99.0	59.0	81.1
835	Lewiston	109.3	78.3	95.4
836-837	Boise	99.5	73.3	87.8
838	Coeur d'Alene	108.5	75.4	93.7
ILLINOIS				
600-603	North Suburban	99.3	131.0	113.4
604	Joliet	99.2	137.7	116.4
605	South Suburban	99.3	131.0	113.4
606-608	Chicago	99.7	138.9	117.2
609	Kankakee	95.2	125.3	108.7
610-611	Rockford	98.1	122.7	109.1
612	Rock Island	95.7	99.0	97.2
613	La Salle	96.9	121.4	107.8
614	Galesburg	96.8	104.8	100.4
615-616	Peoria	99.2	107.3	102.8
617	Bloomington	96.1	108.7	101.7
618-619	Champaign	99.8	107.3	103.1
620-622	East St. Louis	94.6	107.5	100.4
623	Quincy	95.8	99.6	97.5
624	Effingham	95.1	103.6	98.9
625	Decatur	97.0	105.3	100.7
626-627	Springfield	97.5	106.2	101.4
628	Centralia	93.2	106.9	99.3
629	Carbondale	92.9	102.3	97.1
INDIANA				
460	Anderson	95.3	82.8	89.7
461-462	Indianapolis	98.3	85.7	92.7
463-464	Gary	96.7	108.1	101.8
465-466	South Bend	95.3	83.3	89.9
467-468	Fort Wayne	95.9	79.2	88.5
469	Kokomo	93.1	81.7	88.0
470	Lawrenceburg	92.5	78.9	86.4
471	New Albany	93.6	76.4	85.9
472	Columbus	95.9	81.6	89.5
473	Muncie	96.8	81.7	90.1
474	Bloomington	98.1	81.5	90.7
475	Washington	94.6	83.3	89.5
476-477	Evansville	95.9	86.0	91.5
478	Terre Haute	96.6	86.7	92.2
479	Lafayette	95.6	82.8	89.9
IOWA				
500-503,509	Des Moines	98.9	86.2	93.2
504	Mason City	97.5	68.9	84.7
505	Fort Dodge	97.7	60.6	81.1
506-507	Waterloo	99.4	73.4	87.8
508	Creston	98.0	80.2	90.0
510-511	Sioux City	100.1	71.9	87.5
512	Sibley	98.9	57.3	80.3
513	Spencer	100.6	57.3	81.3
514	Carroll	97.3	73.6	86.7
515	Council Bluffs	101.4	80.3	91.9
516	Shenandoah	97.9	73.6	87.1
520	Dubuque	99.7	81.2	91.4
521	Decorah	98.4	66.7	84.3
522-524	Cedar Rapids	100.5	85.5	93.8
525	Ottumwa	98.4	71.9	86.6
526	Burlington	97.5	78.6	89.1
527-528	Davenport	99.4	92.0	96.1
KANSAS				
660-662	Kansas City	99.2	97.3	98.3
664-666	Topeka	99.2	65.2	84.0
667	Fort Scott	97.6	75.2	87.6
668	Emporia	97.8	70.8	85.7
669	Belleville	99.6	66.5	84.8
670-672	Wichita	98.3	65.1	83.4
673	Independence	99.2	70.6	86.4
674	Salina	99.7	68.0	85.5
675	Hutchinson	94.5	66.4	82.0
676	Hays	98.7	66.8	84.4
677	Colby	99.5	66.8	84.9

STATE/ZIP	CITY	MAT.	INST.	TOTAL
KANSAS (CONT'D)				
678	Dodge City	101.0	67.8	86.2
679	Liberal	98.5	66.8	84.3
KENTUCKY				
400-402	Louisville	97.7	83.4	91.3
403-405	Lexington	96.3	75.3	86.9
406	Frankfort	97.1	74.3	86.9
407-409	Corbin	93.5	62.0	79.4
410	Covington	94.4	96.0	95.1
411-412	Ashland	93.2	98.5	95.6
413-414	Campton	94.4	62.6	80.2
415-416	Pikeville	95.7	80.3	88.8
417-418	Hazard	93.7	55.7	76.7
420	Paducah	92.6	86.0	89.7
421-422	Bowling Green	94.7	82.8	89.4
423	Owensboro	95.1	80.2	88.4
424	Henderson	92.3	86.3	89.6
425-426	Somerset	91.6	66.6	80.4
427	Elizabethtown	91.0	83.4	87.6
LOUISIANA				
700-701	New Orleans	102.0	69.2	87.3
703	Thibodaux	100.3	64.3	84.2
704	Hammond	97.4	56.1	78.9
705	Lafayette	99.7	60.5	82.2
706	Lake Charles	100.0	65.0	84.3
707-708	Baton Rouge	101.5	63.6	84.5
710-711	Shreveport	96.9	57.1	79.1
712	Monroe	97.2	54.3	78.0
713-714	Alexandria	97.2	53.8	77.8
MAINE				
039	Kittery	95.0	80.2	88.4
040-041	Portland	99.8	79.0	90.5
042	Lewiston	98.2	79.0	89.6
043	Augusta	97.9	80.0	89.9
044	Bangor	97.9	79.0	89.4
045	Bath	96.3	79.6	88.8
046	Machias	95.8	78.9	88.2
047	Houlton	95.9	79.1	88.4
048	Rockland	94.9	79.5	88.0
049	Waterville	96.3	80.0	89.0
MARYLAND				
206	Waldorf	99.1	68.5	85.4
207-208	College Park	99.2	83.0	92.0
209	Silver Spring	98.3	74.5	87.7
210-212	Baltimore	98.9	84.8	92.6
214	Annapolis	99.1	78.5	89.9
215	Cumberland	94.7	80.1	88.1
216	Easton	96.4	41.9	72.0
217	Hagerstown	95.5	82.2	89.5
218	Salisbury	96.8	51.0	76.3
219	Elkton	93.6	64.9	80.7
MASSACHUSETTS				
010-011	Springfield	100.0	109.5	104.2
012	Pittsfield	99.6	105.3	102.2
013	Greenfield	97.4	108.6	102.4
014	Fitchburg	96.2	121.1	107.3
015-016	Worcester	100.0	122.7	110.1
017	Framingham	95.6	129.4	110.7
018	Lowell	99.4	129.0	112.6
019	Lawrence	100.6	127.8	112.8
020-022, 024	Boston	102.3	136.8	117.7
023	Brockton	100.9	124.5	111.5
025	Buzzards Bay	95.0	122.2	107.1
026	Hyannis	98.0	122.2	108.8
027	New Bedford	100.0	122.3	110.0
MICHIGAN				
480,483	Royal Oak	92.8	103.6	97.6
481	Ann Arbor	95.2	107.7	100.8
482	Detroit	96.7	110.2	102.8
484-485	Flint	94.9	98.2	96.4
486	Saginaw	94.5	94.0	94.3
487	Bay City	94.6	93.9	94.3
488-489	Lansing	95.7	96.5	96.1
490	Battle Creek	95.8	90.4	93.4
491	Kalamazoo	96.1	88.4	92.7
492	Jackson	94.0	96.8	95.2
493,495	Grand Rapids	97.3	76.8	88.1
494	Muskegon	94.7	85.3	90.5

Location Factors

STATE/ZIP	CITY	MAT.	INST.	TOTAL
MICHIGAN (CONT'D)				
496	Traverse City	93.4	79.4	87.2
497	Gaylord	94.5	73.4	85.1
498-499	Iron mountain	96.7	88.8	93.2
MINNESOTA				
550-551	Saint Paul	101.6	123.0	111.2
553-555	Minneapolis	102.3	125.0	112.5
556-558	Duluth	100.6	112.2	105.8
559	Rochester	101.0	109.2	104.7
560	Mankato	97.6	106.4	101.5
561	Windom	96.2	95.3	95.8
562	Willmar	95.9	107.2	100.9
563	St. Cloud	97.3	123.2	108.9
564	Brainerd	97.6	107.3	101.9
565	Detroit Lakes	99.3	100.7	99.9
566	Bemidji	98.5	102.2	100.2
567	Thief River Falls	98.1	97.1	97.7
MISSISSIPPI				
386	Clarksdale	95.7	56.3	78.1
387	Greenville	99.3	66.1	84.5
388	Tupelo	97.1	58.8	80.0
389	Greenwood	97.0	55.8	78.6
390-392	Jackson	96.9	65.3	82.8
393	Meridian	95.9	67.0	83.0
394	Laurel	97.0	59.3	80.2
395	Biloxi	97.6	60.8	81.2
396	Mccomb	95.4	55.3	77.5
397	Columbus	97.0	57.7	79.4
MISSOURI				
630-631	St. Louis	99.4	106.6	102.6
633	Bowling Green	98.1	92.2	95.5
634	Hannibal	96.9	88.8	93.3
635	Kirksville	100.1	85.4	93.5
636	Flat River	99.1	94.7	97.1
637	Cape Girardeau	99.2	91.3	95.7
638	Sikeston	97.1	87.1	92.6
639	Poplar Bluff	96.6	88.4	92.9
640-641	Kansas City	99.0	107.3	102.7
644-645	St. Joseph	98.3	93.2	96.0
646	Chillicothe	95.2	71.1	84.4
647	Harrisonville	94.8	104.8	99.3
648	Joplin	97.1	71.4	85.6
650-651	Jefferson City	95.9	92.1	94.2
652	Columbia	97.4	93.7	95.7
653	Sedalia	96.5	94.6	95.7
654-655	Rolla	95.0	89.5	92.5
656-658	Springfield	98.2	80.9	90.5
MONTANA				
590-591	Billings	103.2	75.9	91.0
592	Wolf Point	103.2	68.6	87.7
593	Miles City	100.6	70.2	87.0
594	Great Falls	104.7	76.5	92.1
595	Havre	101.9	71.3	88.2
596	Helena	102.5	73.2	89.4
597	Butte	103.2	74.6	90.4
598	Missoula	100.5	73.4	88.4
599	Kalispell	99.4	71.3	86.8
NEBRASKA				
680-681	Omaha	99.8	80.6	91.2
683-685	Lincoln	97.7	74.9	87.5
686	Columbus	96.7	72.9	86.1
687	Norfolk	98.4	76.4	88.6
688	Grand Island	98.5	79.8	90.1
689	Hastings	97.7	80.7	90.1
690	Mccook	97.5	71.9	86.0
691	North Platte	97.8	80.9	90.3
692	Valentine	99.5	68.7	85.7
693	Alliance	99.6	66.8	85.0
NEVADA				
889-891	Las Vegas	100.2	112.8	105.9
893	Ely	99.1	99.6	99.3
894-895	Reno	99.2	91.1	95.6
897	Carson City	98.6	90.8	95.1
898	Elko	97.7	84.4	91.8
NEW HAMPSHIRE				
030	Nashua	100.2	86.1	93.9
031	Manchester	100.8	86.1	94.2

STATE/ZIP	CITY	MAT.	INST.	TOTAL
NEW HAMPSHIRE (CONT'D)				
032-033	Concord	98.7	82.5	91.5
034	Keene	97.0	49.9	76.0
035	Littleton	96.7	57.4	79.1
036	Charleston	96.5	47.6	74.6
037	Claremont	95.4	47.6	74.0
038	Portsmouth	97.7	89.9	94.2
NEW JERSEY				
070-071	Newark	102.3	125.6	112.7
072	Elizabeth	100.6	125.6	111.8
073	Jersey City	99.3	125.3	110.9
074-075	Paterson	101.4	125.5	112.2
076	Hackensack	99.3	125.4	111.0
077	Long Branch	98.9	124.1	110.2
078	Dover	99.6	125.5	111.2
079	Summit	99.6	125.6	111.2
080,083	Vineland	97.5	122.8	108.8
081	Camden	99.4	124.0	110.4
082,084	Atlantic City	98.1	122.7	109.1
085-086	Trenton	99.4	122.7	109.8
087	Point Pleasant	99.6	123.3	110.2
088-089	New Brunswick	100.1	125.0	111.2
NEW MEXICO				
870-872	Albuquerque	99.1	75.0	88.3
873	Gallup	99.3	75.0	88.5
874	Farmington	99.8	75.0	88.7
875	Santa Fe	100.1	75.0	88.9
877	Las Vegas	97.7	75.0	87.6
878	Socorro	97.2	75.0	87.3
879	Truth/Consequences	97.5	71.6	85.9
880	Las Cruces	96.3	70.9	85.0
881	Clovis	97.9	74.9	87.6
882	Roswell	99.6	75.0	88.6
883	Carrizozo	100.0	75.0	88.8
884	Tucumcari	98.6	74.9	88.0
NEW YORK				
100-102	New York	106.0	166.3	133.0
103	Staten Island	102.1	160.3	128.1
104	Bronx	99.9	160.3	126.9
105	Mount Vernon	100.0	134.3	115.3
106	White Plains	99.9	134.3	115.3
107	Yonkers	104.8	134.5	118.0
108	New Rochelle	100.3	134.3	115.5
109	Suffern	100.1	123.1	110.4
110	Queens	101.9	160.1	127.9
111	Long Island City	103.6	160.1	128.9
112	Brooklyn	104.0	160.1	129.1
113	Flushing	103.8	160.1	129.0
114	Jamaica	101.9	160.1	127.9
115,117,118	Hicksville	101.7	145.2	121.2
116	Far Rockaway	103.9	160.1	129.1
119	Riverhead	102.5	144.7	121.4
120-122	Albany	97.2	97.9	97.5
123	Schenectady	98.2	97.5	97.9
124	Kingston	101.6	118.7	109.2
125-126	Poughkeepsie	100.7	130.5	114.0
127	Monticello	100.0	119.2	108.6
128	Glens Falls	92.6	93.2	92.8
129	Plattsburgh	97.5	85.0	91.9
130-132	Syracuse	99.4	93.5	96.7
133-135	Utica	97.1	88.7	93.4
136	Watertown	98.9	92.6	96.1
137-139	Binghamton	98.8	90.2	95.0
140-142	Buffalo	100.5	102.9	101.6
143	Niagara Falls	98.3	98.0	98.2
144-146	Rochester	100.3	95.6	98.2
147	Jamestown	97.2	84.4	91.5
148-149	Elmira	96.9	92.0	94.7
NORTH CAROLINA				
270,272-274	Greensboro	99.2	47.9	76.2
271	Winston-Salem	99.1	46.5	75.6
275-276	Raleigh	99.6	48.1	76.6
277	Durham	100.3	48.4	77.1
278	Rocky Mount	96.6	39.8	71.2
279	Elizabeth City	97.5	41.0	72.3
280	Gastonia	98.2	46.0	74.8
281-282	Charlotte	99.2	48.4	76.5
283	Fayetteville	100.0	50.0	77.6
284	Wilmington	96.9	46.3	74.3
285	Kinston	94.9	40.6	70.6

Location Factors

STATE/ZIP	CITY	MAT.	INST.	TOTAL
NORTH CAROLINA (CONT'D)				
286	Hickory	95.2	42.8	71.8
287-288	Asheville	97.4	46.0	74.4
289	Murphy	96.2	33.4	68.1
NORTH DAKOTA				
580-581	Fargo	102.4	64.3	85.4
582	Grand Forks	102.6	53.8	80.8
583	Devils Lake	101.7	56.0	81.3
584	Jamestown	101.9	46.8	77.3
585	Bismarck	100.8	63.0	83.9
586	Dickinson	102.8	58.2	82.9
587	Minot	102.5	70.0	88.0
588	Williston	101.0	58.2	81.9
OHIO				
430-432	Columbus	97.9	90.3	94.5
433	Marion	94.1	80.8	88.2
434-436	Toledo	98.1	95.9	97.1
437-438	Zanesville	94.6	81.1	88.6
439	Steubenville	96.4	90.5	93.8
440	Lorain	98.7	91.5	95.5
441	Cleveland	99.0	100.5	99.7
442-443	Akron	99.8	92.6	96.6
444-445	Youngstown	99.1	87.5	93.9
446-447	Canton	99.2	82.5	91.8
448-449	Mansfield	96.2	87.0	92.1
450	Hamilton	95.7	83.8	90.4
451-452	Cincinnati	96.0	86.3	91.7
453-454	Dayton	95.8	83.6	90.3
455	Springfield	95.7	83.8	90.4
456	Chillicothe	94.5	88.5	91.8
457	Athens	97.0	81.0	89.8
458	Lima	97.4	85.0	91.9
OKLAHOMA				
730-731	Oklahoma City	98.2	61.5	81.8
734	Ardmore	95.2	61.0	79.9
735	Lawton	97.6	62.3	81.8
736	Clinton	96.7	59.6	80.2
737	Enid	97.4	59.6	80.5
738	Woodward	95.4	59.7	79.4
739	Guymon	96.5	31.1	67.3
740-741	Tulsa	97.4	54.2	78.1
743	Miami	94.0	69.0	82.8
744	Muskogee	96.7	39.6	71.2
745	Mcalester	93.7	51.2	74.7
746	Ponca City	94.3	59.6	78.8
747	Durant	94.3	58.5	78.3
748	Shawnee	96.0	57.3	78.7
749	Poteau	93.3	62.3	79.5
OREGON				
970-972	Portland	100.0	99.7	99.9
973	Salem	99.8	98.6	99.3
974	Eugene	99.8	98.3	99.1
975	Medford	101.5	96.7	99.3
976	Klamath Falls	101.5	96.7	99.3
977	Bend	100.3	98.4	99.5
978	Pendleton	94.9	100.3	97.3
979	Vale	92.6	90.4	91.6
PENNSYLVANIA				
150-152	Pittsburgh	99.3	102.9	100.9
153	Washington	96.2	100.8	98.3
154	Uniontown	96.6	99.9	98.0
155	Bedford	97.6	91.1	94.7
156	Greensburg	97.5	99.4	98.3
157	Indiana	96.3	98.2	97.1
158	Dubois	98.0	94.2	96.3
159	Johnstown	97.6	94.9	96.4
160	Butler	92.6	100.3	96.0
161	New Castle	92.6	97.2	94.7
162	Kittanning	93.2	101.4	96.9
163	Oil City	92.6	94.9	93.6
164-165	Erie	94.8	93.5	94.2
166	Altoona	95.0	90.5	93.0
167	Bradford	95.9	93.1	94.7
168	State College	95.5	91.4	93.7
169	Wellsboro	96.7	91.0	94.2
170-171	Harrisburg	98.0	93.8	96.1
172	Chambersburg	96.1	88.6	92.7
173-174	York	96.5	94.0	95.4
175-176	Lancaster	94.7	88.3	91.8

STATE/ZIP	CITY	MAT.	INST.	TOTAL
PENNSYLVANIA (CONT'D)				
177	Williamsport	93.3	79.7	87.2
178	Sunbury	95.5	94.0	94.8
179	Pottsville	94.6	95.6	95.1
180	Lehigh Valley	96.0	114.2	104.2
181	Allentown	98.3	108.6	102.9
182	Hazleton	95.5	97.3	96.3
183	Stroudsburg	95.3	102.3	98.4
184-185	Scranton	99.0	98.7	98.9
186-187	Wilkes-Barre	95.2	97.7	96.3
188	Montrose	94.9	96.3	95.5
189	Doylestown	95.1	121.9	107.1
190-191	Philadelphia	99.7	132.2	114.3
193	Westchester	96.1	124.7	108.9
194	Norristown	95.1	131.2	111.3
195-196	Reading	97.4	99.6	98.4
PUERTO RICO				
009	San Juan	121.8	24.0	78.1
RHODE ISLAND				
028	Newport	99.3	111.5	104.7
029	Providence	100.3	111.5	105.3
SOUTH CAROLINA				
290-292	Columbia	97.6	50.0	76.3
293	Spartanburg	96.4	49.6	75.5
294	Charleston	98.0	58.3	80.2
295	Florence	96.2	50.0	75.5
296	Greenville	96.2	49.6	75.3
297	Rock Hill	95.7	47.7	74.3
298	Aiken	96.7	71.9	85.6
299	Beaufort	97.5	43.0	73.1
SOUTH DAKOTA				
570-571	Sioux Falls	100.6	57.9	81.5
572	Watertown	99.7	52.2	78.4
573	Mitchell	98.3	51.6	77.4
574	Aberdeen	101.2	52.8	79.6
575	Pierre	99.7	54.5	79.5
576	Mobridge	99.1	52.0	78.0
577	Rapid City	101.0	55.3	80.6
TENNESSEE				
370-372	Nashville	96.7	72.7	85.9
373-374	Chattanooga	98.2	66.9	84.2
375,380-381	Memphis	96.6	70.3	84.9
376	Johnson City	97.6	57.4	79.6
377-379	Knoxville	94.5	61.7	79.9
382	Mckenzie	96.3	62.1	81.1
383	Jackson	98.2	63.8	82.8
384	Columbia	94.9	67.3	82.6
385	Cookeville	96.2	60.3	80.1
TEXAS				
750	Mckinney	99.5	50.8	77.7
751	Waxahackie	99.4	58.2	81.0
752-753	Dallas	99.9	67.5	85.4
754	Greenville	99.6	43.7	74.6
755	Texarkana	99.1	51.5	77.8
756	Longview	99.7	40.7	73.4
757	Tyler	100.2	54.9	80.0
758	Palestine	96.1	57.4	78.8
759	Lufkin	96.7	59.1	79.9
760-761	Fort Worth	97.7	63.6	82.4
762	Denton	97.3	48.4	75.4
763	Wichita Falls	98.1	55.5	79.1
764	Eastland	96.8	50.3	76.0
765	Temple	95.3	50.0	75.1
766-767	Waco	97.5	57.1	79.4
768	Brownwood	98.0	49.6	76.3
769	San Angelo	97.6	50.1	76.3
770-772	Houston	100.0	70.4	86.8
773	Huntsville	98.6	56.7	79.9
774	Wharton	99.8	54.8	79.7
775	Galveston	97.6	69.2	84.9
776-777	Beaumont	98.2	62.9	82.4
778	Bryan	95.1	62.1	80.3
779	Victoria	99.9	44.4	75.1
780	Laredo	94.7	53.1	76.1
781-782	San Antonio	95.0	65.1	81.6
783-784	Corpus Christi	97.6	52.8	77.6
785	Mc Allen	97.5	46.9	74.9
786-787	Austin	95.1	59.4	79.1

Location Factors

STATE/ZIP	CITY	MAT.	INST.	TOTAL
TEXAS (CONT'D)				
788	Del Rio	97.0	47.0	74.6
789	Giddings	94.1	57.0	77.5
790-791	Amarillo	98.1	61.9	81.9
792	Childress	97.5	58.1	79.9
793-794	Lubbock	99.6	57.5	80.8
795-796	Abilene	97.9	54.8	78.6
797	Midland	99.7	51.9	78.3
798-799,885	El Paso	97.0	51.7	76.8
UTAH				
840-841	Salt Lake City	102.0	69.1	87.3
842,844	Ogden	97.5	70.7	85.5
843	Logan	99.5	70.7	86.6
845	Price	100.0	67.1	85.3
846-847	Provo	100.0	68.9	86.1
VERMONT				
050	White River Jct.	98.3	55.1	79.0
051	Bellows Falls	96.6	73.7	86.4
052	Bennington	97.0	68.9	84.4
053	Brattleboro	97.4	72.9	86.5
054	Burlington	100.9	66.1	85.4
056	Montpelier	97.7	66.2	83.6
057	Rutland	99.2	66.2	84.4
058	St. Johnsbury	98.3	55.9	79.3
059	Guildhall	96.9	55.7	78.5
VIRGINIA				
220-221	Fairfax	99.9	82.3	92.0
222	Arlington	101.2	80.1	91.8
223	Alexandria	100.2	85.5	93.6
224-225	Fredericksburg	98.5	72.2	86.7
226	Winchester	99.2	63.2	83.1
227	Culpeper	99.1	67.8	85.1
228	Harrisonburg	99.3	64.0	83.5
229	Charlottesville	99.8	63.4	83.5
230-232	Richmond	100.2	66.8	85.3
233-235	Norfolk	101.0	68.2	86.3
236	Newport News	100.0	67.6	85.5
237	Portsmouth	99.4	65.5	84.2
238	Petersburg	99.6	66.8	84.9
239	Farmville	98.7	53.0	78.3
240-241	Roanoke	100.7	60.7	82.8
242	Bristol	98.4	55.1	79.0
243	Pulaski	98.0	50.9	77.0
244	Staunton	98.9	59.1	81.1
245	Lynchburg	99.0	64.6	83.6
246	Grundy	98.3	50.3	76.9
WASHINGTON				
980-981,987	Seattle	103.8	106.6	105.0
982	Everett	103.4	97.1	100.6
983-984	Tacoma	103.5	98.9	101.4
985	Olympia	101.3	98.9	100.2
986	Vancouver	104.8	93.8	99.9
988	Wenatchee	103.3	83.7	94.6
989	Yakima	103.7	93.1	99.0
990-992	Spokane	103.7	80.8	93.5
993	Richland	103.4	88.0	96.5
994	Clarkston	101.9	79.8	92.0
WEST VIRGINIA				
247-248	Bluefield	97.1	77.7	88.4
249	Lewisburg	98.9	83.2	91.9
250-253	Charleston	99.0	90.1	95.0
254	Martinsburg	98.6	79.1	89.9
255-257	Huntington	100.2	91.6	96.3
258-259	Beckley	97.0	87.2	92.6
260	Wheeling	100.3	89.9	95.7
261	Parkersburg	99.1	89.4	94.8
262	Buckhannon	98.5	89.7	94.6
263-264	Clarksburg	99.2	89.7	94.9
265	Morgantown	99.2	88.8	94.6
266	Gassaway	98.4	89.9	94.6
267	Romney	98.4	85.6	92.7
268	Petersburg	98.2	85.0	92.3
WISCONSIN				
530,532	Milwaukee	100.2	107.6	103.5
531	Kenosha	100.0	102.7	101.2
534	Racine	99.4	102.8	100.9
535	Beloit	99.3	97.8	98.6
537	Madison	98.9	99.1	99.0

STATE/ZIP	CITY	MAT.	INST.	TOTAL
WISCONSIN (CONT'D)				
538	Lancaster	97.0	93.2	95.3
539	Portage	95.4	96.3	95.8
540	New Richmond	97.3	96.2	96.8
541-543	Green Bay	101.7	93.3	98.0
544	Wausau	96.2	93.4	95.0
545	Rhinelander	99.8	92.2	96.4
546	La Crosse	97.6	94.9	96.4
547	Eau Claire	99.3	96.1	97.9
548	Superior	97.1	98.1	97.6
549	Oshkosh	97.0	84.1	91.2
WYOMING				
820	Cheyenne	99.2	64.7	83.8
821	Yellowstone Nat'l Park	96.8	60.0	80.4
822	Wheatland	98.0	59.2	80.6
823	Rawlins	99.6	58.0	81.0
824	Worland	97.5	58.0	79.8
825	Riverton	98.6	58.0	80.4
826	Casper	99.5	58.5	81.2
827	Newcastle	97.3	58.0	79.7
828	Sheridan	100.1	61.5	82.8
829-831	Rock Springs	101.5	58.1	82.1
CANADIAN FACTORS (reflect Canadian currency)				
ALBERTA				
	Calgary	130.9	94.8	114.8
	Edmonton	131.1	94.8	114.9
	Fort McMurray	127.0	97.7	113.9
	Lethbridge	120.9	97.1	110.3
	Lloydminster	116.2	93.6	106.1
	Medicine Hat	116.3	92.8	105.8
	Red Deer	116.8	92.8	106.1
BRITISH COLUMBIA				
	Kamloops	117.2	96.2	107.8
	Prince George	118.5	95.4	108.2
	Vancouver	129.1	90.7	111.9
	Victoria	118.5	88.7	105.2
MANITOBA				
	Brandon	116.1	80.3	100.1
	Portage la Prairie	116.1	78.5	99.3
	Winnipeg	128.8	73.5	104.1
NEW BRUNSWICK				
	Bathurst	114.2	71.2	94.9
	Dalhousie	114.2	71.4	95.1
	Fredericton	116.8	76.8	98.9
	Moncton	114.5	72.8	95.8
	Newcastle	114.2	71.9	95.3
	St. John	116.9	76.5	98.8
NEWFOUNDLAND				
	Corner Brook	119.6	71.5	98.1
	St Johns	119.7	76.9	100.6
NORTHWEST TERRITORIES				
	Yellowknife	121.5	90.5	107.7
NOVA SCOTIA				
	Bridgewater	115.8	79.1	99.4
	Dartmouth	117.2	79.1	100.1
	Halifax	118.1	81.7	101.8
	New Glasgow	115.3	79.1	99.1
	Sydney	112.6	79.1	97.6
	Truro	115.3	79.1	99.1
	Yarmouth	115.1	79.1	99.0
ONTARIO				
	Barrie	118.9	98.8	109.9
	Brantford	118.1	103.0	111.3
	Cornwall	118.0	99.1	109.5
	Hamilton	122.6	95.4	110.5
	Kingston	118.9	99.2	110.1
	Kitchener	114.0	92.1	104.2
	London	121.4	91.9	108.2
	North Bay	118.1	96.7	108.5
	Oshawa	117.7	100.0	109.8
	Ottawa	123.0	92.3	109.3
	Owen Sound	119.1	96.8	109.1
	Peterborough	118.1	98.8	109.5
	Sarnia	117.8	103.6	111.4

Location Factors

STATE/ZIP	CITY	MAT.	INST.	TOTAL
ONTARIO (CONT'D)	Sault Ste Marie	112.7	96.3	105.4
	St. Catharines	112.1	93.1	103.6
	Sudbury	112.5	92.1	103.4
	Thunder Bay	113.6	91.9	103.9
	Timmins	118.3	96.7	108.6
	Toronto	123.6	100.7	113.4
	Windsor	112.7	91.6	103.3
PRINCE EDWARD ISLAND				
	Charlottetown	117.6	66.4	94.7
	Summerside	117.1	66.4	94.5
QUEBEC				
	Cap-de-la-Madeleine	114.9	93.0	105.1
	Charlesbourg	114.9	93.0	105.1
	Chicoutimi	113.9	93.2	104.7
	Gatineau	114.6	92.7	104.8
	Granby	114.8	92.7	104.9
	Hull	114.6	92.7	104.8
	Joliette	115.1	93.0	105.2
	Laval	114.3	92.7	104.6
	Montreal	122.0	93.5	109.3
	Quebec	120.1	94.0	108.4
	Rimouski	114.4	93.2	104.9
	Rouyn-Noranda	114.5	92.7	104.7
	Saint Hyacinthe	113.9	92.7	104.4
	Sherbrooke	114.9	92.7	105.0
	Sorel	115.1	93.0	105.2
	St Jerome	114.5	92.7	104.8
	Trois Rivieres	115.1	93.0	105.2
SASKATCHEWAN				
	Moose Jaw	113.4	73.6	95.6
	Prince Albert	112.3	71.6	94.1
	Regina	114.9	87.7	102.7
	Saskatoon	113.4	87.7	101.9
YUKON				
	Whitehorse	113.2	72.4	95.0

R031113-10 Wall Form Materials

Aluminum Forms

Approximate weight is 3 lbs. per S.F.C.A. Standard widths are available from 4″ to 36″ with 36″ most common. Standard lengths of 2′, 4′, 6′ to 8′ are available. Forms are lightweight and fewer ties are needed with the wider widths. The form face is either smooth or textured.

Metal Framed Plywood Forms

Manufacturers claim over 75 reuses of plywood and over 300 reuses of steel frames. Many specials such as corners, fillers, pilasters, etc. are available. Monthly rental is generally about 15% of purchase price for first month and 9% per month thereafter with 90% of rental applied to purchase for the first month and decreasing percentages thereafter. Aluminum framed forms cost 25% to 30% more than steel framed.

After the first month, extra days may be prorated from the monthly charge. Rental rates do not include ties, accessories, cleaning, loss of hardware or freight in and out. Approximate weight is 5 lbs. per S.F. for steel; 3 lbs. per S.F. for aluminum.

Forms can be rented with option to buy.

Plywood Forms, Job Fabricated

There are two types of plywood used for concrete forms.

1. Exterior plyform which is completely waterproof. This is face oiled to facilitate stripping. Ten reuses can be expected with this type with 25 reuses possible.
2. An overlaid type consists of a resin fiber fused to exterior plyform. No oiling is required except to facilitate cleaning. This is available in both high density (HDO) and medium density overlaid (MDO). Using HDO, 50 reuses can be expected with 200 possible.

Plyform is available in 5/8″ and 3/4″ thickness. High density overlaid is available in 3/8″, 1/2″, 5/8″ and 3/4″ thickness.

5/8″ thick is sufficient for most building forms, while 3/4″ is best on heavy construction.

Plywood Forms, Modular, Prefabricated

There are many plywood forming systems without frames. Most of these are manufactured from 1-1/8″ (HDO) plywood and have some hardware attached. These are used principally for foundation walls 8′ or less high. With care and maintenance, 100 reuses can be attained with decreasing quality of surface finish.

Steel Forms

Approximate weight is 6-1/2 lbs. per S.F.C.A. including accessories. Standard widths are available from 2″ to 24″, with 24″ most common. Standard lengths are from 2′ to 8′, with 4′ the most common. Forms are easily ganged into modular units.

Forms are usually leased for 15% of the purchase price per month prorated daily over 30 days.

Rental may be applied to sale price, and usually rental forms are bought. With careful handling and cleaning 200 to 400 reuses are possible.

Straight wall gang forms up to 12′ x 20′ or 8′ x 30′ can be fabricated. These crane handled forms usually lease for approx. 9% per month.

Individual job analysis is available from the manufacturer at no charge.

R031113-40 Forms for Reinforced Concrete

Design Economy

Avoid many sizes in proportioning beams and columns.

From story to story avoid changing column dimensions. Gain strength by adding steel or using a richer mix. If a change in size of column is necessary, vary one dimension only to minimize form alterations. Keep beams and columns the same width.

From floor to floor in a multi-story building vary beam depth, not width, as that will leave slab panel form unchanged. It is cheaper to vary the strength of a beam from floor to floor by means of steel area than by 2″ changes in either width or depth.

Cost Factors

Material includes the cost of lumber, cost of rent for metal pans or forms if used, nails, form ties, form oil, bolts and accessories.

Labor includes the cost of carpenters to make up, erect, remove and repair, plus common labor to clean and move. Having carpenters remove forms minimizes repairs.

Improper alignment and condition of forms will increase finishing cost. When forms are heavily oiled, concrete surfaces must be neutralized before finishing. Special curing compounds will cause spillages to spall off in first frost. Gang forming methods will reduce costs on large projects.

Materials Used

Boards are seldom used unless their architectural finish is required. Generally, steel, fiberglass and plywood are used for contact surfaces. Labor on plywood is 10% less than with boards. The plywood is backed up with 2 x 4′s at 12″ to 32″ O.C. Walers are generally 2 - 2 x 4′s. Column forms are held together with steel yokes or bands. Shoring is with adjustable shoring or scaffolding for high ceilings.

Reuse

Floor and column forms can be reused four or possibly five times without excessive repair. Remember to allow for 10% waste on each reuse.

When modular sized wall forms are made, up to twenty uses can be expected with exterior plyform.

When forms are reused, the cost to erect, strip, clean and move will not be affected. 10% replacement of lumber should be included and about one hour of carpenter time for repairs on each reuse per 100 S.F.

The reuse cost for certain accessory items normally rented on a monthly basis will be lower than the cost for the first use.

After fifth use, new material required plus time needed for repair prevent form cost from dropping further and it may go up. Much depends on care in stripping, the number of special bays, changes in beam or column sizes and other factors.

Costs for multiple use of formwork may be developed as follows:

2 Uses

$$\frac{\text{(1st Use + Reuse)}}{2} = \text{avg. cost/2 uses}$$

3 Uses

$$\frac{\text{(1st Use + 2 Reuse)}}{3} = \text{avg. cost/3 uses}$$

4 Uses

$$\frac{\text{(1st use + 3 Reuse)}}{4} = \text{avg. cost/4 uses}$$

R031113-60 Formwork Labor-Hours

Item	Unit	Hours Required			Total Hours	Multiple Use		
		Fabricate	Erect & Strip	Clean & Move	1 Use	2 Use	3 Use	4 Use
Beam and Girder, interior beams, 12″ wide	100 S.F.	6.4	8.3	1.3	16.0	13.3	12.4	12.0
Hung from steel beams		5.8	7.7	1.3	14.8	12.4	11.6	11.2
Beam sides only, 36″ high		5.8	7.2	1.3	14.3	11.9	11.1	10.7
Beam bottoms only, 24″ wide		6.6	13.0	1.3	20.9	18.1	17.2	16.7
Box out for openings		9.9	10.0	1.1	21.0	16.6	15.1	14.3
Buttress forms, to 8′ high		6.0	6.5	1.2	13.7	11.2	10.4	10.0
Centering, steel, 3/4″ rib lath			1.0		1.0			
3/8″ rib lath or slab form			0.9		0.9			
Chamfer strip or keyway	100 L.F.		1.5		1.5	1.5	1.5	1.5
Columns, fiber tube 8″ diameter			20.6		20.6			
12″			21.3		21.3			
16″			22.9		22.9			
20″			23.7		23.7			
24″			24.6		24.6			
30″			25.6		25.6			
Columns, round steel, 12″ diameter			22.0		22.0	22.0	22.0	22.0
16″			25.6		25.6	25.6	25.6	25.6
20″			30.5		30.5	30.5	30.5	30.5
24″			37.7		37.7	37.7	37.7	37.7
Columns, plywood 8″ x 8″	100 S.F.	7.0	11.0	1.2	19.2	16.2	15.2	14.7
12″ x 12″		6.0	10.5	1.2	17.7	15.2	14.4	14.0
16″ x 16″		5.9	10.0	1.2	17.1	14.7	13.8	13.4
24″ x 24″		5.8	9.8	1.2	16.8	14.4	13.6	13.2
Columns, steel framed plywood 8″ x 8″			10.0	1.0	11.0	11.0	11.0	11.0
12″ x 12″			9.3	1.0	10.3	10.3	10.3	10.3
16″ x 16″			8.5	1.0	9.5	9.5	9.5	9.5
24″ x 24″			7.8	1.0	8.8	8.8	8.8	8.8
Drop head forms, plywood		9.0	12.5	1.5	23.0	19.0	17.7	17.0
Coping forms		8.5	15.0	1.5	25.0	21.3	20.0	19.4
Culvert, box			14.5	4.3	18.8	18.8	18.8	18.8
Curb forms, 6″ to 12″ high, on grade		5.0	8.5	1.2	14.7	12.7	12.1	11.7
On elevated slabs		6.0	10.8	1.2	18.0	15.5	14.7	14.3
Edge forms to 6″ high, on grade	100 L.F.	2.0	3.5	0.6	6.1	5.6	5.4	5.3
7″ to 12″ high	100 S.F.	2.5	5.0	1.0	8.5	7.8	7.5	7.4
Equipment foundations		10.0	18.0	2.0	30.0	25.5	24.0	23.3
Flat slabs, including drops		3.5	6.0	1.2	10.7	9.5	9.0	8.8
Hung from steel		3.0	5.5	1.2	9.7	8.7	8.4	8.2
Closed deck for domes		3.0	5.8	1.2	10.0	9.0	8.7	8.5
Open deck for pans		2.2	5.3	1.0	8.5	7.9	7.7	7.6
Footings, continuous, 12″ high		3.5	3.5	1.5	8.5	7.3	6.8	6.6
Spread, 12″ high		4.7	4.2	1.6	10.5	8.7	8.0	7.7
Pile caps, square or rectangular		4.5	5.0	1.5	11.0	9.3	8.7	8.4
Grade beams, 24″ deep		2.5	5.3	1.2	9.0	8.3	8.0	7.9
Lintel or Sill forms		8.0	17.0	2.0	27.0	23.5	22.3	21.8
Spandrel beams, 12″ wide		9.0	11.2	1.3	21.5	17.5	16.2	15.5
Stairs			25.0	4.0	29.0	29.0	29.0	29.0
Trench forms in floor		4.5	14.0	1.5	20.0	18.3	17.7	17.4
Walls, Plywood, at grade, to 8′ high		5.0	6.5	1.5	13.0	11.0	9.7	9.5
8′ to 16′		7.5	8.0	1.5	17.0	13.8	12.7	12.1
16′ to 20′		9.0	10.0	1.5	20.5	16.5	15.2	14.5
Foundation walls, to 8′ high		4.5	6.5	1.0	12.0	10.3	9.7	9.4
8′ to 16′ high		5.5	7.5	1.0	14.0	11.8	11.0	10.6
Retaining wall to 12′ high, battered		6.0	8.5	1.5	16.0	13.5	12.7	12.3
Radial walls to 12′ high, smooth		8.0	9.5	2.0	19.5	16.0	14.8	14.3
2′ chords		7.0	8.0	1.5	16.5	13.5	12.5	12.0
Prefabricated modular, to 8′ high		—	4.3	1.0	5.3	5.3	5.3	5.3
Steel, to 8′ high		—	6.8	1.2	8.0	8.0	8.0	8.0
8′ to 16′ high		—	9.1	1.5	10.6	10.3	10.2	10.2
Steel framed plywood to 8′ high		—	6.8	1.2	8.0	7.5	7.3	7.2
8′ to 16′ high		—	9.3	1.2	10.5	9.5	9.2	9.0

R032110-10 Reinforcing Steel Weights and Measures

Bar Designation No.**	U.S. Customary Units					SI Units			
	Nominal Weight Lb./Ft.	Nominal Dimensions*				Nominal Dimensions*			
		Diameter in.	Cross Sectional Area, in.^{2}	Perimeter in.	Nominal Weight kg/m	Diameter mm	Cross Sectional Area, cm^2	Perimeter mm
3	.376	.375	.11	1.178	.560	9.52	.71	29.9
4	.668	.500	.20	1.571	.994	12.70	1.29	39.9
5	1.043	.625	.31	1.963	1.552	15.88	2.00	49.9
6	1.502	.750	.44	2.356	2.235	19.05	2.84	59.8
7	2.044	.875	.60	2.749	3.042	22.22	3.87	69.8
8	2.670	1.000	.79	3.142	3.973	25.40	5.10	79.8
9	3.400	1.128	1.00	3.544	5.059	28.65	6.45	90.0
10	4.303	1.270	1.27	3.990	6.403	32.26	8.19	101.4
11	5.313	1.410	1.56	4.430	7.906	35.81	10.06	112.5
14	7.650	1.693	2.25	5.320	11.384	43.00	14.52	135.1
18	13.600	2.257	4.00	7.090	20.238	57.33	25.81	180.1

* The nominal dimensions of a deformed bar are equivalent to those of a plain round bar having the same weight per foot as the deformed bar.
** Bar numbers are based on the number of eighths of an inch included in the nominal diameter of the bars.

R032110-20 Metric Rebar Specification - ASTM A615-81

Grade 300 (300 MPa* = 43,560 psi; +8.7% vs. Grade 40) Grade 400 (400 MPa* = 58,000 psi; –3.4% vs. Grade 60)				
Bar No.	Diameter mm	Area mm^2	Equivalent in.^{2}	Comparison with U.S. Customary Bars
10M	11.3	100	.16	Between #3 & #4
15M	16.0	200	.31	#5 (.31 in.^{2})
20M	19.5	300	.47	#6 (.44 in.^{2})
25M	25.2	500	.78	#8 (.79 in.^{2})
30M	29.9	700	1.09	#9 (1.00 in.^{2})
35M	35.7	1000	1.55	#11 (1.56 in.^{2})
45M	43.7	1500	2.33	#14 (2.25 in.^{2})
55M	56.4	2500	3.88	#18 (4.00 in.^{2})

* MPa = megapascals

R032110-25 Comparison of U.S. Customary Units and SI Units for Reinforcing Bars

U.S. Customary Units								
		Nominal Dimensions[a]			Deformation Requirements, in.			
Bar Designation No.[b]	Nominal Weight, lb/ft	Diameter in.	Cross Sectional Area, in.^{2}	Perimeter in.	Maximum Average Spacing	Minimum Average Height	Maximum Gap (Chord of 12-1/2% of Nominal Perimeter)
3	0.376	0.375	0.11	1.178	0.262	0.015	0.143
4	0.668	0.500	0.20	1.571	0.350	0.020	0.191
5	1.043	0.625	0.31	1.963	0.437	0.028	0.239
6	1.502	0.750	0.44	2.356	0.525	0.038	0.286
7	2.044	0.875	0.60	2.749	0.612	0.044	0.334
8	2.670	1.000	0.79	3.142	0.700	0.050	0.383
9	3.400	1.128	1.00	3.544	0.790	0.056	0.431
10	4.303	1.270	1.27	3.990	0.889	0.064	0.487
11	5.313	1.410	1.56	4.430	0.987	0.071	0.540
14	7.65	1.693	2.25	5.32	1.185	0.085	0.648
18	13.60	2.257	4.00	7.09	1.58	0.102	0.864

SI UNITS							
		Nominal Dimensions[a]			Deformation Requirements, mm		
Bar Designation No.[b]	Nominal Weight kg/m	Diameter, mm	Cross Sectional Area, cm^2	Perimeter, mm	Maximum Average Spacing	Minimum Average Height	Maximum Gap (Chord of 12-1/2% of Nominal Perimeter)
3	0.560	9.52	0.71	29.9	6.7	0.38	3.5
4	0.994	12.70	1.29	39.9	8.9	0.51	4.9
5	1.552	15.88	2.00	49.9	11.1	0.71	6.1
6	2.235	19.05	2.84	59.8	13.3	0.96	7.3
7	3.042	22.22	3.87	69.8	15.5	1.11	8.5
8	3.973	25.40	5.10	79.8	17.8	1.27	9.7
9	5.059	28.65	6.45	90.0	20.1	1.42	10.9
10	6.403	32.26	8.19	101.4	22.6	1.62	11.4
11	7.906	35.81	10.06	112.5	25.1	1.80	13.6
14	11.384	43.00	14.52	135.1	30.1	2.16	16.5
18	20.238	57.33	25.81	180.1	40.1	2.59	21.9

[a]Nominal dimensions of a deformed bar are equivalent to those of a plain round bar having the same weight per foot as the deformed bar.

[b]Bar numbers are based on the number of eighths of an inch included in the nominal diameter of the bars.

R032110-40 Weight of Steel Reinforcing Per Square Foot of Wall (PSF)

Reinforced Weights: The table below suggests the weights per square foot for reinforcing steel in walls. Weights are approximate and will be the same for all grades of steel bars. For bars in two directions, add weights for each size and spacing.

C/C Spacing in Inches	Bar Size								
	#3 Wt. (PSF)	#4 Wt. (PSF)	#5 Wt. (PSF)	#6 Wt. (PSF)	#7 Wt. (PSF)	#8 Wt. (PSF)	#9 Wt. (PSF)	#10 Wt. (PSF)	#11 Wt. (PSF)
2"	2.26	4.01	6.26	9.01	12.27				
3"	1.50	2.67	4.17	6.01	8.18	10.68	13.60	17.21	21.25
4"	1.13	2.01	3.13	4.51	6.13	8.10	10.20	12.91	15.94
5"	.90	1.60	2.50	3.60	4.91	6.41	8.16	10.33	12.75
6"	.752	1.34	2.09	3.00	4.09	5.34	6.80	8.61	10.63
8"	.564	1.00	1.57	2.25	3.07	4.01	5.10	6.46	7.97
10"	.451	.802	1.25	1.80	2.45	3.20	4.08	5.16	6.38
12"	.376	.668	1.04	1.50	2.04	2.67	3.40	4.30	5.31
18"	.251	.445	.695	1.00	1.32	1.78	2.27	2.86	3.54
24"	.188	.334	.522	.751	1.02	1.34	1.70	2.15	2.66
30"	.150	.267	.417	.600	.817	1.07	1.36	1.72	2.13
36"	.125	.223	.348	.501	.681	.890	1.13	1.43	1.77
42"	.107	.191	.298	.429	.584	.753	.97	1.17	1.52
48"	.094	.167	.261	.376	.511	.668	.85	1.08	1.33

R032110-50 Minimum Wall Reinforcement Weight (PSF)

This table lists the approximate minimum wall reinforcement weights per S.F. according to the specification of .12% of gross area for vertical bars and .20% of gross area for horizontal bars.

Location	Wall Thickness	Bar Size	Horizontal Steel Spacing C/C	Sq. In. Req'd per S.F.	Total Wt. per S.F.	Bar Size	Vertical Steel Spacing C/C	Sq. In. Req'd per S.F.	Total Wt. per S.F.	Horizontal & Vertical Steel Total Weight per S.F.
Both Faces	10"	#4	18"	.24	.89#	#3	18"	.14	.50#	1.39#
	12"	#4	16"	.29	1.00	#3	16"	.17	.60	1.60
	14"	#4	14"	.34	1.14	#3	13"	.20	.69	1.84
	16"	#4	12"	.38	1.34	#3	11"	.23	.82	2.16
	18"	#5	17"	.43	1.47	#4	18"	.26	.89	2.36
One Face	6"	#3	9"	.15	.50	#3	18"	.09	.25	.75
	8"	#4	12"	.19	.67	#3	11"	.12	.41	1.08
	10"	#5	15"	.24	.83	#4	16"	.14	.50	1.34

R032110-70 Bend, Place and Tie Reinforcing

Placing and tying by rodmen for footings and slabs runs from nine hrs. per ton for heavy bars to fifteen hrs. per ton for light bars. For beams, columns, and walls, production runs from eight hrs. per ton for heavy bars to twenty hrs. per ton for light bars. Overall average for typical reinforced concrete buildings is about fourteen hrs. per ton. These production figures include the time for placing of accessories and usual inserts, but not their material cost (allow 15% of the cost of delivered bent rods). Equipment handling is necessary for the larger-sized bars so that installation costs for the very heavy bars will not decrease proportionately.

Installation costs for splicing reinforcing bars include allowance for equipment to hold the bars in place while splicing as well as necessary scaffolding for iron workers.

R032110-80 Shop-Fabricated Reinforcing Steel

The material prices for reinforcing, shown in the unit cost sections of the book, are for 50 tons or more of shop-fabricated reinforcing steel and include:

1. Mill base price of reinforcing steel
2. Mill grade/size/length extras
3. Mill delivery to the fabrication shop
4. Shop storage and handling
5. Shop drafting/detailing
6. Shop shearing and bending
7. Shop listing
8. Shop delivery to the job site

Both material and installation costs can be considerably higher for small jobs consisting primarily of smaller bars, while material costs may be slightly lower for larger jobs.

R0322 Welded Wire Fabric Reinforcing

R032205-30 Common Stock Styles of Welded Wire Fabric

This table provides some of the basic specifications, sizes, and weights of welded wire fabric used for reinforcing concrete.

New Designation		Old Designation	Steel Area per Foot				Approximate Weight per 100 S.F.	
Spacing — Cross Sectional Area (in.) — (Sq. in. 100)		Spacing — Wire Gauge (in.) — (AS & W)	Longitudinal		Transverse			
			in.	cm	in.	cm	lbs	kg
Rolls	6 x 6 — W1.4 x W1.4	6 x 6 — 10 x 10	.028	.071	.028	.071	21	9.53
	6 x 6 — W2.0 x W2.0	6 x 6 — 8 x 8 [1]	.040	.102	.040	.102	29	13.15
	6 x 6 — W2.9 x W2.9	6 x 6 — 6 x 6	.058	.147	.058	.147	42	19.05
	6 x 6 — W4.0 x W4.0	6 x 6 — 4 x 4	.080	.203	.080	.203	58	26.91
	4 x 4 — W1.4 x W1.4	4 x 4 — 10 x 10	.042	.107	.042	.107	31	14.06
	4 x 4 — W2.0 x W2.0	4 x 4 — 8 x 8 [1]	.060	.152	.060	.152	43	19.50
	4 x 4 — W2.9 x W2.9	4 x 4 — 6 x 6	.087	.227	.087	.227	62	28.12
	4 x 4 — W4.0 x W4.0	4 x 4 — 4 x 4	.120	.305	.120	.305	85	38.56
Sheets	6 x 6 — W2.9 x W2.9	6 x 6 — 6 x 6	.058	.147	.058	.147	42	19.05
	6 x 6 — W4.0 x W4.0	6 x 6 — 4 x 4	.080	.203	.080	.203	58	26.31
	6 x 6 — W5.5 x W5.5	6 x 6 — 2 x 2 [2]	.110	.279	.110	.279	80	36.29
	4 x 4 — W1.4 x W1.4	4 x 4 — 4 x 4	.120	.305	.120	.305	85	38.56

NOTES: 1. Exact W—number size for 8 gauge is W2.1
2. Exact W—number size for 2 gauge is W5.4

R033053-10 Spread Footings

General: A spread footing is used to convert a concentrated load (from one superstructure column, or substructure grade beams) into an allowable area load on supporting soil.

Because of punching action from the column load, a spread footing is usually thicker than strip footings which support wall loads. One or two story commercial or residential buildings should have no less than 1′ thick spread footings. Heavier loads require no less than 2′ thick. Spread footings may be square, rectangular or octagonal in plan.

Spread footings tend to minimize excavation and foundation materials, as well as labor and equipment. Another advantage is that footings and soil conditions can be readily examined. They are the most widely used type of footing, especially in mild climates and for buildings of four stories or under. This is because they are usually more economical than other types, if suitable soil and site conditions exist.

They are used when suitable supporting soil is located within several feet of the surface or line of subsurface excavation. Suitable soil types include sands and gravels, gravels with a small amount of clay or silt, hardpan, chalk, and rock. Pedestals may be used to bring the column base load down to the top of footing. Alternately, undesirable soil between underside of footing and top of bearing level can be removed and replaced with lean concrete mix or compacted granular material.

Depth of footing should be below topsoil, uncompacted fill, muck, etc. It must be lower than frost penetration but should be above the water table. It must not be at the ground surface because of potential surface erosion. If the ground slopes, approximately three horizontal feet of edge protection must remain. Differential footing elevations may overlap soil stresses or cause excavation problems if clear spacing between footings is less than the difference in depth.

Other footing types are usually used for the following reasons:

A. Bearing capacity of soil is low.

B. Very large footings are required, at a cost disadvantage.

C. Soil under footing (shallow or deep) is very compressible, with probability of causing excessive or differential settlement.

D. Good bearing soil is deep.

E. Potential for scour action exists.

F. Varying subsoil conditions within building perimeter.

Cost of spread footings for a building is determined by:

1. The soil bearing capacity.
2. Typical bay size.
3. Total load (live plus dead) per S.F. for roof and elevated floor levels.
4. The size and shape of the building.
5. Footing configuration. Does the building utilize outer spread footings or are there continuous perimeter footings only or a combination of spread footings plus continuous footings?

Soil Bearing Capacity in Kips per S.F.

Bearing Material	Typical Allowable Bearing Capacity
Hard sound rock	120 KSF
Medium hard rock	80
Hardpan overlaying rock	24
Compact gravel and boulder-gravel; very compact sandy gravel	20
Soft rock	16
Loose gravel; sandy gravel; compact sand; very compact sand-inorganic silt	12
Hard dry consolidated clay	10
Loose coarse to medium sand; medium compact fine sand	8
Compact sand-clay	6
Loose fine sand; medium compact sand-inorganic silts	4
Firm or stiff clay	3
Loose saturated sand-clay; medium soft clay	2

Concrete — R0330 Cast-In-Place Concrete

R033053-60 Maximum Depth of Frost Penetration in Inches

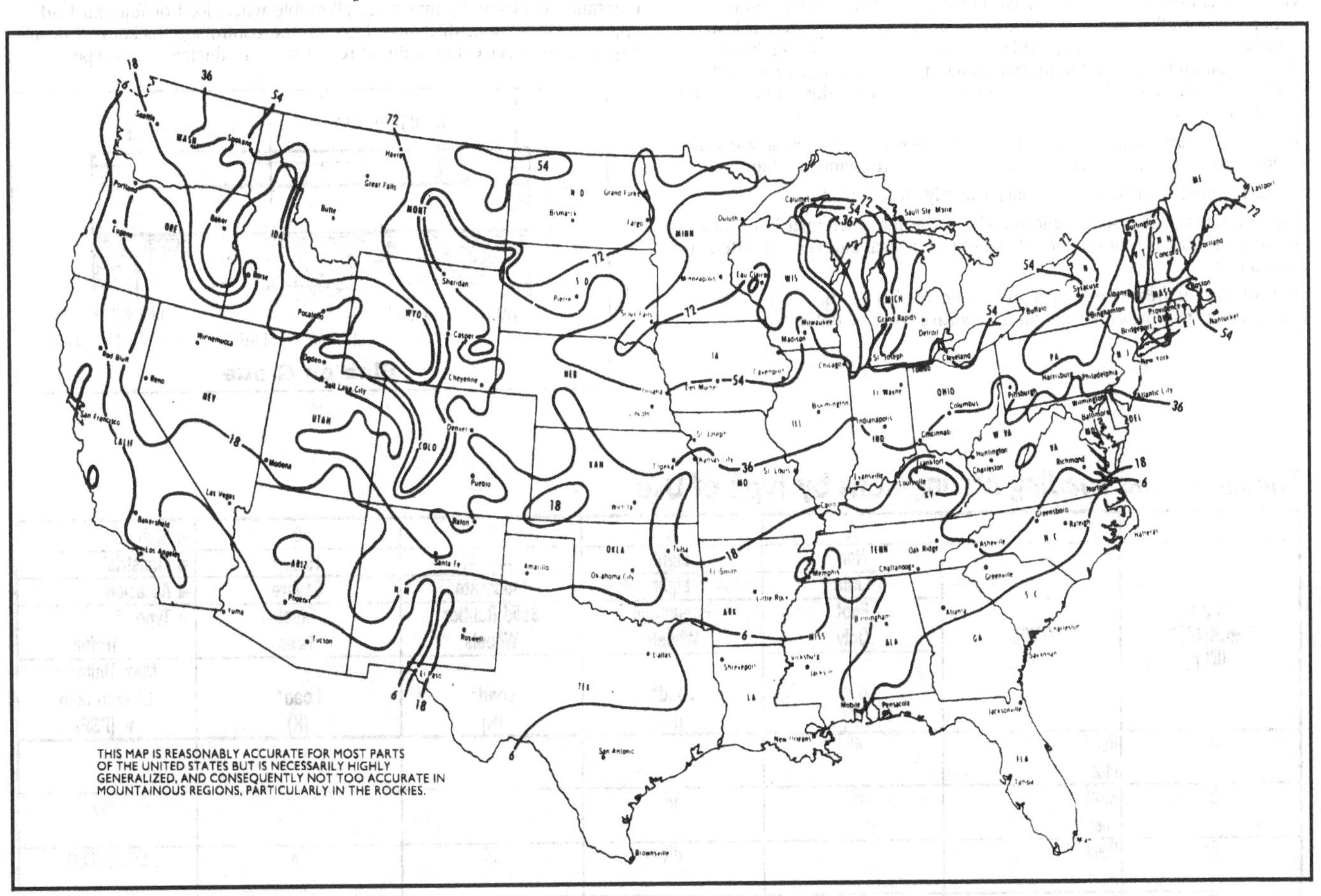

R033105-80 Slab on Grade

General: Ground slabs are classified on the basis of use. Thickness is generally controlled by the heaviest concentrated load supported. If load area is greater than 80 sq. in., soil bearing may be important. The base granular fill must be a uniformly compacted material of limited capillarity, such as gravel or crushed rock. Concrete is placed on this surface of the vapor barrier on top of base.

Ground slabs are either single or two course floors. Single course are widely used. Two course floors have a subsequent wear resistant topping.

Reinforcement is provided to maintain tightly closed cracks.

Control joints limit crack locations and provide for differential horizontal movement only. Isolation joints allow both horizontal and vertical differential movement.

Use of Table: Determine appropriate type of slab (A, B, C, or D) by considering type of use or amount of abrasive wear of traffic type.

Determine thickness by maximum allowable wheel load or uniform load, opposite 1st column, thickness. Increase the controlling thickness if details require, and select either plain or reinforced slab thickness and type.

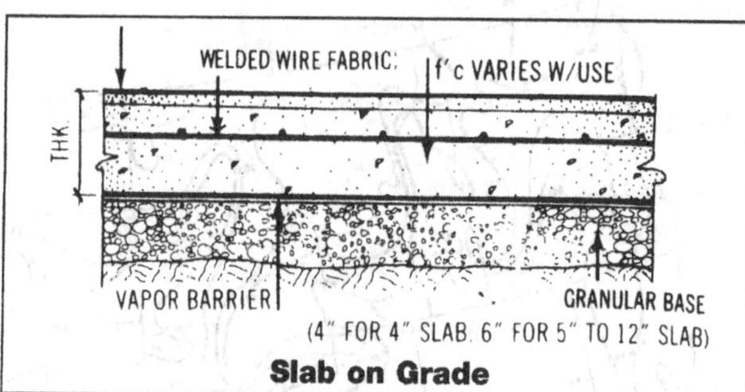

Slab on Grade

Thickness and Loading Assumptions by Type of Use

SLAB THICKNESS (IN.)	TYPE	A	B	C	D	◄ Slab I.D.
		Non	Light	Normal	Heavy	◄ Industrial
		Little	Light	Moderate	Severe	◄ Abrasion
		Foot Only	Pneumatic Wheels	Solid Rubber Wheels	Steel Tires	◄ Type of Traffic
		Load* (K)	Load* (K)	Load* (K)	Load* (K)	Max. Uniform Load to Slab ▼ (PSF)
4"	Reinf. Plain	4K				100
5"	Reinf. Plain	6K	4K			200
6"	Reinf. Plain		8K	6K	6K	500 to 800
7"	Reinf. Plain			9K	8K	1,500
8"	Reinf. Plain				11K	* Max. Wheel Load in Kips (incl. impact)
10"	Reinf. Plain				14K	
12"	Reinf. Plain					
DESIGN ASSUMPTIONS	Concrete, Chuted	f'c = 3.5 KSI	4 KSI	4.5 KSI	Slab @ 3.5 KSI	ASSUMPTIONS BY SLAB TYPE
	Toppings			1" Integral	1" Bonded	
	Finish	Steel Trowel	Steel Trowel	Steel Trowel	Screed & Steel Trowel	
	Compacted Granular Base	4" deep for 4" slab thickness 6" deep for 5" slab thickness & greater				ASSUMPTIONS FOR ALL SLAB TYPES
	Vapor Barrier	6 mil polyethylene				
	Forms & Joints	Allowances included				
	Rein-forcement	WWF as required ≥ 60,000 psi				

Concrete — R0331 Structural Concrete

R033105-85 Lift Slabs

The cost advantage of the lift slab method is due to placing all concrete, reinforcing steel, inserts and electrical conduit at ground level and in reduction of formwork. Minimum economical project size is about 30,000 S.F. Slabs may be tilted for parking garage ramps.

It is now used in all types of buildings and has gone up to 22 stories high in apartment buildings. Current trend is to use post-tensioned flat plate slabs with spans from 22′ to 35′. Cylindrical void forms are used when deep slabs are required. One pound of prestressing steel is about equal to seven pounds of conventional reinforcing.

To be considered cured for stressing and lifting, a slab must have attained 75% of design strength. Seven days are usually sufficient with four to five days possible if high early strength cement is used. Slabs can be stacked using two coats of a non-bonding agent to insure that slabs do not stick to each other. Lifting is done by companies specializing in this work. Lift rate is 5′ to 15′ per hour with an average of 10′ per hour. Total areas up to 33,000 S.F. have been lifted at one time. 24 to 36 jacking columns are common. Most economical bay sizes are 24′ to 28′ with four to fourteen stories most efficient. Continuous design reduces reinforcing steel cost. Use of post-tensioned slabs allows larger bay sizes.

Concrete — R0341 Precast Structural Concrete

R034105-30 Prestressed Precast Concrete Structural Units

Type	Location	Depth	Span in Ft.		Live Load Lb. per S.F.
Double Tee (8′ to 10′)	Floor	28″ to 34″	60 to 80		50 to 80
	Roof	12″ to 24″	30 to 50		40
	Wall	Width 8′	Up to 55′ high		Wind
Multiple Tee (8′)	Roof	8″ to 12″	15 to 40		40
	Floor	8″ to 12″	15 to 30		100
Plank or	Roof or Floor		Roof	Floor	
		4″	13	12	40 for Roof
		6″	22	18	
		8″	26	25	
		10″	33	29	100 for Floor
		12″	42	32	
Single Tee (8′ to 10′)	Roof	28″	40		40
		32″	80		
		36″	100		
		48″	120		
AASHO Girder	Bridges	Type 4	100		Highway
		5	110		
		6	125		
Box Beam (4′)	Bridges	15″	40 to 100		Highway
		27″			
		33″			

The majority of precast projects today utilize double tees rather than single tees because of speed and ease of installation. As a result casting beds at manufacturing plants are normally formed for double tees. Single tee projects will therefore require an initial set up charge to be spread over the individual single tee costs.

For floors, a 2″ to 3″ topping is field cast over the shapes. For roofs, insulating concrete or rigid insulation is placed over the shapes.

Member lengths up to 40′ are standard haul, 40′ to 60′ require special permits and lengths over 60′ must be escorted. Over width and/or over length can add up to 100% on hauling costs.

Large heavy members may require two cranes for lifting which would increase erection costs by about 45%. An eight man crew can install 12 to 20 double tees, or 45 to 70 quad tees or planks per day.

Grouting of connections must also be included.

Several system buildings utilizing precast members are available. Heights can go up to 22 stories for apartment buildings. Optimum design ratio is 3 S.F. of surface to 1 S.F. of floor area.

Masonry — R0401 Maintenance of Masonry

R040130-10 Cleaning Face Brick

On smooth brick a person can clean 70 S.F. an hour; on rough brick 50 S.F. per hour. Use one gallon muriatic acid to 20 gallons of water for 1000 S.F. Do not use acid solution until wall is at least seven days old, but a mild soap solution may be used after two days.

Time has been allowed for clean-up in brick prices.

Masonry — R0405 Common Work Results for Masonry

R040513-10 Cement Mortar (material only)

Type N - 1:1:6 mix by volume. Use everywhere above grade except as noted below. - 1:3 mix using conventional masonry cement which saves handling two separate bagged materials.

Type M - 1:1/4:3 mix by volume, or 1 part cement, 1/4 (10% by wt.) lime, 3 parts sand. Use for heavy loads and where earthquakes or hurricanes may occur. Also for reinforced brick, sewers, manholes and everywhere below grade.

Mix Proportions by Volume and Compressive Strength of Mortar

Where Used	Allowable Proportions by Volume					Compressive Strength @ 28 days
	Mortar Type	Portland Cement	Masonry Cement	Hydrated Lime	Masonry Sand	
Plain Masonry	M	1	1	—	6	
		1	—	1/4	3	2500 psi
	S	1/2	1	—	4	
		1	—	1/4 to 1/2	4	1800 psi
	N	—	1	—	3	
		1	—	1/2 to 1-1/4	6	750 psi
	O	—	1	—	3	
		1	—	1-1/4 to 2-1/2	9	350 psi
	K	1	—	2-1/2 to 4	12	75 psi
Reinforced Masonry	PM	1	1	—	6	2500 psi
	PL	1	—	1/4 to 1/2	4	2500 psi

Note: The total aggregate should be between 2.25 to 3 times the sum of the cement and lime used.

The labor cost to mix the mortar is included in the productivity and labor cost of unit price lines in unit cost sections for brickwork, blockwork and stonework.

The material cost of mixed mortar is included in the material cost of those same unit price lines and includes the cost of renting and operating a 10 C.F. mixer at the rate of 200 C.F. per day.

There are two types of mortar color used. One type is the inert additive type with about 100 lbs. per M brick as the typical quantity required. These colors are also available in smaller-batch-sized bags (1 lb. to 15 lb.) which can be placed directly into the mixer without measuring. The other type is premixed and replaces the masonry cement. Dark green color has the highest cost.

R040519-50 Masonry Reinforcing

Horizontal joint reinforcing helps prevent wall cracks where wall movement may occur and in many locations is required by code. Horizontal joint reinforcing is generally not considered to be structural reinforcing and an unreinforced wall may still contain joint reinforcing.

Reinforcing strips come in 10′ and 12′ lengths and in truss and ladder shapes, with and without drips. Field labor runs between 2.7 to 5.3 hours per 1000 L.F. for wall thicknesses up to 12″.

The wire meets ASTM A82 for cold drawn steel wire and the typical size is 9 ga. sides and ties with 3/16″ diameter also available. Typical finish is mill galvanized with zinc coating at .10 oz. per S.F. Class I (.40 oz. per S.F.) and Class III (.80 oz per S.F.) are also available, as is hot dipped galvanizing at 1.50 oz. per S.F.

Masonry R0421 Clay Unit Masonry

R042110-10 Economy in Bricklaying

Have adequate supervision. Be sure bricklayers are always supplied with materials so there is no waiting. Place best bricklayers at corners and openings.

Use only screened sand for mortar. Otherwise, labor time will be wasted picking out pebbles. Use seamless metal tubs for mortar as they do not leak or catch the trowel. Locate stack and mortar for easy wheeling.

Have brick delivered for stacking. This makes for faster handling, reduces chipping and breakage, and requires less storage space. Many dealers will deliver select common in 2′ x 3′ x 4′ pallets or face brick packaged. This affords quick handling with a crane or forklift and easy tonging in units of ten, which reduces waste.

Use wider bricks for one wythe wall construction. Keep scaffolding away from wall to allow mortar to fall clear and not stain wall.

On large jobs develop specialized crews for each type of masonry unit.

Consider designing for prefabricated panel construction on high rise projects.

Avoid excessive corners or openings. Each opening adds about 50% to labor cost for area of opening.

Bolting stone panels and using window frames as stops reduces labor costs and speeds up erection.

R042110-20 Common and Face Brick

Common building brick manufactured according to ASTM C62 and facing brick manufactured according to ASTM C216 are the two standard bricks available for general building use.

Building brick is made in three grades; SW, where high resistance to damage caused by cyclic freezing is required; MW, where moderate resistance to cyclic freezing is needed; and NW, where little resistance to cyclic freezing is needed. Facing brick is made in only the two grades SW and MW. Additionally, facing brick is available in three types; FBS, for general use; FBX, for general use where a higher degree of precision and lower permissible variation in size than FBS is needed; and FBA, for general use to produce characteristic architectural effects resulting from non-uniformity in size and texture of the units.

In figuring the material cost of brickwork, an allowance of 25% mortar waste and 3% brick breakage was included. If bricks are delivered palletized with 280 to 300 per pallet, or packaged, allow only 1-1/2% for breakage. Packaged or palletized delivery is practical when a job is big enough to have a crane or other equipment available to handle a package of brick. This is so on all industrial work but not always true on small commercial buildings.

The use of buff and gray face is increasing, and there is a continuing trend to the Norman, Roman, Jumbo and SCR brick.

Common red clay brick for backup is not used that often. Concrete block is the most usual backup material with occasional use of sand lime or cement brick. Building brick is commonly used in solid walls for strength and as a fire stop.

Brick panels built on the ground and then crane erected to the upper floors have proven to be economical. This allows the work to be done under cover and without scaffolding.

R042110-50 Brick, Block & Mortar Quantities

Running Bond						
Number of Brick per S.F. of Wall - Single Wythe with 3/8″ Joints				C.F. of Mortar per M Bricks, Waste Included	
Type Brick	Nominal Size (incl. mortar) L H W	Modular Coursing	Number of Brick per S.F.	3/8″ Joint	1/2″ Joint
Standard	8 x 2-2/3 x 4	3C=8″	6.75	10.3	12.9
Economy	8 x 4 x 4	1C=4″	4.50	11.4	14.6
Engineer	8 x 3-1/5 x 4	5C=16″	5.63	10.6	13.6
Fire	9 x 2-1/2 x 4-1/2	2C=5″	6.40	550 # Fireclay	—
Jumbo	12 x 4 x 6 or 8	1C=4″	3.00	23.8	30.8
Norman	12 x 2-2/3 x 4	3C=8″	4.50	14.0	17.9
Norwegian	12 x 3-1/5 x 4	5C=16″	3.75	14.6	18.6
Roman	12 x 2 x 4	2C=4″	6.00	13.4	17.0
SCR	12 x 2-2/3 x 6	3C=8″	4.50	21.8	28.0
Utility	12 x 4 x 4	1C=4″	3.00	15.4	19.6

For Other Bonds Standard Size Add to S.F. Quantities in Table to Left		
Bond Type	Description	Factor
Common	full header every fifth course	+20%
	full header every sixth course	+16.7%
English	full header every second course	+50%
Flemish	alternate headers every course	+33.3%
	every sixth course	+5.6%
Header = W x H exposed		+100%
Rowlock = H x W exposed		+100%
Rowlock stretcher = L x W exposed		+33.3%
Soldier = H x L exposed		—
Sailor = W x L exposed		-33.3%

Concrete Blocks Nominal Size	Approximate Weight per S.F. Standard	Approximate Weight per S.F. Lightweight	Blocks per 100 S.F.	Mortar per M block, waste included Partitions	Mortar per M block, waste included Back up
2″ x 8″ x 16″	20 PSF	15 PSF	113	27 C.F.	36 C.F.
4″ ↓	30 ↓	20 ↓	↓	41 ↓	51 ↓
6″	42	30		56	66
8″	55	38		72	82
10″	70	47		87	97
12″	85	55		102	112

Masonry — R0422 Concrete Unit Masonry

R042210-20 Concrete Block

The material cost of special block such as corner, jamb and head block can be figured at the same price as ordinary block of same size. Labor on specials is about the same as equal-sized regular block.

Bond beam and 16″ high lintel blocks are more expensive than regular units of equal size. Lintel blocks are 8″ long and either 8″ or 16″ high.

Use of motorized mortar spreader box will speed construction of continuous walls.

Hollow non-load-bearing units are made according to ASTM C129 and hollow load-bearing units according to ASTM C90.

Metals — R0505 Common Work Results for Metals

R050516-30 Coating Structural Steel

On field-welded jobs, the shop-applied primer coat is necessarily omitted. All painting must be done in the field and usually consists of red oxide rust inhibitive paint or an aluminum paint. The table below shows paint coverage and daily production for field painting.

See Division 09 97 13.23 for hot-dipped galvanizing and for field-applied cold galvanizing and other paints and protective coatings.

See Division 05 01 10.51 for steel surface preparation treatments such as wire brushing, pressure washing and sand blasting.

Type Construction	Surface Area per Ton	Coat	One Gallon Covers		In 8 Hrs. Person Covers		Average per Ton Spray	
			Brush	Spray	Brush	Spray	Gallons	Labor-hours
Light Structural	300 S.F. to 500 S.F.	1st	500 S.F.	455 S.F.	640 S.F.	2000 S.F.	0.9 gals.	1.6 L.H.
		2nd	450	410	800	2400	1.0	1.3
		3rd	450	410	960	3200	1.0	1.0
Medium	150 S.F. to 300 S.F.	All	400	365	1600	3200	0.6	0.6
Heavy Structural	50 S.F. to 150 S.F.	1st	400	365	1920	4000	0.2	0.2
		2nd	400	365	2000	4000	0.2	0.2
		3rd	400	365	2000	4000	0.2	0.2
Weighted Average	225 S.F.	All	400	365	1350	3000	0.6	0.6

R050521-20 Welded Structural Steel

Usual weight reductions with welded design run 10% to 20% compared with bolted or riveted connections. This amounts to about the same total cost compared with bolted structures since field welding is more expensive than bolts. For normal spans of 18′ to 24′ figure 6 to 7 connections per ton.

Trusses — For welded trusses add 4% to weight of main members for connections. Up to 15% less steel can be expected in a welded truss compared to one that is shop bolted. Cost of erection is the same whether shop bolted or welded.

General — Typical electrodes for structural steel welding are E6010, E6011, E60T and E70T. Typical buildings vary between 2# to 8# of weld rod per ton of steel. Buildings utilizing continuous design require about three times as much welding as conventional welded structures. In estimating field erection by welding, it is best to use the average linear feet of weld per ton to arrive at the welding cost per ton. The type, size and position of the weld will have a direct bearing on the cost per linear foot. A typical field welder will deposit 1.8# to 2# of weld rod per hour manually. Using semiautomatic methods can increase production by as much as 50% to 75%.

R050523-10 High Strength Bolts

Common bolts (A307) are usually used in secondary connections (see Division 05 05 23.10).

High strength bolts (A325 and A490) are usually specified for primary connections such as column splices, beam and girder connections to columns, column bracing, connections for supports of operating equipment or of other live loads which produce impact or reversal of stress, and in structures carrying cranes of over 5-ton capacity.

Allow 20 field bolts per ton of steel for a 6 story office building, apartment house or light industrial building. For 6 to 12 stories allow 25 bolts per ton, and above 12 stories, 30 bolts per ton. On power stations, 20 to 25 bolts per ton are needed.

Metals — R0512 Structural Steel Framing

R051223-10 Structural Steel

The bare material prices for structural steel, shown in the unit cost sections of the book, are for 100 tons of shop-fabricated structural steel and include:

1. Mill base price of structural steel
2. Mill scrap/grade/size/length extras
3. Mill delivery to a metals service center (warehouse)
4. Service center storage and handling
5. Service center delivery to a fabrication shop
6. Shop storage and handling
7. Shop drafting/detailing
8. Shop fabrication
9. Shop coat of primer paint
10. Shop listing
11. Shop delivery to the job site

In unit cost sections of the book that contain items for field fabrication of steel components, the bare material cost of steel includes:

1. Mill base price of structural steel
2. Mill scrap/grade/size/length extras
3. Mill delivery to a metals service center (warehouse)
4. Service center storage and handling
5. Service center delivery to the job site

R051223-20 Steel Estimating Quantities

One estimate on erection is that a crane can handle 35 to 60 pieces per day. Say the average is 45. With usual sizes of beams, girders, and columns, this would amount to about 20 tons per day. The type of connection greatly affects the speed of erection. Moment connections for continuous design slow down production and increase erection costs.

Short open web bar joists can be set at the rate of 75 to 80 per day, with 50 per day being the average for setting long span joists.

After main members are calculated, add the following for usual allowances: base plates 2% to 3%; column splices 4% to 5%; and miscellaneous details 4% to 5%, for a total of 10% to 13% in addition to main members.

The ratio of column to beam tonnage varies depending on type of steels used, typical spans, story heights and live loads.

It is more economical to keep the column size constant and to vary the strength of the column by using high strength steels. This also saves floor space. Buildings have recently gone as high as ten stories with 8″ high strength columns. For light columns under W8X31 lb. sections, concrete filled steel columns are economical.

High strength steels may be used in columns and beams to save floor space and to meet head room requirements. High strength steels in some sizes sometimes require long lead times.

Round, square and rectangular columns, both plain and concrete filled, are readily available and save floor area, but are higher in cost per pound than rolled columns. For high unbraced columns, tube columns may be less expensive.

Below are average minimum figures for the weights of the structural steel frame for different types of buildings using A36 steel, rolled shapes and simple joints. For economy in domes, rise to span ratio = .13. Open web joist framing systems will reduce weights by 10% to 40%. Composite design can reduce steel weight by up to 25% but additional concrete floor slab thickness may be required. Continuous design can reduce the weights up to 20%. There are many building codes with different live load requirements and different structural requirements, such as hurricane and earthquake loadings which can alter the figures.

Structural Steel Weights per S.F. of Floor Area

Type of Building	No. of Stories	Avg. Spans	L.L. #/S.F.	Lbs. Per S.F.	Type of Building	No. of Stories	Avg. Spans	L.L. #/S.F.	Lbs. Per S.F.
		20′x20′		8	Apartments	2-8	20′x20′	40	8
Steel Frame Mfg.	1	30′x30′	40	13		9-25			14
		40′x40′		18		to 10			10
Parking garage	4	Various	80	8.5	Office	20	Various	80	18
Domes	1	200′	30	10		30			26
(Schwedler)*		300′		15		over 50			35

R051223-25 Common Structural Steel Specifications

ASTM A992 (formerly A36, then A572 Grade 50) is the all-purpose carbon grade steel widely used in building and bridge construction.

The other high-strength steels listed below may each have certain advantages over ASTM A992 stuctural carbon steel, depending on the application. They have proven to be economical choices where, due to lighter members, the reduction of dead load and the associated savings in shipping cost can be significant.

ASTM A588 atmospheric weathering, high-strength low-alloy steels can be used in the bare (uncoated) condition, where exposure to normal atmosphere causes a tightly adherant oxide to form on the surface protecting the steel from further oxidation. ASTM A242 corrosion-resistant, high-strength low-alloy steels have enhanced atmospheric corrosion resistance of at least two times that of carbon structural steels with copper, or four times that of carbon structural steels without copper. The reduction or elimination of maintenance resulting from the use of these steels often offsets their higher initial cost.

Steel Type	ASTM Designation	Minimum Yield Stress in KSI	Shapes Available
Carbon	A36	36	All structural shape groups, and plates & bars up thru 8" thick
	A529	50	Structural shape group 1, and plates & bars up thru 2" thick
High-Strength Low-Alloy Quenched & Self-Tempered	A913	50 60 65 70	All structural shape groups
High-Strength Low-Alloy Columbium-Vanadium	A572	42	All structural shape groups, and plates & bars up thru 6" thick
		50	All structural shape groups, and plates & bars up thru 4" thick
		55	Structural shape groups 1 & 2, and plates & bars up thru 2" thick
		60	Structural shape groups 1 & 2, and plates & bars up thru 1-1/4" thick
		65	Structural shape group 1, and plates & bars up thru 1-1/4" thick
High-Strength Low-Alloy Columbium-Vanadium	A992	50	All structural shape groups
Weathering High-Strength Low-Alloy	A242	42	Structural shape groups 4 & 5, and plates & bars over 1-1/2" up thru 4" thick
		46	Structural shape group 3, and plates & bars over 3/4" up thru 1-1/2" thick
		50	Structural shape groups 1 & 2, and plates & bars up thru 3/4" thick
Weathering High-Strength Low-Alloy	A588	42	Plates & bars over 5" up thru 8" thick
		46	Plates & bars over 4" up thru 5" thick
		50	All structural shape groups, and plates & bars up thru 4" thick
Quenched and Tempered Low-Alloy	A852	70	Plates & bars up thru 4" thick
Quenched and Tempered Alloy	A514	90	Plates & bars over 2-1/2" up thru 6" thick
		100	Plates & bars up thru 2-1/2" thick

R051223-30 High Strength Steels

The mill price of high strength steels may be higher than A992 carbon steel but their proper use can achieve overall savings thru total reduced weights. For columns with L/r over 100, A992 steel is best; under 100, high strength steels are economical. For heavy columns, high strength steels are economical when cover plates are eliminated. There is no economy using high strength steels for clip angles or supports or for beams where deflection governs. Thinner members are more economical than thick.

The per ton erection and fabricating costs of the high strength steels will be higher than for A992 since the same number of pieces, but less weight, will be installed.

Metals — R0531 Steel Decking

R053100-10 Decking Descriptions

General - All Deck Products

Steel deck is made by cold forming structural grade sheet steel into a repeating pattern of parallel ribs. The strength and stiffness of the panels are the result of the ribs and the material properties of the steel. Deck lengths can be varied to suit job conditions, but because of shipping considerations, are usually less than 40 feet. Standard deck width varies with the product used but full sheets are usually 12″, 18″, 24″, 30″, or 36″. Deck is typically furnished in a standard width with the ends cut square. Any cutting for width, such as at openings or for angular fit, is done at the job site.

Deck is typically attached to the building frame with arc puddle welds, self-drilling screws, or powder or pneumatically driven pins. Sheet to sheet fastening is done with screws, button punching (crimping), or welds.

Composite Floor Deck

After installation and adequate fastening, floor deck serves several purposes. It (a) acts as a working platform, (b) stabilizes the frame, (c) serves as a concrete form for the slab, and (d) reinforces the slab to carry the design loads applied during the life of the building. Composite decks are distinguished by the presence of shear connector devices as part of the deck. These devices are designed to mechanically lock the concrete and deck together so that the concrete and the deck work together to carry subsequent floor loads. These shear connector devices can be rolled-in embossments, lugs, holes, or wires welded to the panels. The deck profile can also be used to interlock concrete and steel.

Composite deck finishes are either galvanized (zinc coated) or phosphatized/painted. Galvanized deck has a zinc coating on both the top and bottom surfaces. The phosphatized/painted deck has a bare (phosphatized) top surface that will come into contact with the concrete. This bare top surface can be expected to develop rust before the concrete is placed. The bottom side of the deck has a primer coat of paint.

Composite floor deck is normally installed so the panel ends do not overlap on the supporting beams. Shear lugs or panel profile shape often prevent a tight metal to metal fit if the panel ends overlap; the air gap caused by overlapping will prevent proper fusion with the structural steel supports when the panel end laps are shear stud welded.

Adequate end bearing of the deck must be obtained as shown on the drawings. If bearing is actually less in the field than shown on the drawings, further investigation is required.

Roof Deck

Roof deck is not designed to act compositely with other materials. Roof deck acts alone in transferring horizontal and vertical loads into the building frame. Roof deck rib openings are usually narrower than floor deck rib openings. This provides adequate support of rigid thermal insulation board.

Roof deck is typically installed to endlap approximately 2″ over supports. However, it can be butted (or lapped more than 2″) to solve field fit problems. Since designers frequently use the installed deck system as part of the horizontal bracing system (the deck as a diaphragm), any fastening substitution or change should be approved by the designer. Continuous perimeter support of the deck is necessary to limit edge deflection in the finished roof and may be required for diaphragm shear transfer.

Standard roof deck finishes are galvanized or primer painted. The standard factory applied paint for roof deck is a primer paint and is not intended to weather for extended periods of time. Field painting or touching up of abrasions and deterioration of the primer coat or other protective finishes is the responsibility of the contractor.

Cellular Deck

Cellular deck is made by attaching a bottom steel sheet to a roof deck or composite floor deck panel. Cellular deck can be used in the same manner as floor deck. Electrical, telephone, and data wires are easily run through the chase created between the deck panel and the bottom sheet.

When used as part of the electrical distribution system, the cellular deck must be installed so that the ribs line up and create a smooth cell transition at abutting ends. The joint that occurs at butting cell ends must be taped or otherwise sealed to prevent wet concrete from seeping into the cell. Cell interiors must be free of welding burrs, or other sharp intrusions, to prevent damage to wires.

When used as a roof deck, the bottom flat plate is usually left exposed to view. Care must be maintained during erection to keep good alignment and prevent damage.

Cellular deck is sometimes used with the flat plate on the top side to provide a flat working surface. Installation of the deck for this purpose requires special methods for attachment to the frame because the flat plate, now on the top, can prevent direct access to the deck material that is bearing on the structural steel. It may be advisable to treat the flat top surface to prevent slipping.

Cellular deck is always furnished galvanized or painted over galvanized.

Form Deck

Form deck can be any floor or roof deck product used as a concrete form. Connections to the frame are by the same methods used to anchor floor and roof deck. Welding washers are recommended when welding deck that is less than 20 gauge thickness.

Form deck is furnished galvanized, prime painted, or uncoated. Galvanized deck must be used for those roof deck systems where form deck is used to carry a lightweight insulating concrete fill.

Wood, Plastics & Comp. R0611 Wood Framing

R061110-30 Lumber Product Material Prices

The price of forest products fluctuates widely from location to location and from season to season depending upon economic conditions. The bare material prices in the unit cost sections of the book show the National Average material prices in effect Jan. 1 of this book year. It must be noted that lumber prices in general may change significantly during the year.

Availability of certain items depends upon geographic location and must be checked prior to firm-price bidding.

Wood, Plastics & Comp. R0616 Sheathing

R061636-20 Plywood

There are two types of plywood used in construction: interior, which is moisture-resistant but not waterproofed, and exterior, which is waterproofed.

The grade of the exterior surface of the plywood sheets is designated by the first letter: A, for smooth surface with patches allowed; B, for solid surface with patches and plugs allowed; C, which may be surface plugged or may have knot holes up to 1″ wide; and D, which is used only for interior type plywood and may have knot holes up to 2-1/2″ wide. "Structural Grade" is specifically designed for engineered applications such as box beams. All CC & DD grades have roof and floor spans marked on them.

Underlayment-grade plywood runs from 1/4″ to 1-1/4″ thick. Thicknesses 5/8″ and over have optional tongue and groove joints which eliminate the need for blocking the edges. Underlayment 19/32″ and over may be referred to as Sturd-i-Floor.

The price of plywood can fluctuate widely due to geographic and economic conditions.

Typical uses for various plywood grades are as follows:

AA-AD Interior — cupboards, shelving, paneling, furniture

BB Plyform — concrete form plywood

CDX — wall and roof sheathing

Structural — box beams, girders, stressed skin panels

AA-AC Exterior — fences, signs, siding, soffits, etc.

Underlayment — base for resilient floor coverings

Overlaid HDO — high density for concrete forms & highway signs

Overlaid MDO — medium density for painting, siding, soffits & signs

303 Siding — exterior siding, textured, striated, embossed, etc.

Thermal & Moist. Protec. R0731 Shingles & Shakes

R073126-20 Roof Slate

16″, 18″ and 20″ are standard lengths, and slate usually comes in random widths. For standard 3/16″ thickness use 1-1/2″ copper nails. Allow for 3% breakage.

Thermal & Moist. Protec. R0751 Built-Up Bituminous Roofing

R075113-20 Built-Up Roofing

Asphalt is available in kegs of 100 lbs. each; coal tar pitch in 560 lb. kegs. Prepared roofing felts are available in a wide range of sizes, weights and characteristics. However, the most commonly used are #15 (432 S.F. per roll, 13 lbs. per square) and #30 (216 S.F. per roll, 27 lbs. per square).

Inter-ply bitumen varies from 24 lbs. per sq. (asphalt) to 30 lbs. per sq. (coal tar) per ply, MF4@ 25%. Flood coat bitumen also varies from 60 lbs. per sq. (asphalt) to 75 lbs. per sq. (coal tar), MF4@ 25%. Expendable equipment (mops, brooms, screeds, etc.) runs about 16% of the bitumen cost. For new, inexperienced crews this factor may be much higher.

Rigid insulation board is typically applied in two layers. The first is mechanically attached to nailable decks or spot or solid mopped to non-nailable decks; the second layer is then spot or solid mopped to the first layer. Membrane application follows the insulation, except in protected membrane roofs, where the membrane goes down first and the insulation on top, followed with ballast (stone or concrete pavers). Insulation and related labor costs are NOT included in prices for built-up roofing.

Thermal & Moist. Protec. R0752 Modified Bituminous Membrane Roofing

R075213-30 Modified Bitumen Roofing

The cost of modified bitumen roofing is highly dependent on the type of installation that is planned. Installation is based on the type of modifier used in the bitumen. The two most popular modifiers are atactic polypropylene (APP) and styrene butadiene styrene (SBS). The modifiers are added to heated bitumen during the manufacturing process to change its characteristics. A polyethylene, polyester or fiberglass reinforcing sheet is then sandwiched between layers of this bitumen. When completed, the result is a pre-assembled, built-up roof that has increased elasticity and weatherablility. Some manufacturers include a surfacing material such as ceramic or mineral granules, metal particles or sand.

The preferred method of adhering SBS-modified bitumen roofing to the substrate is with hot-mopped asphalt (much the same as built-up roofing). This installation method requires a tar kettle/pot to heat the asphalt, as well as the labor, tools and equipment necessary to distribute and spread the hot asphalt.

The alternative method for applying APP and SBS modified bitumen is as follows. A skilled installer uses a torch to melt a small pool of bitumen off the membrane. This pool must form across the entire roll for proper adhesion. The installer must unroll the roofing at a pace slow enough to melt the bitumen, but fast enough to prevent damage to the rest of the membrane.

Modified bitumen roofing provides the advantages of both built-up and single-ply roofing. Labor costs are reduced over those of built-up roofing because only a single ply is necessary. The elasticity of single-ply roofing is attained with the reinforcing sheet and polymer modifiers. Modifieds have some self-healing characteristics and because of their multi-layer construction, they offer the reliability and safety of built-up roofing.

Thermal & Moist. Protec. R0784 Firestopping

R078413-30 Firestopping

Firestopping is the sealing of structural, mechanical, electrical and other penetrations through fire-rated assemblies. The basic components of firestop systems are safing insulation and firestop sealant on both sides of wall penetrations and the top side of floor penetrations.

Pipe penetrations are assumed to be through concrete, grout, or joint compound and can be sleeved or unsleeved. Costs for the penetrations and sleeves are not included. An annular space of 1″ is assumed. Escutcheons are not included.

Metallic pipe is assumed to be copper, aluminum, cast iron or similar metallic material. Insulated metallic pipe is assumed to be covered with a thermal insulating jacket of varying thickness and materials.

Non-metallic pipe is assumed to be PVC, CPVC, FR Polypropylene or similar plastic piping material. Intumescent firestop sealant or wrap strips are included. Collars on both sides of wall penetrations and a sheet metal plate on the underside of floor penetrations are included.

Ductwork is assumed to be sheet metal, stainless steel or similar metallic material. Duct penetrations are assumed to be through concrete, grout or joint compound. Costs for penetrations and sleeves are not included. An annular space of 1/2″ is assumed.

Multi-trade openings include costs for sheet metal forms, firestop mortar, wrap strips, collars and sealants as necessary.

Structural penetrations joints are assumed to be 1/2″ or less. CMU walls are assumed to be within 1-1/2″ of metal deck. Drywall walls are assumed to be tight to the underside of metal decking.

Metal panel, glass or curtain wall systems include a spandrel area of 5′ filled with mineral wool foil-faced insulation. Fasteners and stiffeners are included.

Openings R0813 Metal Doors

R081313-20 Steel Door Selection Guide

Standard steel doors are classified into four levels, as recommended by the Steel Door Institute in the chart below. Each of the four levels offers a range of construction models and designs, to meet architectural requirements for preference and appearance, including full flush, seamless, and stile & rail. Recommended minimum gauge requirements are also included.

For complete standard steel door construction specifications and available sizes, refer to the Steel Door Institute Technical Data Series, ANSI A250.8-98 (SDI-100), and ANSI A250.4-94 Test Procedure and Acceptance Criteria for Physical Endurance of Steel Door and Hardware Reinforcements.

Level		Model	Construction	For Full Flush or Seamless		
				Min. Gauge	Thickness (in)	Thickness (mm)
I	Standard Duty	1 2	Full Flush Seamless	20	0.032	0.8
II	Heavy Duty	1 2	Full Flush Seamless	18	0.042	1.0
III	Extra Heavy Duty	1 2 3	Full Flush Seamless *Stile & Rail	16	0.053	1.3
IV	Maximum Duty	1 2	Full Flush Seamless	14	0.067	1.6

*Stiles & rails are 16 gauge; flush panels, when specified, are 18 gauge

Openings R0853 Plastic Windows

R085313-20 Replacement Windows

Replacement windows are typically measured per United Inch.

United Inches are calculated by rounding the width and height of the window opening up to the nearest inch, then adding the two figures.

The labor cost for replacement windows includes removal of sash, existing sash balance or weights, parting bead where necessary and installation of new window.

Debris hauling and dump fees are not included.

Openings R0881 Glass Glazing

R088110-10 Glazing Productivity

Some glass sizes are estimated by the "united inch" (height + width). The table below shows the number of lights glazed in an eight-hour period by the crew size indicated, for glass up to 1/4" thick. Square or nearly square lights are more economical on a S.F. basis. Long slender lights will have a high S.F. installation cost. For insulated glass reduce production by 33%. For 1/2" float glass reduce production by 50%. Production time for glazing with two glaziers per day averages: 1/4" float glass 120 S.F.; 1/2" float glass 55 S.F.; 1/2" insulated glass 95 S.F.; 3/4" insulated glass 75 S.F.

Glazing Method	United Inches per Light							
	40"	60"	80"	100"	135"	165"	200"	240"
Number of Men in Crew	1	1	1	1	2	3	3	4
Industrial sash, putty	60	45	24	15	18	—	—	—
With stops, putty bed	50	36	21	12	16	8	4	3
Wood stops, rubber	40	27	15	9	11	6	3	2
Metal stops, rubber	30	24	14	9	9	6	3	2
Structural glass	10	7	4	3	—	—	—	—
Corrugated glass	12	9	7	4	4	4	3	—
Storefronts	16	15	13	11	7	6	4	4
Skylights, putty glass	60	36	21	12	16	—	—	—
Thiokol set	15	15	11	9	9	6	3	2
Vinyl set, snap on	18	18	13	12	12	7	5	4
Maximum area per light	2.8 S.F.	6.3 S.F.	11.1 S.F.	17.4 S.F.	31.6 S.F.	47 S.F.	69 S.F.	100 S.F.

R092000-50 Lath, Plaster and Gypsum Board

Gypsum board lath is available in 3/8″ thick x 16″ wide x 4′ long sheets as a base material for multi-layer plaster applications. It is also available as a base for either multi-layer or veneer plaster applications in 1/2″ and 5/8″ thick–4′ wide x 8′, 10′ or 12′ long sheets. Fasteners are screws or blued ring shank nails for wood framing and screws for metal framing.

Metal lath is available in diamond mesh pattern with flat or self-furring profiles. Paper backing is available for applications where excessive plaster waste needs to be avoided. A slotted mesh ribbed lath should be used in areas where the span between structural supports is greater than normal. Most metal lath comes in 27″ x 96″ sheets. Diamond mesh weighs 1.75, 2.5 or 3.4 pounds per square yard, slotted mesh lath weighs 2.75 or 3.4 pounds per square yard. Metal lath can be nailed, screwed or tied in place.

Many **accessories** are available. Corner beads, flat reinforcing strips, casing beads, control and expansion joints, furring brackets and channels are some examples. Note that accessories are not included in plaster or stucco line items.

Plaster is defined as a material or combination of materials that when mixed with a suitable amount of water, forms a plastic mass or paste. When applied to a surface, the paste adheres to it and subsequently hardens, preserving in a rigid state the form or texture imposed during the period of elasticity.

Gypsum plaster is made from ground calcined gypsum. It is mixed with aggregates and water for use as a base coat plaster.

Vermiculite plaster is a fire-retardant plaster covering used on steel beams, concrete slabs and other heavy construction materials. Vermiculite is a group name for certain clay minerals, hydrous silicates or aluminum, magnesium and iron that have been expanded by heat.

Perlite plaster is a plaster using perlite as an aggregate instead of sand. Perlite is a volcanic glass that has been expanded by heat.

Gauging plaster is a mix of gypsum plaster and lime putty that when applied produces a quick drying finish coat.

Veneer plaster is a one or two component gypsum plaster used as a thin finish coat over special gypsum board.

Keenes cement is a white cementitious material manufactured from gypsum that has been burned at a high temperature and ground to a fine powder. Alum is added to accelerate the set. The resulting plaster is hard and strong and accepts and maintains a high polish, hence it is used as a finishing plaster.

Stucco is a Portland cement based plaster used primarily as an exterior finish.

Plaster is used on both interior and exterior surfaces. Generally it is applied in multiple-coat systems. A three-coat system uses the terms scratch, brown and finish to identify each coat. A two-coat system uses base and finish to describe each coat. Each type of plaster and application system has attributes that are chosen by the designer to best fit the intended use.

Gypsum Plaster Quantities for 100 S.Y.	2 Coat, 5/8″ Thick		3 Coat, 3/4″ Thick		
	Base 1:3 Mix	Finish 2:1 Mix	Scratch 1:2 Mix	Brown 1:3 Mix	Finish 2:1 Mix
Gypsum plaster	1,300 lb.		1,350 lb.	650 lb.	
Sand	1.75 C.Y.		1.85 C.Y.	1.35 C.Y.	
Finish hydrated lime		340 lb.			340 lb.
Gauging plaster		170 lb.			170 lb.

Vermiculite or Perlite Plaster Quantities for 100 S.Y.	2 Coat, 5/8″ Thick		3 Coat, 3/4″ Thick		
	Base	Finish	Scratch	Brown	Finish
Gypsum plaster	1,250 lb.		1,450 lb.	800 lb.	
Vermiculite or perlite	7.8 bags		8.0 bags	3.3 bags	
Finish hydrated lime		340 lb.			340 lb.
Gauging plaster		170 lb.			170 lb.

Stucco-Three-Coat System Quantities for 100 S.Y.	On Wood Frame	On Masonry
Portland cement	29 bags	21 bags
Sand	2.6 C.Y.	2.0 C.Y.
Hydrated lime	180 lb.	120 lb.

R092910-10 Levels of Gypsum Drywall Finish

In the past, contract documents often used phrases such as "industry standard" and "workmanlike finish" to specify the expected quality of gypsum board wall and ceiling installations. The vagueness of these descriptions led to unacceptable work and disputes.

In order to resolve this problem, four major trade associations concerned with the manufacture, erection, finish, and decoration of gypsum board wall and ceiling systems have developed an industry-wide *Recommended Levels of Gypsum Board Finish.*

The finish of gypsum board walls and ceilings for specific final decoration is dependent on a number of factors. A primary consideration is the location of the surface and the degree of decorative treatment desired. Painted and unpainted surfaces in warehouses and other areas where appearance is normally not critical may simply require the taping of wallboard joints and 'spotting' of fastener heads. Blemish-free, smooth, monolithic surfaces often intended for painted and decorated walls and ceilings in habitated structures, ranging from single-family dwellings through monumental buildings, require additional finishing prior to the application of the final decoration.

Other factors to be considered in determining the level of finish of the gypsum board surface are (1) the type of angle of surface illumination (both natural and artificial lighting), and (2) the paint and method of application or the type and finish of wallcovering specified as the final decoration. Critical lighting conditions, gloss paints, and thin wall coverings require a higher level of gypsum board finish than do heavily textured surfaces which are subsequently painted or surfaces which are to be decorated with heavy grade wall coverings.

The following descriptions were developed jointly by the Association of the Wall and Ceiling Industries-International (AWCI), Ceiling & Interior Systems Construction Association (CISCA), Gypsum Association (GA), and Painting and Decorating Contractors of America (PDCA) as a guide.

Level 0: Used in temporary construction or wherever the final decoration has not been determined. Unfinished. No taping, finishing or corner beads are required. Also could be used where non-predecorated panels will be used in demountable-type partitions that are to be painted as a final finish.

Level 1: Frequently used in plenum areas above ceilings, in attics, in areas where the assembly would generally be concealed, or in building service corridors and other areas not normally open to public view. Some degree of sound and smoke control is provided; in some geographic areas, this level is referred to as "fire-taping," although this level of finish does not typically meet fire-resistant assembly requirements. Where a fire resistance rating is required for the gypsum board assembly, details of construction should be in accordance with reports of fire tests of assemblies that have met the requirements of the fire rating acceptable.

All joints and interior angles shall have tape embedded in joint compound. Accessories are optional at specifier discretion in corridors and other areas with pedestrian traffic. Tape and fastener heads need not be covered with joint compound. Surface shall be free of excess joint compound. Tool marks and ridges are acceptable.

Level 2: It may be specified for standard gypsum board surfaces in garages, warehouse storage, or other similar areas where surface appearance is not of primary importance.

All joints and interior angles shall have tape embedded in joint compound and shall be immediately wiped with a joint knife or trowel, leaving a thin coating of joint compound over all joints and interior angles. Fastener heads and accessories shall be covered with a coat of joint compound. Surface shall be free of excess joint compound. Tool marks and ridges are acceptable.

Level 3: Typically used in areas that are to receive heavy texture (spray or hand applied) finishes before final painting, or where commercial-grade (heavy duty) wall coverings are to be applied as the final decoration. This level of finish should not be used where smooth painted surfaces or where lighter weight wall coverings are specified. The prepared surface shall be coated with a drywall primer prior to the application of final finishes.

All joints and interior angles shall have tape embedded in joint compound and shall be immediately wiped with a joint knife or trowel, leaving a thin coating of joint compound over all joints and interior angles. One additional coat of joint compound shall be applied over all joints and interior angles. Fastener heads and accessories shall be covered with two separate coats of joint compound. All joint compounds shall be smooth and free of tool marks and ridges. The prepared surface shall be covered with a drywall primer prior to the application of the final decoration.

Level 4: This level should be used where residential grade (light duty) wall coverings, flat paints, or light textures are to be applied. The prepared surface shall be coated with a drywall primer prior to the application of final finishes. Release agents for wall coverings are specifically formulated to minimize damage if coverings are subsequently removed.

The weight, texture, and sheen level of the wall covering material selected should be taken into consideration when specifying wall coverings over this level of drywall treatment. Joints and fasteners must be sufficiently concealed if the wall covering material is lightweight, contains limited pattern, has a glossy finish, or has any combination of these features. In critical lighting areas, flat paints applied over light textures tend to reduce joint photographing. Gloss, semi-gloss, and enamel paints are not recommended over this level of finish.

All joints and interior angles shall have tape embedded in joint compound and shall be immediately wiped with a joint knife or trowel, leaving a thin coating of joint compound over all joints and interior angles. In addition, two separate coats of joint compound shall be applied over all flat joints and one separate coat of joint compound applied over interior angles. Fastener heads and accessories shall be covered with three separate coats of joint compound. All joint compounds shall be smooth and free of tool marks and ridges. The prepared surface shall be covered with a drywall primer like Sheetrock® First Coat prior to the application of the final decoration.

Level 5: The highest quality finish is the most effective method to provide a uniform surface and minimize the possibility of joint photographing and of fasteners showing through the final decoration. This level of finish is required where gloss, semi-gloss, or enamel is specified; when flat joints are specified over an untextured surface; or where critical lighting conditions occur. The prepared surface shall be coated with a drywall primer prior to the application of final decoration.

All joints and interior angles shall have tape embedded in joint compound and be immediately wiped with a joint knife or trowel, leaving a thin coating of joint compound over all joints and interior angles. Two separate coats of joint compound shall be applied over all flat joints and one separate coat of joint compound applied over interior angles. Fastener heads and accessories shall be covered with three separate coats of joint compound.

A thin skim coat of joint compound shall be trowel applied to the entire surface. Excess compound is immediately troweled off, leaving a film or skim coating of compound completely covering the paper. As an alternative to a skim coat, a material manufactured especially for this purpose may be applied such as Sheetrock® Tuff-Hide primer surfacer. The surface must be smooth and free of tool marks and ridges. The prepared surface shall be covered with a drywall primer prior to the application of the final decoration.

Finishes — R0966 Terrazzo Flooring

R096613-10 Terrazzo Floor

The table below lists quantities required for 100 S.F. of 5/8″ terrazzo topping, either bonded or not bonded.

Description	Bonded to Concrete 1-1/8″ Bed, 1:4 Mix	Not Bonded 2-1/8″ Bed and 1/4″ Sand
Portland cement, 94 lb. Bag	6 bags	8 bags
Sand	10 C.F.	20 C.F.
Divider strips, 4′ squares	50 L.F.	50 L.F.
Terrazzo fill, 50 lb. Bag	12 bags	12 bags
15 Lb. tarred felt		1 C.S.F.
Mesh 2 x 2 #14 galvanized		1 C.S.F.
Crew J-3	0.77 days	0.87 days

2′ x 2′ panels require 1.00 L.F. divider strip per S.F.
3′ x 3′ panels require 0.67 L.F. divider strip per S.F.
4′ x 4′ panels require 0.50 L.F. divider strip per S.F.
5′ x 5′ panels require 0.40 L.F. divider strip per S.F.
6′ x 6′ panels require 0.33 L.F. divider strip per S.F.

Finishes — R0972 Wall Coverings

R097223-10 Wall Covering

The table below lists the quantities required for 100 S.F. of wall covering.

Description	Medium-Priced Paper	Expensive Paper
Paper	1.6 dbl. rolls	1.6 dbl. rolls
Wall sizing	0.25 gallon	0.25 gallon
Vinyl wall paste	0.6 gallon	0.6 gallon
Apply sizing	0.3 hour	0.3 hour
Apply paper	1.2 hours	1.5 hours

Most wallpapers now come in double rolls only.
To remove old paper, allow 1.3 hours per 100 S.F.

R099100-10 Painting Estimating Techniques

Proper estimating methodology is needed to obtain an accurate painting estimate. There is no known reliable shortcut or square foot method. The following steps should be followed:

- List all surfaces to be painted, with an accurate quantity (area) of each. Items having similar surface condition, finish, application method and accessibility may be grouped together.
- List all the tasks required for each surface to be painted, including surface preparation, masking, and protection of adjacent surfaces. Surface preparation may include minor repairs, washing, sanding and puttying.
- Select the proper Means line for each task. Review and consider all adjustments to labor and materials for type of paint and location of work. Apply the height adjustment carefully. For instance, when applying the adjustment for work over 8' high to a wall that is 12' high, apply the adjustment only to the area between 8' and 12' high, and not to the entire wall.

When applying more than one percent (%) adjustment, apply each to the base cost of the data, rather than applying one percentage adjustment on top of the other.

When estimating the cost of painting walls and ceilings remember to add the brushwork for all cut-ins at inside corners and around windows and doors as a LF measure. One linear foot of cut-in with brush equals one square foot of painting.

All items for spray painting include the labor for roll-back.

Deduct for openings greater than 100 SF, or openings that extend from floor to ceiling and are greater than 5' wide. Do not deduct small openings.

The cost of brushes, rollers, ladders and spray equipment are considered to be part of a painting contractor's overhead, and should not be added to the estimate. The cost of rented equipment such as scaffolding and swing staging should be added to the estimate.

R099100-20 Painting

Item	Coat	One Gallon Covers			In 8 Hours a Laborer Covers			Labor-Hours per 100 S.F.		
		Brush	Roller	Spray	Brush	Roller	Spray	Brush	Roller	Spray
Paint wood siding	prime	250 S.F.	225 S.F.	290 S.F.	1150 S.F.	1300 S.F.	2275 S.F.	.695	.615	.351
	others	270	250	290	1300	1625	2600	.615	.492	.307
Paint exterior trim	prime	400	—	—	650	—	—	1.230	—	—
	1st	475	—	—	800	—	—	1.000	—	—
	2nd	520	—	—	975	—	—	.820	—	—
Paint shingle siding	prime	270	255	300	650	975	1950	1.230	.820	.410
	others	360	340	380	800	1150	2275	1.000	.695	.351
Stain shingle siding	1st	180	170	200	750	1125	2250	1.068	.711	.355
	2nd	270	250	290	900	1325	2600	.888	.603	.307
Paint brick masonry	prime	180	135	160	750	800	1800	1.066	1.000	.444
	1st	270	225	290	815	975	2275	.981	.820	.351
	2nd	340	305	360	815	1150	2925	.981	.695	.273
Paint interior plaster or drywall	prime	400	380	495	1150	2000	3250	.695	.400	.246
	others	450	425	495	1300	2300	4000	.615	.347	.200
Paint interior doors and windows	prime	400	—	—	650	—	—	1.230	—	—
	1st	425	—	—	800	—	—	1.000	—	—
	2nd	450	—	—	975	—	—	.820	—	—

R312316-40 Excavating

The selection of equipment used for structural excavation and bulk excavation or for grading is determined by the following factors.

1. Quantity of material
2. Type of material
3. Depth or height of cut
4. Length of haul
5. Condition of haul road
6. Accessibility of site
7. Moisture content and dewatering requirements
8. Availability of excavating and hauling equipment

Some additional costs must be allowed for hand trimming the sides and bottom of concrete pours and other excavation below the general excavation.

When planning excavation and fill, the following should also be considered.

1. Swell factor
2. Compaction factor
3. Moisture content
4. Density requirements

A typical example for scheduling and estimating the cost of excavation of a 15′ deep basement on a dry site when the material must be hauled off the site is outlined below.

Assumptions:

1. Swell factor, 18%
2. No mobilization or demobilization
3. Allowance included for idle time and moving on job
4. No dewatering, sheeting, or bracing
5. No truck spotter or hand trimming

Number of B.C.Y. per truck = 1.5 C.Y. bucket x 8 passes = 12 loose C.Y.

$$= 12 \times \frac{100}{118} = 10.2 \text{ B.C.Y. per truck}$$

Truck Haul Cycle:

Load truck, 8 passes	=	4 minutes
Haul distance, 1 mile	=	9 minutes
Dump time	=	2 minutes
Return, 1 mile	=	7 minutes
Spot under machine	=	1 minute
		23 minute cycle

Fleet Haul Production per day in B.C.Y.

$$4 \text{ trucks} \times \frac{50 \text{ min. hour}}{23 \text{ min. haul cycle}} \times 8 \text{ hrs.} \times 10.2 \text{ B.C.Y.}$$

$$= 4 \times 2.2 \times 8 \times 10.2 = 718 \text{ B.C.Y./day}$$

Add the mobilization and demobilization costs to the total excavation costs. When equipment is rented for more than three days, there is often no mobilization charge by the equipment dealer. On larger jobs outside of urban areas, scrapers can move earth economically provided a dump site or fill area and adequate haul roads are available. Excavation within sheeting bracing or cofferdam bracing is usually done with a clamshell and production is low, since the clamshell may have to be guided by hand between the bracing. When excavating or filling an area enclosed with a wellpoint system, add 10% to 15% to the cost to allow for restricted access. When estimating earth excavation quantities for structures, allow work space outside the building footprint for construction of the foundation and a slope of 1:1 unless sheeting is used.

R312316-45 Excavating Equipment

The table below lists THEORETICAL hourly production in C.Y./hr. bank measure for some typical excavation equipment. Figures assume 50 minute hours, 83% job efficiency, 100% operator efficiency, 90° swing and properly sized hauling units, which must be modified for adverse digging and loading conditions. Actual production costs in the front of the book average about 50% of the theoretical values listed here.

Equipment	Soil Type	B.C.Y. Weight	% Swell	1 C.Y.	1-1/2 C.Y.	2 C.Y.	2-1/2 C.Y.	3 C.Y.	3-1/2 C.Y.	4 C.Y.
Hydraulic Excavator "Backhoe" 15′ Deep Cut	Moist loam, sandy clay	3400 lb.	40%	85	125	175	220	275	330	380
	Sand and gravel	3100	18	80	120	160	205	260	310	365
	Common earth	2800	30	70	105	150	190	240	280	330
	Clay, hard, dense	3000	33	65	100	130	170	210	255	300
Power Shovel Optimum Cut (Ft.)	Moist loam, sandy clay	3400	40	170 (6.0)	245 (7.0)	295 (7.8)	335 (8.4)	385 (8.8)	435 (9.1)	475 (9.4)
	Sand and gravel	3100	18	165 (6.0)	225 (7.0)	275 (7.8)	325 (8.4)	375 (8.8)	420 (9.1)	460 (9.4)
	Common earth	2800	30	145 (7.8)	200 (9.2)	250 (10.2)	295 (11.2)	335 (12.1)	375 (13.0)	425 (13.8)
	Clay, hard, dense	3000	33	120 (9.0)	175 (10.7)	220 (12.2)	255 (13.3)	300 (14.2)	335 (15.1)	375 (16.0)
Drag Line Optimum Cut (Ft.)	Moist loam, sandy clay	3400	40	130 (6.6)	180 (7.4)	220 (8.0)	250 (8.5)	290 (9.0)	325 (9.5)	385 (10.0)
	Sand and gravel	3100	18	130 (6.6)	175 (7.4)	210 (8.0)	245 (8.5)	280 (9.0)	315 (9.5)	375 (10.0)
	Common earth	2800	30	110 (8.0)	160 (9.0)	190 (9.9)	220 (10.5)	250 (11.0)	280 (11.5)	310 (12.0)
	Clay, hard, dense	3000	33	90 (9.3)	130 (10.7)	160 (11.8)	190 (12.3)	225 (12.8)	250 (13.3)	280 (12.0)
				Wheel Loaders				Track Loaders		
				3 C.Y.	4 C.Y.	6 C.Y.	8 C.Y.	2-1/4 C.Y.	3 C.Y.	4 C.Y.
Loading Tractors	Moist loam, sandy clay	3400	40	260	340	510	690	135	180	250
	Sand and gravel	3100	18	245	320	480	650	130	170	235
	Common earth	2800	30	230	300	460	620	120	155	220
	Clay, hard, dense	3000	33	200	270	415	560	110	145	200
	Rock, well-blasted	4000	50	180	245	380	520	100	130	180

R312323-30 Compacting Backfill

Compaction of fill in embankments, around structures, in trenches, and under slabs is important to control settlement. Factors affecting compaction are:

1. Soil gradation
2. Moisture content
3. Equipment used
4. Depth of fill per lift
5. Density required

Production Rate:

$$\frac{1.75' \text{ plate width} \times 50 \text{ F.P.M.} \times 50 \text{ min./hr.} \times .67' \text{ lift}}{27 \text{ C.F. per C.Y.}} = 108.5 \text{ C.Y./hr.}$$

Production Rate for 4 Passes:

$$\frac{108.5 \text{ C.Y.}}{4 \text{ passes}} = 27.125 \text{ C.Y./hr.} \times 8 \text{ hrs.} = 217 \text{ C.Y./day}$$

Example:

Compact granular fill around a building foundation using a 21″ wide x 24″ vibratory plate in 8″ lifts. Operator moves at 50 F.P.M. working a 50 minute hour to develop 95% Modified Proctor Density with 4 passes.

Estimating Tips

- The cost figures in this Square Foot Cost section were derived from approximately 11,200 projects contained in the RSMeans database of completed construction projects. They include the contractor's overhead and profit, but do not generally include architectural fees or land costs. The figures have been adjusted to January of the current year. New projects are added to our files each year, and outdated projects are discarded. For this reason, certain costs may not show a uniform annual progression. In no case are all subdivisions of a project listed.
- These projects were located throughout the U.S. and reflect a tremendous variation in square foot (S.F.) and cubic foot (C.F.) costs. This is due to differences, not only in labor and material costs, but also in individual owners' requirements. For instance, a bank in a large city would have different features than one in a rural area. This is true of all the different types of buildings analyzed. Therefore, caution should be exercised when using these Square Foot costs. For example, for court houses, costs in the database are local court house costs and will not apply to the larger, more elaborate federal court houses. As a general rule, the projects in the 1/4 column do not include any site work or equipment, while the projects in the 3/4 column may include both equipment and site work. The median figures do not generally include site work.
- None of the figures "go with" any others. All individual cost items were computed and tabulated separately. Thus, the sum of the median figures for Plumbing, HVAC and Electrical will not normally total up to the total Mechanical and Electrical costs arrived at by separate analysis and tabulation of the projects.
- Each building was analyzed as to total and component costs and percentages. The figures were arranged in ascending order with the results tabulated as shown. The 1/4 column shows that 25% of the projects had lower costs and 75% had higher. The 3/4 column shows that 75% of the projects had lower costs and 25% had higher. The median column shows that 50% of the projects had lower costs and 50% had higher.
- There are two times when square foot costs are useful. The first is in the conceptual stage when no details are available. Then square foot costs make a useful starting point. The second is after the bids are in and the costs can be worked back into their appropriate units for information purposes. As soon as details become available in the project design, the square foot approach should be discontinued and the project priced as to its particular components. When more precision is required, or for estimating the replacement cost of specific buildings, the current edition of *RSMeans Square Foot Costs* should be used.
- In using the figures in this section, it is recommended that the median column be used for preliminary figures if no additional information is available. The median figures, when multiplied by the total city construction cost index figures (see City Cost Indexes) and then multiplied by the project size modifier at the end of this section, should present a fairly accurate base figure, which would then have to be adjusted in view of the estimator's experience, local economic conditions, code requirements, and the owner's particular requirements. There is no need to factor the percentage figures, as these should remain constant from city to city. All tabulations mentioning air conditioning had at least partial air conditioning.
- The editors of this book would greatly appreciate receiving cost figures on one or more of your recent projects, which would then be included in the averages for next year. All cost figures received will be kept confidential, except that they will be averaged with other similar projects to arrive at Square Foot cost figures for next year's book. See the last page of the book for details and the discount available for submitting one or more of your projects.

50 17 | Square Foot Costs

	50 17 00 | S.F. Costs		UNIT	UNIT COSTS 1/4	MEDIAN	3/4	% OF TOTAL 1/4	MEDIAN	3/4	
01	0010	**APARTMENTS Low Rise (1 to 3 story)**	S.F.	67	84	112				01
	0020	Total project cost	C.F.	6	7.95	9.80				
	0100	Site work	S.F.	5.70	7.80	13.70	6.05%	10.55%	14.05%	
	0500	Masonry		1.31	3.05	5.30	1.54%	3.67%	6.35%	
	1500	Finishes		7.10	9.70	12	9.05%	10.75%	12.85%	
	1800	Equipment		2.18	3.31	4.92	2.73%	4.03%	5.95%	
	2720	Plumbing		5.20	6.70	8.50	6.65%	8.95%	10.05%	
	2770	Heating, ventilating, air conditioning		3.31	4.08	6	4.20%	5.60%	7.60%	
	2900	Electrical		3.86	5.15	6.90	5.20%	6.65%	8.40%	
	3100	Total: Mechanical & Electrical	↓	13.40	17.05	21.50	15.90%	18.05%	23%	
	9000	Per apartment unit, total cost	Apt.	62,000	95,000	140,000				
	9500	Total: Mechanical & Electrical	"	11,800	18,500	24,200				
02	0010	**APARTMENTS Mid Rise (4 to 7 story)**	S.F.	88.50	107	132				02
	0020	Total project costs	C.F.	6.90	9.55	13.05				
	0100	Site work	S.F.	3.54	7	12.60	5.25%	6.70%	9.15%	
	0500	Masonry		5.90	8.10	11.10	5.10%	7.25%	10.50%	
	1500	Finishes		11.15	14.50	18.30	10.55%	13.45%	17.70%	
	1800	Equipment		2.57	3.85	5.05	2.54%	3.48%	4.31%	
	2500	Conveying equipment		1.91	2.43	2.94	1.94%	2.27%	2.69%	
	2720	Plumbing		5.20	8.30	8.80	5.70%	7.20%	8.95%	
	2900	Electrical		5.85	7.90	9.60	6.65%	7.20%	8.95%	
	3100	Total: Mechanical & Electrical	↓	18.70	23.50	28.50	18.50%	21%	23%	
	9000	Per apartment unit, total cost	Apt.	100,000	118,000	195,500				
	9500	Total: Mechanical & Electrical	"	18,900	21,900	27,600				
03	0010	**APARTMENTS High Rise (8 to 24 story)**	S.F.	100	116	139				03
	0020	Total project costs	C.F.	9.75	11.35	14.50				
	0100	Site work	S.F.	3.64	5.90	8.25	2.58%	4.84%	6.15%	
	0500	Masonry		5.80	10.55	13.15	4.74%	9.65%	11.05%	
	1500	Finishes		11.15	13.90	16.45	9.75%	11.80%	13.70%	
	1800	Equipment		3.23	3.97	5.25	2.78%	3.49%	4.35%	
	2500	Conveying equipment		2.28	3.46	4.71	2.23%	2.78%	3.37%	
	2720	Plumbing		7.40	8.70	12.15	6.80%	7.20%	10.45%	
	2900	Electrical		6.90	8.70	11.75	6.45%	7.65%	8.80%	
	3100	Total: Mechanical & Electrical	↓	20.50	26.50	31.50	17.95%	22.50%	24.50%	
	9000	Per apartment unit, total cost	Apt.	104,500	115,000	159,500				
	9500	Total: Mechanical & Electrical	"	22,600	25,800	27,300				
04	0010	**AUDITORIUMS**	S.F.	104	142	204				04
	0020	Total project costs	C.F.	6.55	9.10	13.05				
	2720	Plumbing	S.F.	6.30	9.15	11.10	5.85%	7.20%	8.70%	
	2900	Electrical		8.05	11.75	19.30	6.80%	9.05%	11.35%	
	3100	Total: Mechanical & Electrical	↓	50.50	65.50	90.50	24.50%	29%	31.50%	
05	0010	**AUTOMOTIVE SALES**	S.F.	77	105	130				05
	0020	Total project costs	C.F.	5.10	6.10	7.90				
	2720	Plumbing	S.F.	3.52	6.10	6.65	2.89%	6.05%	6.50%	
	2770	Heating, ventilating, air conditioning		5.45	8.30	8.95	4.61%	10%	10.35%	
	2900	Electrical		6.20	9.75	13.25	7.25%	8.80%	12.15%	
	3100	Total: Mechanical & Electrical	↓	19.50	27.50	33.50	18.90%	20.50%	22%	
06	0010	**BANKS**	S.F.	152	189	239				06
	0020	Total project costs	C.F.	10.80	14.70	19.40				
	0100	Site work	S.F.	17.35	26.50	38.50	7.85%	12.95%	16.95%	
	0500	Masonry		8.15	15.60	27.50	3.36%	6.95%	10.05%	
	1500	Finishes		14	20.50	25.50	5.85%	8.60%	11.55%	
	1800	Equipment		5.70	12.55	25.50	1%	5.55%	10.50%	
	2720	Plumbing		4.74	6.80	9.90	2.82%	3.90%	4.93%	
	2770	Heating, ventilating, air conditioning		9	12.05	16.05	4.86%	7.15%	8.50%	
	2900	Electrical		14.40	19.15	25	8.20%	10.20%	12.20%	
	3100	Total: Mechanical & Electrical	↓	34.50	46.50	55.50	16.25%	19.45%	23%	
	3500	See also division 11 16 00 & 11 17 00								

50 17 | Square Foot Costs

	50 17 00 | S.F. Costs		UNIT	UNIT COSTS			% OF TOTAL			
				1/4	MEDIAN	3/4	1/4	MEDIAN	3/4	
13	0010	**CHURCHES**	S.F.	102	129	168				13
	0020	Total project costs	C.F.	6.30	8	10.50				
	1800	Equipment	S.F.	1.22	2.91	6.20	.95%	2.11%	4.31%	
	2720	Plumbing		3.97	5.55	8.15	3.51%	4.96%	6.25%	
	2770	Heating, ventilating, air conditioning		9.30	12.10	17.10	7.50%	10%	12%	
	2900	Electrical		8.60	11.80	15.85	7.30%	8.80%	10.95%	
	3100	Total: Mechanical & Electrical	↓	27	35.50	47.50	18.30%	22%	25%	
	3500	See also division 11 91 00								
15	0010	**CLUBS, COUNTRY**	S.F.	109	132	166				15
	0020	Total project costs	C.F.	8.80	10.75	14.85				
	2720	Plumbing	S.F.	6.60	9.80	22.50	5.60%	7.90%	10%	
	2900	Electrical		8.60	11.80	15.35	7%	8.95%	11%	
	3100	Total: Mechanical & Electrical	↓	45.50	57	60.50	19%	26.50%	29.50%	
17	0010	**CLUBS, SOCIAL Fraternal**	S.F.	87	125	168				17
	0020	Total project costs	C.F.	5.45	8.25	9.85				
	2720	Plumbing	S.F.	5.50	6.85	10.35	5.60%	6.90%	8.55%	
	2770	Heating, ventilating, air conditioning		7.90	9.55	12.30	8.20%	9.25%	14.40%	
	2900	Electrical		6.55	10.80	12.35	6.50%	9.50%	10.55%	
	3100	Total: Mechanical & Electrical	↓	34	36.50	46.50	21%	23%	23.50%	
18	0010	**CLUBS, Y.M.C.A.**	S.F.	114	146	181				18
	0020	Total project costs	C.F.	5.05	8.45	12.60				
	2720	Plumbing	S.F.	6.90	13.80	15.45	5.65%	7.60%	10.85%	
	2900	Electrical		8.95	10.95	16.40	6.25%	7.95%	9.25%	
	3100	Total: Mechanical & Electrical	↓	33.50	40.50	68	20.50%	22.50%	28.50%	
19	0010	**COLLEGES Classrooms & Administration**	S.F.	111	151	201				19
	0020	Total project costs	C.F.	8.10	11.75	18.10				
	0500	Masonry	S.F.	8.10	15.30	18.80	5.10%	7.45%	10.50%	
	2720	Plumbing		5.65	11.55	21.50	5.10%	6.60%	8.95%	
	2900	Electrical		9.20	14	19.45	7.80%	9.85%	12%	
	3100	Total: Mechanical & Electrical	↓	36.50	51.50	69	23.50%	28%	32%	
21	0010	**COLLEGES Science, Engineering, Laboratories**	S.F.	206	241	279				21
	0020	Total project costs	C.F.	11.80	17.20	19.55				
	1800	Equipment	S.F.	11.45	26	28.50	2%	6.45%	12.65%	
	2900	Electrical		16.95	24	37	7.10%	9.40%	12.10%	
	3100	Total: Mechanical & Electrical	↓	63	74.50	116	28.50%	31.50%	41%	
	3500	See also division 11 53 00								
23	0010	**COLLEGES Student Unions**	S.F.	131	178	215				23
	0020	Total project costs	C.F.	7.30	9.60	11.85				
	3100	Total: Mechanical & Electrical	S.F.	49.50	53.50	63	23.50%	26%	29%	
25	0010	**COMMUNITY CENTERS**	S.F.	108	134	181				25
	0020	Total project costs	C.F.	7.05	10.10	13.10				
	1800	Equipment	S.F.	2.62	4.44	7.05	1.50%	3.04%	5.45%	
	2720	Plumbing		5.15	9	12.25	4.85%	7%	8.95%	
	2770	Heating, ventilating, air conditioning		8.20	12.05	17.15	6.80%	10.35%	12.90%	
	2900	Electrical		8.85	11.80	17	7.20%	8.95%	10.45%	
	3100	Total: Mechanical & Electrical	↓	30.50	38	55	20%	25%	31%	
28	0010	**COURT HOUSES**	S.F.	156	184	254				28
	0020	Total project costs	C.F.	12	14.35	18.05				
	2720	Plumbing	S.F.	7.45	10.40	14.90	5.95%	7.45%	8.20%	
	2900	Electrical		16.60	18.45	27	8.90%	10.45%	11.85%	
	3100	Total: Mechanical & Electrical	↓	46.50	61	66.50	22.50%	27.50%	30%	
30	0010	**DEPARTMENT STORES**	S.F.	58	78	98.50				30
	0020	Total project costs	C.F.	3.10	4.01	5.45				
	2720	Plumbing	S.F.	1.80	2.27	3.45	1.82%	4.21%	5.90%	
	2770	Heating, ventilating, air conditioning	↓	5.25	8.10	12.20	8.20%	9.10%	14.80%	

Reference Section

50 17 | Square Foot Costs

	50 17 00 | S.F. Costs		UNIT	UNIT COSTS 1/4	UNIT COSTS MEDIAN	UNIT COSTS 3/4	% OF TOTAL 1/4	% OF TOTAL MEDIAN	% OF TOTAL 3/4	
30	2900	Electrical	S.F.	6.65	9.10	10.75	9.05%	12.15%	14.95%	30
	3100	Total: Mechanical & Electrical	↓	11.65	14.95	26	13.20%	21.50%	50%	
31	0010	**DORMITORIES Low Rise (1 to 3 story)**	S.F.	110	143	190				31
	0020	Total project costs	C.F.	6.20	10	15				
	2720	Plumbing	S.F.	6.60	8.80	11.15	8.05%	9%	9.65%	
	2770	Heating, ventilating, air conditioning		6.95	8.35	11.10	4.61%	8.05%	10%	
	2900	Electrical		7.25	11.05	15.10	6.40%	8.65%	9.50%	
	3100	Total: Mechanical & Electrical	↓	38	40.50	63	22%	25%	27%	
	9000	Per bed, total cost	Bed	46,400	51,500	110,500				
32	0010	**DORMITORIES Mid Rise (4 to 8 story)**	S.F.	135	176	217				32
	0020	Total project costs	C.F.	14.85	16.30	19.50				
	2900	Electrical	S.F.	14.30	16.25	22	8.20%	10.20%	11.95%	
	3100	Total: Mechanical & Electrical	"	40	80	81	25.50%	34.50%	37.50%	
	9000	Per bed, total cost	Bed	19,200	43,700	253,000				
34	0010	**FACTORIES**	S.F.	51	76	117				34
	0020	Total project costs	C.F.	3.27	4.87	8.10				
	0100	Site work	S.F.	5.80	10.60	16.75	6.95%	11.45%	17.95%	
	2720	Plumbing		2.74	5.10	8.45	3.73%	6.05%	8.10%	
	2770	Heating, ventilating, air conditioning		5.35	7.65	10.35	5.25%	8.45%	11.35%	
	2900	Electrical		6.35	10	15.30	8.10%	10.50%	14.20%	
	3100	Total: Mechanical & Electrical	↓	18.10	27.50	36.50	21%	28.50%	35.50%	
36	0010	**FIRE STATIONS**	S.F.	101	139	187				36
	0020	Total project costs	C.F.	5.90	8.10	10.80				
	0500	Masonry	S.F.	14.95	26.50	35	8.35%	11.55%	16.15%	
	1140	Roofing		3.29	8.90	10.15	1.90%	4.94%	5.05%	
	1580	Painting		2.56	3.82	3.91	1.37%	1.57%	2.07%	
	1800	Equipment		1.26	2.41	4.45	.62%	1.86%	3.54%	
	2720	Plumbing		5.65	9.10	13.15	5.85%	7.35%	9.45%	
	2770	Heating, ventilating, air conditioning		5.60	9.05	14	5.15%	7.40%	9.40%	
	2900	Electrical		7.20	12.65	17.15	6.80%	8.60%	10.60%	
	3100	Total: Mechanical & Electrical	↓	37	47.50	53.50	18.40%	23%	26%	
37	0010	**FRATERNITY HOUSES & Sorority Houses**	S.F.	101	130	178				37
	0020	Total project costs	C.F.	10.05	10.50	12.65				
	2720	Plumbing	S.F.	7.65	8.75	16	6.80%	8%	10.85%	
	2900	Electrical		6.65	14.40	17.65	6.60%	9.90%	10.65%	
	3100	Total: Mechanical & Electrical	↓	7.85	25.50	30.50		15.10%	15.90%	
38	0010	**FUNERAL HOMES**	S.F.	107	145	264				38
	0020	Total project costs	C.F.	10.90	12.10	23.50				
	2900	Electrical	S.F.	4.71	8.65	9.45	3.58%	4.44%	5.95%	
	3100	Total: Mechanical & Electrical	↓	16.75	24.50	34	12.90%	12.90%	12.90%	
39	0010	**GARAGES, COMMERCIAL (Service)**	S.F.	60.50	93.50	129				39
	0020	Total project costs	C.F.	3.97	5.85	8.50				
	1800	Equipment	S.F.	3.40	7.65	11.90	2.21%	4.62%	6.80%	
	2720	Plumbing		4.18	6.45	11.75	5.45%	7.85%	10.65%	
	2730	Heating & ventilating		5.50	7.30	9.85	5.25%	6.85%	8.20%	
	2900	Electrical		5.75	8.75	12.60	7.15%	9.25%	10.85%	
	3100	Total: Mechanical & Electrical	↓	12.65	24.50	36	12.35%	17.40%	26%	
40	0010	**GARAGES, MUNICIPAL (Repair)**	S.F.	94	119	169				40
	0020	Total project costs	C.F.	5.55	7.05	12.10				
	0500	Masonry	S.F.	8.30	16.25	25.50	5.60%	9.15%	12.50%	
	2720	Plumbing		3.98	7.65	14.40	3.59%	6.70%	7.95%	
	2730	Heating & ventilating		6.80	9.85	19	6.15%	7.45%	13.50%	
	2900	Electrical		6.55	10.50	14.95	6.65%	9.25%	11.15%	
	3100	Total: Mechanical & Electrical	↓	30.50	42	61	21.50%	23%	28.50%	

50 17 | Square Foot Costs

	50 17 00 | S.F. Costs		UNIT	UNIT COSTS 1/4	UNIT COSTS MEDIAN	UNIT COSTS 3/4	% OF TOTAL 1/4	% OF TOTAL MEDIAN	% OF TOTAL 3/4	
41	0010	**GARAGES, PARKING**	S.F.	34.50	50.50	86.50				41
	0020	Total project costs	C.F.	3.24	4.40	6.40				
	2720	Plumbing	S.F.	.98	1.52	2.34	1.72%	2.70%	3.85%	
	2900	Electrical		1.89	2.32	3.65	4.33%	5.20%	6.30%	
	3100	Total: Mechanical & Electrical	↓	3.87	5.40	6.70	7%	8.90%	11.05%	
	3200									
	9000	Per car, total cost	Car	14,600	18,300	23,300				
43	0010	**GYMNASIUMS**	S.F.	96	128	173				43
	0020	Total project costs	C.F.	4.79	6.50	8				
	1800	Equipment	S.F.	2.28	4.28	8.20	1.98%	3.30%	6.70%	
	2720	Plumbing		6.05	7.20	9.30	4.65%	6.40%	7.75%	
	2770	Heating, ventilating, air conditioning		6.55	9.95	19.95	5.15%	9.05%	11.10%	
	2900	Electrical		7.40	10.05	13.25	6.60%	8.30%	10.30%	
	3100	Total: Mechanical & Electrical	↓	27	36.50	43.50	19.60%	23.50%	28%	
	3500	See also division 11 67 00								
46	0010	**HOSPITALS**	S.F.	184	230	315				46
	0020	Total project costs	C.F.	13.95	17.40	25				
	1800	Equipment	S.F.	4.68	9	15.50	.80%	2.53%	4.80%	
	2720	Plumbing		15.85	22.50	29	7.60%	9.10%	10.85%	
	2770	Heating, ventilating, air conditioning		23.50	30	41.50	7.80%	12.95%	16.65%	
	2900	Electrical		20.50	27.50	39	9.90%	11.75%	14.10%	
	3100	Total: Mechanical & Electrical	↓	59.50	82.50	123	28%	33.50%	37%	
	9000	Per bed or person, total cost	Bed	213,500	294,000	338,500				
	9900	See also division 11 71 00								
48	0010	**HOUSING For the Elderly**	S.F.	90.50	115	141				48
	0020	Total project costs	C.F.	6.45	8.95	11.45				
	0100	Site work	S.F.	6.30	9.80	14.35	5.05%	7.90%	12.10%	
	0500	Masonry		2.76	10.30	15.05	1.30%	6.05%	11%	
	1800	Equipment		2.19	3.02	4.80	1.88%	3.23%	4.43%	
	2510	Conveying systems		2.20	2.96	4.02	1.78%	2.20%	2.81%	
	2720	Plumbing		6.75	8.60	10.80	8.15%	9.55%	10.50%	
	2730	Heating, ventilating, air conditioning		3.45	4.89	7.30	3.30%	5.60%	7.25%	
	2900	Electrical		6.75	9.15	11.75	7.30%	8.50%	10.25%	
	3100	Total: Mechanical & Electrical	↓	24.50	29	37	18.10%	22.50%	29%	
	9000	Per rental unit, total cost	Unit	84,500	98,500	110,000				
	9500	Total: Mechanical & Electrical	"	18,800	21,600	25,200				
50	0010	**HOUSING Public (Low Rise)**	S.F.	76	106	138				50
	0020	Total project costs	C.F.	6.80	8.45	10.50				
	0100	Site work	S.F.	9.70	13.95	22.50	8.35%	11.75%	16.50%	
	1800	Equipment		2.07	3.38	5.15	2.26%	3.03%	4.24%	
	2720	Plumbing		5.50	7.25	9.20	7.15%	9.05%	11.60%	
	2730	Heating, ventilating, air conditioning		2.76	5.35	5.85	4.26%	6.05%	6.45%	
	2900	Electrical		4.61	6.85	9.50	5.10%	6.55%	8.25%	
	3100	Total: Mechanical & Electrical	↓	22	28	31.50	14.50%	17.55%	26.50%	
	9000	Per apartment, total cost	Apt.	83,500	95,000	119,500				
	9500	Total: Mechanical & Electrical	"	17,800	22,000	24,400				
51	0010	**ICE SKATING RINKS**	S.F.	65	152	167				51
	0020	Total project costs	C.F.	4.79	4.90	5.65				
	2720	Plumbing	S.F.	2.43	4.56	4.66	3.12%	3.23%	5.65%	
	2900	Electrical		6.95	10.70	11.30	6.30%	10.15%	15.05%	
	3100	Total: Mechanical & Electrical	↓	11.55	16.35	20	18.95%	18.95%	18.95%	
52	0010	**JAILS**	S.F.	198	256	330				52
	0020	Total project costs	C.F.	17.85	25	29				
	1800	Equipment	S.F.	7.75	23	39	2.77%	5.55%	10.35%	
	2720	Plumbing		19.10	25.50	34	7%	8.90%	13.35%	
	2770	Heating, ventilating, air conditioning		17.90	24	46	7.50%	9.45%	17.75%	
	2900	Electrical	↓	21	28	35	8.80%	11.55%	14.95%	

50 17 | Square Foot Costs

	50 17 00 | S.F. Costs		UNIT	UNIT COSTS 1/4	UNIT COSTS MEDIAN	UNIT COSTS 3/4	% OF TOTAL 1/4	% OF TOTAL MEDIAN	% OF TOTAL 3/4	
52	3100	Total: Mechanical & Electrical	S.F.	54	98.50	117	27.50%	30%	34%	52
53	0010	**LIBRARIES**	S.F.	127	162	209				53
	0020	Total project costs	C.F.	8.55	10.75	13.70				
	0500	Masonry	S.F.	10	17.50	29.50	5.80%	7.60%	11.60%	
	1800	Equipment		1.71	4.60	6.95	.37%	1.50%	4.07%	
	2720	Plumbing		4.63	6.75	9.10	3.38%	4.60%	5.70%	
	2770	Heating, ventilating, air conditioning		10.25	17.35	22.50	7.80%	10.95%	12.80%	
	2900	Electrical		12.85	16.65	21.50	8.35%	10.30%	11.95%	
	3100	Total: Mechanical & Electrical		37.50	48	59.50	20.50%	23%	26.50%	
54	0010	**LIVING, ASSISTED**	S.F.	116	137	161				54
	0020	Total project costs	C.F.	9.75	11.40	12.95				
	0500	Masonry	S.F.	3.41	4.06	4.77	2.36%	3.16%	3.86%	
	1800	Equipment		2.63	3.06	3.92	2.12%	2.47%	2.87%	
	2720	Plumbing		9.70	13	13.45	6.05%	8.15%	10.60%	
	2770	Heating, ventilating, air conditioning		11.50	12.05	13.20	7.95%	9.35%	9.70%	
	2900	Electrical		11.35	12.55	14.50	9%	10%	10.70%	
	3100	Total: Mechanical & Electrical		31.50	37	43	25%	28.50%	31.50%	
55	0010	**MEDICAL CLINICS**	S.F.	118	145	184				55
	0020	Total project costs	C.F.	8.65	11.20	14.85				
	1800	Equipment	S.F.	3.19	6.70	10.40	1.05%	2.94%	6.35%	
	2720	Plumbing		7.85	11.05	14.75	6.15%	8.40%	10.10%	
	2770	Heating, ventilating, air conditioning		9.35	12.25	18	6.65%	8.85%	11.35%	
	2900	Electrical		10.15	14.40	18.80	8.10%	10%	12.25%	
	3100	Total: Mechanical & Electrical		32.50	44	60	22.50%	27%	33.50%	
	3500	See also division 11 71 00								
57	0010	**MEDICAL OFFICES**	S.F.	111	137	168				57
	0020	Total project costs	C.F.	8.30	11.20	15.15				
	1800	Equipment	S.F.	3.66	7.25	10.30	.70%	5.10%	7.05%	
	2720	Plumbing		6.15	9.45	12.75	5.60%	6.80%	8.50%	
	2770	Heating, ventilating, air conditioning		7.40	10.70	14.10	6.10%	8%	9.70%	
	2900	Electrical		9.05	12.85	17.90	7.60%	9.80%	11.40%	
	3100	Total: Mechanical & Electrical		24	34.50	50	19.30%	22.50%	28.50%	
59	0010	**MOTELS**	S.F.	70	101	132				59
	0020	Total project costs	C.F.	6.25	8.35	13.65				
	2720	Plumbing	S.F.	7.10	9.05	10.75	9.45%	10.60%	12.55%	
	2770	Heating, ventilating, air conditioning		4.32	6.45	11.55	5.60%	5.60%	10%	
	2900	Electrical		6.60	8.35	10.40	7.45%	9.05%	10.45%	
	3100	Total: Mechanical & Electrical		21	28	48.50	18.50%	24%	25.50%	
	5000									
	9000	Per rental unit, total cost	Unit	35,600	67,500	73,000				
	9500	Total: Mechanical & Electrical	″	6,950	10,500	12,200				
60	0010	**NURSING HOMES**	S.F.	110	142	173				60
	0020	Total project costs	C.F.	8.60	10.75	14.70				
	1800	Equipment	S.F.	3.46	4.58	7.65	2.02%	3.62%	4.99%	
	2720	Plumbing		9.40	14.25	17.15	8.75%	10.10%	12.70%	
	2770	Heating, ventilating, air conditioning		9.90	15.05	19.90	9.70%	11.45%	11.80%	
	2900	Electrical		10.85	13.55	18.45	9.40%	10.55%	12.45%	
	3100	Total: Mechanical & Electrical		26	36.50	60.50	26%	29.50%	30.50%	
	9000	Per bed or person, total cost	Bed	48,700	61,000	78,500				
61	0010	**OFFICES Low Rise (1 to 4 story)**	S.F.	92.50	120	156				61
	0020	Total project costs	C.F.	6.60	9.10	12				
	0100	Site work	S.F.	7.40	13.15	19.35	6.20%	9.70%	13.55%	
	0500	Masonry		3.58	7.35	13.15	2.62%	5.45%	8.45%	
	1800	Equipment		.98	1.92	5.25	.60%	1.50%	3.50%	
	2720	Plumbing		3.31	5.10	7.50	3.66%	4.50%	6.10%	
	2770	Heating, ventilating, air conditioning		7.30	10.20	14.90	7.20%	10.30%	11.70%	
	2900	Electrical		7.55	10.75	15.25	7.45%	9.65%	11.35%	

50 17 | Square Foot Costs

	50 17 00 | S.F. Costs		UNIT	UNIT COSTS 1/4	UNIT COSTS MEDIAN	UNIT COSTS 3/4	% OF TOTAL 1/4	% OF TOTAL MEDIAN	% OF TOTAL 3/4	
61	3100	Total: Mechanical & Electrical	S.F.	21	28.50	42.50	18.20%	22%	27%	61
62	0010	**OFFICES Mid Rise (5 to 10 story)**	S.F.	98	119	161				62
	0020	Total project costs	C.F.	6.95	8.85	12.55				
	2720	Plumbing	S.F.	2.96	4.58	6.60	2.83%	3.74%	4.50%	
	2770	Heating, ventilating, air conditioning	↓	7.45	10.60	16.95	7.65%	9.40%	11%	
	2900	Electrical	↓	7.25	9.40	14.05	6.35%	7.80%	10%	
	3100	Total: Mechanical & Electrical	↓	18.80	24.50	46	18.95%	21%	27.50%	
63	0010	**OFFICES High Rise (11 to 20 story)**	S.F.	120	152	187				63
	0020	Total project costs	C.F.	8.40	10.50	15.05				
	2900	Electrical	S.F.	7.30	8.90	13.20	5.80%	7.85%	10.50%	
	3100	Total: Mechanical & Electrical	↓	23.50	31.50	53	16.90%	23.50%	34%	
64	0010	**POLICE STATIONS**	S.F.	144	189	239				64
	0020	Total project costs	C.F.	11.50	14.05	19.25				
	0500	Masonry	S.F.	13.55	24	30	6.70%	9.10%	11.35%	
	1800	Equipment	↓	2.29	10.05	15.95	.98%	3.35%	6.70%	
	2720	Plumbing	↓	8.05	16.05	20	5.65%	6.90%	10.75%	
	2770	Heating, ventilating, air conditioning	↓	12.60	16.75	23.50	5.85%	10.55%	11.70%	
	2900	Electrical	↓	15.75	22.50	30	9.80%	11.85%	14.80%	
	3100	Total: Mechanical & Electrical	↓	51.50	62	84	25%	31.50%	32.50%	
65	0010	**POST OFFICES**	S.F.	114	141	179				65
	0020	Total project costs	C.F.	6.85	8.65	9.85				
	2720	Plumbing	S.F.	5.15	6.35	8	4.24%	5.30%	5.60%	
	2770	Heating, ventilating, air conditioning	↓	8	9.90	11	6.65%	7.15%	9.35%	
	2900	Electrical	↓	9.40	13.20	15.70	7.25%	9%	11%	
	3100	Total: Mechanical & Electrical	↓	27	35.50	40	16.25%	18.80%	22%	
66	0010	**POWER PLANTS**	S.F.	790	1,050	1,925				66
	0020	Total project costs	C.F.	21.50	47.50	101				
	2900	Electrical	S.F.	56	118	176	9.30%	12.75%	21.50%	
	8100	Total: Mechanical & Electrical	↓	139	450	1,025	32.50%	32.50%	52.50%	
67	0010	**RELIGIOUS EDUCATION**	S.F.	91.50	120	147				67
	0020	Total project costs	C.F.	5.10	7.30	9.10				
	2720	Plumbing	S.F.	3.83	5.45	7.70	4.40%	5.30%	7.10%	
	2770	Heating, ventilating, air conditioning	↓	9.70	10.95	15.50	10.05%	11.45%	12.35%	
	2900	Electrical	↓	7.25	10.25	13.55	7.60%	9.05%	10.35%	
	3100	Total: Mechanical & Electrical	↓	29.50	39	47.50	22%	23%	27%	
69	0010	**RESEARCH Laboratories & Facilities**	S.F.	137	197	287				69
	0020	Total project costs	C.F.	10.35	20	24				
	1800	Equipment	S.F.	6.10	12	29.50	.94%	4.58%	9.10%	
	2720	Plumbing	↓	13.75	17.60	28.50	6.15%	8.30%	10.80%	
	2770	Heating, ventilating, air conditioning	↓	12.35	41.50	49	7.25%	16.50%	17.50%	
	2900	Electrical	↓	16.05	26.50	43.50	9.45%	11.15%	15.40%	
	3100	Total: Mechanical & Electrical	↓	52.50	91.50	132	31.50%	37.50%	42%	
70	0010	**RESTAURANTS**	S.F.	133	172	223				70
	0020	Total project costs	C.F.	11.20	14.70	19.25				
	1800	Equipment	S.F.	8.60	21.50	32	6.10%	13%	15.65%	
	2720	Plumbing	↓	10.55	12.80	16.80	6.10%	8.15%	9%	
	2770	Heating, ventilating, air conditioning	↓	13.40	18.50	22.50	9.20%	12%	12.40%	
	2900	Electrical	↓	14.10	17.35	22.50	8.35%	10.55%	11.55%	
	3100	Total: Mechanical & Electrical	↓	43.50	46.50	60.50	21%	24.50%	29.50%	
	9000	Per seat unit, total cost	Seat	4,875	6,500	7,700				
	9500	Total: Mechanical & Electrical	"	1,225	1,625	1,925				
72	0010	**RETAIL STORES**	S.F.	62	83.50	111				72
	0020	Total project costs	C.F.	4.21	6	8.35				
	2720	Plumbing	S.F.	2.25	3.76	6.40	3.26%	4.60%	6.80%	
	2770	Heating, ventilating, air conditioning	↓	4.87	6.65	10	6.75%	8.75%	10.15%	
	2900	Electrical	↓	5.60	7.65	11.05	7.25%	9.90%	11.60%	
	3100	Total: Mechanical & Electrical	↓	14.90	19.05	26.50	17.05%	21%	23.50%	

50 17 | Square Foot Costs

		50 17 00 | S.F. Costs	UNIT	UNIT COSTS 1/4	UNIT COSTS MEDIAN	UNIT COSTS 3/4	% OF TOTAL 1/4	% OF TOTAL MEDIAN	% OF TOTAL 3/4	
74	0010	**SCHOOLS Elementary**	S.F.	101	125	154				74
	0020	Total project costs	C.F.	6.65	8.50	11				
	0500	Masonry	S.F.	9.10	15.60	23	5.45%	10.65%	14.10%	
	1800	Equipment		2.77	4.73	8.70	1.90%	3.20%	4.70%	
	2720	Plumbing		5.85	8.25	11	5.70%	7.15%	9.35%	
	2730	Heating, ventilating, air conditioning		8.75	13.90	19.45	8.15%	10.80%	14.90%	
	2900	Electrical		9.55	12.65	16.10	8.40%	10.05%	11.70%	
	3100	Total: Mechanical & Electrical	↓	34.50	43	53	25%	27.50%	30%	
	9000	Per pupil, total cost	Ea.	11,600	17,300	38,700				
	9500	Total: Mechanical & Electrical	"	3,275	4,150	10,500				
76	0010	**SCHOOLS Junior High & Middle**	S.F.	105	128	157				76
	0020	Total project costs	C.F.	6.65	8.60	9.65				
	0500	Masonry	S.F.	13.40	17.60	20.50	8.60%	11.60%	14.35%	
	1800	Equipment		3.34	5.40	8.10	1.79%	3.09%	4.86%	
	2720	Plumbing		6.10	7.50	9.30	5.30%	6.80%	7.25%	
	2770	Heating, ventilating, air conditioning		12.15	14.80	26	8.90%	11.55%	14.20%	
	2900	Electrical		10.25	12.35	15.95	7.90%	9.35%	10.60%	
	3100	Total: Mechanical & Electrical	↓	33.50	43	53	23.50%	27%	29.50%	
	9000	Per pupil, total cost	Ea.	13,300	17,400	23,400				
78	0010	**SCHOOLS Senior High**	S.F.	108	133	167				78
	0020	Total project costs	C.F.	6.55	9.65	15.55				
	1800	Equipment	S.F.	2.86	6.75	9.50	1.88%	2.91%	4.56%	
	2720	Plumbing		6.10	9.15	16.75	5.60%	6.90%	8.30%	
	2770	Heating, ventilating, air conditioning		12.45	14.25	27	8.95%	11.60%	15%	
	2900	Electrical		11	14.25	21.50	8.65%	10.20%	12.25%	
	3100	Total: Mechanical & Electrical	↓	36	42.50	71	23.50%	26.50%	28.50%	
	9000	Per pupil, total cost	Ea.	10,300	20,900	26,100				
80	0010	**SCHOOLS Vocational**	S.F.	88	128	158				80
	0020	Total project costs	C.F.	5.45	7.85	10.85				
	0500	Masonry	S.F.	5.15	12.75	19.50	3.20%	4.61%	10.95%	
	1800	Equipment		2.75	6.85	9.55	1.24%	3.10%	4.26%	
	2720	Plumbing		5.65	8.40	12.35	5.40%	6.90%	8.55%	
	2770	Heating, ventilating, air conditioning		7.90	14.65	24.50	8.60%	11.90%	14.65%	
	2900	Electrical		9.15	12.55	17.15	8.45%	11%	13.20%	
	3100	Total: Mechanical & Electrical	↓	34.50	35	60.50	27.50%	29.50%	31%	
	9000	Per pupil, total cost	Ea.	12,300	32,800	49,000				
83	0010	**SPORTS ARENAS**	S.F.	77	103	158				83
	0020	Total project costs	C.F.	4.18	7.50	9.65				
	2720	Plumbing	S.F.	4.46	6.75	14.30	4.35%	6.35%	9.40%	
	2770	Heating, ventilating, air conditioning		9.60	11.35	15.75	8.80%	10.20%	13.55%	
	2900	Electrical		8	10.90	14.05	8.60%	9.90%	12.25%	
	3100	Total: Mechanical & Electrical	↓	19.95	35	47	21.50%	25%	27.50%	
85	0010	**SUPERMARKETS**	S.F.	71	82.50	96.50				85
	0020	Total project costs	C.F.	3.96	4.79	7.25				
	2720	Plumbing	S.F.	3.97	5	5.80	5.40%	6%	7.45%	
	2770	Heating, ventilating, air conditioning		5.85	7.75	9.45	8.60%	8.65%	9.60%	
	2900	Electrical		8.90	10.25	12.10	10.40%	12.45%	13.60%	
	3100	Total: Mechanical & Electrical	↓	23	25	32.50	20.50%	26.50%	31%	
86	0010	**SWIMMING POOLS**	S.F.	115	193	410				86
	0020	Total project costs	C.F.	9.25	11.50	12.55				
	2720	Plumbing	S.F.	10.65	12.15	16.45	4.80%	9.70%	20.50%	
	2900	Electrical		8.65	14	20.50	5.75%	6.95%	7.60%	
	3100	Total: Mechanical & Electrical	↓	21	54.50	73	11.15%	14.10%	23.50%	
87	0010	**TELEPHONE EXCHANGES**	S.F.	153	224	284				87
	0020	Total project costs	C.F.	9.50	15.25	21				
	2720	Plumbing	S.F.	6.45	9.95	14.60	4.52%	5.80%	6.90%	
	2770	Heating, ventilating, air conditioning	↓	14.95	30	37.50	11.80%	16.05%	18.40%	

50 17 | Square Foot Costs

	50 17 00 | S.F. Costs		UNIT	UNIT COSTS 1/4	UNIT COSTS MEDIAN	UNIT COSTS 3/4	% OF TOTAL 1/4	% OF TOTAL MEDIAN	% OF TOTAL 3/4	
87	2900	Electrical	S.F.	15.55	24.50	43.50	10.90%	14%	17.85%	87
	3100	Total: Mechanical & Electrical	↓	46	87	123	29.50%	33.50%	44.50%	
91	0010	**THEATERS**	S.F.	96	120	182				91
	0020	Total project costs	C.F.	4.44	6.55	9.65				
	2720	Plumbing	S.F.	3.20	3.47	14.20	2.92%	4.70%	6.80%	
	2770	Heating, ventilating, air conditioning		9.35	11.30	14	8%	12.25%	13.40%	
	2900	Electrical		8.40	11.35	23	8.05%	9.35%	12.25%	
	3100	Total: Mechanical & Electrical	↓	21.50	29	34.50	21.50%	25.50%	27.50%	
94	0010	**TOWN HALLS City Halls & Municipal Buildings**	S.F.	107	136	178				94
	0020	Total project costs	C.F.	9.80	11.75	16.50				
	2720	Plumbing	S.F.	4.48	8.40	15.40	4.31%	5.95%	7.95%	
	2770	Heating, ventilating, air conditioning		8.10	16.10	23.50	7.05%	9.05%	13.45%	
	2900	Electrical		10.20	14.55	19.85	8.05%	9.45%	11.65%	
	3100	Total: Mechanical & Electrical	↓	35.50	45	68.50	22%	26.50%	31%	
97	0010	**WAREHOUSES & Storage Buildings**	S.F.	40	59.50	85.50				97
	0020	Total project costs	C.F.	2.10	3.28	5.45				
	0100	Site work	S.F.	4.13	8.20	12.40	6.05%	12.95%	19.85%	
	0500	Masonry		2.41	5.70	12.30	3.73%	7.40%	12.30%	
	1800	Equipment		.64	1.39	7.80	.91%	1.82%	5.55%	
	2720	Plumbing		1.34	2.39	4.48	2.90%	4.80%	6.55%	
	2730	Heating, ventilating, air conditioning		1.53	4.30	5.75	2.41%	5%	8.90%	
	2900	Electrical		2.37	4.46	7.40	5.15%	7.20%	10.10%	
	3100	Total: Mechanical & Electrical	↓	6.60	10.15	20	12.75%	18.90%	26%	
99	0010	**WAREHOUSE & OFFICES Combination**	S.F.	49	65.50	90				99
	0020	Total project costs	C.F.	2.52	3.66	5.40				
	1800	Equipment	S.F.	.86	1.65	2.46	.52%	1.20%	2.40%	
	2720	Plumbing		1.90	3.37	4.93	3.74%	4.76%	6.30%	
	2770	Heating, ventilating, air conditioning		3	4.69	6.55	5%	5.65%	10.05%	
	2900	Electrical		3.30	4.91	7.75	5.75%	8%	10%	
	3100	Total: Mechanical & Electrical	↓	9.20	14.20	22.50	14.40%	19.95%	24.50%	

Square Foot Project Size Modifier

One factor that affects the S.F. cost of a particular building is the size. In general, for buildings built to the same specifications in the same locality, the larger building will have the lower S.F. cost. This is due mainly to the decreasing contribution of the exterior walls plus the economy of scale usually achievable in larger buildings. The Area Conversion Scale shown below will give a factor to convert costs for the typical size building to an adjusted cost for the particular project.

The Square Foot Base Size lists the median costs, most typical project size in our accumulated data, and the range in size of the projects.

The Size Factor for your project is determined by dividing your project area in S.F. by the typical project size for the particular Building Type. With this factor, enter the Area Conversion Scale at the appropriate Size Factor and determine the appropriate cost multiplier for your building size.

Example: Determine the cost per S.F. for a 100,000 S.F. Mid-rise apartment building.

$$\frac{\text{Proposed building area} = 100{,}000\text{ S.F.}}{\text{Typical size from below} = 50{,}000\text{ S.F.}} = 2.00$$

Enter Area Conversion scale at 2.0, intersect curve, read horizontally the appropriate cost multiplier of .94. Size adjusted cost becomes .94 x \$107.00 = \$101.00 based on national average costs.

Note: For Size Factors less than .50, the Cost Multiplier is 1.1
For Size Factors greater than 3.5, the Cost Multiplier is .90

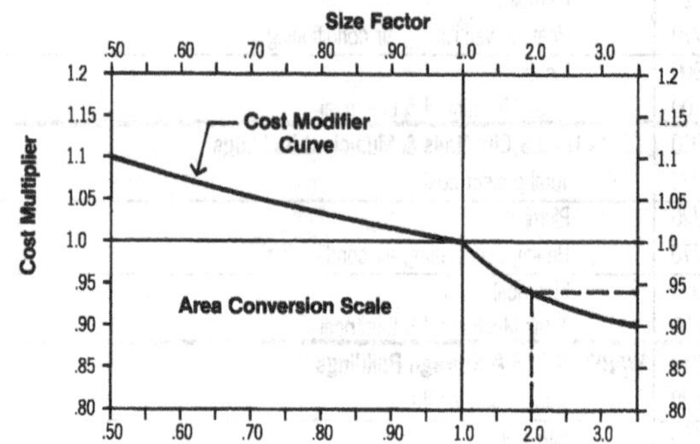

Square Foot Base Size

Building Type	Median Cost per S.F.	Typical Size Gross S.F.	Typical Range Gross S.F.	Building Type	Median Cost per S.F.	Typical Size Gross S.F.	Typical Range Gross S.F.
Apartments, Low Rise	\$ 84.00	21,000	9,700 - 37,200	Jails	\$ 256.00	40,000	5,500 - 145,000
Apartments, Mid Rise	107.00	50,000	32,000 - 100,000	Libraries	162.00	12,000	7,000 - 31,000
Apartments, High Rise	116.00	145,000	95,000 - 600,000	Living, Assisted	137.00	32,300	23,500 - 50,300
Auditoriums	142.00	25,000	7,600 - 39,000	Medical Clinics	145.00	7,200	4,200 - 15,700
Auto Sales	105.00	20,000	10,800 - 28,600	Medical Offices	137.00	6,000	4,000 - 15,000
Banks	189.00	4,200	2,500 - 7,500	Motels	101.00	40,000	15,800 - 120,000
Churches	129.00	17,000	2,000 - 42,000	Nursing Homes	142.00	23,000	15,000 - 37,000
Clubs, Country	132.00	6,500	4,500 - 15,000	Offices, Low Rise	120.00	20,000	5,000 - 80,000
Clubs, Social	125.00	10,000	6,000 - 13,500	Offices, Mid Rise	119.00	120,000	20,000 - 300,000
Clubs, YMCA	146.00	28,300	12,800 - 39,400	Offices, High Rise	152.00	260,000	120,000 - 800,000
Colleges (Class)	151.00	50,000	15,000 - 150,000	Police Stations	189.00	10,500	4,000 - 19,000
Colleges (Science Lab)	241.00	45,600	16,600 - 80,000	Post Offices	141.00	12,400	6,800 - 30,000
College (Student Union)	178.00	33,400	16,000 - 85,000	Power Plants	1050.00	7,500	1,000 - 20,000
Community Center	134.00	9,400	5,300 - 16,700	Religious Education	120.00	9,000	6,000 - 12,000
Court Houses	184.00	32,400	17,800 - 106,000	Research	197.00	19,000	6,300 - 45,000
Dept. Stores	78.00	90,000	44,000 - 122,000	Restaurants	172.00	4,400	2,800 - 6,000
Dormitories, Low Rise	143.00	25,000	10,000 - 95,000	Retail Stores	83.50	7,200	4,000 - 17,600
Dormitories, Mid Rise	176.00	85,000	20,000 - 200,000	Schools, Elementary	125.00	41,000	24,500 - 55,000
Factories	76.00	26,400	12,900 - 50,000	Schools, Jr. High	128.00	92,000	52,000 - 119,000
Fire Stations	139.00	5,800	4,000 - 8,700	Schools, Sr. High	133.00	101,000	50,500 - 175,000
Fraternity Houses	130.00	12,500	8,200 - 14,800	Schools, Vocational	128.00	37,000	20,500 - 82,000
Funeral Homes	145.00	10,000	4,000 - 20,000	Sports Arenas	103.00	15,000	5,000 - 40,000
Garages, Commercial	93.50	9,300	5,000 - 13,600	Supermarkets	82.50	44,000	12,000 - 60,000
Garages, Municipal	119.00	8,300	4,500 - 12,600	Swimming Pools	193.00	20,000	10,000 - 32,000
Garages, Parking	50.50	163,000	76,400 - 225,300	Telephone Exchange	224.00	4,500	1,200 - 10,600
Gymnasiums	128.00	19,200	11,600 - 41,000	Theaters	120.00	10,500	8,800 - 17,500
Hospitals	230.00	55,000	27,200 - 125,000	Town Halls	136.00	10,800	4,800 - 23,400
House (Elderly)	115.00	37,000	21,000 - 66,000	Warehouses	59.50	25,000	8,000 - 72,000
Housing (Public)	106.00	36,000	14,400 - 74,400	Warehouse & Office	65.50	25,000	8,000 - 72,000
Ice Rinks	152.00	29,000	27,200 - 33,600				

Abbreviations

A	Area Square Feet; Ampere
ABS	Acrylonitrile Butadiene Stryrene; Asbestos Bonded Steel
A.C.	Alternating Current; Air-Conditioning; Asbestos Cement; Plywood Grade A & C
A.C.I.	American Concrete Institute
AD	Plywood, Grade A & D
Addit.	Additional
Adj.	Adjustable
af	Audio-frequency
A.G.A.	American Gas Association
Agg.	Aggregate
A.H.	Ampere Hours
A hr.	Ampere-hour
A.H.U.	Air Handling Unit
A.I.A.	American Institute of Architects
AIC	Ampere Interrupting Capacity
Allow.	Allowance
alt.	Altitude
Alum.	Aluminum
a.m.	Ante Meridiem
Amp.	Ampere
Anod.	Anodized
Approx.	Approximate
Apt.	Apartment
Asb.	Asbestos
A.S.B.C.	American Standard Building Code
Asbe.	Asbestos Worker
ASCE.	American Society of Civil Engineers
A.S.H.R.A.E.	American Society of Heating, Refrig. & AC Engineers
A.S.M.E.	American Society of Mechanical Engineers
A.S.T.M.	American Society for Testing and Materials
Attchmt.	Attachment
Avg., Ave.	Average
A.W.G.	American Wire Gauge
AWWA	American Water Works Assoc.
Bbl.	Barrel
B&B	Grade B and Better; Balled & Burlapped
B.&S.	Bell and Spigot
B.&W.	Black and White
b.c.c.	Body-centered Cubic
B.C.Y.	Bank Cubic Yards
BE	Bevel End
B.F.	Board Feet
Bg. cem.	Bag of Cement
BHP	Boiler Horsepower; Brake Horsepower
B.I.	Black Iron
Bit., Bitum.	Bituminous
Bit., Conc.	Bituminous Concrete
Bk.	Backed
Bkrs.	Breakers
Bldg.	Building
Blk.	Block
Bm.	Beam
Boil.	Boilermaker
B.P.M.	Blows per Minute
BR	Bedroom
Brg.	Bearing
Brhe.	Bricklayer Helper
Bric.	Bricklayer
Brk.	Brick
Brng.	Bearing
Brs.	Brass
Brz.	Bronze
Bsn.	Basin
Btr.	Better
Btu	British Thermal Unit
BTUH	BTU per Hour
B.U.R.	Built-up Roofing
BX	Interlocked Armored Cable
°C	degree centegrade
c	Conductivity, Copper Sweat
C	Hundred; Centigrade
C/C	Center to Center, Cedar on Cedar
C-C	Center to Center
Cab.	Cabinet
Cair.	Air Tool Laborer
Calc	Calculated
Cap.	Capacity
Carp.	Carpenter
C.B.	Circuit Breaker
C.C.A.	Chromate Copper Arsenate
C.C.F.	Hundred Cubic Feet
cd	Candela
cd/sf	Candela per Square Foot
CD	Grade of Plywood Face & Back
CDX	Plywood, Grade C & D, exterior glue
Cefi.	Cement Finisher
Cem.	Cement
CF	Hundred Feet
C.F.	Cubic Feet
CFM	Cubic Feet per Minute
c.g.	Center of Gravity
CHW	Chilled Water; Commercial Hot Water
C.I.	Cast Iron
C.I.P.	Cast in Place
Circ.	Circuit
C.L.	Carload Lot
Clab.	Common Laborer
Clam	Common maintenance laborer
C.L.F.	Hundred Linear Feet
CLF	Current Limiting Fuse
CLP	Cross Linked Polyethylene
cm	Centimeter
CMP	Corr. Metal Pipe
C.M.U.	Concrete Masonry Unit
CN	Change Notice
Col.	Column
CO_2	Carbon Dioxide
Comb.	Combination
Compr.	Compressor
Conc.	Concrete
Cont.	Continuous; Continued, Container
Corr.	Corrugated
Cos	Cosine
Cot	Cotangent
Cov.	Cover
C/P	Cedar on Paneling
CPA	Control Point Adjustment
Cplg.	Coupling
C.P.M.	Critical Path Method
CPVC	Chlorinated Polyvinyl Chloride
C.Pr.	Hundred Pair
CRC	Cold Rolled Channel
Creos.	Creosote
Crpt.	Carpet & Linoleum Layer
CRT	Cathode-ray Tube
CS	Carbon Steel, Constant Shear Bar Joist
Csc	Cosecant
C.S.F.	Hundred Square Feet
CSI	Construction Specifications Institute
C.T.	Current Transformer
CTS	Copper Tube Size
Cu	Copper, Cubic
Cu. Ft.	Cubic Foot
cw	Continuous Wave
C.W.	Cool White; Cold Water
Cwt.	100 Pounds
C.W.X.	Cool White Deluxe
C.Y.	Cubic Yard (27 cubic feet)
C.Y./Hr.	Cubic Yard per Hour
Cyl.	Cylinder
d	Penny (nail size)
D	Deep; Depth; Discharge
Dis., Disch.	Discharge
Db.	Decibel
Dbl.	Double
DC	Direct Current
DDC	Direct Digital Control
Demob.	Demobilization
d.f.u.	Drainage Fixture Units
D.H.	Double Hung
DHW	Domestic Hot Water
DI	Ductile Iron
Diag.	Diagonal
Diam., Dia	Diameter
Distrib.	Distribution
Div.	Division
Dk.	Deck
D.L.	Dead Load; Diesel
DLH	Deep Long Span Bar Joist
Do.	Ditto
Dp.	Depth
D.P.S.T.	Double Pole, Single Throw
Dr.	Drive
Drink.	Drinking
D.S.	Double Strength
D.S.A.	Double Strength A Grade
D.S.B.	Double Strength B Grade
Dty.	Duty
DWV	Drain Waste Vent
DX	Deluxe White, Direct Expansion
dyn	Dyne
e	Eccentricity
E	Equipment Only; East
Ea.	Each
E.B.	Encased Burial
Econ.	Economy
E.C.Y	Embankment Cubic Yards
EDP	Electronic Data Processing
EIFS	Exterior Insulation Finish System
E.D.R.	Equiv. Direct Radiation
Eq.	Equation
EL	elevation
Elec.	Electrician; Electrical
Elev.	Elevator; Elevating
EMT	Electrical Metallic Conduit; Thin Wall Conduit
Eng.	Engine, Engineered
EPDM	Ethylene Propylene Diene Monomer
EPS	Expanded Polystyrene
Eqhv.	Equip. Oper., Heavy
Eqlt.	Equip. Oper., Light
Eqmd.	Equip. Oper., Medium
Eqmm.	Equip. Oper., Master Mechanic
Eqol.	Equip. Oper., Oilers
Equip.	Equipment
ERW	Electric Resistance Welded
E.S.	Energy Saver
Est.	Estimated
esu	Electrostatic Units
E.W.	Each Way
EWT	Entering Water Temperature
Excav.	Excavation
Exp.	Expansion, Exposure
Ext.	Exterior
Extru.	Extrusion
f.	Fiber stress
F	Fahrenheit; Female; Fill
Fab.	Fabricated

Abbreviations

FBGS	Fiberglass
F.C.	Footcandles
f.c.c.	Face-centered Cubic
f'c.	Compressive Stress in Concrete; Extreme Compressive Stress
F.E.	Front End
FEP	Fluorinated Ethylene Propylene (Teflon)
F.G.	Flat Grain
F.H.A.	Federal Housing Administration
Fig.	Figure
Fin.	Finished
Fixt.	Fixture
Fl. Oz.	Fluid Ounces
Flr.	Floor
F.M.	Frequency Modulation; Factory Mutual
Fmg.	Framing
Fdn.	Foundation
Fori.	Foreman, Inside
Foro.	Foreman, Outside
Fount.	Fountain
fpm	Feet per Minute
FPT	Female Pipe Thread
Fr.	Frame
F.R.	Fire Rating
FRK	Foil Reinforced Kraft
FRP	Fiberglass Reinforced Plastic
FS	Forged Steel
FSC	Cast Body; Cast Switch Box
Ft.	Foot; Feet
Ftng.	Fitting
Ftg.	Footing
Ft lb.	Foot Pound
Furn.	Furniture
FVNR	Full Voltage Non-Reversing
FXM	Female by Male
Fy.	Minimum Yield Stress of Steel
g	Gram
G	Gauss
Ga.	Gauge
Gal., gal.	Gallon
gpm, GPM	Gallon per Minute
Galv.	Galvanized
Gen.	General
G.F.I.	Ground Fault Interrupter
Glaz.	Glazier
GPD	Gallons per Day
GPH	Gallons per Hour
GPM	Gallons per Minute
GR	Grade
Gran.	Granular
Grnd.	Ground
H	High Henry
H.C.	High Capacity
H.D.	Heavy Duty; High Density
H.D.O.	High Density Overlaid
H.D.P.E.	high density polyethelene
Hdr.	Header
Hdwe.	Hardware
Help.	Helper Average
HEPA	High Efficiency Particulate Air Filter
Hg	Mercury
HIC	High Interrupting Capacity
HM	Hollow Metal
HMWPE	high molecular weight polyethylene
H.O.	High Output
Horiz.	Horizontal
H.P.	Horsepower; High Pressure
H.P.F.	High Power Factor
Hr.	Hour
Hrs./Day	Hours per Day
HSC	High Short Circuit
Ht.	Height
Htg.	Heating
Htrs.	Heaters
HVAC	Heating, Ventilation & Air-Conditioning
Hvy.	Heavy
HW	Hot Water
Hyd.;Hydr.	Hydraulic
Hz.	Hertz (cycles)
I.	Moment of Inertia
IBC	International Building Code
I.C.	Interrupting Capacity
ID	Inside Diameter
I.D.	Inside Dimension; Identification
I.F.	Inside Frosted
I.M.C.	Intermediate Metal Conduit
In.	Inch
Incan.	Incandescent
Incl.	Included; Including
Int.	Interior
Inst.	Installation
Insul.	Insulation/Insulated
I.P.	Iron Pipe
I.P.S.	Iron Pipe Size
I.P.T.	Iron Pipe Threaded
I.W.	Indirect Waste
J	Joule
J.I.C.	Joint Industrial Council
K	Thousand; Thousand Pounds; Heavy Wall Copper Tubing, Kelvin
K.A.H.	Thousand Amp. Hours
KCMIL	Thousand Circular Mils
KD	Knock Down
K.D.A.T.	Kiln Dried After Treatment
kg	Kilogram
kG	Kilogauss
kgf	Kilogram Force
kHz	Kilohertz
Kip.	1000 Pounds
KJ	Kiljoule
K.L.	Effective Length Factor
K.L.F.	Kips per Linear Foot
Km	Kilometer
K.S.F.	Kips per Square Foot
K.S.I.	Kips per Square Inch
kV	Kilovolt
kVA	Kilovolt Ampere
K.V.A.R.	Kilovar (Reactance)
KW	Kilowatt
KWh	Kilowatt-hour
L	Labor Only; Length; Long; Medium Wall Copper Tubing
Lab.	Labor
lat	Latitude
Lath.	Lather
Lav.	Lavatory
lb.; #	Pound
L.B.	Load Bearing; L Conduit Body
L. & E.	Labor & Equipment
lb./hr.	Pounds per Hour
lb./L.F.	Pounds per Linear Foot
lbf/sq.in.	Pound-force per Square Inch
L.C.L.	Less than Carload Lot
L.C.Y.	Loose Cubic Yard
Ld.	Load
LE	Lead Equivalent
LED	Light Emitting Diode
L.F.	Linear Foot
L.F. Nose	Linear Foot of Stair Nosing
L.F. Rsr	Linear Foot of Stair Riser
Lg.	Long; Length; Large
L & H	Light and Heat
LH	Long Span Bar Joist
L.H.	Labor Hours
L.L.	Live Load
L.L.D.	Lamp Lumen Depreciation
lm	Lumen
lm/sf	Lumen per Square Foot
lm/W	Lumen per Watt
L.O.A.	Length Over All
log	Logarithm
L-O-L	Lateralolet
long.	longitude
L.P.	Liquefied Petroleum; Low Pressure
L.P.F.	Low Power Factor
LR	Long Radius
L.S.	Lump Sum
Lt.	Light
Lt. Ga.	Light Gauge
L.T.L.	Less than Truckload Lot
Lt. Wt.	Lightweight
L.V.	Low Voltage
M	Thousand; Material; Male; Light Wall Copper Tubing
M^2CA	Meters Squared Contact Area
m/hr.; M.H.	Man-hour
mA	Milliampere
Mach.	Machine
Mag. Str.	Magnetic Starter
Maint.	Maintenance
Marb.	Marble Setter
Mat; Mat'l.	Material
Max.	Maximum
MBF	Thousand Board Feet
MBH	Thousand BTU's per hr.
MC	Metal Clad Cable
M.C.F.	Thousand Cubic Feet
M.C.F.M.	Thousand Cubic Feet per Minute
M.C.M.	Thousand Circular Mils
M.C.P.	Motor Circuit Protector
MD	Medium Duty
M.D.O.	Medium Density Overlaid
Med.	Medium
MF	Thousand Feet
M.F.B.M.	Thousand Feet Board Measure
Mfg.	Manufacturing
Mfrs.	Manufacturers
mg	Milligram
MGD	Million Gallons per Day
MGPH	Thousand Gallons per Hour
MH, M.H.	Manhole; Metal Halide; Man-Hour
MHz	Megahertz
Mi.	Mile
MI	Malleable Iron; Mineral Insulated
mm	Millimeter
Mill.	Millwright
Min., min.	Minimum, minute
Misc.	Miscellaneous
ml	Milliliter, Mainline
M.L.F.	Thousand Linear Feet
Mo.	Month
Mobil.	Mobilization
Mog.	Mogul Base
MPH	Miles per Hour
MPT	Male Pipe Thread

Abbreviations

MRT	Mile Round Trip
ms	Millisecond
M.S.F.	Thousand Square Feet
Mstz.	Mosaic & Terrazzo Worker
M.S.Y.	Thousand Square Yards
Mtd., mtd.	Mounted
Mthe.	Mosaic & Terrazzo Helper
Mtng.	Mounting
Mult.	Multi; Multiply
M.V.A.	Million Volt Amperes
M.V.A.R.	Million Volt Amperes Reactance
MV	Megavolt
MW	Megawatt
MXM	Male by Male
MYD	Thousand Yards
N	Natural; North
nA	Nanoampere
NA	Not Available; Not Applicable
N.B.C.	National Building Code
NC	Normally Closed
N.E.M.A.	National Electrical Manufacturers Assoc.
NEHB	Bolted Circuit Breaker to 600V.
N.L.B.	Non-Load-Bearing
NM	Non-Metallic Cable
nm	Nanometer
No.	Number
NO	Normally Open
N.O.C.	Not Otherwise Classified
Nose.	Nosing
N.P.T.	National Pipe Thread
NQOD	Combination Plug-on/Bolt on Circuit Breaker to 240V.
N.R.C.	Noise Reduction Coefficient/ Nuclear Regulator Commission
N.R.S.	Non Rising Stem
ns	Nanosecond
nW	Nanowatt
OB	Opposing Blade
OC	On Center
OD	Outside Diameter
O.D.	Outside Dimension
ODS	Overhead Distribution System
O.G.	Ogee
O.H.	Overhead
O&P	Overhead and Profit
Oper.	Operator
Opng.	Opening
Orna.	Ornamental
OSB	Oriented Strand Board
O.S.&Y.	Outside Screw and Yoke
Ovhd.	Overhead
OWG	Oil, Water or Gas
Oz.	Ounce
P.	Pole; Applied Load; Projection
p.	Page
Pape.	Paperhanger
P.A.P.R.	Powered Air Purifying Respirator
PAR	Parabolic Reflector
Pc., Pcs.	Piece, Pieces
P.C.	Portland Cement; Power Connector
P.C.F.	Pounds per Cubic Foot
P.C.M.	Phase Contrast Microscopy
P.E.	Professional Engineer; Porcelain Enamel; Polyethylene; Plain End
Perf.	Perforated
PEX	Cross linked polyethylene
Ph.	Phase
P.I.	Pressure Injected
Pile.	Pile Driver
Pkg.	Package
Pl.	Plate
Plah.	Plasterer Helper
Plas.	Plasterer
Pluh.	Plumbers Helper
Plum.	Plumber
Ply.	Plywood
p.m.	Post Meridiem
Pntd.	Painted
Pord.	Painter, Ordinary
pp	Pages
PP, PPL	Polypropylene
P.P.M.	Parts per Million
Pr.	Pair
P.E.S.B.	Pre-engineered Steel Building
Prefab.	Prefabricated
Prefin.	Prefinished
Prop.	Propelled
PSF, psf	Pounds per Square Foot
PSI, psi	Pounds per Square Inch
PSIG	Pounds per Square Inch Gauge
PSP	Plastic Sewer Pipe
Pspr.	Painter, Spray
Psst.	Painter, Structural Steel
P.T.	Potential Transformer
P. & T.	Pressure & Temperature
Ptd.	Painted
Ptns.	Partitions
Pu	Ultimate Load
PVC	Polyvinyl Chloride
Pvmt.	Pavement
Pwr.	Power
Q	Quantity Heat Flow
Qt.	Quart
Quan., Qty.	Quantity
Q.C.	Quick Coupling
r	Radius of Gyration
R	Resistance
R.C.P.	Reinforced Concrete Pipe
Rect.	Rectangle
Reg.	Regular
Reinf.	Reinforced
Req'd.	Required
Res.	Resistant
Resi.	Residential
Rgh.	Rough
RGS	Rigid Galvanized Steel
R.H.W.	Rubber, Heat & Water Resistant; Residential Hot Water
rms	Root Mean Square
Rnd.	Round
Rodm.	Rodman
Rofc.	Roofer, Composition
Rofp.	Roofer, Precast
Rohe.	Roofer Helpers (Composition)
Rots.	Roofer, Tile & Slate
R.O.W.	Right of Way
RPM	Revolutions per Minute
R.S.	Rapid Start
Rsr	Riser
RT	Round Trip
S.	Suction; Single Entrance; South
SC	Screw Cover
SCFM	Standard Cubic Feet per Minute
Scaf.	Scaffold
Sch., Sched.	Schedule
S.C.R.	Modular Brick
S.D.	Sound Deadening
S.D.R.	Standard Dimension Ratio
S.E.	Surfaced Edge
Sel.	Select
S.E.R., S.E.U.	Service Entrance Cable
S.F.	Square Foot
S.F.C.A.	Square Foot Contact Area
S.F. Flr.	Square Foot of Floor
S.F.G.	Square Foot of Ground
S.F. Hor.	Square Foot Horizontal
S.F.R.	Square Feet of Radiation
S.F. Shlf.	Square Foot of Shelf
S4S	Surface 4 Sides
Shee.	Sheet Metal Worker
Sin.	Sine
Skwk.	Skilled Worker
SL	Saran Lined
S.L.	Slimline
Sldr.	Solder
SLH	Super Long Span Bar Joist
S.N.	Solid Neutral
S-O-L	Socketolet
sp	Standpipe
S.P.	Static Pressure; Single Pole; Self-Propelled
Spri.	Sprinkler Installer
spwg	Static Pressure Water Gauge
S.P.D.T.	Single Pole, Double Throw
SPF	Spruce Pine Fir
S.P.S.T.	Single Pole, Single Throw
SPT	Standard Pipe Thread
Sq.	Square; 100 Square Feet
Sq. Hd.	Square Head
Sq. In.	Square Inch
S.S.	Single Strength; Stainless Steel
S.S.B.	Single Strength B Grade
sst, ss	Stainless Steel
Sswk.	Structural Steel Worker
Sswl.	Structural Steel Welder
St.;Stl.	Steel
S.T.C.	Sound Transmission Coefficient
Std.	Standard
Stg.	Staging
STK	Select Tight Knot
STP	Standard Temperature & Pressure
Stpi.	Steamfitter, Pipefitter
Str.	Strength; Starter; Straight
Strd.	Stranded
Struct.	Structural
Sty.	Story
Subj.	Subject
Subs.	Subcontractors
Surf.	Surface
Sw.	Switch
Swbd.	Switchboard
S.Y.	Square Yard
Syn.	Synthetic
S.Y.P.	Southern Yellow Pine
Sys.	System
t.	Thickness
T	Temperature; Ton
Tan	Tangent
T.C.	Terra Cotta
T & C	Threaded and Coupled
T.D.	Temperature Difference
Tdd	Telecommunications Device for the Deaf
T.E.M.	Transmission Electron Microscopy
TFE	Tetrafluoroethylene (Teflon)
T. & G.	Tongue & Groove; Tar & Gravel
Th., Thk.	Thick
Thn.	Thin
Thrded	Threaded
Tilf.	Tile Layer, Floor
Tilh.	Tile Layer, Helper
THHN	Nylon Jacketed Wire
THW.	Insulated Strand Wire

Abbreviations

THWN	Nylon Jacketed Wire
T.L.	Truckload
T.M.	Track Mounted
Tot.	Total
T-O-L	Threadolet
T.S.	Trigger Start
Tr.	Trade
Transf.	Transformer
Trhv.	Truck Driver, Heavy
Trlr	Trailer
Trlt.	Truck Driver, Light
TTY	Teletypewriter
TV	Television
T.W.	Thermoplastic Water Resistant Wire
UCI	Uniform Construction Index
UF	Underground Feeder
UGND	Underground Feeder
U.H.F.	Ultra High Frequency
U.I.	United Inch
U.L.	Underwriters Laboratory
Uld.	unloading
Unfin.	Unfinished
URD	Underground Residential Distribution
US	United States
USP	United States Primed
UTP	Unshielded Twisted Pair
V	Volt
V.A.	Volt Amperes
V.C.T.	Vinyl Composition Tile
VAV	Variable Air Volume
VC	Veneer Core
Vent.	Ventilation
Vert.	Vertical
V.F.	Vinyl Faced
V.G.	Vertical Grain
V.H.F.	Very High Frequency
VHO	Very High Output
Vib.	Vibrating
V.L.F.	Vertical Linear Foot
Vol.	Volume
VRP	Vinyl Reinforced Polyester
W	Wire; Watt; Wide; West
w/	With
W.C.	Water Column; Water Closet
W.F.	Wide Flange
W.G.	Water Gauge
Wldg.	Welding
W. Mile	Wire Mile
W-O-L	Weldolet
W.R.	Water Resistant
Wrck.	Wrecker
W.S.P.	Water, Steam, Petroleum
WT., Wt.	Weight
WWF	Welded Wire Fabric
XFER	Transfer
XFMR	Transformer
XHD	Extra Heavy Duty
XHHW, XLPE	Cross-Linked Polyethylene Wire Insulation
XLP	Cross-linked Polyethylene
Y	Wye
yd	Yard
yr	Year
Δ	Delta
%	Percent
~	Approximately
Ø	Phase; diameter
@	At
#	Pound; Number
<	Less Than
>	Greater Than
Z	zone